Molecular Virology of Human Pathogenic Viruses

Molecular Virology of Human Pathogenic Viruses

Wang-Shick Ryu

Department of Biochemistry, Yonsei University,
Seoul, Korea

ELSEVIER

AMSTERDAM • BOSTON • HEIDELBERG • LONDON
NEW YORK • OXFORD • PARIS • SAN DIEGO
SAN FRANCISCO • SINGAPORE • SYDNEY • TOKYO

Academic Press is an imprint of Elsevier

Academic Press is an imprint of Elsevier
125 London Wall, London EC2Y 5AS, UK
525 B Street, Suite 1800, San Diego, CA 92101-4495, USA
50 Hampshire Street, 5th Floor, Cambridge, MA 02139, USA
The Boulevard, Langford Lane, Kidlington, Oxford OX5 1GB, UK

British Library Cataloguing-in-Publication Data
A catalogue record for this book is available from the British Library.

Library of Congress Cataloging-in-Publication Data
A catalog record for this book is available from the Library of Congress.

ISBN: 978-0-12-800838-6

Publisher: Sara Tenney
Acquisition Editor: Jill Leonard
Editorial Project Manager: Fenton Coulthurst
Production Project Manager: Julia Haynes
Designer: Victoria Pearson

Typeset by MPS Limited, Chennai, India

Short Contents

Contents

Part III
RNA Viruses

Part III-1
Positive-Strand RNA Viruses

Part III-2
Negative-Strand RNA Viruses

Part IV
RT Viruses

Preface

Since viruses are agents of life-threatening infectious diseases, viruses have become of interest to people from all walks of life. More recently, this field has gained increased attention because of the 2009 H1N1 influenza pandemic, avian influenza, Ebola outbreak, SARS outbreak, and more recently MERS outbreak and Zika virus outbreak. Besides these newly emerging viruses, the existing viruses, such as HIV/AIDS and influenza virus, nonetheless, represent a significant disease burden on the public health worldwide. Why do viruses cause diseases to their host organisms to whom they are indebted? When and where did viruses come from in the first place? Why are new viruses frequently emerging? Why does modern medicine fail to control or conquer these culprits? These are the kinds of questions that people might wonder about viruses. The aim of this book is to provide fundamental knowledge on the "virus" from which students could find their own answers.

The motivation of writing this book fulfills a self-serving purpose in teaching *virology* to undergraduate and graduate students. My experience in teaching *virology* convinced me that the conveyance of lecture contents to students is a challenging task. Students can become overwhelmed by the information pouring out of numerous virus families, and they can easily get lost by the unfamiliar nomenclature and terminology. Another important element that makes the task even more difficult is the lack of textbooks that can be embraced by undergraduate and graduate students. Some of the existing virology textbooks are either too comprehensive for students to grasp or too sketchy to engage. My intent is to make a book, which is "concise" but "informative." In line with this aim, this book mainly focuses on human pathogenic viruses.

This book is primarily written in the context of virus families, as opposed to principles (mechanism). The description of virus families allows students to focus on one virus family at a time. Although many features of each virus family are distinct, some features are shared by diverse viruses. These common principles are described in Part I Principle that includes *Classification*, *Structure*, *Virus Life Cycle*, *Diagnosis and Methods*, and *Host Immune Response*. From Part II to Part IV, each chapter is dedicated to the individual virus families. Specifically, 10 major human virus families are covered, in order of *DNA viruses*, *RNA viruses* (*positive-strand RNA viruses* and *other negative-strand RNA viruses*), and *reverse transcribing viruses*. Other miscellaneous viruses belonging to these three virus groups are covered only briefly in the following chapters: *Other DNA Viruses*, *Other Positive-Strand RNA Viruses*, and *Other Negative-Strand RNA Viruses*. Inclusion of these chapters lets the students learn at least some aspects of the neglected miscellaneous viruses. In my view, it is critical to arrange chapters in a logical manner, for example, according to Baltimore classification. My 20 years of teaching experience convinced me that the chapters for DNA viruses, which are more familiar to students, should appear before the chapters for RNA viruses. Less familiar reverse transcribing viruses come afterward. Other related viruses are described in Part V including *Viral vectors*, *Subviral agents*, and *New emerging viruses*. Finally, Part VI Viruses and Disease features medically related content, such as *HIV and AIDS*, *Vaccines*, and *Antivirals*.

Virology as a discipline is inherently diverse and cannot be readily contained in a single volume format. This apparent "mission impossible" can be bravely accomplished only by a virologist, who has a limited knowledge on diverse virus families, except for the ones that he has encountered in his career. Being a single-authored book, the book has a uniform organization of its individual chapters. Consistency throughout the book is an important virtue that students might appreciate most. In each chapter, the viral life cycle is narratively written in a consistent manner, in order of classification, virion structure, genome structure, viral proteins, life cycle, and effects on host. In particular, throughout the text, an emphasis is borne on the genome replication and virus–host interaction. The writing style is engaging, but narrative without compromising rigor. Lastly, the book is heavily illustrated. This reflects my conviction that students are visual learners.

MAIN FEATURES

To limit its volume to less than 500 pages, which is suitable for a one-semester undergraduate course, the main text is written in a brief tone. To make up the conciseness of the main text, special features are included to enrich students' learning. These features include boxes, perspectives, summaries, study questions, suggested readings, and journal club.

- *Boxes*: The boxes are to provide interesting side topics. This part covers information that is relevant but may be inappropriate in the main text. Some of them contain fundamental knowledge on molecular and cellular biology that is relevant to the chapter. Others contain more detailed information that would be of interest to some advanced students. We owe our knowledge to the accomplishments made by the leading scientists who at times showed exceptional insights. The stories of some of these legendary virologists are recounted in the boxes. These boxes could be skipped without compromising the readability of the main text or they could be read separately.
- *Perspectives*: The major advances in the field over decades are highlighted. I have attempted to point out the key questions that remain to be answered and the tasks that represent unmet medical needs and public health concerns. This feature is to provoke students intellectually to engage in scientific endeavors beyond the classroom.
- *Summary*: The main texts are summarized by five to seven short paragraphs with an emphasis on keywords. This part is to refresh what we have learned in the text in a brief tone.
- *Study Questions*: This is to help students evaluate what they have learned in the chapter and to learn more about the viruses in an experimental setting. To be concise and consistent, only two to three questions are included. These features will expose students to scientific inquiries.
- *Suggested Readings*: Recent articles that made primary discoveries as well as review articles to provide an overview are listed. For brevity, only five papers in each chapter are included.
- *Journal Club*: Journal Club makes this book appropriate for a graduate course. One recent article that reflects the recent progress is carefully chosen and the highlight of the article is described.

The intended readers of this book are undergraduate and graduate students majoring in life sciences. As the book mainly deals with human pathogenic viruses, it is suitable for medical students as well. After studying virology with this book, it is hoped that students are inspired by the "intellectual challenge" posed by viruses and become more interested in virology.

Acknowledgments

Writing a textbook involves many people behind the scenes. My special thanks goes to the crew at Life Science Publishing Company in Korea, who did almost all the illustrations on my behalf. The Graphic staffs led by Mr Yoon painstakingly responded to my demanding requests.

The book greatly benefits from internet resources such as Wikipedia, Viralzone, and Public Health Image Library/CDC (PHIL). Numerous photos/images from Wikipedia and PHIL that are freely available as public domain are used in this book. In addition, many illustrations of virus particles and diagrams of the virus genome available in Viralzone are used with permission. Without their generosity, this book would have not been completed.

Peer evaluation was critical in revising and improving the accuracy of the book. I am grateful to these colleagues for their time and expertise and for their insights. My special thanks goes to Drs Young-Min Lee (Utah State University), Hans Netter (Monash University), Marc Windisch (Institute Pasteur Korea), Seungtaek Kim (Yonsei University College of Medicine), Eui-Chul Shin (KAIST), Choongho Lee (Dong-Kuk University), Byung-Yoon Ahn (Korea University), Soon-Bong Hwang (Hallym University), Jin-Hyun Ahn (Sungkyunkwan University), Kyun-Hwan Kim (Kon-Kuk University), Sung-Key Jang (POSTECH), Sang-Joon Ha (Yonsei University), Ji-Young Min (Institute Pasteur Korea), Myung-Kyun Shin (University of Wisconsin-Madison), Jong-Hwa Kim (Samsung Medical Center), and Carroll Brooks (Yonsei University).

I am greatly indebted to Dr Janet Mertz, a supervisor of my graduate study at University of Wisconsin-Madison, and to Dr John Taylor, a supervisor of postdoctoral training at Fox Chase Cancer Center-Philadelphia. During my time there, I was deeply inspired by their professional integrity as well as their intellectual discipline. Without their continuing encouragement, this book would not be completed. Lastly, I am also grateful to Dr Howard Temin (University of Wisconsin-Madison), who led me to the world of virology via his engaging lecture at the University of Wisconsin-Madison. Last, but not the least, I should admit that his lecture note served as a keystone of this book.

Part I

Principles

A virus is an "obligate intracellular parasite." A virus can reproduce only inside host organisms. Viruses can be found in all living organisms on Earth, ranging from bacteria, fungi, and amoeba, to plants and animals. Despite the diversity of host organisms, there are common principles that are shared by diverse viruses. In Part I, the principles that underlie diverse viruses are described. Chapter "Discovery and Classification" covers the discovery and classification of viruses. In chapter "Virus Structure," the structural features of viral capsids and the principles of capsid assembly are covered. The principles of the viral life cycle are discussed in chapter "Virus Life Cycle." The methods used for viral diagnostics and virus research are covered in chapter "Diagnosis and Methods." Finally, the host immune response to viral infection is covered in chapter "Host Immune Response."

Chapter 1

Discovery and Classification

Chapter Outline

The earliest discoveries of viruses and how they have been woven into the history of virology will be recounted first. Then, the elementary facts on virus including the definition and classification will be described. Although this book mainly deals with human viruses, viruses are found in almost all living organisms on this planet and we will take a glimpse of diverse viruses found in other living organisms. Finally, the subviral agents such as prions and viroids will be covered.

1.1 DISCOVERY OF VIRUS

1.1.1 Virus in Ancient History

How long ago did human viruses first appear on Earth? Although human beings are believed to have originated about 3–4 million years ago, the oldest record of virus in history was found only 4000 years ago in ancient Egypt (Fig. 1.1). The victim of poliovirus was inscribed in a stele. In addition, evidence of *smallpox*[1] was found in Egyptian mummies. The earliest physical evidence of it is probably the pustular rash found on the mummified body of Pharaoh Ramses V of Egypt (1149–1145 BC). Smallpox was described in the literature of ancient China (700 BC) as well. It is believed that the collapse of the Inca and Aztec cultures in South America can be attributed to smallpox and *measles*[2] that were brought by European explorers. After all, viruses have greatly influenced the fates of ancient cultures extinguished in human history.

1.1.2 Discovery of Virus in Plants and Animals

By the end of 19th century, during a time when all the transmissible agents were believed to be microbes, the existence of transmissible agents, which were smaller than a microbe, had begun to be perceived. In 1892, Dimitri Ivanowski, a Russian scientist, reported an unexpected observation during his study on the Tobacco mosaic disease of a plant (Fig. 1.2). He found that the filtrates of the transmissible agent caused the disease. He modestly stated that "according to my experiments, the filtered extract introduced into healthy plants produces the symptoms of the disease just as surely as does the unfiltered sap." In the year 1898, Martinus Beijerinck, a Dutch scientist, independently made similar observations in his studies on Tobacco mosaic disease of a plant. Further, he speculated that the pathogen exists only in

1. **Smallpox** A fatal infectious disease of human that is caused by a pox virus.

2. **Measles** Measles is an infection of the respiratory system caused by a virus belonging to family *Paramyxoviridae* (see chapter: Other Negative-Strand RNA Viruses).

Molecular Virology of Human Pathogenic Viruses. DOI: http://dx.doi.org/10.1016/B978-0-12-800838-6.00001-1

FIGURE 1.1 The historical record of virus. The oldest record of a virus was found in a stele from 13th-century BC Egypt. A man (priest) standing with a stick is believed to be a victim of poliomyelitis.

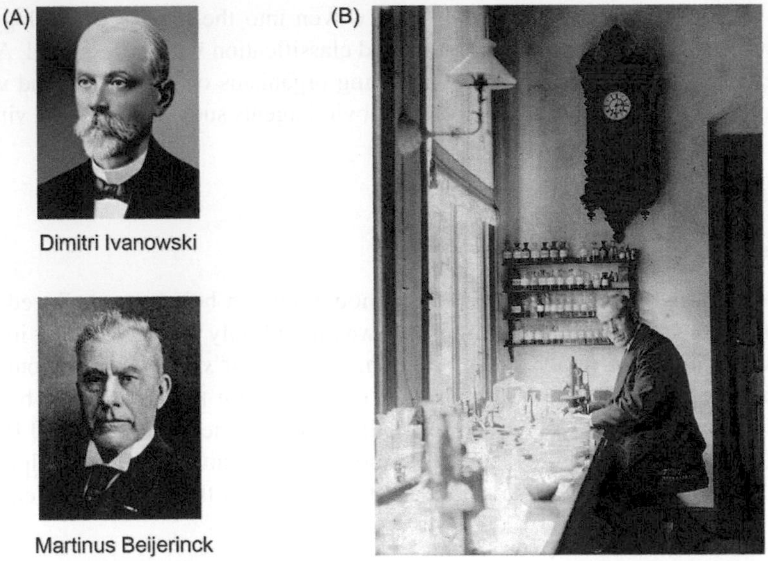

FIGURE 1.2 The photos of two pioneers, who discovered 'virus' as a filterable agent. (A) Dimitri Ivanowski (1864–1920), a Russian botanist, the first man to discover virus in 1892 and thus one of the founders of virology (above). In 1898, Martinus Beijerinck (1851–1931), independently reproduced Ivanowski's filtration experiments and then showed that the infectious agent was able to reproduce and multiply in the host cells of the tobacco plant (below). (B) A photo that captured Martinus Beijerinck in his laboratory in 1921. Martinus Beijerinck coined "virus" to articulate the nonbacterial nature of the causal agent of Tobacco mosaic disease.

living tissues. He named the new pathogen *virus*[3] to highlight its nonbacterial nature. Importantly, he articulated two experimental definitions of viruses as the following: the ability to pass through a porcelain filter, and the need for living cells on which to grow. Subsequently, a similar observation was made of an animal virus as well. Loeffler and Frosch, German scientists, found a filterable agent in their studies on *foot-and-mouth disease*[4] in cows in 1897.

---

3. **Virus** The word is from the Latin *virus* referring to poison.
4. **Foot-and-mouth disease** Animal disease that is caused by foot-and-mouth disease virus (FMDV), which belongs to family *Picornaviridae*.

Although it was clear that viruses are small entities, significantly smaller than bacteria, the physical identity of viruses remained unclear until the virus particle was crystallized. In the 1930s, Wendel Stanley successfully made crystals of Tobacco mosaic virus (TMV), a finding that implicated that the virus particle constitutes a simple structure, parallel to proteins, because crystallization can be achievable only from molecules or particles having a simple structure. Accordingly, he then speculated that TMV is principally composed of protein only.

1.1.3 Discovery of the Human Viruses

As stated above, viruses were discovered from plants and animals just before the turn of the 20th century. Nonetheless, a human virus had not yet been discovered. As a matter of fact, many human pathogenic microbes had been discovered since 1884, when *Koch's postulates*[5] for identification of the agent responsible for a specific disease prevailed. Many attempts were made to search for the cause of scourges that considerably threatened human life, including yellow fever, *rabies*[6], and poliomyelitis. An etiological agent for these three plagues was discovered soon after the turn of the 20th century. Historical accounts for the discovery of the culprits are revealing, as the following.

Yellow fever was one whose etiology was uncovered first among the human pathogenic viruses (Fig. 1.3). The work of Loeffler and Frosch on animal viruses encouraged Walter Reed and his colleagues, who were working on yellow fever, a terrifying human disease. Walter Reed, who led the U.S. Army Yellow Fever Commission residing in Cuba, was able to demonstrate that an inoculum from an infected individual can infect healthy volunteers even after filtration. It was the moment of discovery of the yellow fever virus (YFV) in 1902—the first human virus ever isolated. In fact, the discovery of YFV involved human volunteers including colleagues of Walter Reed, some of whom unfortunately succumbed to the virus they discovered. In retrospect, the lack of animal models for transmission, which was established almost three decades later, subjected human volunteers to an experimental infection.

A year after the discovery of YFV, an etiologic agent for rabies was discovered (see Fig. 1.3). Rabies spread in the 1880s in Europe as dogs became a popular pet animal. Louis Pasteur had earlier used rabbits for transmission of rabies for the development of rabies vaccine (see Box 25.1). Therefore, unlike yellow fever, an animal model for transmission was already available before the discovery of the etiologic agent. Nonetheless, Louis Pasteur failed to isolate and cultivate a rabies microbe in a media, in which bacterium are expected to grow. Almost two decades later in 1903, Paul Remlinger demonstrated that the filtrates of rabies inoculum transmit the disease to animals. Thus, rabies virus, a filterable agent that causes rabies in animals, was discovered. In 1908, Karl Landsteiner, an Austrian scientist, reported the

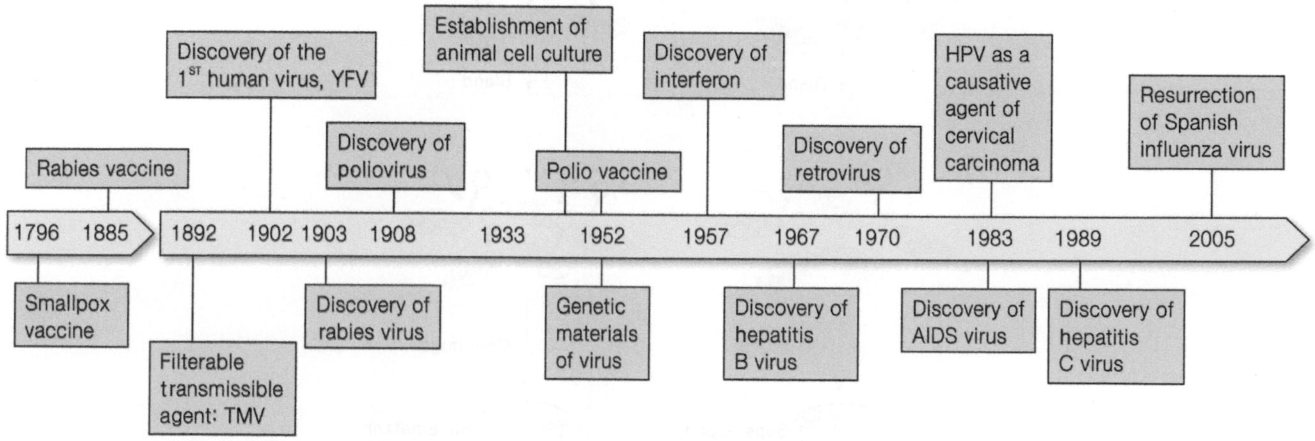

FIGURE 1.3 Chronicle of the major discoveries in virus research. The major milestones in virus research are shown. Small pox vaccine (by Edward Jenner) and rabies vaccine (by Louis Pasteur) were developed even prior to the discovery of virus by Ivanowski in 1892, as demarcated by a separate arrow bar. In the first decade of the 20th century, three human viruses (YFV, Rabies virus, and Poliovirus) were discovered. YFV, yellow fever virus; HPV, human papillomavirus.

5. **Koch's postulates** It refers to a set of criteria that Koch developed to establish a causative relationship between a microbe and a disease. Koch applied the postulates to describe the etiology of cholera and tuberculosis.

6. **Rabies** It refers to a viral disease that causes acute encephalitis in animals. Rabies virus belongs to a family *Rhabdoviridae* (see chapter: Rhabdovirus). The rabies is a Latin word for "madness."

transmission of poliomyelitis to monkeys, revealing the etiological agent of poliomyelitis. Overall, during the first decade of the 20th century, the etiological agents of three major viral scourages, YFV, rabies virus, and poliovirus, were discovered (see Fig. 1.3).

1.1.4 Discovery of Genetic Materials

One big question that remained unanswered until the 1950s was what is the genetic material of life? A prevailing view was that protein rather than nucleic acid was more likely to be the genetic materials responsible for inheritance due to its higher diversity (ie, 20 amino acids versus 4 bases). In fact, the experiment that proved nucleic acids to be the genetic material was carried out by an experiment using *bacteriophages*[7] (ie, a virus of bacteria). Bacteriophages, which propagate rapidly in bacterial hosts, became a favorite experimental subject during 1940–1960. In 1952, when the identity of the genetic material was still in debate, Hershey clearly demonstrated that the nucleic acid component of bacteriophage T2 was the genetic material (Fig. 1.4). In his experiment, nucleic acids and proteins were radiolabeled distinctively either with ^{32}P and ^{35}S, respectively. Such a prepared T2 phage was used to infect a bacterium, and examined to see whether ^{32}P or ^{35}S was detectable in the cell pellet after centrifugation. Since T2 phage was known to inject the genetic material into the host cell, what is detected in the cell pellet represents the genetic materials. Indeed, ^{32}P (nucleic acid) was abundantly detected in the pellet, while ^{35}S (protein) was detected in the supernatant, revealing that

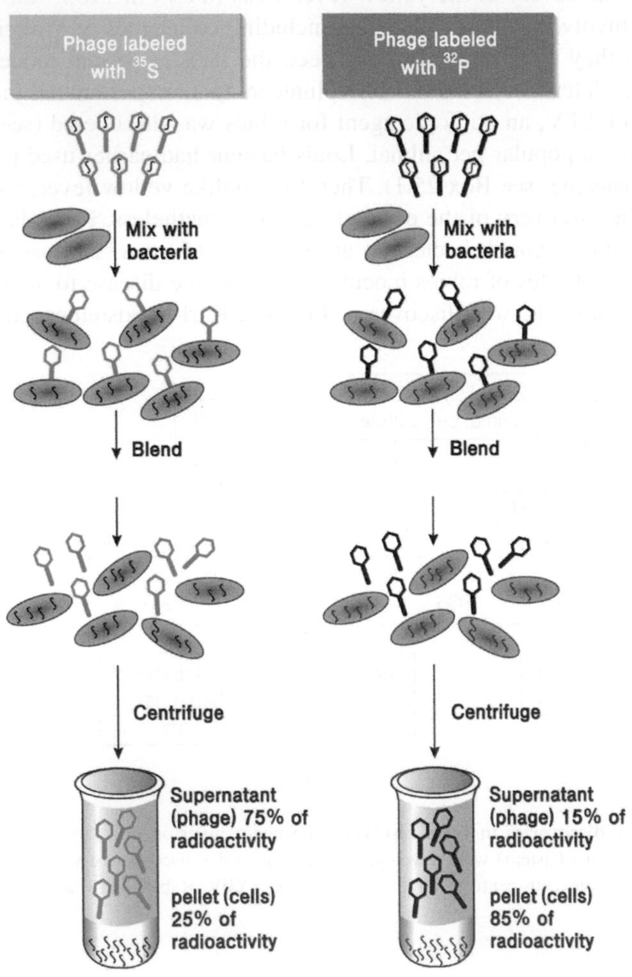

FIGURE 1.4 A seminal experiment, which demonstrated that DNA is the genetic material. T2 phage, that was propagated either in the presence of ^{32}P (purple) or ^{35}S (green), was used to infect *Escherichia coli*, a host. The mixture was subjected to blending to detach the phage from the host cells. Following centrifugation, the amount of ^{32}P and ^{35}S was measured from cell pellet (*E. coli*) and supernatants (T2 phage).

7. **Bacteriophage** A bacteriophage (informally, *phage*) is a virus that infects and replicates within bacteria. The term is derived from Greek word *phagein* "to eat."

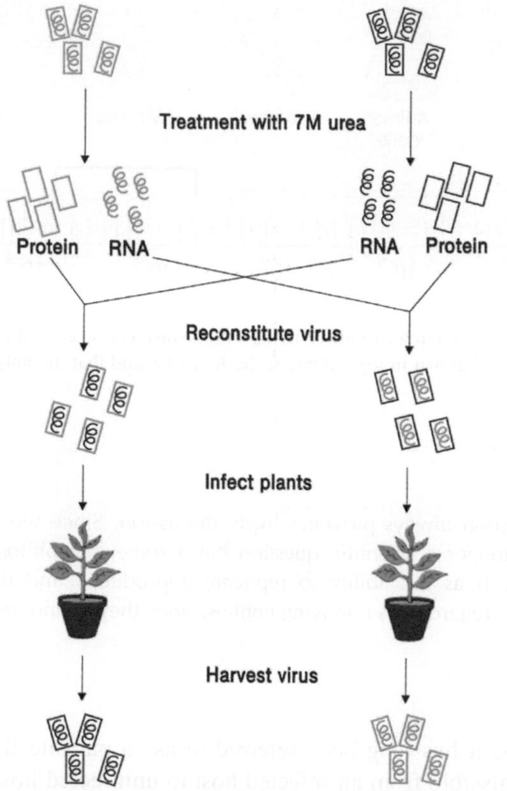

FIGURE 1.5 A reconstitution experiment, which proved that RNA is the genetic material of TMV. This experiment exploited the fact that TMV can be reconstituted in vitro, following fractionation into proteins and RNA by 7M urea treatment. Two strains of TMV were colored differently by either green or purple.

the nucleic acids are the genetic materials. This work represents a seminal discovery that clearly proved that "DNA is the genetic material" (see Fig. 1.3).

A few years later, in 1957, Fraenkel-Conrat and Singer confirmed that the nucleic acid (RNA) is the genetic material by an experiment using TMV. In their experiment, they explored the fact that TMV can be reconstituted in vitro by mixing proteins and RNA components following fractionation of TMV particles (Fig. 1.5). Briefly, two strains of TMV, that exhibits distinct lesions on the leaf, were employed. After separating into proteins and RNA, two hybrid viruses were reconstituted by mixing two components. These hybrid viruses were used to infect the host plant. Remarkably, the phenotype of the reconstituted hybrid viruses was determined by the RNA, but not the proteins. This reconstitution experiment confirmed that "nucleic acid (RNA) is the genetic material."

As stated above, RNA as well as DNA was discovered to be the genetic material of viruses.

1.2 DEFINITION OF VIRUS

What are the defining features of a "virus"? As stated above, unlike the other pathogens known in those days, a "virus" is a filterable transmissible agent. In addition, it is submicroscopic and the physical size of the most of animal viruses ranges from 30 to 300 nm in diameter (Fig. 1.6). Hence, a virus can be aptly said to be a "nanoparticle" in nature. On the other hand, a "virus" can be viewed as a molecular complex constituted of the nucleic acids and the protein shells that encompass the nucleic acids. The nucleic acids can be either RNA or DNA. Although all living organisms on the globe have a DNA genome, viruses are the only organisms, if you may, that still employs RNA as genome. The role of protein shells, also called capsids, is to protect the viral genome from biochemical damage. In addition, some viruses have an envelope (ie, a lipid bilayer) that coats the capsids.

What are the common features that are shared by viruses? Because viruses are found in almost all living organisms on earth, their biological properties should be as diverse as their host organisms. Nonetheless, five common features of viruses, regardless of their hosts, are perceived. A defining feature is that a "virus" replicates only inside living cells. In other words, a "virus" is simply a physical entity outside of cells, as it cannot reproduce outside of cells. More precisely speaking, what a virus really belongs to is an interface between living and nonliving matter (Box 1.1). Since a virus

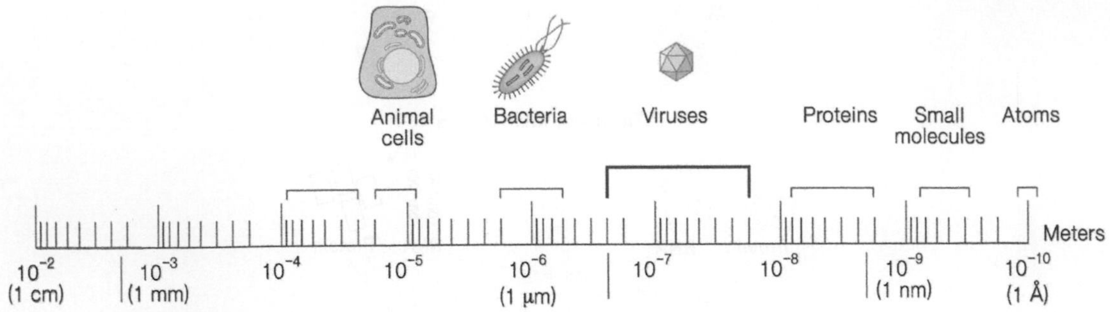

FIGURE 1.6 Comparison of the size of virus particles to other living organisms. The size of most animal viruses ranges from 30 to 300 nm in diameter, as marked by red bracket. The ruler is drawn in logarithmic scale. It can be said that the animal viruses are about 100–1000 times smaller than their host cell.

BOX 1.1 Are Viruses Alive?

"Are viruses alive or not?" This question always provokes lively discussion. Since we are now well aware of the molecular details of virus, this question is no longer a scientific question but a somewhat philosophical question. Viruses share many features with a living organisms, such as the ability to replicate (reproduce), and the possession of inheritable materials (genome). Nonetheless, viruses can be regarded as nonliving entities, since they cannot replicate outside of host cells.

can propagate only inside host cells, it has long been referred to as "a parasite living in cells." Second, a virus is "an infectious agent," in that it is transmissible from an infected host to uninfected hosts. Third, a virus propagates itself via assembly. In other words, the assembly of its components in infected cell is the way of multiplication, not by division as per cells. Fourth, a virus could rapidly cope with environmental changes (eg, host cell, drug, and antibodies), a property that is attributable to its higher mutation rate (see Box 1.3). Fifth, a virus is an unique organism that delivers its genome to the host cells via the process called "infection." This special feature of viruses is exploited as vehicles in gene therapy (see chapter: Virus Vectors).

The vast majority of viruses are associated with diseases, because they were discovered as etiological agents for infectious diseases, such as yellow fever, rabies, and poliomyelitis. One might wonder whether the disease-causing properties are common property of all viruses. This is not the case. There are some viruses that are nonpathogenic to their hosts, for example, adeno-associated virus (AAV) is not pathogenic to its host, human (see chapter: Other DNA Viruses). A widely accepted view is that the aim of virus evolution is not to cause disease in its host, but is to maximize its spread (see Box PVI.1).

1.3 ADVANCES IN VIROLOGY

Let's consider the major discoveries made in the past century, including the early discoveries of viruses (see Fig. 1.3). In fact, it was Louis Pasteur, who first started virus research in the laboratory setting. He successfully developed a rabies vaccine, presuming the microbial cause in 1885, which is even before the virus was officially discovered as a filterable agent. Following the discovery of TMV as a filterable agent, the first human virus, YFV, was discovered. Soon after, rabies virus was discovered as a cause of rabies, and poliovirus was discovered as a cause of poliomyelitis. It is worth noting that these discoveries were made even before DNA was discovered to be the genetic material in 1952 (see Fig. 1.3).

Until 1950s, animal viruses could not be experimentally studied in a laboratory due to the lack of animal cell culture. Instead, bacteriophages, a virus of bacteria, had greatly advanced "virology" as a discipline of experimental science, primarily because bacteriophages can be readily propagated in bacterial culture (see Fig. 1.4). A breakthrough of that advanced animal virus research was the successful establishment of animal cell culture in the early 1930s (see Fig. 1.3). Needless to say, animal cell culture was instrumental for the development of the poliovirus vaccine by Salk in 1956 (see Box 25.2), which saved thousands of lives. A cell line derived from monkey kidney was used to propagate poliovirus.

Another breakthrough that greatly contributed to the advance of the discipline of virology was the advance of recombinant DNA technology that was established in the early 1970s. Recombinant DNA technology has drastically changed the way of studying viruses from classical experimental science to modern biology. It was exemplified by the discovery of the retrovirus in 1970 by Howard Temin and David Baltimore that led down the road to the molecular era (see Fig. 1.3 and Box 17.1). Undoubtedly, the discovery of retrovirus was the cornerstone for the discovery of the AIDS virus, HIV, in 1983.

As stated above, virology, as a discipline that studies the diverse aspects of viral infection of host cells and its consequence, became established during the early 20th century. The discipline of virology can be divided into a few subdisciplines such as viral epidemiology, clinical virology, viral immunology, and molecular virology. Viral epidemiology investigates the mode of viral transmission, and the risk factors for disease. Clinical virology develops the diagnostic methods for detecting viral infection. Viral immunology studies the consequence of host immune response to viral infection. Molecular virology studies the molecular mechanism of viral replication in the context of virus life cycle. In fact, the division into four subdisciplines is somewhat vague, and the breadths of each subdiscipline inevitably overlap. This book, as the title implies, is inclined toward molecular virology.

What are the aims of virus research? One of the important reasons is to gain knowledge that is instrumental in controlling viral diseases. As stated above, viruses have made a great impact on human life throughout history. The Spanish flu pandemic is the best example that clearly shows the magnitude of the impact of viral diseases on human life (Fig. 1.7). It was one of the deadliest natural disasters in human history, which killed 30–50 million people between 1918 and 1919. The HIV epidemic is another example of a viral disease that has significantly impacted our life. It has killed more than 30 million people in the past three decades. Thanks to intensive research, the HIV epidemic is more or less under control, at least in the Western hemisphere (see chapter: HIV and AIDS).

The second aim of virus research is to exploit viruses as tools for academic research. In the early period of molecular biology in the 1970s, the virus was a favorite experimental model, because it is easier to manipulate in laboratory due to its small genome size. Consequently, many important findings on eukaryotic molecular biology were made by using virus, such as the discovery of introns, enhancers, and so on. Moreover, oncogenes and tumor suppressor genes were mainly discovered through studies on RNA tumor viruses and DNA tumor viruses, respectively (see chapter: Tumor Viruses). In fact, the field of molecular biology itself owes much to the earlier discoveries made by using animal viruses as experimental models.

Viruses are also of interest to pharmaceutical industries. Diagnostics, vaccines, and therapeutic drugs are just few kinds of products that drug industries are interested in that are related to virus research. We will see how these products are made by pharmaceutical industries and clinically used in chapter "Diagnosis and Methods" on viral diagnostics, in chapter "Vaccines" on vaccines, and in chapter "Antiviral Therapy" on antiviral therapy.

Lastly, viruses are often exploited as gene delivery vehicles or gene therapy vectors (see chapter: Virus Vectors). One outstanding feature of viruses is that they have the ability to deliver their genome to target cells. A few animal

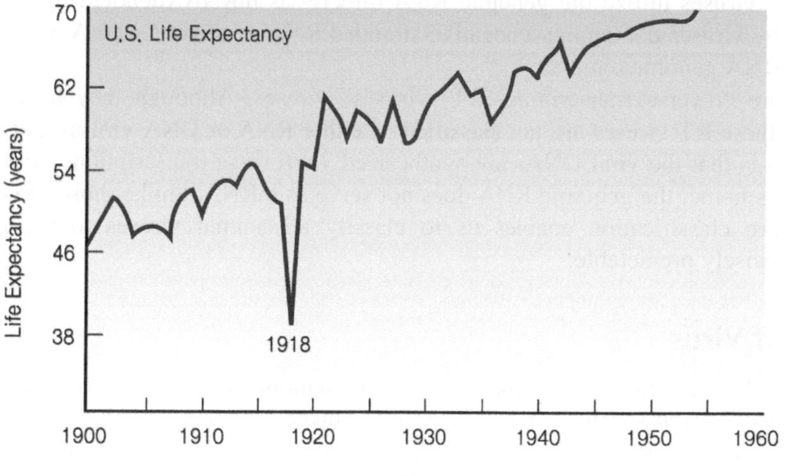

FIGURE 1.7 The impact of viral diseases on the human life expectancy. Life expectancy gradually increased during the past century in U.S.A. Spanish flu during 1918–1919 at the end of World War I was the only episode that adversely affected the incremental increase of life expectancy in the past century.

viruses, such as retrovirus and adenovirus, have been extensively developed for therapeutic purpose. Therapeutic utilization of otherwise pathogenic viruses is a wise strategy to "exploit an enemy to conquer an enemy", a quote from an ancient Chinese literature, although therapeutic application is not yet in practice.

1.4 CLASSIFICATION OF VIRUSES

In the early days, viruses were discovered as the etiological agents of the disease they caused. Then, viruses were named after, and often classified based on, the diseases that they caused, such as yellow fever virus, and rabies virus. Since the advent of molecular technologies, the genome-based classification was established. The nucleotide sequence relatedness allows a more precise classification of virus species. Furthermore, the genome-based classification allows to predict the mode of viral genome replication. For instance, SARS virus, a new emerging virus in 2003, was immediately identified as a new member of the coronavirus family by the nucleotide sequence analysis.

1.4.1 Genome-Based Classification

The nature of the nucleic acids in the genome is the criteria of the genome-based classification. Viruses that have RNA as a genome are called "RNA viruses," whereas viruses that have DNA as a genome are called "DNA viruses" (Fig. 1.8). An exception to this rule comprises viruses that replicate via reverse transcription, which are grouped separately as "reverse transcribing (RT) viruses," regardless of whether the genome is RNA or DNA. As a result, animal viruses can be largely classified into three groups based on the nucleic acids of the virus genome (Table 1.1).

Animal viruses can be further classified into seven groups, based on the genome features, which pertain to the mode of genome replication (Fig. 1.9). In addition to the nucleic acids species (whether DNA or RNA), whether it is a positive-strand or negative-strand (ie, *polarity*[8]) and whether it is a single-strand or double-strand are considered. This genome-based classification is also called the "Baltimore Classification," named after the prominent scientist who envisioned the genome-based classification (Box 1.2).

According to the Baltimore classification, animal viruses are subdivided into seven groups: DNA viruses (Group I and II), RNA viruses (Group III, IV, and V), and RT viruses (Group VI and VII). The schematic diagram in Fig. 1.9 illustrates how viruses in each group differently synthesize their mRNAs. After all, distinct mRNA transcription strategies represent the hallmark of each virus group. Group I is represented by viruses containing a double-stranded DNA genome. Group I viruses synthesize mRNA by transcription from the DNA genome template. Group II is represented by viruses containing a single-stranded DNA genome. Group II viruses first convert their single-stranded DNA genome to double-stranded DNA, which is then used as a template for mRNA transcription. Group III is represented by viruses containing a double-stranded RNA genome. Group III viruses synthesize mRNA by transcription from their double-stranded RNA template. Group IV is represented by viruses containing a positive-stranded RNA genome. Group IV viruses utilize the genomic RNA directly as mRNA (denoted by dotted lines in the figure). Group V is represented by viruses containing a negative-stranded RNA genome. Group V viruses synthesize mRNA by transcription from their RNA genome template.

Group VI and VII are "reverse transcribing (RT) viruses" viruses. Although they have either RNA or double-stranded DNA genome, these RT viruses are not classified as either RNA or DNA viruses. An important feature that is shared by the RT viruses is that the viral DNAs are synthesized via reverse transcription. Note that although Group VI viruses contain an RNA genome, the genomic RNA does not serve as mRNA, unlike those of Group IV.

Overall, the Baltimore classification enables us to classify all animal viruses to the extent that the genome replication strategy is precisely predictable.

1.4.2 Taxonomy of Virus

Viral species are officially classified and named by an international committee, the International Committee on Taxonomy of Virus (ICTV)[9]. Viral species can be placed in a ranked hierarchy, starting with orders, which are divided into families, then genera (singular: genus), and then species (singular: species) (Table 1.2). Species can be further

8. **Polarity** It is a terminology used in referring to one of two strands in nucleic acids. The strand that is the same sense as mRNA is called "positive-strand," while the complement is called "negative-strand," by definition.

9. **International Committee on Taxonomy of Virus (ICTV)** http://www.ictvdb.org/.

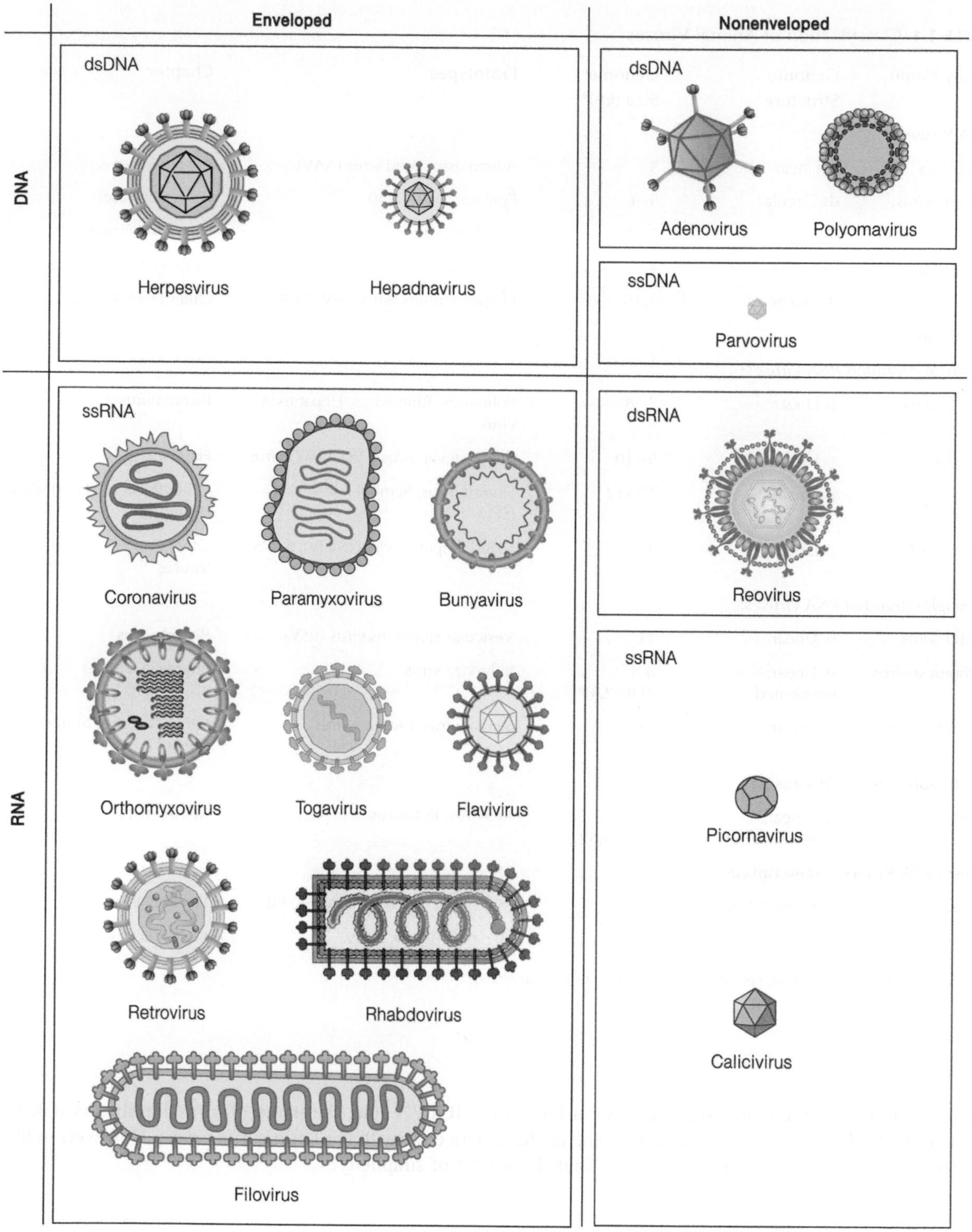

FIGURE 1.8 Classification of viruses. Animal viruses can be classified by the nature of the nucleic genome (ie, DNA or RNA). In addition, viruses are often called by their morphological features. For instance, according to the existence of envelope, it can be called either an enveloped virus or naked (nonenveloped virus). Family names are given below. *dsDNA*, double-strand DNA; *ssDNA*, single-strand DNA; *dsRNA*, double-strand RNA; *ssRNA*, single-strand RNA.

TABLE 1.1 Classification of Animal Viruses

Family Name	Genome Structure	Genome Size (kb)*	Prototypes	Chapter
DNA Viruses				
Parvovirus	ss Linear	5	Adeno-associated virus (AAV)	Adenoviruses
Polyomavirus	ds Circular	5–6	Polyomavirus, SV40	Polyomaviruses
Papillomavirus	ds Circular	7–8	Papillomavirus	Papillomaviruses
Adenovirus	ds Linear	42	Adeonovirus	Herpesviruses
Herpesvirus	ds Linear	120–200	Herpes simplex virus, EBV, VZV	Other DNA Viruses
RNA Viruses				
(+) Single-Stranded RNA Viruses:				
Picornavirus	ss Linear	7–8	Poliovirus, Rhinovirus, Hepatitis A virus	Picornaviruses
Flavivirus	ss Linear	9–10	Yellow fever virus, Hepatitis C virus	Flaviviruses
Togavirus	ss Linear	10–12	Sindbis virus, Semliki Forest virus (SFV)	Other Positive-Strand RNA Viruses
Coronavirus	ss Linear	30	Mouse hepatitis virus (MHV), SARS	Other Positive-Strand RNA Viruses
(–) Single-Stranded RNA Viruses:				
Rhabdovirus	ss Linear	11	Vesicular stomatitis virus (VSV)	Rhabdovirus
Orthomyxovirus	ss Linear, segmented	8 × (0.8–2.3)*	Influenza virus	Influenza Viruses
Paramyxovirus	ss Linear	15	Sendai virus, Measles virus	Other Negative Strand-RNA Viruses
Double-Stranded RNA Viruses:				
Reovirus	ds Linear, segmented	10 × (1.2–3.8)*	Reovirus, Rotavirus	Retroviruses
Viruses with Reverse Transcriptase				
Retrovirus	ss Linear RNA	8–10	Murine leukemia virus (MLV), HIV	Hepadnaviruses
Hepadnavirus	Circular, ds DNA	3.2	Hepatitis B virus (HBV)	Virus Vectors

**The range of genome sizes of segment genome are denoted within parenthesis.*

subdivided into "genotype" or "subtype." According to the ICTV's rule, the name of virus families is italicized and ends with the Latin suffix *-viridae*, and the name of the genera ends with the Latin suffix *-virus*. However, in this book, virus family names are often referred in plain English for sake of simplicity.

1.4.3 Mutation and Evolution

All living organisms on earth employ DNA as a genome. In contrast, viruses employ either DNA or RNA as a genome. Having a more flexible RNA molecule as a genome, RNA viruses exhibit a higher mutation rate (Box 1.3). Consequently, RNA viruses can mutate or evolve rapidly upon antiviral stresses, such as host immune response and treatment of antiviral drugs. Not surprisingly, the majority of newly emerging viruses are RNA viruses (see chapter: New Emerging Viruses).

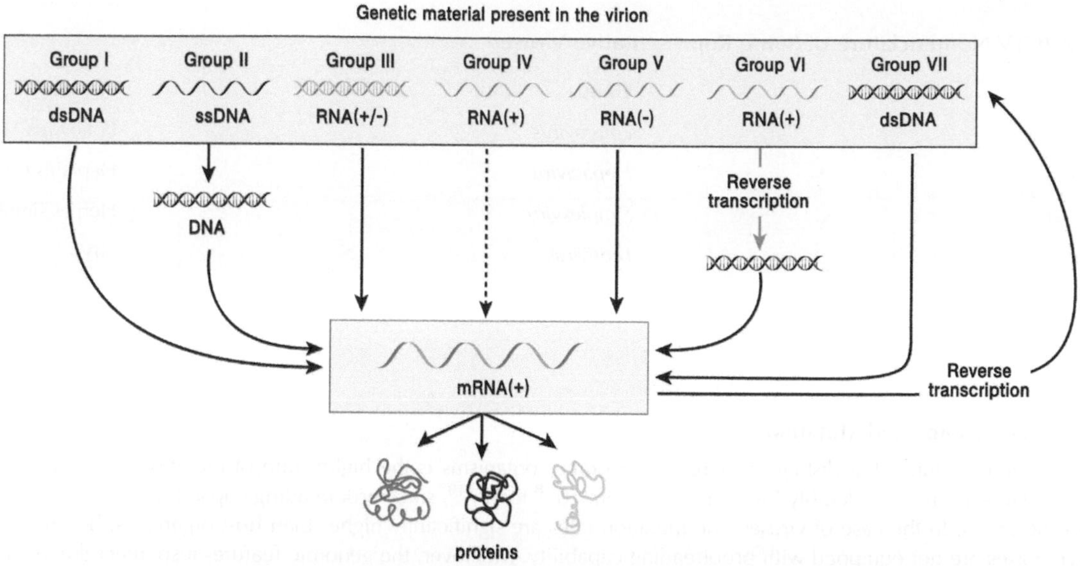

FIGURE 1.9 Classification of viruses based on the genome structure. Animal viruses are divided into three groups: DNA viruses (Group I and II), RNA viruses (Group III, IV, and V), and RT viruses (Group VI and VII). The box on the top contains the viral genome structure in each group, highlighting the nature of nucleic acids (DNA or RNA), the strandness (*ss*: single-strand or *ds*: double-strand), and the polarity (plus or minus). The relationship of the genome to mRNA is indicated either by solid line (transcription) or dotted line (no transcription). *RT*: reverse transcription.

BOX 1.2 Baltimore Classification

David Baltimore was the one who proposed the genome-based classification of animal viruses. It was a truly visionary insight that foretold the era of molecular virology. Undoubtedly, the Baltimore classification serves as a gold standard of virus classification. In addition to his contribution as a virologist, Baltimore has profoundly influenced international science, including key contributions to immunology, virology, cancer research, biotechnology, and recombinant DNA research, through his accomplishments as a researcher, administrator, educator, and public advocate for science and engineering. He shared the Nobel Prize in 1975 with Howard Temin for the discovery of reverse transcriptase (see Box 17.1) at the age of only 37. Even afterward, he continued to be productive as a research scientist not only in virology but also in the field of immunology. He is also well known as the discoverer of transcription factor NF-kB. In addition, his contribution extended to his role as a university administrator. He has served as the president of California Institute of Technology (Caltech) from 1997 to 2006. He is currently the President Emeritus and Robert Andrews Millikan Professor of Biology at Caltech.

Photo of David Baltimore (1938–).

1.5 VIRUSES IN OTHER ORGANISMS

As described earlier, this book deals only with human viruses. In fact, viruses are discovered in almost all organisms on earth, including animals, plants, insects, amoeba, plankton, and bacteria. Here, it is worth noting what kinds of viruses are found in such diverse organisms. Table 1.3 includes only one or two viral species per organism for brevity, which is only the tip of the iceberg. For instance, plant viruses are found in many agricultural products, such as rice, corns, potato, and tobacco (Fig. 1.10). Pathogenic plant viruses damage the crops, resulting in significant economic loss in the farming industry. Interestingly, viruses do not always cause disease in host organisms. For instance, tulip breaking virus

TABLE 1.2 ICTV Nomenclature of Some Representative Viruses

Family	Genus	Species
Picornaviridae	*Enterovirus*	Poliovirus 1
Flaviviridae	*Hepacivirus*	Hepatitis C virus
Herpesviridae	*Simplexvirus*	Herpes simplex virus 1
Retroviridae	*Lentivirus*	HIV

BOX 1.3 Viral Genome and Mutation

One of the salient features that distinguish viruses from other organisms is the higher rate of mutation. The mutation rates of eukaryotic organisms are considerably lower, ranging from 10^{-8} to 10^{-10}, since proofreading capability (10^{-3}) takes largely care of most of the errors. In the case of viruses, the mutation rates are significantly higher than host organisms, because viral DNA/RNA polymerases are not equipped with proofreading capability. Moreover, the genomic features also affect the mutation rates. For instance, the mutation rates of RNA viruses (10^{-3} to 10^{-5}) are significantly higher than those of DNA viruses (10^{-6} to 10^{-8}). In other words, the RNA virus with 10 kb genome size has at least one or more mutation per genome per replication cycle. RNA being more flexible than DNA, RNA viruses have intrinsically higher mutation rates. On the other hand, in the case of DNA virus, single-stranded DNA viruses have higher mutation rates than double-stranded DNA viruses. Note that the mutation rate of retroviruses are lower than that of single-stranded RNA viruses, but comparable to that of single-stranded DNA viruses.

What is the biological implication of having a higher mutation rate? A living organism has to cope with environmental changes via a process called "adaptation" or "evolution." Mutation is the driving force for evolution. In other words, mutations are not always harmful, but can be beneficial to an organism. Being random events, the vast majority of mutations lead to the loss of gene function. Some of the mutations, at least, can be advantageous for the virus to survive in challenging environments. The process of outgrowth of a virus having such advantageous mutations is termed "selection" or "adaptation." Therefore, the emergence of viral mutants (ie, variants) is the consequence of not only mutation but also selection. Selection constitutes an important concept in understanding viral evolution.

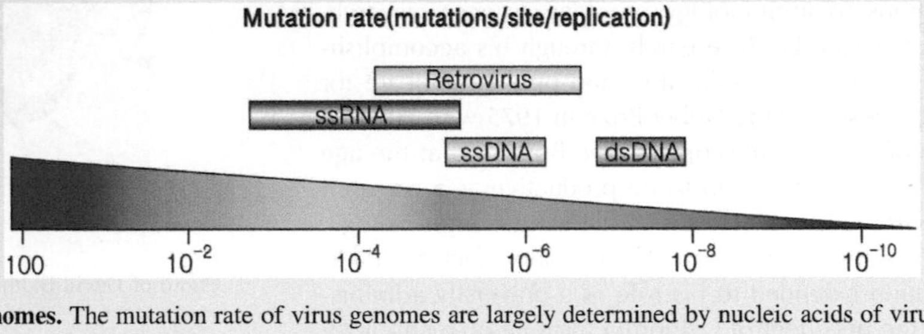

Mutation rate of virus genomes. The mutation rate of virus genomes are largely determined by nucleic acids of virus genomes: whether single- or double-stranded, and whether DNA or RNA.

does not cause pathogenic lesion on the host plant but leaves a beautiful stripe pattern on the flower that is appreciated by people (Box 1.4). Notably, many plant viruses constitute "families" that do not have a counterpart in animal viruses. For instance, TMV is classified in the "Virgavirus family," and cauliflower mosaic virus is classified in the "Caulimovirus family," neither of which have any animal members (Table 1.3).

In addition to higher eukaryotes, viruses are also found in unicellular organisms such as amoeba, yeasts, and bacteria. Recently, *giant viruses*[10], which are an extraordinary size (400 nm in diameter), were discovered in amoeba (Box 1.5). The genome size of "Mimivirus," the first giant virus discovered, is about 1200 kb, which is five times larger than any other known virus (ie, 230 kb of cytomegalovirus). It even rivals some bacteria in the genome size. After all, the discovery of giant viruses in amoeba makes the distinction between virus and organism blurred.

10. **Giant virus** A novel virus found in amoeba, which is bigger than any other known viruses (400 nm in diameter, 1200 kb in genome).

TABLE 1.3 Viruses in Other Organisms

Host Range		Major Virus	Infectious Disease	Family Name	Genome	Chapter
Vertebrates						
Primates	Chimpanzee	Simian immunodeficiency virus	AIDS-like	Retrovirus	ssRNA(RT)	Retroviruses
	Gorilla	Ebola virus	Fatal	Filovirus	(−)RNA	Other Negative-Strand RNA Viruses
	Monkeys	Simian virus 5	–	Paramyxovirus	(−)RNA	Other Negative-Strand RNA Viruses
Mammalians	Cows, Pigs	Foot-and-mouth disease virus	Blisters	Picornavirus	(+)RNA	Picornavirus
	Rabbits	Rabbit hemorrhagic disease virus	Fatal	Calicivirus	(+)RNA	Other Postive-strand RNA Viruses
	Cats	Feline parvovirus	Fatal	Parvovirus	ssDNA	Adenoviruses
	Dogs	Canine parvovirus	Diarrhea	Parvovirus	ssDNA	Other DNA Viruses
	Bats	Bat coronavirus	–	Coronavirus	(+)RNA	Other Positive-Strand RNA Viruses
Rodents	Mouse	Mouse hepatitis virus	Fatal	Coronavirus	(+)RNA	Other Positive-Strand RNA Viruses
	Mouse	Minute virus of mice	–	Parvovirus	ssDNA	Other DNA Viruses
	Chicken	Newcastle disease virus	Fatal	Paramyxovirus	(−)RNA	Other Negative-Strand RNA Viruses
	Pigeons	Pigeon coronavirus	–	Coronavirus	(+)RNA	Other Positive-Strand RNA Viruses
Fishes	Salmons	Infectious salmon anemia virus	Fatal	Orthomyxovirus	(−)RNA	Influenza Viruses
	Trouts	Infectious hematopoietic necrosis virus	Fatal	Rhabdovirus	(−)RNA	Rhabdovirus
	Breams	Red sea bream iridovirus	Fatal	Iridovirus	dsDNA	–
	Flounders	Viral hemorrhagic septicemia virus	Sepsis	Rhabdovirus	(−)RNA	Rhabdovirus
Reptiles	Turtles	Turtle herpes virus	Wart	Herpesvirus	dsDNA	Herpesviruses
	Snakes	Snake adenovirus	–	Adenovirus	dsDNA	Adenoviruses
	Lizards	Lizard adenovirus	–	Adenovirus	dsDNA	Adenoviruses
Amphibians	Frogs	Frog virus 3	Fatal	Iridovirus	dsDNA	–
Invertebrates						
Arthropods	Shrimps	White spot syndrome virus	Fatal	Nimavirus	dsDNA	–
Shellfishes	Oysters, Clams	Ostreid herpesvirus-1	–	Herpesvirus	dsDNA	Herpesviruses
Insects	Mosquitoes	Sindbis virus	–	Togavirus	(+)RNA	Other Positive-Strand RNA Viruses

(Continued)

TABLE 1.3 (Continued)

Host Range		Major Virus	Infectious Disease	Family Name	Genome	Chapter
	Moths	Nuclear polyhedrosis virus	–	Baculovirus	dsDNA	Virus Vectors
	Honey bees	Israeli acute paralysis virus	Fatal	Picornavirus-like	(+)RNA	Picornavirus
	Beetles	Flock house virus	–	Nodavirus	(+)RNA	Other Positive-Strand RNA Viruses
	Cricket	Cricket paralysis virus	Fatal	Picornavirus-like	(+)RNA	Picornavirus
Plants						
	Tobacco	Tobacco mosaic virus	Necrosis	Virgavirus	(+)RNA	–
	Tulip	Tulip breaking virus	Stripes	Potyvirus	(+)RNA	–
	Grass	Brome mosaic virus	Necrosis	Alphavirus-like	(+)RNA	Other Positive-Strand RNA Viruses
	Potato	Potato spindle tuber viroid	–	Viroid	ssRNA	Subviral Agents and Prions
	Rice	Rice dwarf virus	–	Reovirus	dsRNA	Other Negative-Strand RNA Viruses
	Maize	Maize streak virus	–	Geminivirus	ssDNA	–
	Cauliflower	Cauliflower mosaic virus	Necrosis	Caulimovirus	dsDNA(RT)	–
Unicellular Organisms						
	Amoeba	Mimivirus	–	Mimivirus	dsDNA	Discovery and Classification
	Yeasts	Saccharomyces L-A virus	–	Totivirus	dsRNA	–
	Fungi	Botrytis porri RNA virus 1	–	Mycovirus	dsRNA	–
	Bacteria	λ phage, T2 phage	Cell lysis	–	dsDNA	–

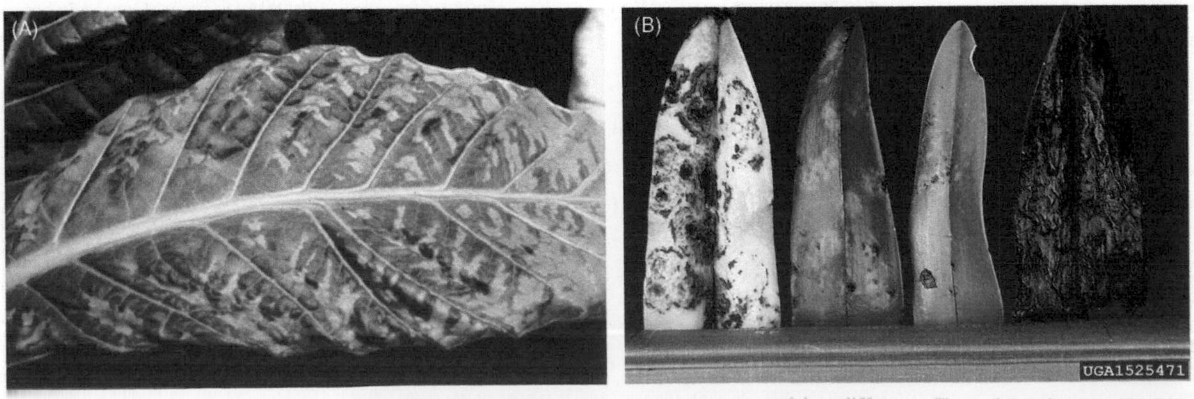

FIGURE 1.10 Tobacco mosaic virus, a plant virus. TMV has a very wide host range and has different effects depending on the host being infected. (A) Tobacco mosaic virus symptoms on tobacco. (B) Tobacco mosaic virus symptoms on orchid.

BOX 1.4 Tulip Breaking Virus

Tulip breaking virus (TBV), also known as tulip mosaic virus, is a plant virus. In peculiar, TBV infection of tulip leaves a stripe pattern without pathogenic lesions on the host. Tulips with the stripe pattern were once sold at extraordinarily high prices, which was about 10 times the annual income of average workers during the so-called Tulip mania period during the 17th century in the Netherlands. Of course, the stripe pattern was highly valued for its artistic beauty, without knowing it was the result of a viral infection. In fact, TBV belongs to the potyvirus family (see Table 1.3). The stripe pattern is believed to be the result of bleaching caused by TBV infection.

The portrait of tulip named "Semper Augutus." The tulip was sold at the higher price in the market during 17th century in the Netherlands under "tulip mania." The effects of the TBV are seen in the striking streaks of white in its red petals.

1.6 SUBVIRAL AGENTS

In addition, some filterable agents, which do not comply with the classical definition of "virus," were discovered. They are so-called subviral agents[11] or virus-like transmissible agents. These subviral agents have long been considered to be "virus," since they are transmissible, pathogenic to their host, and filterable. These subviral agents will be described in more detail in chapter "Subviral Agents and Prions."

Subviral agents comprise three kinds: satellite viruses, viroids, and prions (Table 1.4). A satellite virus is morphologically indistinguishable from a regular virus, and composed of nucleic acids and capsid proteins. However, one important distinction is that its replication depends on another virus (ie, a host virus). Since the satellite virus cannot replicate in the absence of a host virus, it can be said to be "a parasite of parasite." The second kind of subviral agent is the "viroids" that are found only in plants. A viroid contains a small RNA molecule ($\sim$0.3 kb circular RNA) only, but is devoid of proteins. In other words, the RNA itself is the transmissible agent. The third kind of subviral agent is the "prions" that are associated with *TSE* (transmissible spongiform encephalopathy) or scrapie. In contrast to the viroids, prions are composed of proteins only, but devoid of nucleic acids. The "prions hypothesis" which states that protein, devoid of nucleic acids, is the only etiologic agent for TSE, had not been accepted in the science community until recently (see Box 20.1).

11. **Subviral agent** It refers to virus-like transmissible agents, which do not comply with the classical definition of "virus."

BOX 1.5 Giant Virus and Virophage

Attention is paid to the discovery of two giant viruses in amoeba that lie beyond the classical definition of virus. The first giant virus discovered in 2003 was named "mimivirus." The name derived from "mimicking microbes." It was initially mistaken by Jean-Michel Claverie as a microbe and left in his freezer for 10 years, since its size nearly parallels the microbe (~400 nm) under microscope. In fact, its genome size is 1200 kb, which is five times larger than herpesvirus, which is known to have the largest genome among viruses. And it encodes 1200 genes. Considering the genome size and the number of genes it encodes, mimivirus is closer to a microbe than a virus.

Recently, Jean-Michel Claverie discovered another giant virus, named "mamavirus." An interesting twist in this discovery is that another virus was found inside the giant mamavirus. In other word, a satellite virus of giant virus was discovered, which is packaged inside the host virus. Although satellite viruses were previously reported in other organisms, this is the first example where the satellite virus is found packaged inside host virus particles. The satellite virus is now called "Sputnik," named after the first man-made satellite. It has a 18 kb DNA genome, which encodes 21 genes (Table Box 1.5). Interestingly, Sputnik-infected mamavirus was somewhat attenuated in progeny production and morphologically distorted. Briefly put, Sputnik makes its host sick. Moreover, the satellite virus of the giant virus is now dubbed a "virophage," as an analogy to the bacteriophage. Jean-Michel Claverie wittily added that "The discovery of virophage makes 'virus' more living organisms." Paradoxically, "being sick" can be said to be the evidence of "living." It was the first report that a virus particle was found packaged inside a host virus, which is already acting as a parasite of the host organism.

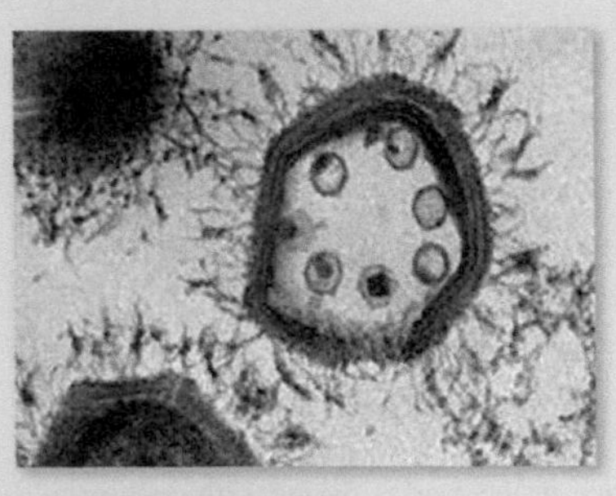

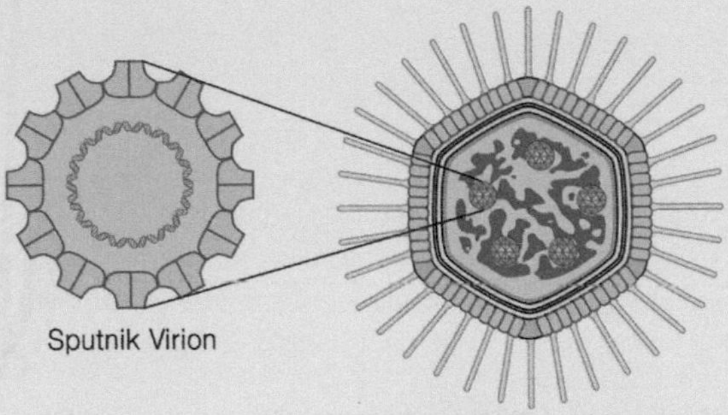

Mamavirus and virophage. (A) Transmission electron microscopy. Sputnik particles (six) are clearly visible inside mamavirus capsids. (B) Illustration of mamavirus and its satellite Sputnik virus.

TABLE BOX 1.5 Giant Virus and Virophage

Host Organism	Giant Virus (Genome Size)	Virophage (Genome Size)
Amoeba	Mimivirus (~1200 kb)	–
Amoeba	Mamavirus (~1200 kb)	Sputnik (18 kb)

1.7 PERSPECTIVES

A virus is physically smaller than a cell, the basic unit of life, and it propagates only inside cells, and often causes diseases. Strenuous efforts were made in the past century to control scourges caused by viruses. Since the discovery of the virus as a filterable agent at the dawn of the 20th century, the study of viruses has been established and has advanced a great deal during the past century. In retrospect, the achievements made in virus research are truly remarkable. Currently, almost all viral etiologic agents of human infectious diseases are believed to be identified, and have been extensively subjected to experimental analysis. For instance, viruses are now precisely classified by genome features rather than the diseases that they cause. The Baltimore classification now enables us to predict the

TABLE 1.4 Outstanding Features of Subviral Agents

Features	Satellite Virus	Viroid	Prion
Genome (Nucleic acid)	o	o	X
Protein coding	o	X	X
Particle protein	o (Capsid)	X	o (PrP)

genome replication mechanism of a novel virus immediately upon the availability of nucleotide sequence. On the other hand, subviral agents, such as viroids and prions, are extraordinary in the sense that RNA or protein itself is an infectious agent, as detailed in chapter "Subviral Agents and Prions." Another extraordinary infectious agents are giant viruses discovered in amoeba. These two extraordinary infectious agents seem to blur and challenge the classical definition of the virus.

1.8 SUMMARY

- Discovery: Historical records on the scourges caused by viruses have been found in ancient relics. The virus was first described as a filterable transmissible agent that causes disease in plants and animals.
- Definition: A virus is a "submicroscopic and intracellular parasite." Viruses are found in almost all living organisms on earth.
- Virology: Virology is a discipline, which studies the diverse aspects of virus replication and its consequences to the host cell. Virology, the study of viruses, has been established and has advanced a great deal over the past century.
- Classification: Animal viruses are classified into three groups: DNA viruses, RNA viruses, and reverse transcribing (RT) viruses. They are further classified into seven groups, according to their genome or by the Baltimore classification.
- Subviral agent: Virus-like transmissible agents, which do not comply with the classical definition of a "virus" are termed "subviral agents." Satellite viruses, viroids, and prions are the tree types of subviral agents.

STUDY QUESTIONS

1.1 "Computer virus" is coined as an analogy to "virus." State in what respects these two seemingly unrelated entities are related.

1.2 According to the Baltimore classification, viruses that belong to two distinct groups contain a positive-stranded RNA genome. List these two groups and describe the differences of the genome replication mechanisms between two groups.

1.3 List three kinds of subviral agents and state why each of these subviral agents does not comply with the classical definition of a "virus."

SUGGESTED READING

Duffy, S., Shackelton, L.A., Holmes, E.C., 2008. Rates of evolutionary change in viruses: patterns and determinants. Nat. Rev. Genet. 9 (4), 267–276.

Enquist, L.W., Editors of the Journal of Virology, 2009. Virology in the 21st century. J. Virol. 83 (11), 5296–5308.

Raoult, D., Forterre, P., 2008. Redefining viruses: lessons from Mimivirus. Nat. Rev. Microbiol. 6 (4), 315–319.

Villarreal, L.P., 2004. Are viruses alive? Sci. Am. 291 (6), 100–105.

JOURNAL CLUB

- Enquist, L.W., 2009. Virology in the 21st century. J. Virol. 83 (11), 5296–308.

  Highlight: An interesting perspective on Virology in the 21st century, which is written by Prof. Enquist, a leading virologist and an Editor-in-Chief of the *Journal of Virology*.

BOOK CLUB

- Boose, J. and August, M.J., 2013. To Catch a Virus, ASM Press, Washington, DC.

  Highlight: Historical accounts of earlier discovery on human pathogenic viruses are vividly described from early work of Louis Pasteur on rabies vaccine to the more recent Barre-Sinoussi and Luc Montagnier's work on the discovery of HIV. It is a must-read for anyone who is interested in the historical account of the virus discovery.
- Crotty, S., 2001. Ahead of the Curve: David Baltimore's Life in Science, University of California Press.

  Highlight: A compelling biography of David Baltimore, which details the life and work of one of the most brilliant, powerful, and controversial scientists of our time.

INTERNET RESOURCES

- International Committee on Taxonomy of Virus (ICTV): http://www.ictvdb.org/
- Viral Zone (http://viralzone.expasy.org/); a Swiss Institute of Bioinformatics web resource for all viral genus and families, providing general molecular and epidemiological information, along with virion and genome figures.
- The official web site of Nobel Prize: http://www.nobelprize.org/

Chapter 2

Virus Structure

Chapter Outline

Viruses are a kind of "nanoparticle" existing in nature. As described in the previous chapter, the viruses in diverse families differ from each other in their morphology. Nonetheless, the virus particles are built on common underlying principles. This chapter will describe the principles of virus particle structure. In addition, modern technologies used for the structural studies of virus particles will be described: electron microscopy (EM), cryo-EM, and X-ray crystallography.

2.1 TERMINOLOGIES

Several terminologies frequently used for description of the virus particles are listed in Table 2.1. First, the viral proteins that constitute the virus particles are called "structural protein," while the viral proteins that are absent in virus particles are called "nonstructural protein." For instance, the proteins that serve as building block of the viral capsid are classified as structural proteins. In particular, "nucleocapsid"[1] is used to refer a viral capsid that is associated with the viral genome. In addition, the envelope proteins are also structural proteins constituting the viral particles. On the other hand, not all virus-coded proteins are found in the virus particles. For instance, many virus-coded enzymes such DNA/RNA polymerase and proteases are frequently not found in the virus particle. These viral proteins are classified as nonstructural proteins, as they do not constitute the virus particle.

2.2 ENVELOPE AND CAPSID

Virus particles can either "enveloped" or "nonenveloped," depending on the presence or absence of the envelope. For instance, poliovirus and norovirus particles do not have an envelope, while influenza virus and yellow fever virus particles do have an envelope (Fig. 2.1). The envelope is composed of a lipid bilayer, which is derived from cell membranes, and virally coded envelope proteins (Fig. 2.2). In addition, in many enveloped particles there is a viral protein, termed matrix protein, that coats the inner leaf of the lipid bilayer. The matrix proteins are involved in the virion assembly of enveloped virus particles. A virus particle that does not have an envelope is aptly called a "naked virus," which hints at the lack of a coat. Inside the envelope, capsid is found that is the protein shell that encloses the viral genome and any other components necessary for to virus structure or function. Importantly, the capsid does play a protective role, sequestering the genome material from physical and chemical damaging agents. In fact, two kinds of capsid structures are found: a spherical capsid and a helical capsid, which will be detailed below.

1. **Nucleocapsid** It refers to a viral capsid that is associated with the viral genome.

Molecular Virology of Human Pathogenic Viruses. DOI: http://dx.doi.org/10.1016/B978-0-12-800838-6.00002-3

TABLE 2.1 Terminologies Used for Description of the Viral Structures

Term	Description
Structural protein	Viral proteins that make up the viral particles
Nonstructural protein	Viral proteins that is absent in the viral particles
Subunit	Basic assembly unit of viral capsids
Capsid	The protein shell that encloses and protects the nucleic acid genome of a virus
Nucleocapsid	A viral capsid and its associated genome
Envelope	Lipid bilayer that surrounds many type of viruses
Virion	A completely assembled virus particle outside its host cell
Subviral particle	A kind of incompletely assembled virus particle

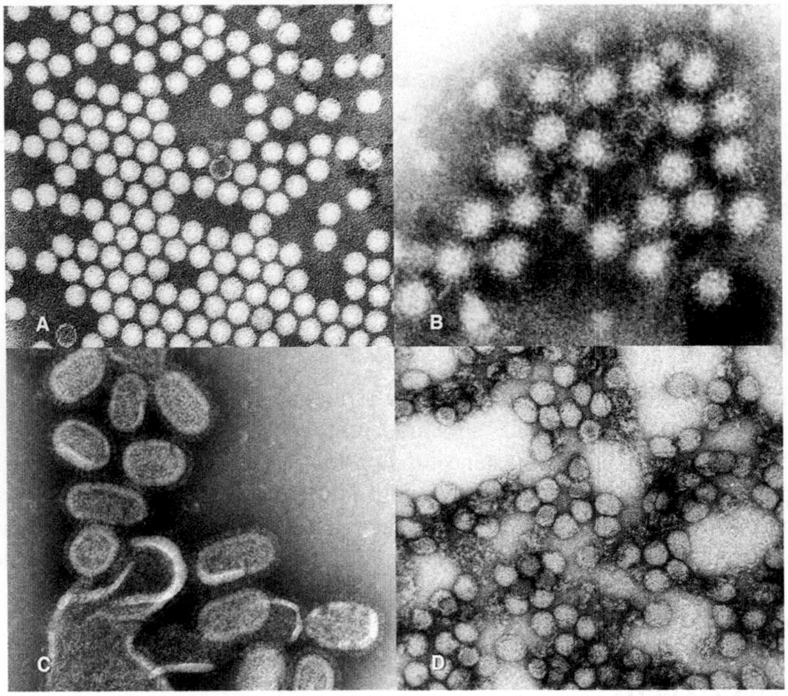

FIGURE 2.1 **Electron microscopic views of the major human pathogenic viruses.** Transmission electron micrographs of poliovirus (~30 nm diameter) (A), norovirus (27–38 nm in diameter) (B), influenza virus (80–120 nm diameter) (C), and yellow fever virus (D). Poliovirus and norovirus particles are naked (nonenveloped) particles, while influenza and yellow fever virus particles are enveloped particles.

2.3 CAPSID STRUCTURE

Spherical capsids possess an icosahedral structure, while helical capsids possess an elongated capsid structure. Capsids are constituted by numerous building blocks (ie, subunits) in that each subunit is made of one or a few molecules of structural proteins. Although each individual structural protein is not symmetrical, the capsids made of numerous structural proteins are symmetrical. A question raised was "how can a symmetrical structure be made by using unsymmetrical building blocks?" A short answer to this question is to implement the principle of "subunit assembly," by arranging unsymmetrical building blocks in a symmetric manner, which can be either icosahedral symmetry or helical symmetry.

Then, what is the advantage of subunit assembly in the capsid assembly? First, the subunit assembly enables the building of the capsids spontaneously without energy expenditure (ie, self-assembly). Numerous repetitive interactions between subunits can readily derive the capsid assembly, if the capsids are made by only one or a few kinds of subunit.

Second, the subunit assembly minimizes the genome expenditure for coding structural proteins, because only a few genes are required. Third, the subunit assembly allows to construct a more stable capsid, because far more molecular interactions than otherwise are involved in the capsid assembly.

Viruses have adopted "subunit assembly" as a strategy to build robust capsid structures. Conversely, the capsid needs to be dismantled eventually upon infection. In other words, the capsid is not a static shell; rather, it is a dynamic structure comprised of flexible subunits.

Let's consider two symmetrical structures: helical capsids and icosahedral capsids.

Helical capsids: The simplest way to build a symmetric structure is to arrange the building blocks around a circle. Ultimately, a tube-shaped structure can be made by putting multiple layers on the top of the circle. This is the way in which a helical capsid can be built by using unsymmetrical building blocks. The best example of this kind of capsid structure is that of TMV (Fig. 2.3). Cooperative binding of capsid subunits to the helical RNA results in a helical nucleocapsid. Notably, the capsid of TMV assembles spontaneously in vitro without energy expenditure. In addition to TMV, a few animal viruses possess helical nucleocapsid structure as well, including influenza virus, rhabdovirus, and Ebola virus.

Icosahedral capsids: Another way to build a symmetric structure is to construct a three dimensional (3D) symmetric structure. In theory, only five kinds of 3D symmetric structure are possible: ones with 4, 6, 8, 12, and 20 facets. In order to build a 3D symmetric structure with a minimal number of subunits, three subunits can be arranged in a triangular facet. In other words, 24 subunits are needed to build an octahedral structure, while 60 subunits are needed to build an icosahedral structure. Consistent with this theory, electron microscopic examination of viral capsids revealed that the vast majority of spherical vial capsids possess a capsid structure of 20 facets, an icosahedral structure (Fig. 2.4).

Then, one wonders about the advantages of having an icosahedral symmetrical structure. Compared to other symmetrical structures, an icosahedral structure is the symmetric structure that can be built using the minimal kinds and the maximal numbers of subunits. An icosahedral structure is the most stable structure that can be built by multiple subunits

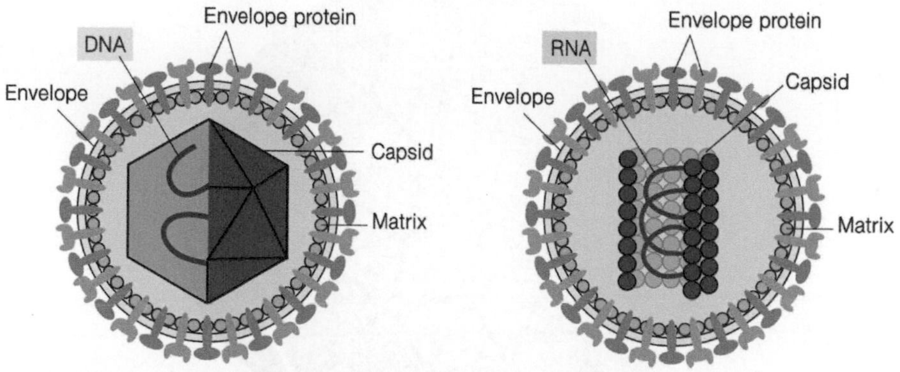

FIGURE 2.2 Schematic diagram of virus particles. (A) A typical envelope virus particle with a spherical capsid. The left side of the capsid is uncovered to show the viral genome inside (eg, DNA). (B) A typical envelope virus particle with a helical capsid. A part of the nucleocapsid is uncoated to show the viral genome inside (eg, RNA). Three kinds of virus structural proteins are denoted: envelope glycoproteins, capsid protein, and matrix protein. The viral structural protein that cover the inner leaf of viral envelope is often referred as matrix protein.

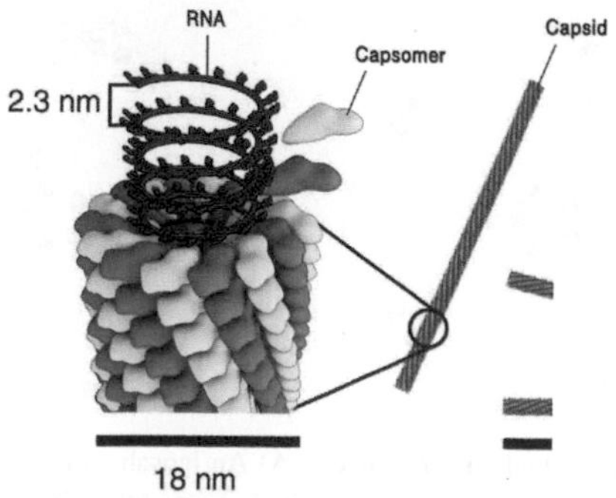

FIGURE 2.3 The helical capsid structure of TMV. The helical structure is made by the cooperative binding of the subunits protein (ie, capsomer) to the RNA genome of TMV. Note that all subunits bind to the RNA in equivalent manner.

(Fig. 2.5). Moreover, the use of a few genes to code for structural proteins is a strategy to minimize the genome expenditure.

How can the subunits be put together to build icosahedral capsid? In fact, the majority of viral capsids are composed by the multiplicity of 60 subunits. The simplest icosahedral structure is one with three subunits in each facet, resulting in 60 subunits (Fig. 2.6A). To build a more complex structure, each facet of an icosahedral structure can be divided into four small triangles, in which three subunits in each triangle are positioned. As such, 240 subunits (12 subunits × 20 facets) can be

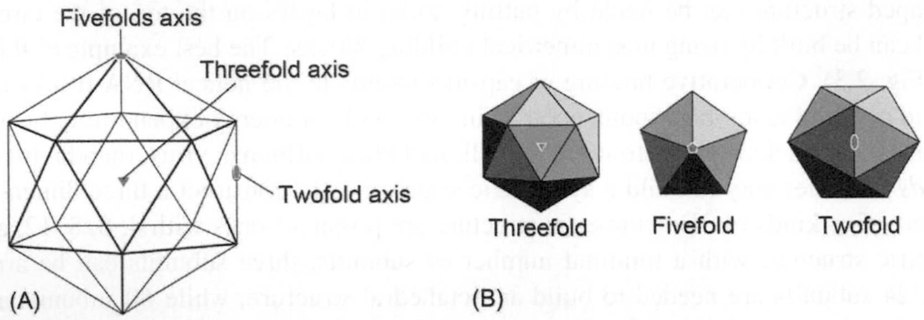

FIGURE 2.4 Schematic diagram of icosahedral structure. (A) An icosahedral structure is composed of 20 triangular facets and 12 vertices. (B) An icosahedral structure has a threefold axis of symmetry in the middle of facet, fivefold axis of symmetry at vertices, and twofold axis of symmetry.

FIGURE 2.5 An icosahedral sculpture. An icosahedral sculpture is often used by architects. A dome structure built in roof of a building at Yonsei University in Korea. A triangular facet is evident by linking three fivefold axes of symmetry (the white dot).

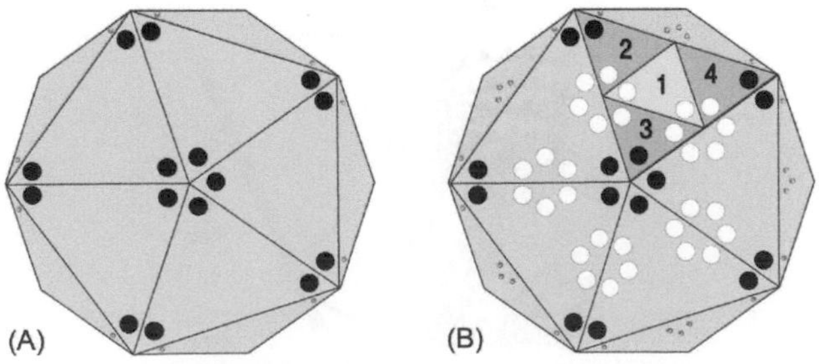

FIGURE 2.6 The arrangement of subunits in icosahedral structures. (A) An icosahedral structure with T number of 1. The structure is made of 60 subunits, having 20 triangular facets (3 subunits per facet). (B) An icosahedral structure with T number of 4. The structure is made of 240 subunits, having 20 facets (12 subunits per facet). In particular, 5 subunits (black dot) constitute the fivefold axis of symmetry (12 vertices), whereas 6 subunits (white dot) constitute newly formed vertices. Although all subunits appear to be equally arranged, some of them are arranged as a pentagon, while others are arranged as a hexagon. In other words, a subunit is said to be "quasi-equivalent" (Box 2.1).

BOX 2.1 Triangulation Number

An icosahedral symmetrical structure constitutes 20 equilateral triangle facets, each of which can be made of 3 subunits. In other words, 3 asymmetric subunits (ASU) can make up the facet. Thus, an icosahedral capsid could be made by at least 60 subunits. In reality, most viral capsids are composed of more than 60 subunits. To account for the principle of viral capsid assembly, Drs Casper and Klug proposed "Quasi-equivalence theory." The theory states that, in the case where the capsids are composed of more than 60 subunits, each subunit is not topologically equivalent; however, if they are considered to be equivalent, they can be assembled into an icosahedral structure under the same principle. According to this theory, each facet (ie, triangle) is constituted by an integer number of triangles, in which the number is dubbed "triangulation (T) number." Furthermore, T number can be obtained by an equation: $T = H^2 + HK + K^2$, where H and K are the position of a fivefold axis of symmetry in vector space.

On the other hand, the numbers of subunits that constitute the capsid are calculated by the following equation: the subunit number $= 12 \times 5 + 10(T-1) \times 6 = 60$ T. For instance, a capsid having T = 4 symmetry is composed of 240 subunits. T number can be obtained by high resolution EM image of capsids. Therefore, the number of subunits that make up a capsid from an EM image of the capsid can be estimated without biochemical analysis.

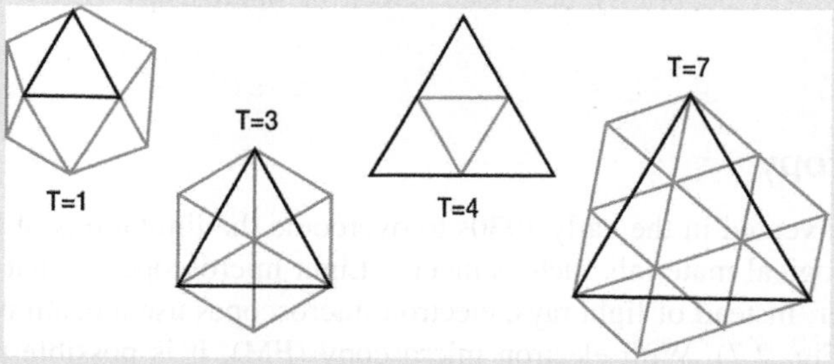

T numbers of icosahedral structures. The bold lines, linking the fivefold axis of symmetry of each structure, make up one triangle (facet). For T = 4, 4 subunits make up the facet. For T = 7, it is considered that seven subunits make up a facet, although it is not evident in the diagram shown here.

TABLE 2.2 T Number of the Major Animal Viruses

Virus Species	Family	T Number	Subunit Copies
Parvovirus	Parvovirus	1	60
Poliovirus, Rhinovirus	Picornavirus	3	180
Hepatitis B virus	Hepadnavirus	4	240
SV40	Polyomavirus	7	420
Reovirus	Reovirus	13	780
HSV-1	Herpesvirus	16	960
Adenovirus type 2	Adenovirus	25	1500

arranged (Fig. 2.6B). Likewise, an icosahedral structure having an increasing number of subunits can be made. *Triangulation number*[2] (T), a parameter that pertains to the topological features, can estimate the precise number of subunits constituting an icosahedral structure (Box 2.1). T numbers of the majority of animal viruses are known (Table 2.2). In general, the bigger the capsids are, the higher T numbers are. T numbers can be acquired from a high-resolution electron microscope image of a viral capsid structure.

2. **Triangulation number** It refers to the number of subunits that constitute a triangular facet of icosahedral capsid.

2.4 ROLES OF VIRUS STRUCTURE

What are the roles of virus structure? First, one of the most important roles of virus structure is a protective role. The capsid structure protects the viral genome from physicochemical damage, such as nucleases, and radiation (eg, ultraviolet). Second, the role of the virus structure is to recognize the cellular receptor for the entry. Specifically, one of the viral structural proteins (either an envelope glycoprotein for the enveloped viruses or a capsid protein for the nonenveloped viruses) directly binds to the cellular receptor, for the viral entry. Third, the viral capsids play a role in delivering the viral genome to the site of genome replication. For instance, for the viruses that replicate in the nucleus, the viral capsids play a critical role in the nuclear entry of the viral genomes (see Fig. 3.7).

2.5 TOOLS USED FOR VIRUS STRUCTURE RESEARCH

As stated above, virus particles represent a "nanoparticle" existing in nature, mostly ranging from 30 to 300 nm in size, which can be viewed only by an electron microscope. Thus, optical instruments, such as the electron microscope, are essential to examine the morphology of virus particles. Three kinds of modern technologies that are utilized for the visualization of virus particles will be briefly described: electron microscopy (EM), cryo-electron microscopy, and X-ray crystallography.

2.5.1 Electron Microscopy

The electron microscope was invented in the early 1930s to overcome the limitations of light microscopes. It was originally invented to view nonbiological materials such as metals. Light microscopes at that time could magnify specimens as high as 1000 times. However, instead of light rays, electron microscopes use a beam of electrons focused by magnets to resolve minute structures (Fig. 2.7). With electron microscopy (EM), it is possible to magnify a structure 100,000 times. Two kinds of electron microscopic technologies have been developed: transmission electron microscopy (*TEM*) and scanning electron microscopy (*SEM*). TEM is a microscopic technique in which a beam of electrons is transmitted

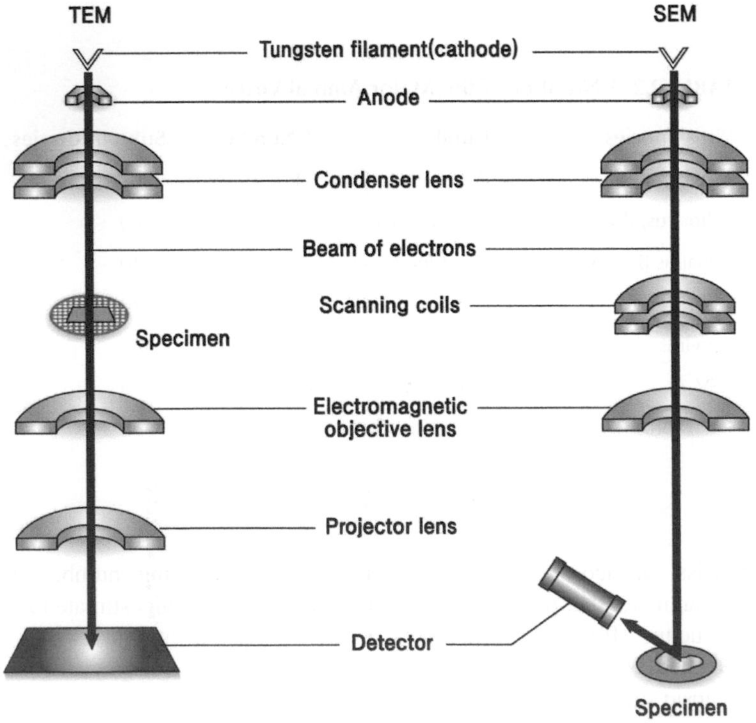

FIGURE 2.7 Principles of electron microscope. The principle of TEM and SEM is illustrated for comparison. Electron beam lines are indicated by red line. TEM generates the image by electrons that transmit the specimen, while SEM generates the image by electrons that diffract from the specimen.

through an ultra-thin specimen, interacting with it as the beam passes through. TEM relies on negative staining of purified virus particles with an electron-dense material, such as uranyl acetate or phosphotungstate. An image is formed from the interaction of the electrons transmitted through the specimen; the image is magnified and focused onto a layer of photographic film (Fig. 2.8). What can be learned from TEM images of virus particles? TEM reveals the overall morphology (whether viruses are spherical or elongated particles and whether they are naked or enveloped). In addition, even the symmetric parameter (ie, T number) of an icosahedral capsid could be obtained from a high resolution image.

On the other hand, SEM is a type of electron microscopic technique that produces images of a sample by scanning it with a focused beam of electrons (see Fig. 2.7). The electrons interact with atoms in the sample, producing various signals that can be detected and that contain information about the topography and composition of the sample's surface. As you can see in Fig. 2.8, more realistic 3D images are obtained.

2.5.2 Cryo-Electron Microscopy

Cryo-electron microscopy (Cryo-EM) is a kind of TEM where the sample is examined at cryogenic temperatures (generally lower than −160°C). A feature of cryo-EM is the ability to freeze the specimen rapidly so that water molecules in the specimen turn into transparent ice crystals. The ice crystal of the solvent transforms the biomolecules (ie, virus particle) into a rigid state so that a high resolution image can be acquired. Numerous images are captured from the specimen on grid that is kept frozen. Then, such acquired images are digitally processed to reconstruct a high resolution image (Fig. 2.9). One outstanding feature of cryo-EM is that it examines the specimen in its intact state without artificial treatments such as fixing, dehydration, and staining. In other words, the cryo-EM images are more likely to represent the native state of biological structures. More importantly, the analysis of cryo-EM image yields even interior structure of capsids as well as exterior structure (see Fig. 2.9).

What kinds of objects (ie, viral particles) are suitable for the analysis with cryo-EM? Primarily, the symmetry of the particles is an important element to acquire a high resolution image. In addition, the rigidity and conformational homogeneity of the objects are also important for obtaining a high resolution. Notably, the resolution of cryo-EM images has been greatly improved by recent advances in image reconstruction technology. For instance, some virus structures

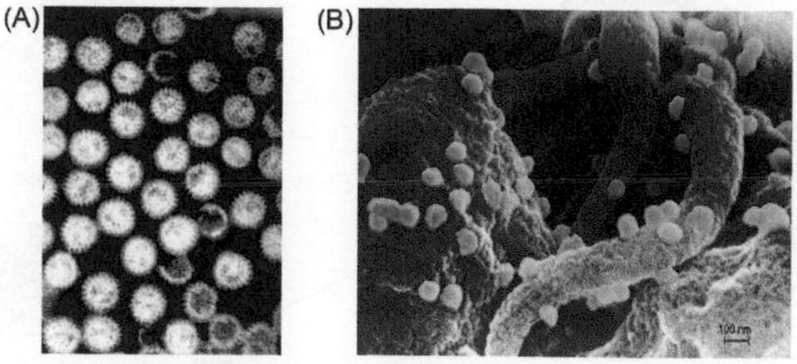

FIGURE 2.8 Electron microscopic image of virus particles. (A) Transmission electron micrograph of rotavirus particles (~70 nm in diameter). (B) Scanning electron microscope image of HIV particles (green) budding from the infected T lymphocytes (pink and blue).

FIGURE 2.9 High-resolution image obtained by cryo-electron microscope. Cross-section and two cut away views of a hepatitis B virus (HBV) virion was obtained by cryo-electron microscope. HBV virion is comprised of a T = 4 icosahedral capsid (blue) with 120 spikes and an outer envelope with projections of envelope glycoproteins (yellow). The image analysis of the capsid shell yields an interior dodecahedral cage of density, which is ascribed to ordered mature double-strand DNA.

obtained by cryo-EM are already at a resolution that can be interpreted in terms of an atomic model. For instance, a high resolution (~3.5 angstrom) image of adenovirus particles, which provides sufficient resolution to trace polypeptide chains in the capsid structure, was recently achieved by cryo-EM. It should be noted that the structural analysis of adenovirus particles represents a formidable challenge, since it has a gigantic capsid particle having a 150-megadalton capsid containing nearly 1 million amino acid residues (ie, equivalent to having over 3000 molecules of a protein composed of 300 amino acids residues). Now, it can be said that cryo-EM rivals X-ray crystallography, when applied to large, homogenous, and highly symmetric objects.

2.5.3 X-ray Crystallography

X-ray crystallography is a tool used for determining the atomic and molecular structure of a crystal. The underlying principle is that the crystalline atoms cause a beam of X-rays to diffract into many specific directions (Fig. 2.10). By measuring the angles and intensities of these diffracted beams, a crystallographer can produce a 3D picture of the density of electrons within the crystal. From this electron density image, the mean positions of the atoms in the crystal can be determined, as well as their chemical bonds, their disorder, and various other information. The method revealed the structure and function of many biological molecules, including vitamins, drugs, proteins, and nucleic acids, such as DNA. Note that the double helix structure of DNA discovered by James Watson and Francis Crick was revealed by X-ray crystallography. Recent advances in image reconstruction technology have made X-ray crystallography amenable to the structural analysis of much larger complexes, such as virus particles (Fig. 2.11). The major shortcoming of X-ray crystallography is that it is difficult to obtain a crystal of virus particles, which is a prerequisite for X-ray crystallography. Another shortcoming is that X-ray crystallography generally requires placing the samples in nonphysiological environments, which can occasionally lead to functionally irrelevant conformational changes.

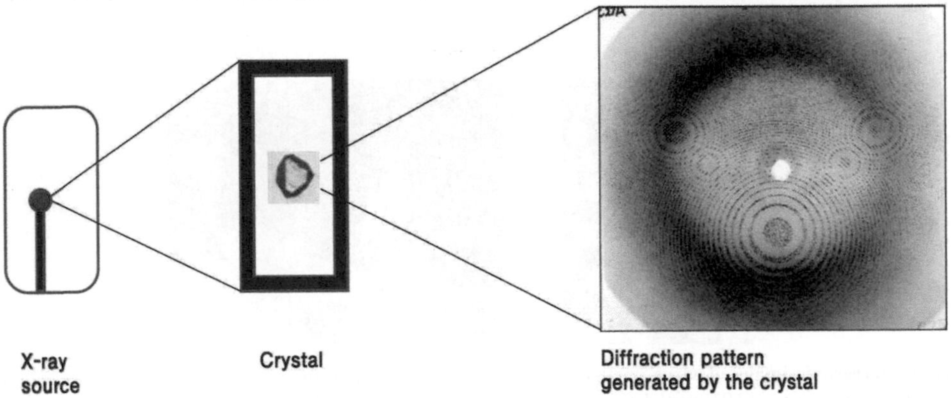

FIGURE 2.10 Principle of X-ray crystallography. High-resolution capsid structure can be obtained from the diffraction pattern generated by crystals of the virus particles (eg, human rhinovirus 14).

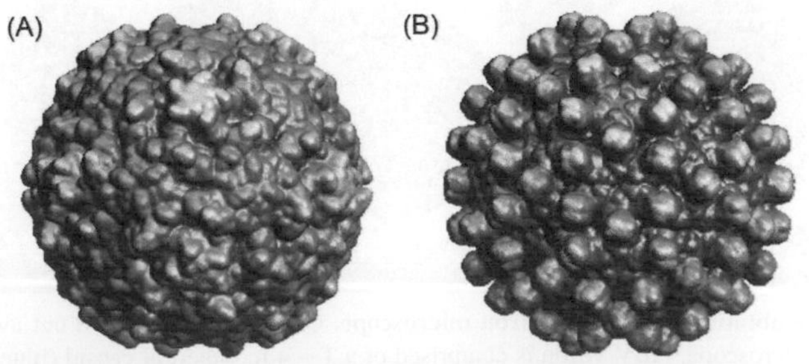

FIGURE 2.11 Viral capsid structure obtained by X-ray crystallography. (A) Poliovirus capsid with T = 3 symmetry. (B) Hepatitis B virus capsid with T = 4 symmetry (http://viperdb.scripps.edu/).

2.6 PERSPECTIVES

The study of viruses involves the microscopic examination and the structural analysis of the virus particles. In fact, the structural analysis of virus particles played an important role in the early advances in virus research and contributed to the establishment of "virology" as a discipline. The crystal structure of TMV had been resolved in 1935, when virology was still in its infancy. Electron microscopy was applied to virus particles soon after the first instrument was constructed in the early 1930s. Importantly, the electron microscope significantly transformed the science of virology. It provided a powerful approach for rapid diagnosis, and became an exquisitely sensitive tool to define new viral agents, especially when other methods failed. Recent advances in digital image processing have provoked the renaissance of "structural virology." The advances have allowed the structural analysis of larger virus particles, such as adenovirus. In particular, the advances in cryo-EM made the analysis of virus particles more feasible. Notably, not only the shell structure but also the structure of molecules present inside the capsid particles such as the viral genomes (or noncapsid viral proteins) has been obtained. It is expected that tools for structural virology will make a greater contribution to advance our knowledge on many aspect of virology, and eventually help us to design better strategies to combat the pathogens. Remember that "seeing is believing."

2.7 SUMMARY

- *Virus structure:* It is a kind of nanoparticle found in nature. It can be visualized only by an electron microscope. Virus particles are composed of the capsid that encompasses the virus genome and the viral envelope. The viral envelope is absent in some viruses, known as nonenveloped viruses.
- *Capsid structures:* Two kinds of the viral capsid structure are found: spherical icosahedral capsid and cylinder-shaped helical capsid. The icosahedral capsid has 20 facets, each of which represents a triangular shape.
- *T number:* T number defines the topological feature of the icosahedral structure. It indicates the number of subunits that constitute a triangular facet made of three fivefold axes of symmetry.
- *Tools for the virus structure analysis:* Three kinds of technologies that are utilized for the visualization of virus particles are the following: electron microscopy (EM), cryo-EM, and X-ray crystallography.

STUDY QUESTIONS

2.1 Describe the biological roles of the viral capsids in the virus life cycle.
2.2 Estimate the T number of the sculpture shown in Fig. 2.5. Estimate how many sheets of glass are needed to assemble the sculpture.
2.3 Describe two methods that are utilized to obtain a high resolution image of a virus particle structure. And compare the pros and cons of the two methods.

SUGGESTED READING

Bostina, M., Levy, H., Filman, D.J., Hogle, J.M., 2011. Poliovirus RNA is released from the capsid near a twofold symmetry axis. J. Virol. 85 (2), 776–783.

Grigorieff, N., Harrison, S.C., 2011. Near-atomic resolution reconstructions of icosahedral viruses from electron cryo-microscopy. Curr. Opin. Struct. Biol. 21 (2), 265–273.

Kuhn, R.J., Rossmann, M.G., 2005. Structure and assembly of icosahedral enveloped RNA viruses. Adv. Virus Res. 64, 263–284.

Liu, H., Jin, L., Koh, S.B., Atanasov, I., Schein, S., Wu, L., et al., 2010. Atomic structure of human adenovirus by cryo-EM reveals interactions among protein networks. Science. 329 (5995), 1038–1043.

Wang, J.C., Nickens, D.G., Lentz, T.B., Loeb, D.D., Zlotnick, A., 2014. Encapsidated hepatitis B virus reverse transcriptase is poised on an ordered RNA lattice. Proc. Natl. Acad. Sci. U. S. A. 111 (31), 11329–11334.

JOURNAL CLUB

- Bostina, M., Levy, H., Filman, D.J., Hogle, J.M., 2011. Poliovirus RNA is released from the capsid near a twofold symmetry axis. J. Virol. 85, 776–783.

  Highlight: Single particle cryo-EM showed that the RNA, which has never been observed by any other technology, is clearly visible both inside and outside the capsids. Further, the RNA being released from the capsid during uncoating was caught in the act of exiting. Advances in imaging technology have allowed acquiring the dynamic movement of subjects in action, as opposed to still images of the subjects acquired from fixed specimen.

Chapter 3

Virus Life Cycle

Chapter Outline

Viruses are obligate intracellular parasites. Therefore, viruses must gain entry into target cells and usurp the host cellular machinery to propagate and to produce progeny viruses. The multiple steps involved in the virus propagation occurring inside cells are collectively termed the "virus life cycle." The virus life cycle can be divided into three stages—entry, genome replication, and exit. Here, we focus on entry and exit, in which the commonality of mechanisms among viruses prevails. On the other hand, the genome replication, which is a step that is distinct for each of the virus families, is described in Part II to Part IV, where virus families are individually considered.

3.1 STEPS IN VIRUS LIFE CYCLE

A virus encounters multiple obstacles during its journey to enter the host cells. Cellular membranes pose as barriers for the invaders. The plasma membrane represents the first barrier that all animal viruses have to penetrate. The nuclear membrane represents the second barrier to some viruses that replicate their genome in the nucleus. Let's see how viruses obviate the barriers.

The virus life cycle can be divided into three stages—entry, genome replication, and exit (Fig. 3.1). The virus life cycle can be described in analogy with a businessman's life; the entry to his way to work, the genome replication to his task at work, and the exit to his way home. The first stage is entry. Entry involves *attachment*, in which a virus particle encounters the host cell and attaches to the cell surface, *penetration*, in which a virus particle reaches the cytoplasm, and *uncoating*, in which the virus sheds its capsid. Following the uncoating, the naked viral genome is utilized for gene expression and viral genome replication. Finally, when the viral proteins and viral genomes are accumulated, they are assembled to form a progeny virion particle and then released extracellularly. Virion assembly and the release from the cell constitute the *exit*.

3.2 VIRAL ENTRY

Entry, the first step of virus infection, involves the recognition of viral receptor by a virus particle. The viral entry can be divided into four steps: attachment, penetration, cytoplasmic trafficking, and uncoating. These steps are often linked to each other so that the division into four steps is obscure, but serves the explanation purpose.

3.2.1 Attachment

The attachment refers to the first encounter of virus particles with host cells, which involves two kinds of host proteins on the plasma membrane: (1) attachment factors and (2) viral receptors. The attachment factor on the cell surface recruits and holds the virus particles, thereby facilitating the interaction of the viral particle with the entry receptor.

Molecular Virology of Human Pathogenic Viruses. DOI: http://dx.doi.org/10.1016/B978-0-12-800838-6.00003-5

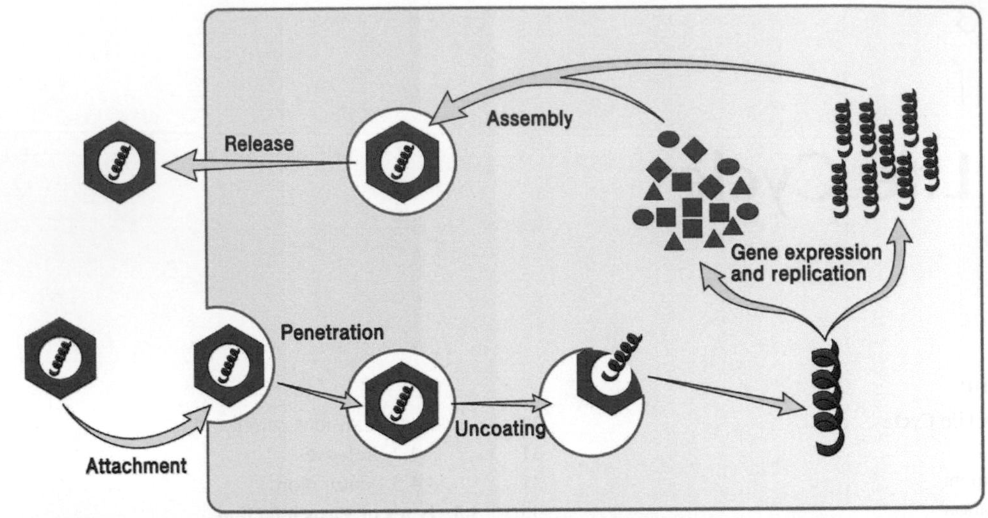

FIGURE 3.1 The life cycle of virus. The virus life cycle could be divided into six steps: attachment, penetration, uncoating, gene expression and replication, assembly, and release. The viral capsid (blue) and genome (brown) are schematically drawn for the purpose of explanation. The nucleus is omitted for clarity.

TABLE 3.1 Viral Receptors for Major Human Viruses

Family: Prototype Virus	Receptor (Coreceptor)	Attachment Factor	Viral Antireceptor
DNA Virus			
Parvovirus: AAV	HSPG (FGFR, integrin)	HSPG	CAP
Polyoma: SV40	GM1 gangliosides	–	VP1
Adeonovirus: Ad5	Integrin	CAR	Fiber protein
Herpesvirus: HSV-1	Nectin-1/HVEM, PILRα	HSPG	gD, gB
Herpesvirus: CMV	EGFR	HSPG	
Herpesvirus: EBV	CD21	HSPG	gp350
RNA Virus			
Picornavirus: Poliovirus	PVR/CD155	–	VP1, VP2, VP3
Picornavirus: Coxsackie virus	CAR	–	
Picornavirus: Rhinovirus	ICAM-1 or LDL receptor	DC-SIGN, L-SIGN	VP1, VP2, VP3
Flavivirus: Hepatitis C virus	CD81, Claudin-1, Occludin	SR-B1, LDL receptor	E2
Coronavirus: SARS virus	ACE2	–	Spike
Orthomyxovirus: Influenza virus	Sialic acid	–	HA
Rhabdovirus: VSV	Phosphatidyl serine	–	G protein
Rhabdovirus: Rabies	NCAM-1/CD56	–	G protein
Reovirus: Reovirus	JAM-A	–	Spike protein s1
RT Virus			
Retrovirus: HIV-1	CD4 (CXCR4 or CCR5)	DC-SIGN, L-SIGN	gp120
Hepatitis B virus	NTCP	HSPG	Pre-S1

Note: ACE, angiotensin-converting enzyme; CAR, coxsackie-adenovirus receptor; EGFR, epidermal growth factor receptor; HSPG, heparin sulfate proteoglycan; HVEM, herpes virus entry mediator; ICAM, intercellular adhesion molecule; NTCP, sodium taurocholate cotransporting polypeptide; PVR, poliovirus receptor.

In fact, glycoaminoglycans, such as heparins, serve as the attachment factor for diverse viruses, revealing the broader specificity of the attachment factors (Table 3.1). Unlike attachment factors, viral receptors, upon binding to the virus particles, promote the penetration of virus particles into cells. Further, the viral receptors are virus-specific and more importantly, determine cell tropism. For example, *CD4*[1] is specifically recognized by HIV, which infects CD4-expressing T lymphocytes.

BOX 3.1 Strategies to Discovery Viral Receptor

What are the experimental approaches to unveil viral receptors on the cell surface of the susceptible cells? Three experimental methods have been successfully used. The first approach is to identify the receptors by biochemical purification of cellular proteins on the cell surface that bind to the viral antireceptors (ie, viral structural proteins). Affinity purification of plasma membrane proteins using the viral structural proteins as a ligand is feasible. Alternatively, immunoprecipitation of plasma membrane proteins that bind to the viral structural protein could lead to many candidate proteins in SDS–PAGE gel. Subsequently, each protein band on the gel is subjected to mass spectroscopy (eg, MALDI-TOF) for identification. The second approach is to use *monoclonal antibodies*[2] (Mab) against plasma membrane proteins. It exploits the fact that an antibody specific to the receptor could block the entry. A set of the Mab is prepared against the plasma membrane of the susceptible cell in the first place. Then, each individual Mab is tested for its ability to block the viral infection. If the virus infection is blocked by a specific Mab, the ligand for the Mab is a potential candidate for the viral receptor. Ultimately, the ligand can be identified by immunoprecipitation, followed by mass spectroscopy. The third approach is to identify the receptor by *functional cloning,* which exploits the cDNA expression library. An individual cDNA of the library is transfected into a nonsusceptible cell, then tested as to whether the cell becomes infected. This method necessarily involves high-throughput robotics, as the human cDNA library is composed of over 30,000 genes. Recently, this modern technology has been successfully implemented to clone the HCV receptor. Once the potential molecules for the viral receptors are identified by one of above three approaches, the functionality of the receptors needs to be validated. This validation is to confirm whether the cDNA transfection to a nonsusceptible cell is necessary and sufficient to convert the nonsusceptible cell to a susceptible cell.

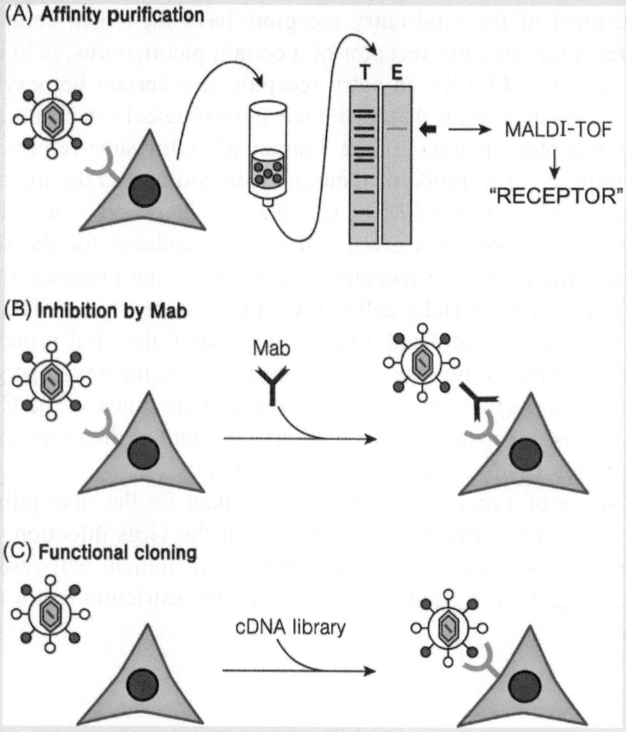

(A) Affinity purification. (B) Inhibition by Monoclonal antibodies (Mab). (C) Functional cloning.

1. **CD4** A glycoprotein found on the surface of immune cells such as T helper cells, monocytes, macrophages, and dendritic cells. CD4 is best known as a cellular marker for T helper lymphocyte.

2. **Monoclonal antibody** Monoclonal antibodies (mAb) are monospecific antibodies that are made by identical immune cells that are all clones of a unique parent cell.

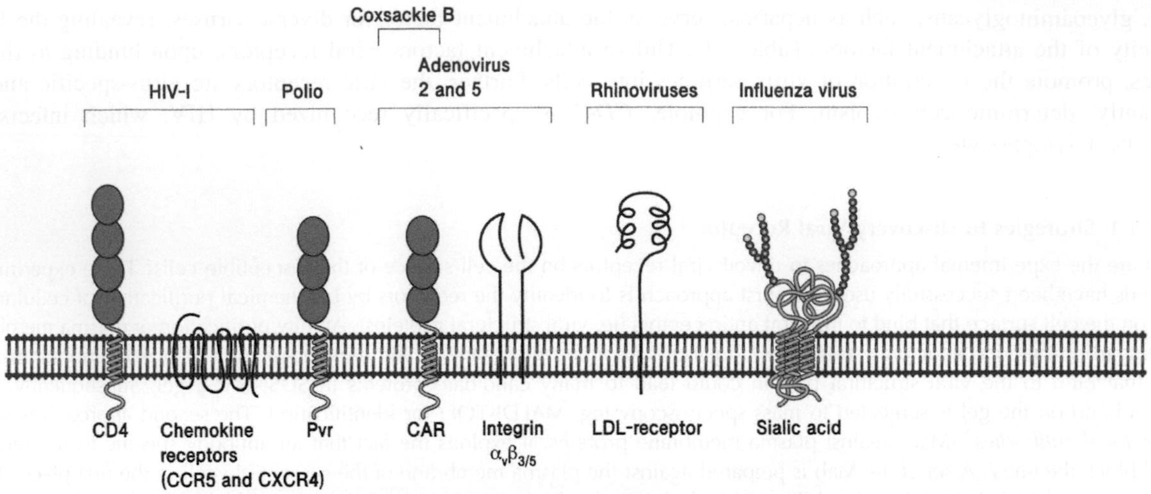

FIGURE 3.2 The viral receptors of representative human pathogenic viruses. The ectodomain of the receptor molecules is recognized by virus particles. Immunoglobulin (IgG) fold, which is shared by members of immunoglobulin superfamily (ie, CD4, Pvr, and CAR), is drawn in a *circle*. The sialic acid moiety of glycan in membrane glycoproteins is marked by yellow. Pvr, poliovirus receptor; CAR, coxsackie virus and adenovirus receptor.

Almost all viral receptors for the major human pathogenic viruses have been identified during the past three decades (Box 3.1). Advances in molecular biology made since the 1970s played a critical role for the discovery. Importantly, a few intriguing points stand out regarding the attributes of viral receptors. First, the molecular nature of the viral receptors is quite diverse, ranging from glycoproteins to phospholipids (see Table 3.1). Even the carbohydrate moiety of membrane glycoproteins is utilized as the viral receptor. For instance, *sialic acid*[3] residue of glycans is the entry receptor for influenza virus. Second, most of the viral entry receptors have their own cellular functions. For instance, the physiological function of LDL receptor, an entry receptor of a certain picornavirus, is to uptake LDL particles into cells, while epidermal growth factor receptor (EGFR), an entry receptor of a certain herpesvirus, is an EGFR (Fig. 3.2). In other words, viruses subvert the cellular proteins that have their physiological functions, and utilize them as entry receptors. Third, many of the viral receptors belong to the immunoglobulin superfamily, such as CD4 and CAR (see Fig. 3.2). Fourth, some viruses require coreceptors for their entry, in addition to the main receptors for entry. For example, HIV requires chemokine receptors, such as *CCR5* or *CXCR4*, as a coreceptor for efficient entry (see Fig. 17.4).

Importantly, the presence of the receptor in a given cell is a determinant for the susceptibility to a certain virus. Thus, cell *tropism*[4] is largely determined by the receptor. For instance, the presence of CD4 in T helper lymphocyte confers the susceptibility to HIV infection. A HeLa cell that is otherwise resistant to HIV infection becomes susceptible to HIV infection if CD4 is experimentally expressed. On the other hand, the viral proteins that recognize the receptors are ones on the surface of virus particles. In the case of enveloped viruses, the envelope glycoproteins on the viral envelope bind to the receptors. For instance, *gp120* of the HIV virion particle binds to the CD4 molecule of the target cells (see Fig. 17.4). In case of naked viruses, capsid proteins may directly bind to the receptor. For instance, the fiber protein of the adenovirus particle binds to the *CAR*[5] molecule on the target cells.

Although the presence or absence of a receptor is a key determinant for the susceptibility of a certain virus, cell tropism can also be determined by *host restriction factors*, which limit the virus infection. For instance, *TRIM5α* of monkey cell restricts HIV infection of monkey cell, whereas TRIM5α of human cell restricts simian immunodeficiency virus infection of human cell (see Fig. 17.17). Thus, TRIM5α is a host restriction factor that is responsible for determining host range of primate lentivirus.

---

3. **Sialic acid** It is a generic term that refers to the N-acetyl neuraminic acid, an amino sugar, which is terminally linked to glycans on the cell membrane.

4. **Tropism** The term *tropism* is derived from Greek word for "a turning"—*tropos*—indicating growth or turning movement of a biological organism. It refers to the cell specificity of viral infection in this context.

5. **CAR (Coxsackie-Adenovirus Receptor)** CAR is exploited by two unrelated viruses for entry (ie, coxsackie type B3 and adenovirus type 5).

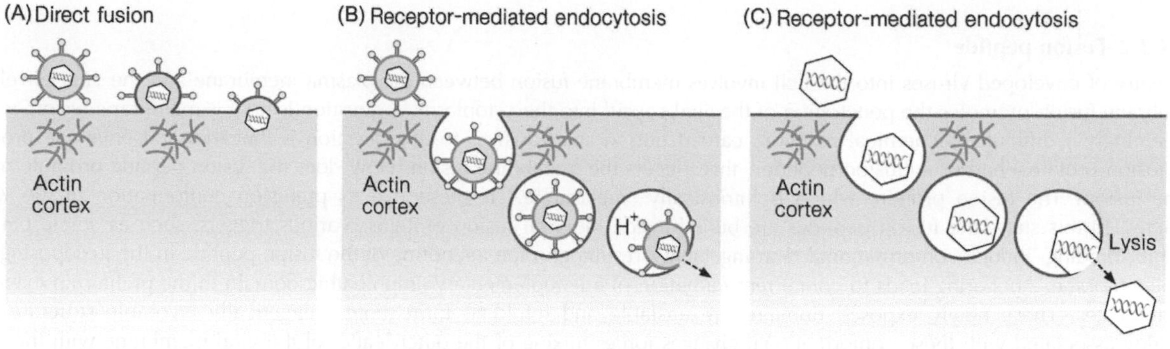

FIGURE 3.3 Three modes of viral penetration. (A) Direct fusion. The viral nucleocapsids enter the cell by the fusion between viral envelope and plasma membrane. (B) Receptor-mediated endocytosis for enveloped viruses. The virus particle acts as a ligand for the endocytosis. The virus particle is located inside endosomes. The fusion takes place between the viral envelope and endosomal membrane. Note that actin cortex (a mechanical support of the plasma membrane), which serves a physical barrier for the entry, is bypassed through the endocytosis. (C) Receptor-mediated endocytosis for nonenveloped naked viruses. A component of viral nucleocapsid triggers the lysis of endosomal membrane necessary for the release of the viral genome to the cytoplasm.

3.2.2 Penetration

Following attachment of the virus particle on the target cells, the next step is the penetration into the cytoplasm. The mechanism for the penetration differs, whether enveloped or not. For enveloped viruses, one of the following two mechanisms is used: *direct fusion* and *receptor-mediated endocytosis*. For nonenveloped naked viruses, receptor-mediated endocytosis is used for penetration.

Direct fusion: Direct fusion, as its name implies, is a mechanism in which two membranes (ie, the viral envelope and cell membrane) fuse (Fig. 3.3A). In this case, the viral nucleocapsid is directly delivered to the cytoplasm, leaving the viral envelope behind on the plasma membrane. Retrovirus is a representative that penetrates by direct fusion (see Fig. 17.4).

Receptor-mediated endocytosis: Although some viruses, as described above, penetrate into the cytosol directly through the plasma membrane, most viruses depend on endocytic uptakes, a process termed *receptor-mediated endocytosis* (Fig. 3.3B). Following the engagement of viral particles on the receptor, the virus particle-receptor complex triggers the endocytosis by forming a coated pit on the plasma membrane, leading to endosome formation. As a result, the virus particle becomes located inside the endosome. The next step is to breakdown the endosome to penetrate to the cytoplasm. The process of endosome breakdown differs whether enveloped or not. For enveloped viruses, the membrane fusion between the viral envelope and the endosomal membrane triggered by acidic pH at early endosome causes the endosome breakdown. More precisely, the fusion peptide embedded on the envelope glycoprotein becomes exposed (ie, activated) as a consequence of conformational change upon low pH; then the fusion is triggered by the fusion peptide (Box 3.2). For nonenveloped naked viruses, the endosome lysis is induced by one of the capsid proteins. In other words, membrane fusion is the mechanism of penetration for envelope viruses (see Fig. 3.3B), while membrane lysis is the mechanism of penetration for nonenveloped viruses (Fig. 3.3C).

As stated above, most viruses enter the cell via receptor-mediated endocytosis. What would be the advantages of receptor-mediated endocytosis, as opposed to direct fusion? Unlike direct fusion, evidently, receptor-mediated endocytosis bypasses the actin cortex or the meshwork of microfilaments in the cortex that presents an obstacle for the penetration (see Fig. 3.3). Moreover, by being taken up by endocytosis, animal viruses can avoid leaving the viral envelope glycoprotein on the plasma membrane, thus likely causing a delay in detection by immune system.

Typically, receptor-mediated endocytosis proceeds via a clathrin-dependent manner (Fig. 3.4). Receptor-mediated endocytosis is the mechanism intrinsic to the cells, which is utilized to take extracellular molecules into the cells. *Clathrin*-mediated[6] endocytosis, which is also the pathway utilized for uptake of LDL, is employed by many viruses, such as influenza virus and adenovirus. Upon the binding of the virus particle with the receptor, a clathrin-coated pit is formed, as clathrins are recruited near the plasma membrane. Following the formation of an endocytic vesicle, the vesicles are fused with early endosomes. The virus particles are now located inside the early endosomes.

6. **Clathrin** A protein that plays a major role in the formation of coated vesicles.

BOX 3.2 Fusion peptide

The entry of enveloped viruses into the cell involves membrane fusion between the plasma membrane and the viral envelope. Membrane fusion promotes the penetration of the viral capsid into the cytoplasm. A question is how is membrane fusion, which is seemingly a difficult biochemical reaction, carried out? A short answer to this question is that the viral envelope proteins (eg, fusion proteins) harbor a "fusion peptide," that triggers the membrane fusion. How does the fusion peptide promote membrane fusion? The fusion protein, which is intrinsically "metastable," is present in its prefusion conformation in the virion particles. In this state, the fusion peptides are buried inside the viral fusion proteins. Various triggers, such as acidic pH and receptor binding, induce conformational rearrangements, resulting in the anchoring of the fusion peptide in the juxtaposing cellular membrane. Anchoring leads to concurrent formation of a complementary amphipathic domain in the prehairpin extended intermediates. These newly exposed domains are unstable and refold to form more energetically favorable structure. The enthalpy associated with these conformational changes forces mixing of the outer leaflet of the viral membrane with the outer layer of the cellular membrane, resulting in formation of the "hemifusion" stalk. The inner leaflet of the lipid bilayers then come into contact and begin mixing, opening a pore (fusion pore) between viral and cellular membranes as the trimeric structures refold into a highly stable postfusion conformation.

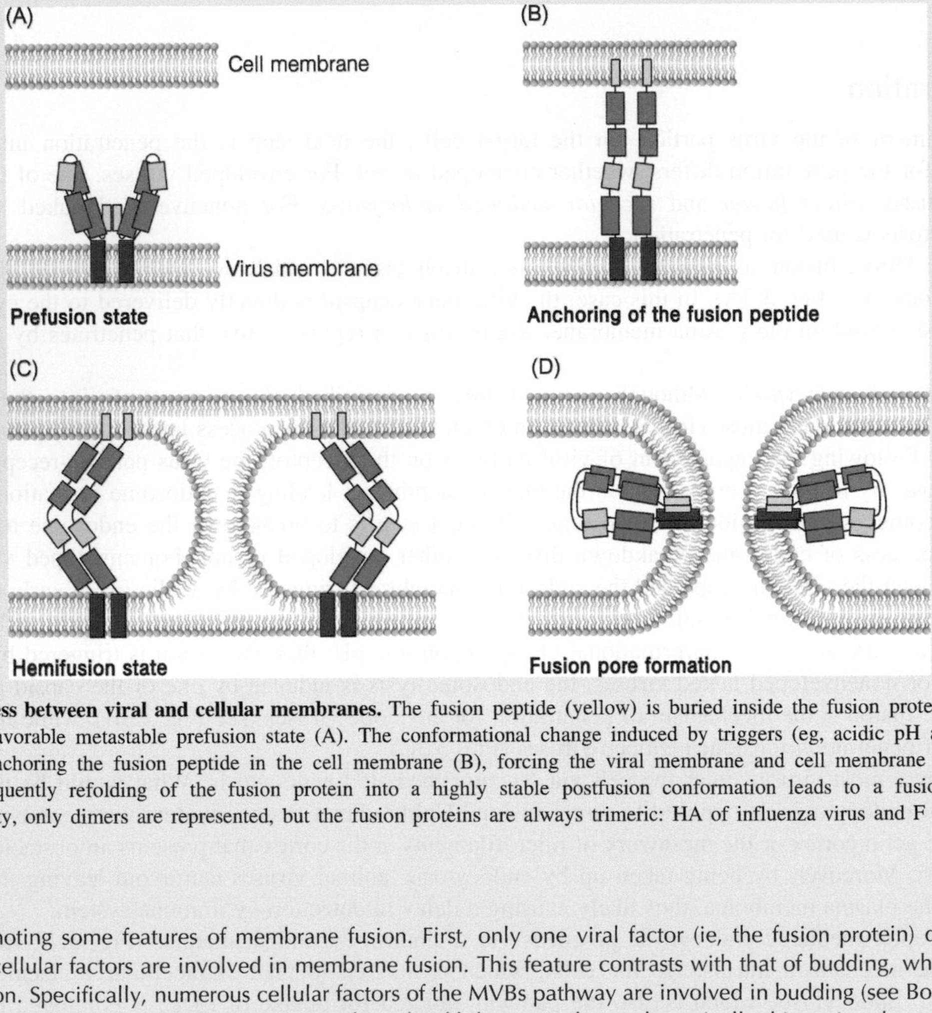

The fusion process between viral and cellular membranes. The fusion peptide (yellow) is buried inside the fusion protein, which is in an energetically unfavorable metastable prefusion state (A). The conformational change induced by triggers (eg, acidic pH and receptor binding) results in anchoring the fusion peptide in the cell membrane (B), forcing the viral membrane and cell membrane into a hemifusion state (C). Subsequently refolding of the fusion protein into a highly stable postfusion conformation leads to a fusion pore formation (D). For simplicity, only dimers are represented, but the fusion proteins are always trimeric: HA of influenza virus and F (fusion) protein of paramyxovirus.

It is worth noting some features of membrane fusion. First, only one viral factor (ie, the fusion protein) drives membrane fusion, but no cellular factors are involved in membrane fusion. This feature contrasts with that of budding, which also involves membrane fusion. Specifically, numerous cellular factors of the MVBs pathway are involved in budding (see Box 3.4). Secondly, membrane fusion, which is essentially the mixing of two lipid bilayers, is thermodynamically driven. In other words, membrane fusion is accompanied with the conformational changes of the fusion protein that is poised to change from a metastable prefusion state to a highly stable postfusion state.

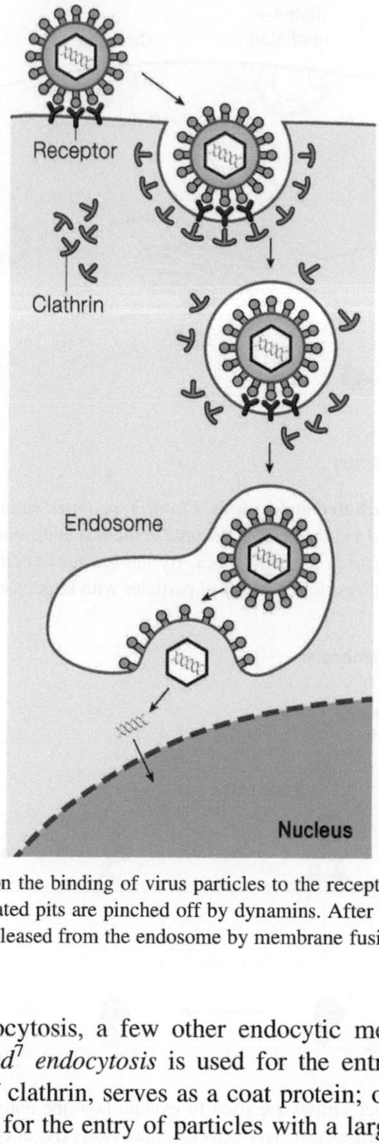

FIGURE 3.4 **Clathrin-mediated endocytosis.** Upon the binding of virus particles to the receptor, clathrins are recruited to form the clathrin-coated pit via its interaction with AP-2 adapter. Clathrin-coated pits are pinched off by dynamins. After the vesicle coat is shed, the uncoated endocytic vesicle fuses with the early endosome. The capsids are released from the endosome by membrane fusion between viral envelope and endosome that is triggered by low pH inside the endosome.

In addition to receptor-mediated endocytosis, a few other endocytic mechanisms are utilized by animal viruses (Fig. 3.5). For instance, caveolin-*mediated*[7] *endocytosis* is used for the entry of polyomaviruses, such as SV40 (see Fig. 6.3). In this case, caveolin, instead of clathrin, serves as a coat protein; otherwise it is similar to *clathrin-mediated endocytosis*. *Macropinocytosis*[8] is utilized for the entry of particles with a larger size, such as vaccinia virus and herpes viruses. The virus particle first activates the signaling pathways that trigger actin-mediated membrane ruffling and blebbing. The formation of large vacuoles (macropinosomes) at the plasma membrane is followed by the internalization of virus particles and penetration into the cytosol by the viruses or their capsids.

3.2.3 Intracellular Trafficking

Following successful penetration inside cells, the virus particles need to get to an appropriate site in the cell for genome replication. This process is termed *intracellular trafficking*. In fact, the biological importance of the cytoplasmic trafficking was not realized until the invention of live cell imaging technology. For viruses that replicate in the cytoplasm, the viral nucleocapsids need to be routed to the site for replication. In fact, *microtubule-mediated transport* coupled with receptor-mediated endocytosis is the mechanism for the transport (Fig. 3.6). In addition, for viruses that replicate in the nucleus, the viral nucleocapsids need to enter the nucleus. For many DNA viruses, the viral nucleocapsids are routed to the perinuclear area via microtubule-mediated transport. In this process, a *dynein motor* powers the movement of virus particles. As an analogy, the viral nucleocapsids can be envisioned as a train in a railroad.

7. **Caveolins** A family of integral membrane proteins which are the principal components of caveolae membranes.

8. **Macropinocytosis** An endocytic mechanism normally involved in fluid uptake.

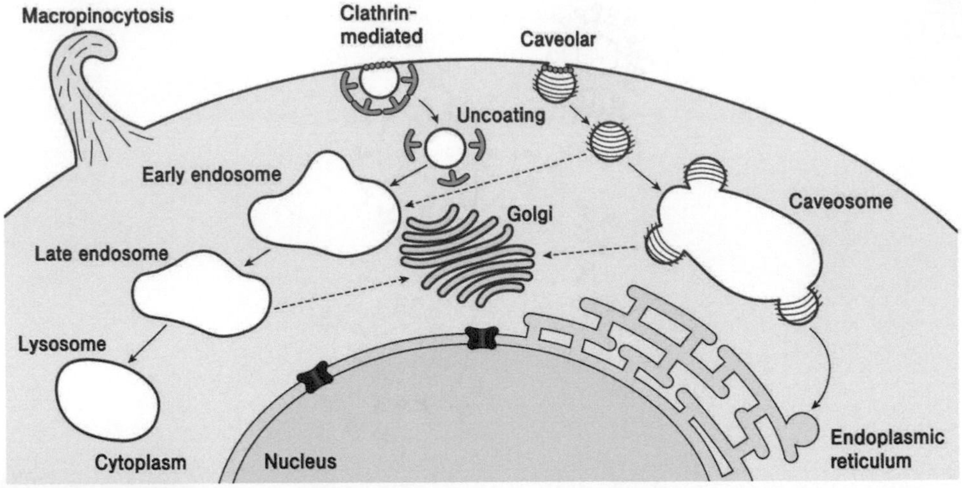

FIGURE 3.5 Diverse pathways for the receptor-mediated endocytosis. *Clathrin-mediated endocytosis.* This pathway is the most commonly observed uptake pathway for viruses. The viruses are transported via the early endosome to the late endosome and eventually to the lysosome. *Caveolin/lipid-raft-mediated endocytosis.* The caveola pathway brings viruses to caveosomes. By the second vesicle transport step, viruses are transported to Golgi, and then to ER. *Macropinocytosis.* Macropinocytosis is utilized for the entry of particles with larger size, such as vaccinia viruses and herpes viruses.

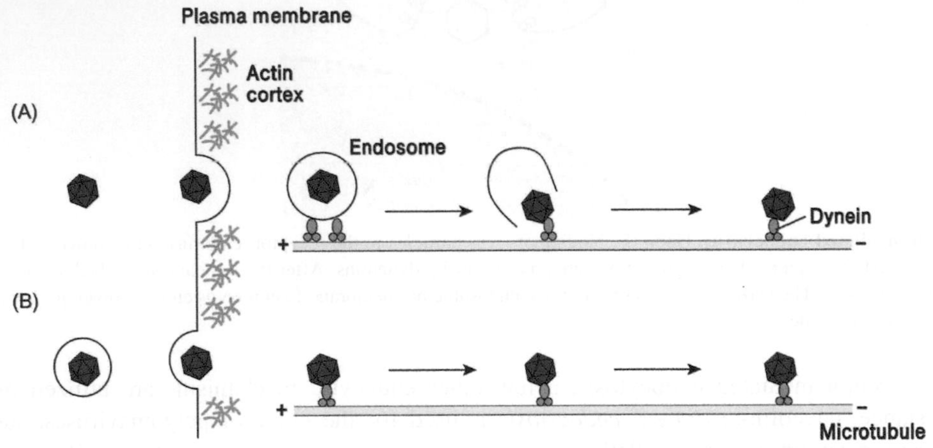

FIGURE 3.6 Cytoplasmic trafficking. Two distinct viruses are used to explain how the entry is linked to cytoplasmic trafficking: (A) adenovirus (naked) and (B) herpes virus (enveloped). Incoming viruses can enter cells by endocytosis (A) or direct fusion (B). Following penetration into cytoplasm, either endocytic vesicles or viral capsids exploit dynein motors to traffic toward the minus ends of microtubules. Either the endocytic vesicles (A) or the capsids (B) interact directly with the microtubules. The virus can also lyse the endocytic membrane, releasing the capsid into the cytosol (A).

3.2.4 Uncoating

As the virus particles approach to the site of replication, from the cell periphery to the perinuclear space, the viral genome becomes exposed to cellular machinery for viral gene expression, a process termed uncoating. Uncoating is often linked with the endocytic route or cytoplasmic trafficking (see Fig. 3.6).

For viruses that replicate in the nucleus, the viral genome needs to enter the nucleus via a nuclear pore. Multiple distinct strategies are utilized, largely depending on their genome size (Fig. 3.7). For the virus with a smaller genome, such as polyomavirus, the viral capsid itself enters the nucleus. For viruses with a larger genome, the docking of nucleocapsids to a nuclear pore complex causes a partial disruption of the capsid (eg, adenovirus) or induces a minimal change in the viral capsid (eg, herpes virus), allowing the transit of DNA genome into the nucleus.

3.3 VIRAL GENE EXPRESSION AND GENOME REPLICATION

The viral genome replication strategies are distinct from each other among the virus families. In fact, the genome replication mechanism is the one that defines the identity of each virus family. Furthermore, the extent to which each virus family relies on host machinery is also diverse, ranging from one that entirely depends on host machinery to one that is

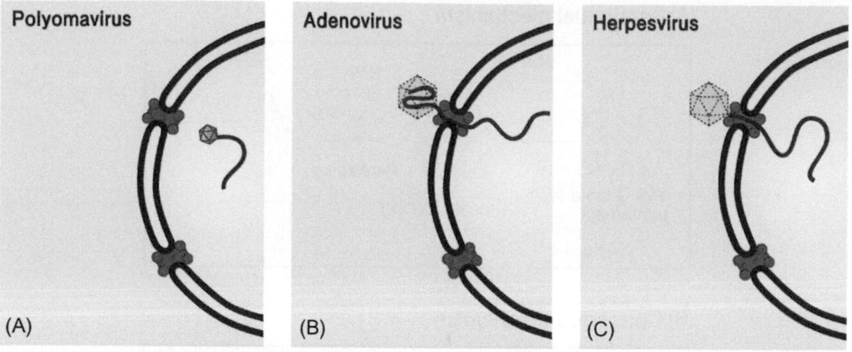

FIGURE 3.7 **Strategies for the nuclear entry of DNA viruses.** (A) Polyomavirus capsids are small enough to enter the nucleus directly via the nuclear pore complex without disassembly. Uncoating of the polyomavirus genome takes place in the nucleus. (B) The adenovirus capsids are partially disrupted upon binding to the nuclear pore complex, allowing the transit of the DNA genome into the nucleus. (C) For herpesvirus, the nucleocapsids are minimally disassembled to allow transit of the DNA genome into the nucleus.

quite independent. However, all viruses, without exception, entirely rely on host translation machinery, ribosomes, for their protein synthesis. This stage of the virus life cycle will be covered in some detail from Part II to Part IV.

3.4 EXIT

Exit can be divided into three steps: capsid assembly, release, and maturation.

3.4.1 Capsid Assembly

The capsid assembly follows as the viral genome as well as the viral proteins abundantly accumulates. The capsid assembly can be divided into two processes: capsid assembly and genome packaging. Depending on viruses, these two processes can occur sequentially or simultaneously in a coupled manner. Picornavirus is an example of the former, while adenovirus is an example of the latter (Fig. 3.8). In the case of picornavirus, the capsids (ie, immature capsid or procapsid) are assembled first without the RNA genome. Subsequently, the RNA genome is packaged or inserted via a pore formed in the procapsid structure. By contrast, in the case of adenovirus, the capsid assembly is coupled with the DNA genome packaging. Then, a question that arises is how does the virus selectively package the viral genome? A *packaging signal*,[9] a *cis*-acting element present in the viral genome, is specifically recognized by the viral capsid proteins, which selectively package either RNA or DNA.

3.4.2 Release

For naked viruses, the virus particles are released via cell lysis of the infected cells. Thus, no specific exit mechanism is necessary, because the cell membrane that traps the assembled virus particles are dismantled. Examples of naked viruses are polyomavirus (ie, SV40) and adenovirus. By contrast, in cases of enveloped viruses, *envelopment*, a process in which the capsids become surrounded by lipid bilayer, takes place prior to the release. With respect to the relatedness of the capsid assembly to the envelopment, two mechanisms exist. First, the envelopment can proceed after the completion of capsid assembly (Fig. 3.9A). In this sequential mechanism, the fully assembled capsids are recruited to the membrane by interaction of the viral capsids with viral envelope glycoprotein. Examples of this include herpesvirus and hepatitis B virus. Alternatively, the envelopment can occur simultaneously with the capsid assembly (Fig. 3.9B). Retrovirus is the representative of this coupled mechanism.

On the other hand, regarding the membrane for envelopment, two cellular membranes are exploited. The plasma membrane is the site of envelopment for some viruses, such as retrovirus and influenza virus, whereas endosomes, such as endoplasmic reticulum (ER) and Golgi bodies, are the site of envelopment for others, such as herpesvirus (see Fig. 9.11) and hepatitis B virus (see Fig. 18.4).

Then, how are the viruses released from the infected cells? Most enveloped viruses are released extracellularly via *exocytosis*[10]; often, this process is also called *budding*, as an analogy of buds in plants. Via budding, the envelopment

9. **Packaging signal** A sequence element in the viral genome that is essential for the genome packaging.

10. **Exocytosis** The process in which a cell directs the contents of secretory vesicles out of the cell membrane into the extracellular space.

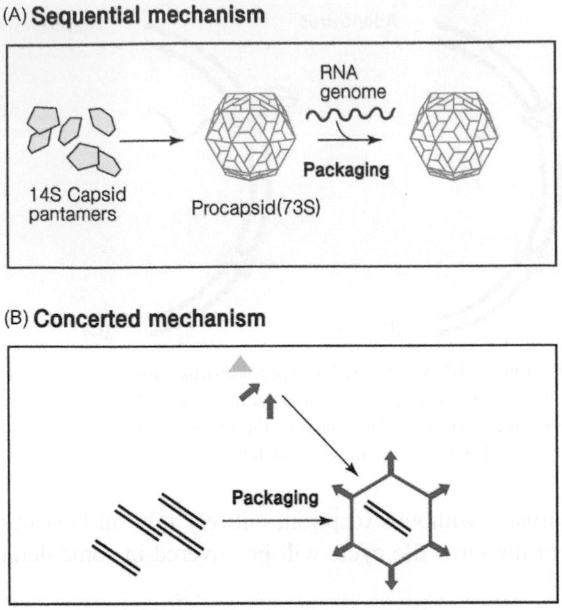

FIGURE 3.8 Relationship between capsid assembly and genome packaging. (A) Sequential mechanism. For picornavirus, the procapsid, a precursor of the capsids, is preassembled without RNA genome. Subsequently, the RNA genome penetrates into the procapsid via a pore. (B) Coupled mechanism. For adenovirus, the DNA genome is packaged into the capsid during capsid assembly.

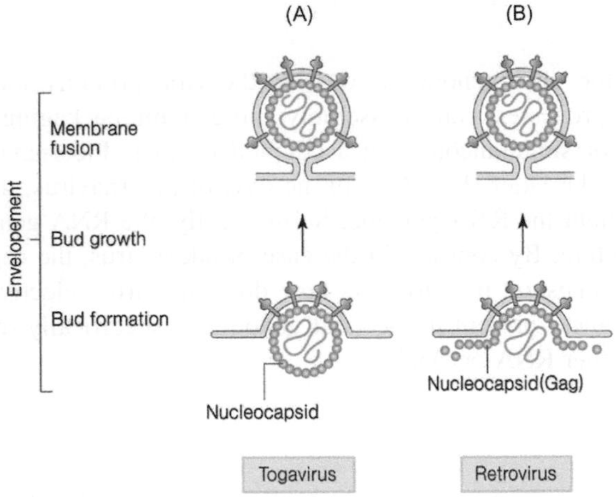

FIGURE 3.9 Relationship between capsid assembly and envelopment. (A) Sequential mechanism. The capsid assembly occurs prior to the envelopment. The assembled capsid is then targeted to the membrane for envelopment. Togavirus constitutes a family of positive-strand RNA viruses (see Table 13.2). (B) Coupled mechanism. Capsid proteins and the viral genome are recruited together to the budding site on the membrane. Capsid assembly and the envelopment of the capsid proceeds simultaneously. The envelopment process can be divided into three steps: a bud formation, a bud growth, and finally membrane fusion.

proceeds in a linked manner with extracellular release. Then a question that arises is how mechanistically is budding triggered? The clue for this was revealed by the identification of a peptide motif termed *late (L) domain*,[11] which is instrumental in triggering the budding process (Box 3.3). For instance, the retroviral Gag protein encodes "PTAP" motif as a late domain. Briefly, Gag protein, via its late domain, recruits cellular factors involved in the *multivesicular bodies*[12] (MVBs) pathway and subverts the MVB pathway for budding. After all, it is intriguing to learn how viruses exploit cellular mechanisms to produce their own progeny extracellularly.

11. **Late domain** A peptide motif (four amino acid), that involves in the budding of enveloped viruses. It is composed of four amino acids such as "PTAP" or "PPXY" residues (see Box 3.3).
12. **Multivesicular bodies (MVBs)** An intracellular structure that is generated by the inward vesiculation in late endosomes. MBV plays a large role in the transport of ubiquitinated proteins and receptors to a lysosome.

BOX 3.3 Late Domain

Late domain, which was first discovered in the Gag polyproteins of retroviruses and M (matrix) proteins of rhabdoviruses, is involved in the budding process of the enveloped viruses. It is composed of four amino acid residues (ie, PTAP, PPPY, and PPEY) encoded in either the rhabdovirus matrix protein or the p6 subdomain of HIV Gag polyprotein (panel A). Intriguingly, a substitution mutation of the late domain motif results in virions attached to the plasma membrane without being released, as if the viral release is blocked at a late stage of budding, a phenotype that is reflected in the nomenclature. It was found that the late domains are involved in recruiting cellular factors involved in the MVBs pathway (panel B). For instance, PTAP motif is critical for the interaction of Gag polyprotein with *Tsg101*,[13] which is a component of *ESCRT-I*.[14] Importantly, the budding of retrovirus particles is topologically equivalent to MVB biogenesis: in both cases budding is directed away from the cytoplasm. In other words, the retroviral late domains mediate the viral egress from the plasma membrane by coopting the cellular machinery for MVB biogenesis.

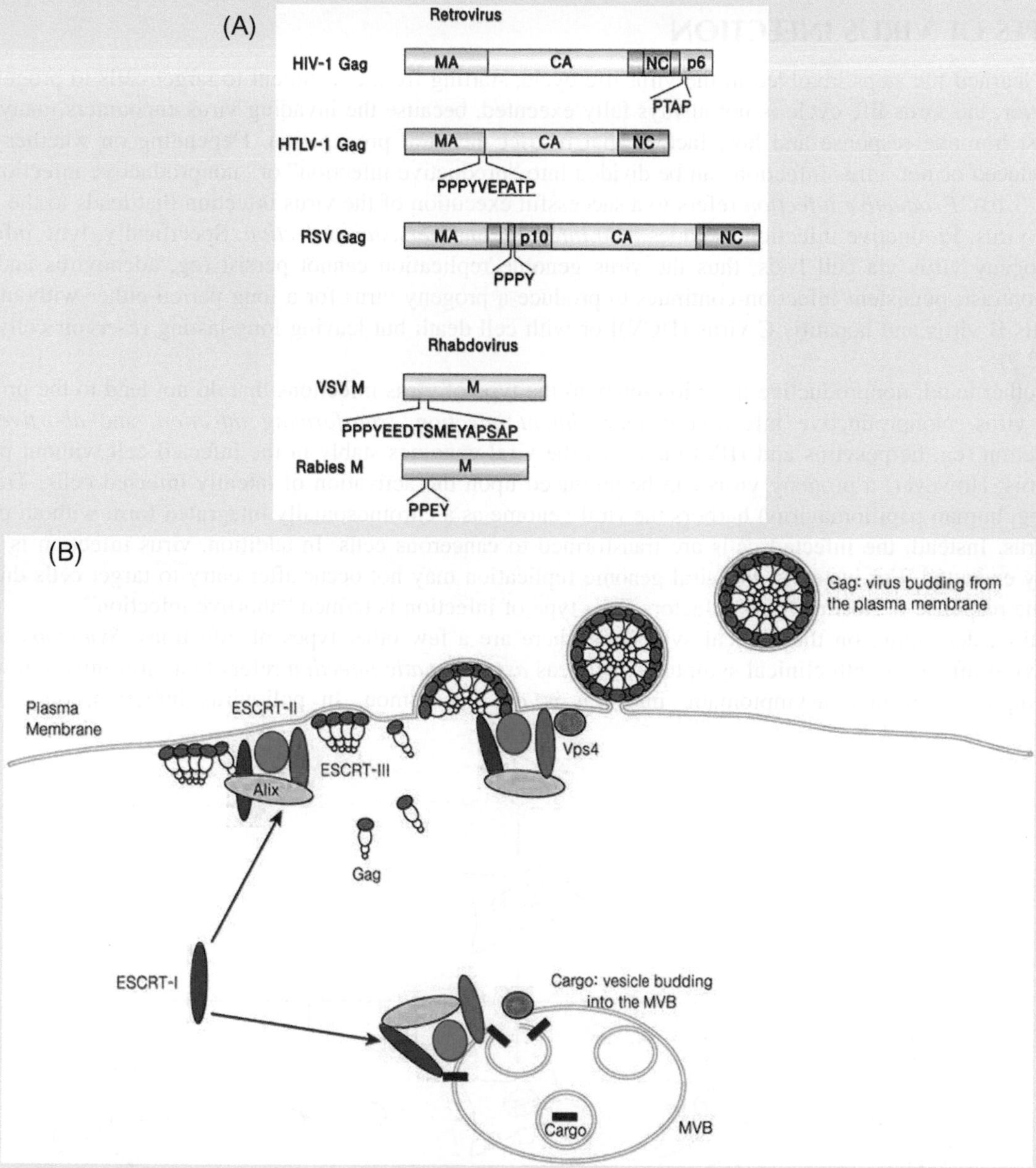

Late domain and its function in MVBs pathway. (A) Three kinds of L domain found in enveloped viruses such as retroviruses (ie, HIV, HTLV-1, and RSV) and rhabdoviruses (ie, vesicular stomatitis virus and rabies virus): PTAP motif, PPPY motif, and PPEY motif. A (alanine), E (glutamic acid), P (proline), T (threonine), and Y (tyrosine). (B) L domain and ESCRT complexes involved in MVBs pathway. The budding of retroviral Gag is facilitated by ESCRT complexes, which are normally involved in the MVB pathway. In HIV, Gag interacts with Tsg101 and Alix, an adapter protein, leading to recruitment of the additional components of the MVB pathway, that is, ESCRT-II (green), ESCRT-III (purple), to assemble into a functional complex. Vps4 (red) is involved in recycling the MVB machinery.

13. **Tsg101 (tumor suppressor gene)** Vps23p, the yeast ortholog of Tsg101, is a component of the ESCRT-I complex, which is involved in the MVB pathway.

14. **ESCRT (endosomal sorting complex required for transport)** ESCRT machinery is made up of cytosolic protein complexes referred to as ESCRT-0, -I,-II, and -III. Together with a number of accessory proteins, these ESCRT complexes enable a unique mode of membrane remodeling that results in membranes bending/budding away from the cytoplasm.

3.4.3 Maturation

The last step of the virus particle assembly is "maturation," a process that occurs extracellularly following release. For picornavirus and retrovirus, maturation is an essential step to acquire infectivity. In case of retrovirus, the cleavage of the Gag polyprotein by the viral PR protein (aspartate protease) occurring in the released virion is accompanied with a considerable morphological transition such as the condensation of the capsid structure (see Fig. 17.10). Importantly, such a maturation process confers the particle its infectivity.

3.5 TYPES OF VIRUS INFECTION

Above, we learned the steps involved in the virus life cycle, starting from attachment to target cells to progeny production. However, the virus life cycle is not always fully executed, because the invading virus encounters many obstacles, such as host immune response and host factors, that restrict the viral propagation. Depending on whether a progeny virus is produced or not, virus infection can be divided into "productive infection" or "nonproductive infection," respectively (Fig. 3.10). *Productive infection* refers to a successful execution of the virus infection that leads to the production of progeny virus. Productive infection includes *lytic infection* and *persistent infection*. Specifically, lytic infection produces a progeny virus via cell lysis, thus the virus genome replication cannot persist (eg, adenovirus and influenza virus). In contrast, persistent infection continues to produce a progeny virus for a long period either without cell death [eg, hepatitis B virus and hepatitis C virus (HCV)] or with cell death but leaving long-lasting reservoir cells (eg, HIV) (see Fig. 22.9).

On the other hand, nonproductive infection refers to the type of virus infections that do not lead to the production of a progeny virus. Nonproductive infection includes *latent infection*, *transforming infection*, and *abortive infection*. Latent infection (eg, herpesvirus and HIV) maintains the viral genomes stably in the infected cell without producing a progeny virus. However, a progeny virus can be produced upon the activation of latently infected cells. Transforming infection (eg, human papillomavirus) harbors the viral genome as a chromosomally integrated form without producing a progeny virus. Instead, the infected cells are transformed to cancerous cells. In addition, virus infection is not always successfully executed. For instance, the viral genome replication may not occur after entry to target cells due to strong host immune response or host restriction factors. This type of infection is termed "abortive infection".

In addition, depending on the clinical symptoms, there are a few other types of infections. *Symptomatic infection* refers to a viral infection with clinical symptoms, whereas *asymptomatic infection* refers to a viral infection without any clinical symptoms. In fact, asymptomatic infection is not uncommon. In poliovirus infection, only one out of

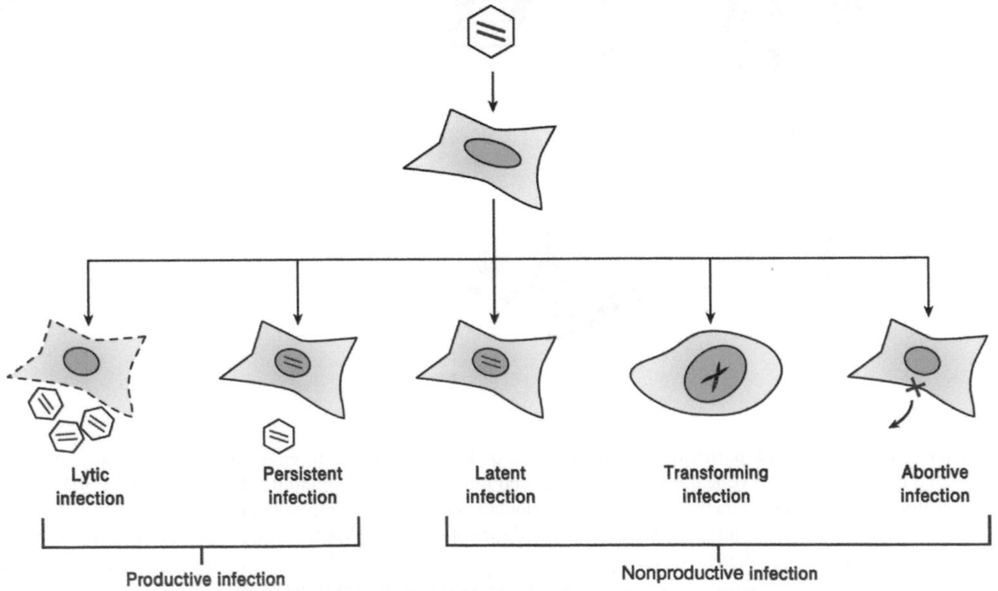

FIGURE 3.10 Types of virus infection. Five types of virus infections are illustrated with emphasis on the progeny virus production and the state of the viral genome (red). The virus life cycle including viral genome replication is fully executed in productive infection, while the virus life cycle is not fully executed in nonproductive infection. The type of virus infection is determined by the intricate interplay between virus and host interaction.

approximately 200 infected people manifests clinical symptoms. In this case, the viral life cycle is executed in a limited way, but an infected individual often ends up having antibodies. Thus, this type of infection is also called *inapparent infection*.

3.6 PERSPECTIVES

Viruses, being intracellular parasites, rely on hosts for their propagation. Through evolution, viruses have acquired the abilities to subvert host functions to comply with their needs. In this regard, similarities are notable with respect to entry, penetration, assembly, and exit stages of the virus life cycle. In fact, many steps in the virus life cycle have been extensively studied in the past three decades. Nevertheless, some novel steps in virus life cycles have only begun to be unraveled. In particular, cell–cell transmission (Box 3.4) is one novel mechanism that draws significant attention,

BOX 3.4 Cell–Cell Transmission

It was believed that viruses could infect neighboring cells only by being released extracellularly from the infected cells. In contrast to this belief, a novel mechanism has been described, in which a virus could infect the neighboring cell without being released. This new mode of infection is termed cell–cell transmission or cell-to-cell spread. Four distinct modes of cell–cell transmission mechanisms have been described. First, cell–cell transmission is mediated by plasma membrane fusion between two cells. The viral capsids are transmitted from an infected cell to uninfected cells without being enveloped. This mode of cell–cell transmission is described in retrovirus and herpesvirus. Second, cell–cell transmission occurs across a *tight junction*.[15] The virus exits basolaterally from an infected cell and is trapped between the infected and uninfected cell membranes at the tight junctions. Using viral entry receptors on the target cell, virions enter the uninfected target cells. This mode of cell–cell transmission is described in herpesvirus and HCV. Third, cell-to-cell spread occurs across a neural synapse. Virions, either mature or incomplete (naked core), assemble in either the postsynaptic or presynaptic cell depending on the virus, and either bud through the membrane into the synaptic space or are released from synaptic vesicles into the cleft. Virions then either fuse directly with the opposing synaptic cells or are endocytosed. Rhabdovirus, herpes viruses, and paramyxoviruses move across neural synapses. Fourth, cell–cell transmission occurs across a virological synapse. Immune cells can be polarized via cell contact, which is termed an *immunological synapse*.[16] Likewise, an infected cell can polarize viral budding toward the receptor-expressing target cell in a structure called a "virological synapse." Virions bud from the infected cell into a synaptic cleft, from which they fuse with the target cell. HIV and HTLV-1, lymphotropic retrovirus, are examples for this mode of viral transmission.

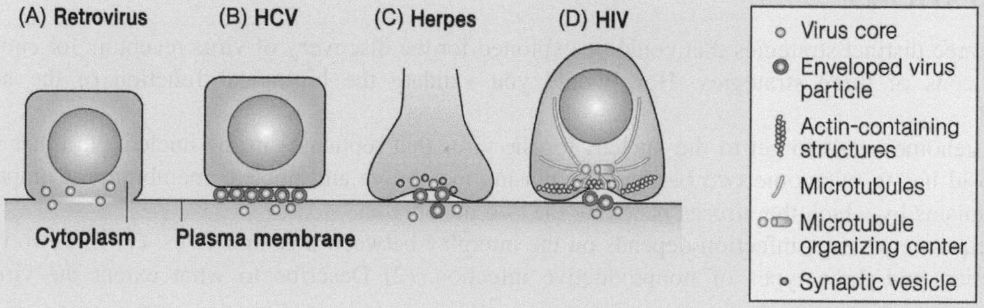

Four modes of cell-to-cell spread. (A) Via plasma membrane fusion. (B) Across tight junction. (C) Across a neural synapse. (D) Across a virological synapse.

(Continued)

15. **Tight junction** Tight junctions are the closely associated areas of two cells whose membranes join together forming a virtually impermeable barrier to fluid. They help to maintain the polarity of cells by preventing the lateral diffusion of integral membrane proteins between the apical and lateral/basal surfaces, allowing the maintenance of specialized functions of each surface.

16. **Immunological synapse** An immunological synapse is the interface between an antigen-presenting cell or target cell and a lymphocyte, such as an effector T cell, which is named as an analogy to a neural synapse.

BOX 3.4 (Continued)

What is the advantage of cell–cell transmission? It facilitates the viral spread from infected cell to uninfected neighboring cells via direct contact without diffusion. More importantly, the viruses associated with cells are physically protected from neutralizing antibodies. In fact, cell–cell transmission was uncovered by an experiment characterizing the neutralizing antibodies-resistant viral transmission. Notably, cell–cell transmission was found only in enveloped viruses, but not in naked viruses. Perhaps, cell–cell transmission is useless for naked viruses, where a bulk of virion is abruptly released upon cell lysis.

because of its implication in viral pathogenesis. It is hoped that our better understanding on cell-to-cell spread could be exploited for the treatment of chronic viral diseases, such as AIDS and viral hepatitis.

3.7 SUMMARY

- *Virus life cycle*: Virus life cycle can be divided into three stages: entry, genome replication, and exit. Entry can be subdivided into attachment, penetration, and uncoating. Exit can be subdivided into virion assembly and release.
- *Attachment*: Two kinds of molecules on cell surface are involved: attachment factors and viral receptors. Glycoaminoglycans, such as heparins, act as attachment factors for many viruses, while membrane proteins that belong to immunoglobulin superfamily act as cellular receptors for the viral entry.
- *Penetration*: Receptor-mediated endocytosis is exploited for the entry of most viruses. Alternatively, direct fusion is used for some viruses.
- *Cytoplasmic trafficking*: Following penetration, microtubule-mediated transport is used to deliver the virus particle to appropriate sites in the cell.
- *Exit*: Naked viruses exit cells via cell lysis, while enveloped viruses exit cells via budding through cellular membranes.

STUDY QUESTIONS

3.1 Describe three distinct strategies that could be exploited for the discovery of virus receptors for entry. Compare the pros and cons of three strategies. How would you validate the biological function of the newly identified receptors?

3.2 The viral genome needs to get to the nucleus for the virus that replicates in the nucleus. In other words, the viral nucleocapsid has to overcome two barriers (ie, plasma membrane and nuclear membrane). Compare and contrast the mechanisms by which the viruses penetrate the two membranes.

3.3 The consequence of virus infection depends on the interplay between host and virus. (1) List two types of productive infection and three types of nonproductive infection. (2) Describe to what extent the virus life cycle is executed.

SUGGESTED READING

Connolly, S.A., Jackson, J.O., Jardetzky, T.S., Longnecker, R., 2011. Fusing structure and function: a structural view of the herpesvirus entry machinery. Nat. Rev. Microbiol. 9 (5), 369–381.

Mercer, J., Schelhaas, M., Helenius, A., 2010. Virus entry by endocytosis. Annu. Rev. Biochem. 79, 803–833.

Sattentau, Q., 2008. Avoiding the void: cell-to-cell spread of human viruses. Nat. Rev. Microbiol. 6 (11), 815–826.

Vigant, F., Santos, N.C., Lee, B., 2015. Broad-spectrum antivirals against viral fusion. Nat. Rev. Microbiol. 13 (7), 426–437.

Yamauchi, Y., Helenius, A., 2013. Virus entry at a glance. J. Cell Sci. 126 (Pt 6), 1289–1295.

JOURNAL CLUB

- Banerjee, I., Miyake Y., Nobs S.P., Schneider C., Horvath P., Kopf M., Matthias P., Helenius A., Yamauchi Y., 2014. Influenza A virus uses the aggresome processing machinery for host cell entry. Science 346 (6208), 473–477.

  Highlight: During cell entry, capsids of incoming influenza viruses must be uncoated before viral ribonucleoproteins (vRNPs) can enter the nucleus for replication. After membrane fusion in late endocytic vacuoles, the vRNPs and the matrix proteins dissociate from each other and disperse within the cytosol. A question is how the vRNPs that disperse in the cytoplasm could make it to the nucleus. This paper revealed that influenza virus subverts "aggresome" formation machinery by mimicking misfolded protein aggregates by carrying unanchored ubiquitin chains.

Chapter 4

Diagnosis and Methods

Chapter Outline

To control the viral transmission, the identification of the culprit is far more important than anything else. Historically, immunological methods have been used extensively for virus diagnosis until recently. As molecular methods became implemented in viral diagnosis since the 1990s, polymerase chain reaction (PCR)-based molecular methods have become a standard in clinical laboratories, because it is sensitive and readily applicable to diverse pathogens. In addition to molecular and immunological tools, some classical methods such as plaque assay are still used for some viruses. Here, the methods used for viral diagnosis as well as the experimental tools used for virus research will be described. Electron microscopy, which is also utilized for viral diagnosis as well as basic research is described in chapter "Virus Structure."

4.1 VIRUS DIAGNOSIS

For the diagnosis of virus infection, diverse experimental tools are utilized to detect, measure, and quantify the viral products (ie, protein and nucleic acids). These experimental approaches can be largely divided into two groups: immunological methods to detect viral proteins (antigens) and molecular methods to detect viral nucleic acids. In addition, hemagglutination is often used for some enveloped viruses, such as influenza viruses.

4.1.1 Protein Detection

Proteins are one of the principal components of virus particles. Historically, the viral proteins have been utilized as viral markers, since they are readily detectable by antibodies. Three immunological methods used for the detection of viral proteins (ie, antigens) will be covered here: ELISA, immunofluorescence assay (IFA), and immunoblotting (IB). All three methods rely on the specificity of antigen—antibody binding. Either polyclonal or *monoclonal antibodies*[1] (Mab) can be used; however, the use of Mab is becoming a standard, as a large number of Mab with their defined epitopes become commercially available.

1. **Monoclonal antibodies (Mab)** Monospecific antibodies that are made by identical immune cells that are all clones of a unique parent cell. Monoclonal antibodies have monovalent affinity, in that they bind to the same epitope.

Molecular Virology of Human Pathogenic Viruses. DOI: http://dx.doi.org/10.1016/B978-0-12-800838-6.00004-7

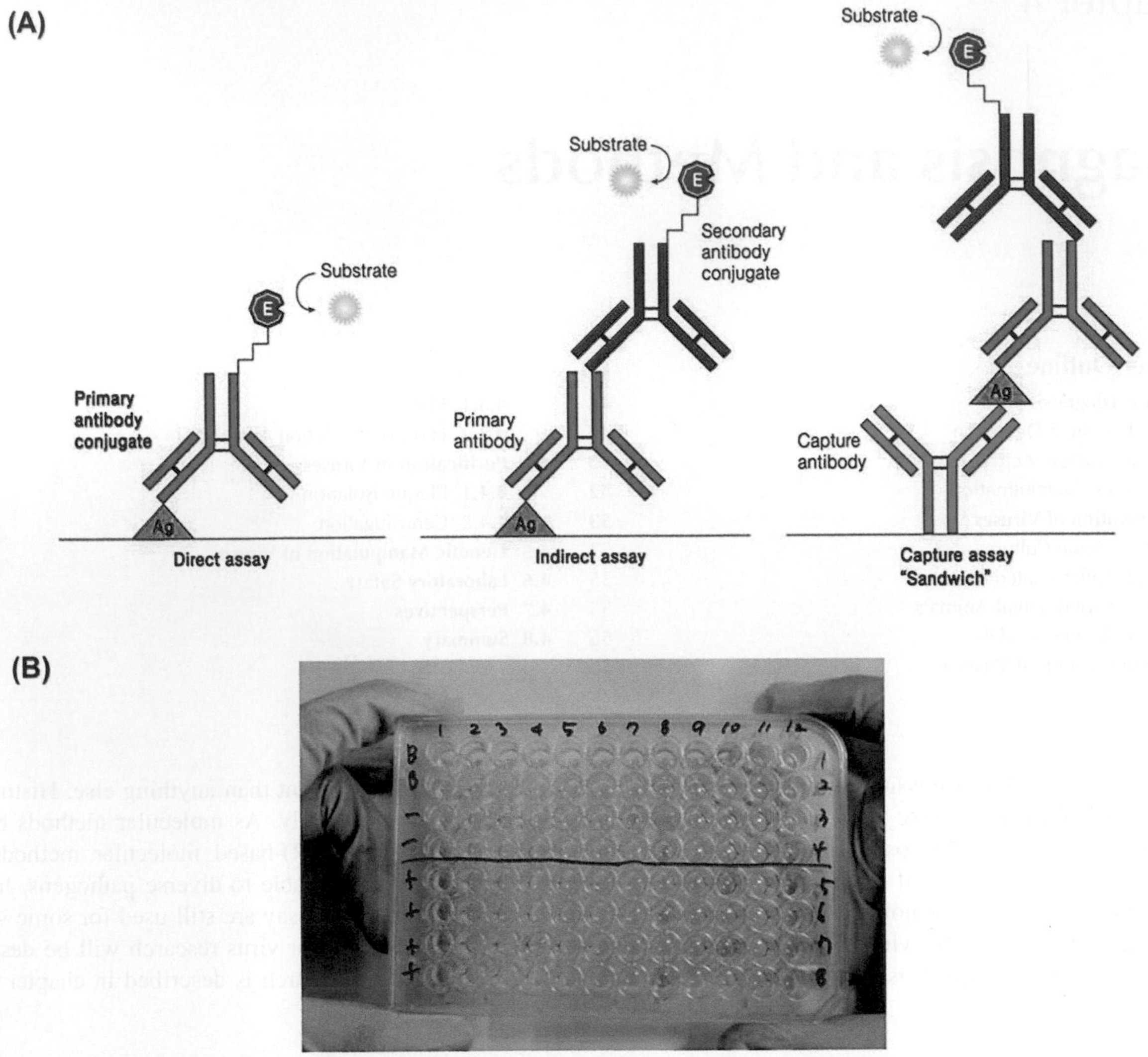

FIGURE 4.1 **ELISA.** (A) Three common formats of ELISA. *Direct ELISA*. First, the plate (ie, well) is coated with antigens derived from specimens (eg, patient serum). Then, the primary antibody conjugate is loaded. The enzyme (ie, AP or HRP) linked to the primary antibody converts the substrate to the colored product. *Indirect ELISA*. The only difference from the direct ELISA is that the enzyme-linked secondary antibody is employed. The antibody specific for the Fc region of the primary antibody is chosen for the secondary antibody. Indirect ELISA is commonly used when the diverse kinds of secondary antibody conjugates are commercially available. *Sandwich ELISA*. This format is similar to that of indirect ELISA, except that the plate is coated with antibody (ie, capture antibody), instead of antigen. (B) A photo of 96-well microtiter plate. The 96 wells are arranged in an 8 × 12 format. Note that the plate is made of polystyrene, to which proteins are adherent. Wells in yellow represent the antigen or antibody-positivity of the specimens, resulting from a typical ELISA reaction.

4.1.1.1 ELISA

ELISA[2] is the most frequently used diagnostic tool for virus detection that combines the exquisite specificity of antigen-antibody binding and the sensitivity of enzyme reaction. In principle, ELISA is a format of analytic biochemistry assay that uses a solid-phase enzyme immunoassay to detect the presence of an antigen in a liquid sample. Commercial ELISA kits are currently used for the diagnosis of human pathogenic viruses including human immunodeficiency virus (HIV), hepatitis B virus (HBV), and hepatitis C virus (HCV) infection in clinical laboratories. It can be formatted to detect antigen as well as antibody. Three formats are typically used: (1) direct ELISA, (2) indirect ELISA, and (3) sandwich ELISA.

Direct ELISA: Direct ELISA is used for the quantitation of antigens present in specimens (Fig. 4.1A).

2. **ELISA** (enzyme-linked immunosorbent assay).

Typically, a 96-well microtiter plate is used as a solid support (Fig. 4.1B). Antigens from the specimens, which are being measured, are attached to the surface of the plate. Then, the primary antibody (ie, antigen-specific) that is conjugated to an enzyme [ie, alkaline phosphatase (AP) or horseradish peroxidase (HRP)] is applied to the well so that it can bind to the antigen. The enzyme linked to the primary antibody converts a chromogenic substrate to a colored product (yellow), when a chromogenic substrate is added. In this format, the antigen attached to the plate is diluted to be limiting, the extent of enzyme reaction (ie, color change) is related to the amount of antigen attached to the plate. As a result, the amount of antigen present in the specimens can be quantitatively estimated. One flaw of the direct ELISA is that the primary antibody conjugates need to be prepared individually for each antigen. Indirect ELISA is designed to obviate this flaw by employing so called secondary antibody conjugate.

Indirect ELISA: Indirect ELISA is fundamentally similar to the direct ELISA, except that one additional antibody (ie, secondary antibody) is employed (see Fig. 4.1A). The secondary antibody is specific to the *Fc fragment* of IgG molecule (see Fig. 5.12) of the primary antibody so that the secondary antibody conjugate can be broadly used. For instance, the secondary antibody specific for the Fc fragment of rabbit primary antibody can be broadly used for any rabbit antibodies, regardless of their antigen specificity. This procedure is dubbed "indirect ELISA," as antigens are indirectly measured by the use of secondary antibodies. Indirect ELISA can be used for the quantitation of antibodies as well as antigens present in specimens. For the quantitation of antibodies, the primary antibody from specimens is diluted to be limiting so that the extent of enzyme reaction (ie, color change) is related to the amount of antibodies.

Sandwich ELISA: Sandwich ELISA or capture assay is used for the quantitation of antigens. Unlike indirect ELISA, two specific antibodies (A and B), instead of one, are required (see Fig. 4.1A). To begin with, a specific antibody A is attached to a 96-well microtiter plate. Then, antigens in the sample are applied to the plate so that antigens could bind to the antibody A attached on a solid support. Then, a specific antibody B is applied so that it could bind to antigens captured on a solid support. Next, the enzyme-linked secondary antibody is applied, as above. Finally, a chromogenic substrate is added for measurement. As the extent of enzyme reaction (ie, color change) is also correlated to the amount of antigen bound to the antibody A, the amount of antigen present in the sample can be quantitatively estimated. This procedure is dubbed "sandwich ELISA," as the antigen is sandwiched between two antibodies. It is important that two antigen-specific antibodies should bind to two distinct sites on the antigen. Two Mab having distinct binding specificities (ie, epitope) on a given antigen are conventionally used (see Fig. 5.13).

Overall, two formats are practically used: (1) indirect ELISA for the detection of antibodies and (2) sandwich ELISA for the detection of antigens (Fig. 4.1).

1.1.1.1 Immunofluorescence Assay

IFA is a microscopic method that can detect and visualize the viral proteins expressed in cells via antigen—antibody reaction (Fig. 4.2). First of all, the virus infected cells are grown on cover glass. To visualize the antigen, the cells are then fixed by paraformaldehyde so that the proteins in the cells are poised to antibody binding. Cells are then permeabilized by a proper detergent such as triton X-100. A specific antibody (eg, patient sera) is applied on the surface so that the viral antigen is recognized by the antibody. Then, the secondary antibody, which is specific to Fc fragment of IgG molecule of the primary antibody, is applied. Since a fluorescence dye is linked to the secondary antibody, the antigen can be visualized under a fluorescence microscope. Subcellular localization of the viral antigen can be revealed by IFA. Moreover, two or even three antigens can be readily visualized simultaneously by using two to three distinct antibodies. Currently, IFA is used for research rather than diagnostic purposes. Nonetheless, in the early days when ELISA kit was not yet commercially available, IFA was used for diagnostic purposes by using patient sera as primary antibody.

1.1.1.2 Immunoblotting

IB is a convenient tool extensively used in laboratories. IB, also called *Western blot*, can readily detect proteins transferred onto a solid support (ie, nitrocellulose membrane). It combines the protein resolving capability of SDS gel electrophoresis with the specificity of antigen—antibody binding (Fig. 4.3). First, the proteins in mixtures (eg, cell lysates) are resolved with *SDS–PAGE*.[3] Next, the proteins on the gel are transferred to a membrane by electric current; they are literally "blotted" to the membrane, as the name implies. Subsequently, the proteins on the membrane are detected by a series of antibody reactions: first with primary antibody for the specific detection of antigens, then with the secondary

3. **SDS–PAGE** Polyacrylamide gel electrophoresis (PAGE), is a technique widely used to separate biological macromolecules, usually proteins or nucleic acids, according to their electrophoretic mobility. SDS–PAGE is a method to separate proteins following denaturation by SDS, a detergent.

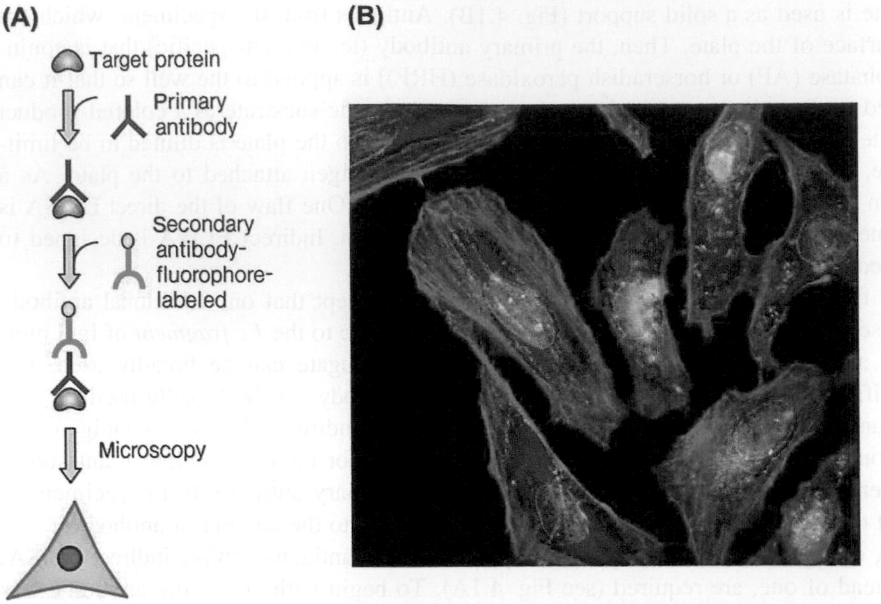

FIGURE 4.2 Immunofluorescence assay. (A) A diagram illustrating the procedure involved in IFA. After cells grown on cover glass are fixed and permeabilized, a primary antibody is added to detect specific antigen, and the secondary antibody conjugated to fluorescence tag (eg, FITC) is sequentially added to bind to the primary antibody. The viral protein in a fixed cell is microscopically detected under fluorescence microscope. (B) Image obtained by IFA. HeLa cells grown in tissue culture and stained with antibody to actin (green), vimentin (red), and DNA (blue).

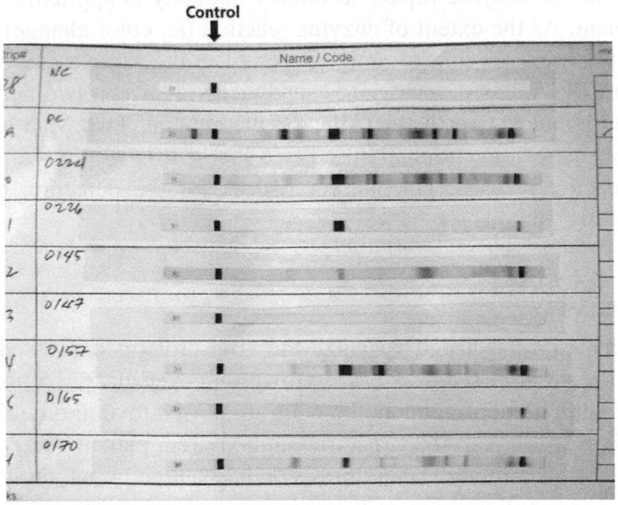

FIGURE 4.3 Immunoblot for viral diagnosis. Immunoblot used for detection of anti-HIV antibodies is shown. A strip of nitrocellulose membrane, onto which HIV antigens had been transferred, was used for detection of anti-HIV antibodies present in serums. Each band on the strip corresponds to different viral antigens separated by SDS–PAGE. A strip with NC (negative control) shows that a serum is negative for the viral antibodies, while a strip with PC (positive control) shows that a serum is positive for the viral antibodies. The band detected in all strips, denoted by an *arrow*, serves as a control for the assay.

antibody, which is linked to an enzyme (AP or HRP) that allows visualization. Finally, the protein bands on the membrane are visualized by the chemiluminescence method, which is better known as *ECL*.[4]

4.1.2 Nucleic Acids Detection

Besides proteins, the nucleic acid component of the virus particles could be readily detectable for the diagnostic purpose. Two methods are frequently used for the detection of the nucleic acids: PCR and real-time PCR (RT-PCR).

4. **ECL (enhanced chemiluminescence)** ECL is a common technique for a variety of detection assays in biology. An antibody that is conjugated to an enzyme (either HRP or AP) is used. The enzyme catalyzes the conversion of the enhanced chemiluminescent substrate into a sensitized reagent, which emits light. ECL allows detection of minute quantities of a biomolecule.

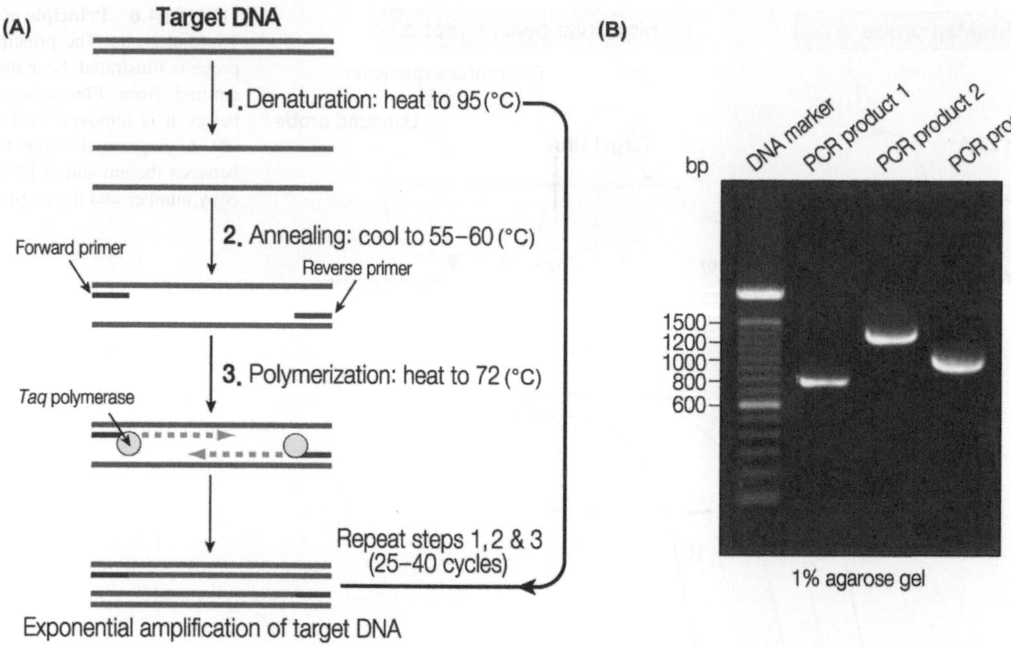

FIGURE 4.4 Principle of PCR. (A) A diagram illustrating steps involved in one cycle of PCR reaction: denaturation, annealing, and polymerization. A PCR reaction typically constitutes 25–40 cycles for the amplification. (B) Agarose gel electrophoresis showing the result of PCR.

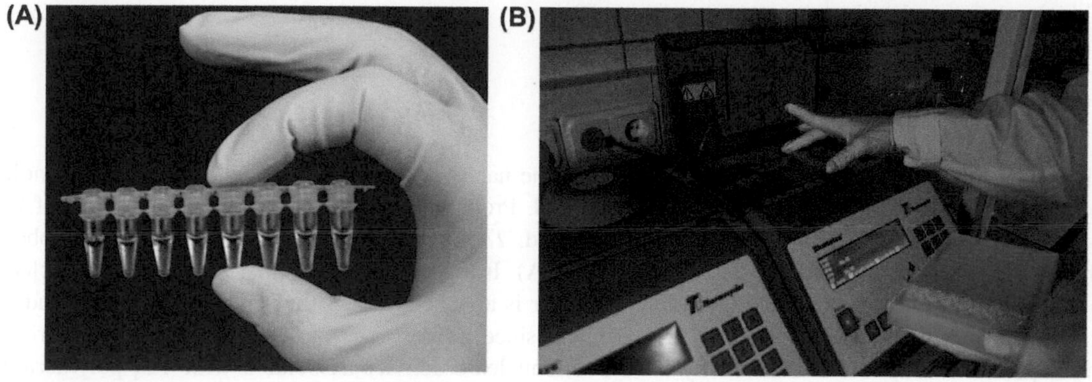

FIGURE 4.5 Thermocycler. (A) A strip of eight PCR tubes, each containing a 100 μL reaction mixture. (B) An eight-tube strip is inserted into heat block of thermocycler.

4.1.2.1 PCR

PCR has dramatically revolutionized viral diagnosis. Considering its versatility, PCR is a truly handy technology, because it requires only one instrument (ie, thermocycler) and only few reagents (Fig. 4.4). More importantly, PCR allows the diagnosis of the viral infection at very low titer (eg, $<10^6$ particle per mL). For instance, clinical diagnosis of early phase of HIV infection, when neither the viral antigens nor antibodies are detectable, can be accomplished by PCR.

The method relies on thermal cycling, consisting of cycles of heating and cooling of the reaction for DNA melting and enzymatic amplification of the DNA (Fig. 4.5). As PCR progresses, the DNA serves as a template for amplification, setting in motion a chain reaction in which the DNA is exponentially amplified. Primers containing sequences complementary to the target region enable specific and repeated amplification. A heat-stable *Taq polymerase* is another key component for the repeated amplification. One shortcoming is that the result of PCR is only semiquantitative, as it inherently involves amplification. To overcome this shortcoming, RT-PCR was developed.

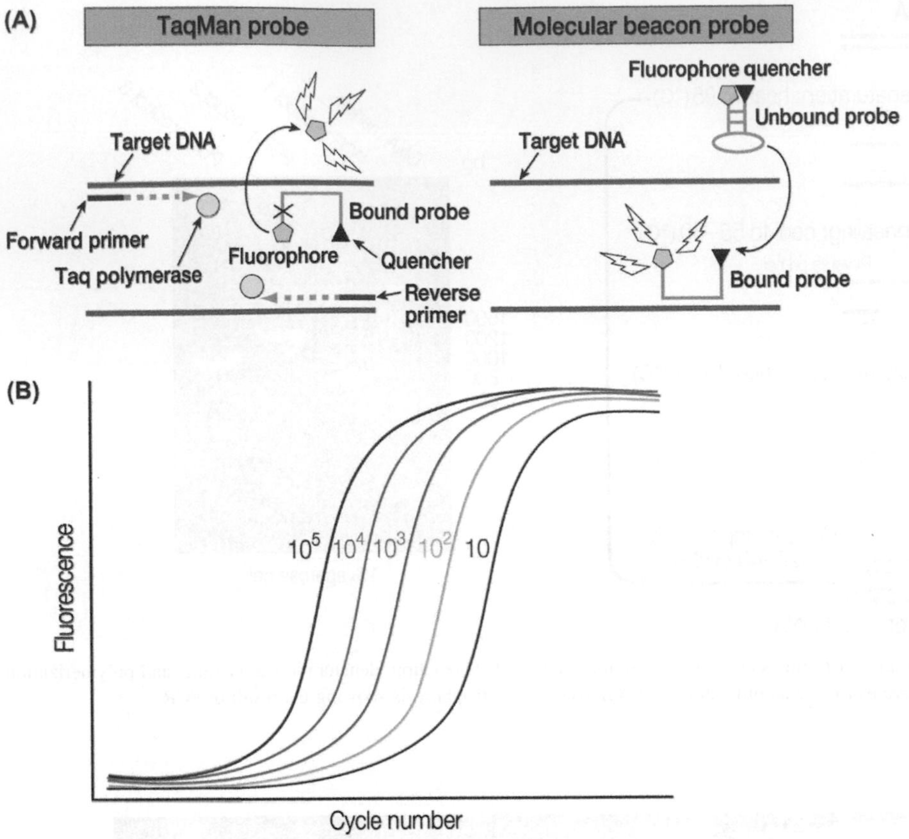

FIGURE 4.6 Principle of RT-PCR. (A) TaqMan probe. The principle of TaqMan probe is illustrated. Note that fluorescence emitted from Fluorophore is quenched before it is removed by Taq polymerase. (B) A diagram showing the relationship between the amount of DNA measured by copy number and the amplification cycle.

4.1.2.2 Real-Time PCR

RT-PCR is an innovative advance of PCR technology. As the name implies, RT-PCR is built with a technology that is capable of monitoring PCR product as it is being amplified. Probing technology is the critical feature of RT-PCR. A few kinds of probing technology have been commercialized: *TaqMan probe* and Molecular Beacon probe. Here, for brevity, the TaqMan technology will be described (Fig. 4.6A). It represents a specially designed oligonucleotide probe that has a fluorophore attached to one end, while a quencher is attached at the other end. When it is bound to the template during the annealing step, no fluorescence is emitted, since fluorescence is quenched. However, during the polymerization step, the fluorophore is cleaved by 5′ to 3′ exonuclease activity associated with Taq polymerase, then the fluorescence can be detected. As the amplification proceeds, fluorescence emitted correspondingly increases. The real-time monitoring capability has transformed the PCR from qualitative to quantitative. For instance, the use of a standard DNA with known copy number in parallel allowed quantitation within a less than twofold error range (Fig. 4.6B).

4.1.3 Hemagglutination

Hemagglutination is used for the diagnosis of some enveloped viruses such as influenza viruses. This method relies on the specific feature of some enveloped viruses that can adsorb to red blood cells (RBCs). Specifically, *hemagglutinin*[5] (HA), an envelope glycoprotein of some enveloped viruses, imparts this property. In the absence of virus particles, RBCs precipitate by gravity to the bottom of the well, giving rise to a distinct red-colored dot in a conical shaped well (Fig. 4.7A). In the presence of virus particles, RBCs clump together as a result of interaction between HA proteins of virus particles and RBC, leading to a lattice formation. In this case, as RBCs are dispersed as a clump, a red dot is not formed. In a given sample, a red dot will appear beyond a certain dilution fold. To carry out a hemagglutination assay,

5. **Hemagglutinin (HA)** Hemagglutinin refers to a substance that causes RBCs to agglutinate or to clump together. This process is called hemagglutination.

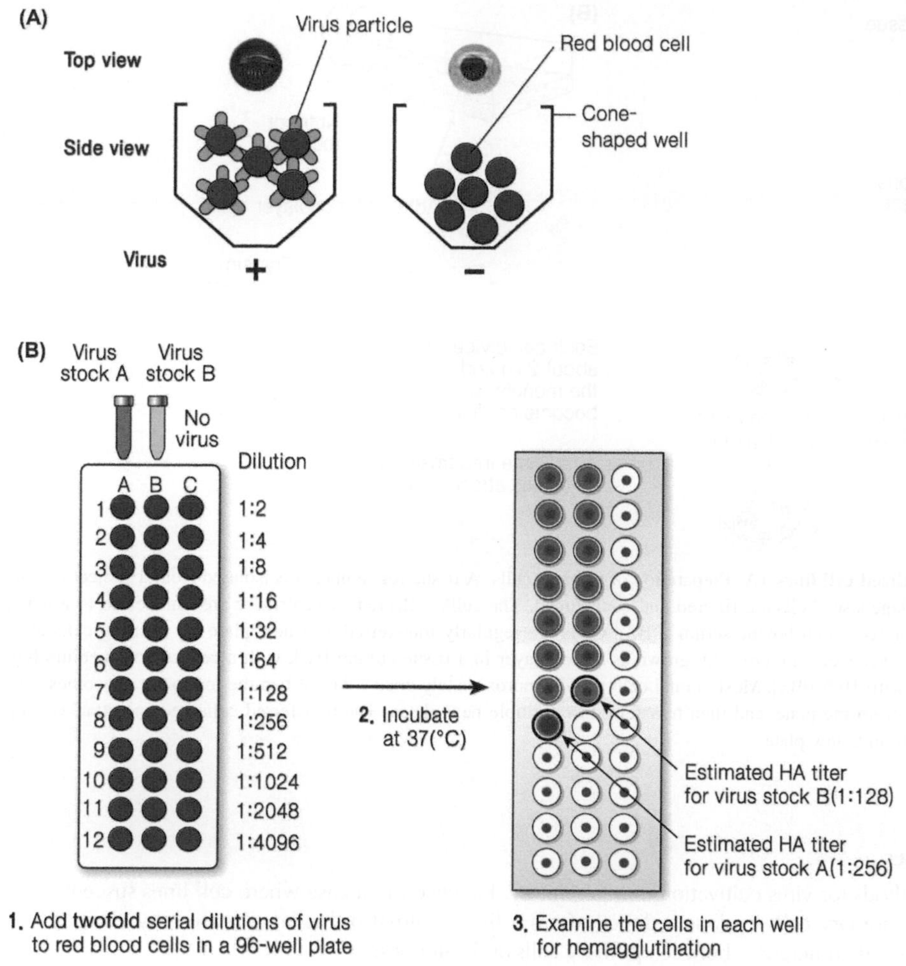

FIGURE 4.7 Hemagglutination assay. (A) A diagram illustrating the principle of hemagglutination. In a positive reaction, RBC becomes agglutinated by virus particles, showing a lattice formation. In a negative reaction, RBC precipitates to the bottom of the well, forming a distinct red dot in a cone-shaped bottom. (B) Titration of virus stocks by hemagglutination assay. The wells denoted by arrows represent the highest dilution that exhibits hemagglutination. This dilution corresponds to HA titer.

a twofold serial dilution of virus-containing samples is dispensed into individual wells of a 96-well microtiter plate (Fig. 4.7B). Then, aliquots of RBC are added to each well. The highest dilution at which clumping is observed is regarded as the *HA titer* of the sample. The virus titer in a sample can be estimated by multiplying the dilution fold. In a standard condition, 1 HA unit corresponds to 10^4 particles per mL.

Hemagglutination is a classical method for viral diagnosis, but is still used for diagnosis of influenza virus today. One outstanding advantage of this method is that it does not require any equipment. Moreover, it is a robust and rapid diagnostic tool, but the sensitivity is somewhat limited.

Now, we will discuss experimental methods used in virus laboratories for research purposes, including virus cultivation, quantitation, purification, and genetic analysis.

4.2 CULTIVATION OF VIRUSES

To investigate viruses and their life cycles, one should be able to propagate the virus in a laboratory. Animal viruses, with a few exceptions, are cultivable in the laboratory by using tissue culture. A variety of animal and human cell lines are now available and used for virus cultivation. In addition to tissue culture, embryonated eggs and experimental animals are used for virus cultivation in certain circumstance.

4.2.1 Tissue Culture

Animal viruses are typically grown by using tissue culture in laboratories (Fig. 4.8). In most cases, cell lines, instead of tissue, are used. Cell lines refer to immortalized cells that have acquired the ability to proliferate indefinitely (Box 4.1).

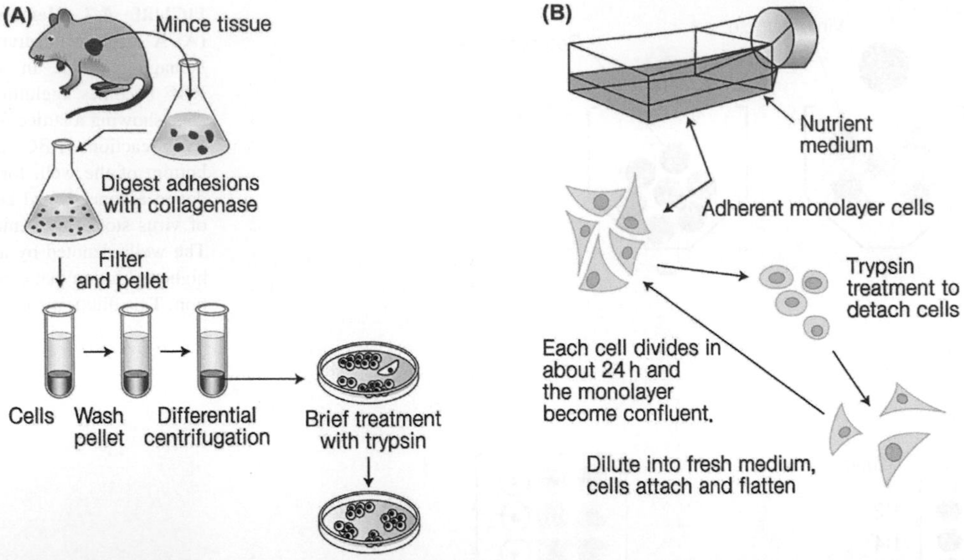

FIGURE 4.8 Primary cells versus immortalized cell lines. (A) Preparation of primary cells. A tissue (eg, embryo) is isolated from a subject by surgery, minced by a knife, and treated with collagenase. Cells are filtered and centrifuged. The cells collected by centrifuge are transferred to a plate, and cultivated by adding culture medium with 10% fetal bovine serum (FBS). Cells are regularly transferred to a new plate by splitting cells after treatment with trypsin. (B) Immortalized cell lines. Cells are typically grown as a monolayer in a tissue culture flask with a cell culture medium [eg, Dulbecco's modified eagle medium (DMEM) with 10% FBS]. Most animal cells divide approximately once a day. When the monolayer becomes confluent, cells are treated with trypsin to detach from the plate, and then transferred to multiple new plates. The transferred cells become attached to a new plate within a day and resume their growth in a new plate.

BOX 4.1 Primary Cells and Cell Lines

Cell lines are the most convenient methods for virus cultivation in a laboratory. However, in a case where cell lines susceptible to the virus infection are unavailable, primary cells explanted directly from a living animal or human are the only choice for virus cultivation. What are the characteristic differences between primary cells and cell lines?

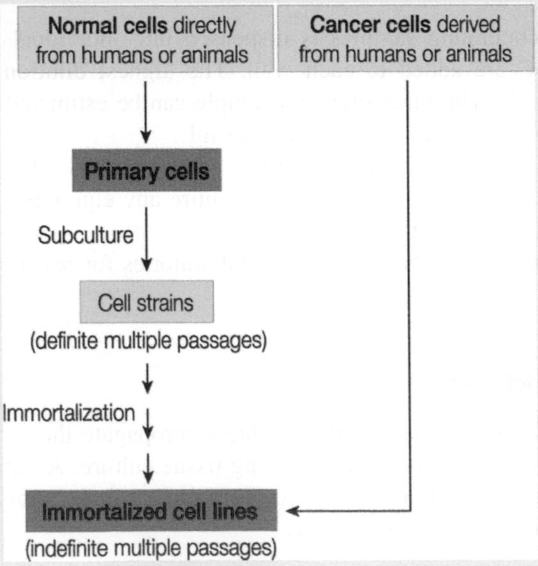

A diagram depicting steps involved in preparation of primary cell versus immortalized cell line. Cell strains are cells that have a limited proliferation capability and can be immortalized. Cell line can be directly prepared from cancerous cells that are already immortalized during cancer development

(Continued)

BOX 4.1 (Continued)

Primary cells: Cells that are isolated directly from a subject are known as primary cells. With the exception of some derived from tumors, most primary cell cultures have a limited lifespan. After a certain number of population doublings (called the "Hayflick limit"), cells undergo the process of senescence and stop dividing, while generally retaining viability. Hence, primary cells can be maintained only for a month or so. Therefore, primary cells need to be prepared each time from a subject. Thus, it is impractical to use primary cells routinely for virus infection in most laboratories. One outstanding feature is, nonetheless, that primary cells maintain almost all properties of tissues (eg, hepatocytes). For instance, none of hepatoma (liver cancer) cell lines available are susceptible to HBV infection, while primary hepatocytes isolated from an individual are susceptible to HBV infection. It is believed that the viral receptor essential for HBV infection is not expressed in hepatoma cell lines, but expressed in primary cells.

Cell lines: An established or immortalized cell line has acquired the ability to proliferate indefinitely via multiple genetic mutations. The genetic changes parallel those found in cancers, including the activation of protooncogenes and the inactivation of tumor suppressor genes (see chapter: "Tumor Viruses"). Furthermore, most cell lines have chromosomal abnormality. One practical advantage of the immortalized cell lines is that they can be cultivated for long periods (ie, months) by splitting cells regularly to a new tissue culture plate.

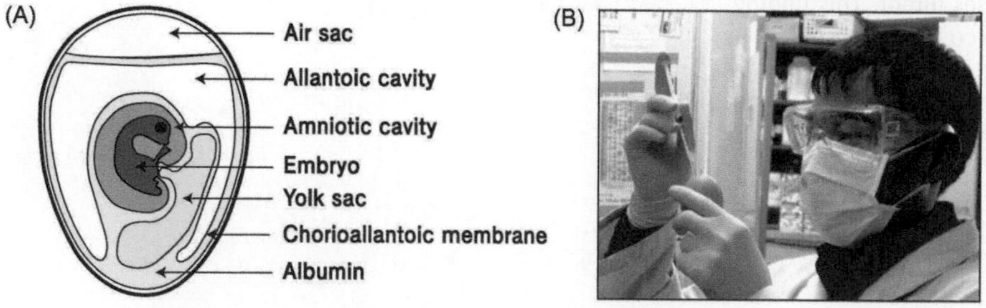

FIGURE 4.9 Embryonated eggs for virus cultivation. (A) A diagram showing the anatomic details of embryonated eggs. Diversely differentiated tissues at day 10–14 after fertilization are denoted. (B) A virus inoculum is being injected to an embryonated egg.

In fact, cell lines from diverse tissue origins (human cancer) were established. For instance, HeLa cells are derived from tissue of human cervical carcinoma (see Box 7.4). For most human viruses, cell lines that support the viral replication are available, with a few exceptions. In the latter case, primary cells are instead used, albeit it is cumbersome to prepare the primary cells taken directly from a living animal.

4.2.2 Embryonated Eggs

Before tissue culture technology were established in the 1950s, embryonated eggs had been used for cultivation of some viruses such as influenza virus during the 1930s–1950s. Even today, embryonated eggs are utilized for the propagation of influenza virus; in particular, for vaccine production (see Fig. 25.10). A chick embryo at 10–14 days after fertilization constitutes diversely differentiated tissues including amnion, allantois, chorion, and yolk sac (Fig. 4.9). Viral inoculum can be administered into the allantoic cavity of embryonated eggs, and a few days later, the progeny viruses can be harvested from the amniotic fluids. Although embryonated eggs have been largely replaced by tissue culture these days, it is still used in a laboratory to obtain high viral titer for some viruses such as influenza virus and Sendai virus.

4.2.3 Experimental Animals

According to Animal Welfare Acts, animal usage for any experiments should be minimized. Nonetheless, animals are often used in virus research laboratories under a certain circumstance. For a given virus, both natural and nonnatural hosts are used as experimental animals. For instance, mice are often used as an experimental animal for the infection of influenza virus, although mice are not the natural host for influenza virus. Moreover, experimental animals are indispensable for antiviral drug development and vaccine development (see chapters: Vaccines and Antiviral Therapy). For instance, cell culture experiment is inappropriate for studying infection pathology, since only one type of cells is cultured. By contrast, in animals, all kinds of cell types including immune cells are present in physiological circumstance.

BOX 4.2 Plaque Assay and Renato Dulbecco

Renato Dulbecco greatly contributed to the advances in cell biology as well as virology by developing important tools that are still being used. First, he developed the *DMEM*, a tissue culture medium, by adding vitamins and amino acids, iron, and high dose glucose to Eagle's Medium. It is remarkable that DMEM, which was developed in the 1950s, is still used these days. More relevantly, in fact, Renato Dulbecco first established the plaque assay for animal viruses. He extended an idea of plaque assay that was then established for counting bacteriophages. Specifically, he infected chicken embryo fibroblasts with Western equine encephalitis virus and observed plaque formation. Establishment of the plaque assay made animal viruses amenable to genetic analysis.

Another important contribution is his work on polyomavirus (SV40), an animal tumor virus (see chapter "Polyomaviruses: SV40"). He discovered that SV40 causes a tumor by integration of its DNA into the host chromosome. This finding established a notion that the viral gene causes tumor. This finding represents a keystone that led to the discovery of the underlying mechanism of viral oncogenesis. For his discovery of viral oncogenes, he shared the 1975 Nobel Prize in Physiology and Medicine with his former student Howard Temin (see Box 17.1) and David Baltimore (see Box 1.2).

A photo of Renato Dulbecco (1914–2012).

Consequently, when the discrepancies of results between the studies of cell culture and those of experimental animal are noted, the results from animal studies are respected.

4.2.4 Animal Models

Humans are the only natural host for some human pathogenic viruses such as HIV. Since human beings cannot be the subject of experimentation, it is crucial to find alternatives that can reflect features manifested in human infection. Such alternatives are collectively called *animal models*. Although animal models are not the natural host of human viruses, they can be experimentally infected by human viruses, and are amenable to experimental analysis. In practice, mice are the foremost choice, as they are small, affordable, and genetically well characterized. Although mice are not the natural host of influenza virus, mice can be experimentally infected, for instance, by the human influenza virus. On the other hand, for some viruses, such as HIV, primates (ie, chimpanzee) are the only choice, as they are the only animal permissive of HIV infection.

4.3 QUANTIFICATION OF VIRUSES

The quantification of viruses involves counting the number of virus particles in a sample. It is an essential method in virus research laboratories. In principle, two methods are used for the quantitation of virus particles or virus infectivity: (1) molecular methods such as PCR, and RT-PCR and (2) virological methods such as plaque assay, end-point dilution assay, and hemagglutination assay. Although the molecular methods have largely replaced the classical virological methods, they cannot measure the infectivity, a defining feature of the virus particle.

4.3.1 Plaque Assay

Plaque[6] assays are the standard method that have long been used to determine the virus titer (ie, infectious dose) (Box 4.2). It determines the number of *plaque forming units* (pfu) in a sample. Typically, 10-fold serial dilutions of the

6. **Plaque** It refers to an area of empty hole in the monolayer of cells in plate, resulting from the cell lysis induced by virus infection.

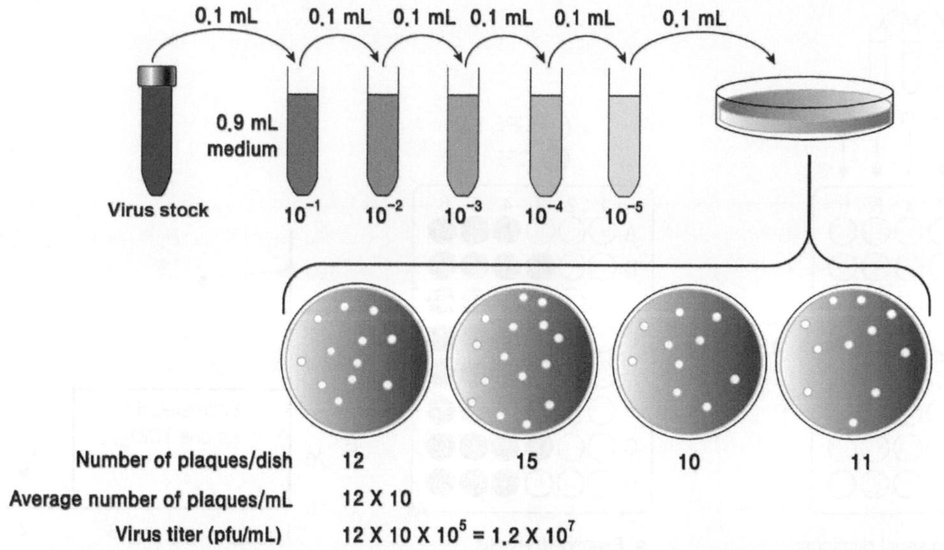

FIGURE 4.10 Virus quantification by plaque assay. This assay is based on a microbiological method conducted in a plate. Specifically, a confluent monolayer of host cells is infected with the virus at varying dilutions and covered with a semisolid medium, such as agar, to prevent the virus infection from spreading indiscriminately. Plaque formation can take about 10–14 days, depending on the virus being analyzed. Plaques are generally counted manually and the results, in combination with the dilution factor used to prepare the plate, can be used to calculate the number of pfu per sample unit volume (pfu/mL).

virus stock are inoculated into each plate (Fig. 4.10). The virus infected cell will lyse and spread the infection to adjacent cells. As the infection-to-lysis cycle is repeated, the infected cell area will form a plaque which can be seen visually. The pfu/mL result represents the number of infectious particles in the sample, assuming that each plaque is caused by a single infectious virus particle. Note that plaque assay is restricted to the viruses that induce cell lysis or death, such as picornavirus, influenza virus, and herpesvirus.

In addition to pfu, *MOI*[7] is frequently used as the unit of virus titer. It refers to the number of virus particles loaded per cell for infection experiment. Low MOI (eg, 0.1) is used for the experiment where only a subset of cells needs to be infected. In contrast, high MOI (eg, >10) is used for the experiment where the vast majority of cells need to be infected.

4.3.2 End-Point Dilution Assay

Plaque assay is limited to only a subset of animal viruses that can lead to cell lysis, forming plaques on the monolayer of cells in a cell culture plate. In fact, many animal viruses do not form plaques on the monolayer, but nonetheless induce a discernible *CPE*.[8] These morphological changes that are observable under microscope can be exploited for quantitation, a method termed "end-point dilution assay." Likewise, 10-fold dilutions of virus stock are individually added to each well of the cell monolayer (Fig. 4.11). A few days later, wells are examined to see whether CPE is present. Then, the fold of dilutions that led to the CPE in 50% of wells seeded can be estimated. This value of virus titer is termed, $TCID_{50}$.[9] For instance, $TCID_{50} = 100$ means that 100-fold dilution of a given virus stock is estimated to induce CPE in 50% of wells seeded. Note that the specific CPE can be diverse, depending on viruses, ranging from vacuole formation to cell lysis.

4.4 PURIFICATION OF VIRUSES

Isolation and purification of virus particles from specimens (ie, tissues or bloods) is essential for the characterization of the virus of interest. For this purpose, virus particles are routinely purified by two methods: either by biological

7. **MOI (multiplicity of infection)** It refers to as the number of virus particles imposed to one cell.
8. **CPE (cytopathic effect)** It refers to any pathological changes (or lesion) in the host cells that are caused by virus infection.
9. **$TCID_{50}$ (tissue culture infectious dose)** It refers to the dilution fold of virus stock that could lead to CPE in 50% of wells seeded. It is a unit of virus titer.

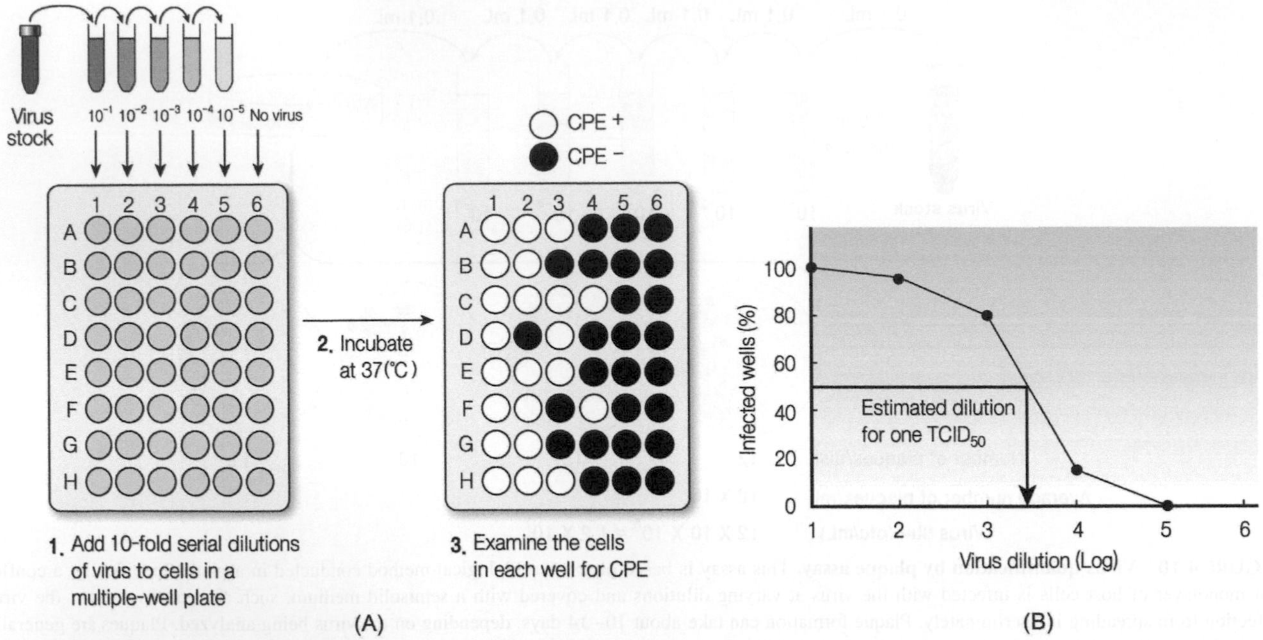

FIGURE 4.11 End-point dilution assay. (A) A diagram illustrating the steps involved in the end-point dilution assay. The virus inoculums in each dilution are seeded into multiple cells. The presence of CPE in each well is denoted in dark blue. (B) Estimation of $TCID_{50}$ value. The percentages of CPE in each dilution are plotted against the logarithmic value of the dilutions in *X*-axis. The arrowed line shows how $TCID_{50}$ can be obtained from the graph.

methods, such as plaque isolation, or by physical methods, such as centrifugation. Importantly, plaque isolation is used for the qualitative purification of (genetically identical) virus particle, while centrifugation is used for the quantitative purification of (biochemical purity) virus particles.

4.4.1 Plaque Isolation

Virus particles can be purely isolated from a single plaque by plaque assay. For instance, a sample can be taken from a single plaque by pipette tip. In fact, the amount of virus particles taken by a tip represents only a trace amount of virus particles. This sample is used as an inoculum to propagate some quantity of virus particles by using cell culture. Plaque-purified virus is considered to be genetically identical. Note that this is only for plaque forming viruses.

4.4.2 Centrifugation

Viruses can be biochemically purified from specimens by physical means such as centrifugation. Centrifugation is an experimental method used to separate objects (ie, virus particles) from specimens by centrifugal force. For instance, a tube containing a sample in a density medium such as sucrose is subjected to centrifugal forces by spinning. Centrifugal force applied to the sample is termed *RCF* (relative centrifugal force). For instance, $1000 \times g$ (gravity) means that 1000-fold of gravity force is applied by centrifugation. The migration speed of objects by centrifugation is determined by size, shape, and density of particles as well as RCF.

A method called "*density gradient centrifugation*" is used for isolation of virus particles. The density gradient of sucrose is prepared from top to bottom of the centrifuge tube. Commonly, sucrose or cesium chloride is used as a density medium. Then, two methods of separation are used: (1) rate-zonal separation and (2) isopycnic (equilibrium) separation (Fig. 4.12).

Rate-zonal separation. The objects in the density medium migrate depending on its size, even though their density is identical. The bigger the size of the particle, the faster it migrates. Specifically, a sample containing viral specimen is loaded onto the top of the preformed density gradient (eg, 10–50% sucrose), and then subjected to centrifugation for a few hours (Fig. 4.12A).

The virus particles migrate or sediment depending on their size, and form a band along the density medium (Fig. 4.12B). By taking a fraction corresponding to the virus-containing band, the virus particles can be biochemically

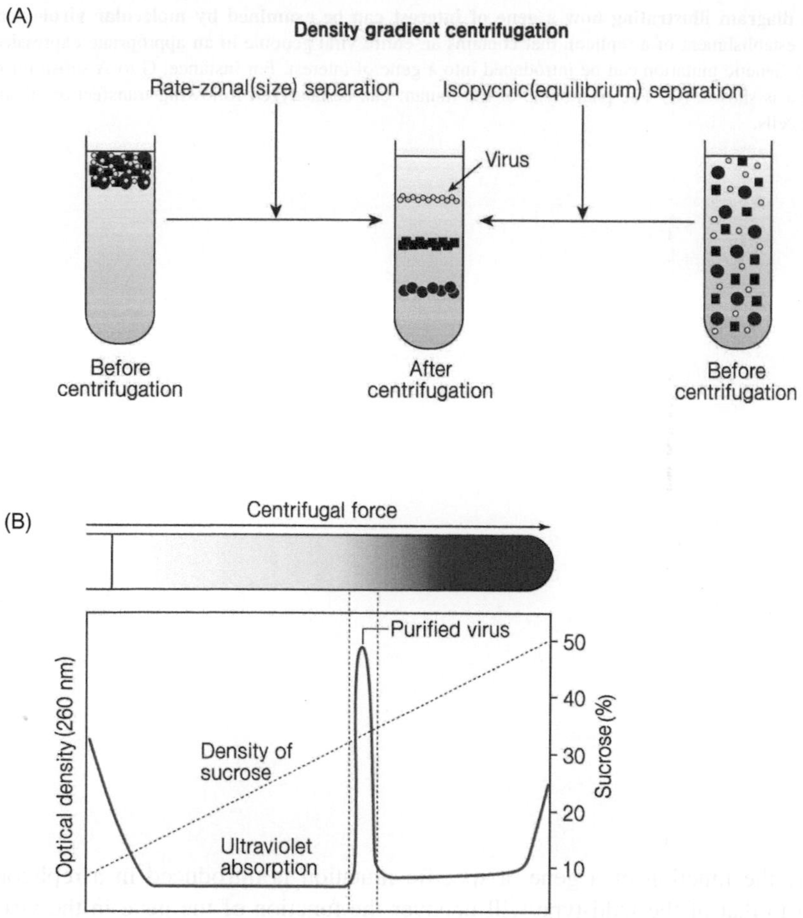

FIGURE 4.12 Density gradient analysis. (A) Rate-zonal separation versus isopycnic (equilibrium) separation. To highlight the difference of principles of separation between two methods, the distribution of objects in the centrifuge tube is shown from one before and one after centrifugation. In rate-zonal separation, the sample is loaded on the top of the density medium, and the particles in the sample migrate from top to bottom, depending on its size. In isopycnic (equilibrium) separation, the sample is mixed with density medium, and the objects in the sample migrate to a position where the density of medium is equal to their density. (B) A diagram illustrating the result of density gradient analysis. The optical density (ie, proteins and nucleic acids) and sucrose concentration (10–50%) are plotted against each fraction from top to bottom of the centrifuge tube, as shown above. The direction of centrifugal force is denoted along with the concentration of density medium (sucrose).

purified. The sedimentation behavior of an object is often referred to as the *sedimentation coefficient*[10] (S), which reflects the relative sedimentation properties of a particle (or molecule) in a standard condition. Practically, it can be considered to simply reflect the size of an object. For instance, an eukaryotic ribosome has a value of 80S.

Isopycnic (equilibrium) separation. When subjected to centrifugal force, an object having a certain density migrates until the density of the surrounding medium is equal to its own density (ie, equilibrium). This method of separation is termed "isopycnic separation."

What is a practical difference between rate-zonal separation and isopycnic separation? In rate-zonal centrifugation, the density gradient is premade before centrifugation, while in isopycnic separation, the density gradient is made during centrifugation. As a result, the former takes only a few hours (1–4 h), while the latter takes much longer (>16 h). More importantly, the former is used to separate an object by its physical size, while the latter is used to separate an object by its density.

4.5 GENETIC MANIPULATION OF VIRUSES

To gain insights into the molecular mechanisms of virus genome replication, the effect of virus gene mutations on virus replication needs to be examined. Recombinant DNA technology had been implemented in virology in the 1970s, and as a result, "*molecular virology*" was established as a subdiscipline of virology. To get started, one needs to obtain the full-length DNA of the viral genome (or the cDNA copy of RNA virus genome), and insert the cDNA into an appropriate plasmid vector to express the viral genes. The plasmid, that is capable of inducing the viral genome replication, is

10. **Sedimentation coefficient** The unit of sedimentation coefficient represents the size of the particle that precipitates upon centrifugation. It is also called Svedberg number (S).

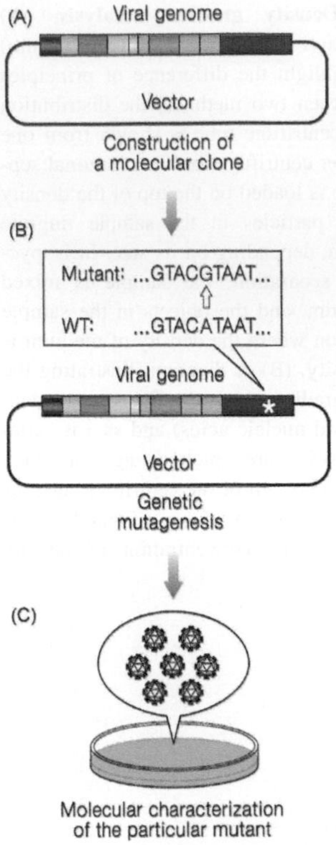

FIGURE 4.13 A diagram illustrating how a gene of interest can be examined by molecular virological approach. (A) An establishment of a replicon that contains an entire viral genome in an appropriate expression plasmid vector. (B) Genetic mutation can be introduced into a gene of interest. For instance, G to A substitution in a gene of interest is shown. (C) The phenotype of the mutant can be analyzed following transfection of the replicon DNA into cells.

termed a "*replicon*."[11] For instance, to study the function of a gene, a specific mutation is introduced in a replicon. Comparison of the phenotypes of the mutant to that of the wild-type will uncover the function of the gene in the virus life cycle (Fig. 4.13).

Molecular virology is a powerful tool for genetic analysis of viral genes. In particular, PCR-mediated mutagenesis enables us to generate all kind of mutations on the replicon plasmid. For instance, one could make a deletion or insertion or substitution mutation in a gene. Molecular virology is also amenable to RNA viruses by inserting cDNA of a full-length RNA genome into appropriate expression plasmid vectors (see Box 11.1). By using a replicon, the mutation analysis of RNA viruses as well as DNA viruses can be carried out.

4.6 LABORATORY SAFETY

Precautions are necessary to prevent laboratory hazards. In particular, workers in clinical laboratories handle human serum and other body fluids that may contain HIV, or HBV or HCV. Proper laboratory practices involve wearing gloves and laboratory coats. On the other hand, research laboratories handling human pathogens, such as AIDS virus and Ebola virus, need to have a biosafety facility. The biosafety facility is essential not only to prevent laboratory infection of investigators but also to preclude spread of laboratory strains to community. A biosafety facility can be classified by the relative danger as *biological safety levels (BSL)*. The levels of containment range from the lowest biosafety level 1 (BSL-1) to the highest level 4 (BSL-4) (Fig. 4.14). Higher numbers indicate a greater risk to the external environment. At the lowest level of biocontainment, the containment zone may only be a chemical fume hood. At the highest level the containment involves isolation of an organism by means of building systems, sealed rooms, sealed containers, positive pressure personnel suits (sometimes referred to as "space suits") and elaborate procedures for entering the room, and decontamination procedures for leaving the room. In most cases this also includes high levels of security for access to the facility, ensuring that only authorized personnel may be admitted to any area that may have some effect on the quality of the containment zone.

11. **Replicon** A replicon in genetics is a region of DNA or RNA, that replicates from a single origin of replication. Here, it refers to a plasmid construct that can induced the viral genome replication, when transfected into cells.

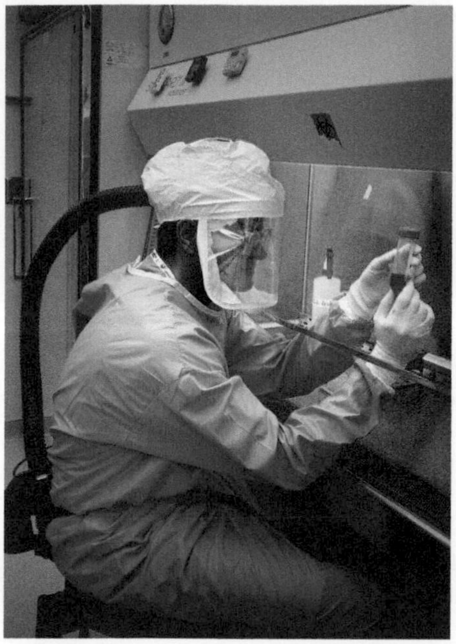

FIGURE 4.14 **Biosafety facility.** Researcher at US Centers for Disease Control working with the influenza virus under biosafety level 3 (BSL-3) conditions, with a respirator inside a biosafety cabinet.

The biosafety level for a given pathogen is stipulated by government authorities (eg, Centers for Disease Control in the United States). For instance, BSL-2 facility is required for studies of pathogens that cause only mild disease to humans or are difficult to contract via aerosol in a laboratory setting, such as HIV, HBV, and influenza virus. BSL-3 facility is required for studies of pathogens that cause severe to fatal disease in humans but for which treatments exist, such as SARS-coronavirus and West Nile virus. On the other hand, BSL-4 facility is required for work with dangerous and exotic agents that pose a high individual risk of aerosol-transmitted laboratory infections, agents which cause severe to fatal disease in humans for which vaccines or other treatments are *not* available, such as Ebola virus, Marburg virus, and various other hemorrhagic diseases.

4.7 PERSPECTIVES

Experimental tools for viral diagnosis have dramatically evolved from morphological and immunological methods to current molecular methods. Nowadays, molecular diagnosis (ie, PCR-based methods) is in a mainstay of viral diagnosis, owing to its sensitivity, reliability, and lower cost. Nonetheless, classical methods such as ELISA and hemagglutination assay are still used in many clinical settings, because of their convenience. On the other hand, advances in recombinant DNA technology have established "molecular virology," which studies the virus life cycle and its impact on the host at the molecular level. In fact, the introduction of molecular technology led to the establishment of molecular virology that has prevailed in the field of virology discipline for past decades. As a result, many details of the virus life cycles of human pathogenic viruses are unraveled. Despite these advances, preventive vaccines and effective antiviral treatment for many human viral diseases are not yet available, including vaccines for HIV, HCV, and Ebola virus. To address these unmet medical needs, several gaps in viral methods need to be addressed. In particular, the importance of animal models for human pathogenic viruses should not be underestimated, because animal models are critically important for studies involving vaccine development and antiviral drug discovery as well as for studies of the infection pathology. Lastly, a biosafety facility becomes essential for virus laboratories handling human pathogenic viruses such as HIV and Ebola virus.

4.8 SUMMARY

- *Nucleic acid diagnosis*: Nucleic acid diagnosis such as PCR, and RT-PCR are mainly used for the detection of viral DNA or viral RNA.
- *Immunological methods*: ELISA, IFA, and immunoblotting are used for the detection of viral antigens.

- *Virus cultivation*: Tissue culture by using cell lines is a convenient method for virus cultivation for most animal viruses.
- *Virus quantitation*: Plaque assay and end-point dilution assays are historically used for virus quantitation. Hemagglutination assay is used for some enveloped viruses such as influenza virus. These classical methods are now largely replaced by molecular methods such as PCR.
- *Molecular virology*: A viral replicon, the plasmid construct capable of inducing viral genome replication, needs to be established for molecular studies. Introduction of mutations in a replicon construct enable us to examine the function of a gene of interest.

STUDY QUESTIONS

4.1 Indirect ELISA can be used for the quantitation of either antigen or antibody. How differently do the ELISA procedures proceed?

4.2 Immunological methods utilized for viral diagnosis involve a solid support onto which the viral antigens are bound or presented. What kinds of solid supports are utilized for three immunological methods? (1) ELISA, (2) IFA, (3) Immunoblotting.

4.3 Transfection of viral replicons is frequently used for the investigation of certain human viruses. Please state the advantages of using replicons for the examination of virus genome replication.

BOOK CLUB

- Boose, J., August, M.J., 2013. To Catch a Virus. ASM Press, Washington, DC, 364 p.
  Highlight: A historical account of virus discovery is vividly captured by authors. From the discovery of Yellow fever virus, the first human virus discovered, to the discovery of HIV, this is a wonderful reading for anyone interested in early day of virus discovery.

INTERNET RESOURCES

- CDC (Centers for Disease Control) site for biosafety: http://www.cdc.gov/biosafety/

Chapter 5

Host Immune Response

Chapter Outline

Upon the infection of microbes, the host organisms recognize the invading microbes and trigger an immune response to eliminate them. The resulting activated immune cells execute a diverse antiviral function to block the virus infection. It is impossible to cover all the details of host immune response to the virus infection in a single book chapter in a comprehensive manner, because of its complexity (ie, numerous immune cells and molecules are involved). Instead, we focus on innate immunity with an emphasis on the recognition of pathogens and the antiviral defense mechanisms.

5.1 DEFENSE SYSTEM OF HUMAN

The front line protection from virus infection is to avoid contact with viruses. For instance, either wearing a face mask or using an air filter is the front line protection from, for example, the respiratory infection such as influenza virus. Daily sanitation practices such as washing hands are also important for protection from the virus infection. Human defense systems against infectious pathogens are divided into two kinds (Fig. 5.1). Natural defense physically protects the infection externally, while immunological defense protects the infection internally. The skin and mucous membrane constitute natural defense. Skin, which is dry, acidic, and partly coated with bacteria, is an inappropriate environment for the viral infection. Mucous secretions from respiratory tracts, urinary organ, and eyes also effectively block the virus infection. Recent appreciation of the importance of the *mucous membrane*[1] in human virus infections such as HIV and herpesvirus infection has drawn attention to *mucosal immunity*.[2] Once the virus successfully invades the body interior, the immune system is the only defense system for the protection.

5.1.1 Innate Immunity Versus Adaptive Immunity

Immune response to viral infection includes innate immunity, which ensues immediately after infection, and an adaptive immunity, that comes after a delay. The adaptive immunity begins to occur when the innate immune response becomes

1. **Mucous membrane** The mucous membrane (or mucosa) is the lining covered in epithelium, which is involved in absorption and secretion. It lines cavities that are exposed to the external environment and internal organs.

2. **Mucosal immunity** It refers to immune response pertaining to the mucous membrane, which is distinct in that IgA, instead of IgG, is the type of immunoglobulin that acts.

Molecular Virology of Human Pathogenic Viruses. DOI: http://dx.doi.org/10.1016/B978-0-12-800838-6.00005-9

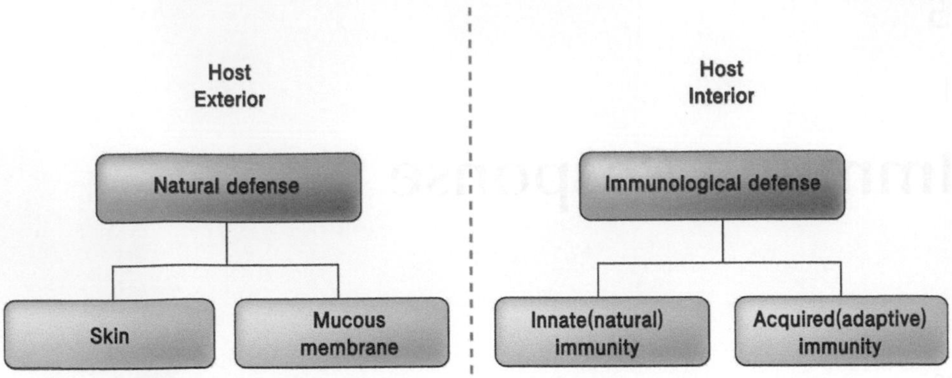

FIGURE 5.1 Host defense mechanism against pathogens. Host exterior such as skin and mucous membrane serve as a natural defense or barrier against invading pathogens. Once inside the human body (host interior), the invading pathogens encounter immunological defense, such as innate immunity and adaptive immunity.

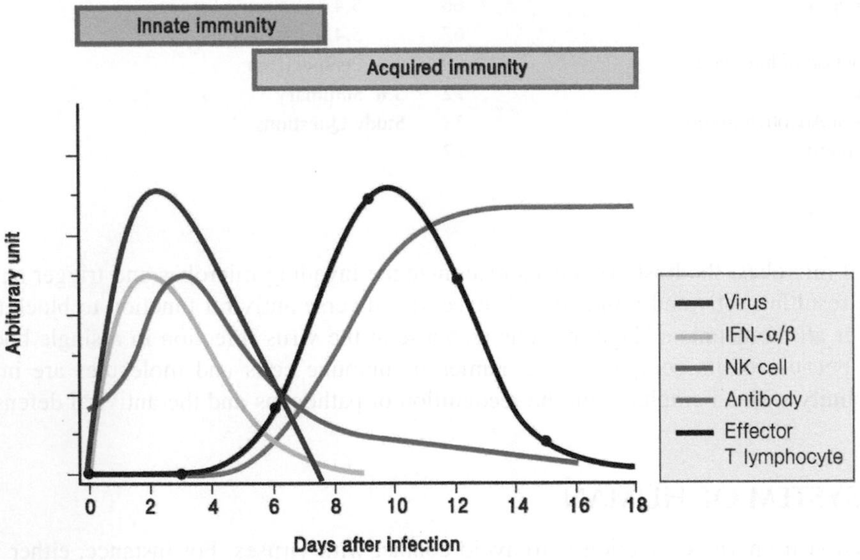

FIGURE 5.2 Time courses of the innate and adaptive immune response. Immediately after infection, the innate immune response, such as interferon production, and NK cell activity, acts to suppress the viral replication. The viral clearance is achieved by the adaptive immune response, which initiates upon the decrease of the innate immune response. Among the antiviral functions of the adaptive immunity, the CTL activity is gradually decreasing, while the antibody level is maintained for a long period.

attenuated (Fig. 5.2). In short, the innate immunity controls the early phase of virus infection, while the adaptive immunity controls the later phase of virus infection.

Let's consider the difference between innate immunity and adaptive immunity. (1) First of all, distinct kinds of immune cells are involved (Table 5.1). The innate immunity involves dendritic cells (DC), macrophages (Mϕ), and NK cells, while the adaptive immunity involves B lymphocytes and T lymphocytes. (2) The difference in the response kinetics stands out. The innate immune response is a rapid response, in which dendritic cells are ready to act, while the adaptive immune response is a delayed response, in which B or T lymphocytes need to be activated, and differentiated. (3) The diversity of pathogenic molecules recognized is characteristically different. The diversity of molecules recognized by innate immunity is limited, because the common molecular patterns of pathogens are recognized. In contrast, the diversity of molecules recognized by adaptive immunity is unlimited in principle. (4) The antiviral function of innate immunity is nonspecific, while that of the adaptive immunity is specific. For instance, *interferon (IFN)-α*,[3] an

3 **IFN** IFN are a group of cytokines, which trigger the induction of a broad array of antiviral proteins. Interferons are named for their ability to "interfere" with viral replication by protecting cells from virus infections

TABLE 5.1 Features of Innate Immunity and Adaptive Immunity

	Innate Immunity	Adaptive Immunity
Recognition	Molecular pattern (PAMP)	Antigen, peptide epitope
Recognition receptor	TLR, RIG-I, MDA5	BCR, TCR
Response	Rapid	Delayed
Diversity	Limited	Unlimited
Duration	Transient	Prolonged
Cells	Dendritic cell (DC)	Dendritic cell (DC)
	Macrophage (MΦ)	B lymphocyte
	Natural killer cell (NK)	T lymphocyte
Antiviral function	Cytokines (IFN-α/β)	Antibodies, CTL

antiviral cytokine resulting from innate immune response from one particular virus, suppresses other unrelated viruses as well. In contrast, antibodies induced by a particular antigen (viral protein) could not bind to other antigens. (5) Lastly, the duration of immunity is different. The innate immunity is only transient, while the adaptive immunity is prolonged due to *immunological memory*.[4]

Importantly, although the antigen recognition by the innate immunity is distinct from that of adaptive immunity, both have the capability of distinguishing self from nonself. Further, innate immunity is closely linked to adaptive immunity, as you will see below (see Fig. 5.11).

5.2 INNATE IMMUNITY

Innate immune response is a rapid response occurring within only a few hours after the invasion of pathogens. One of the main outcomes is the production of IFN-α, which is the hallmark of antiviral response induced by innate immunity.

5.2.1 Cells Constituting Innate Immunity

As stated above, the innate immune system comprises many types of immune cells, including DCs, macrophages, and NK cells (Fig. 5.3). These cells react to the pathogens via distinct mechanisms. (1) Dendritic cells: DCs recognize the pathogens, thereby leading to the production of IFN-α/β. (2) Macrophages: Macrophages engulf and then digest the invading pathogens. They also stimulate lymphocytes and other immune cells in respond to the pathogens. (3) NK cells: NK cells are a type of cytotoxic T lymphocyte pertaining to the innate immunity. The role of NK cells is analogous to that of cytotoxic T cells in adaptive immune response. NK cells are unique, however, because they have the ability to recognize the cells in the absence of an *MHC*[5] molecule, allowing a faster immune reaction.

Next, we will consider how innate immune cells recognize the pathogens.

5.2.2 Toll-Like Receptors

How do cells constituting the innate immunity recognize the invading microbes? It was speculated early on that a yet-to-be-known cellular receptor would recognize the invading pathogens. In support of this prediction, the cellular receptor was indeed identified as *toll*-like[6] receptor (TLR) (Box 5.1). Fifteen TLRs have now been discovered in mammals (Fig. 5.4). In fact, TLRs are expressed not only in immune cells, such as dendritic cells and macrophages, but also in

4. **Immunological memory** It refers to the function of memory lymphocytes that are differentiated from the activated lymphocytes, and can swiftly respond to the antigens.

5. **MHC (major histocompatibility complex)** An MHC molecule displays an oligopeptide, called epitope, of antigens on the cell surface of antigen-presenting cells (APCs) such as DCs and macrophages.

6. **Toll** a gene that was discovered in Drosophila. Toll was derived from her acclamation in German, "Das war ja toll (That is great!)."

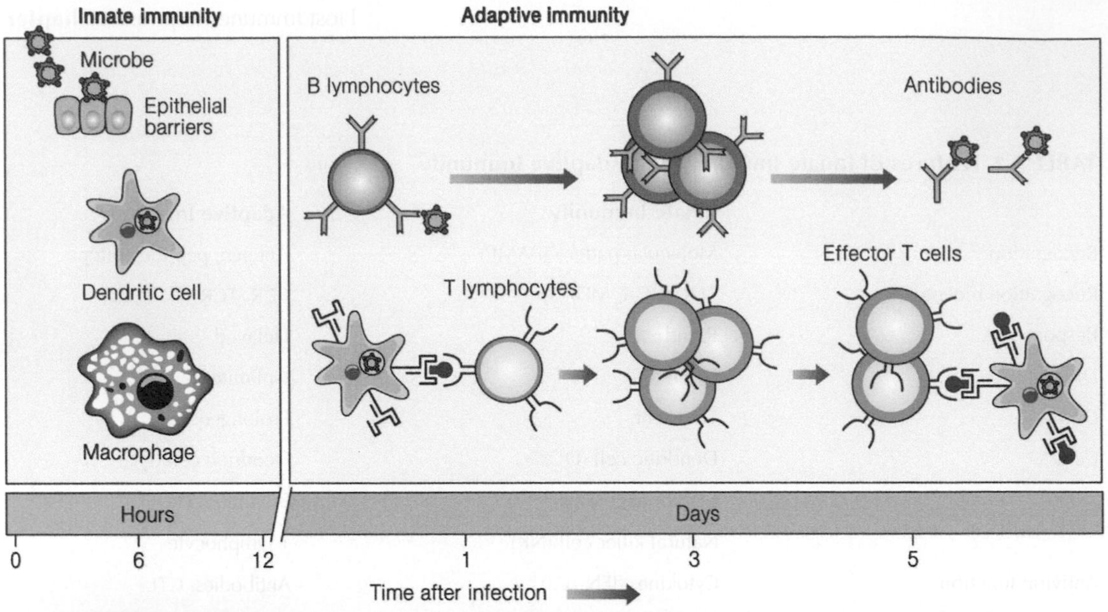

FIGURE 5.3 Immune cells involved in the innate immunity and adaptive immunity. Innate immunity is the front-line defense once the microbes bypass the epithelial barriers. Phagocytes such as dendritic cells, and macrophages engulf the invading microbes, whereas NK cells kill the infected cells. Adaptive immunity ensues. B lymphocytes drive humoral immunity, while T lymphocytes drive cellular immunity. Antibodies, the product of humoral immunity, directly bind to the virus particles and block the infection. Cytotoxic T lymphocyte (CTL), the product of cellular immunity, kills the infected cells.

BOX 5.1 Toll-Like Receptor and Dendritic Cells

One of the enigmas in immunology was how the innate immune system recognizes the invading pathogens, which are diverse in nature, ranging from bacteria, fungi, yeast, viruses, etc. Charles Janeway, an eminent immunologist, hypothesized in 1989 that characteristic molecules of the microbes are recognized by cellular receptors, which he coined "pathogen recognition receptor." A clue for his hypothetical receptor was obtained in a *Drosophila* experiment by Jules Hoffman. Specifically, he found that a fly lacking a *Toll* gene is vulnerable to fungal infection, hinting that a sensor for the recognition of the invading fungi was lacking in the mutant fly. Subsequently, a mammalian homolog of Toll (Toll-like receptor or TLR) was discovered in rodents, implicating the conservation from invertebrates to vertebrates.

In addition, another important finding in innate immunity was the discovery of the dendritic cell, which is instrumental in sensing the invading pathogens. In fact, Ralph Steinman found a star-shaped cell that looks like dendrites in the lymph node, which he called the dendritic cell (DC) based on its cellular morphology. Subsequently, he and others showed that DC plays a central role in sensing the invading pathogens via TLRs. The discovery of TLR and DC paved the road to our current understanding of innate immunity. For their discovery of TLR and DC cells, Ralph Steinman and Jules Hoffman shared the 2011 Nobel Prize.

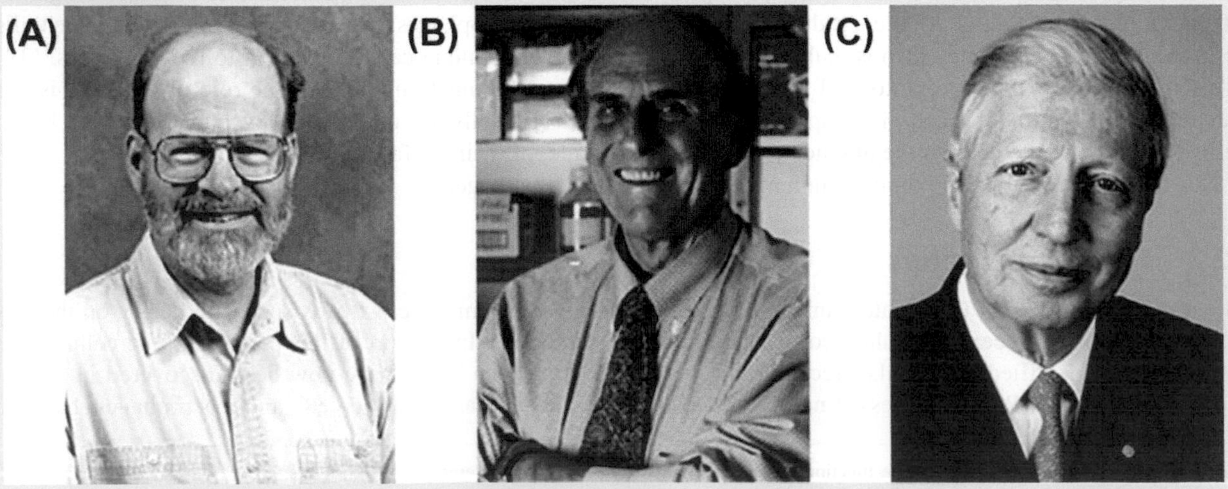

Photos of three leading immunologists. (A) Charles Janeway (1943–2003) of Yale University. He is also well known as an author of "Janeway's Immunobiology," a textbook for immunology. (B) Ralph M. Steinman (1943–2011) of Rockefeller University. (C) Jules Hoffman of University of Strasbourg in France (1941–).

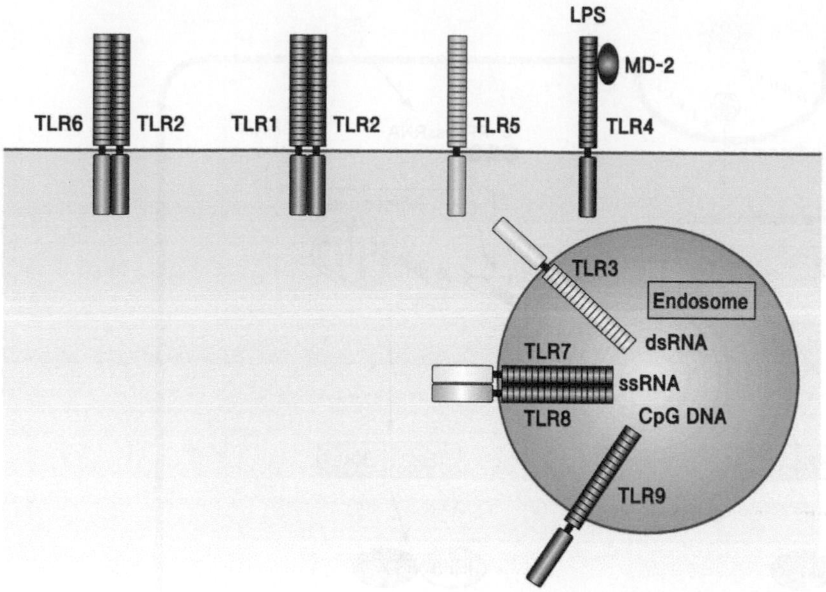

FIGURE 5.4 Toll-like receptors and their ligands. Some TLRs (TLR1, 2, 4, 5, and 6) are located on the plasma membrane, while other TLRs (TLR3, TLR7/8, TLR9) are located on endosomes. The PAMPs that are recognized by each TLR are denoted. Note that the endosomal TLRs recognize viral nucleic acids such as ssRNA, dsRNA, and CpG DNA. In fact, many TLRs act as either homo- or heterodimers such as TLR1/2, TLR2/6, and TLR7/8.

nonimmune cells, such as epithelial cells. How does TLR selectively recognize diverse invading pathogens? In fact, TLRs recognize pathogen-associated molecular patterns (PAMP), which are found on the invading pathogens. In general, double-strand RNA, single-strand RNA, and CpG dinucleotide DNA of the invading viruses are recognized by TLRs that are located at endosomes. Specific TLRs that recognize the many human viruses have been uncovered. For instance, TLR7/8 recognizes the genomic RNA of influenza virus. Often, these endosomal TLRs (ie, TLR3, 7, 8, and 9) are called "nucleotide-sensing TLR." On the other hand, bacterial cell wall components such as LPS and diacyl lipopeptide are typically recognized by TLR4 and TLR2/6, which are located on the plasma membrane.

How does TLR receptor trigger the innate immune response? In brief, so-called TLR signaling, which culminates to induce IFN expression, is activated upon the engagement of ligands to the receptors (Fig. 5.5). Specifically, TLR3, upon the ligand binding, recruits *TRIF*[7] to the receptor, which then activates *TBK1*[8]*/IKKε*, a serine/threonine kinase. The activated TBK1/IKKε phosphorylates *IRF3/7*,[9] which then becomes translocated to the nucleus. Subsequently, the IRF3/7 binds to the promoter of the IFN gene, thereby inducing transcription. IRF3/7, which is normally expressed in the cytoplasm, traffics to the nucleus upon the phosphorylation and acts as a transcription factor.

5.2.3 RIG-I

TLR signaling is critical for the immune defense against invading microbes. Unexpectedly, however, it was found that a transgenic mouse lacking TLRs remains resistant against viral infection, implicating the existence of another receptor that is critical in sensing the invading viruses. Subsequently, in addition to TLRs, *RIG-I*[10] and *MDA-5*[11] were identified as receptors sensing the viral RNAs (see Fig. 5.5). Interestingly, RIG-I/MDA-5 belong to the DEAD-box RNA helicase family. These RNA helicases serve as an "RNA sensor" in the cytoplasm of the virus infected cells. They are also known as "intracellular receptor," as opposed to TLRs that detect extracellularly located PAMPs. Since both TLRs and RIG-I/MDA5 recognize the molecular pattern of the invading pathogens, these receptors are collectively called "pattern recognition receptors" (*PRR*[12]) (Table 5.2).

7. **TRIF** (TIR domain-containing adaptor-inducing interferon).
8. **TBK** (TRAF-associated NF-kB binding kinase 1)/IKKε (ikB kinase).
9. **IRF** (Interferon regulatory factor).
10. **RIG-I (retinoic acid inducible gene)** It serves as an RNA sensor for the detection of viral RNAs and belongs to the DEAD-box RNA helicase family. It was first discovered as the retinoic acid-inducible gene, as the name implies.
11. **MDA-5 (melanoma differentiation associated gene)** It serves as an RNA sensor for the detection of viral RNAs and belongs to the DEAD-box RNA helicase family. It was first discovered as melanoma differentiation associated gene, as the name implies.
12. **PRR** (pattern recognition receptor).

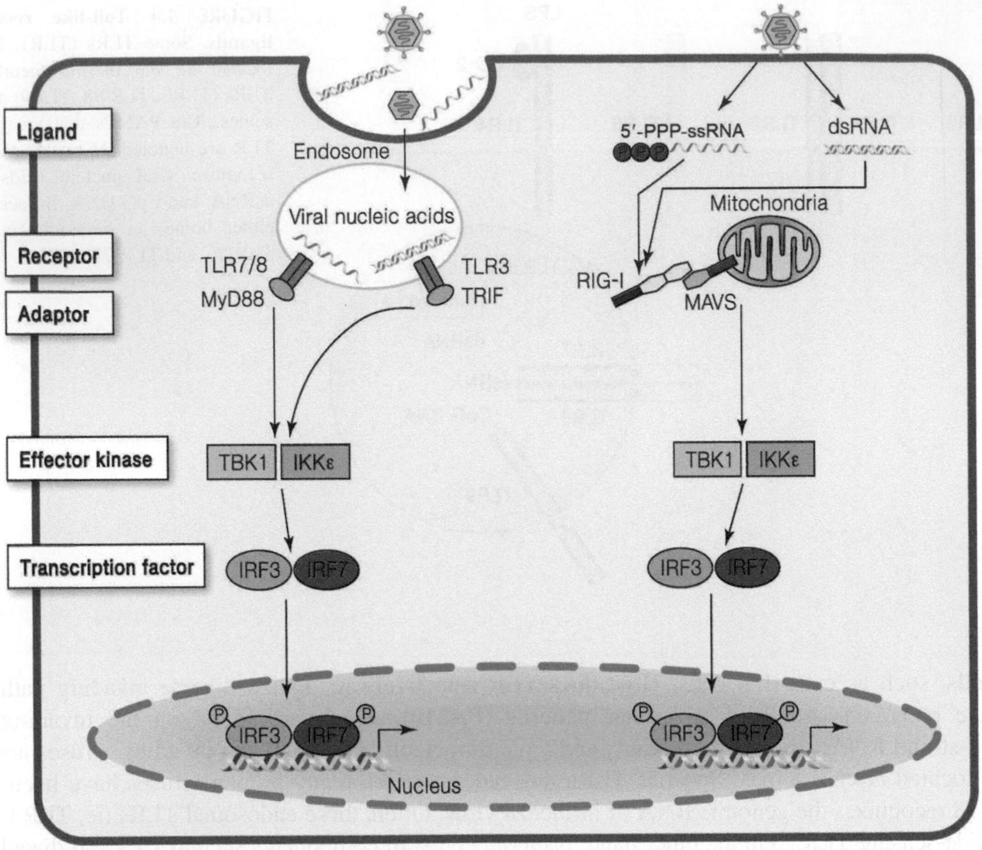

FIGURE 5.5 **Toll-like receptor signaling versus RIG-I signaling.** TLR3 and TLR7/8 are shown as the representative of TLR signaling, while RIG-I is shown as representative of RIG-I/MDA5 signaling. Viral RNAs in endosome are recognized by TLR3 or TLR7/8, whereas the viral RNAs in the cytoplasm are recognized by RIG-I/MDA5. Note that five steps of the signaling, from ligand to transcription factor, are highlighted on the left. On left, common nomenclatures for the molecules involved in each step of TLR/RIG-I signaling are highlighted: in particular, the adaptor refers to a protein linking the receptor and an effector kinase via protein–protein interaction.

TABLE 5.2 TLR Signaling Versus RIG-I Signaling

	TLR Signaling	RIG-I Signaling
Pattern Recognition Receptor	TLR (extracellular receptor)	RIG-I, MDA-5 (intracellular receptor)
Ligand	Diverse	5′-PPP-RNA, dsRNA
Adaptor	TRIF, MyD88	MAVS (mitochondrial)
Effector kinase	TBK1, TAK1	TBK1, TAK1
Outcome	IFN-α	IFN-α

A question is how does RIG-I/MDA5 specifically recognize viral RNAs? It was thought that double-strand RNAs present as an intermediate during viral RNA replication are recognized as motifs that represent nonself. Intriguingly, RIG-I specifically recognizes *5′-PPP-RNA*,[13] a characteristic nucleotide present at the 5′ end of the viral RNA genome, as well as double-strand RNA (Fig. 5.6). RIG-I prefers to detect smaller double-strand RNA (<1 kb), while MDA5 prefers to detect larger double-strand RNA (>1 kb).

13. **5′-pppRNA** a characteristic nucleotide present at the 5′ end of the viral RNA genome such as influenza virus. Note that cellular mRNA has a cap structure at 5′ end, while tRNA and rRNA have a monophosphate at 5′ end.

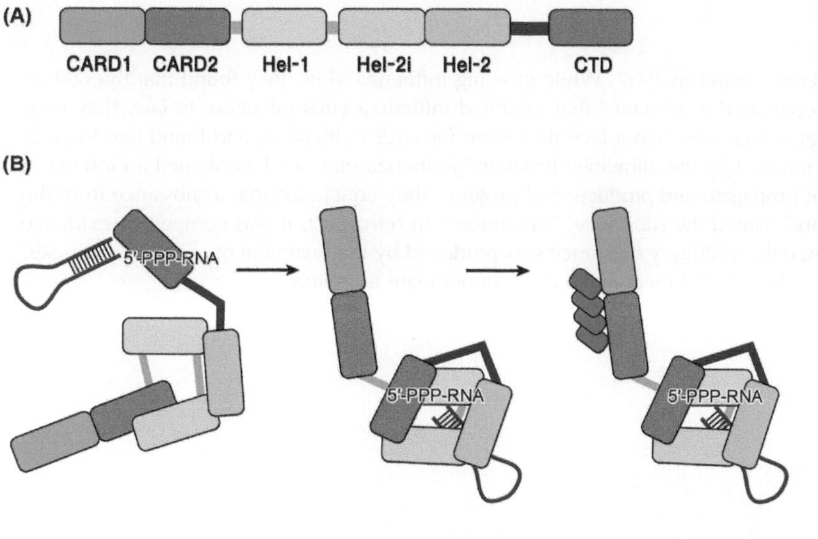

FIGURE 5.6 RIG-I and its ligands. (A) Domains structure of RIG-I. RIG-I contains two CARD domains at N-terminus interacts with the CARD domain of MAVS, and the CTD domain binds to the 5′-PPP-RNA, the ligand. CTD (C-terminus domain). (B) The conformation change of RIG-I upon the engagement of ligands. The CARD domain becomes exposed upon the ligand binding (ie, 5′-PPP-RNA) and the activated RIG-I is stabilized by ubiquitination (blue).

How then does RIG-I/MDA5 receptor trigger the cellular signaling to induce innate immunity? RIG-I/MDA5 acts as a sensor in detecting double-strand RNA in the cytoplasm. Upon the engagement of double-strand RNA, the activation of multiple signaling molecules that are mediated via protein–protein interactions leads to the phosphorylation of transcription factors (IRF3/7). RIG-I, a member of DEAD-box RNA helicase, contains two *CARD*[14] domains (see Fig. 5.6). In particular, the engagement of ligands induces conformational changes of RIG-I such that the CARD domains become exposed or activated to interact with the CARD domain of MAVS, an adaptor. Such activated *MAVS*[15] activates TBK1/IKKε, a serine-threonine protein kinase, to phosphorylate IRF3/7 (see Fig. 5.5).

Intriguingly, the signal triggered by RIG-I parallels with that of TLR in many respects: (1) the recruitment of adaptor protein, (2) the activation of effector kinase, (3) the phosphorylation, nuclear translocation of IRFs by the effector kinases, (4) and finally, the induction of IFN expression (see Fig. 5.5). It should be noted that while the receptors and the adaptors are distinct, both TLR- and RIG-I signaling culminate in inducing IFN, antiviral cytokines.

It is important to emphasize that RIG-I/MDA5 is not redundant with TLR in sensing the invading pathogen. First, TLRs recognize the pathogen-associated molecules extracellularly, while RIG-I/MDA5 recognizes the pathogen-associated molecules intracellularly. Hence, RIG-I/MDA5 is often called an "intracellular receptor." Secondly, TLRs recognize diverse kinds of the pathogen-associated molecules, such as lipopeptides, proteins, and nucleic acids, while RIG-I/MDA5 recognizes the specific features of RNA molecules (ie, double-strand RNA and 5′-PPP-RNA). Overall, TLR and RIG-I/MDA5 are complementary in sensing the invading pathogens.

5.2.4 Antiviral Function of Interferon

The first line of defense against viral infection is the IFN response, which triggers the induction of a broad array of antiviral proteins. Cells mounting innate immune response produce IFNs. IFN was discovered in 1957 by Alick Isaacs and Jean Lindenmann as an antiviral cytokine that blocks the infection of influenza virus (Box 5.2). In fact, there are three types: IFN-α, β, and γ. IFN-α is mainly produced in immune cells, such as dendritic cells, while IFN-β is produced in most types of cells. On the other hand, IFN-γ is produced in activated T lymphocytes and NK cells. IFN-α and IFN-β share a receptor, while IFN-γ uses a distinct receptor. The former are called "type I IFN," while the latter is called "type II IFN."

How then does IFN block the viral infection? IFN induces the antiviral state of the infected cell via its binding to the IFN receptor (Fig. 5.7). It is worth noting two points. First, IFN, a cytokine that is extracellularly secreted, acts on not only the virus infected cells, but also neighboring cells, a phenomenon called the "paracrine effect." In other words, IFN renders neighboring cells into an antiviral state so that the virus spread is prevented. Second, the antiviral functions

14. **CARD (caspase recruitment domain)** CARDs are "interaction motifs" found in a wide array of proteins, typically those involved in processes related to inflammation and apoptosis. These domains mediate the formation of larger protein complexes via direct interactions between individual CARDs.

15. **MAVS (mitochondrial antiviral signaling)** It was discovered as mitochondrial protein essential for antiviral signaling (ie, IFN induction). It is also termed as IPS (interferon promoter stimulator) or VISA or Cardif.

BOX 5.2 Discovery of Interferons

IFN was discovered by Alick Isaacs and Jean Lindenmann in 1957. While growing influenza virus, they found that the culture medium derived from the virus infected cells contained a substance that inhibited influenza virus infection. In fact, they used chorioallantoic membrane of embryonated eggs, which was then a favorite system for virus cultivation, and found that the culture medium derived from cells that had been infected by the ultraviolet-irradiated influenza viral stock contained an inhibitory substance. Since the inactivated virus could not propagate and produce viral proteins, they concluded that a substance from the host cell "interfered" with the virus infection and named the substance, "interferon." In retrospect, it was compelling evidence that the inhibitory substance is a host factor, since the inhibitory substance was produced by the treatment of inactivated viruses. More importantly, the discovery of IFNs inspired the establishment of the concept of innate immunity.

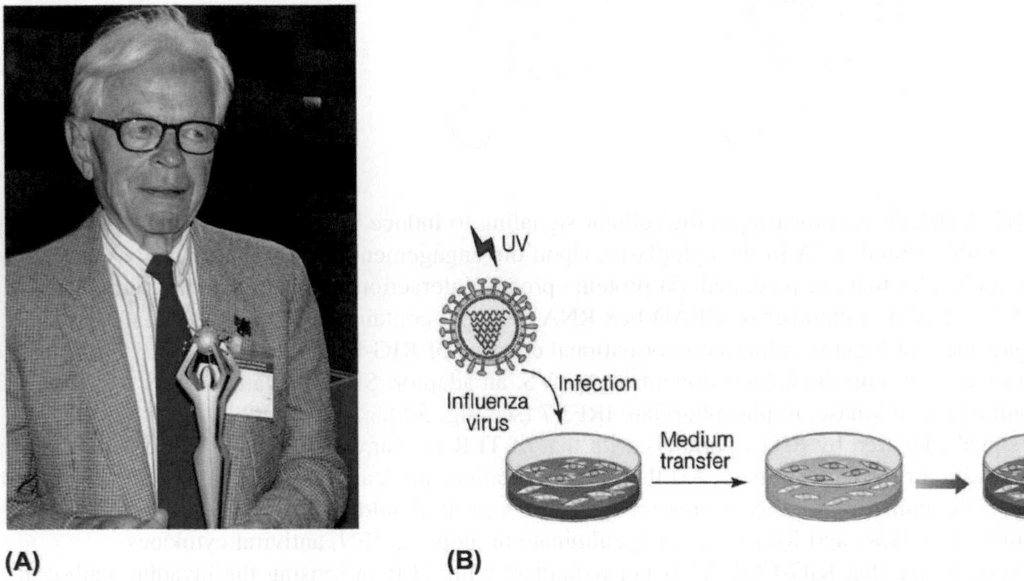

(A) A recent photo of Jean Lindenmann (1924–2015). (B) A schematic showing an experiment that leads to the discovery of interferons. The culture medium taken from cells that had been infected by the UV-inactivated influenza virus blocked the infection of influenza virus.

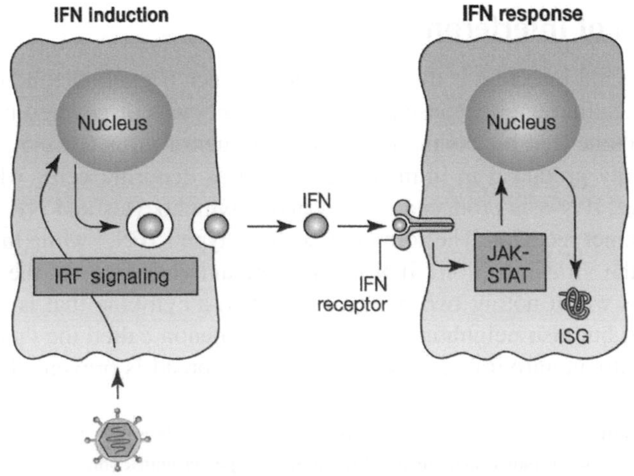

FIGURE 5.7 IFN induction versus IFN response. Note that differences between IFN induction and IFN response in followings; (1) ligands: PAMP versus IFN, (2) receptors: TLR or RIG-I versus IFN receptor, (3) signaling: IRF signaling versus JAK-STAT signaling, and (4) outcomes: IFN versus ISGs.

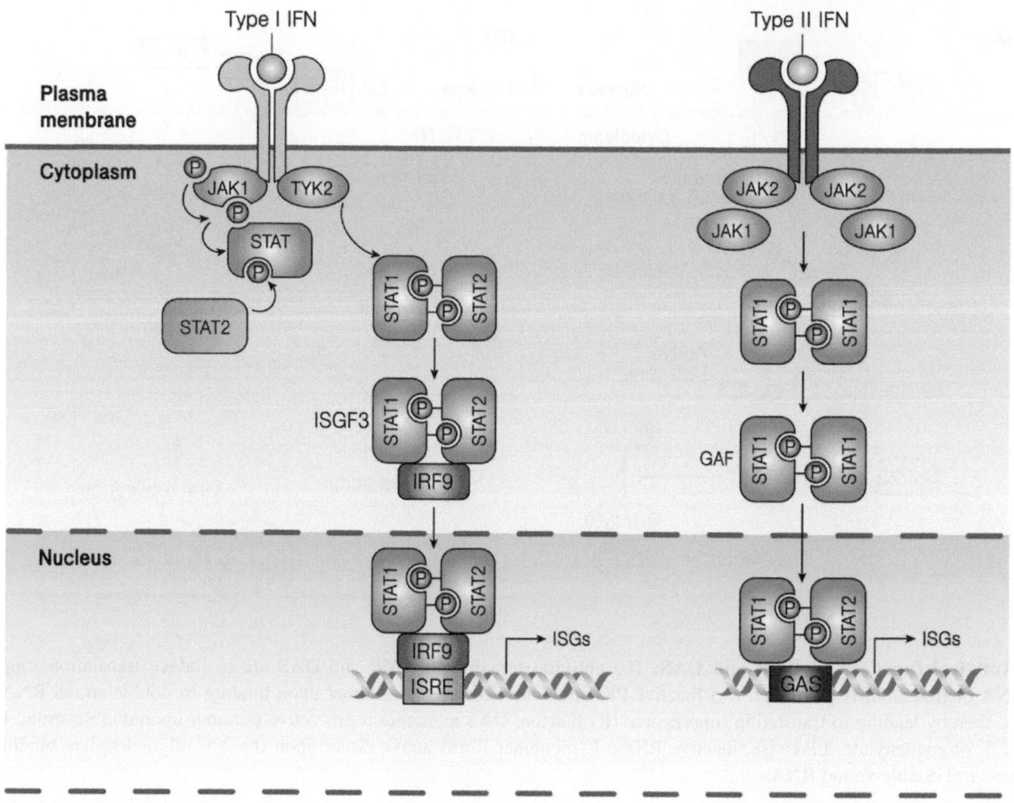

FIGURE 5.8 IFN triggers JAK-STAT signaling. Upon the binding of type I IFN (ie, IFN-α/ β), the IFN receptor recruits, activates TYK2/JAK1, a serine/threonine kinase. The phosphorylated STAT1/STAT2, complexed with IRF9, resulting in the formation of IFN-stimulated gene factor 3 (ISGF3). Then, the ISGF3 moves to the nucleus, binds to ISRE element, and induces transcription of ISGs such as OAS and PKR. Likewise, type II IFN (ie, IFN-γ) induces GAF (IFN-γ activation factor) via JAK-STAT signaling, which binds to GAS promoter and induce transcription of ISGs.

of IFN are not virus-specific. Therefore, IFN produced by a certain virus could block the infection of other unrelated viruses. The lack of specificity is a hallmark of innate immunity, which is a feature distinct from adaptive immunity (see Table 5.1).

Upon the engagement of IFN, the activated IFN receptor triggers *JAK-STAT signaling*, leading to the induction of many IFN-stimulated genes (*ISG*[16]) (Fig. 5.8). *PKR*[17] and *2′-5′ OAS*[18] are two representatives of over 300 ISGs that are induced upon IFN stimulation.

Among over 300 ISGs, two of them are well known (Fig. 5.9). First, PKR, a serine/threonine kinase, becomes activated upon the engagement of double-strand RNA. The activated PKR phosphorylates eIF2α, thereby leading to translation suppression. Second, OAS is an enzyme that synthesizes the *2′-5′ oligoadenylate (2′-5′)*, an uncommon kind of nucleotide. The inactive monomeric OAS becomes an active tetramer upon double-strand RNA binding, and produces the 2′-5′. In turn, the inactive *RNase L*[19] is activated upon the binding of the 2′-5′, and degrades double-stand RNA. Overall, two representative ISGs block viral infection either via the translation suppression or the degradation of viral RNA at the early infection, when the viral protein synthesis is critical for the successful establishment of infection.

It should be emphasized that the innate immune response is induced not only by immune cells such as dendritic cells and macrophages, but also by nonimmune cells, such as epithelial cells. Moreover, since the vast majority of animal viruses infect nonimmune cells rather than immune cells, the innate immune response occurring in nonimmune cells is more relevant for the protection from the viral infection.

16. **ISG** (interferon-stimulated gene).
17. **PKR (protein kinase RNA-activated)** a protein kinase that is activated by double-stranded (dsRNA). PKR phosphorylate eIF2α, thereby leading to translation suppression.
18. **2′-5′ OAS (oligoadenylate synthetase)** an enzyme that synthesizes the 2′5′ oligoadenylate (2′-5′).
19. **RNase L (for latent)** RNase that is activated by 2′-5′. It is also called 2′-5′-dependent ribonuclease.

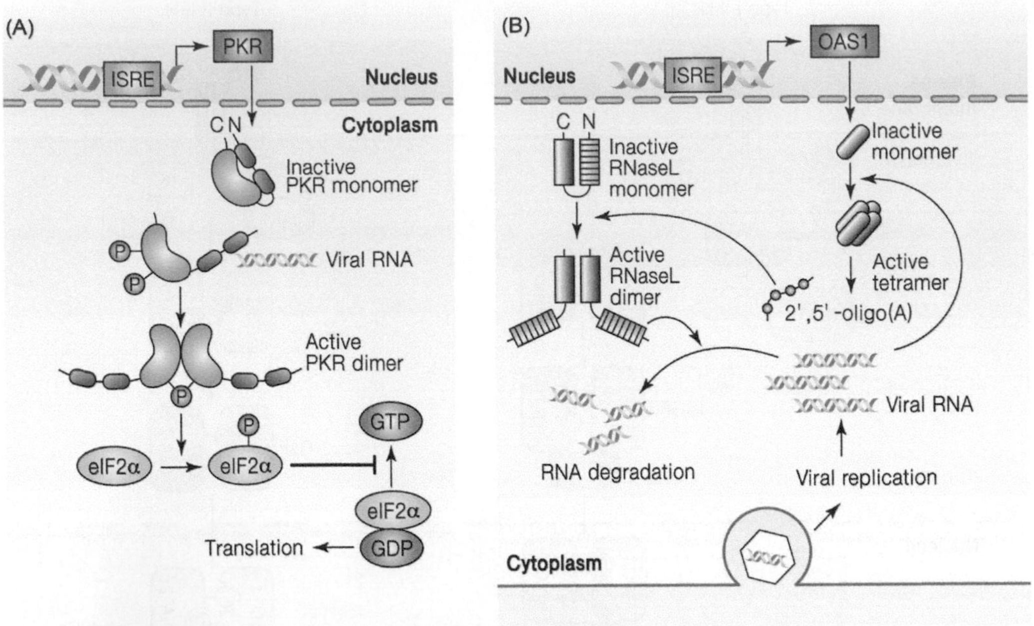

FIGURE 5.9 Antiviral functions of PKR and OAS. The antiviral functions of PKR and OAS are to induce translation suppression and viral double-strand RNA degradation, respectively. (A) Inactive PKR monomer forms active dimer upon binding to double-strand RNA. The PKR phosphorylates eIF2α, thereby leading to translation suppression. (B) Inactive OAS monomer forms active tetramer upon double-strand RNA binding, and synthesizes the 2′5′ oligoadenylate. Likewise, inactive RNase L monomer forms active dimer upon the 2′5′ oligoadenylate binding. Such activated RNase L degrades viral double-strand RNAs.

So far, we have learned about the innate immunity that is induced in the early phase of virus infection. The adaptive immune response that follows the innate immune response is driven by immune cells (ie, B and T lymphocytes) that are distinct from those for innate immunity (DC and macrophage); nonetheless, the adaptive immune response depends considerably on innate immunity, as you will see below.

5.3 ADAPTIVE IMMUNITY

Adaptive immunity embraces two arms: *humoral immunity* and *cellular immunity* (Fig. 5.10). Humoral immunity is executed largely by antibodies (ie, immunoglobulins) that are produced by plasma cells, a differentiated B lymphocyte, while cellular immunity is executed by *helper T lymphocyte*[20] (Th) and *cytotoxic T lymphocyte*[21] (CTL), a differentiated $CD4^+$ and $CD8^+$ T lymphocyte, respectively. Antibodies block viral infection by neutralizing the virions via direct binding, while CTL kills the virus infected cells by introducing toxic substances. In addition, the adaptive immune response results in memory lymphocytes so that the host can respond quickly and effectively to subsequent reinfection. In other words, the persistency and the memory of immune response are the defining features of the adaptive immunity (Table 5.3).

To what extent is the adaptive immunity influenced by innate immunity? The roles of DCs reflect the intimacy between the innate and adaptive immunity. For instance, DC is critical in sensing the invading pathogens not only in the context of innate immunity, but also in the context of adaptive immunity. In particular, DC presents antigenic peptide (ie, *epitope*[22]) in association with *MHC molecules*, to T cell receptors, serving as an *antigen-presenting cell*[23] *(APC)* (Fig. 5.11). Thus, DCs serve as a "sentinel," in both the innate and adaptive immune response.

20. **Helper T lymphocyte (Th)** a subset of $CD4^+$ T lymphocyte that helps B or $CD8^+$ T lymphocyte to proliferate, and differentiate to effector cells.

21. **Cytotoxic T cell (CTL)** a subset of T lymphocyte that expresses CD8 marker and is capable of killing target cells.

22. **Epitope** An epitope represents the part of an antigen that is recognized by the adaptive immune system, specifically by antibodies, B cells, or T cells.

23. **Antigen-presenting cell (APC)** an immune cell that presents antigenic peptides (epitope) in context of MHC to T lymphocytes. DC cell, macrophage, and B lymphocyte are said to be "professional APC."

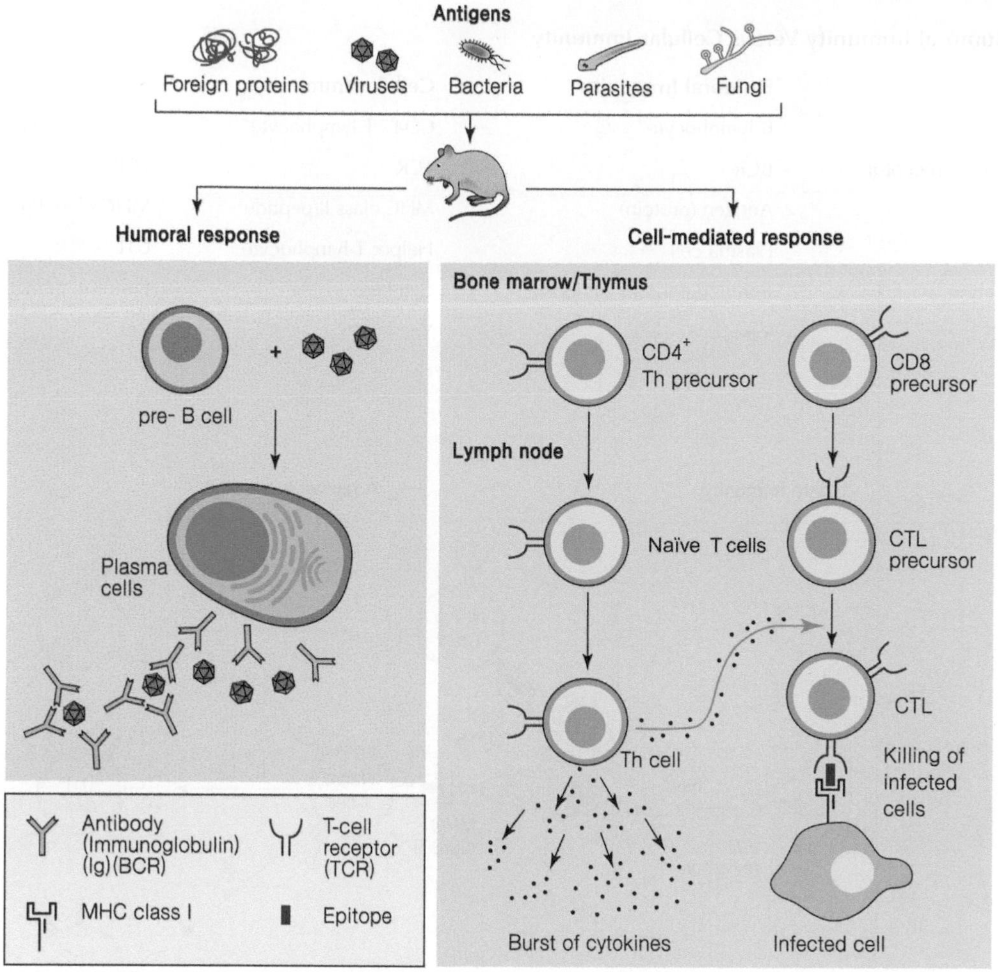

FIGURE 5.10 **Humoral immunity versus cellular immunity.** (A) Humoral immune response. The recognition of antigens by a B-cell receptor (BCR) activates a specific pre-B cell. The activated pre-B cell differentiates into a plasma cell that produces immunoglobulins. (B) Cellular immune response. T-cell precursors ($CD4^+$ or $CD8^+$) derived from bone marrow are educated in the thymus and released as naïve T cells. In the lymph node, naïve T cells become activated by recognition of the antigen epitope presented by DCs. $CD4^+$ T cell differentiates into T helper cell (Th), while $CD8^+$ T cell differentiates into cytotoxic T cell (CTL).

5.3.1 Components of Adaptive Immunity

The first step of inducing adaptive immunity is the recognition of foreign antigens by B or T lymphocytes. B or T lymphocytes are the central players of adaptive immunity and are derived from *hematopoietic stem cells (HSC)* (Box 5.3).

How do B or T lymphocytes recognize antigens? *B-cell receptor* (BCR), an immunoglobulin molecule expressed on the surface of B lymphocyte, recognizes antigens (Fig. 5.12), while *T-cell receptor* (TCR) expressed on the surface of T lymphocyte recognizes antigens. An important difference is that the BCR binds to the antigen directly, while the TCR recognizes the peptide presented by the MHC class molecules on the APCs (Fig. 15.13). As a result, the BCR recognizes either a linear or a discontinuous epitope of native antigens, while the TCR recognizes the linearized peptide derived from antigens (Fig. 5.13).

How is the MHC/peptide complex formed? This process, termed "antigen presentation," occurs mainly in APCs such as DCs and macrophages (Fig. 5.14). Two classes of MHC molecules exist, which are subtly different in their antigen presenting pathways.

In the case of MHC class I, endogenous antigens are processed by proteasomes to the peptides (mainly 9–11 mers), and these peptides are transported into the lumen of ER by TAP transporter, loaded onto MHC class I molecules, and are presented on the cell surface. In the case of MHC class II, exogenous antigens engulfed by endocytosis are transported in an endocytic vesicle, processed to the peptides (10–30 mers), loaded onto MHC class II molecules, and presented on the cell surface. In brief, the MHC class I molecule presents the peptides derived from endogenous antigens, while the MHC class II molecule presents the peptides derived from exogenous antigens.

TABLE 5.3 Humoral Immunity Versus Cellular Immunity

	Humoral Immunity	Cellular Immunity	
Precursor cells	B lymphocyte	$CD4^+$ T lymphocyte	$CD8^+$ T lymphocyte
Antigen recognition receptor	BCR	TCR	TCR
Ligands	Antigen (protein)	MHC class II/peptide	MHC class I/peptide
Activated cells	Plasma cell	Helper T lymphocyte	CTL
Molecules in action	Immunoglobulin (antibody)	Cytokine	Perforin, granzymes, cytokines
Memory	Yes	Yes	Yes

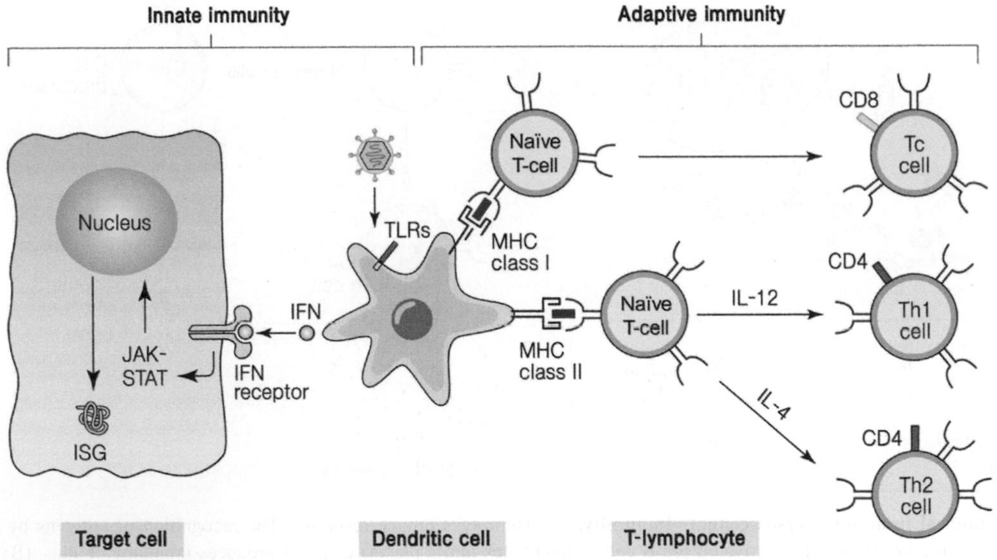

FIGURE 5.11 The roles of DC in linking the innate immunity to adaptive immunity. The central function of DC in both innate immunity and adaptive immunity is to recognize the invading pathogens. DC recognizes the invading pathogens by TLR and RIG-I receptors in the innate immune response. On the other hand, in adaptive immunity, DC present foreign epitope in the context of MHC molecules to naïve T cell, thereby leading to the activation and the differentiation of T cell.

BOX 5.3 Immune cells

How are cells involved in immune response produced, and how are they related to each other? Immune cells are produced and differentiated in the lymphoid organs, and transported to peripheral tissues via blood vessels and lymphatic vessels. There are two kinds of lymphoid organs: (1) primary lymphoid organs (bone marrow and thymus), in which immune cells are produced, and (2) secondary lymphoid organs (lymph node and spleen), in which immune cells reside. All immune cells are derived from HSC. HSC differentiate to two progenitor cells: common myeloid progenitor and common lymphoid progenitor. Myeloid progenitor cell differentiates to erythrocytes, platelets, basophils, eosinophils, neutrophils, and monocytes. Monocytes differentiate into macrophage and dendritic cells in tissues. Lymphoid progenitor cells differentiate into B lymphocyte, T lymphocyte, and natural killer (NK) cell. These differentiated blood cells are present in blood via circulation. All blood cells except erythrocytes are involved in immune response. Hence, blood cells are often conveniently classified by red blood cell (erythrocytes) or white blood cells (leukocytes), which comprise all immune cells.

(Continued)

BOX 5.3 (Continued)

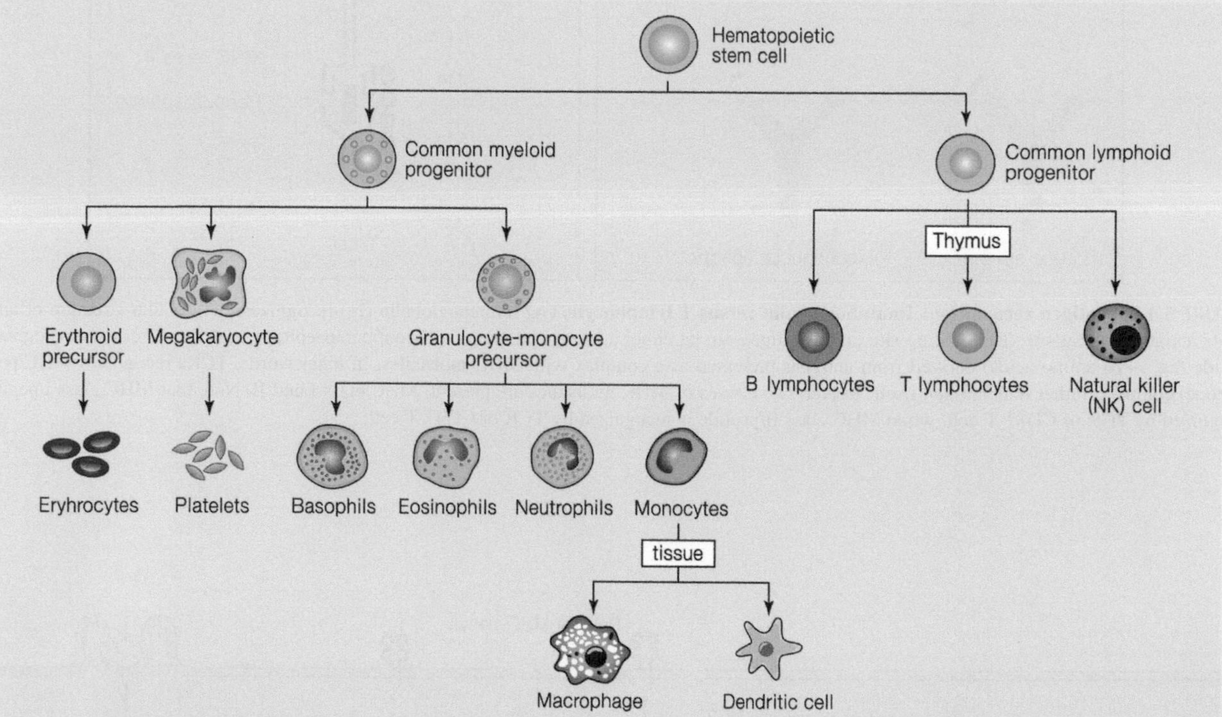

Hematopoiesis. Diverse bloods cells are derived from hematopoietic stem cells (HSCs). HSC, a stem cell of blood cells, differentiates into two progenitor cells: common myeloid progenitor and common lymphoid progenitor. Then, the myeloid progenitor and lymphoid progenitor cells differentiate into diverse blood cells.

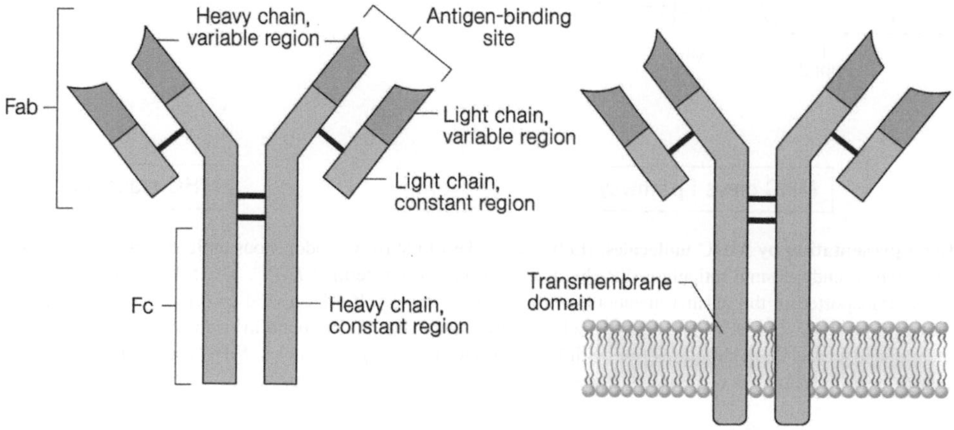

FIGURE 5.12 **Schematic structures of immunoglobulin and B-cell receptor.** (A) Immunoglobulin is composed of two heavy chains and two light chains that are linked by disulfide bonds. The variable region at N-terminus constitutes the antigen-binding site. Immunoglobulin could be divided into two fragments by trypsin treatment: Fab (the antigen-binding fragment), and Fc (the constant fragment). (B) B-cell receptor. B-cell receptor is an isoform of immunoglobulin with the C-terminal extension that contains the transmembrane domain.

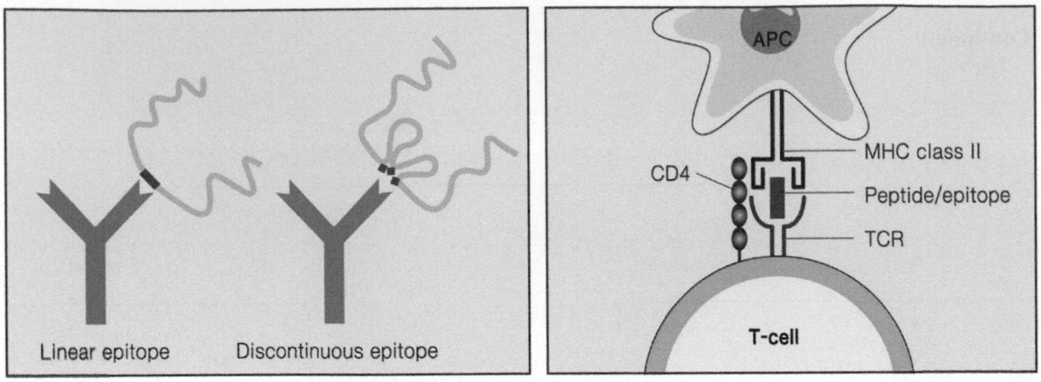

FIGURE 5.13 **Antigen recognition: Immunoglobulin versus T lymphocyte.** (A) Immunoglobulin (Ig) recognizes the globular structure of antigen via its antigen-binding site. The binding site on the antigen can be either a linear epitope or a discontinuous epitope. (B) T-cell receptor recognize the peptide (eg, ~10 amino acids) derived from antigens presented as a complex with MHC molecules. In other words, TCRs recognizes MHC/peptide (epitope) complex, rather than antigen itself. In fact, two classes of MHC molecules are present: MHC class I and II. Note that MHC class I/peptide is recognized by TCR of $CD8^+$ T cell, while MHC class II/peptide is recognized by TCR of $CD4^+$ T cell.

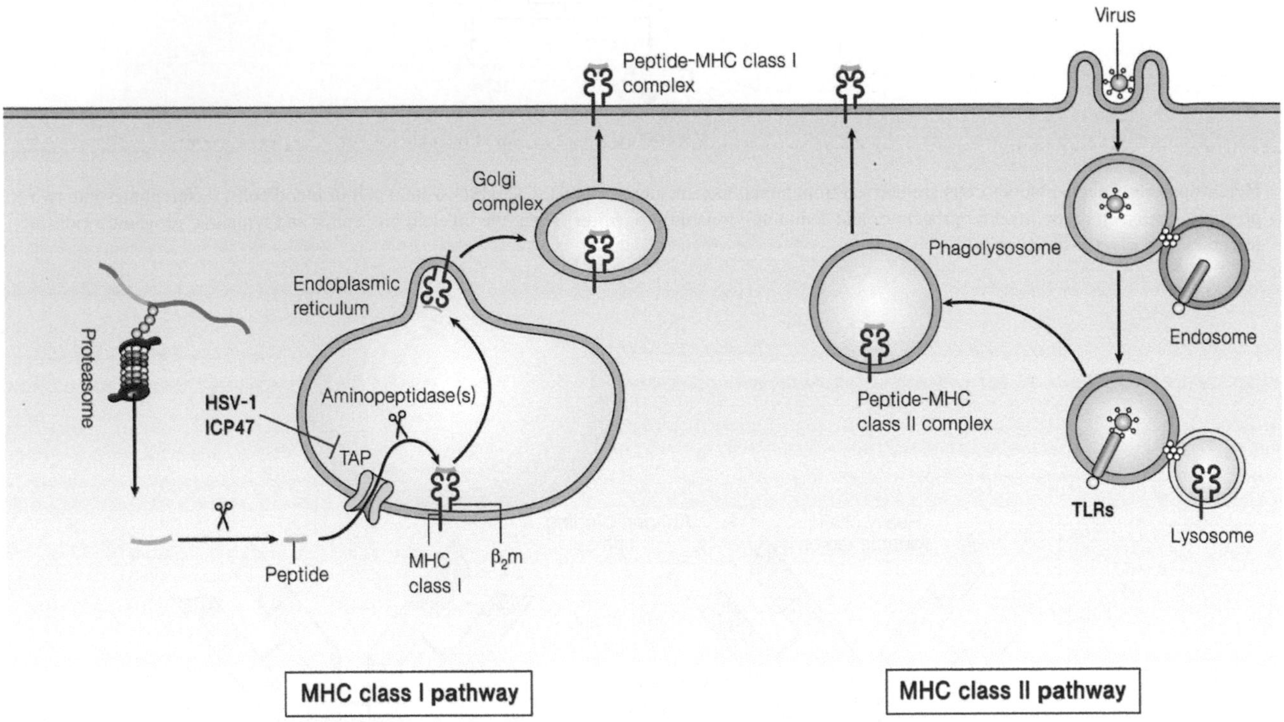

FIGURE 5.14 **Antigen presentation by MHC molecules.** (Left) MHC class I pathway. Endogenous antigens are first processed by proteasome and then, transported into lumen of endoplasmic reticulum (ER) by a peptide transporter, termed *TAP*[24]. The MHC class I/peptide complex is formed in lumen of ER and then, is transported to the plasma membrane via vesicular transport. ICP47, a viral protein of herpes simplex virus type I (HSV-I), blocks the transport function of TAP. (Right) MHC class II pathway. Viruses engage TLRs found in endocytic vesicle that recognizes nucleic acids. Following maturation of phagosomes, these structures fuse with lysosomes to form phagolysosomes. MHC class II molecules that are contained in the lysosomes are loaded with peptide fragments of viruses that are processed by lysosomal proteases.

24. **TAP (transporter associated with antigen presentation)** A membrane protein located in endoplasmic reticulum that functions to uptake peptides into lumen side.

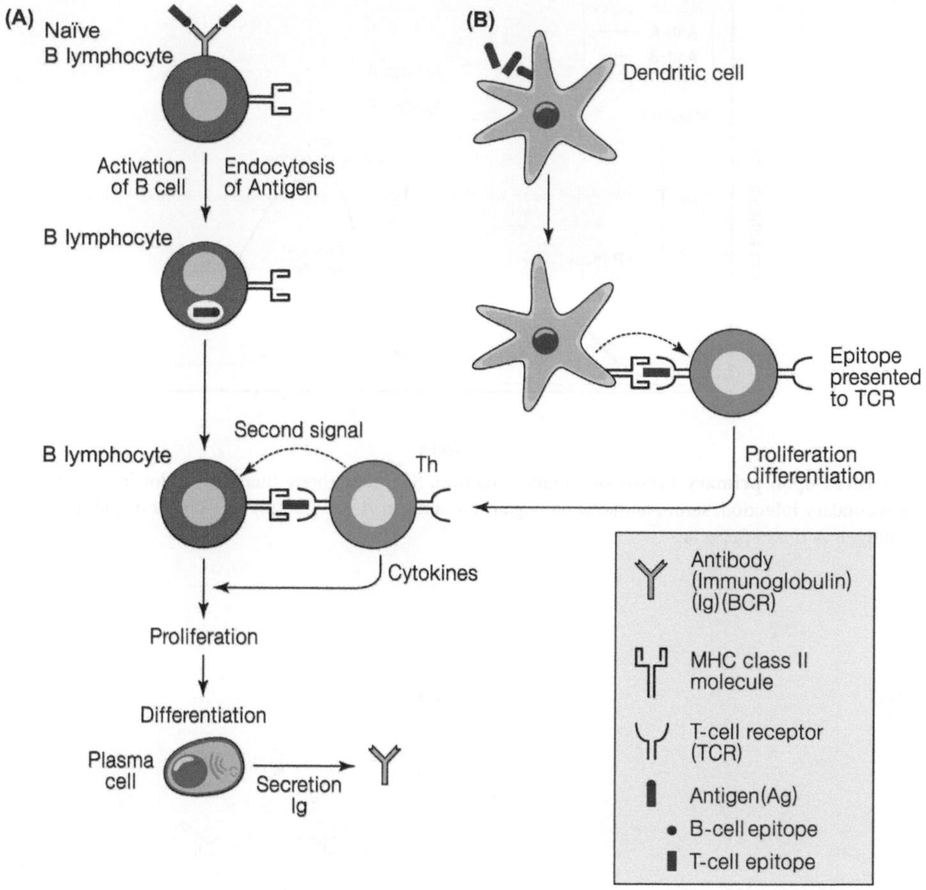

FIGURE 5.15 Activation of humoral immunity. (A) The activation of B lymphocyte. (B) The activation of helper T lymphocyte. Note that the MHC/peptide complex on the surface of activated B lymphocyte that is recognized by the helper T lymphocyte is identical to that on the surface of dendritic cells recognized by naïve T cell.

5.3.2 Humoral Immunity

How is the immunoglobulin molecule (antibody), which is an effector molecule in humoral immunity, produced? Humoral immune response is initiated by a B lymphocyte that has been selected in the bone marrow, and activated by antigen binding to BCR (Fig. 5.15). Upon activation, a specific clone of B cells engulfs the antigen, and digests the antigen to peptides, which are presented as a peptide epitope by an MHC class II molecule to the helper T lymphocyte cell (Th). Upon the second signal from the Th, the B cell proliferates and differentiates to *plasma cell.* The plasma cell produces a secretory immunoglobulin. Note that the Th is activated by a dendritic cell that presents the T cell epitope loaded onto an MHC class II molecule.

A defining feature of the adaptive immunity is "immunologic memory." How is immunologic memory produced? During the proliferation and differentiation of B lymphocytes, *memory B cells* are produced. Plasma cells continually produce antibodies even after clearing a virus infection. Upon subsequent infection by the same virus, the secondary antibody response elicited by the memory B cell is rapid and much higher than the first antibody response (Fig. 5.16). Thus, the invading pathogens are rapidly cleared by the circulating antibodies upon the secondary infection by the same virus. Antibodies that are capable of blocking virus infections are called *neutralizing antibodies*.

How does the neutralizing antibody block virus infection? The antibody can block the virus infection by impeding the receptor binding of the virus particles (Fig. 5.17). In addition, the antibody can clear the virus particles via a process called *opsonization*, in which a macrophage engulfs the virus-antibody complex via its Fc receptor via phagocytosis.

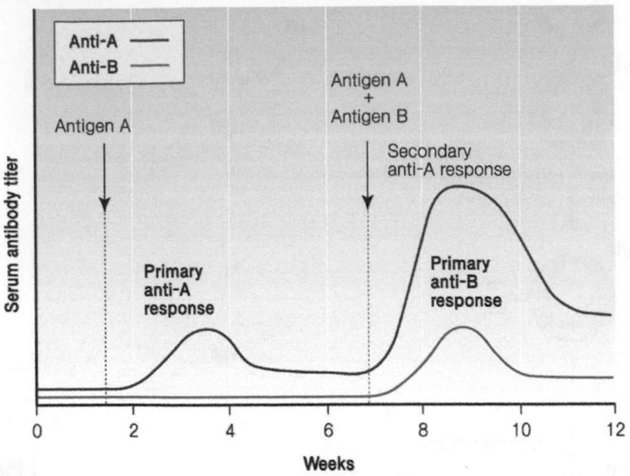

FIGURE 5.16 Antibody titers upon primary versus secondary infection. Serum antibody titer rises following a certain lag period (1–2 weeks) after infection. Upon the secondary infection, serum antibody titer against to antigen A rises rapidly in a higher magnitude without the lag phase, while antibody titer against to antigen A rises after a lag.

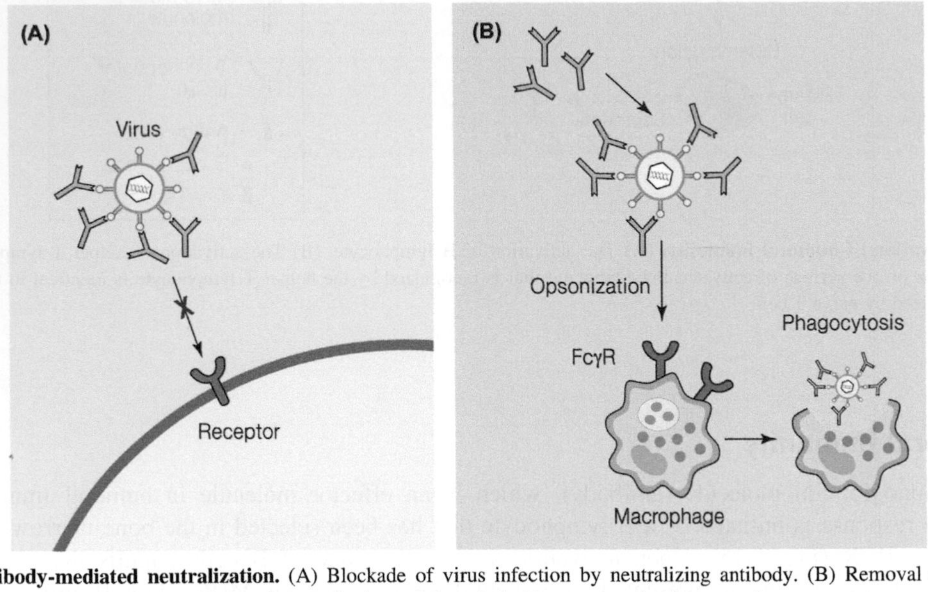

FIGURE 5.17 Antibody-mediated neutralization. (A) Blockade of virus infection by neutralizing antibody. (B) Removal of the virus-antibody complex by opsonization. Macrophage engulfs the virus-antibody complex via its Fc receptor via phagocytosis.

5.3.3 Cellular Immunity

How is cellular immunity induced? If viruses infect directly APC, such as DC and macrophage, in the infection site, the viral antigens expressed in APC are processed and presented to an MHC class I molecule (*endogenous pathway*) (see Fig. 5.14). Alternatively, the viral antigens are abundantly present in the infected tissues, since the infected cells are killed by NK cells activated at an early phase of immune response. APC such as DCs in the tissue engulf the viral antigens, process them to peptides, and present the peptides complexed with MHC class I molecules (*exogenous pathway*) (Fig. 5.18). In both cases, the APC that present antigenic peptide loaded onto MHC class I molecules migrate to a nearby lymph node, activating naïve $CD8^+$ T cells. Naïve $CD4^+$ T cells can be activated by the APC that present the peptides loaded onto MHC class II molecules on the surface (see Fig. 5.11). Then, the differentiated Th helps the differentiation of $CD8^+$ T cells to CTL by producing cytokines such as IL-2 and IFN-γ. Such activated CTL recognize the

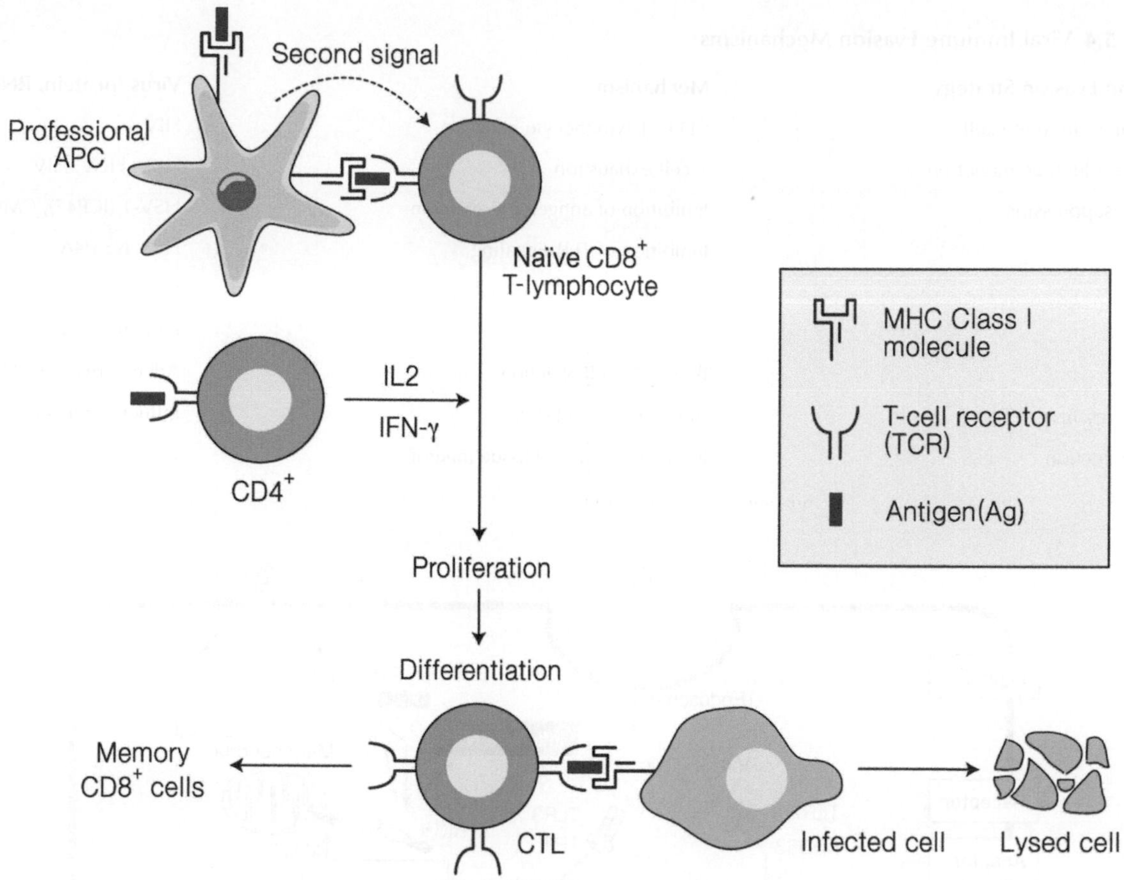

FIGURE 5.18 Activation of cellular immunity. Naïve $CD8^+$ T lymphocytes are activated by DC cells, which present viral epitope complex with MHC class I molecule. Such activated T lymphocytes proliferate, and differentiate to become CTL. The virus infected cells are recognized by TCR on the activated CTL, and killed.

MHC class I/peptide complexes presented on the surface of the infected cells via a TCR, and lyse the cell by introducing toxic substances such as *perforin*[25] and *granzyme*[26]. As a result, the virus infected cell is killed and eliminated.

How is the exquisite specificity of CTL acquired? Note that the activation of a naïve T lymphocyte is triggered by the interaction between TCR and the MHC class I/peptide complex. Conversely, the recognition of the virus infected cell by the CTL is mediated by the precisely identical molecular interaction between TCR and the MHC class I/peptide (see Fig. 5.18). In other words, out of an astronomical number of naïve T lymphocytes, only a few T lymphocytes that can recognize the virus infected cells are activated in the first place.

5.4 IMMUNE EVASION

So far, we have learned how effectively the host immune system responds to the virus infection. Nonetheless, to establish an infection successfully, viruses counteract the host immune response. The final outcome of virus infection depends on the extent of viral countermeasures to the host response. In fact, many viruses evade host immunity effectively, thereby successfully establishing their infection. Viruses acquire diverse strategies to evade the host immune response during their evolution. Such strategies can be grouped into five categories (Table 5.4).

---

25. **Perforin** Perforin is a cytolytic protein found in the granules of cytotoxic T lymphocytes (CTLs) and NK cells.
26. **Granzyme** It refers to the serine proteases that are released by cytoplasmic granules within cytotoxic T cells and NK cells.

TABLE 5.4 Viral Immune Evasion Mechanisms

Immune Evasion Strategy	Mechanism	Virus (protein, RNA)
Infection of Immune cell	$CD4^+$ T-lymphocyte infection	HIV
Immune tolerance induction	T-cell exhaustion	HBV, HCV, HIV
Immunosuppression	Inhibition of antigen presentation	HSV-1 (ICP47), CMV (US3)
	Inhibition of TLR signaling	HCV NS3/4A
	Inhibition of RNG-I signaling	HCV NS3/4A
		Influenza virus NS1
	Blockade of IFN action	Adenovirus (VA RNA)
Immune escape	Antigen drift and shift	Influenza virus HA
Latent infection	Immuno-privileged tissue (neuron)	Herpesvirus

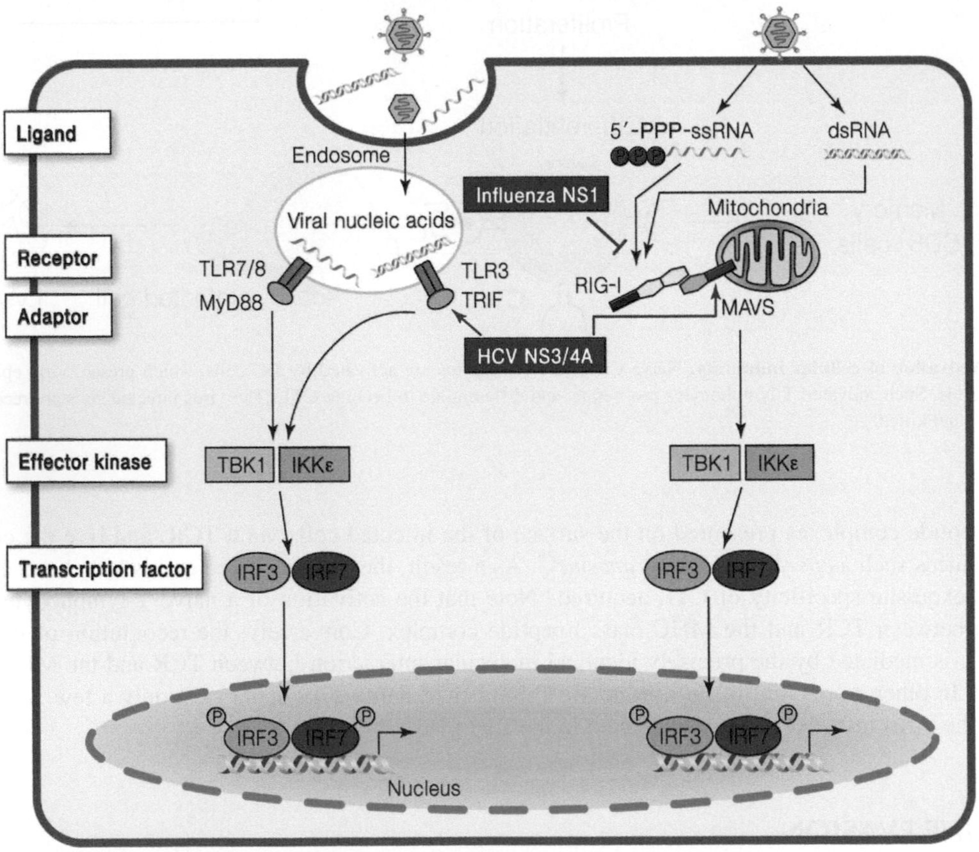

FIGURE 5.19 Blockade of innate immunity by viruses. Two representative examples are depicted here. HCV NS3/4A serine protease cleaves TRIF and MAVS, which are the adaptor molecules in TLR- and RIG-I signaling, respectively. Influenza virus NS1 protein blocks the TRIM25-mediated ubiquitination of RIG-I, thereby suppressing RIG-I signaling.

5.4.1 Immune Suppression

Suppression of innate immunity: IFN, the main product of innate immune response, is induced by either TLR signaling or RIG-I signaling. Not surprisingly, many viruses have the ability to suppress IFN induction by blocking TLR signaling or RIG-I signaling. For instance, hepatitis C virus (HCV) blocks TLR signaling and RIG-I signaling via the cleavage of TRIF and MAVS by the viral NS3/4A serine protease, respectively (Fig. 5.19). Another well-studied example is that the NS1 protein of influenza virus blocks RIG-I signaling by inhibiting RIG-I ubiquitination (see Fig. 15.13).

On the other hand, some viruses suppress the IFN action. Intriguingly, PKR is inhibited by multiple viruses. For instance, the NS5A protein of HCV inhibits PKR activity via its binding. Likewise, the NS1 protein of influenza virus inhibits PKR activity via it binding. Moreover, PKR is inhibited by adenovirus VA RNA (see Fig. 8.10). Overall, almost all viruses, if not all, similarly adopt diverse strategies to block innate immunity; the examples mentioned above are ones that have been studied extensively, and therefore represent only the tip of iceberg.

Suppression of adaptive immunity: Many viruses suppress adaptive immunity to establish their infections. Intriguingly, viruses frequently inhibit adaptive immunity by impairing antigen presentation. Some viruses impair antigen presentation by downregulation of MHC class I molecule expression (see Table 5.4). For example, adenoviruses, herpesviruses, and HIV suppress CTL induction by downregulating the expression of MHC class I molecule in the infected cells. In addition, herpesviruses, including herpes simplex virus (HSV)-1, impair antigen presentation that is mediated by MHC class I molecule. For instance, ICP47 of HSV-1 inhibits TAP-mediated peptide translocation by competing with antigen peptides for TAP binding site (see Fig. 5.14). In the case of chronic persistent hepatitis B virus (HBV) and HCV infection (see Fig. 3.10), virus-specific CTL activity is functionally impaired. In particular, persistent infection proceeds in the absence of cell killing by CTL in chronic HBV and HCV infection, a phenomena called as "T-cell exhaustion" (see Table 5.4). Under the persistence of virus, virus-specific Th and CTL progressively lose their effector functions such as cytokine production and cytolytic activity. T cell exhaustion is initiated by repeated interaction between TCR on T cell and peptide-MHC class I or II complex on APC (see Fig. 18.13).

5.4.2 Infection of Immune Cells

Among many distinct strategies of evading host immune response, the most direct way is to infect immune cells, thereby impairing their functions (see Table 5.4). One example that stands out is HIV. HIV infects $CD4^+$ T lymphocytes (Th), and induces apoptosis of the infected cells. Abrupt decline of $CD4^+$ T lymphocyte in the blood is the hallmark of HIV infection (see Fig. 22.4). The reduction of the helper T lymphocytes compromises overall host immunity, thereby ushering *opportunistic infection*[27] by other pathogenic microbes.

5.4.3 Immune Escape

Some viruses frequently provoke antigenic variations (ie, mutations) so that viral antigens could be no longer recognized by the host immune system. Immune evasion by genetic mutations is also termed "immune escape." In particular, for RNA viruses, which manifest inherently higher mutation rates (see Box 1.3), the immune escape is the principal mechanism for their immune evasion. For influenza virus and HIV, the mutant virus harboring mutations in epitope peptide are frequently selected during viral infection. It is established that the mutant viruses could escape not only from neutralizing antibodies but also from cell killing by CTL. For instance, a novel antigenic variant of the influenza virus, which escapes from neutralization by circulating antibodies, causes a seasonal epidemic every year (see Fig. 15.15). Another example is the case of HIV, where viral mutants harboring a mutation in CTL epitope accumulate over long-term infection (see Fig. 22.10).

5.4.4 Latent Infection

To avoid the immune system, some viruses hide themselves in the immune-privileged tissues of the human body (ie, neuronal tissue), where immune cells are modest or absent. One example of this kind of immune evasion is the neuronal infection by herpesviruses infection. Although the primary infection by HSV-1 occurs in the epithelial cells, it reaches the ganglia, which is a so-called immuno-privileged area lacking immune cells, and establishes a latent infection (see Fig. 9.12). During latent infection, HSV-1 resides in the body without being recognized by the host immune system.

5.5 PERSPECTIVES

Our current knowledge on immune response to virus infection has been greatly advanced recently and, not surprisingly, "viral immunology" has become established as a subdiscipline of immunology specializing in the immunological aspect of virus infection. A notion that innate immunity simply represents a fast and transient immune response has been challenged in the past decade. In fact, innate immunity is more tightly linked to adaptive immunity than it was previously

27. **Opportunistic infection** An opportunistic infection is an infection caused by pathogens, particularly "opportunistic pathogens" that usually do not cause disease in a healthy host.

appreciated. In other words, adaptive immunity is significantly influenced by the robustness of innate immunity. Recently, diverse cellular receptors that sense the invading pathogens have been revealed, including 15 members of the TLR family and RIG-I/MDA5. In particular, RIG-I/MDA5 senses the viral RNAs in the cytoplasm. By contrast to a wealth of new information on RNA sensing, little is known about DNA sensing. However, recently, an enzyme called cGAMP synthase (cGAS) was identified as a DNA sensor (see Journal Club). To what extent cGAS serves as a DNA sensor of diverse DNA viruses has just begun to be elucidated. Not surprisingly, almost all viruses are built with countermeasures to suppress innate immune response either via blockade of IFN induction or IFN action. Viral evasion of innate immunity represents attractive antiviral targets that need to be exploited in the near future.

5.6 SUMMARY

- *Immune response*: Upon the recognition of the invading pathogens, the host immune system provokes an immune response to eliminate the pathogens itself or the infected cells.
- *Innate immunity*: Innate immunity is a fast, nonspecific immune response to the invading pathogens. Upon recognition of pathogens, TLR and RIG-I activates the cytoplasmic signaling, leading to IFN production. IFN is the main outcome of the host innate response to the invading pathogens.
- *Adaptive immunity*: Adaptive immunity is a delayed but specific immune response to the invading pathogens. The specific recognition of viral antigen by B lymphocytes and T lymphocytes lead to humoral immunity (ie, antibody) and cell-mediated immunity (ie, CTL), respectively.
- IFN*s*: IFNs, antiviral cytokines, induce the expression of over 300 cellular genes (ie, ISGs) including PKR and 2′5′-OAS
- *Immune evasion*: Viruses have diverse strategies to evade the host immune response, such as immune cell infection, immune suppression, immune escape, and latent infection.

STUDY QUESTIONS

5.1 Compare and contrast the TLR signaling to RIG-I signaling.

5.2 Compare the IFN induction versus IFN action, regarding molecules shown below. (1) ligand, (2) receptor, (3) transcription factor, (4) genes induced.

5.3 Almost all animal viruses have adapted strategies during evolution to evade host immune response to establish viral infection. State and compare how animal viruses listed below evade host immunity. (1) Influenza virus, (2) Hepatitis C virus, (3) Herpesvirus, (4) HIV.

SUGGESTED READING

Ablasser, A., Bauernfeind, F., Hartmann, G., Latz, E., Fitzgerald, K.A., Hornung, V., 2009. RIG-I-dependent sensing of poly(dA:dT) through the induction of an RNA polymerase III-transcribed RNA intermediate. Nat. Immunol. 10 (10), 1065–1072.

Bowie, A.G., Unterholzner, L., 2008. Viral evasion and subversion of pattern-recognition receptor signalling. Nat. Rev. Immunol. 8 (12), 911–922.

Jensen, S., Thomsen, A.R., 2012. Sensing of RNA viruses: a review of innate immune receptors involved in recognizing RNA virus invasion. J. Virol. 86 (6), 2900–2910.

Sadler, A.J., Williams, B.R., 2008. Interferon-inducible antiviral effectors. Nat. Rev. Immunol. 8 (7), 559–568.

Wu, J., Sun, L., Chen, X., Du, F., Shi, H., Chen, C., et al., 2013. Cyclic GMP-AMP is an endogenous second messenger in innate immune signaling by cytosolic DNA. Science. 339 (6121), 826–830.

JOURNAL CLUB

- Sun, L., Wu, J., Du, F., Chen, X., Chen, ZJ., Cyclic GMP-AMP synthase is a cytosolic DNA sensor that activates the type I interferon pathway. Science. 2013. 339 (6121), 786–791.
- Highlight: Unlike viral RNA, little is known about how viral DNA is sensed by the host immune system. This paper describes an enzyme called cGAMP synthase (cGAS), which detects cytoplasmic DNA and triggers interferon induction. It is implied that cGAS could be the receptor for sensing DNA virus.

Part II

DNA Viruses

DNA viruses can be divided into two classes: single-strand DNA viruses and double-strand DNA viruses. Parvoviruses are the only virus family with a single-strand DNA genome among the animal viruses. In contrast, double-strand DNA viruses can be subdivided into three groups: (1) those with a small size DNA genome (<10 kb), such as polyomaviruses and papillomaviruses; (2) those with a medium size DNA genome (ca., 35 kb), such as adenoviruses; and (3) those with a large size DNA genome (ca., 150–250 kb), such as herpesviruses.

II.1 Historical Accounts

Research performed on bacteriophages during the 1950s and 1960s greatly advanced prokaryotic molecular biology. Similarly, the research on animal virology during the 1970s and 1980s greatly advanced eukaryotic molecular biology. In fact, many molecular biologists during that period worked with small DNA viruses, such as SV40. Unlike the host genome, the viral genome is small enough to be easily handled in laboratories, and the establishment of animal cell cultures allowed virus propagation. Thus, small DNA viruses were employed rather as a model to study eukaryotic molecular biology. Not surprisingly, many discoveries on eukaryotic molecular biology were obtained from DNA viruses (Table PII.1). For instance, Rb, protein, a tumor suppressor gene, was discovered as a SV40 T-antigen binding protein. Considering the importance of Rb in cell biology, research on DNA viruses has greatly contributed to cancer biology and cell biology as well as virology.

II.2 Features Shared by DNA Viruses

Before an individual DNA virus family is described by each chapter, let's consider some common properties shared by these DNA viruses. First, unlike RNA viruses, DNA viruses could rely on host DNA polymerase for their genome synthesis. Interestingly, the extent of their host dependence correlates with their genome size. DNA viruses having a small genome (<10 kb), such as polyomaviruses, papillomaviruses, and parvoviruses, do not encode a gene for the viral DNA polymerase but instead use host DNA polymerase for their genome synthesis (Table PII.2). On the other hand, DNA viruses having a larger genome, such as adenoviruses and herpesviruses, encode their own viral polymerases. Second, the viral genome replication of almost all DNA viruses occurs in the nucleus, regardless of their host dependence. The viral genome replication of DNA viruses that rely on host DNA polymerase should occur in the nucleus, where cellular DNA synthesis is carried out. Even DNA viruses who have their own DNA polymerase, such as adenoviruses and herpesviruses, replicate their genome in the nucleus, implicating that cellular factors other than DNA polymerase are utilized for their viral genome synthesis.

Another important feature that should be noted is that many DNA viruses, including polyomaviruses, papillomaviruses, adenoviruses, and a subset of herpesviruses, are tumor viruses. Notably, the viral oncoproteins of three DNA viruses with a small sized genome—polyomaviruses, papillomaviruses, adenoviruses—dysregulate

TABLE PII.1 Major Discoveries in DNA Viruses

Year	Discovery	Scientists (Inst.)	Remark
1953	Discovery of Polyomavirus	Gross	
1960	Discovery of SV40	Hilleman	
1977	Discovery of Introns from Adenovirus	Sharp (MIT)	1993 Nobel Prize
1983	HPV-16 as a causative agent of cervical carcinoma	Zur Hausen (Germany)	2008 Nobel Prize
1988	SV40 T-antigen as an Rb-binding protein	Livingston (Harvard Univ)	
1988	Adenovirus E1A as an Rb-binding protein	Harlow (Cold Spring Harbor Lab)	
1993	Ubiquitin E3 ligase activity of HPV-16 E6 and E6AP complex	Howley (Harvard Univ)	

TABLE PII.2 Features of DNA Viruses

Family	Prototype	Genome	Genome Size	DNA Polymerase	Tumor
Parvovirus	AAV	ssDNA	5 kb	Host	–
Polyomavirus	SV40	dsDNA	5 kb	Host	Yes
Papillomavirus	HPV	dsDNA	8 kb	Host	Yes
Adenovirus	Ad2	dsDNA	32–35 kb	Virus	Yes
Herpesvirus	HSV-1	dsDNA	120–235 kb	Virus	Yes

the function of tumor suppressor proteins, such as p53 and Rb, via direct protein–protein interaction. For instance, SV40 T-antigen interacts with both p53 and Rb proteins, and the interactions are essential for the viral oncogenesis. Likewise, E1A and E1B-55K proteins of adenoviruses interact with Rb and p53, respectively, and their interactions are essential for the viral oncogenesis.

Chapter 6

Polyomaviruses: SV40

Chapter Outline

Simian virus 40 (SV40) had been extensively studied during the 1970s–1980s. A reason for that is, perhaps, because it was easier to grow SV40 than other animal viruses in tissue culture, and it was relatively easier to handle SV40 genome experimentally due to its small size. Further, SV40 has been employed as a model to study cell transformation and viral oncogenesis and became a prototype DNA tumor virus. As a consequence, SV40 contributed greatly to our current knowledge of cancer biology.

6.1 DISCOVERY OF POLYOMAVIRUSES

Discovery: *Polyomaviruses*[1] were first discovered in the 1950s in the rodent tissues as a tumor-associated virus (Table 6.1). As its name implies, it causes diverse kinds of tumors in rodents. On the other hand, SV40 was discovered in a monkey cell line that was utilized in the 1960s for poliovirus vaccine production. In fact, it is believed that a considerable number of people had been unknowingly infected by SV40 that was inadvertently contained in the vaccine preparation. The observation that SV40 causes tumors in hamsters led to a realization that SV40 is related to the rodent polyomavirus. Moreover, the discovery of a primate polyomavirus (ie, SV40) strongly supported a notion that a related virus could be found in humans as well.

Human polyomaviruses: Two human polyomaviruses—*JCV*[2] and *BKV*[3]—were discovered in 1965 and 1971, respectively (see Table 6.1). Although their relatedness to SV40 is nearly 70%, evidence for their association with human cancers has not been reported. They are transmitted via respiration during infancy, and nearly all adults (40–90%) are persistently infected without any apparent clinical symptoms. Since these viruses coexist with humans, it is believed that these polyomaviruses somehow establish a symbiotic relationship with the human body. Nonetheless, JCV can rarely cause progressive multifocal leukoencephalopathy (*PML*[4]) in immune compromised individuals. Recently, a novel human polyomavirus was discovered in Merkel cell carcinoma (MCC), a rare kind of skin cancer. MCV is of interest, since it is the first polyomavirus that has been found to cause cancer in humans (Box 6.1).

Since the discovery of human polyomaviruses, and in particular since the discovery of MCV, more attention is being paid to human polyomaviruses. However, this chapter mainly focuses on SV40, since our current knowledge on polyomaviruses has been largely acquired from earlier studies on SV40 (Table 6.2).

1. **Polyomaviruses** The term *polyoma* is derived from Greek word for "many"-*poly* and "tumor" -*oma*.
2. **JCV** JCV was discovered in 1965 from a patient with progressive multifocal leukoencephalopathy (PML). It was named using two initials of a patient.
3. **BKV** BKV was discovered in 1971 from the urine of a renal transplant patient, initials B.K.
4. **PML** PML is a rare and usually fatal viral disease characterized by progressive damage (*-pathy*) or inflammation of the white matter (*leuko-*) of the brain (*-encephalo-*) at multiple locations (*multifocal*). It occurs almost exclusively in people with severe immune deficiency, such as transplant patients on immunosuppressive medications.

Molecular Virology of Human Pathogenic Viruses. DOI: http://dx.doi.org/10.1016/B978-0-12-800838-6.00006-0

TABLE 6.1 The Classification of Human Polyomaviruses

Virus	Host	Sero-Prevalence	Disease
Simian virus 40 (SV40)	Monkey	–	Tumor in rodents
Polyomavirus (PyV)	Mouse	–	Tumor
JC virus (JCV)	Human	~81%	Progressive multifocal leukoencephalopathy
BK virus (BKV)	Human	82–99%	Cystitis, nephropathy
Merkel cell polyomavirus (MCV)	Human	60–81%	Merkel cell carcinoma

BOX 6.1 Merkel Cell Polyomavirus

A novel human polyomavirus—Merkel cell polyomavirus (MCV)—was discovered in 2008 from Merkel cell carcinoma, a rare but aggressive form of skin cancer. Merkel cells are oval receptor cells found in the dermis of skin that are associated with the sense of light touch discrimination. MCV is suspected to cause the majority of cases of Merkel cell carcinoma. Approximately 80% of Merkel cell carcinoma (MCC) tumors have been found to be infected with MCV. MCV appears to be a common—if not universal—infection of older children and adults (60–81% sero-prevalence). Most MCV viruses found in MCC tumors have at least two mutations that render the virus nontransmissible: (1) The virus is integrated into the host genome in a monoclonal fashion and (2) The viral T-antigen has truncation mutations that make T-antigen defective in initiating DNA replication. Importantly, the integration into the host chromosome precedes the viral oncogenesis, a feature that is not obligatory in the virus life cycle.

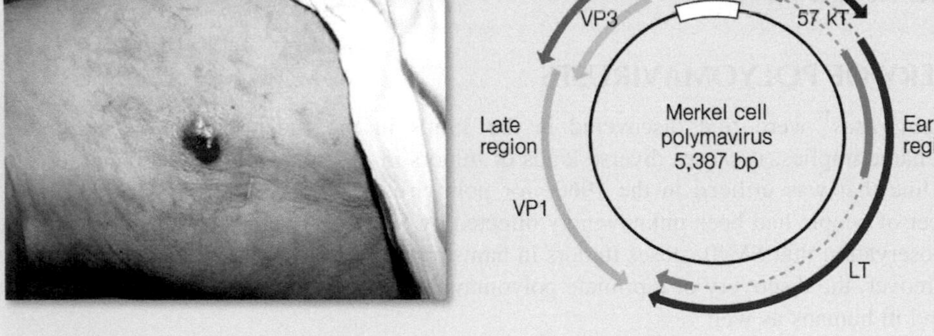

Merkel cell carcinoma (MCC) lesion and MCV genome. (left) Clinical appearance of Merkel cell carcinoma (MCC) lesion. Red, raised, nodular MCC lesion on the arm of a patient. (right) Genomic organization of Merkel cell polyomavirus. NCRR (noncoding regulatory region) contains a bidirectional promoter and the origin of replication. LT (large T-antigen), sT (small T-antigen), and 57-kT antigen constitute the early gene products.

TABLE 6.2 The Defining Features of Polyomaviruses

Genome	Virion Structure	Genome Replication
Circular dsDNA (5 kb)	Nonenveloped	Host DNA polymerase
Minichromosome	Icosahedral symmetry	T-antigen
Life Cycle	**Interaction with Host**	**Diseases**
Early-to-late switch	T-antigen-p53 interaction	Tumor
Cell cycle deregulation	T-antigen-Rb interaction	Merkel cell carcinoma

6.2 THE GENOME STRUCTURE OF SV40

Virion structure: SV40 virions are small, nonenveloped capsid particles with T = 7 icosahedral symmetry (Fig. 6.1). VP1 pentamers constitute the vertices of icosahedron, in which one molecule of VP2/VP3 underlies. In particular, VP2 is modified by *myristate*, a saturated fatty acid with 14 carbons. The fatty acid moiety facilitates an interaction with cell membrane. Inside the capsid, the viral DNA genome is found in association with histones, forming a nucleosome-like structure called "minichromosome." Unlike cellular nucleosome, notably, H1 linker histone is absent in the minichromosome. The nucleosome-like structure of the SV40 genome is one of the features that makes SV40 suitable for a model for eukaryotic DNA replication.

Genome structure: The viral genome is a small double-stranded circular DNA of approximately 5.2 kb in length (Fig. 6.2). All regulatory elements, including the *replication origin*[5] (Ori), promoter of early genes, and promoter of late genes, are clustered within a segment called the *regulatory region*. The viral mRNAs are transcribed from both strands of the DNA genome, as two promoters lie in the opposite direction (ie, bidirectional promoter). Both early and late RNAs are alternatively spliced and processed into multiple mRNAs for the synthesis of viral proteins.

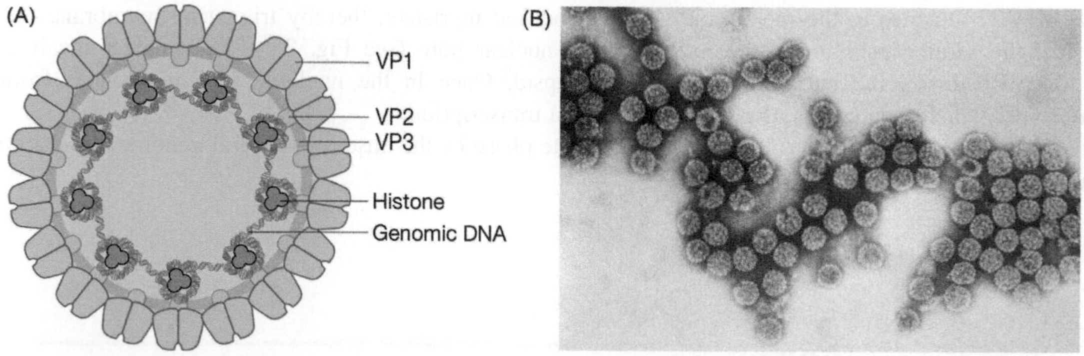

FIGURE 6.1 **Virion structure of SV40.** (A) Diagram depicting the cross section of the SV40 particle. It is an icosahedral capsid made of 72 capsomers, each composed of VP1 pentamers, with T = 7 symmetry. Each VP1 pentamer is underlined by one molecule of VP2/VP3. The viral genomic DNA is associated with histones. (B) Electron micrograph of polyomavirus particles.

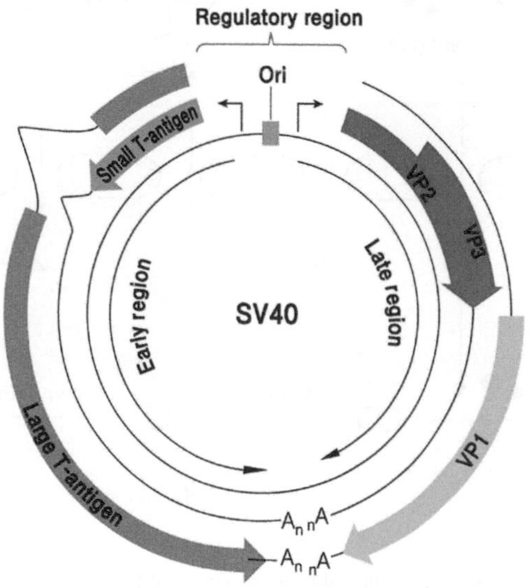

FIGURE 6.2 **Genome structure of SV40.** Early and late genes are transcribed from respective promoters embedded into "regulatory region" in opposite direction. The origin of replication (Ori) is also positioned in the regulatory region. The ORFs for early genes (ie, small and large T-antigens) and late genes (ie, VP1, VP2, and VP3) are depicted by arrowed boxes. Polyadenylation sites for early and late mRNAs are denoted by A with an arrow, respectively.

5. **Replication origin (Ori)** a *cis*-acting element where the DNA synthesis begins.

Protein coding: Early mRNAs are translated into two early proteins: small T-antigen and large *T-antigen*,[6] while late mRNAs are translated into three capsids proteins, VP1, VP2, and VP3.

6.3 THE LIFE CYCLE OF SV40

Cell culture: SV40 lytically infects monkey cells, such as CV-1 cell, a cell line derived from the kidney of an African green monkey.

Entry: Diverse glycolipid molecules are utilized as a receptor for the entry of polyomaviruses. In particular, SV40 utilizes *GM1 ganglioside*[7] as a receptor for the entry (Fig. 6.3). The VP1 of SV40 capsid binds to GM1 ganglioside via the galactose and sialic acid residues. The capsid enters cells via a receptor-mediated endocytosis. SV40 particles are subsequently internalized inside on endosome which is called "caveolae."[8] In other words, SV40 enters the cell via *caveolin/lipid raft-mediated endocytosis*, as opposed to clathrin-mediated endocytosis (see Fig. 3.5). Virions are then transported inside these vesicles through the cytosol to the lumen of the endoplasmic reticulum, near the nuclear membrane. The lower pH inside the endosome induces conformational change of VP2 such that the now exposed N-terminus of VP2 attaches to the membrane via the modified myristate, thereby triggering membrane fusion. Being small capsids, the viral capsids enter the nucleus via a nuclear pore (see Fig. 3.7). The nuclear localization signal (ie, NLS) on VP1 drives the nuclear import of the capsid. Once in the nucleus, VP1 dissociates from the viral minichromosome, which then serves as a template for viral transcription.

The life cycle of SV40 can be divided into early and late phase by the onset of the viral genome replication, a feature that is shared by all DNA viruses.

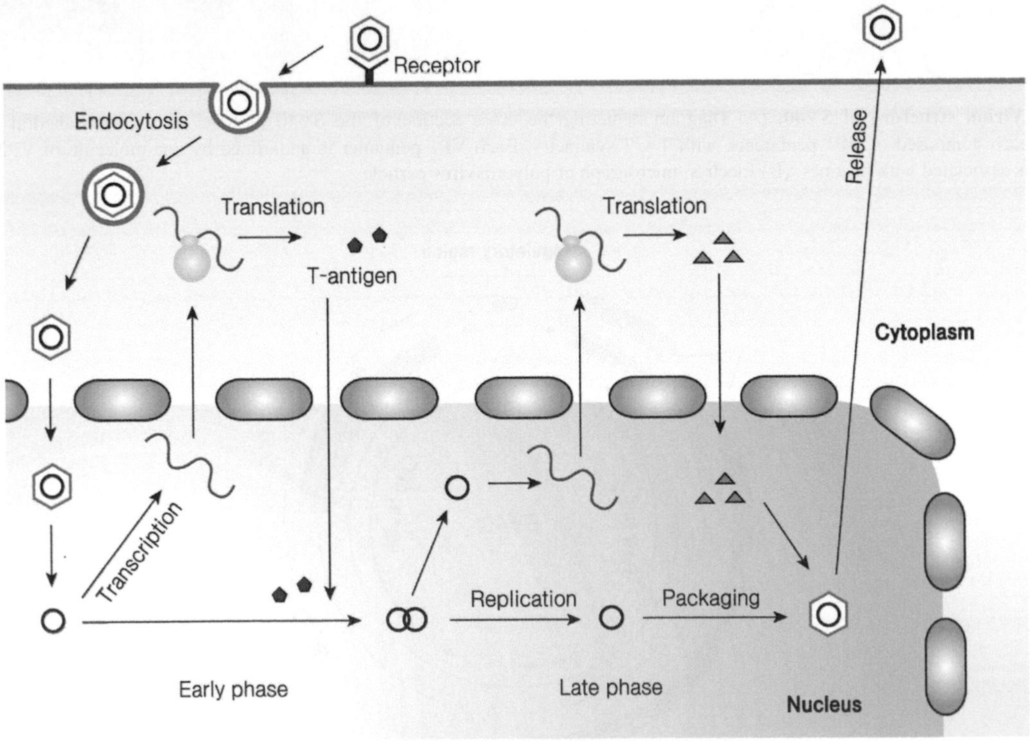

FIGURE 6.3 The life cycle of SV40. SV40 enters the cell via endocytosis. The viral DNA is uncoated in the nucleus following nuclear entry of the viral capsid via nuclear pore. The viral genome replication occurs in the nucleus largely by host DNA replication machinery. Early and late phase are divided by the onset of viral genome replication.

---

6. **T-antigen** It was discovered as a tumor-specific antigen (ie, tumor antigen) in SV40-induced tumors. thus named as "T-antigen."
7. **GM1 ganglioside** a kind of glycolipid (ie, glycosphingolipid) that was further modified with sialic acid.
8. **Caveolae** Caveolae are a special type of lipid raft, which are small (50–100 nm) invaginations of the plasma membrane found in many cell types.

6.3.1 Early Phase

Immediately following the nuclear entry of the viral DNA, the viral transcription starts.

Transcription: Early RNAs are transcribed by the cellular RNA polymerase II from early promoter (Fig. 6.4). Early promoter represents a typical eukaryotic promoter having a "TATA box" and upstream elements, such as SP1 binding sites. Early transcripts are alternatively spliced into two mRNAs, which are translated to large T-antigen and small T-antigen (see Fig. 6.2). Early RNAs encode two proteins from one transcript via alternative splicing, resulting in an extension of the coding capacity.

Switching to late phase: T-antigen begins to bind to the T-antigen binding sites, starting from the T-antigen binding site I to III, as it accumulates in the nucleus (see Fig. 6.4). T-antigen binding to the viral genome triggers the switch from early to late phase. Specifically, the engagement of T-antigen to the binding site I leads to transcription suppression from the early promoter. The suppression is also called "autoregulation," because T-antigen regulates its own expression (Fig. 6.5). The engagement of T-antigen to the binding site II triggers the initiation of viral genome replication. Lastly, the engagement of T-antigen to the binding site III activates transcription from the late promoter, a process also called "transactivation."

6.3.2 Late Phase

The late phase begins with the initiation of viral DNA replication.

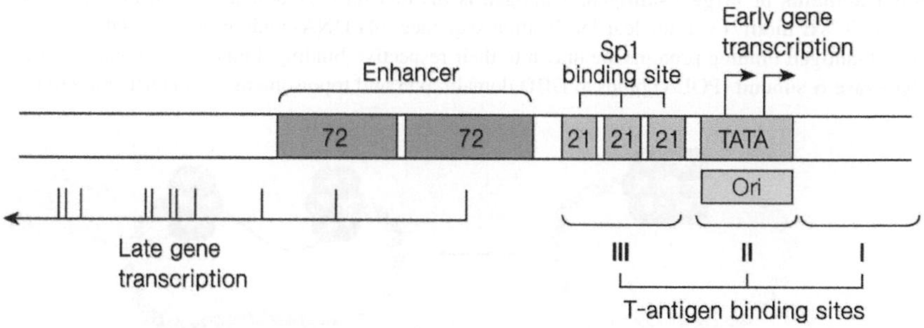

FIGURE 6.4 Regulatory region of SV40. Almost all *cis*-acting elements essential for the viral genome replication are embedded into the 0.5 kb so-called "regulatory region": (1) early promoter, (i2) late promoter, and (3) origin of replication. Early and late transcripts are transcribed in the opposite direction, as denoted by arrows. Early promoter constitutes TATA box, SP1 binding sites, and 72-bp repeat enhancer region. Late promoter does not have TATA box, but have numerous 5′ initiation site. Three T-antigen binding sites (ie, I, II, and III) are denoted. The origin of replication (Ori) overlaps with T-antigen binding site II.

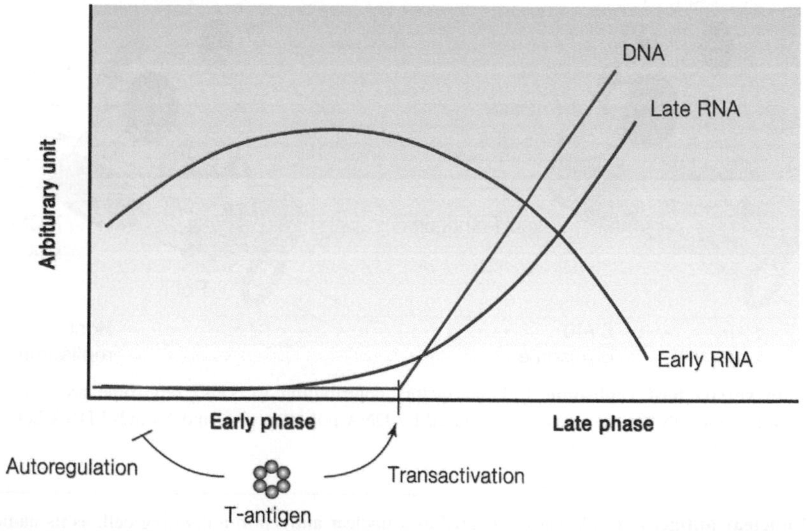

FIGURE 6.5 Switch from early to late phase. T-antigen plays a pivotal role in the switch from early phase to late phase. Autoregulation of early gene transcription and transactivation of late gene transcription by T-antigen is schematically depicted for the purpose of explanation.

Genome replication: SV40 genome replication heavily relies on the cellular replication machinery, including not only DNA polymerase but also other accessory factors such as topoisomerase and *PCNA*.[9] In fact, T-antigen is sufficient to orchestrate the viral genome replication in infected cells. As a matter of fact, the diverse biochemical activities of T-antigen contribute to DNA synthesis: (1) DNA helicase, (2) ATPase, and (3) DNA polymerase binding activity (Fig. 6.6).

SV40 has served as a model in studying eukaryotic DNA replication, as it heavily relies on cellular DNA replication machinery. The DNA synthesis had been demonstrated in vitro using a template DNA containing SV40 origin (Ori), when cellular extracts (ie, DNA replication machinery) was complemented by the purified T-antigen. Note that T-antigen is the only viral protein required to execute the DNA synthesis. Establishment of the SV40 DNA replication in vitro was instrumental for subsequent identification of the cellular factors contributing to eukaryotic DNA replication. The DNA synthesis is initiated by binding of two hexamers of T-antigen to the origin (Ori) on the SV40 genome, thereby melting the duplex DNA (Fig. 6.7). Subsequently, cellular DNA replication machineries are recruited to constitute the so-called SV40 replisome[10] that can execute DNA synthesis (Fig. 6.8). Importantly, all cellular factors

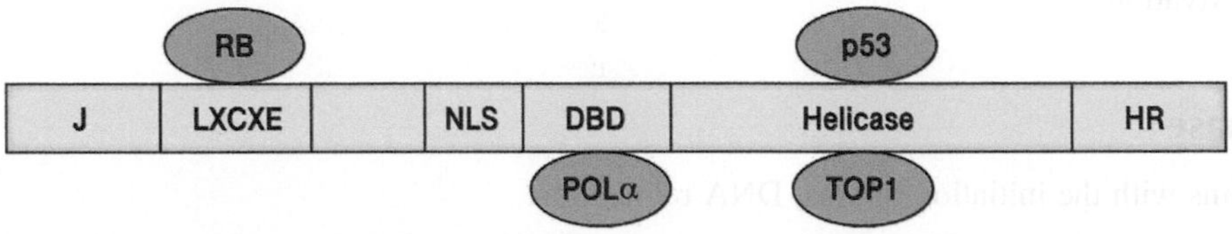

FIGURE 6.6 The functional domains of large T-antigen. T-antigen is divided into six domains, each of which has a biological activity: from N-terminus, (1) J domain, (2) LXCXE motif, (3) a nuclear localization sequence, (4) DNA-binding domain (DBD), (5) a helicase domain, and (6) a host range domain (HR). Four T-antigen binding proteins are drawn to their respective binding domains: retinoblastoma-associated protein (RB) binds to LXCXE motif; DNA polymerase α subunit (POLα) binds to DBD domain; p53 and topoisomerase 1 (TOP1) bind to the helicase domain.

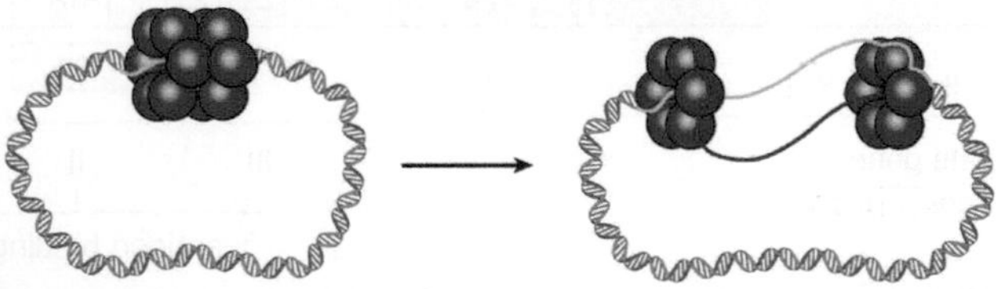

FIGURE 6.7 Initiation of SV40 DNA replication by T-antigen. SV40 DNA replication begins when the large T-antigen binds to the origin of replication (Ori). Two hexamers of T-antigen form a head-to-head orientation at the origin, unwinding the viral DNA followed by bidirectional replication.

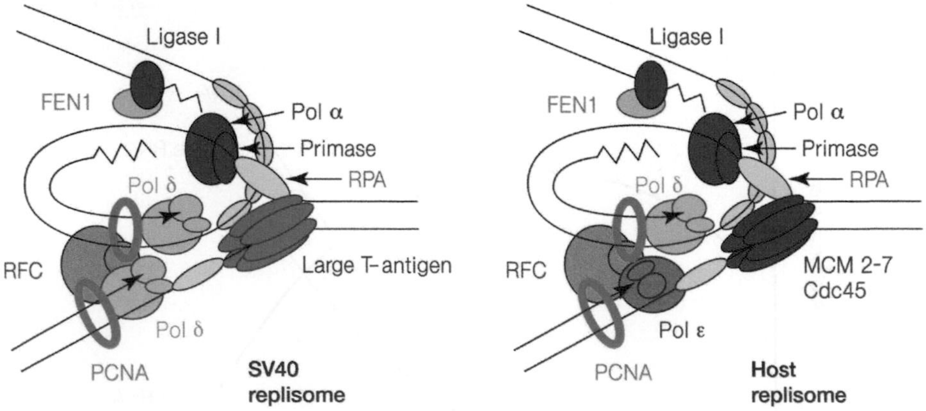

FIGURE 6.8 SV40 replisome versus host replisome. Cellular factors contributing to SV40 replisome are almost identical to those in host replisome, except that the leading strand DNA polymerase ε is replaced by DNA polymerase δ, and Mcm2-7 DNA helicase is replaced by T-antigen. RPA (replication protein A), PCNA (proliferating cell nuclear antigen).

9. **PCNA (proliferating cell nuclear antigen)** PCNA was discovered as a nuclear antigen in a dividing cell, as its name implies. It acts as a sliding clamp at the replication fork, leading to increase of the "processivity" of DNA polymerase.

10. **Replisome** a complex molecular machine that carries out replication of DNA.

involved in the SV40 replisome are essentially identical with those involved in the "host replisome," except that Mcm2-7/Cdc45 DNA helicase is replaced by T-antigen (ie, viral DNA helicase).

Deregulation of cell cycle: SV40 deregulates cell cycle control to achieve multiple round of the viral DNA replication. SV40 overrides cell cycle control in two distinct mechanisms. First, T-antigen impedes the progress of S phase in the cell cycle, thereby prolonging the S phase (ie, S phase arrest). How does T-antigen hold the progress of S phase? T-antigen induces the *ATR*[11] /*ATM*[12]-mediated *DNA damage response (DDR*[13]*)* pathway (Box 6.2). Such activated ATR directly activates the p53 isoform Δp53, leading to upregulation of the Cdk inhibitor p21, which forces the host cell to stay in the replicative S phase. This pathway is called *ATR-Δp53-p21 pathway*[14] (see Journal Club). Intriguingly, ultraviolet (UV) irradiation of cells at G1/S transit also induces cell cycle arrest at S phase via ATR-Δp53-p21 pathway. Similar to UV irradiation, SV40 infection induces the DDR pathway to prolong the S phase. Then, what is the benefit for SV40 of inducing the DDR pathway? Since SV40 replication depends on the host factors, such as DNA polymerase α plus δ, topoisomerases, and other factors that are functional only in S phase, the prolonged S phase facilitates the progeny virus production.

Second, T-antigen overrides the "re-replication block," which is the central regulatory mechanism to maintain the integrity of the host chromosome. The re-replication block is to ensure that origins are utilized only once per cell cycle so that all chromosome DNA are equally duplicated. Intriguingly, although SV40 heavily relies on host DNA machinery, SV40 overrides the re-replication block so that it induces multiple rounds of cellular DNA synthesis, giving rise to polyploid cells. How does the T-antigen override the re-replication block? A recent report showed that T-antigen binds to *Nbs1*[15] and blocks the function of Nbs1, which is involved in the re-replication block. The deregulation of the re-replication block by SV40 T-antigen represents the viral strategy to coopt the host's cell cycle control for its own benefit.

Late Gene Transcription: The onset of the viral genome replication cues the late gene transcription from the late promoter (see Fig. 6.5). The late RNAs are alternatively spliced into multiple mRNAs, which are translated into VP1 to VP3 proteins. These capsid proteins are translocated to the nucleus, where the viral capsid assembly occurs.

Assembly and Release: The viral capsids are assembled in the nucleus, and then extracellularly released via cell lysis (see Fig. 6.3). It remained uncertain as to how cell lysis is triggered. Recently, a novel viral capsid protein, called VP4, was identified. It is translated from the third AUG codon of VP2 ORF, sharing the C-terminus with VP2 and VP3. Importantly, VP4 was shown to exert "viroporin" function, which induces pore formation in the membrane. It is VP4 that triggers cell lysis by disrupting the cell membrane.

6.4 EFFECTS ON HOST

For a permissive host, such as primates (ie, monkeys and human), the viral life cycle is fully executed in that viral genome replication occurs, and the progeny virus are produced (Fig. 6.9). SV40 induces cell lysis and forms plaques on monolayer cells of the permissive cells. In contrast, for *nonpermissive hosts,*[16] such as rodents (mouse and hamster), only the early phase of viral life cycle is executed, whereas the late phase (ie, the viral genome replication) is blocked. Instead, however, *cell transformation*[17] is observed (see Fig. 6.9). Thus, the consequence of SV40 infection is different depending upon whether the host is permissive or nonpermissive for SV40 infection.

Tumor formation: SV40 infection leads to tumor formation in rodents, a *nonpermissive host*. Upon infection, T-antigen is continually expressed, although the late gene expression is not permitted. Which function of T-antigen contributes to cell transformation? As described above, T-antigen has multiple biochemical activities such as DNA binding

11. **ATR (ataxia telangiectasia and RAD3-related)** ATM-related gene. It is a serine/threonine kinase that is involved in sensing DNA damage and activating the DNA damage checkpoint, leading to cell cycle arrest.

12. **ATM (ataxia telangiectasia-mutated)** a tumor suppressor gene identified as a mutated gene in ataxia telangiectasia. Ataxia telangiectasia is a rare, neurodegenerative, inherited disease causing severe disability. ATM, a serine/threonine protein kinase, is activated by double-strand break (DSB), acting as a sensor of the DNA damage.

13. **DNA damage response (DDR)** it refers to the signaling pathway that is induced by DNA damage (see Box 6.2).

14. **ATR-Δp53-p21 pathway** it refers to a signaling pathway that is triggered by ATR. The activated ATR activates an isoform of p53 (ie, Δp53) via phosphorylation, which then leads to induction of p21 expression. Such induced p21 induces cell cycle arrest. Δp53, an isoform of p53, resulted from alternative splicing. Unlike p53, it does not bind to T-antigen.

15. **Nbs1** a component of MRN complex (Mre11, Rad50, and Nbs1) that is involved in initial processing of double-strand DNA break repair.

16. **Nonpermissive host** a host that permits only the early phase of the viral life cycle.

17. **Cell transformation** It refers to a process by which cells acquires the properties of cancer cells.

BOX 6.2 DNA Damage Response and DNA Viruses

Genome stability is an essential feature for cell survival. Eukaryotic cells have evolved a complex set of pathways to repair DNA and ensure the faithful duplication of the genome. Diverse kinds of genotoxic stresses, such as UV irradiation and reactive oxygen species, cause DNA damage. In addition to genotoxic stresses, double-strand break (DSB), a fatal DNA damage, takes place inevitably during DNA replication. To overcome such DNA damages, cells are equipped with diverse DNA repair mechanisms. In this process, the signaling pathway that senses the DNA damage and activates the DNA repair mechanisms is collectively called *DNA damage response (DDR)*.

To what extent is DDR related to DNA viruses? Being a parasite, a virus co-opts diverse cellular functions. In fact, DDR is the one that is most extensively exploited by DNA viruses. Intriguingly, some viruses trigger DDR to induce the viral DNA synthesis in resting cells they infect, while other viruses suppress DDR. For instance, SV40 and HPV induce ATM/ATR-mediated DDR, in that the activation of S phase checkpoint results in S phase delay or arrest, which prolongs the ensuing viral genome replication (Table 6.B1). In contrast, adenovirus blocks DDR for efficient viral genome replication. In fact, DDR is induced by adenovirus infection, since the adenoviral DNA itself is recognized as double-strand break damage (see Box 8.1). Consequently, the concatemers of the viral genome are formed, unless the DSB repair is blocked. Thus, the inhibition of DSB repair by the virally coded proteins is critical for the efficient adenoviral genome replication. In the case of herpesviruses, the replication intermediates of linear DNA genomes are recognized by ATM/ATR without invoking the DDR signaling. In cells infected by herpes simplex virus (HSV-1), ATM/ATR-mediated DDR is also blocked by the virally coded proteins. It is apparent that DNA viruses have evolved to acquire functions that block DDR signaling in order to avoid unwanted DNA products, as the linear DNA genomes are inevitably recognized by ATM/ATR as DNA damages.

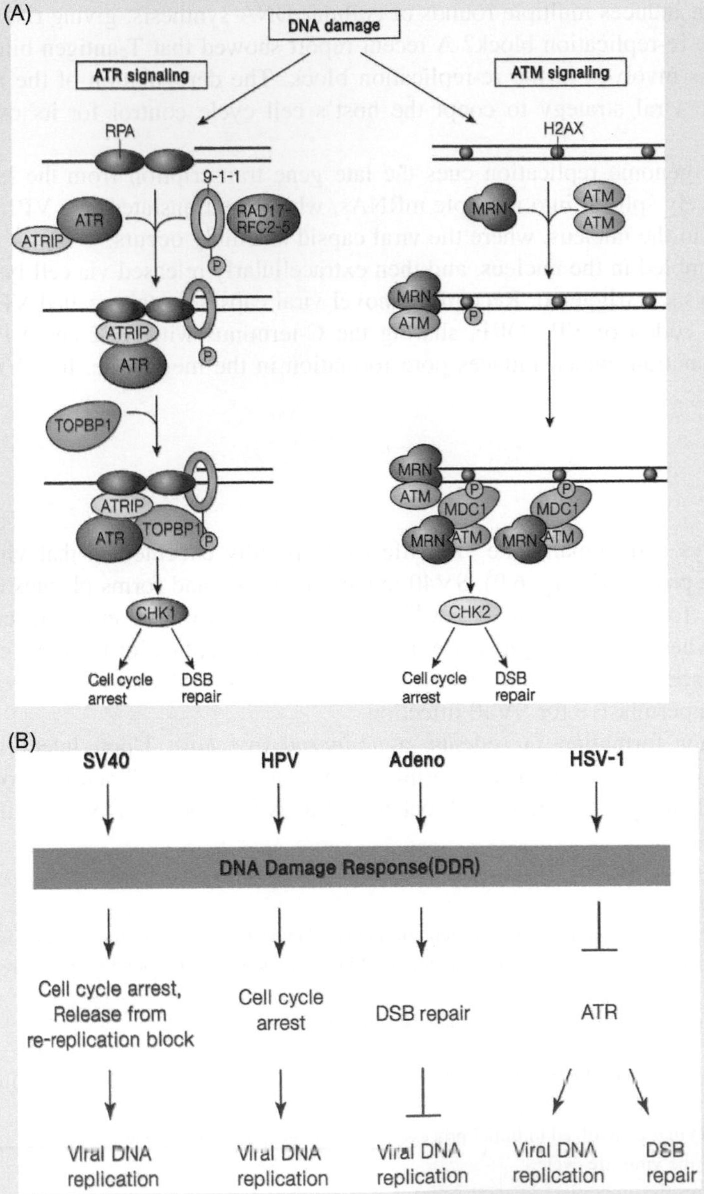

DNA damage response signaling pathway in DNA viruses. (A) ATM signaling and ATR signaling. ATM kinase is activated when the double-strand break DNA damage is sensed via a mechanism involving MRN (Mre11/Rad50/Nbs1). The activated ATM then phosphorylates the histone variant H2AX. The phosphorylated H2AX (γH2AX) binds to the mediator of DNA damage checkpoint protein-1 (MDC1), leading to recruitment of additional ATM-MRN complexes and further phosphorylates H2AX. The ATM also phosphorylates downstream targets, CHK2 kinase, leading to cell cycle arrest and DSB repair. ATR kinase is activated by sensing DNA damage via a mechanism involving Replication protein A (RPA). ATR/ATRIP and Rad9-Rad1-Hus1 (also known as 9-1-1) are independently recruited to the damaged sites. Then, ATR activator topoisomerase-binding protein-1 (TOPBP1) is recruited via interaction with Rad9 of 9-1-1. TOPBP1 binds and activates ATR, leading to phosphorylation of CHK1. In response to DNA damage or replication stress, cell cycle arrest is induced. (B) The roles of DNA damage response (DDR) signaling pathway in DNA viruses. The activation of DDR signaling is essential for cells to enter S phase that ensure the viral DNA synthesis. For some viruses such as SV40 and HPV, virus-encoded proteins, T-antigen and E7, induce the DDR, while linear DNA genome itself could induce DDR in adenovirus and HSV-1 infected cells. On the other hand, DNA viruses having the linear DNA genome, the activated DDR needs to be blocked for efficient viral DNA replication.

(Continued)

BOX 6.2 (Continued)

TABLE BOX 6.2 DNA Damage Response and DNA Viruses

Family	Virus	DNA Damage Response Signaling	Host Factors Targeted by	Viral Proteins	Function
Polyomavirus	SV40	ATM signaling activated	–	T-antigen	S phase delay
Papillomavirus	HPV-31	ATM signaling activated	–	E7	S phase arrest
Adenovirus	Ad5	ATM signaling inhibited	MRN complex	E1B-55/E4 orf6	Blockade of DSB repair
Herpesvirus	HSV-1	ATM signaling inhibited	DNA-PK	ICP0	Blockade of DSB repair
		ATR signaling inhibited	ATR/ATRIP/RPA complex	ICP8/UL8/UL5/UL52 complex	Blockade of DSB repair

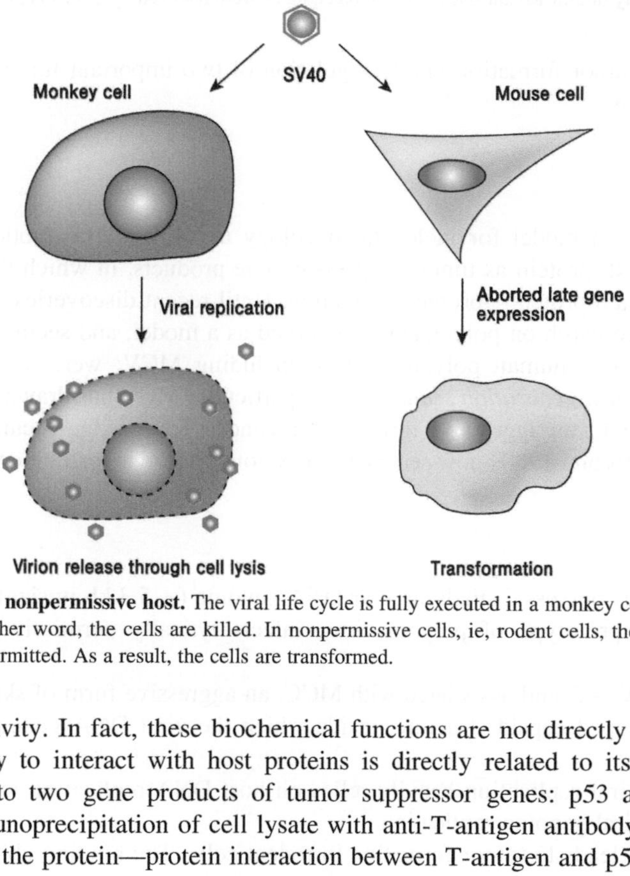

FIGURE 6.9 **Permissive host versus nonpermissive host.** The viral life cycle is fully executed in a monkey cell, a permissive host, and the progeny virions are released by cell lysis. In other word, the cells are killed. In nonpermissive cells, ie, rodent cells, the early phase of the virus life cycle is executed, while the late phase is not permitted. As a result, the cells are transformed.

activity and DNA helicase activity. In fact, these biochemical functions are not directly related to cell transformation. Instead, the T-antigen's ability to interact with host proteins is directly related to its cell transformation function. Specifically, T-antigen binds to two gene products of tumor suppressor genes: p53 and *Rb*[18] protein (ie, pRB[105]) (see Fig. 6.6). Indeed, co-immunoprecipitation of cell lysate with anti-T-antigen antibody brought down a protein having a size of 53 kDa, revealing the protein—protein interaction between T-antigen and p53 (Fig. 6.10). Mechanistically, T-antigen blocks the tumor suppressor function of p53 via the protein—protein interaction, thereby contributing to cell transformation (see Fig. 24.4). Likewise, co-immunoprecipitation of cell lysate with anti-Rb antibody brought down T-antigen as well as pRB[105] or vice versa, revealing the protein—protein interaction between T-antigen and pRB[105] (Fig. 6.10). After all, T-antigen blocks the tumor suppressor function of pRB[105] via protein—protein interaction, thereby contributing to cell transformation (see Fig. 24.3).

18. **Rb (retinoblastoma)** a tumor suppressor gene that was first identified as a defective gene in retinoblastoma.

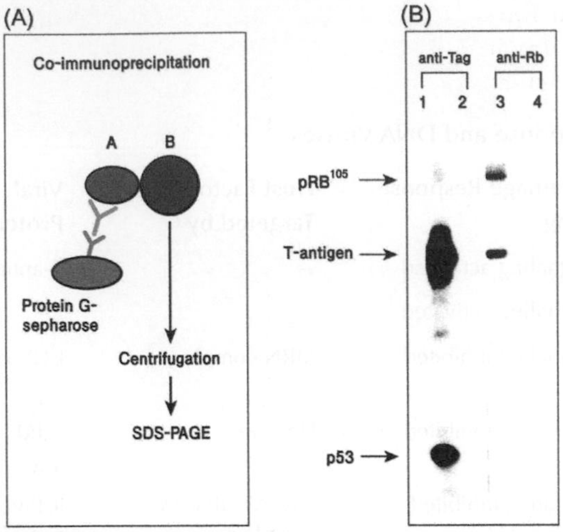

FIGURE 6.10 Evidence for the interactions between T-antigen and p53 and between T-antigen and Rb protein. (A) A diagram illustrating the strategy of co-immunoprecipitation to examine the protein—protein interaction. Antibody specific for the protein A can bring down the protein B, if the protein B binds to the protein A in cells. The protein G, which has the ability to bind to the Fc fragment of IgG, linked to Sepharose bead is utilized to precipitate the antigen—antibody complex. The protein A-interacting proteins can be revealed by SDS-PAGE. (B) Cell lysates were precipitated with either anti-T-antigen antibody or anti-Rb antibody, and the precipitates were resolved by SDS-PAGE.

Altogether, SV40 induces tumor formation via dysregulation of two important tumor suppressor proteins—p53 and Rb—by T-antigen (see Fig. 24.8).

6.5 PERSPECTIVES

Polyomaviruses have served as a model for molecular oncology as well as eukaryotic molecular biology. A payoff was the discovery of p53 and Rb protein as tumor suppressor gene products, in which their tumor suppressor functions are suppressed by T-antigen via a protein—protein interaction. Until recent discoveries of human polyomaviruses, such as BK virus and JC virus, the research on polyomaviruses served as a model, and seemed to be unrelated to human diseases. More recently, quite a few human polyomaviruses, including MCV, were discovered by the advent of new sequencing technology termed *next generation sequencing*. In particular, MCV has drawn a lot of attention, as it is clearly found to be associated with MCC, an aggressive form of skin cancer. Knowledge accumulated from studies on animal polyomaviruses during the 1970s and 1980s now serves as a keystone for the investigation of human polyomaviruses.

6.6 SUMMARY

- *Polyomavirus*: Polyomavirus possess a small circular DNA genome (~5.3 kb in size).
- *SV40 genome*: SV40, the prototype of polyomavirus, infects lytically permissive monkey cells but transforms nonpermissive rodent cells.
- *Human polyomavirus*: MCV is found associated with MCC, an aggressive form of skin cancer.
- *Virion structure*: Nonenveloped capsid structure, inside which the viral DNA genome is found in association with histone molecules.
- *Genome replication*: The viral replication heavily relies on host DNA replication machinery. T-antigen is the only viral protein essential for viral genome replication.
- *Host effect*: SV40 infection leads to tumor formation in rodent cells. The binding ability of T-antigen to p53 and Rb protein is critical for tumorigenesis.

STUDY QUESTIONS

6.1 The life cycle of SV40 can be divided into early and late phase by the onset of the viral genome replication, where a switch from early to late phase is facilitated by T-antigen. Explain the molecular mechanisms underlying the early to late switch.

6.2 Describe to what extent the viral life cycle proceeds in the following SV40 mutants, when these mutants DNA were transfected into CV-1 cell (monkey cell) or COS cell (T-antigen expressing monkey cell). (1) mutant B (T-antigen helicase mutant), and (2) mutant B (Origin-defective mutant)

6.3 SV40 mutants (ie, Ori-defective) are frequently used in establishing human cell lines from a primary cell. Explain why the Ori-defective mutants are used, instead of wild-type SV40.

SUGGESTED READING

Chang, Y., Moore, P.S., 2012. Merkel cell carcinoma: a virus-induced human cancer. Annu. Rev. Pathol. 7, 123–144.

DeCaprio, J.A., Garcea, R.L., 2013. A cornucopia of human polyomaviruses. Nat. Rev. Microbiol. 11 (4), 264–276.

Jiang, M., Zhao, L., Gamez, M., Imperiale, M.J., 2012. Roles of ATM and ATR-mediated DNA damage responses during lytic BK polyomavirus infection. PLoS Pathog. 8 (8), e1002898.

Lilley, C.E., Schwartz, R.A., Weitzman, M.D., 2007. Using or abusing: viruses and the cellular DNA damage response. Trends Microbiol. 15 (3), 119–126.

Sowd, G.A., Fanning, E., 2012. A wolf in sheep's clothing: SV40 co-opts host genome maintenance proteins to replicate viral DNA. PLoS Pathog. 8 (11), e1002994.

JOURNAL CLUB

- Rohaly, G., Korf, K., Dehde, S., Dornreiter, I., 2010. Simian virus 40 activates ATR-Δp53 signaling to override cell cycle and DNA replication control. J. Virol. 84 (20), 10727–10747.

  Highlight: SV40 overrides cell cycle control, thereby prolonging the S phase. This paper demonstrated that SV40 T-antigen induces S phase arrest by activating ATR-Δp53-p21 pathway.

Chapter 7

Papillomaviruses

Chapter Outline

Papillomaviruses (family Papillomaviridae) are small, nonenveloped, icosahedral viruses that possess a circular double-strand DNA genome of 8 kb. While the majority of human papillomaviruses (HPVs[1]) infections remain subclinical or cause benign lesions only, infections by a subset of HPVs, known as high-risk types, can lead to cancer. This chapter will focus on the molecular aspects of HPV with an emphasis on its genome replication and its effect on host cells, in particular HPV-associated carcinoma (Table 7.1).

7.1 DISCOVERY OF PAPILLOMAVIRUSES

Papillomaviruses were first discovered in 1933 in the cotton-tail rabbit as a wart-causing virus. A similar wart-causing virus, bovine papillomaviruses (BPV), was soon discovered in cows. Subsequently after its discovery HPV became the research focus, in particular the discovery of cervical carcinoma-associated high-risk HPV subtypes.

TABLE 7.1 The Defining Features of Papillomaviruses

Genome	Virion Structure	Genome Replication
Circular dsDNA (8 kb)	Nonenveloped	Host DNA polymerase
Minichromosome	Icosahedral symmetry	Episomal and vegetative replication
Life Cycle	**Interaction with Host**	**Diseases**
Early-to-late switch	E6-p53 interaction	Cervical carcinoma
Differentiation-linked	E7-Rb interaction	Warts

1. **HPVs (human papillomaviruses)** The term *papilloma* is derived from Greek word for "nipple" *-papilla* and Greek word for "tumor" *-oma*. Papilloma refers to a benign epithelial tumor, growing outwardly projecting in finger-like fronds.

Molecular Virology of Human Pathogenic Viruses. DOI: http://dx.doi.org/10.1016/B978-0-12-800838-6.00007-2

BOX 7.1 High-Risk HPVs and Harold zur Hausen

The nomination of Harold zur Hausen as the 2008 Nobel Prize winner represents the milestones of papillomavirus research and reflects the impact of his discovery of linking high-risk HPVs to cervical carcinoma. Earlier epidemiology study in the 1960s had shown that cervical carcinoma is a sexually transmitted disease (STD). In fact, it has been thought that herpes simplex virus (HSV-2) is linked to cervical carcinoma. During his own endeavor to search for the culprit, Harold zur Hausen paid attention to papillomavirus, since papillomavirus is a tumor virus that causes genital warts. He first cloned out HPV DNAs from genital warts and used them as a probe to search for one in cervical carcinoma. After over 10 year's effort, he was able to clone out certain genotypes of HPVs that are consistently found in cervical carcinoma, which are now known as high-risk HPVs: HPV-16 and 18. In retrospect, his thought turned out to be visionary. In his interview by a reporter asking his response to his nomination to Nobel award, he softly said that "he is glad that his thought is now vindicated." According to recent reports, HPV-16 and 18 are found in more than 70% of cervical carcinoma, and the number becomes close to 100% (ie, 99.7%), if some low-risk HPV genotypes are included. More importantly, Harold zur Hausen's seminal discovery paved the way for the HPV vaccine development and toward eventual eradication of the culprit.

A photo of Harold zur Hausen (1936–).

7.1.1 Classification and Epidemiology

Among over 150 HPV subtypes currently reported, most of the HPVs are associated with warts, a benign tumor. Importantly, HPV-16 and 18 are associated with cervical carcinoma, while HPV-6 and 11 are associated with genital warts (Box 7.1). Approximately half a million people worldwide are newly diagnosed every year with cervical cancer, which is mostly attributable to HPV infection. More than half of cervical cancer patients died due to the disease. Fortunately, HPV vaccines recently became available making HPV-associated cervical carcinoma preventable (Box 7.2). A recent survey of HPV prevalence using a shotgun sequencing approach revealed that the overall HPV prevalence among healthy people is significantly high (68.9%), and is highest in the skin (61.3%), followed by the vagina (41.5%), mouth (30%), and gut (17.3%). However, the significance of high HPV prevalence in healthy human subjects is unknown.

7.1.2 Transmission

human papillomaviruses infect the stratified epithelium of the skin and mucosa. They cause cutaneous, anogenital, or cervical infection depending on subtypes (Table 7.2). Transmission via the latter two routes occurs via sexual contact. Although the epidermis of the skin serves as a physical barrier for cutaneous infection, cuts in skin provide access to the underlying cells. Cell-mediated immunity in the skin tissue plays a significant role in preventing the skin infection. However, more or less, HPV manages to establish skin infection and causes wart.

7.2 THE GENOME STRUCTURE OF HPV

Virion structure: HPV virions are small (55 nm), nonenveloped particles with $T = 7$ icosahedral symmetry (Fig. 7.1). Major and minor capsids proteins—L1 and L2, respectively—constitute viral capsids. Inside viral capsids, the viral

BOX 7.2 HPV Vaccines

Worldwide, cervical cancer is the second most common and the fifth deadliest cancer for women. It affects about 16 per 100,000 women per year and kills about 9 per 100,000 per year. Approximately 80% of cervical cancers occur in developing countries. Worldwide, in 2008, it was estimated that there were 473,000 cases of cervical cancer, and 253,500 deaths per year. For men, HPV infection causes only genital warts, a benign tumor, which is not fatal. In contrast, for women, high-risk HPV infection in women causes cervical carcinoma, which is a malignant tumor and fatal. Even in developed countries, over 20% of women are infected by HPV. Most HPV infections in young females are temporary and have little long-term significance. Seventy percent of infections are cleared in 1 year and 90% in 2 years. However, when the infection persists—in 5% to 10% of infected women—there is high risk of developing precancerous lesions of the cervix, which can progress to an invasive cervical cancer. This process usually takes 10–15 years, providing many opportunities for the detection and treatment of precancerous lesion. Recent development of HPV vaccines has made cervical carcinoma preventable. For instance, Gardasil is made of recombinant *VLPs*[2] of the major capsid protein, L1, expressed in yeasts. It is recommended for youngsters before the onset of sexual life.

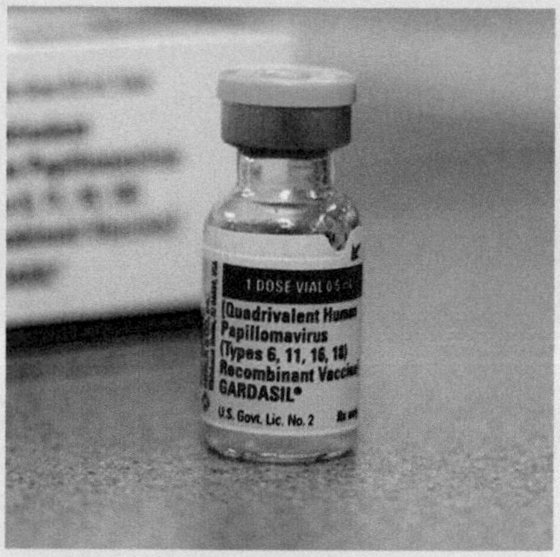

A photo of HPV vaccines. Gardasil is a quadrivalent HPV vaccine composed of capsid proteins from four subtypes of high-risk HPVs (ie, HPV-6, 11, 16, and 18).

TABLE 7.2 The Classification of Human Papillomaviruses

HPV Genus	HPV Genotype	Tissue	Disease
HPV-α (mucosal type)	HPV-6, 11	Mucosal epithelium of genital tract Mucosal epithelium of respiratory tract	Genital wart papilloma of larynx
	HPV-16, 18	Mucosal epithelium of genital tract	Cervical carcinoma oropharyngeal cancer
HPV-β (cutaneous type)	HPV-5, 8	Skin epithelium	Skin cancer
HPV-μ (cutaneous type)	HPV-1	Skin epithelium	Wart
	HPV-38	Skin epithelium	Wart, skin cancer

genomic DNA is associated with histones. Hence, the viral genome holds a minichromosome structure, a property that is shared with polyomavirus.

Genome structure: The viral genome is an 8 kb long double-stranded circular DNA (Fig. 7.2). Like SV40, all regulatory elements, including the *replication origin*[3] (Ori), early promoter (PE), and late promoter (PL), are clustered within a segment called the *long control region*[4] (LCR). However, unlike SV40, the viral mRNAs are transcribed only from one strand of the DNA genome, because two promoters lie in the same direction: P_E and P_L. Early and late viral RNAs are alternatively spliced, polyadenylated by using distinct polyadenylation signals, and processed into multiple mRNAs for the synthesis of viral proteins. Early mRNAs are translated into six early proteins: E1 to E7 protein, while late

2. **VLP (virus-like particle)** It refers to viral particles devoid of the viral genome.
3. **Replication origin (Ori)** A *cis*-acting element, where the DNA synthesis begins.
4. **Long control region (LCR)** A segment of HPV genome, where the origin and viral promoters are clustered.

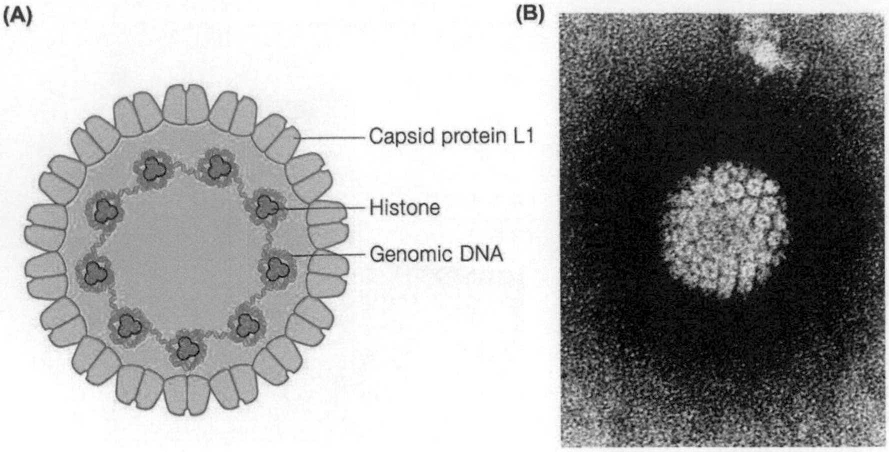

FIGURE 7.1 Virion structure of HPV. (A) Diagram depicting the cross section of HPV particle. The viral capsids are an icosahedral capsids made of 72 molecules of L1 pentamer. For clarity, L2 protein, which is present inside of L1 pentamer with ratio of one copy per L1 pentamer, is not drawn. The viral genomic DNA, which is associated with histones, is found inside the nucleocapsid. (B) Transmission electron micrograph of negatively stained papillomavirus particle. The viral capsid is a relatively small particle with a diameter of 55 nm

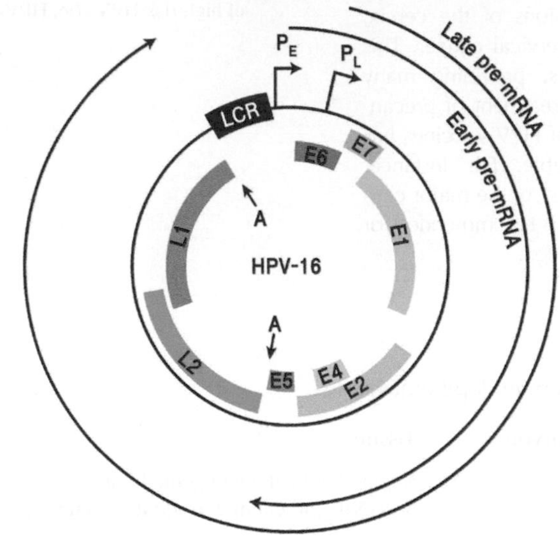

FIGURE 7.2 Genome structure of HPV. Early and late mRNAs transcribed clockwise from the respective promoters are denoted by the solid line. Polyadenylation sites for early and late mRNAs are denoted by A with an arrow. Early and late pre-mRNAs are alternatively spliced in multiple ways, which is not shown for brevity. Open reading frames of six early (E) genes and two late (L) genes are shown in boxes. LCR; long control region.

mRNAs are translated into two capsid proteins, L1 and L2 (Table 7.3). Some viral ORFs overlap each other, reflecting the compact nature of the viral genome organization. Being a small DNA virus, the viral genome does not encode its own DNA polymerase.

7.3 THE LIFE CYCLE OF HPV

Cell culture: A cell line that fully supports HPV propagation is not yet available. The lack of proper cell culture has hampered the advance of HPV research, in particular, on the late phase of the life cycle. In fact, papillomaviruses can propagate only in differentiated epithelial cell; however, the differentiation of epithelial cell has become achievable only recently.

Cell tropism: HPVs infect *keratinocytes*,[5] which constitute the epidermis of stratified *epithelium*[6] in human body such as skin. Narrow tissue tropism is a defining feature of HPV infection. In particular, HPV infects cells in the basal layer of the epithelium, which becomes exposed through microwounds (Fig. 7.3). However, the viral genome replication and virion assembly could occur only in differentiated cells at the upper layer.

5. **Keratinocytes** Cells that constitute epidermis of epithelium and express abundant keratin, a cytoskeleton protein, as the name implies.
6. **Epithelium** A type of tissue that covers the surface of various organs in human body.

TABLE 7.3 The List of HPV Proteins and Their Functions

HPV Proteins	Functions
E1	DNA helicases, origin-binding protein
E2	Viral transcription/replication regulator
E4	Essential for E2 function
E5	Membrane-associated protein, cell proliferation
E6	Transforming protein, p53 binding
E7	Transforming protein, pRB binding
L1	Major capsid protein
L2	Minor capsid protein

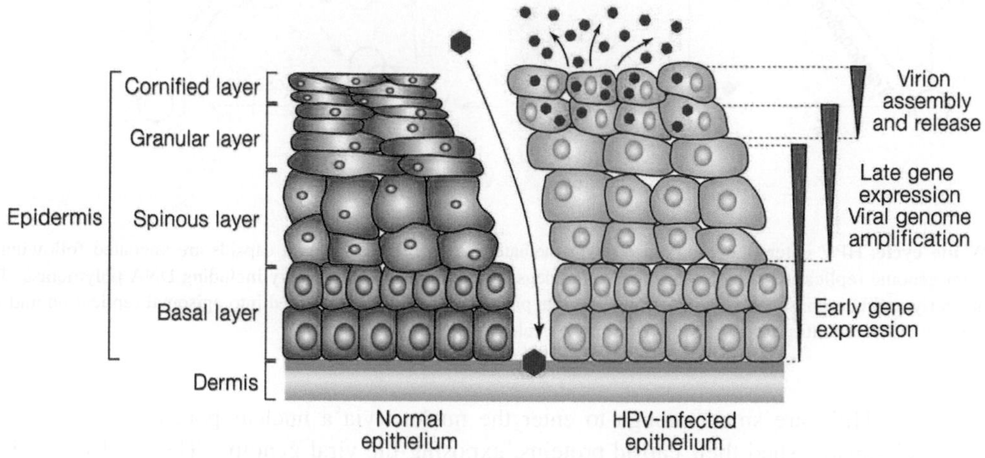

FIGURE 7.3 HPV infection in skin epidermis. HPV replication is linked to cell differentiation. Epidermis, which covers the underlying dermis, can be divided into three parts depending on differentiation status: basal layer (stem cells), upper layer, and uppermost cornified layer (sloughing cells). The virus infects cells in the basal layer of the epithelium, which are the only proliferating cells in normal epithelia, as differentiated cells in the upper layers have exited the cell cycle. The viral DNA is maintained at a low copy number in these cells. When the basal cells divide, some daughter cells move up to the epithelium and begin the process of differentiation. Differentiation of HPV-infected cells induces the productive phase of the viral life cycle. Progeny virions (red) are assembled and shed from the uppermost layers of the epithelium.

The life cycle of HPV can be divided into early and late phase by onset of the viral genome replication (Fig. 7.4). Importantly, the early-to-late transition of the HPV life cycle is linked to epithelial cell differentiation.

7.3.1 Early Phase

Entry: HPV virion attaches to cells through the *extracellular matrix*[7] (ECM) and then binds to cells via *heparin sulfate proteoglycan*[8] (HSPG) as a receptor and enters cells via receptor-mediated endocytosis (see Fig. 7.4). A recent study has reported that HPSG as well as growth factor receptors, such as epidermal growth factor receptor (EGFR), plays an important role for the entry. Furthermore, evidence implicates that the activation of EGFR signaling upon engagement of the viral capsids is critical for the viral entry. Following entry into cytoplasm, the viral capsids are located inside endosomes. Upon acidification (low pH) of endosomes, the L2 capsid protein becomes exposed and triggers the lysis of the endosomal membrane.

7. **Extracellular matrix (ECM)** The extracellular part of animal tissue that usually provides structural support to the animal cells. The ECM is the defining feature of connective tissue in animals, which is largely composed of fibrous protein, collagen, and glycoaminoglycan (GAG).

8. **Heparin sulfate proteoglycan (HSPG)** A kind of proteoglycans present abundantly on cell surface, in which heparin sulfate represents the glycan moiety.

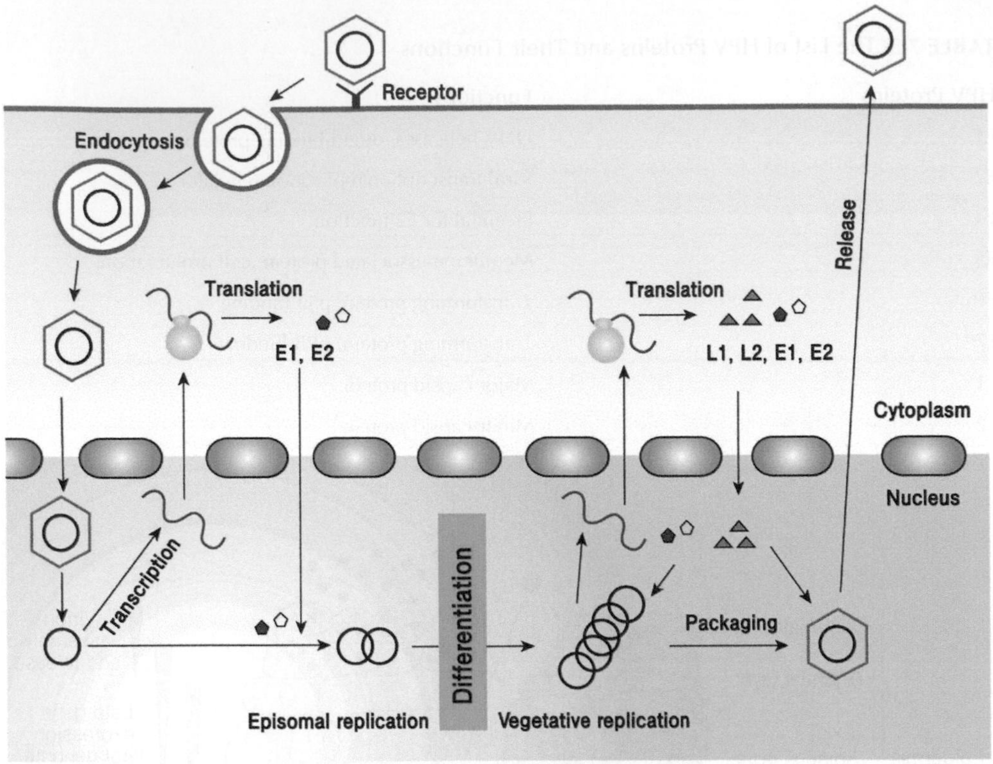

FIGURE 7.4 HPV life cycle. HPV enters the cells via receptor-mediated endocytosis. The viral capsids are uncoated following nuclear entry via nuclear pore. The viral genome replication, which occurs in the nucleus, depends on host machinery including DNA polymerase. The viral life cycle could be divided into early and late phase, likewise to SV40. The late phase could be further divided into episomal replication and vegetative replication with respect to the cell differentiation status. Progeny virions are released by cell lysis.

The viral capsids of HPV are small enough to enter the nucleus via a nuclear pore (see Fig. 3.7). Following the nuclear entry, the viral capsids shed their capsid proteins, exposing the viral genome. The viral DNA derived from the viral capsids possesses a nucleosome structure associated with core histones molecules.

Transcription: Early RNAs are transcribed from the early promoter located within LCR by cellular RNA polymerase II (see Fig. 7.2). The viral RNAs are alternatively spliced into multiple mRNAs, which are translated to viral early proteins from E1 to E7 proteins.

7.3.2 Late Phase

Transcription: Late RNAs are transcribed from the late promoter (ie, P_L) positioned in LCR (see Fig. 7.2). The viral RNAs are alternatively spliced, forming multiple mRNAs, which are translated into viral late proteins, L1 and L2 proteins.

Genome replication: The viral genome replication can be divided into three phases—(1) establishment, (2) maintenance, and (3) amplification—which are closely related to the extent of cell differentiation (Fig. 7.5).

(1) Establishment: Upon infection of basal keratinocytes, the viral genome is amplified to approximately 50 to 100 copies per cell.
(2) Maintenance: Then, the viral genomes or *episomes*[9] are replicated in synchrony with the cellular DNA replication, a process dubbed "episomal replication." How are viral episomes equally segregated into two daughter cells upon cell division? The answer to this question became apparent by the discovery of the interaction between E2 and *Brd4*[10] proteins. Intriguingly, the viral genome is attached to mitotic chromosomes via interaction with Brd4

9. **Episome** A DNA that is stably present in the cell, excluding chromosomal DNA. The term "epi" is derived from Greek word for "above."
10. **Brd4 (bromodomain-containing protein 4)** Brd4 is a chromatin adaptor protein that binds to interphase chromatin and mitotic chromosomes through bromodomain, which specifically recognize acetylated histones.

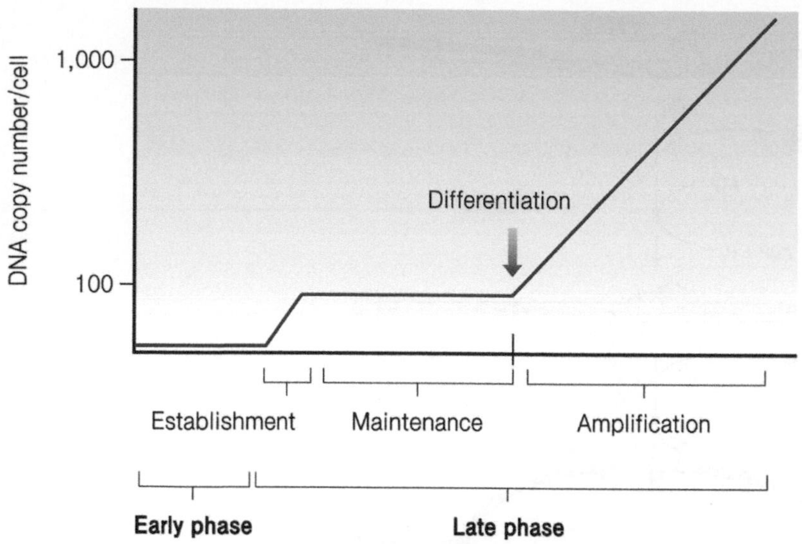

FIGURE 7.5 Three phases of HPV genome replication. Viral life cycle could be divided into early and late phase, which begins by onset of viral DNA synthesis. In the late phase, the viral DNA copy number is increased up to 100 copies per cell during "establishment" phase. Afterwards, the viral genome replicates in synchrony with cellular DNA, a process called "episomal replication." In this phase, the copy number is maintained at a constant level in undifferentiated cells. Differentiation of HPV-infected cells allows productive replication, a process called "amplification phase." In this phase, the copy number per cell is amplified up to 1000 copies per cell.

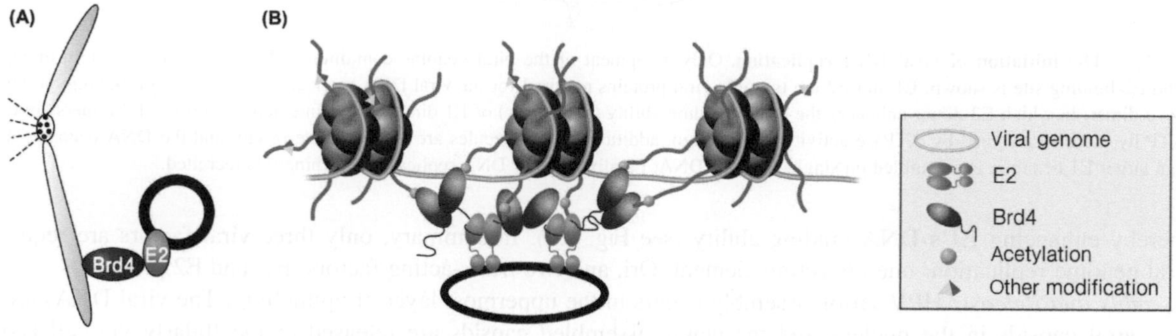

FIGURE 7.6 A model illustrating the E2-mediated tethering of the viral genome to host chromatin. (A) The E2 protein tethers the HPV genome to the mitotic chromosome via interaction with Brd4. (B) The Brd4 interacts with acetylated lysine residues on histone tails protruding from the host nucleosomes. E2 protein is composed of two independently folded domains: the N-terminal transactivation domain and the C-terminal DNA-binding domain.

(Fig. 7.6). In fact, E2 tethers the viral genome to the mitotic chromosomes via an E2-Brd4 interaction, which ensures faithful partitioning of replicated viral DNAs to the nuclei of both daughter cells, a mechanism often referred to as a "piggy-back mechanism." The E2-Brd4 interaction contributes to the viral episome maintenance during latent infection.

(3) Amplification: As the infected cells migrate toward the upper layers of the epithelium and become highly differentiated, the viral genome is further replicated, reaching up to 1000 copies per cell. It is also at this productive stage of the life cycle that the capsid protein L1 and L2 are synthesized. Accumulation of viral capsid proteins allows the packaging of amplified viral DNAs into new viral particles that are shed by desquamation of the terminally differentiated keratinocytes. In this phase, E6 and E7 protein deregulates cell cycle control, pushing differentiating cells into S phase, allowing viral genome amplification in cells that normally would have exited the cell cycle.

Genome replication mechanism: Like SV40, HPV genome replication occurs via a mechanism that entirely depends on the cellular DNA replication machinery, including DNA polymerase. Only two viral early proteins, E1 and E2, are required for viral genome replication. First, E1 dimer binds to the Ori element located in LCR (Fig. 7.7). E1 protein, an initiator, exhibits multiple biochemical activities, including DNA helicases, ATPase, and DNA-binding ability (see Table 7.3). Its role is to bind the Ori element and recruit all cellular DNA replication machinery that can execute DNA replication of the viral genome. In this regard, it can be compared to the SV40 T-antigen. On the other hand, E2, which is better known as a transcriptional regulator, also acts in the viral genome replication. Specifically, E2 interacts with

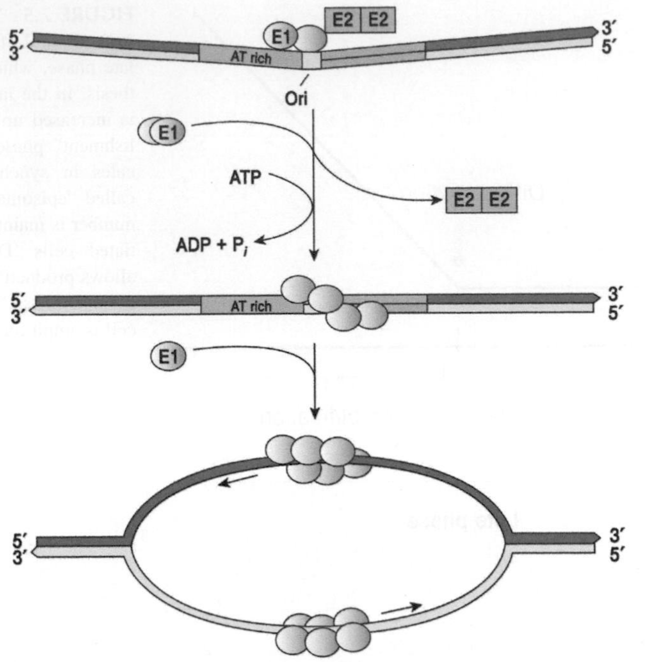

FIGURE 7.7 The initiation of viral DNA replication. Only a segment of the viral genome containing AT-rich region, the origin of replication (Ori), and E2-binding site is shown. E1 and E2 are the only viral proteins required for the viral DNA replication. E1, the origin-binding protein, binds to Ori as a dimer, in which E2 dimer enhances the origin-binding ability (specificity) of E1 dimer via an interaction with E1. E2 dimers are displaced upon ATP hydrolysis catalyzed by ATPase activity of E1. Then, additional E1 molecules are loaded at the origin, and the DNA becomes distorted. Finally, a larger E1 hexamer is assembled on single-stranded DNA. Finally, cellular DNA replication machinery is recruited.

E1, thereby enhancing E1's DNA-binding ability (see Fig. 7.7). In summary, only three viral factors are required for the viral genome replication: one *cis*-acting element, Ori, and two *trans*-acting factors, E1, and E2 proteins.

Assembly and Release: HPV virion assembly occurs in the uppermost layer of epithelium. The viral DNAs are packaged by viral capsids in the nucleus and the newly assembled capsids are released extracellularly via cell lysis (see Fig. 7.4). It remains unclear how the cell lysis occurs.

7.4 EFFECTS ON HOST

Tumor formation: HPV is a DNA tumor virus that causes cervical carcinoma. Among more than 120 genotypes of HPVs, most of them cause a *benign tumor*,[11] wart. By contrast, some 15 genotypes that infect genital mucosa cause a *malignant tumor*,[12] cervical carcinoma. These are termed so-called high-risk HPVs. Which HPV genes are responsible for HPV-associated carcinogenesis? Two viral genes, E6 and E7, were identified as viral oncogenes responsible for the carcinogenesis. The expression of two viral proteins is necessary and sufficient for cellular transformation in culture. Then, the question is how these viral proteins work concertedly to transform normal cells into cancerous states.

E7 protein: Cells in the upper layer of epithelium (ie, suprabasal layer cells), where HPV genome replication is carried out, are differentiated cells, in which the DNA synthesis does not normally occur. In HPV infected cells, however, it is the E7 protein that allows the otherwise resting cells to reenter S phase of the cell cycle. In fact, E7 protein binds to Rb protein, which is a product of tumor suppressor gene, *Rb*.[13] Specifically, Rb protein binds to *E2F*,[14] thereby regulating E2F's role in G1 to S phase transition. The binding of E7 protein to Rb protein releases E2F from the Rb-E2F complex (Fig. 7.8). Then, the released E2F, S phase-specific transcription factor, directs the transcription of S phase-specific genes such as DNA polymerase, which leads to S phase.

11. **Benign tumor** A tumor that lacks the ability to metastasize.
12. **Malignant tumor** A tumor that has the ability to metastasize, and also called cancer.
13. **Rb (retinoblastoma)** A tumor suppressor gene that was first identified as a defective gene in retinoblastoma.
14. **E2F** A transcription factor that drives the entry to S phase. It was first known as a transcription factor that binds to adenovirus E2 promoter, as its name implies.

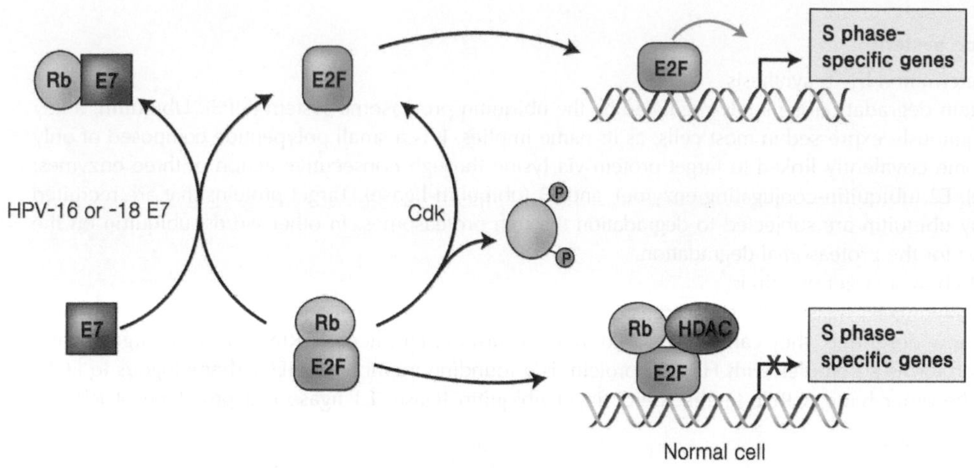

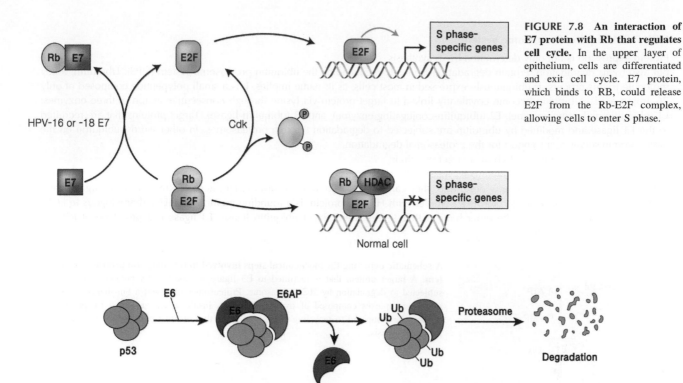

FIGURE 7.8 An interaction of E7 protein with Rb that regulates cell cycle. In the upper layer of epithelium, cells are differentiated and exit cell cycle. E7 protein, which binds to RB, could release E2F from the Rb-E2F complex, allowing cells to enter S phase.

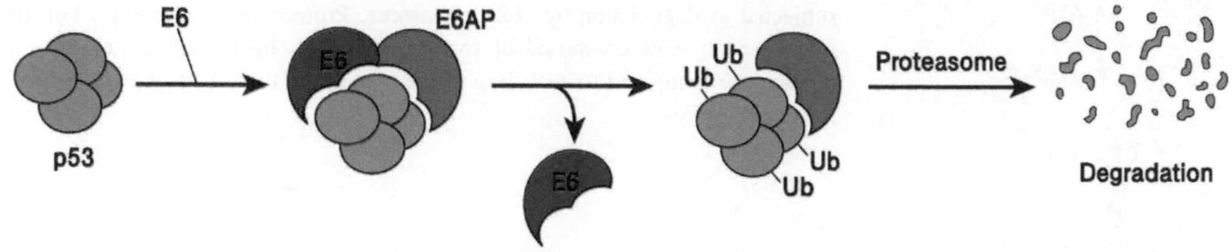

FIGURE 7.9 E6 protein-mediated p53 degradation. In normal cell, p53 protein, a tumor suppressor protein, exhibits a short half-life, and is expressed at low level. Stress stimuli, such as DNA damage, and virus infection, stabilize p53 and accumulates to high level. In HPV-infected cells, E6 recruits E6AP and forms E6AP complex, a novel ubiquitin E3 ligase that targets p53.

How do cells that reenter S phase remain supportive to the viral genome replication without exiting S phase of cell cycle? The answer to this question was recently reported. It is also the E7 protein that triggers DNA damage response (*DDR*[15]) via a mechanism involving *ATM*,[16] thereby inducing arrest in S phase (see Journal Club).

In summary, HPV E7 protein subverts two tumor suppressor proteins—Rb and ATM—in order to support the sustained viral genome replication in the upper layer cells, in which DNA synthesis does not normally occur otherwise. Not surprisingly, via interactions with the tumor suppressor proteins, E7 protein also deregulates their tumor suppressor function, thereby inducing tumor formation. For instance, the Rb-binding abilities of high-risk HPVs, such as HPV-16 and 18, are stronger than those of low-risk HPVs, an observation related to the Rb-binding ability of E7 to oncogenic potency.

E6 protein: As stated above, in HPV infected cell, DNA synthesis and unscheduled cell division do occur in differentiated cells. p53-mediated *apoptosis*[17] is induced to compensate for the excessive cell division occurring in the upper layer of the epithelium. To sustain its propagation, the virus needs a mechanism to block the apoptosis of HPV-infected cells. It is the E6 protein that downregulates p53 via ubiquitin-mediated proteolysis (Fig. 7.9). How does HPV E6 protein accomplish this task? E6 protein binds to *E6AP*[18] and forms an E6-E6AP complex, which facilitates ubiquitination of p53. The ubiquitinated p53 is subjected to proteolysis by proteasomes (Box 7.3). Thus, the E6-E6AP complex acts as a novel ubiquitin E3 ligase. Intriguingly, E6 proteins of high-risk HPVs (ie, HPV-16 and 18), but not those of low-risk HPVs, are capable of p53 ubiquitination. This observation relates the p53-binding ability of E6 to its oncogenic potency.

---

15. **DDR** It refers to signaling pathway that is active by DNA damage (see Box 6-2).

16. **ATM (ataxia telangiectasia-mutated)** A tumor suppressor gene identified as a mutated gene in ataxia telangiectasia. Ataxia telangiectasia is a rare, neurodegenerative, inherited disease causing severe disability. ATM, a serine/threonine protein kinase, is activated by double-strand break (DSB), acting as a sensor of the DNA damage.

17. **Apoptosis** It refers to the programmed cell death that may occur in multicellular organisms.

18. **E6AP (E6-associated protein)** A host factor that was first identified as HPV E6-binding protein. It is now classified as a member of HECT (homologous to E6AP carboxyl terminus)-type ubiquitin E3 ligase family.

BOX 7.3 Ubiquitin Proteasome System

Protein expression is not only determined by its synthesis but also by its degradation. Protein degradation is mainly regulated by the ubiquitin proteasome system (UPS). Ubiquitin, a key molecule in this process, is ubiquitously expressed in most cells, as its name implies. It is a small polypeptide composed of only 76 amino acids. Ubiquitin become covalently linked to target protein via lysine through consecutive action of three enzymes: E1 (ubiquitin-activating enzyme), E2 (ubiquitin-conjugating enzyme), and E3 (ubiquitin-ligase). Target proteins that are recruited to the E3 ligase and modified by ubiquitin are subjected to degradation through proteasomes. In other words, ubiquitin on the target protein serves as a signpost for the proteasomal degradation.

Then, the question arises as to how a target protein is specifically recognized and recruited to E3 ligase? The substrate specificity for ubiquitination is conferred by E3 ligases. In fact, although diverse E3 ligases are now described, they can be classified into two distinct families: (1) RING finger E3 ligases, and (2) HECT E3 ligases. E6AP, which forms a complex with HPV E6 protein, is a founding member of HECT (homologous to E6AP carboxyl-terminus) family. On the other hand, CRUL (cullin RING finger ubiquitin ligase) E3 ligase is a prototype of *RING*[19] finger E3 ligases (see Box 8.1).

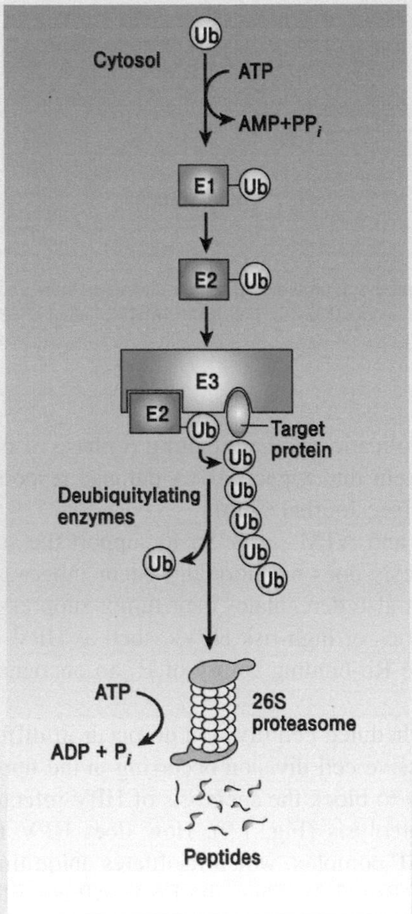

A schematic depicting the biochemical steps involved in the ubiquitin proteasome system. A target protein that is recruited to E3 ligase is modified by polyubiquitin and is subjected to degradation by 26S proteasome. Proteasomes represent a bin-shaped multiple protein complexes composed of over 30 proteins, inside which are filled by proteases. It could be compared to recycling bin in the sense that residual peptides are recycled for protein synthesis.

On the other hand, E6 protein is also essential for the establishment of the successful HPV infection. Prevention of the premature apoptosis of HPV-infected cells by E6 protein is essential for the completion of the virus life cycle.

Cervical carcinoma: Cells isolated from cervical carcinoma, which is a malignant tumor, can be cultivated in vitro. Importantly, HPV DNA is found in almost all cervical carcinoma cells. Moreover, it is found in an integrated state in

19. **RING (really interesting new gene)** A protein motif found in RING finger family proteins.

BOX 7.4 HeLa Cell

Despite strenuous effort, it had been proving difficult to cultivate human cells. It was only in the early 1950s that scientists could cultivate human cells in laboratories. HeLa cells that are grown in many laboratories are the ones that paved the way. In 1951, George Gey, a professor at Johns Hopkins Medical School, was able to successfully cultivate cells isolated from tissues of a cervical carcinoma patient. The cancer cells of this patient grew well under the culture condition, in which other cancer cells failed to grow. It was named after a patient donor, Henrietta Lacks, a 31-year-old Black woman who later died of cervical cancer. Not surprisingly, HPV infection was the cause of her cancer, although she could have not known.

The impact of the HeLa cell on the advancement of biomedical research has been enormous. Successful cultivation of the HeLa cell has enabled us to cultivate many other cancer cells. It has been utilized for the preparation of polio vaccine, which saved many people's life. On the other hand, no other cells have been publically debated as much as the HeLa cell. Apparently, the cells were isolated without the consent of the donor, which is not acceptable by today's bioethics standard. Moreover, her family was unaware of the existence of the HeLa cell, until recently. The story of the HeLa cell reminds us of medical practice in the past. Although Henrietta Lack never had a chance to travel outside of Baltimore, the HeLa cell traveled even to space. The HeLa cell, derived from the cancer of a woman, who died at the age of only 31, has greatly contributed to biomedical research.

(A) (B)

HeLa cell and its donor. (A) Photograph of Henrietta Lack. (B) HeLa cells grown in tissue culture and stained with antibody to cytoskeletal microtubules (magenta) and DNA (cyan).

the chromosome. The integration site on chromosome appears randomly, and, frequently, more than one copy of HPV DNA is integrated. Although the chromosomal integration is not an obligate step in the viral life cycle, it does happen in HPV infected cells (see Fig. 24.5). Intriguingly, E6 and E7 genes remain intact in the integrated HPV genome, while other HPV genes such as E1 and E2 are frequently lost or mutated. Although HPV genome replication is not supported in these carcinoma cells, E6 and E7 proteins remain to be expressed. Thus, persistent expression of E6 and E7 proteins appears to be critical for tumorigenesis. Note that E6 and E7 proteins are also expressed in HeLa cells, which are derived from cervical carcinoma (Box 7.4).

7.5 PERSPECTIVES

Discovery of high-risk HPVs as etiologic agents for cervical carcinoma in 1983 drew much attention to papillomaviruses over the three decades since then. Worldwide, approximately half a million women are newly diagnosed every year with HPV-associated cervical cancer. Pre-neoplastic stages (ie, carcinoma in site or dysplasia) of cervical cancer can be readily screened by histological examination of cervix tissue by a procedure termed *PAP test*[20] (Box 7.5). An HPV DNA test, as well as the PAP test of cervical tissues, allows early diagnosis of cervical cancer, which has saved

20. **PAP test** The Papanicolaou test (abbreviated as Pap test, known earlier as Pap smear) is a method of screening used to detect potentially precancerous and cancerous processes in the cervix. The test was invented by and named after the Greek doctor Georgios Papanicolaou.

BOX 7.5 HPV and Cervical Carcinoma

Out of 150 HPV subtypes, most of them cause a benign tumor such as warts and refers to as low-risk HPVs. High-risk HPV subtypes, which include about 15 subtypes, are associated with cervical carcinoma. Prominent examples include HPV-16 and 18. High-risk HPV subtypes are transmitted via sexual contact. Prevalence of HPV infection in the United States reaches 30% in younger women in their 20s. The HPV prevalence is much higher in people in South East Asia, South America, and Africa. Although HPV-induced cervical cancer is confined to women, men are not safe either. Importantly, HPV is also causally associated with other types of cancers as well, for example, anus and vagina. Fortunately, preneoplastic stages (ie, carcinoma in site or dysplasia) of cervical cancer can be readily screened by histological examination of cervix tissue by a procedure termed *PAP test*. Moreover, recently developed HPV vaccines effectively prevent HPV infection (see Box 7.2). Thus, HPV vaccination is highly recommended to be given prior to the onset of their sexual life. HPV vaccine is regarded as a "cancer vaccine," because it can prevent not only HPV infection but also HPV-induced cancer. It is hoped that HPV vaccines will save women from cervical carcinoma.

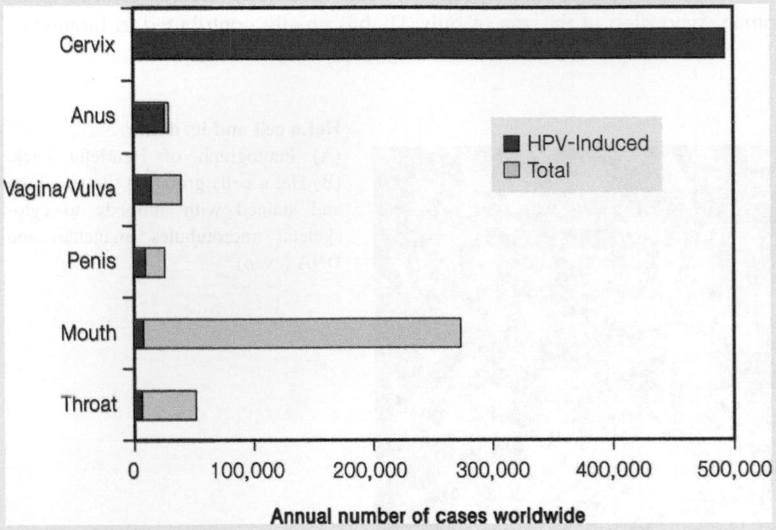

The graph shows the annual number of cases of various cancers worldwide. The fraction of cancers estimated to be induced by HPV is shown in red. HPV is associated not only with cervical carcinoma but also with other organs such as mouth and throat.

millions of women. Furthermore, an HPV vaccine that could effectively prevent HPV-related cervical carcinoma in women was developed. In fact, it is a "cancer vaccine." Importantly, it shows how basic science could benefit public health. In retrospect, two viral oncogenes of HPV—E6 and E7—have played a crucial role in the advance of cancer biology. E6 and E7 proteins interact with the two most important tumor suppressor proteins (ie, p53 and Rb), thereby deregulating their roles in apoptosis and cell cycle control, respectively. Furthermore, the identification of E6AP, as an E6-binding protein, led to the discovery of a novel HECT-type family of ubiquitin E3 ligase. On the other hand, research on the late phase of viral life cycle has been hampered by the lack of proper cell lines that support the viral genome replication. It has been proven difficult to propagate epithelial cells in a differentiated state that supports viral genome replication. A recent advance in the differentiation of epithelial cells, which was successful only recently in a few laboratories, is expected to provide new insights into the regulatory mechanism as to how a switch from episomal replication to vegetative replication is regulated. Recent approval of "Imiquimod" (see Fig. 26.3), as an anti-HPV drug for the treatment of warts, is expected to draw much attention to antiviral therapy.

7.6 SUMMARY

- *Taxonomy:* Papillomaviruses have been discovered in humans (HPV) as well as in animals (BPV). A subset of anogenital HPVs are associated with cervical cancers. HPV is a DNA tumor virus that causes a fatal tumor in humans.
- *HPV genome:* HPV possesses a small double-stranded circular DNA of 8 kb.
- *Virion structure:* HPV particles are small, nonenveloped icosahedral capsid particles.
- *Genome replication:* HPV genome replication heavily depends on the host machinery in that E1 and E2 proteins are the only viral proteins essential for the viral genome replication. A salient feature of HPV replication is that the differentiation status of the epithelial cell is linked to the viral genome replication.

- *Host effect:* High-risk HPVs cause cervical cancer. E7 protein dysregulates Rb's function in cell cycle control, while E6 protein degrades p53.
- *HPV vaccines:* HPV vaccines are developed and proven to be effective for the prevention of cervical carcinoma.

STUDY QUESTIONS

7.1. HPV genome replication does occur in two distinct modes. List and compare the differences between the two distinct modes of the viral genome replication.

7.2. In more than 50% of human cancer, p53 mutation is found. By contrast, p53 mutation is not found in HPV-associated cervical carcinoma. Explain why? Justify your answer.

7.3. HeLa cell is a favorite cell line that is used in many laboratories. (1) State the tissue from which HeLa cell was originally isolated. (2) List the genes that were responsible for the immortalization of HeLa cell. (3) It is thought that HeLa cell is not appropriate for the study of apoptosis and cell cycle regulation. Explain why?

SUGGESTED READING

Gillespie, K.A., Mehta, K.P., Laimins, L.A., Moody, C.A., 2012. Human papillomaviruses recruit cellular DNA repair and homologous recombination factors to viral replication centers. J. Virol. 86 (17), 9520–9526.

Ma, Y., Madupu, R., Karaoz, U., Nossa, C.W., Yang, L., Yooseph, S., et al., 2014. Human papillomavirus community in healthy persons, defined by metagenomics analysis of human microbiome project shotgun sequencing data sets. J. Virol. 88 (9), 4786–4797.

Moody, C.A., Laimins, L.A., 2010. Human papillomavirus oncoproteins: pathways to transformation. Nat. Rev. Cancer. 10 (8), 550–560.

Raff, A.B., Woodham, A.W., Raff, L.M., Skeate, J.G., Yan, L., Da Silva, D.M., et al., 2013. The evolving field of human papillomavirus receptor research: a review of binding and entry. J. Virol. 87 (11), 6062–6072.

Schelhaas, M., Shah, B., Holzer, M., Blattmann, P., Kuhling, L., Day, P.M., et al., 2012. Entry of human papillomavirus type 16 by actin-dependent, clathrin- and lipid raft-independent endocytosis. PLoS Pathog. 8 (4), e1002657.

JOURNAL CLUB

- Moody, C. et al., 2009. HPV activates the ATM DNA damage pathway for viral genome amplification upon differentiation. PLoS Pathog. 5:e1000605.

  Highlight: The article clearly shows that HPV-induced ATM-mediated DNA damage signaling is essential for HPV genome replication. Note that an interesting epithelial cell culture system, called "organotypic raft culture," was employed to induce epithelial cell differentiation.

Chapter 8

Adenoviruses

Chapter Outline

Adenoviruses[1] cause only mild diseases in human such as respiratory infection, gastroenteritis, and epidemic conjunctivitis. On the other hand, adenoviruses cause tumor in rodents such as hamsters. Thus, adenovirus is the *tumor virus* that was isolated from human. Being the first tumor virus isolated from human, adenovirus has served as a model in studying the underlying mechanism of tumorigenesis as well as viral tumorigenesis. Ironically, however, adenoviruses do not cause tumor in human. This chapter will focus on molecular aspects of adenovirus with emphasis on virus life cycle (Table 8.1). In addition, adenoviruses are utilized as a gene therapy vector; this aspect will be described in some detail in chapter "Virus Vectors."

8.1 DISCOVERY AND CLASSIFICATION

Discovery: Adenoviruses were first isolated in the *adenoid*[2] of a person suffering from respiratory infection. In addition to respiratory infection, adenovirus is also known to cause enterogastritis, and epidemic conjunctivitis in human. Although these diseases are self-limiting in healthy persons, they cause significant morbidity in AIDS, cancer, and organ transplant patients with compromised immune systems.

Classification: Adenoviruses are found not only in human but also in diverse vertebrates including reptiles (see Table 1.4). Over 60 *subtypes*[3] of human adenoviruses are currently reported, which are divided into six subgroups, which are denoted as A-G (Table 8.2). Among these, the subtype 5 (HAdV-C5), which causes respiratory infection in human, has been studied extensively as a prototype.

8.2 THE VIRION AND GENOME STRUCTURE

Virion Structure: The adenovirus particle is a spherical capsid with 12 fibers, which project outward (Fig. 8.1). It is a nonenveloped capsid particle with $T = 25$ icosahedral symmetry. It is a rather sophisticated structure constituted of up to 15 structural proteins. The capsid is made of two building blocks, termed hexons and pentons. Twenty facets are constituted of 240 hexons, and the vertex is constituted of pentamers. Fibers are attached to the vertex. Inside viral capsids, the viral genomic DNA is associated with three kinds of viral proteins (V, VII, and Mu). An electron micrograph image of an adenovirus particles is shown in Fig. 10.1.

Genome Structure: The viral genome is a 35 kb molecule of double-stranded linear DNA (Fig. 8.2A). It has an inverted terminal repeat (*ITR*) with 100–180 bp in length at both ends. The nucleotide sequence in ITR is involved in

1. **Adenovirus** The term "adeno" is derived from Greek word for "spring"-*adeno*.
2. **Adenoid** The adenoid, also known as a pharyngeal tonsil or nasopharyngeal tonsil, is a mass of lymphatic tissue located posterior to the nasal cavity, in the roof of the nasopharynx, where the nose blends into the throat.
3. **Subtype** It refers to a rank that below "species" in taxonomic rank.

Molecular Virology of Human Pathogenic Viruses. DOI: http://dx.doi.org/10.1016/B978-0-12-800838-6.00008-4

TABLE 8.1 The Defining Features of Adenoviruses

Genome	Particle Structure	Replication Mechanism
Linear dsDNA (35 kb)	Nonenveloped	Viral DNA polymerase
Protein-linked	Icosahedral nucleocapsid	Protein-priming
Life Cycle	**Host Effect**	**Disease**
Early-to-late switch	E1A-Rb interaction	Tumor
Cell cycle deregulation	E1B-p53 interaction	Respiratory infection

TABLE 8.2 The Classification of Human Adenoviruses

Human Adenovirus		Disease
Subgroup A	type 12	Enteritis
Subgroup C	type 1, 2, 5	Respiratory infection
Subgroup E	type 4	Respiratory infection

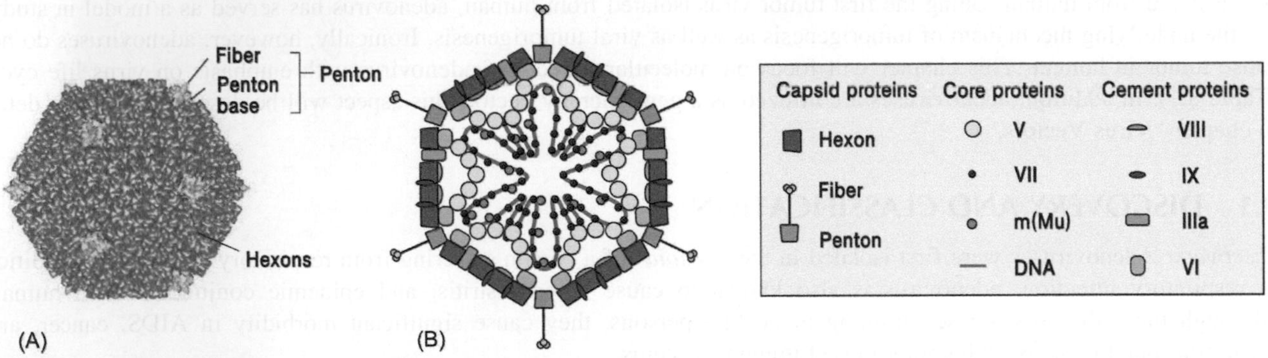

FIGURE 8.1 Structure of adenovirus capsids. (A) The adenovirus capsid with a diameter of ~150 nm. The capsid structure is highlighted with hexons constituting facets and pentons constituting the vertex, where the fiber is attached. (B) A diagram depicting the cross section of the adenovirus particle. Six out of 12 fibers projecting from the vertex are schematically shown. The viral genomic DNA is associated with the viral core proteins (V, VII, and Mu) inside the capsid. Cement protein VI is associated with cement protein VIII and core protein V, which together glue hexons beneath the vertex region and connect them to the viral DNA genome.

forming a panhandle structure, an intermediate structure observed during viral genome replication (Fig. 8.2B). Notably, a viral protein termed terminal protein (TP) is covalently linked to the 5′ terminus of the viral genome. TP protein serves as a *protein primer* for viral DNA replication.

Viral RNAs and Their Protein Coding: Adenovirus encodes a large number of proteins, as the viral proteins are coded by both strands of the genome (Fig. 8.3). First, early genes are expressed from four distinct promoters from E1 to E4, from both strands. Early gene products contribute to the establishment of viral infection and viral genome replication. The late genes (ie, L1 to L5) are transcribed from a single promoter, termed as a major late promoter (MLP). Each transcript is alternatively spliced and polyadenylated to encode a variety of structural proteins, ranging from capsid proteins (penton, hexon, and fiber) and genome binding proteins (V, VII, and Mu).

8.3 THE LIFE CYCLE OF ADENOVIRUS

Cell Tropism: Adenovirus subtype 5 (HAdV-C5) infects the epithelium of the upper respiratory tract in human and causes respiratory infection.

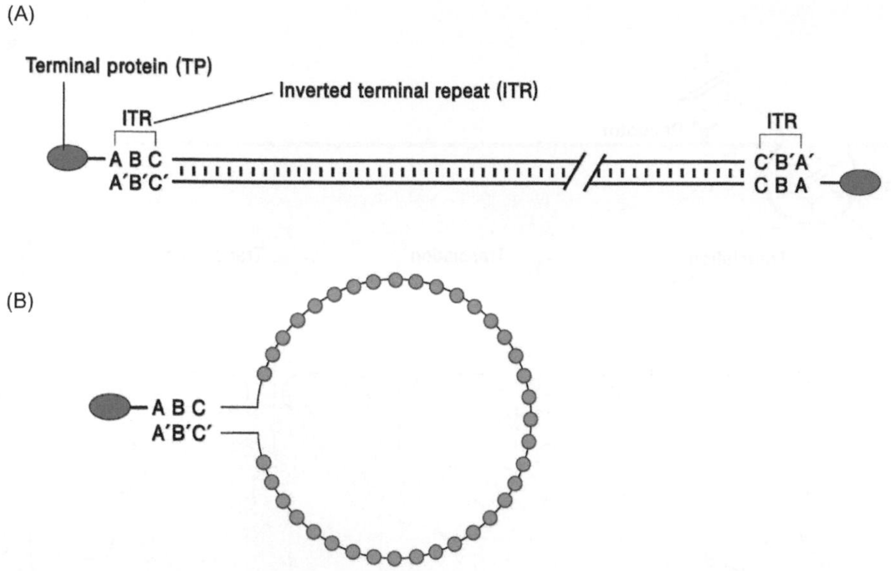

FIGURE 8.2 Genome structure of adenovirus. (A) ITR elements located at both ends of the DNA genome are highlighted: ABC is complementary to A′B′C′. TP (terminal protein) is linked to both 5′ termini of double-strand DNA genome. *ITR*, inverted terminal repeat. (B) A panhandle structure of a single-strand DNA formed via a complementarity of ITR sequence.

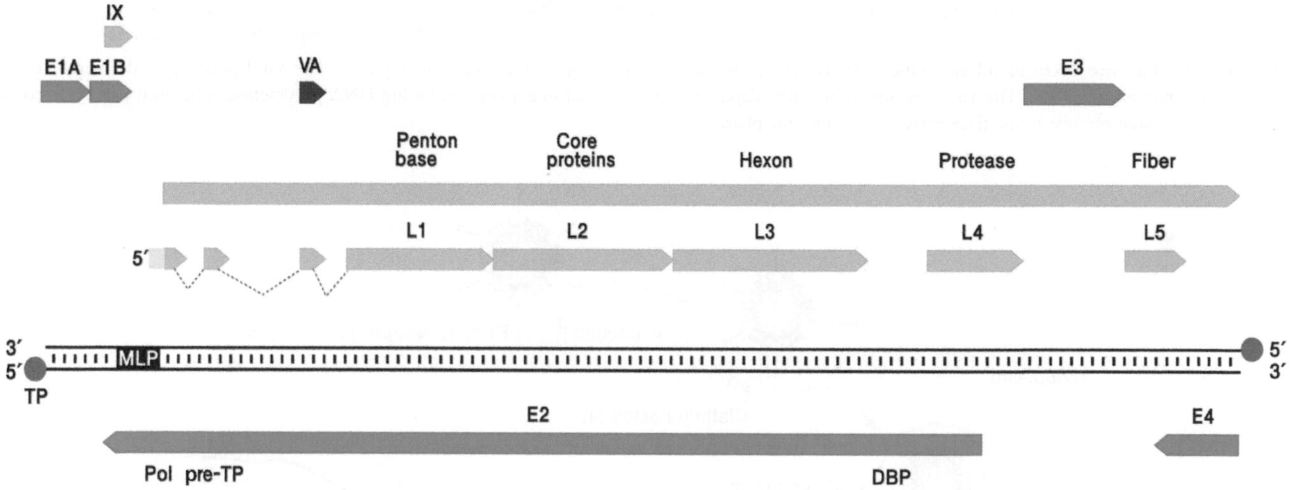

FIGURE 8.3 Adenoviral transcripts and their protein codings. Early mRNAs (blue) and late mRNAs (green) are denoted by arrowed bars. Early (E1–E4) genes and late (L1–L5) genes are indicated above the ORFs. Both strands are utilized for transcription. *DBP*, DNA-binding protein; *MLP*, major late promoter; *VA*, virus-associated RNA.

Similar to other DNA viruses, the virus life cycle can be divided into two phases, with respect to the viral genome replication: early phase and late phase. The viral genome replication occurs in the nucleus of the infected cell (Fig. 8.4).

Entry: Adenovirus enters cells via receptor-mediated endocytosis. The fiber of the viral capsids binds to a *CAR*[4] molecule on the cell surface as an attachment factor (Fig. 8.5). The engagement of *integrin*[5] is needed for efficient internalization as the major entry receptor. The viral capsid enters the cell via clathrin-mediated endocytosis, and

4. **CAR (Coxsackie-Adenovirus Receptor)** CAR is exploited by two unrelated viruses for entry (ie, coxsackie type B3 and adenovirus type 5) (see Fig. 3.2).

5. **Integrin** Integrins are transmembrane proteins that mediate the attachment between a cell and its surroundings, such as other cells or the extracellular matrix (ECM).

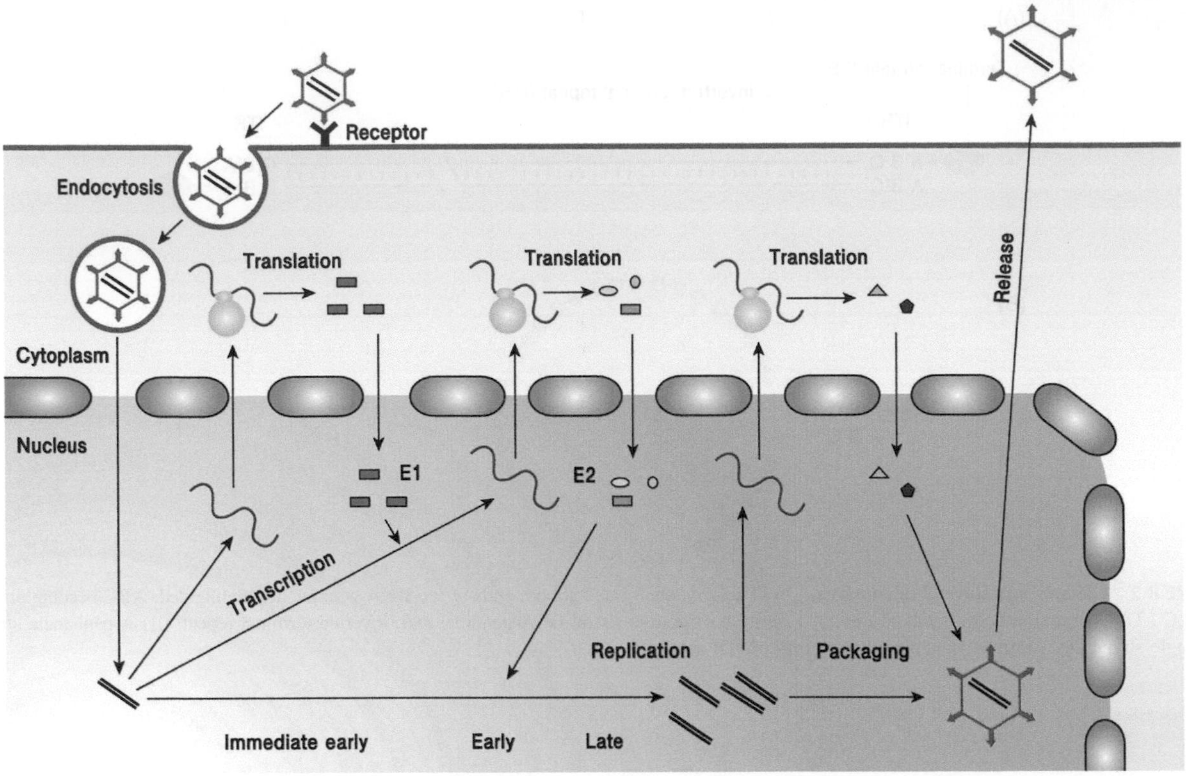

FIGURE 8.4 The life cycle of adenovirus. Adenovirus enters the cell via receptor-mediated endocytosis. The viral genome is delivered to the nucleus for genome replication. The viral genome replication depends on the cellular machinery including DNA polymerase. The viral life cycle could be divided into three phases: immediate early, early, and late phases.

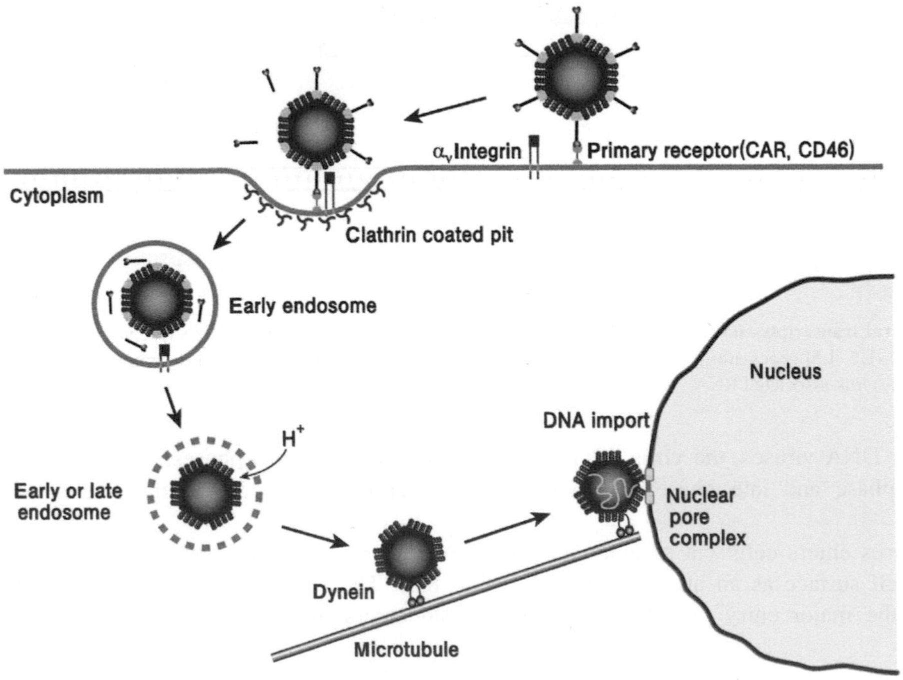

FIGURE 8.5 The steps involved in adenovirus entry. The viral capsid enters the cell following the recognition of CAR as an attachment factor, and integrin as the major receptor. Inside the endosome, fibers are dismantled by acidic pH, and the penton base is exposed. Such exposed penton base triggers membrane lysis, thereby allowing penetration of the capsid to the cytoplasm. The capsid trafficks toward the nucleus via microtubule-mediated transport. A recent study showed that kinesin-1 motors are associated with the viral capsid disrupt, in part, the viral capsid and the NPC, thereby facilitating entry of the viral DNA into the nucleus.

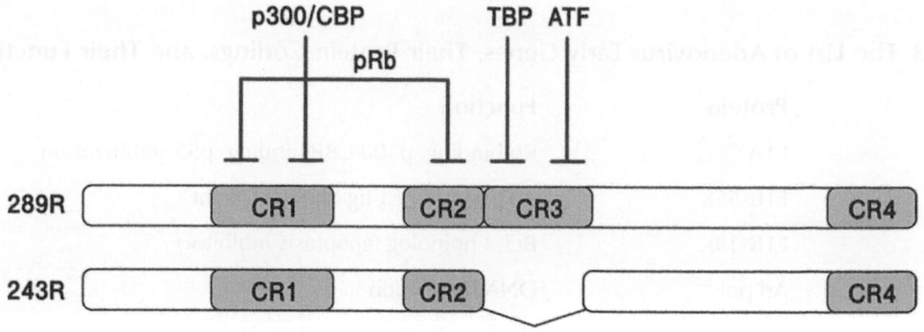

FIGURE 8.6 **The domain structure of E1A proteins.** Two kinds of E1A proteins are expressed due to alternative splicing: 289R and 243R. Four conserved region of E1A protein (ie, CR1 to CR4), which are conserved among serotypes, are denoted by boxes. Note that the CR3 is not present in 243R. E1A binds to numerous cellular factors, some of which are indicated above to respective domains. p300/CBP: transcriptional coactivator. *pRb*, retinoblastoma protein; *TBP*, TATA-binding protein; *ATF*, cAMP-response element/activating transcription factor.

locates itself inside the endosome. Interestingly, the fibers of the capsid are dismantled in the endosome by acidic pH. Such exposed penton base triggers the lysis of endosomal membrane, which allows the capsid to penetrate into the cytoplasm; in fact, a membrane-lysis domain at the N-terminus of the cement protein VI (see Fig. 8.1B) is cleaved, and released to mediate endosome rupture. Next, the partially disrupted viral capsid trafficks toward the nucleus using the microtubule minus end-directed motor complex dynein/dynactin, and attaches to the nuclear pore complex (NPC) via an interaction of hexon with a NPC component. As a result, the viral capsids become partly disrupted near the nuclear pore; the viral genome, instead of the capsid itself, enters the nucleus. Note that unlike polyomavirus and papillomavirus, which have a smaller capsid, the adenoviral capsid cannot transverse nuclear pore (see Fig. 3.7).

8.3.1 Early Phase

Transcription: Early gene transcription takes place, prior to the onset of the viral genome replication. Early gene transcription is executed in two phases: these are appropriately termed "immediate early genes" and "early genes." E1A proteins play a central role in early transcription. The viral transcripts of the E1A gene, which is located at the 5′ terminus of the viral DNA genome, are alternatively spliced into two mRNAs (ie, 12S and 13S RNA). The 12S and 13S RNA are then translated to E1A proteins—243R and 289R, respectively (Fig. 8.6). Although E1A is not a DNA-binding protein, it acts as a transcriptional transactivator via interaction with transcription factors such as p300/CBP, TBP, and ATF. E1A proteins act as transcriptional transactivator of E2 gene transcription. E3 and E4 gene transcription follow, thereafter. These early gene transcripts are processed, transported to cytoplasm, and translated to E2 to E4 proteins (Table 8.3).

In addition to its role in early gene transcription, E1A plays a central role in establishing the viral infection by triggering the entry of cells to S phase, a prerequisite for viral DNA synthesis. Adenovirus infects epithelial cells, which are terminally differentiated to nondividing cells. In order to permit the viral DNA synthesis, cells need to enter the S phase of the cell cycle (dividing). E1A is the viral protein that facilitates the cells to enter S phase, via its interaction with *Rb*[6] protein (see Fig. 24.3).

Switch: Two early proteins—E1A and E2—trigger the switch from early phase and late phase (Fig. 8.7). As stated above, the early phase can be subdivided into two phases: immediate early and early phases. E1A triggers the switch from immediate early to early phase by transactivation of early genes (E2, E3, and E4). E2 gene encodes three viral proteins contributing to the viral genome replication: (1) Ad polymerase, (2) pre-TP, and (3) DBP (see Table 8.3). E2 proteins induce not only the viral DNA replication but also simultaneously the switch from early to late phase.

8.3.2 Late Phase

The viral genome replication, the capsid assembly, and the release of the assembled capsid occur in the late phase (see Fig. 8.4).

Genome Replication: Three viral proteins are involved in the adenoviral DNA replication (Fig. 8.8). These are three E2 gene products: adenovirus DNA polymerase (*Ad Pol*), *pre-TP* (precursor of TP), and *DBP* (DNA-binding protein). Importantly, unlike polyomavirus and papillomavirus, which have a small DNA genome, adenovirus encodes its own

6. **Rb (retinoblastoma)** a tumor suppressor gene that was first identified as a defective gene in retinoblastoma.

TABLE 8.3 The List of Adenovirus Early Genes, Their Protein Codings, and Their Functions

Gene	Protein	Function
E1a	E1A	Rb binding, p300/CBP binding, p53 stabilization
E1b	E1B-55K	Ad ubiquitin E3 ligase component
	E1B-19K	Bcl-2 homolog (apoptosis inhibitor)
E2	Ad pol	DNA replication
	pre-TP	DNA replication
	DBP	DNA replication
E3	E3-gp18K	Protecting cells from CTL lysis
	E3-11.6K	ADP (adenovirus death protein)
E4	E4-orf3	Inhibition of IFN response; inhibition of DNA damage response
	E4-orf4	Induction of apoptosis
	E4-orf6	Ad ubiquitin E3 ligase component
VA RNA	–	Inhibition of IFN-α action

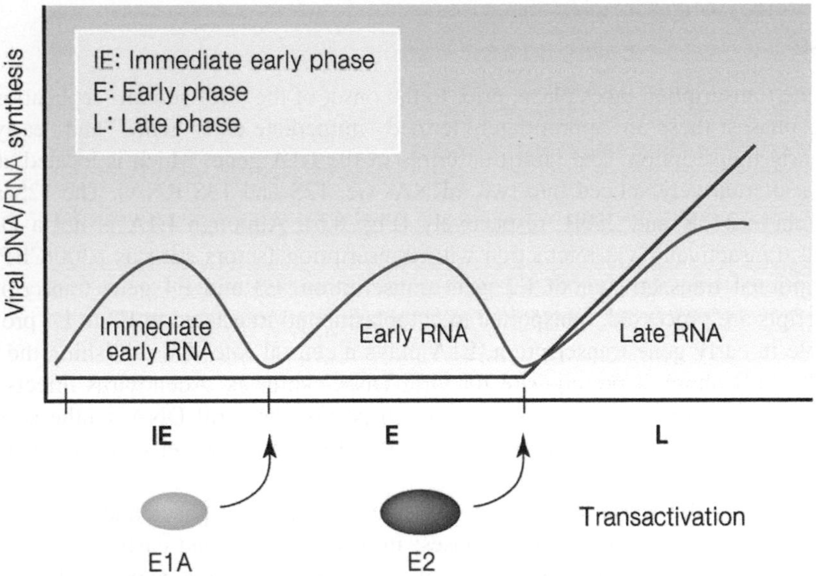

FIGURE 8.7 A graphic illustrating the switch from early to late phase. E1A and E2 proteins play pivotal roles in the switching from immediate early (IE) to early (E) phase and early to late (L) phase, respectively. The level of viral RNA and viral DNA are not drawn in scale but only for the purpose of explanation.

viral DNA polymerase (Ad Pol). Thus, the host dependence of adenoviral DNA replication is minimal, although its DNA replication occurs in the nucleus.

Ad Pol initiates the DNA synthesis using the pre-TP protein as primer. Notably, protein, instead of nucleotide, serves as the primer for the DNA synthesis, a process termed "*protein-priming*." In fact, the first nucleotide being synthesized is linked to the hydroxyl group of serine residue of pre-TP protein. As the nascent DNA synthesis ensues, the nontemplated strand becomes displaced and coated by DBP, a single-strand binding protein. Moreover, the displaced single-strand DNA folds into a *panhandle structure* via complementarity between two ITR elements, when the nascent strand DNA synthesis is completed. The first round of DNA synthesis ends up yielding two DNA molecules (ie, one

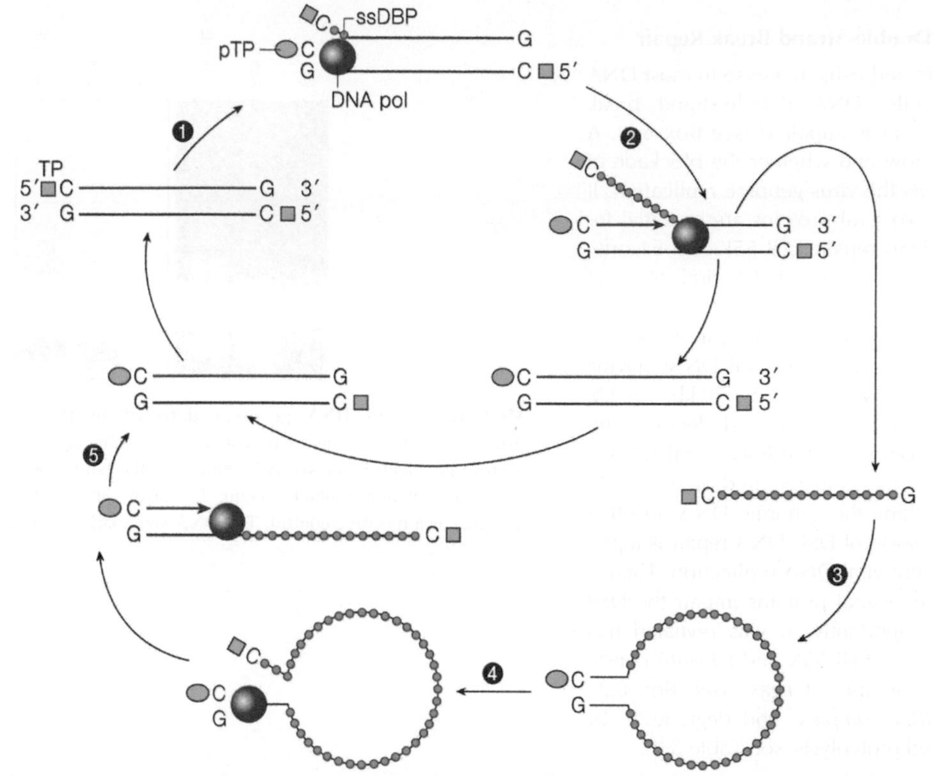

FIGURE 8.8 Genome replication mechanism of adenovirus. The first round of the DNA synthesis: (1) the initiation of DNA synthesis: Ad Pol/pre-TP complex binds to the terminus of the template, and initiates the DNA synthesis via protein-priming. The serine residue of pre-TP protein is linked to the first nucleotide (ie, C). (2) The displaced strand is coated by DBP. (3) The panhandle structure is formed by the displaced strand. (4)(5) The second round of the DNA synthesis: the newly generated end of the panhandle structure is used as a template for the second strand DNA synthesis.

double-strand DNA and one single-strand DNA). The second round of the DNA synthesis starts at the tip of the panhandle structure. In fact, the newly formed double-stranded region of the panhandle is indistinguishable from that of the linear viral DNA template used for the first round of the DNA synthesis. Likewise, the initiation of the DNA synthesis for the second round occurs via protein-priming. Eventually, two molecules of the DNA genome are yielded. Finally, the pre-TP (80 kDa) linked to the DNA is processed to TP (55 kDa) by a viral protease.

One salient feature of adenoviral DNA genome replication is that two strands of DNA synthesis are accomplished in a sequential manner. Only one strand of the viral double-strand DNA is used as a template for the first round of DNA synthesis, and then, the displaced strand is used for the second round of DNA synthesis, a process termed as "*asynchronous replication*."

DSB Repair: The blockade of DNA *double-strand break (DSB) repair* is essential for efficient adenoviral DNA replication; otherwise, concatemers of the viral genome are formed, as the linear viral DNA is recognized by DSB repair machinery as a double-strand DNA break (Box 8.1). Interestingly, a specific type of cullin RING ubiquitin E3 ligase, termed *adenovirus ubiquitin E3 ligase* (Box 8.2), expressed in adenovirus infected cells targets *MRN*[7] *complex* and degrades it via proteasome. Specifically, E1B-55K and E4 orf6 are two viral proteins that constitute the adenovirus ubiquitin E3 ligase (Table 8.4). It is thought that adenovirus has acquired the ability to target and degrade a critical component of DSB repair signaling (ie, MRN complex) throughout evolution by subverting cullin RING E3 ligase.

Transcription: Late genes are transcribed from a single promoter, dubbed "MLP" (Fig. 8.9). The primary transcripts from the late genes are unusually larger RNAs. They are alternatively spliced, and polyadenylated, yielding numerous transcripts. As a result, five major groups of transcripts, from L1 to L5, are accumulated in the cytoplasm, and translated into structural proteins, building blocks for the capsids, including penton, core, hexon, and fiber.

7. **MRN (Mre11-Rad50-Nbs1)** a trimeric complex involved in DSB repair that senses the DNA damage (Box 8.1). It also exhibits endonuclease and exonuclease activity.

BOX 8.1 DNA Double-Strand Break Repair

In adenovirus infected cells, likewise in most DNA virus infected cells, DNA double-strand break (DSB) repair function is impaired (see Box 6.2). A question is then how and whether the blockade of DSB repair benefits the virus genome replication. It was shown that two viral proteins are essential for the blockade of DSB repair: E1B-55K and E4 orf6. Intriguingly, the concatemer of the viral genomic DNA, a head-to-tail multimer, is detected in the cells infected by adenovirus deficient in one of these two genes. In fact, larger viral DNAs having multimeric unit of the viral DNA (ie, 70 kb, 105 kb, and etc.) are observed by agarose gel electrophoresis. An interpretation is that the linear viral DNA is mistaken as DSB DNA and repaired to the multimeric units by linking the genomic DNA. In other words, the suppression of DSB DNA repair is a prerequisite for efficient viral DNA replication. Then, a question is how two viral proteins impair the DSB repair function. Importantly, it was revealed that two viral proteins (ie, E1B-55K and E4 orf6) constitute *adenovirus ubiquitin E3 ligase* (see Box 8.2), which targets *MRN complex*, and degrades it by ubiquitin-mediated proteolysis (see Table 8.4).

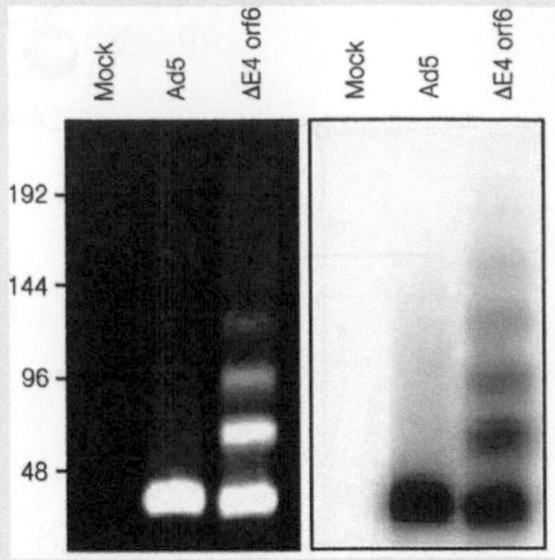

Multimeric viral DNA genomes detected in the adenovirus mutant. Adenovirus DNA isolated from adenovirus infected cells were analyzed by agarose gel electrophoresis. Ad5 represents the wild-type and ΔE4 orf6 represents the deletion mutant lacking E4 orf6. An uninfected cell (mock) is included as a negative control. The DNA size markers are denoted to the left.

Assembly and Release: The viral capsids are assembled in the nucleus, and become released from the cell by cell lysis (see Fig. 8.4). One E3 gene product, namely *E3-11.6K*, is known to be involved in cell death and virion release (see Table 8.3). Although the E3 protein, also known as *ADP* (adenovirus death protein), is an early gene product, it remains to be expressed in late phase. It is a membrane protein expressed in endoplasmic reticulum and Golgi, but its mechanism leading to cell lysis remains uncertain.

8.4 EFFECTS ON HOST

To establish the viral infection, adenovirus suppresses multiple cellular functions, including host immune response and mRNA processing. In addition, adenovirus infection leads to apoptosis, resulting in cell lysis.

Immune Suppression: Adenovirus suppresses the host immune response in diverse ways. First, adenovirus infection blocks the phosphorylation of *STAT1*, thereby inhibiting IFN action (see Fig. 5.6). Second, the virally coded *VA RNA*[8] inhibits the innate immune response (Fig. 8.10). VA RNA binds to *PKR* (protein kinase R), blocking the dimerization of PKR, which is required for activation (see Fig. 5.7). In short, adenovirus blocks innate immunity by inhibiting the activation of two factors critical for IFN action: (1) STAT1 and (2) PKR.

In addition, adenovirus suppresses adaptive immunity. In particular, the E3 gene encodes multiple genes that regulate adaptive immunity. For instance, *gp18K* protects cells from cell killing induced by cytotoxic T lymphocyte (see Table 8.3).

mRNA Processing: Intriguingly, the nuclear export of cellular mRNA is blocked in the adenovirus infected cells, while that of viral mRNA occurs normally. As a result, the cellular mRNA level decreases, while the viral mRNA abundantly accumulates. How is the selective nuclear export of viral mRNA accomplished? Recent reports implicated that *E1B-55K* and *E4 orf6* are essential for the selective nuclear export of viral mRNA (see Table 8.4). A speculation is that a host factor involved in the selective nuclear export is targeted, and degraded by *adenovirus ubiquitin E3 ligase*. However, the identity of the host factor remains uncertain.

8. **VA (virus-associated) RNA** a noncoding RNA transcribed by RNA polymerase III.

BOX 8.2 Cullin RING Ubiquitin E3 Ligase

Cullin RING ubiquitin E3 ligases (CRL) is the largest family of ubiquitin E3 ligases. In fact, CRL is a multiprotein complex, in which cullin serves as a scaffold linking a ubiquitin donor, E2 ubiquitin conjugating enzyme, to a ubiquitin acceptor, substrate (target protein). The C-terminus of cullin scaffold is associated with ROC, a *RING*[9] domain protein, linked to the E2 conjugating enzyme, while the N-terminus of cullin scaffold is bound with adapter/linker proteins that are to recruit diverse substrates. How could then a variety of substrates be recruited to CRL? At least four different sets of adapter/linker complex are known. One example is Skp1/F-box complex in so-called SCF (Skp1-cullin-F box) E3 ligase, where Skp1 serves as a linker and F-box protein serves as an adapter for the substrate.

On the other hand, in adenovirus infected cells, a specific kind of CRL, termed "*adenovirus ubiquitin E3 ligase*," plays a critical role in virus life cycle. In this viral E3 ligase, the viral E1B-55K/E4 orf6 protein complex acts as an adapter for substrate binding. In particular, p53 and MRN complex were identified as target substrate for the ubiquitin-mediated degradation in the virus infected cells (see Table 8.4).

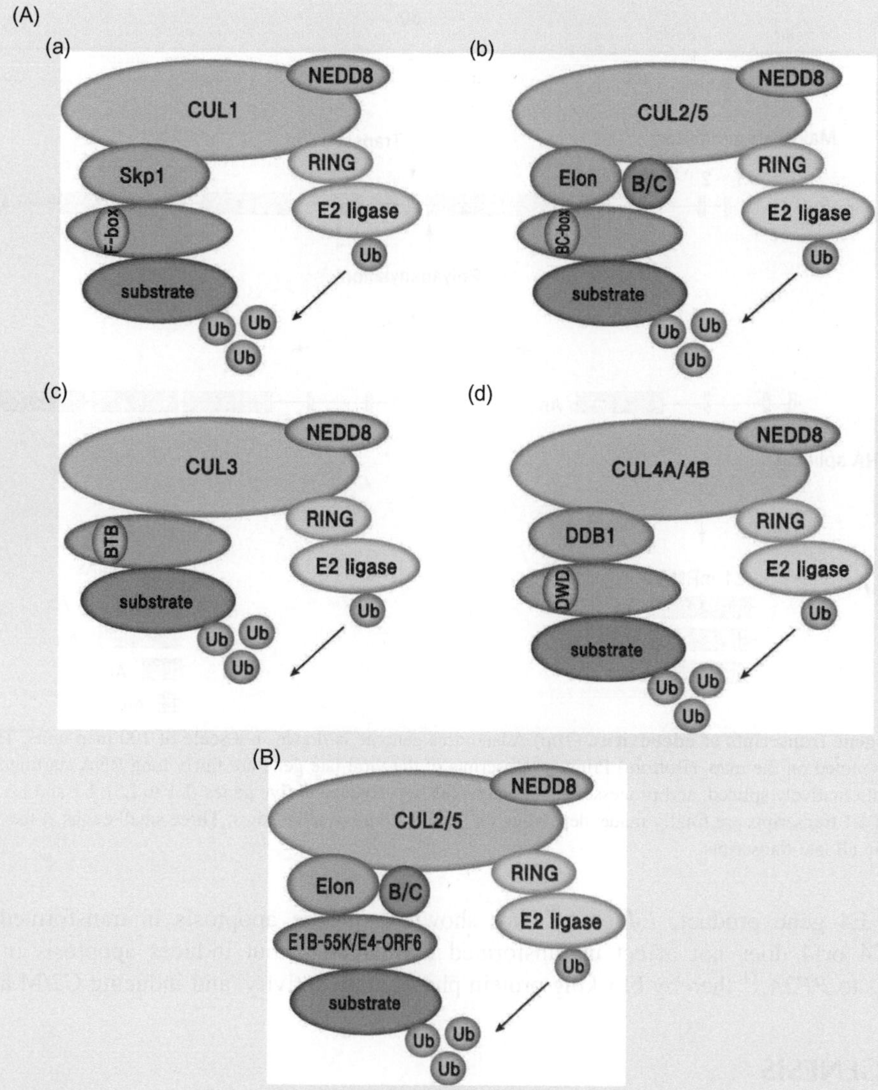

Cullin RING ubiquitin E3 ligase. (A) Cullin RING ubiquitin E3 ligase. (a) SCF E3 ligase, where Skp1 serves as an adapter and F-box protein serves as a substrate receptor. (b) Cullin 2/5-based E3 ligase, where elongin B/C complex serves as an adapter. (c) Cullin 3-based E3 ligase, where BTB protein serves as an adapter. (d) Cullin 4-base E3 ligase, where DDB1 serves as an adapter. (B) Adenovirus ubiquitin E3 ligase. E1B-55K/E4 orf6 protein complex serves as a substrate receptor for substrate binding, while elongin B/C complex serves as an adapter.

---

9. **RING (really interesting new gene)** a protein motif found in RING finger family proteins.

TABLE 8.4 The Function of E1B-55K-E4 orf6 Complex

Target Substrate	Function	Adenovirus Life Cycle	Host Effect
p53	Tumor suppressor	E1A-mediated p53 stabilization	Apoptosis
MRN (Mre11-Rad50-Nbs1) complex	DSB repair	Blockade of viral DNA concatemer formation	DSB repair
Unknown	mRNA export	Viral mRNA export	Host shutoff

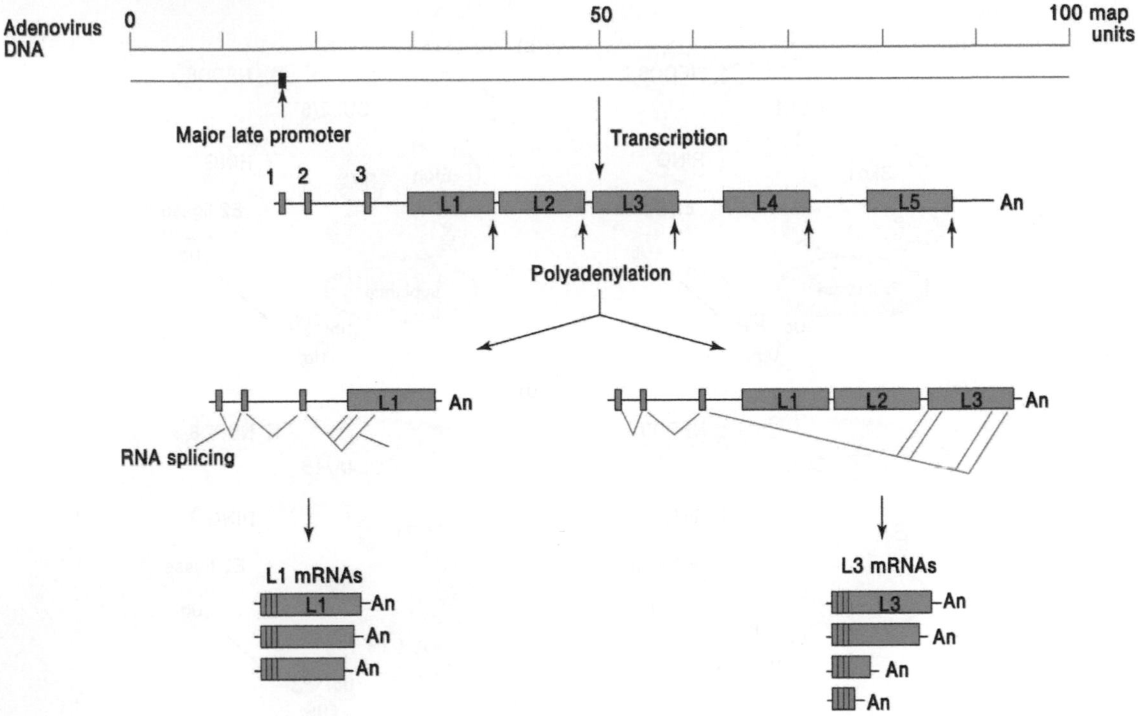

FIGURE 8.9 The late gene transcripts of adenovirus. (Top) Adenovirus genome is drawn in a scale of 100 map units. The position of the major late promoter (MLP) is denoted on the map. (Bottom) Primary transcripts of the viral late gene are fairly long RNA starting from the major late promoter (MLP). They are alternatively spliced, and processed to numerous mRNAs to encode five genes (L1 to L5). L1 and L3 transcripts are shown as examples. Three kinds of L1 transcripts are finally made, depending on three alternative splice sites. Three small exons at the 5′ end, termed "tripartite leader," are represented in all late transcripts.

Apoptosis: One E4 gene product, *E4 orf4*,[10] was shown to induce apoptosis in transformed cells (Table 8.3). Interestingly, the E4 orf4 does not affect untransformed normal cells, but induces apoptosis in transformed cells. Specifically, it binds to *PP2A*,[11] thereby blocking protein phosphatase activity, and inducing G2/M arrest and apoptosis.

8.5 TUMORIGENESIS

Adenovirus infection could lead to cell *transformation*.[12] Ironically, adenovirus infection of the human cell, a natural host, leads to cell lysis, instead of cell transformation. In other words, the human cell is a "permissive cell," in which the virus life cycle—late phase as well as early phase—is fully permitted, and as a result, the progeny virus is produced. In contrast, rodent cells such as mouse and hamsters are "nonpermissive cells," in which late phase of the viral life cycle is not permitted. As a result, rodent cells are transformed by adenovirus infection.

10. **E4 orf4** a viral protein encoded by orf4 (open reading frame) of E4 gene.
11. **PP2A (protein phosphatase)** anti-oncogenic protein exhibiting serine/threonine phosphatase activity.
12. **Transformation** It refers to a process in which cells acquire the properties of cancer (ie, immortalization).

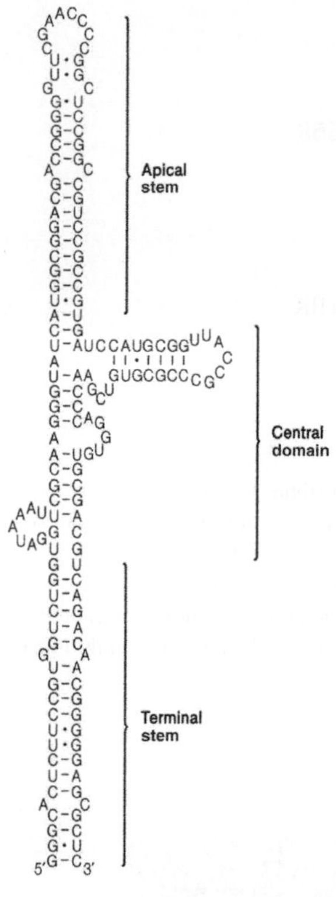

FIGURE 8.10 The secondary structure of VA RNA. VA RNA represents a small noncoding RNA, which is about 160 nt in length. Although the variation of VA RNA sequences among different genotypes is high, their secondary structures are conserved. The abundance in adenovirus infected cells is high—up to 10^8 copies per cell. Three structural domains are denoted by brackets: apical stem, central domain, and terminal stem.

Two gene products of the E1 gene, *E1A* and *E1B*, are essential for cell transformation (see Table 8.3). A question then arises as to how E1A and E1B contribute to cell transformation. The oncogenic property of E1A essentially pertains to its binding ability to Rb protein, a tumor suppressor gene product (see Fig. 24.3). By binding to Rb, E1A releases *E2F*[13] from Rb-E2F complex, thereby leading to entry to S phase in the cell cycle.

In addition, E1A leads to p53 stabilization. Specifically, E1A-mediated induction of *ARF*[14] protein suppresses MDM2-mediated p53 degradation (Fig. 8.11). This signaling pathway is dubbed the "*ARF-MDM2-p53 pathway*." The induction of p53, a tumor suppressor gene product, by E1A sounds counterintuitive. Such accumulated p53's function is regulated by E1B and E4 orf3 (see Fig. 8.11).

Specifically, E1b gene encodes two viral proteins: *E1B-19K* and *E1B-55K* (see Fig. 24.6). E1B-19K forms a complex with *Bax*, a homolog of Bcl-2, thereby blocking apoptosis, and contributing to cell transformation. As stated earlier, E1B-55K, as a component of adenovirus E3 ligase, targets p53 for the degradation via ubiquitin-mediated proteolysis (see Box 8.2). Likewise, E1B-55K is also believed to contribute to cell transformation via targeting the MRN complex (see Box 8.1).

Importantly, neither E1A nor E1B gene products are sufficient for cell transformation. In other words, both E1A and E1B gene products are required for cell transformation (see Fig. 24.7). The collaborative roles of two viral oncogenes will be described in more detail in chapter "Tumor Viruses."

8.6 PERSPECTIVES

Adenovirus research plays a central role in the advancement of molecular biology, molecular oncology, and virology. For instance, *HEK293 cells*, a cell line commonly used in many laboratories, is the cell line that has been transformed by adenovirus (Box 8.3). One of a few anecdotal findings made in studying adenovirus is the discovery of the intron (intervening sequence), which was discovered in the study of the tripartite leader region of the late transcripts. Studies

13. **E2F** A cellular transcription factor that was first discovered as a binding protein to adenovirus E2 promoter, as its name implies. It is best known as S phase-specific transcription factor.

14. **ARF (alternative reading frame)** It refers to a tumor suppressor gene product, whose gene was initially discovered as an alternative reading frame in INK4a gene (CDK inhibitor or $p16^{INK4a}$). It is also dubbed, $p14^{ARF}$, according to its mass.

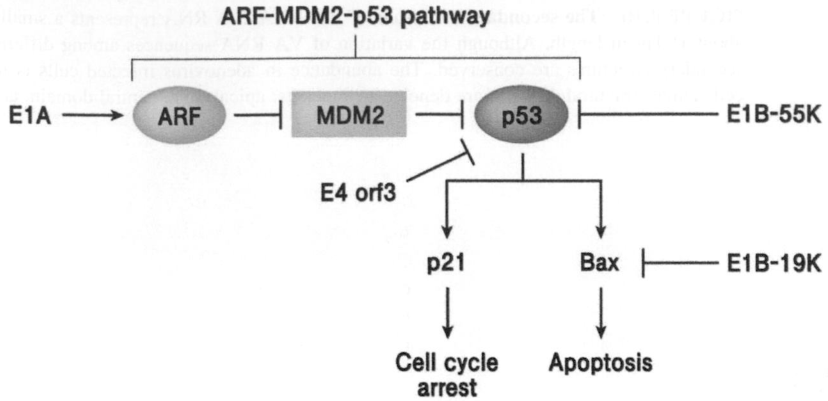

FIGURE 8.11 Oncogenic roles of adenoviral E1A, E1B, and E4 gene products. At least four viral proteins contribute to the viral oncogenesis. E1A induces p53 stabilization via ARF-MDM2-p53 pathway. The stabilized p53 could lead to cell cycle arrest via p21 induction or apoptosis via Bax induction. Such otherwise anti-oncogenic p53 functions are blocked by three viral proteins: E1B-55K, E1B-19K, and E4 orf3. For instance, E1B-55K targets p53 for ubiquitin-mediated degradation, while E1B-19K blocks apoptosis via its binding to Bax. On the other hand, E4 orf3 regulates p53-induced transcription of p21 and Bax via epigenetic regulation (histone methylation). Tumor-related properties of three cellular genes are denoted by circles (tumor suppressor gene) and rectangle (oncogene). Arrows indicate the activation or induction, while the letter T indicates inhibition or suppression.

BOX 8.3 HEK293 Cell

In addition to HeLa cell (Box 7.3), *HEK293* cell is one of the most frequently used cell lines in laboratories. HEK293 is the choice of cell line, as (1) transfection efficiency is reproducibly high, and (2) proteins are highly expressed. Many investigators use HEK293 (also called 293 cell) cells without knowing that 293 cell are transformed by adenovirus. First of all, 293 cell was made by transfecting adenovirus 5 DNA fragments to human primary fetal kidney cells, as HEK stands for human embryonic kidney, while 293 simply stands for the experiment number that Dr. Graham carried out. In fact, 4.5 kb DNA fragment of Ad5 genome is integrated into the chromosome, and expresses E1A as well as E1B. In other words, HEK293 cells are transformed by E1A and E1B viral oncogenes, in which Rb and p53 function are impaired. As a result, HEK293 cells are inappropriate for cell cycle regulation and apoptosis.

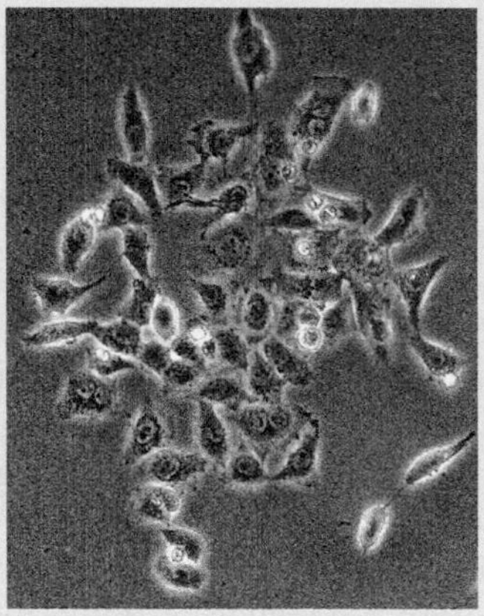

Phase contrast microscopic view of HEK293 cells.

using adenovirus as a model for oncogenesis has truly paid off. In fact, the discovery of E1A-Rb binding and E1B-55K-p53 binding uncovered the secret of cancer biology, representing seminal discoveries in molecular oncology. The identification of two cellular proteins that are regulated by viral oncogene products via protein–protein interaction established that Rb and p53 are the most important tumor suppressor genes that are targeted by viral oncogenes. On the other hand, adenovirus was exploited as a gene therapy vector in the 1990s. Its oncolytic potency is being explored for the cancer therapy (Box 8.4).

8.7 SUMMARY

- *Human adenovirus*: Over 51 subtypes of human adenovirus have been reported. Its infection leads to only mild symptoms, such as respiratory disease.
- *Viral genome*: Adenoviral genome represents a 35 kb linear double-strand DNA, in which TP proteins are covalently linked to the 5′ terminus.
- *Virion structure*: It is a nonenveloped naked virus. The viral capsid harbors 12 fibers protruding from the vertex.

BOX 8.4 Oncolytic Adenovirus

An effort to explore the oncolytic properties of some viruses for cancer therapy has drawn attention. It aims to exploit viruses that are capable of killing cancerous cells, but not normal cells. These are collectively called "oncolytic viruses." They include adenovirus, herpesvirus, poxvirus, reovirus, and measles virus. The selectivity that distinguishes normal cells from cancer cells is the essential feature of oncolytic viruses. Frequently, the viral variants that have lost their ability for replication or attenuated variants are utilized. These viral variants cannot grow in normal cells but can grow in cancer, leading to cell lysis.

Adenovirus is a representative oncolytic virus that has been extensively investigated. The finding that adenovirus lacking the E1b gene (ie, Ad-Δ E1b-55k) selectively kills cancerous cells led to the development of this variant as an oncolytic virus for cancer therapy. Interestingly, adenovirus lacking E1b-55K does not propagate in normal cells but does propagate in cancer, which lacks p53. The adenovirus variant is expected to kill cancer cells selectively, when treated to cancer patients. What is the underlying mechanism for cell killing? An interpretation is that WT adenovirus can propagate in normal cells, as E1B-55K suppresses p53, a cellular guardian, while the mutant cannot. In contrast, the mutant virus can propagate in cancer cells, which lack p53. The adenovirus variant, dubbed *"ONYX-015,"* has been in clinical trials for cancer gene therapy. ONYX-015 has been extensively tested in clinical trials, with the data indicating that it is safe and selective for cancer. However, a limited therapeutic effect has been demonstrated following injection and systemic spread of the virus was not detected. During these trials a plethora of reports emerged challenging the underlying p53-selectivity, with some reports showing that in some cancers with a wild-type p53 ONYX-015 actually did better than in their mutant p53 counterparts. These reports slowed the advancement through Phase III trials in the United States. In contrast, in China, an oncolytic adenovirus similar to ONYX-015 (E1B-55K/E3B-deleted) was approved in 2005 for use in combination with chemotherapy for the treatment of late-stage refractory nasopharyngeal cancer. Outside of China, the push to the clinic for ONYX-015 has been largely been discontinued for financial reasons and until a "real mechanism" can be found. Overall, cancer gene therapy by using oncolytic viruses became a hope for patients whose cancer cells are hopelessly resistant to chemotherapy. Adenovirus, which is a human pathogen, is now being converted to a remedy tool. This anecdote relates to a famous Chinese saying "use the enemy to kill the enemy."

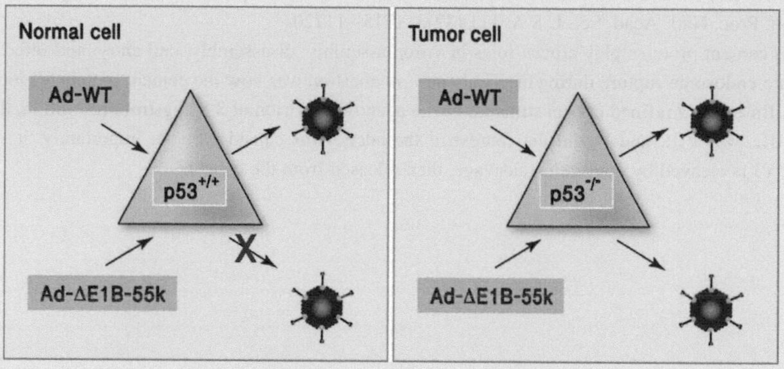

Normal cell versus tumor cell with respect to the adenovirus infection lacking E1B-55K. In normal cells, which contain the wild-type allele of p53, the mutant virus lacking E1B-55K gene cannot replicate. In tumor cells, which contain the defective allele of p53, the mutant virus can replicate.

- *Genome replication*: Adenoviral DNA polymerase carries out DNA synthesis. The viral TP protein serves as a protein-primer for the DNA synthesis. One salient feature of adenoviral DNA genome replication is that two strands of DNA synthesis are accomplished in a sequential manner, a process termed as "asynchronous replication."
- *Host effect*: Adenovirus infection causes oncogenic transformation in rodent, but not in human, a permissive host. Instead, adenovirus infection leads to cell lysis of human cells.

STUDY QUESTIONS

8.1 Adenovirus DNA genome replication is carried out in two steps, in which two strands of the viral DNA are asynchronously copied. (1) State the similarity and difference between the two templates. (2) List viral proteins that are involved in the DNA synthesis. State any difference between the two stages of the DNA replication.

8.2 A cullin RING ubiquitin E3 ligase is formed in adenovirus infected cells. (1) Please list the components of the adenovirus ubiquitin E3 ligase. (2) Please list cellular factors that are known to be targeted by the adenovirus ubiquitin E3 ligase, and discuss its biological impact.

8.3 Adenovirus is a DNA tumor virus, which has been isolated from human. (1) Please list the viral oncoproteins and describe their tumor causing properties. (2) Adenovirus infection causes tumor in rodent, but not in human. Explain why?

SUGGESTED READING

Cassany, A., Ragues, J., Guan, T., Begu, D., Wodrich, H., Kann, M., et al., 2015. Nuclear import of adenovirus DNA involves direct interaction of hexon with an N-terminal domain of the nucleoporin Nup214. J. Virol. 89 (3), 1719–1730.

Lam, E., Stein, S., Falck-Pedersen, E., 2014. Adenovirus detection by the cGAS/STING/TBK1 DNA sensing cascade. J. Virol. 88 (2), 974–981.

Pelka, P., Ablack, J.N., Fonseca, G.J., Yousef, A.F., Mymryk, J.S., 2008. Intrinsic structural disorder in adenovirus E1A: a viral molecular hub linking multiple diverse processes. J. Virol. 82 (15), 7252–7263.

Reddy, V.S., Nemerow, G.R., 2014. Structures and organization of adenovirus cement proteins provide insights into the role of capsid maturation in virus entry and infection. Proc. Natl. Acad. Sci. U. S. A. 111 (32), 11715–11720.

Strunze, S., Engelke, M.F., Wang, I.H., Puntener, D., Boucke, K., Schleich, S., et al., 2011. Kinesin-1-mediated capsid disassembly and disruption of the nuclear pore complex promote virus infection. Cell Host Microbe. 10 (3), 210–223.

JOURNAL CLUB

- Reddy, V.S., Nemerow, G.R., 2014. Structures and organization of adenovirus cement proteins provide insights into the role of capsid maturation in virus entry and infection. Proc. Natl. Acad. Sci. U.S.A. 111 (32), 11715–11720.

  Highlight: Adenovirus cement proteins play crucial roles in virion assembly, disassembly, cell entry, and infection. In particular, cement protein VI is known to mediate endosome rupture during the viral entry. A question was how the cement protein VI located on the capsid interior can induce endosome rupture. Based on a refined crystal structure of the adenovirus virion at 3.8 angstrom resolution, this article uncovered the structure of cement proteins (IIIa, VI, VIII, and IX) in the context of the adenovirus capsid particle. Importantly, it shows that the membrane-lytic domain of cement protein VI is cleaved by maturation cleavage, then released from the capsid.

Chapter 9

Herpesviruses

Chapter Outline

We are familiar with *herpesviruses*,[1] because most of us have suffered their infection. Herpesviruses are associated with multiple human disorders, ranging from mild symptoms such as cold sores to more severe diseases, such as cancers (Table 9.1). For instance, oral herpes is caused by herpes simplex virus type 1 (HSV-1) (Fig. 9.1). Here, HSV-1, a prototype of human alpha-herpesvirus, will be mainly described. Besides HSV-1, human cytomegalovirus (HCMV), a prototype of beta-herpesvirus, and Epstein-Barr virus (EBV), a prototype of gamma-herpesvirus, will be described briefly in separate boxes.

9.1 CLASSIFICATION

Taxonomy: Herpesviruses are found not only in human but also in many vertebrates such as mammals, birds, reptiles, and fishes. Over 130 species of herpesviruses have been reported, and they can be divided into three subgroups depending on their biological properties: alpha-, beta-, and gamma-herpesviruses.

Eight species of human herpesviruses have been discovered. Human herpesviruses are divided into three genera: alpha-, beta-, and gamma-herpesviruses (Table 9.2). In alpha-herpesviruses, three species are reported: (1) herpes simplex virus type 1 (HSV-1), which causes cold sores or oral herpes, (2) herpes simplex virus type 2 (HSV-2), which causes genital herpes, and (3) Varicella-Zoster virus (VZV), which causes chicken pox in children and shingles in adults. Beta-herpesviruses include human cytomegalovirus (HCMV), which causes cytomegalic inclusion disease, and human herpesvirus 6 and 7, which cause roseola. In gamma-herpesviruses, two species are reported: *Epstein-Barr virus (EBV)*[2] and Kaposi's sarcoma-associated herpesvirus (*KSHV*). EBV is associated with two types of human cancers: Burkitt's lymphoma and nasopharyngeal carcinoma. KSHV is associated with Kaposi's sarcoma in AIDS patients. These human gamma-herpesviruses are *tumor viruses*.[3] In addition, human herpesviruses are systemically named by the order of discovery, from HHV-1 to HHV-8 (see Table 9.2).

Of the eight human herpesviruses, HSV-1, a prototype of human alpha-herpesvirus, will be mainly described for brevity.

1. **Herpesvirus** the term "herpes" is derived from Greek word for "creep = latent or chronic"-*herpin*.
2. **Epstein-Barr virus (EBV)** A human gamma-herpesvirus that is associated with Burkitt's lymphoma. The virus was named after two scientists who discovered: Dr. Epstein and Dr. Barr (see Box 9.3).
3. **Tumor virus** Five human viruses, including EBV and KSHV, are known to be an etiologic agent for human cancer (see Table 24.2).

Molecular Virology of Human Pathogenic Viruses. DOI: http://dx.doi.org/10.1016/B978-0-12-800838-6.00009-6

TABLE 9.1 The Defining Features of Herpesvirus

Genome	Particle Structure	Replication Mechanism
Large linear dsDNA (120 ~ 235 kb)	Enveloped	Viral DNA polymerase
Terminal repeats	Icosahedral nucleocapsid	Rolling-circle mechanism
Life Cycle	**Host Effect**	**Disease**
IE-E-L phase	Host shutoff	Cold sore, chicken pox, zoster
Latent infection	Immune evasion	Tumor (lymphoma)

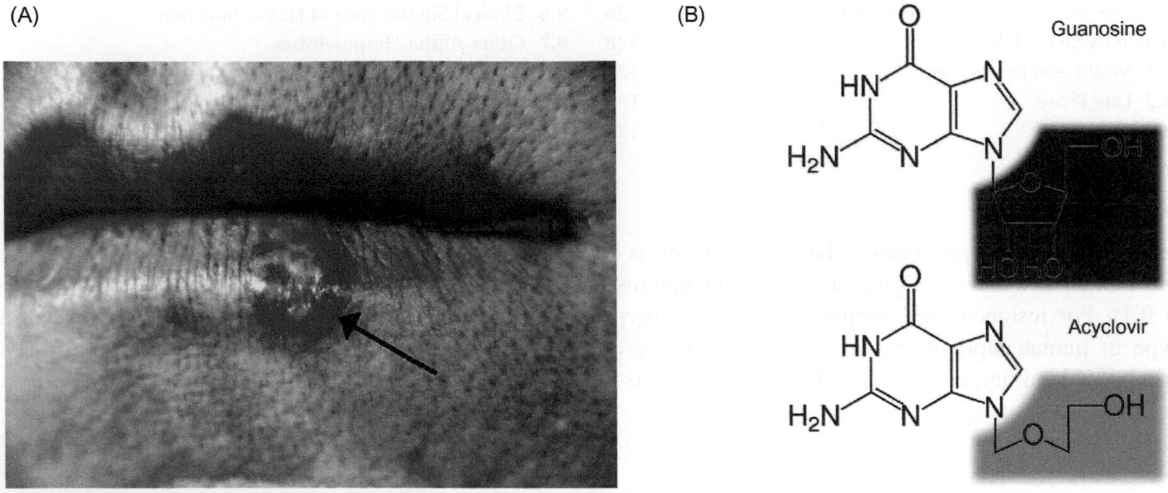

FIGURE 9.1 Oral herpes and antiherpes drugs. (A) Herpes simplex lesion of lower lip, second day after onset. Oral herpes, the visible symptoms of which are colloquially called *cold sores* or *fever blisters*, is the most common form of herpesvirus infection. (B) Comparison of chemical structures of guanosine and guanosine analog acyclovir. Acyclovir is an antiherpes drug prescribed for cold sores.

9.2 THE VIRION AND GENOME STRUCTURE OF HSV-1

Virion Structure: HSV-1 is an enveloped virus with a diameter of 150 nm, in which the capsid with a diameter of 90 nm is encompassed (Fig. 9.2). In the viral envelope, more than nine viral envelope glycoproteins, ranging from gB to gN glycoprotein, are embedded. A peculiar feature of the virion structure of herpesviruses is that the space between the envelope and the capsid, termed *tegument*,[4] is filled with numerous viral proteins. Importantly, the tegument proteins are essential for the establishment of virus infection. Among more than 15 viral tegument proteins identified, three of them are well characterized: VP16, ICP0, and Vhs protein. *VP16* is a tegument protein that acts as a transcriptional transactivator of the viral genes (ie, immediate early genes). *ICP0*, the viral ubiquitin E3 ligase, targets a number of cellular proteins that restrict viral infection for the degradation (see Fig. 9.7). On the other hand, *Vhs*[5] protein counteracts host innate immune response by reducing mRNA stability, in particular, those of interferons and pro-inflammatory cytokines.

Genome Structure: HSV-1 has a large 150 kb linear double-strand DNA (Fig. 9.3). Two kinds of repeat elements are present in the genome: (1) a pair of direct repeat elements at both termini, termed *TR* (terminal repeat), and (2) two pairs of inverted repeat elements in the middle of the genome, termed *IR* (inverted repeat). TR is essential for viral DNA genome packaging. However, the biological function of IR remains uncertain. Obviously, the vast majority of

4. **Tegument** A space between envelope and nucleocapsid in virions of certain viruses, notably herpesviruses. Tegument proteins can be compared with matrix protein in other enveloped virions.

5. **Vhs (virion host shutoff)** It blocks host immune response by degradation of cellular mRNAs via its endonuclease activity.

TABLE 9.2 Classification of Human Herpesviruses

Human Herpesviruses	Common Name	Disease
Alpha-Herpesvirus		
Herpes simplex virus type-1	HSV-1 (HHV-1)	Cold sore
Herpes simplex virus type-2	HSV-2 (HHV-2)	Genital ulcer
Varicella-Zoster virus	VZV (HHV-3)	Chicken pox or varicella, zoster
Beta-Herpesvirus		
Human cytomegalovirus	HCMV (HHV-5)	Mononucleosis
		Cytomegalic inclusion disease
Human herpesvirus 6	HHV-6	Roseola
Human herpesvirus 7	HHV-7	Roseola
Gamma-Herpesvirus		
Epstein-Barr virus	EBV (HHV-4)	Infectious mononucleosis
		Burkitt's lymphoma
		Nasopharyngeal carcinoma
Kapposi's sarcoma	KSHV (HHV-8)	Karposi's sarcoma

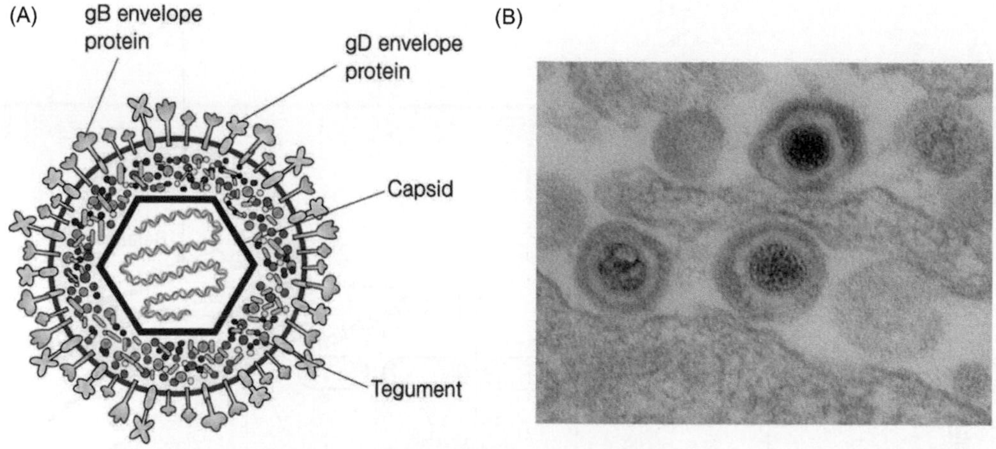

FIGURE 9.2 Virion structure of HSV-1. (A) Multiple envelope glycoproteins, including gB and gD, are present in the viral envelope. Tegument is stuffed with numerous viral proteins such as VP16 and Vhs. The viral capsid is larger, with $T = 16$ symmetry, which packages a linear double-strand DNA with 150 kb in length. (B) Electron micrograph of HSV-1 virion. Tegument, a space between the capsid and the envelope, is clearly visible.

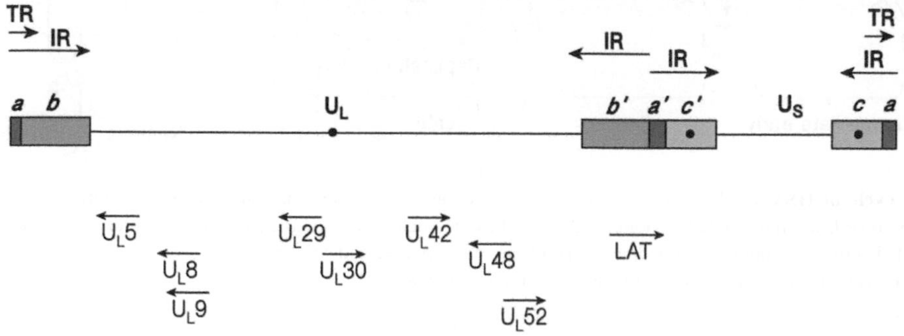

FIGURE 9.3 Genome organization of HSV-1. The linear double strand DNA genome of HSV-1 is drawn with emphasis on repeat elements (box), which are denoted by "a," "b," and "c." U_L and U_S elements are demarcated by two neighboring IR elements ("ab" or "ac," respectively). Note that "**a**′" is complement of "**a**." The terminal repeats ("a") at the both ends of the genome are denoted by TR, while the inverted repeats are denoted by IR. Arrows indicate the directionality of the nucleotide sequence. OriL located in U_L region and OriS located in "c" elements are indicated by red dots. A subset of HSV-1 transcripts is also indicated below. *LAT*, latency-associated transcript.

DNA genome sequence is located between two IRs, and these are aptly named U_L (unique long) and U_S (unique short), as opposed to the "repeat" element, which is reserved for a redundant element. Most of the viral proteins are encoded within U_L and U_S regions. Intriguingly, three origins of DNA replication (Ori) elements are found: one in U_L region (termed *OriL*) and two in U_S region (termed *OriS*).

Protein Coding: Having a large genome, HSV-1 encodes over 90 proteins (see Fig. 9.3). The U_L region encodes 65 proteins, while the U_S region encodes 14 proteins. Interestingly, a few viral proteins are encoded even by the repeat elements. The viral proteins are named after their gene number. For instance, U_L 30 refers to the gene number 30 that is located in the U_L region; the protein encoded by the U_L 30 gene is termed U_L 30 protein.

Multiple nomenclatures of the viral proteins were used in the past. Recently, the nomenclature according to the gene number became widely accepted, as this is systemic and less confusing. For instance, VP16 is called by three different nomenclatures: (1) *VP16* is named after structural proteins constituting virion particle, (2) *UL48* is named after the gene number, and (3) *ICP25*[6] is named after the serial number of viral proteins expressed in the infected cells. For a given protein, one of these is more frequently used than others; in fact, VP16 is more commonly used than U_L48 and ICP25.

9.3 THE LIFE CYCLE OF HSV-1

Cell tropism: HSV-1 infects the *epithelium* of the mucosal layer. Following primary infection, HSV-1 invades the central nerve system (CNS) via sensory neurons and finally establishes latent infection in nerve ganglia (see Fig. 9.12).

Similar to other DNA viruses, the virus life cycle can be divided into two phases by the onset of viral genome replication: early phase and late phase. The viral genome replication occurs in the nucleus of infected cells (Fig. 9.4).

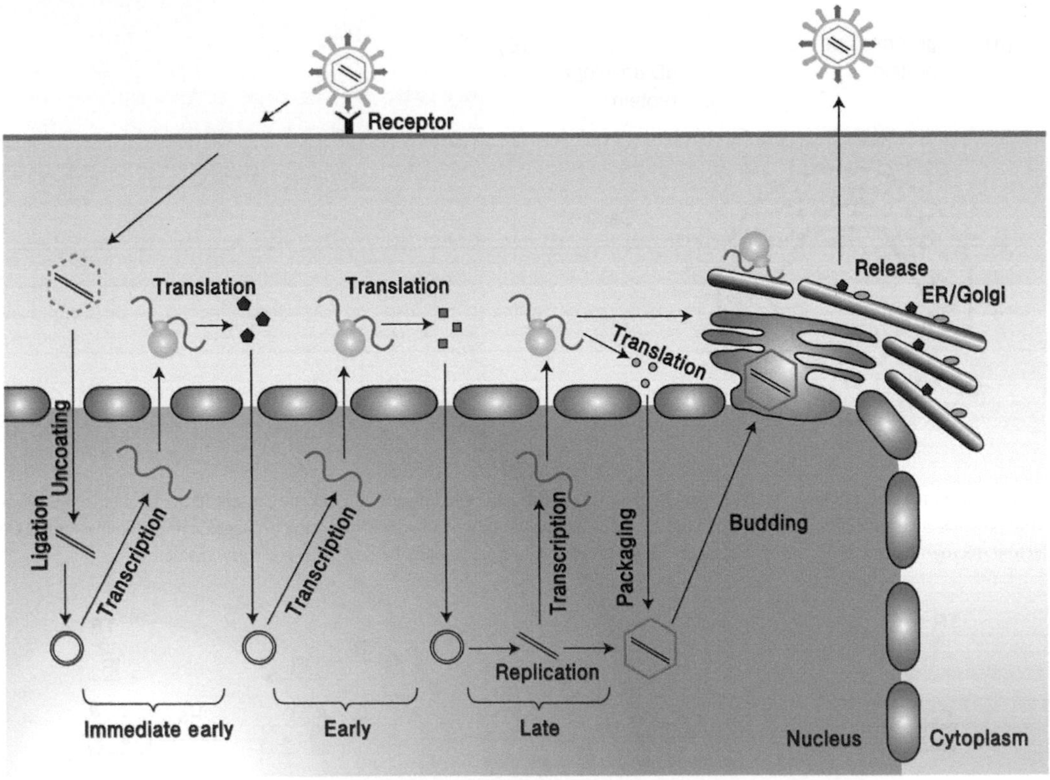

FIGURE 9.4 The life cycle of HSV-1. The virion enters the cell by either direct fusion or endocytosis (for clarity, only entry by direct fusion is shown). Following entry to cell, the linear DNA genome is delivered to the nucleus, and converted into a circular form in the nucleus. The viral gene expression can be divided into three phases: immediate early phase, early phase, and late phase. The capsid assembly occurs in the nucleus. The assembled capsid egresses via nuclear membrane, and then via secretory pathway.

6. **ICP** (infected cell protein)

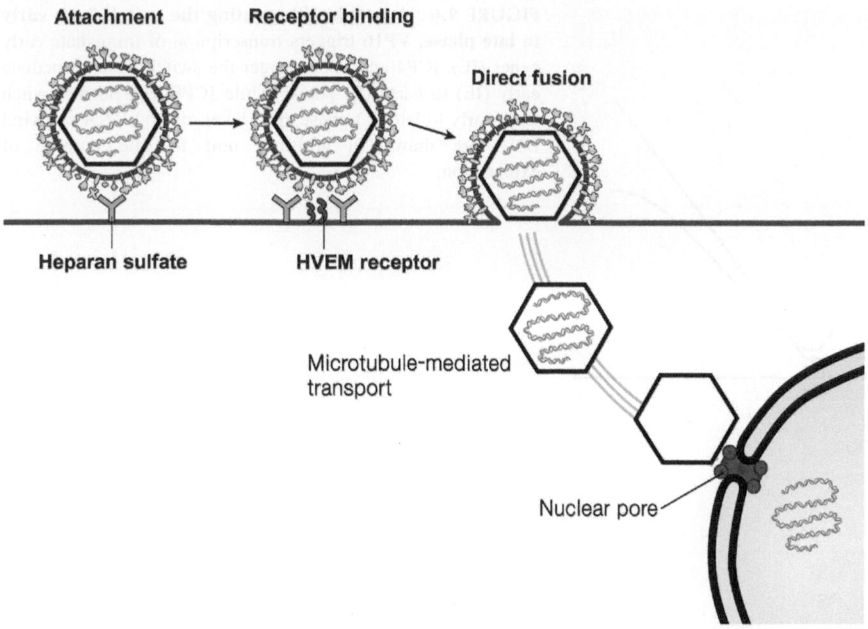

FIGURE 9.5 The HSV-1 entry via direct fusion. The virion attaches to cell surface via heparan sulfate as attachment factors. The gD binds to an entry receptor, HVEM. This gD-HVEM interaction triggers gB, the fusion protein, to mediate membrane fusion. Following the penetration, the viral capsid trafficks toward the nucleus via microtubule-mediated transport. Upon docking to the nuclear pore, the capsids become partially disrupted, and then release their genomes into the nucleus.

Entry: HSV-1 enters cells either via direct fusion or receptor-mediated endocytosis. Virion attaches to cell surface glycosaminoglycans such as heparan sulfate for entry (Fig. 9.5). Several viral glycoproteins, such as gD and gB, are involved in membrane fusion with the host cell. In particular, gD interacts with a cellular receptor, *HVEM*.[7] This subsequently triggers gB, the fusion protein, to mediate membrane fusion. Following penetration to the cytoplasm, the capsids traffic to nuclear membrane via *microtubule-mediated transport*. Upon docking to the nuclear pore, the viral DNA released from the capsids enters the nucleus via a nuclear pore. The linear DNA genome becomes converted to a circular configuration. The circularization of DNA genome is facilitated by ligation of "**a**" elements at the end or by recombination between "**a**" elements (see Fig. 9.3).

9.3.1 Viral Gene Expression

Upon entry to the nucleus, the viral DNA is localized near nuclear bodies known as *nuclear domain 10*[8] (ND10), where the viral chromosome exists in a repressed state. In other words, ND10 represents the host *intrinsic immunity* against the viral invasion. The viral gene transcription takes place only after the ICP0-mediated epigenetic regulation.

Three Phases: Prior to viral genome replication, the viral transcription takes place, and these genes are termed "early genes." Like adenovirus, early gene transcription of herpesvirus is subdivided into two phases: these are termed "immediate early genes (IE)," and "early genes (E)." Immediate early gene expression is stimulated by viral tegument proteins, while early gene expression is stimulated by immediate early gene products (Fig. 9.6). Early gene products are largely involved in the viral genome replication, and trigger the switch from early to late phase. In summary, the viral genes are divided into three groups, depending on the phase of the life cycle; immediate early (IE), early (E), and late (L) genes. Alternatively, they are also called α, β, and γ genes, respectively.

Immediate Early Phase: Prior to the onset of viral genome replication, immediate early and early gene expression occurs in a sequential manner. *VP16*, a tegument protein, acts as the transcriptional transactivator of IE (immediate early) genes. It binds to the promoter region of IE genes and potently transactivates transcription of IE genes. In addition, *Vhs*, a tegument protein, acts to degrade cellular mRNAs for few hours after infection. Because of Vhs function, the viral mRNAs are preferentially accumulated in the cytoplasm, as opposed to host mRNAs. Interestingly, Vhs was shown to degrade even the immediate early mRNA, which is no longer needed, during early phase. In late phase, on the other hand, the nascent made VP16 blocks the RNase activity of Vhs via direct binding.

7. **HVEM (herpes virus entry mediator)** It is a membrane protein that belongs to TNF-α superfamily.

8. **Nuclear domain 10 (ND10)** A small proteineous subnuclear structure that is composed of multiple factors including PML (promyelocytic leukemia protein) and Sp100. It is also called PML nuclear bodies.

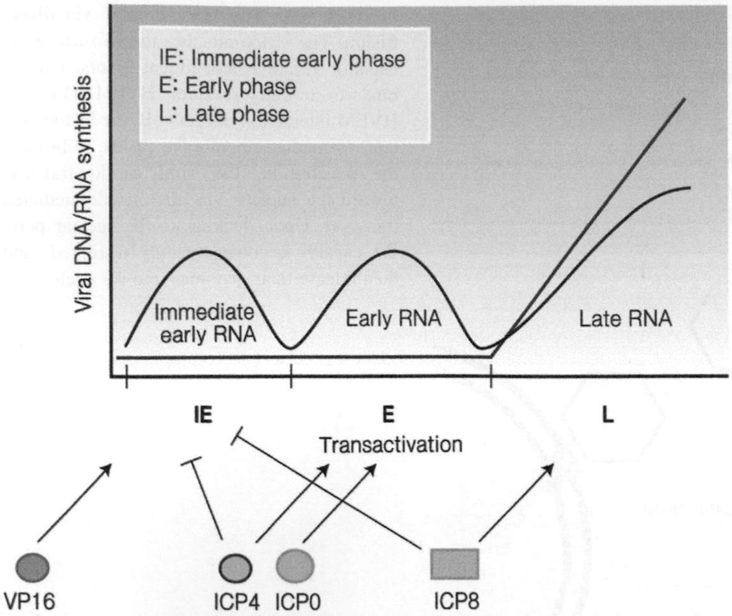

FIGURE 9.6 A graphic illustrating the switch from early to late phase. VP16 triggers transcription of immediate early genes (IE). ICP4 and ICP0 trigger the switch from immediate early (IE) to early (E) phase, while ICP8 triggers the switch from early to late (L) phase. The level of viral RNA and viral DNA are drawn in arbitrary unit for the purpose of explanation.

TABLE 9.3 The Immediate Early Phase Proteins of HSV-1 and Their Functions

Protein	Function
ICP0	Transcriptional transactivator and the viral ubiquitin E3 ligase
ICP4	Transcriptional transactivator
ICP27	Blocks cellular RNA splicing; facilitates viral mRNA export
ICP47	Blocks antigen presentation by binding to TAP transporter

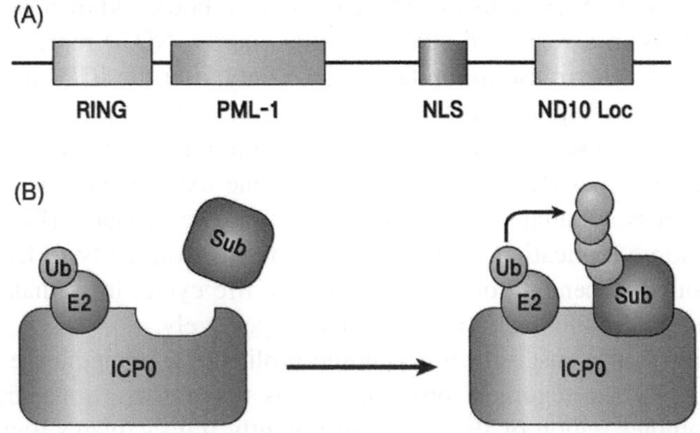

FIGURE 9.7 ICP0 protein: functional domains. (A) The functional domain of ICP0 protein, including RING finger, PML-1 binding region, NLS sequence, and ND10 localization domain. (B) A diagram illustrating substrate-targeting mechanism employed by ICP0 ubiquitin E3 ligase.

Switch to Early Phase: The functions of some IE gene products are well known (Table 9.3). Two IE gene products, *ICP0* and *ICP4*, play a role in switching from immediate early to early phase via transactivation of early gene transcription (see Fig. 9.6). ICP4 is also required for late gene transcription. ICP0 is a multifunctional protein, which interacts with numerous cellular factors including PML (Fig. 9.7). Moreover, ICP0 has a RING finger domain that confers *ubiquitin E3 ligase activity*. ICP0 influences chromatin structure in that it impedes the assembly of a repressed viral chromatin structures. As stated above, ICP0-mediated inactivation of host intrinsic restriction factors is essential for viral transcription. Whether or not ICP0 influences chromatic structure directly through its ubiquitin E3 ligase activity remains unclear. In contrast, ICP4 enhances transcription via recruiting cellular transcription factors to the promoter regions. Conversely, ICP4 can repress transcription of its own gene by directly binding to the promoter region (ie, "auto-regulation").

On the other hand, *ICP27* blocks the nuclear export of cellular mRNAs, but nonetheless facilitates the nuclear export of the viral mRNAs. In other word, ICP27 promotes viral gene expression by regulating mRNA processing (see Table 9.3). Only 4 h after viral infection, IE gene expression begins to decrease. The downregulation of IE gene expression occurs primarily by Vhs and by ICP4, which suppresses IE gene transcription (see Fig. 9.6).

Early Phase: Early gene products are mainly involved in viral genome replication (see Table 9.4). In particular, *ICP8* triggers the switch from early to late phase, by suppressing IE gene expression as well as by promoting late gene expression (see Fig. 9.6).

9.3.2 Late Phase

The viral genome replication, the late gene expression, the capsid assembly, and the release of the assembled capsid occur in the late phase.

Genome Replication: Factors essential for viral DNA synthesis can be divided into two groups: *cis*-acting elements and *trans*-acting factors. First, the origin of replication, the site for initiation of DNA synthesis, represents *cis*-acting elements for DNA synthesis. HSV-1 genome contains three origin of replication (one *OriL* and two *OriS*); these three elements are functionally redundant, as only one of them is sufficient for viral DNA replication.

Regarding *trans*-acting factors, *UL30* serves as a viral DNA polymerase for the HSV-1 genome replication (Fig. 9.8). Besides, six additional viral proteins contribute to the viral genome replication (Table 9.4). These

TABLE 9.4 The DNA Replication Proteins of HSV-1

Protein	Function
U_L9	Origin-binding protein; ATPase & helicase activity
U_L9 (ICP8)	Single-strand DNA-binding protein
$U_L5/U_L8/U_L52$	Helicase-primase complex
U_L30	DNA polymerase
U_L42	Processivity factor

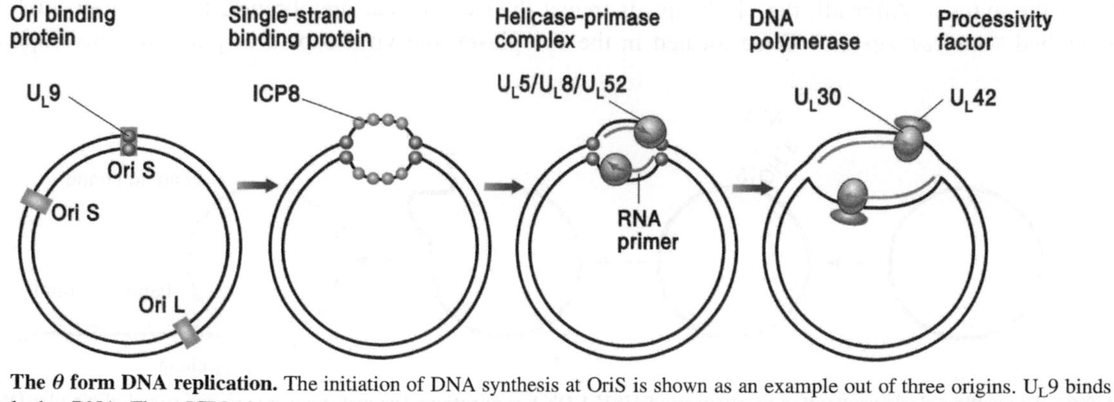

FIGURE 9.8 **The θ form DNA replication.** The initiation of DNA synthesis at OriS is shown as an example out of three origins. U_L9 binds to OriS to unwind the duplex DNA. Then, ICP8 binds to single-strand DNA region, recruiting $U_L5/U_L8/$ U_L52 helicase-primase complex to synthesize RNA primer. Finally, U_L30, the viral DNA polymerase, continues to synthesize the DNA from the RNA primer. U_L42 acts as a processivity factor. This mode of DNA replication is also called "bidirectional DNA replication," as the nascent DNA synthesis occurs simultaneously in both strands in opposite directions.

include: (1) *UL9*, the origin-binding protein, (2) *UL5/UL8/UL52* complex, helicase-primase complex, and (3) *UL48*, a processivity factor. Moreover, *ICP8* (U_L48) is a single strand DNA-binding protein, which stimulates viral DNA synthesis.

The mechanism for HSV-1 genome replication is not fully understood. The reason for this is primarily due to the lack of an in vitro system that recapitulates viral DNA synthesis. Nonetheless, the mechanism of the viral DNA synthesis is expected to be fundamentally similar with that of the cellular genome. The circularized DNA genome, which is formed immediately after its entry into the nucleus, is the template for the viral genome replication. First, the DNA replication begins via bidirectional DNA replication by using the circular template (see Fig. 9.8). This mode of DNA replication is dubbed "*θ form replication*"; it was named after the shape of the DNA intermediates. Nonetheless, the bulk of DNA synthesis is achieved by a so-called *rolling-circle mechanism* (Fig. 9.9). The evidence for this is the *concatemers*, the multimeric form of the DNA genome, found during the viral genome replication. It is believed that viral genome replication starts with the θ form replication, and then, the mode of DNA replication switches from *θ* form replication to the rolling-circle mechanism. It is thought that the nick is made in one strand of DNA prior to the switch.

What is the biological significance of the circularization of linear DNA prior to the genome replication? It is believed that the circularization of linear DNA is the viral strategy to overcome "*the end-replication problem*," the problems inherent to the linear DNA genome (see Box 10.2).

Having a larger genome, herpesviruses encode genes pertaining to nucleic acids metabolism as well. Many of these viral proteins contribute to the viral DNA synthesis. For instance, thymidine kinase (*UL23*), ribonucleotide reductase (*UL39/UL40*), and deoxyuridine triphosphatase (*UL50*) are involved in dNTPs synthesis. *N*-glycosylase (*UL2*) is involved in DNA repair. Interestingly, these metabolic enzymes are dispensable for dividing cells, which carry out DNA synthesis themselves, but indispensable for nondividing cells such as neuronal cells.

Late Gene Expression: Once viral DNA replication has initiated, viral late gene expression is increased. Viral immediate early proteins, such as ICP4, ICP22, and ICP27, and early gene products, such as ICP8, are required for late gene expression (see Fig. 9.6). Late gene transcription and DNA replication as well as virion assembly take place in *replication compartments* in the nucleus.

Assembly and Release: The capsid assembly occurs in the nucleus. Unlike most other DNA viruses, the capsids are preassembled first, and then one unit of viral genome DNA is subsequently packaged into the preassembled capsid (Fig. 9.10). The sequence element termed "**a**" element serves as a *packaging signal*. When one unit of the viral genome, from "a" to "a," is enclosed into the capsid, the DNA is cleaved by the cleavage and packaging proteins. In other words, the DNA genome packaging is completed by the cleavage of the concatemer. The viral factors involved in the cleavage and packaging remains to be identified.

Cell lysis is the way to release the assembled capsid for most DNA viruses that replicate in the nucleus. The DNA viruses we have seen so far are all nonenveloped viruses, including polyomavirus, papillomavirus, and adenovirus. In contrast, being an enveloped virus, herpesviruses need to acquire the envelope. In particular, herpesviruses acquire their envelope from the nuclear membrane in the first place (Fig. 9.11). In fact, herpesviruses bud into the nuclear envelope (*primary envelopment*) before entering the cytoplasm; as a matter of fact, herpesviruses are the only known viruses that bud into the nuclear envelope. The evidence for this exit is that enveloped virions residing inside perinuclear space are observed by electron microscopy. Once the virion resides inside the perinuclear space, the membrane fusion of the virion with the outer nuclear membrane, termed "*de-envelopment*," releases the naked capsid into the cytosol. After all, the viral capsids transit the two nuclear membranes; this peculiar nuclear exit process is dubbed "*nuclear egress*." Once located in the cytoplasm, the viral capsid acquires the envelope again from

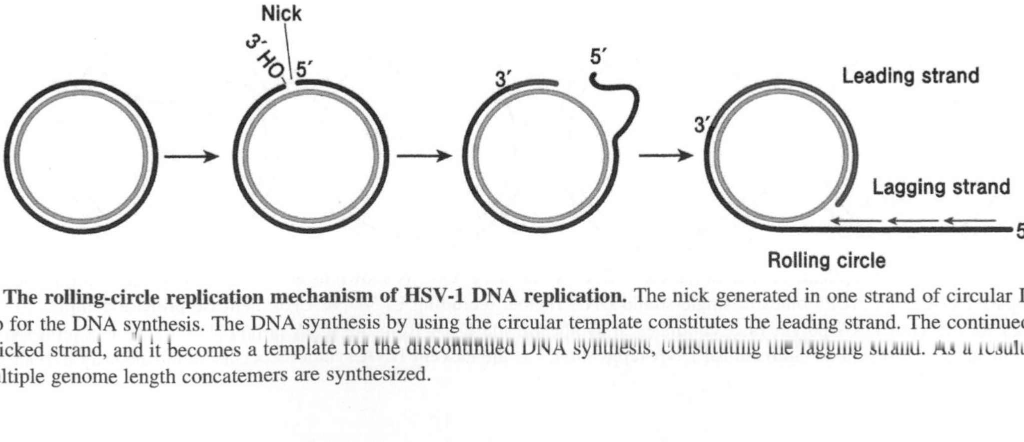

FIGURE 9.9 **The rolling-circle replication mechanism of HSV-1 DNA replication.** The nick generated in one strand of circular DNA provides 3′ hydroxyl group for the DNA synthesis. The DNA synthesis by using the circular template constitutes the leading strand. The continued DNA synthesis displaces the nicked strand, and it becomes a template for the discontinued DNA synthesis, constituting the lagging strand. As a result of rolling circle replication, multiple genome length concatemers are synthesized.

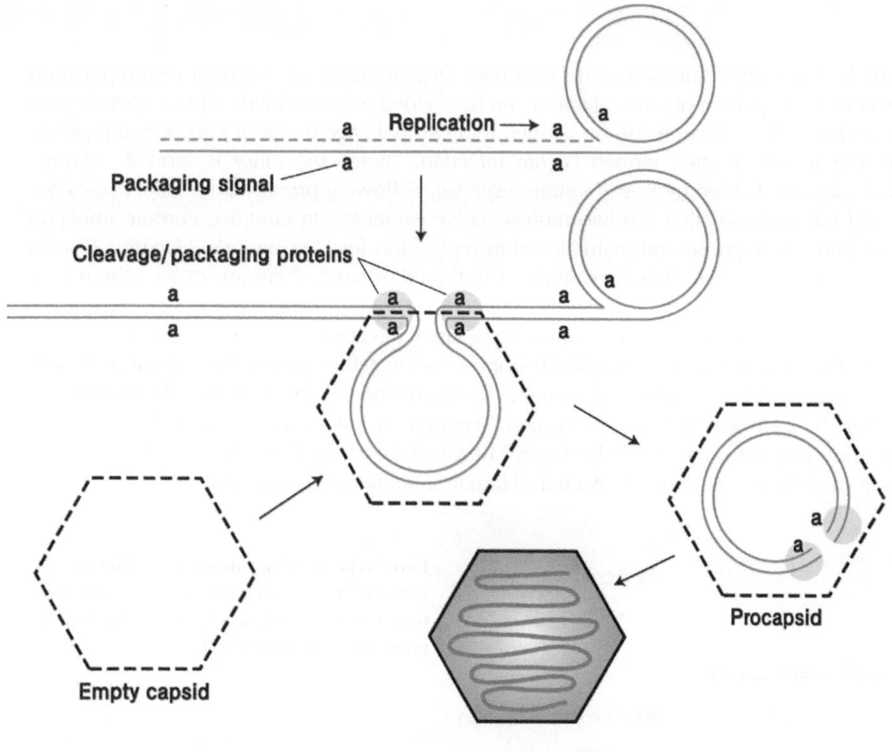

FIGURE 9.10 The steps involved in the viral DNA genome packaging. The viral genome replication is coupled with the genome packaging. The genome packaging is initiated by the recognition of "a" elements by the preassembled capsid. The cleavage of the concatemer by cleavage/packaging proteins is coupled with the DNA genome packaging. The sequence element "a" is denoted.

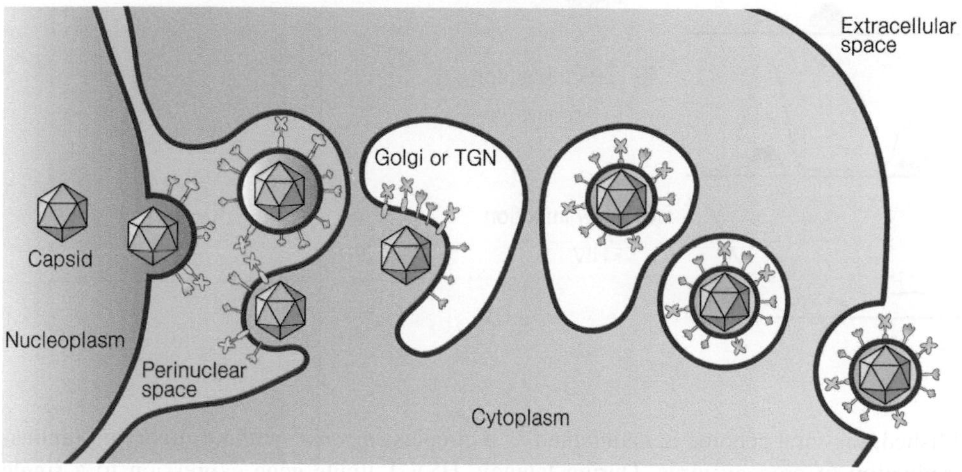

FIGURE 9.11 Nuclear egress and budding of HSV-1 particles. Nuclear egress of HSV-1 particle occurs through two sequential envelopment processes. The viral capsid assembled in the nucleus acquires its envelope via budding into perinuclear space (primary envelopment). Such enveloped particles lose its envelope via fusion with outer nuclear membrane. The capsid entered in cytoplasm acquires its envelope again via budding through Golgi or tans-Golgi network (TGN) (secondary envelopment) and then released.

the Golgi or *trans*-Golgi network (TGN) (*secondary envelopment*). The enveloped virion is then transported in a vesicle to the plasma membrane for extracellular release. Importantly, some tegument proteins (as well as viral envelope glycoproteins) are obtained during secondary envelopment.

9.4 LATENT INFECTION

One salient feature of herpesvirus infection is "*latent infection*" (Box 9.1). In fact, primary infection of HSV-1 occurs in the *mucous layer* of the epithelium, while latent infection occurs mainly in the sensory neuron (Fig. 9.12). After productive primary infection, HSV-1 enters the sensory neuron axon and migrates along the axon to the cell body in a sensory ganglion of the periphery nervous system. In the sensory neuron cell body, the HSV-1 DNA is stably maintained as an episome in the nucleus. Thus, HSV-1 establishes a latent infection in the neuronal system deficient in immune cells. This residence in an immunologically privileged region is a way of *immune evasion* (see Table 5.4).

BOX 9.1 Patterns of Virus Infection

Outcomes of virus infection mainly rely on host immune response to virus infection. Depending on the extent of viral replication and pathogenesis during the course of infection, the pattern of viral infection can be divided into four kinds. First, depending on the viral persistence following primary infection, virus infection can be divided into two patterns: *transient infection* and *persistent infection*. In clinical circumstance, the former is often termed *"acute infection,"* while the latter is termed *"chronic infection."* Acute infection refers as a viral infection limited by host immune response. Following primary infection, viruses are cleared from the body. Influenza virus and rotavirus infection are two representative examples. In contrast, chronic infection refers to a viral infection that evades host immune response and maintains viral replication for long periods. Hepatitis B virus (HBV) and hepatitis C virus (HCV) infection are two representative examples. On the other hand, there are cases, where viral replication discontinuously occurs without clearance after primary infection. Depending on viral persistency, this discontinuous infection can be further divided into *latent infection* or *slow infection*. In the case of latent infection, viral replication occurs transiently during primary infection. Soon after, the virus stops replicating and becomes undetectable even in the circulating bloodstream, a period dubbed *"viral latency."* During latency, the viral genome is maintained without being eliminated and occasionally, the virus resumes its replication by a process termed *"reactivation."* Herpesvirus infection, such as HSV-1, is a representative of latent infection. The pattern of slow infection is similar to that of latent infection, but differs in that infection pathology becomes evident following long periods (ie, decades) of infection. Human immunodeficiency virus (HIV) infection is a representative of slow infection.

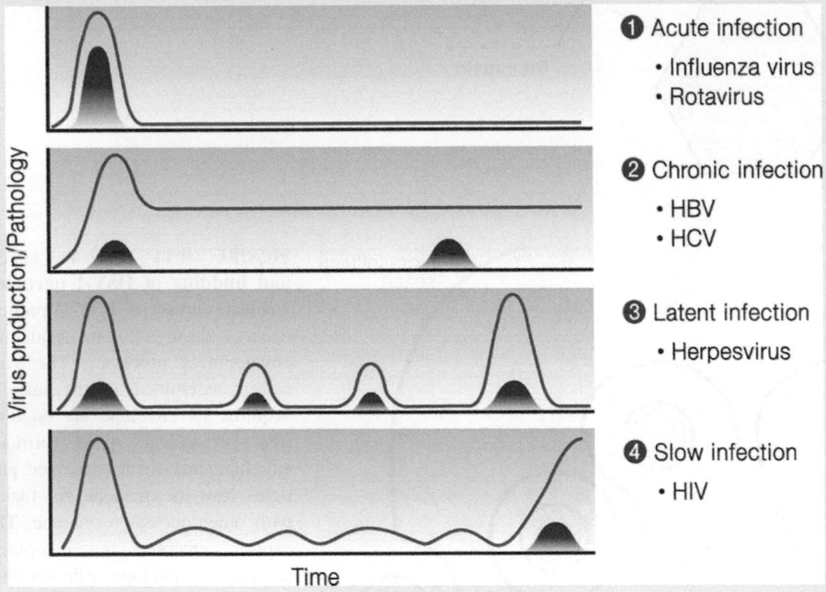

Four type of virus infection patterns. The extent of viral production (line) and the infection pathology (red) are drawn along with the progress of virus infection.

Once latent infection is established, the viral genome is maintained as a circular *episome*[9] without involving replication of the viral genome and production of progeny virus. During latency, HSV-1 limits gene expression to a single locus, the *latency-associated transcript (LAT)*. The role of LAT has been obscure, because of the lack of protein-coding. Four kinds of viral transcripts are detected that are all transcribed from the "**b**" element of the HSV-1 genome (Fig. 9.13). Evidence that LAT expresses any viral proteins is lacking. Intriguingly, LAT are processed to be *microRNA*.[10] Importantly, these miRNAs are shown to be critical for the maintenance of latency. In fact, two LAT-derived miRNAs suppress translation of ICP0 and ICP4 protein (see Fig. 9.13). Note that ICP0 and ICP4 are essential for viral early gene transcription (see Table 9.3). A long-standing puzzle about LAT has been resolved.

On the other hand, the virus in latency gets reactivated occasionally, and reenters the lytic phases of its life cycle, a process termed "*reactivation*." Although it is not clear yet as to what causes the reactivation, stimuli such as hormone, ultraviolet radiation, and stress are the suspects. Upon reactivation, the progeny virus trafficks via anterograde transport to peripheral tissues and induces lytic infection (see Fig. 9.12).

9. **Episome** A DNA that is stably present in the cell, excluding chromosomal DNA. The term "epi" is derived from Greek word for "above."
10. **MicroRNA** A kind of short RNA (ie, 20~22 nt in length) found in eukaryotic cells, that regulates mRNA translation and mRNA stability.

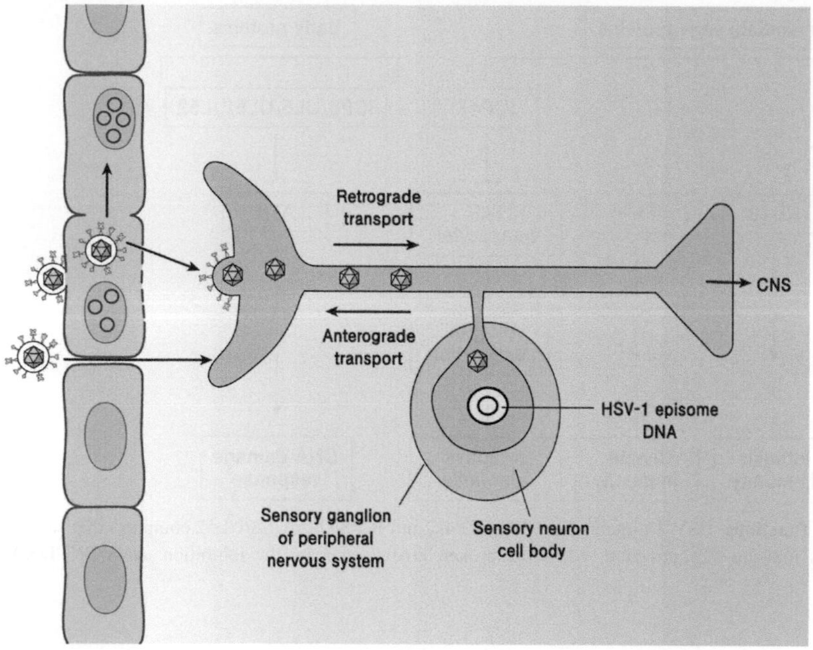

FIGURE 9.12 Viral spread to ganglion cells. Primary infection of HSV-1 occurs in the epithelial cell, where lytic infection is established. The viral particles released invade neuronal tissue by infecting nerve terminals in close contact. These axon terminals can derive from sensory neuron in dorsal root ganglia. The nucleocapsid is transported within axon to the neural cell body by microtubule-mediated transport, a process called "retrograde transport." Then, the viral DNA is delivered to the nucleus, where the viral episome is stably maintained, a hallmark of HSV-1 latency. Upon reactivation, the viral genome replication occurs and the resulting progeny capsid traffics via anterograde transport and gets to epithelial cells for lytic infection. *CNS*, central nervous system.

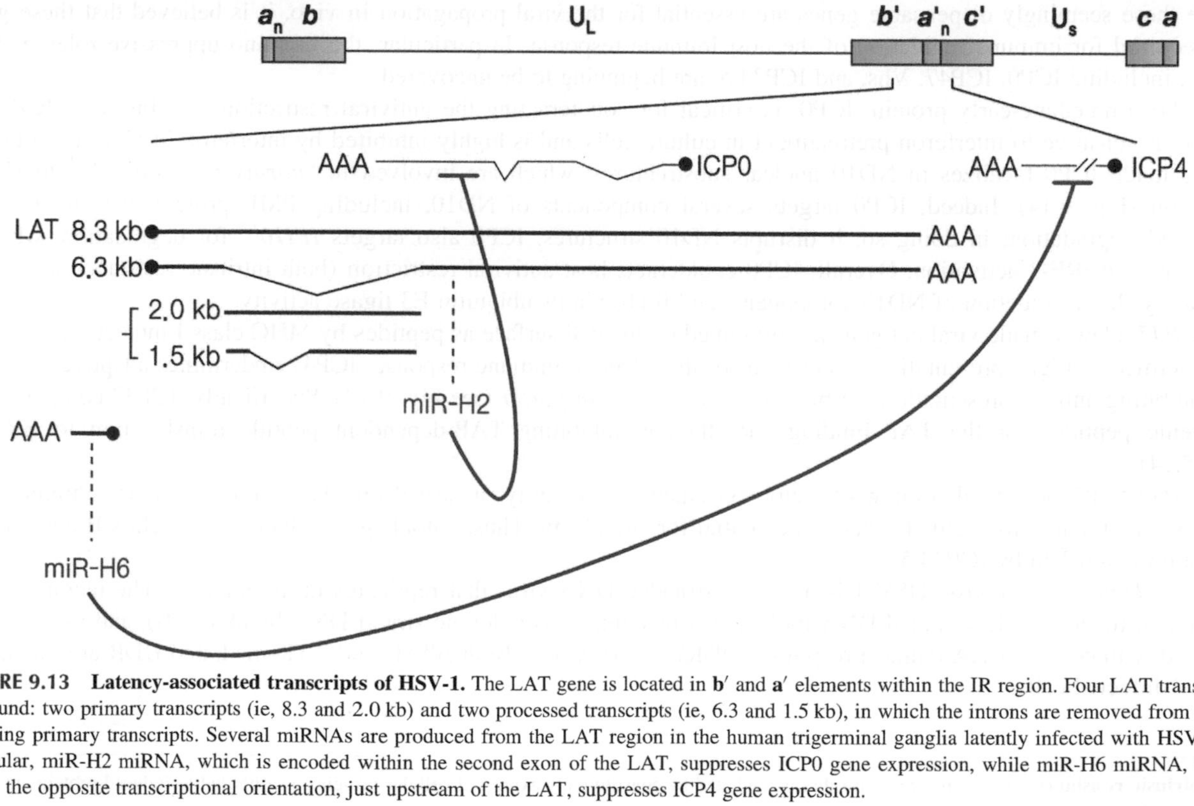

FIGURE 9.13 Latency-associated transcripts of HSV-1. The LAT gene is located in **b′** and **a′** elements within the IR region. Four LAT transcripts are found: two primary transcripts (ie, 8.3 and 2.0 kb) and two processed transcripts (ie, 6.3 and 1.5 kb), in which the introns are removed from corresponding primary transcripts. Several miRNAs are produced from the LAT region in the human trigerminal ganglia latently infected with HSV-1. In particular, miR-H2 miRNA, which is encoded within the second exon of the LAT, suppresses ICP0 gene expression, while miR-H6 miRNA, which lies in the opposite transcriptional orientation, just upstream of the LAT, suppresses ICP4 gene expression.

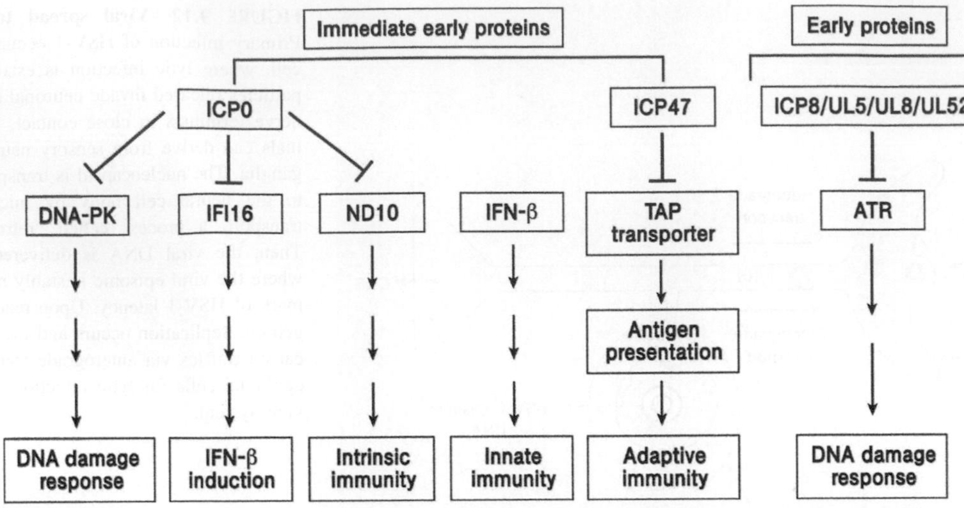

FIGURE 9.14 **The effects of HSV-1 proteins on host functions.** HSV-1 proteins (ie, ICP0, ICP47, and ICP8/UL5/UL8/UL52 complex) that subvert host functions are illustrated (see text). Arrowed lines indicate the activation, while the broken arrows indicate the inhibition. *DNA-PK*, DNA-dependent protein kinase.

9.5 EFFECTS ON HOST

HSV-1 infection significantly affects host cells in that it blocks DNA damage response signaling as well as host immunity.

Immunosuppression. HSV-1 can infect culture cells lytically. Nonetheless, only 37 genes out of 84 genes identified in the HSV-1 genome are essential for viral growth in cell culture, whereas the remaining 47 genes are dispensable. Since these seemingly dispensable genes are essential for the viral propagation in vivo, it is believed that these genes are required for immune regulation of the host immune response. In particular, the immunosuppressive roles of these genes, including ICP0, ICP47, Vhs, and ICP34.5, are beginning to be uncovered.

ICP0. Immediate-early protein, ICP0, is critical for counteracting the antiviral restriction. For instance, ICP0-null mutant is sensitive to interferon pretreatment in culture cells and is highly inhibited by interferon in vivo. Intriguingly, newly made ICP0 localizes in ND10 nuclear substructures, which are involved in "*intrinsic resistance*"[11] to HSV-1 infection (Fig. 9.14). Indeed, ICP0 targets several components of ND10, including PML protein, via an ubiquitin-mediated degradation; in doing so, it disrupts ND10 structures. ICP0 also targets *IFI16*[12] for degradation, which is implicated in IRF-3 activation. Overall, ICP0 counteracts host antiviral restriction (both intrinsic resistance and innate immunity) by degradation of ND10 components and IFI16 via its ubiquitin E3 ligase activity.

ICP47. Newly made viral antigens are presented on the cell surface as peptides by MHC class I molecules. This process, termed "antigen presentation," is critical for the adaptive immune response. ICP47 undermines adaptive immunity by inhibiting antigen presentation via binding to the *TAP transporter* (see Fig. 9.14). Specifically, ICP47 competes with antigenic peptides for the TAP binding site, thereby inhibiting TAP-dependent peptide translocation to ER (see Fig. 5.14).

ICP34.5. ICP34.5 is also known to suppress adaptive immunity. It turned out that ICP34.5 protein inhibits autophagy via its binding to beclin-1, which is essential for autophagy. Thus, autophagy-mediated MHC class II antigen presentation is blocked by ICP34.5.

DNA Damage Response. HSV-1 is a double-stranded DNA virus that replicates in the nucleus. The incoming viral DNA as a linear double-stranded DNA molecule is recognized as a double-strand DNA break (DSB), and as such must contend with cellular DNA damage response (DDR) (see Box 6.2). Both ATM- and ATR-mediated DDR are blocked in HSV-1 infected cells.

11. **Intrinsic resistance** An antiviral response that is mediated by constitutively expressed cellular proteins, as opposed to induced proteins in innate immunity. It is also called intrinsic immunity.

12. **IFI16 (interferon gamma-inducible protein 16)** It is a type of pattern recognition receptors (PRRs) that recognizes the DNA genome of DNA viruses. It is also known as a "nuclear DNA sensor."

ICP0. First, ICP0 blocks *ATM-mediated DDR* by degradation of DNA-PK (protein kinase), which is activated in response to double-strand breaks (see Fig. 9.14).

ICP8. Secondly, the DNA damage response kinase, ATR, is specifically inactivated in HSV-1 infected cells. ATR and other associated factors, including ATRIP, RPA70, and Claspin, are recruited to the viral replication compartments in the nucleus. It appears that the viral DNA replication intermediates are recognized by ATR without concurrent activation. Recently, it was shown that ICP8 binds to ATR complex along with the helicase/primase complex (UL5/UL8/UL52), thereby inactivating ATR kinase. Overall, HSV-1 co-opts cellular DDR in that it utilizes some of ATR-mediated DDR machinery for efficient viral DNA replication without invoking the activation of ATR-mediated DDR.

9.6 CLINICAL SIGNIFICANCE OF HSV-1 INFECTION

Most adults are latently infected with multiple, at least two or three, herpesviruses (ie, HSV-1, VZV, and EBV), and are seropositive to these viral antigens. Herpesvirus infections do not cause serious diseases to healthy individuals. Nonetheless, the latently infected herpesviruses often cause serious sequels under certain circumstances such as in immunocompromised individuals. For instance, HSV-1 causes cold sores or watery blisters in the skin or mucous membranes in mouth and lips following reactivation. Treatment usually involves nucleoside analog antiviral drugs such as *acyclovir* that inhibits viral replication (see Fig. 9.1). Acyclovir is a *prodrug*,[13] which is converted into acyclovir triphosphate that acts as an inhibitor of the viral DNA polymerase (see Fig. 26.3). A preventive vaccine for HSV-1 is not available.

9.7 OTHER ALPHA-HERPESVIRUSES

Besides HSV-1, HSV-2 and VZV are also alpha-herpesviruses (see Table 9.2). HSV-2 is related to HSV-1, exhibiting over 50% genetic similarity. HSV-2 causes disorders in the reproductive organs, such as genital herpes. Genital herpes, known simply as *herpes*, is the second most common form of herpes. Acyclovir is used for the treatment, but a vaccine is not available.

On the other hand, VZV causes chicken pox to youngsters and herpes zoster to adults, which is often referred to as shingles (Fig. 9.15). Being transmitted via an airborne route, VZV is extremely contagious, and most people are infected before or during youth. Primary VZV infection to youngsters results in chicken pox (varicella), which may result in complications including encephalitis or pneumonia. After clinical symptoms of chicken pox are resolved, VZV still remains

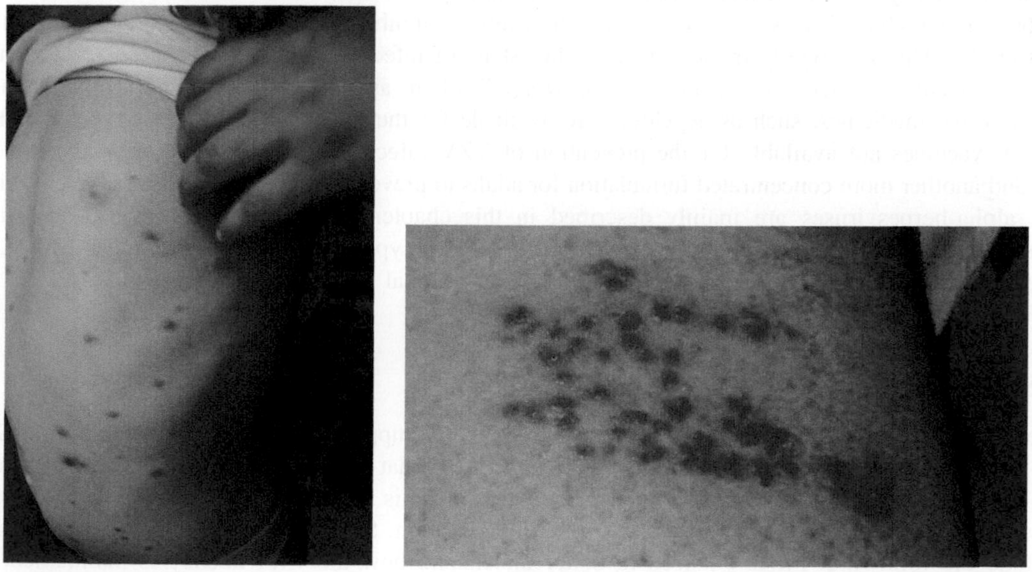

FIGURE 9.15 Photos of patients with varicellar and zoster. (Left) A 3-year-old girl with a chicken pox rash on her torso. (Right) Shingles blisters showing characteristic purple color.

13. **Prodrug** A prodrug is a medication that is administered in an inactive form, and then it becomes converted to its active form through a normal metabolic process.

BOX 9.2 Human Cytomegalovirus

HCMV (human cytomegalovirus) is a prototype of human beta-herpesviruses. Worldwide, over 80% of adults have been infected by HCMV, as indicated by the presence of antibodies. Although they may be found throughout the body, HCMV infections are frequently associated with the salivary glands. HCMV infection is typically unnoticed in healthy people, but can be life-threatening for the immunocompromised, such as HIV-infected persons and organ transplant recipients. In other words, HCMV is an *opportunistic pathogen*.[14] After infection, HCMV has an ability to remain latent within the body for long periods. HCMV is also the most frequently transmitted virus to a developing fetus. HCMV infection is more widespread in developing countries and is the most significant viral cause of birth defects in industrialized countries. Congenital HCMV is the leading infectious cause of deafness, learning disabilities, and mental retardation in children. Nucleoside analog drugs such as ganciclovir (see Fig. 26.1) are used for patients with life-threatening illnesses.

BOX 9.3 Epstein-Barr Virus

EBV (Epstein-Barr virus) is a prototype of human gamma-herpesviruses. Worldwide, over 90% of adults have already been infected by EBV. Infection with EBV occurs by oral transfer of saliva and genital secretions. EBV infects B cells of the immune system and epithelial cells. Once the virus's initial lytic infection is brought under control, EBV latently persists in the individual's B cells for the rest of the individual's life. EBV is best known as the cause of *infectious mononucleosis* (also known as "*mono*,") and sometimes being referred as the *kissing disease*. Most people are exposed to the virus as children, when the disease produces no noticeable or only flu-like symptoms. When infection with EBV occurs during adolescence, it causes infectious mononucleosis with frequency of 35–50%. The disease is manifested by fever, sore throat, and fatigue, along with several other possible signs and symptoms. It is generally a self-limiting disease, and little treatment is normally required. On the other hand, EBV is also associated with particular forms of cancer, such as *Burkitt's lymphoma*, and *nasopharyngeal carcinoma*. Importantly, Burkitt's lymphoma, the cancer of B lymphocytes, is geographically confined to malaria endemic regions in Africa, whereas nasopharyngeal carcinoma is also confined to Southern China. Hence, it is believed that EBV infection is a cofactor for carcinogenesis.

dormant in the nervous system of the infected person (virus latency), including the cranial nerve ganglia, dorsal root ganglia, and autonomic ganglia. VZV is often reactivated and causes a number of neurologic conditions, collectively termed "*Zoster*" or shingles. Unless properly treated at the earlier stage of infection, the consequence of zoster can be severe, resulting in complications such as "*postherpetic neuralgia*" often associated with dreadful pains and disability. Fortunately, effective medicines, such as acyclovir, are available for the treatment of herpes zoster. Two kinds of live attenuated VZV vaccines are available for the prevention of VZV infections: one for children to prevent chicken pox (ie, Varivax) and another more concentrated formulation for adults to prevent shingles (ie, Zostavax) (see Table 25.1)

Although alpha-herpesviruses are mainly described in this chapter for the sake of brevity, beta- and gamma-herpesviruses are equally important clinically. Here, HCMV, a prototype of beta-herpesviruses, and EBV, a prototype of gamma-herpesviruses, are briefly described with emphasis on clinical features (Boxes 9.2 and 9.3).

9.8 PERSPECTIVES

Herpesviruses stand out among human viruses in that they potently suppress the host immune response. Since herpesviruses have coexisted in human body for long time, it is believed that herpesviruses acquired multiple functions that are essential for maintaining latent infection during evolution. In this regard, herpesvirus has been an experimental model to investigate host immunity, in particular, antigen presentation. Some of the current issues relating herpesvirus research include the following. First, we need to better understand how the viral latency is maintained. Second, we need a better understanding of the biological significance of viral miRNAs. In addition to protein-coding genes, in fact, herpesviruses genome encodes quite a number of miRNAs, and intriguingly, some viral miRNAs derived from LAT control the reactivation of viral latency. Apparently, the viral miRNAs are critically important for the maintenance of

14. **Opportunistic pathogens** pathogens that usually do not cause disease in a healthy host, one with a healthy immune system, but do cause disease in compromised immune system.

latency. Any measures that control the LAT expression could be an effective strategy to block reactivation. Third, opportunistic pathogens, such as HSV-1, VZV, HCMV, and EBV, could become a serious threat to elderly persons as the aging population is increasing. Antiviral drugs that can prevent or control the reactivation of latently infected herpesviruses represent an unmet medical need. Finally, Kaposi's sarcoma-associated herpesvirus (KSHV) has been a focus of the last two decades, since KSHV is associated with Kaposi's sarcoma in AIDS patients. We have just begun to uncover some of the viral oncogenes involved in the viral carcinogenesis. Taken together, our better understanding of herpesvirus biology will help to prevent the sequels of the creeping viruses.

9.9 SUMMARY

- *Classification*: human herpesviruses are classified into three genera: alpha-, beta-, and gamma-herpesviruses. HSV-1 is a prototype of human alpha-herpesviruses.
- *Virion structure*: HSV-1 is an enveloped virus, in which the capsid is enclosed. Some tegument proteins, such as VP16, ICP0, and Vhs, are essential for the viral infection.
- *Genome*: HSV-1 genome represents a linear double-strand DNA, 150 kb in length, and it has a direct repeat element at both ends, termed terminal repeat or TR, that is essential for the viral genome replication.
- *Genome replication*: HSV-1 genome replication occurs largely via a rolling-circle mechanism and the DNA synthesis is driven by the virally encoded DNA polymerase.
- *Latency*: HSV-1 replicates primarily in epithelial cells but establishes latency in neuronal cells. The viral genome is maintained as a circular episome in latently infected neuronal cells. During latent infection, a latency-associated transcript or LAT is expressed.
- *Beta-herpesviruses*: HCMV, a prototype of human beta-herpesviruses, causes congenital disease and produces serious complications in immunocompromised individuals.
- *Gamma-herpesviruses*: EBV, a prototype of human gamma-herpesviruses, is a human tumor virus in that it is associated with Burkitt's lymphoma, and nasopharyngeal carcinoma.

STUDY QUESTIONS

9.1 An immediate early protein, ICP0, is important for the regulation of lytic and latent viral infection. ICP0-null mutants could not lead to lytic infection in human diploid fibroblasts (primary cells), while it could lead to lytic infection in certain cell lines such as HeLa cells and U2OS cells. Explain why?

9.2 Compare the similarity and difference between two modes of HSV-1 DNA replication.

9.3 LAT was shown to be processed to miRNAs that regulate reactivation. What are the advantages of using miRNA, as opposed to mRNA (protein-coding genes) in a viral perspective?

SUGGESTED READING

Boutell, C., Everett, R.D., 2013. Regulation of alphaherpesvirus infections by the ICP0 family of proteins. J. Gen. Virol. 94 (Pt. 3), 465–481.

Johnson, D.C., Baines, J.D., 2011. Herpesviruses remodel host membranes for virus egress. Nat. Rev. Microbiol. 9 (5), 382–394.

Kerur, N., Veettil, M.V., Sharma-Walia, N., Bottero, V., Sadagopan, S., Otageri, P., et al., 2011. IFI16 acts as a nuclear pathogen sensor to induce the inflammasome in response to Kaposi Sarcoma-associated herpesvirus infection. Cell Host Microbe. 9 (5), 363–375.

Paludan, S.R., Bowie, A.G., Horan, K.A., Fitzgerald, K.A., 2011. Recognition of herpesviruses by the innate immune system. Nat. Rev. Immunol. 11 (2), 143–154.

Pan, D., Flores, O., Umbach, J.L., Pesola, J.M., Bentley, P., Rosato, P.C., et al., 2014. A neuron-specific host microRNA targets herpes simplex virus-1 ICP0 expression and promotes latency. Cell Host Microbe. 15 (4), 446–456.

JOURNAL CLUB

- Mohni, K.N., Smith, S., Dee, A.R., Schumacher, A.J., Weller, S.K., 2013. Herpes simplex virus type 1 single strand DNA binding protein and helicase/primase complex disable cellular ATR signaling. *PLoS Pathog*. 9(10), e1003652.

  Highlight: HSV-1 has evolved to disable the cellular DNA damage response kinase, ATR. This article demonstrated that the HSV-1 single strand binding protein (ICP8) and the helicase/primase complex (UL8/UL5/UL52) form a nuclear complex in transfected cells that is necessary and sufficient to disable ATR signaling. The data suggested that these four viral proteins prevent ATR activation by binding to the DNA substrate and obscuring the loading of cellular factors. This is the first example of viral DNA replication proteins obscuring access to DNA substrate that would normally trigger DNA damage response.

Chapter 10

Other DNA Viruses

Chapter Outline

In Chapters 6–9, the four major human DNA viruses were described. Here, two additional DNA viruses will be briefly described. They are parvovirus and poxvirus. Parvovirus has the smallest DNA genome (~5 kb), while poxvirus has the largest DNA genome (~200 kb) among animal DNA viruses. Human parvoviruses cause either no symptom or only mild symptoms. Adeno-associated virus (AAV) is a prototype of human parvovirus. Being a satellite virus, AAV replication depends on a helper virus, adenovirus. By contrast, infection by smallpox virus, a human poxvirus, is dreadful and often fatal. Smallpox virus is the only human virus that has been eradicated on Earth. Vaccinia virus is a prototype of poxvirus.

10.1 PARVOVIRUS

Parvoviruses (family Parvoviridae) are small (28 nm in diameter), naked icosahedral viruses that possess a single-strand DNA of 5 kb.

Parvovirus is a unique animal virus that has a single-stranded DNA as a genome. *Adeno-associated virus (AAV)* is a prototype of parvovirus. As its name implies, AAV was first discovered in a sample isolated from an adenovirus infected person. Moreover, AAV replication depends on the presence of adenovirus. In other words, *AAV is a satellite virus*[1] that requires adenovirus as a helper for propagation.

Taxonomy: In fact, parvoviruses are found in most animals, not only in vertebrates but also in invertebrates such as mosquitoes. Parvoviruses can be divided into two groups: one that depends on the helper virus, and another that does not depend on the helper virus. The former is dubbed "*defective parvovirus*," while the latter is dubbed "*autonomous parvovirus*" (Table 10.1). AAV, a prototype of parvovirus, is classified as defective parvovirus, because it depends on the helper virus. AAV does not cause any diseases to human. On the other hand, *B19 virus*, another human parvovirus that infects erythrocytes, causes *erythema infectiosum* in children, a mild disease with rash. Human *bocavirus*,[2] the third human parvovirus recently discovered, is responsible for respiratory infection during the summer. In contrast to human parvoviruses, animal parvoviruses cause more severe diseases (see Table 10.1). For instance, canine parvovirus causes gastroenteritis, a leading cause of death of pet dogs.

Here, AAV, a prototype of human parvovirus, will be described.

Virion Structure: AAV virion is nonenveloped, and it has a small icosahedral capsid 28 nm in diameter (Fig. 10.1).

Genome structure: AAV has a small single-strand DNA genome, 5 kb in length. Notably, both termini have a DNA sequence that is self-complementary, forming a hairpin structure (Fig. 10.2). It is a *palindrome* sequence, 115 nt in length, and is termed *ITR* (inverted terminal repeat). Importantly, ITR serves as a primer for viral DNA replication. The viral genome replication yields both positive- or negative-strand single-strand DNA. AAV capsid particles contain

1. **Satellite virus** It refers to a virus, which depends on the host virus as a helper.
2. **Bocavirus** It is a parvovirus first isolated both in cattle (*bo*vine) and dog (*ca*nine), which is named after the two initials from two animal species.

Molecular Virology of Human Pathogenic Viruses. DOI: http://dx.doi.org/10.1016/B978-0-12-800838-6.00010-2

TABLE 10.1 Classification of Parvoviruses

Genus	Virus Species	Host	Helper Dependence	Disease
Dependovirus	AAV	Human	Defective	–
Densovirus	Culex densovirus	Mosquito	Autonomous	–
Erythrovirus	B19 virus	Human	Autonomous	Erythema infectiosum
Parvovirus	MVM (minute virus of mice)	Mouse	Autonomous	–
	Canine parvovirus	Canine	Autonomous	Gastroenteritis
Bocavirus	Human bocavirus	Human	Autonomous	Respiratory infection

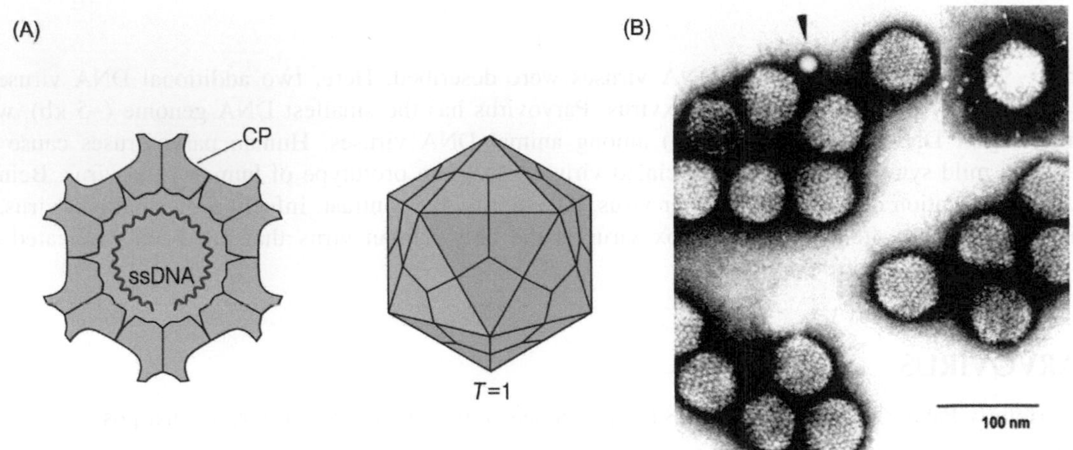

FIGURE 10.1 AAV capsids. (A) A diagram illustrating AAV capsid ($T = 1$ symmetry). Single-strand DNA genome packaged inside is drawn. (B) Electron microscopic image of AAV particles. AAV particle (*arrow head*) is seen in the midst of adenovirus particles, a helper virus.

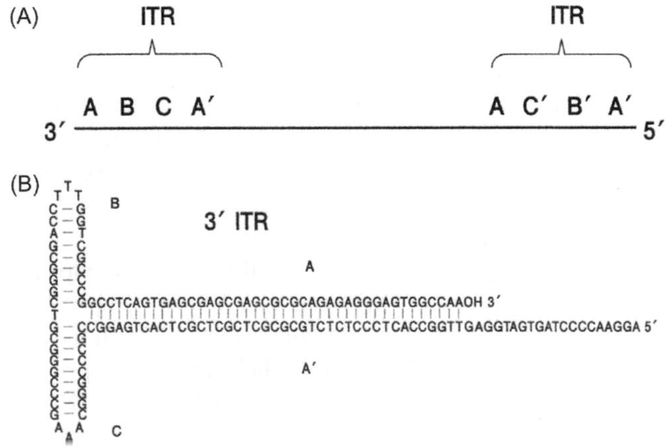

FIGURE 10.2 Genome structure of AAV. (A) Single-strand DNA genome of AAV. Single-strand DNA genome of AAV with a negative polarity is shown, where ITR elements located at both termini is highlighted. Note that the 3′ terminus is positioned to the left, as the single-strand DNA shown represents the negative polarity. (B) The hairpin structure of the ITR element. The nucleotide sequences encoded of the ITR element are drawn. Note that "A" stands for the sequence that is complementary to the sequence in "A′."

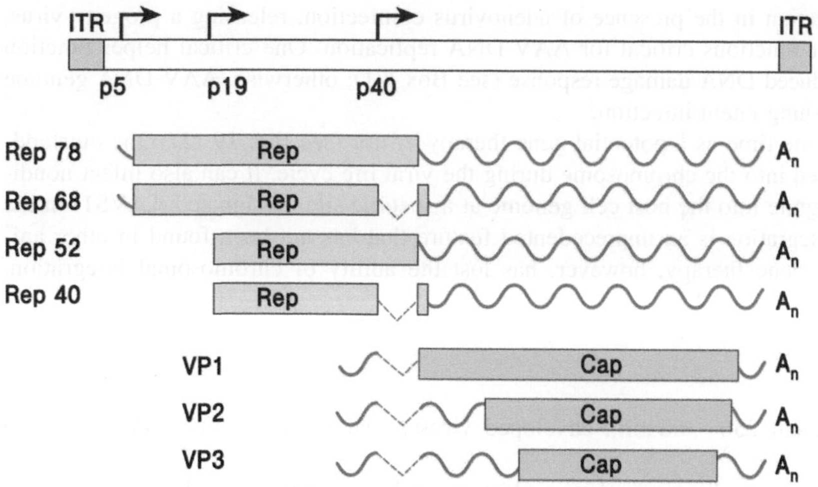

FIGURE 10.3 Protein-coding of AAV. The double-strand DNA genome is shown on top, with ITR elements at both termini. Three viral promoters (ie, p5, p19, and p40) are denoted with *arrows* showing the direction of transcription. The viral proteins derived from Rep and Cap ORFs are drawn on the viral transcripts (boxes).

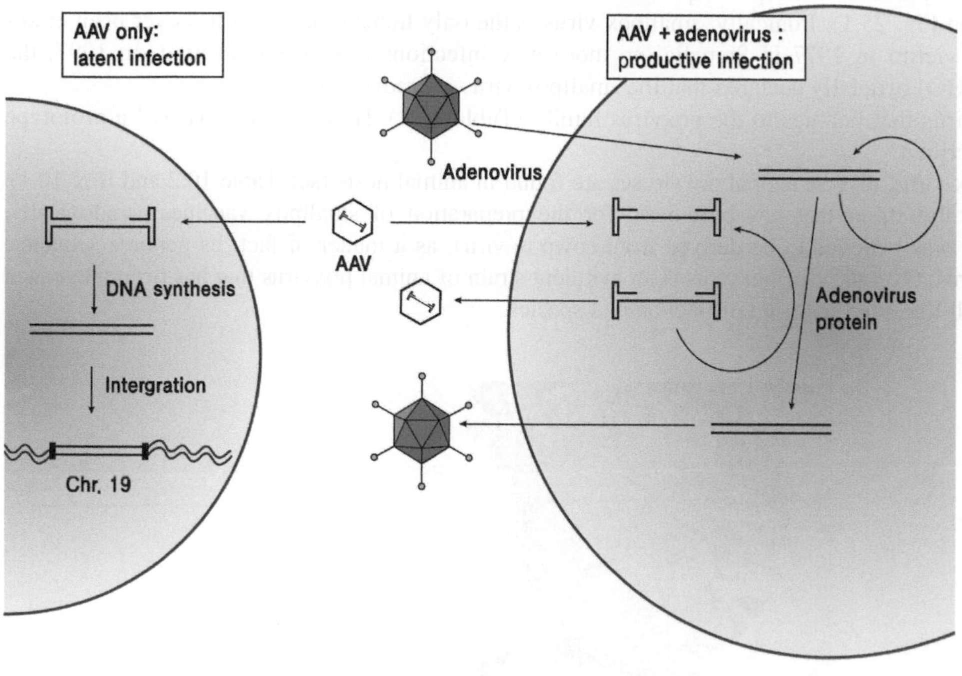

FIGURE 10.4 AAV life cycle. (Left) In the absence of adenovirus coinfection, AAV infection leads to a latent infection. AAV genome becomes integrated into the chromosome at a specific site (ie, chromosome 19, AAVS1 locus). The site-specific integration depends on AAV Rep78/68 protein. Neither the viral genome replication nor progeny virus production occurs. (Right) Adenovirus coinfection leads to a productive infection. AAV as well as adenovirus is produced.

either positive- or negative-strand DNA. As the single-strand DNA is converted to double-strand DNA during the viral DNA replication, the *polarity*[3] of single-strand DNA in the particle does not matter.

Protein-Coding: Parvoviruses encode essentially two ORFs: *Rep* ORF and Cap ORF (Fig. 10.3). Rep ORF expresses four kinds of Rep proteins involved in the viral genome replication. Two RNAs having distinct 5′ ends are transcribed from two distinct promoters (ie, *p5* and *p19*). These transcripts are alternatively spliced to express four related Rep proteins: Rep78, Rep68, Rep 52, and Rep 40. On the other hand, transcripts derived from the p40 promoter are alternatively spliced to express three Cap-related proteins: VP1, VP2, and VP3.

AAV Life Cycle: In the absence of adenovirus coinfection, AAV can establish latency by integrating into the chromosome 19 (Fig. 10.4). Following the conversion of the single-strand genome DNA to double-strand DNA, it becomes integrated into the chromosome. The site-specific integration of AAV genome depends on Rep78 and Rep68 proteins.

3. **Polarity** It refers to the strandness of a single-strand DNA or RNA. The strand that corresponds to that of mRNA is defined as "plus" or "positive." The strand that is complementary to mRNA is defined as "minus" or "negative."

In contrast, AAV undergoes productive infection in the presence of adenovirus coinfection, releasing a progeny virus. A question is what are the adenoviral helper functions critical for AAV DNA replication. One critical helper function of adenovirus is the suppression of AAV-induced DNA damage response (see Box 8.1); otherwise, AAV DNA genome gets integrated into the chromosome, establishing latent infection.

Interest: AAV has drawn attention for some time as a potential gene therapy vector (see Fig. 19.12). One outstanding feature is that the AAV genome is inserted into the chromosome during the viral life cycle. It can also infect nondividing cells and has the ability to stably integrate into the host cell genome at a specific site (designated AAVS1) in the human chromosome 19. The site-specific integration is an unprecedented feature that has not been found in other animal viruses. Note that the AAV vector for gene therapy, however, has lost the ability of chromosomal integration, because the *rep* and *cap* genes are deleted (see Fig. 19.12).

10.2 POXVIRUS

Poxviruses (family Poxviridae) are large (about 250–360 nm), enveloped viruses that possess a linear double-strand DNA genome of over 200 kb in length.

Smallpox remains as the most fatal infectious disease in human history and cost millions of lives (Fig. 10.5). Smallpox virus has not only greatly influenced human life, but also impacted on the human endeavor on fighting infectious diseases. For instance, the proof of concept on preventive vaccine was first demonstrated by *Edward Jenner* (1749–1823) using cowpox (see Fig. 25.1). Ironically, smallpox virus is the only human virus that has ever been eradicated on Earth. Since the last victim in 1977 in Somalia, no more new infections have been reported. In 1997, the World Health Organization (WHO) officially declared that the smallpox virus is eradicated.

Smallpox virus is a DNA virus that belongs to the poxvirus family (Table 10.2). Here, *vaccinia virus*,[4] a prototype of poxvirus family, will be described.

Taxonomy: Besides smallpox virus, diverse animal poxviruses are found in animal hosts (see Table 10.2 and Box 10.1). Even though vaccinia virus is the strain that has been used for the preparation of smallpox vaccine, paradoxically, its origin remained uncertain. It was believed to be derived from cowpox virus; as a matter of fact, its genome sequence is more related to buffalopox virus. Overall, vaccinia virus is an avirulent strain of animal poxvirus that has been attenuated by multiple passages through cell lines derived from distinct animal species.

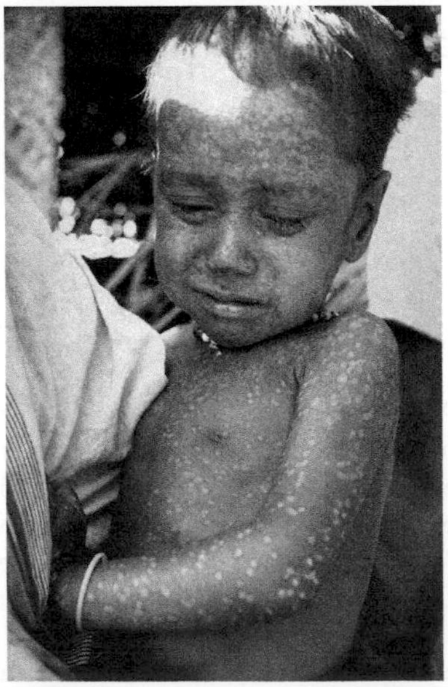

FIGURE 10.5 A picture of smallpox victim. This 1975 photograph depicts a 2-year-old female child by the name of Rahima Banu, who was actually the last known case of naturally occurring smallpox, or variola major in the world. The case occurred in the Bangladesh district of Barisal, in a village named Kuralia, on Bhola Island.

4. **Vaccinia virus** It is a poxvirus strain that has been used to produce smallpox vaccine, as its name was derived from "vaccine."

TABLE 10.2 Classification of Poxviruses

Virus Species	Reservoir (Vector)	Host (Natural)
Variola (smallpox)	–	Human
Vaccinia	Human	Cow, buffalo
Monkeypox	Squirrel	Monkey
Cowpox	Rodents	Cow, cat
Buffalopox	–	Buffalo
Myxoma	Ticks and mosquitoes	Cotton-tail rabbits

BOX 10.1 Rabbit Poxvirus as a Pest Control in Australia

Virus has been used unsuccessfully as a biological pest control. Ironically, this episode has happened in Australia, a country with the reputation of having stringent *quarantine*[5] to prevent the introduction of bugs from other continents. Australia had imported European rabbits for game animals since 1859. The rabbits propagated exponentially in the continent without predators and ate up grasses that were to be fed on by cattle and sheep. Some plant species were endangered and even vanished. To reduce the rabbit density, Australian authorities employed diverse tricks such as fences to contain them. None of the tricks worked. Finally, to control the rabbit numbers they decided to introduce another exotic organism, which was the myxomavirus, a rabbit poxvirus. It sounded like a clever strategy, as an ancient Chinese philosopher said "use your enemy to control your enemy."

Myxomavirus is a poxvirus, whose natural host is a cotton-tail rabbit. It causes myxoma, a benign tumor in the natural host, but it kills European rabbits. Myxomavirus is a vector-borne virus and is transmitted via insects such as ticks and mosquitoes. The impact was rewarding, as the rabbits density was dramatically dampened, as expected. Unexpectedly, the rabbit density recovered soon after. Intriguingly, a rabbit species that was resistant to myxomavirus infection was found to have evolved and became dominant. This anecdote eloquently illustrates that selection pressure derives mutation or evolution.

Rabbits around a waterhole in the myxomatosis trial site on Wardang Island in 1938.

Virion Structure: Poxvirus has an enveloped, brick-shaped or ovoid virion, 220–450 nm long and 140–260 nm wide (Fig. 10.6). In fact, two distinct infectious particles exist: intracellular mature virion (MV) and extracellular virion (EV). The EV is peculiar in that it is enveloped twice. In other words, the EV consists of one form of MV

5. **Quarantine** The word came from an Italian word (17th-century Venetian) "quaranta," meaning forty. It is the number of days for which ships were required to be isolated before passengers and crew could go ashore during the Black Death plague epidemic.

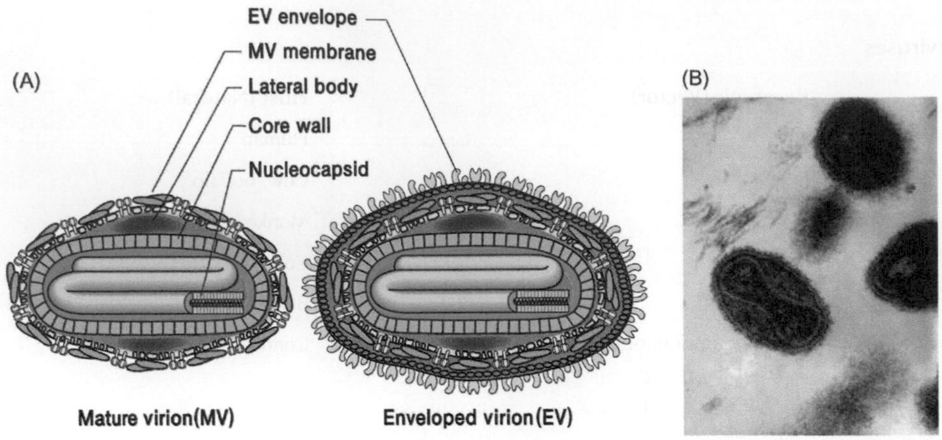

FIGURE 10.6 Poxvirus particles. (A) Enveloped, brick-shaped or ovoid virion, 220–450 nm long and 140–260 nm wide. The surface membrane displays surface tubules or surface filaments. Two distinct infectious virus particles exist: the intracellular mature virus (IMV) and the extracellular enveloped virus (EEV). (B) Transmission electron micrograph (TEM) depicts a number of smallpox virions; Mag—approximately 370,000 ×. The "dumb-bell-shaped" structure inside the smallpox virion is the viral core, which contains the viral DNA.

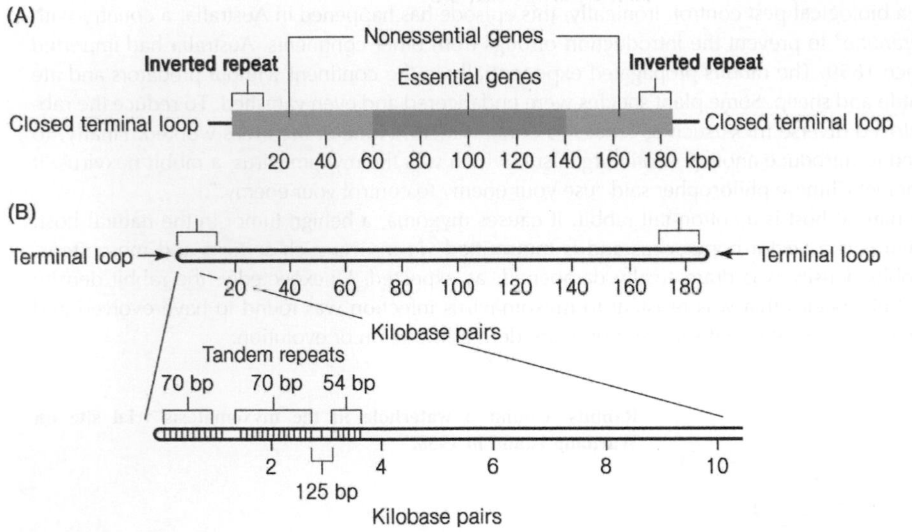

FIGURE 10.7 The genome structure of poxvirus. (A) The linear double-strand DNA genome of poxvirus is schematically drawn with an emphasis on the inverted terminal repeat (ITR). It is long, approximately 200 kb in length, and encodes over 200 proteins. The closed terminal loop is denoted. (B) The structure of inverted terminal repeat. The repeat elements, 50–70 nt in length, at the terminus are shown below.

wrapped within a second lipid membrane. The EVs often comprise <1% of the total progeny virions. While EVs are not essential for virus infection either in vitro or in vivo, viruses lacking the capacity to efficiently produce EVs are usually highly attenuated in vivo. These attributes suggest that EVs have a specialized role distinct from those of infectious MVs in viral replication. It has been suggested that EVs promote their own release from the infected cells, the dissemination of virus in the host, and the suppression of immune response to infection.

Genome Structure: Poxvirus has a larger double-strand DNA genome of 130–375 kb in length (Fig. 10.7). The linear genome is flanked by *inverted terminal repeat (ITR)* sequences which are covalently closed at their extremities. In particular, ITR, which contains 50–70 nt repeats, is believed to act as a telomere. Presumably, the viral telomere is the viral strategy acquired during evolution to resolve "the end-replication problem" that a virus having a linear double-strand DNA genome has to solve during the genome replication (Box 10.2).

10.3 PERSPECTIVES

Human parvoviruses are only modestly pathogenic or nonpathogenic and rarely cause clinical symptoms. AAV, a prototype of human parvovirus, has been extensively studied as a potential gene therapy vector. Despite much effort over the past decades, it has not yet been developed for clinical use, primarily because of its adverse side effects, such as a danger to induce cancer (see chapter: Virus Vectors). In contrast to human parvoviruses that cause only mild symptoms, human poxviruses, such as smallpox virus, cause fatal diseases. Thanks to the effective vaccines, smallpox virus has been officially eradicated on the globe, except a few specimens kept just in case. Whether even the specimens have to

BOX 10.2 End-Replication Problem

DNA replication is a process that precisely duplicates the DNA template. Due to an intrinsic property of the DNA polymerase, it is difficult to copy the very end of the DNA template. This problem is often referred to as the "*end-replication problem.*" The problem is attributable to two features of the DNA replication. First, all DNA polymerases require a primer for the initiation of DNA synthesis, and the primers are often RNA molecules that have to be removed after the completion of DNA synthesis. The removal of the RNA primer will generate a gap region; otherwise filled by another DNA polymerase that proceeds in the opposite direction. Second, the DNA synthesis proceeds only from 5′ to 3′ direction. In other words, no DNA polymerases proceed from 3′ to 5′ direction. Consequently, the gap region cannot be filled. Therefore, a strategy to obviate the end-replication problem is needed for any organism having a DNA genome.

Four disparate strategies are employed by DNA viruses. The first is to have a circular DNA genome so that no end is present in the first place. SV40 and HPV are good examples of circular DNA genomes. The second is to circularize a linear DNA genome prior to DNA replication. This is the case for herpesviruses. The third strategy is to initiate the DNA synthesis form the very end by using a protein primer, instead of an RNA primer. As a result, the protein primer remains linked to the DNA genome, as exemplified in the adenovirus. The fourth strategy is to explore a special end structure (ie, telomere), as demonstrated in poxviruses.

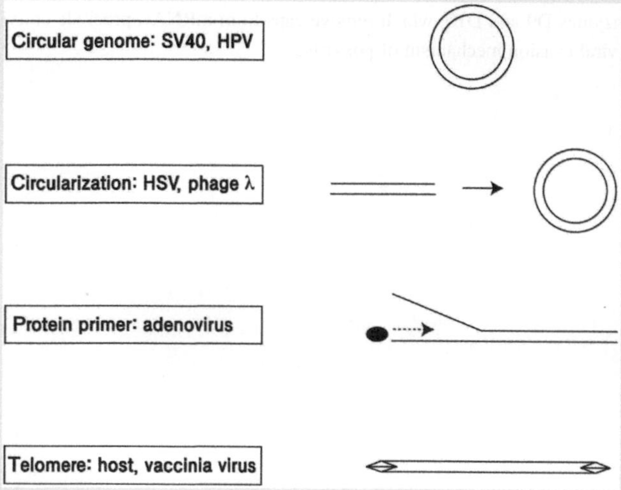

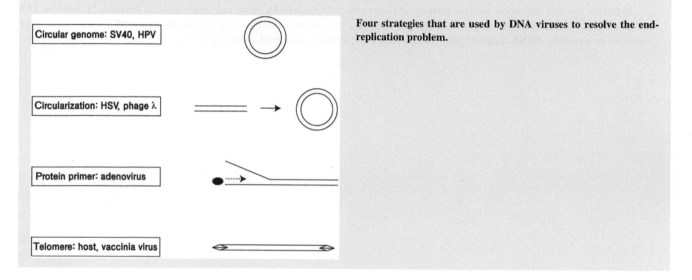

Four strategies that are used by DNA viruses to resolve the end-replication problem.

be destroyed or not is being debated. An argument in support of keeping the specimens is because of the potential use of smallpox virus as a bioterrorism weapon.

10.4 SUMMARY

- *Parvovirus*: Parvoviruses are small, naked icosahedral viruses that possess a single-strand DNA of 5 kb. Human pathogenic parvoviruses including B19 virus and bocavirus cause only mild symptoms. In contrast, canine parvovirus causes severe gastroenteritis, which is often fatal.
- *Adeno-associated virus (AAV)*: AAV is a prototype of parvovirus. AAV has drawn attention for some time as a potential gene therapy vector. AAV replication depends on the presence of adenovirus.
- *Poxvirus*: Smallpox virus is a DNA virus belonging to the poxvirus family. Smallpox remains as the most fatal infectious disease in human history and cost millions of lives. Thanks to global vaccination, the smallpox virus is eradicated on the globe.
- *Vaccinia virus*: Vaccinia virus is the poxvirus strain that was used for smallpox vaccination.

STUDY QUESTIONS

10.1 Depict the genome structure of AAV with respect to the repeat elements. State how these repeat elements contribute to the viral genome replication.

10.2 Depict the genome structure of poxvirus with respect to the repeat elements. State how these repeat elements contribute to the viral genome replication.

SUGGESTED READING

Adeyemi, R.O., Pintel, D.J., 2014. Parvovirus-induced depletion of cyclin B1 prevents mitotic entry of infected cells. PLoS Pathog. 10 (1), e1003891.

Lentz, T.B., Samulski, R.J., 2015. Insight into the mechanism of inhibition of adeno-associated virus by the mre11/rad50/nbs1 complex. J. Virol. 89 (1), 181–194.

Liu, S.W., Katsafanas, G.C., Liu, R., Wyatt, L.S., Moss, B., 2015. Poxvirus decapping enzymes enhance virulence by preventing the accumulation of dsRNA and the induction of innate antiviral responses. Cell. Host. Microbe. 17 (3), 320–331.

Pickup, D.J., 2015. Extracellular virions: the advance guard of poxvirus infections. PLoS Pathog. 11 (7), e1004904.

Schwartz, R.A., Carson, C.T., Schuberth, C., Weitzman, M.D., 2009. Adeno-associated virus replication induces a DNA damage response coordinated by DNA-dependent protein kinase. J. Virol. 83 (12), 6269–6278.

JOURNAL CLUB

- Liu, S.W., Katsafanas, G.C., Liu, R., Wyatt, L.S., Moss, B. 2015. Poxvirus decapping enzymes enhance virulence by preventing the accumulation of dsrna and the induction of innate antiviral responses. Cell. Host. Microbe. 17 (3), 320–331.

  Highlight: Poxvirus replication involves synthesis of double-strand RNA (dsRNA), which can trigger antiviral responses by inducing PKR and stimulating 2′5′OAS. This paper showed that poxvirus decapping enzymes D9 and D10, which remove caps from mRNAs, preclude antiviral responses by preventing dsRNA accumulation. This represents a unique viral evasion mechanism of poxvirus.

Part III

RNA Viruses

All living organisms on the globe employ DNA as their genetic material, apart from RNA viruses, which can exploit RNA as their genetic material. RNA viruses share a few notable features. First, the mutation rate of RNA viruses is much higher than that of DNA viruses. Therefore, variant viruses are more frequently generated in RNA viruses, a property that enables the RNA viruses to cope with challenges such as antibodies. The high mutation rate is the reason for the difficulty in controlling RNA viruses with antiviral drugs and vaccines. Second, all RNA viruses encode a gene for **RdRp** (RNA-dependent RNA polymerase) to synthesize their own RNA genome. This is because RNA viruses cannot exploit cellular RNA polymerase, which employs a solely DNA template. A few exception to this rule (ie, hepatitis delta virus and viroids) will be described in chapter "Subviral Agents and Prions."

RNA viruses are subdivided into three groups, depending on their genome polarity. RNA viruses whose genome is the same strand as mRNA are called "**positive-strand RNA viruses**," while RNA viruses whose genome is the complement strand of mRNA are called "**negative-strand RNA virus**." RNA viruses that contain double-strand RNA as a genome are called "**double-strand RNA virus**."

Depending on the genome polarity, the strategies for RNA genome replication differ. First, the positive-strand RNA virus replicates its genome using negative-strand RNA as an intermediate. Conversely, the negative-strand RNA virus replicates its genome using positive-strand RNA as an intermediate. Likewise, double-strand RNA virus replicates its genome using positive-strand RNA as an intermediate. As far as the genome replication strategy is concerned, a double-strand RNA virus can be considered as a negative-strand RNA virus (Fig. PIII.1).

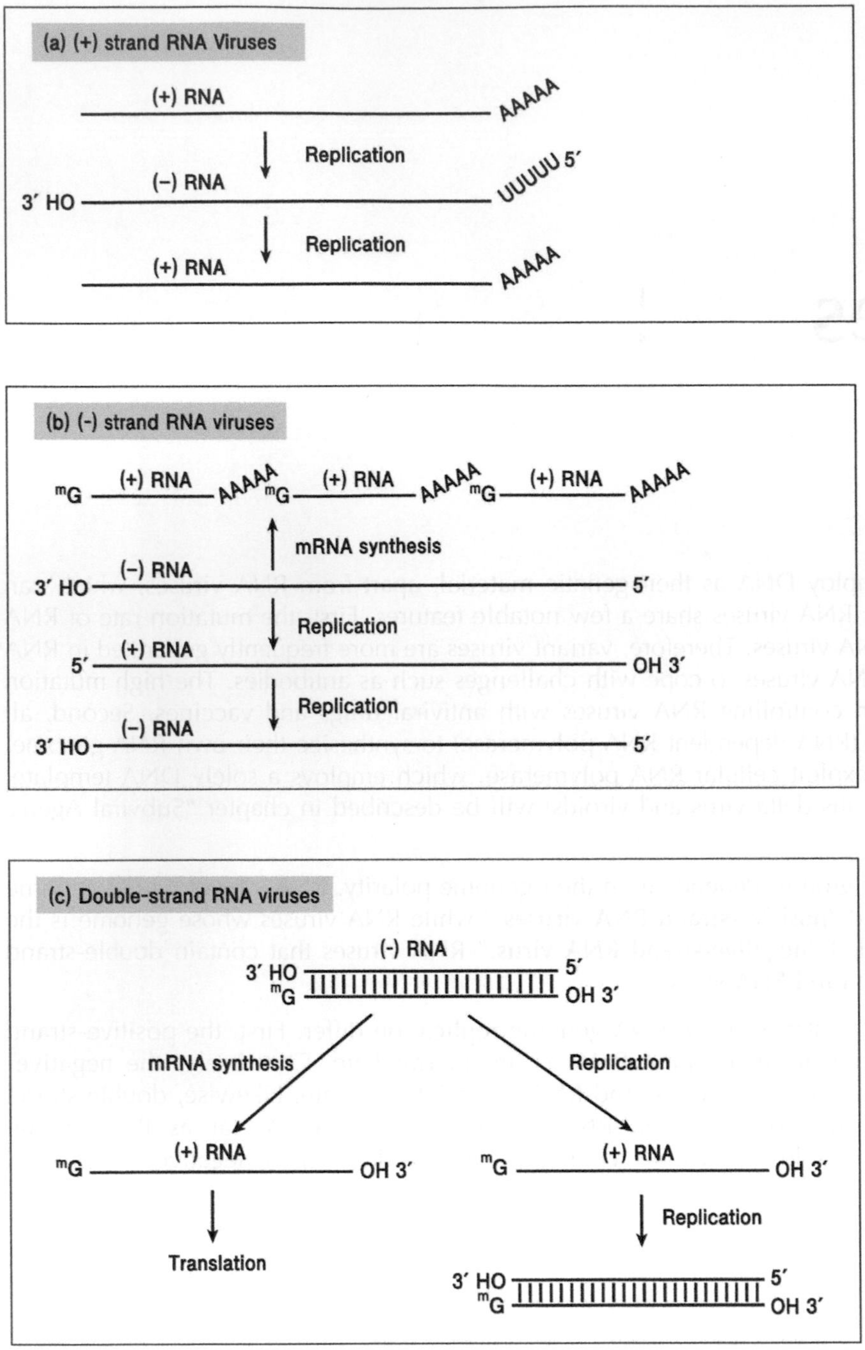

FIGURE PIII.1 Genome replication strategies of RNA viruses. (A) Positive-strand RNA virus first synthesizes the negative-strand RNA, which is utilized as a template for nascent positive-strand RNA. (B) Negative-strand RNA virus synthesizes two kinds of positive-strand RNAs. A few subgenomic RNAs are first synthesized that serve as mRNA for translation. Then, a genomic length positive-strand RNA is synthesized that serves as a template for nascent negative-strand RNA. (C) Double-strand RNA virus utilizes solely the negative strand RNA as a template for viral genome replication. Similar to negative-strand RNA virus, two positive-strand RNAs are synthesized. One serves as mRNA, while other serves as a template for nascent negative-strand RNA genome. ^{m}G, cap structure.

Part III-1

Positive-Strand RNA Viruses

Positive-strand RNA viruses comprise of five families—picornavirus, flavivirus, calicivirus, togavirus, and coronavirus (Table PIII-1). Some commonality is outstanding among these positive-strand RNA viruses. First, the genomic RNA itself is *"infectious."* In other words, a progeny virus is produced when the viral genomic RNA is introduced to the cell by transfection. Second, the genomic RNA itself serves as mRNA for the synthesis of viral proteins. This feature is a hallmark of the positive-strand RNA viruses. Third, the viral RNA genome replicates via double-strand RNA as an intermediate. Thus, double-strand RNA exists, even though only transiently, during the RNA replication. Fourth, the virus life cycle including the viral genome replication is confined to the cytoplasm. The nuclear functions of host cells are largely dispensable for the propagation of positive-strand RNA viruses.

TABLE PIII-1 Classification of Positive-Strand RNA Viruses

Family	Genus	Species
Picornavirus	Enterovirus	Poliovirus
	Rhinovirus	Human rhinovirus
Flavivirus	Flavivirus	Japanese encephalitis virus
	Hepacivirus	Hepatitis C virus
Calicivirus	Norovirus	Norovirus
Togavirus	Alphavirus	Semliki forest virus
	Rubivirus	Rubella virus
Coronavirus	Coronavirus	SARS-coronavirus

Chapter 11

Picornavirus

Chapter Outline

Picornavirus represents the prototype of the positive-strand RNA viruses. In fact, many human pathogenic viruses belong to the family Picornaviridae. This chapter will focus on poliovirus, which is the prototype of the *picornavirus*[1] family (Table 11.1). Poliovirus is the etiologic agent of *poliomyelitis*,[2] an acute flaccid paralysis affecting 1−2% of infected individuals and, on rare occasions, causing death.

11.1 CLASSIFICATION OF PICORNAVIRUSES

Classification: Picornaviruses are associated with diseases of many organs including gastrointestinal tracts, respiratory tracts, neuronal tissue, and muscles. Depending on the tissue tropism, picornaviruses are divided into five genera (Table 11.2). For instance, one that infects gastrointestinal tracts is classified as *Enterovirus*[3] genus. Poliovirus belongs to genus *Enterovirus*. *Rhinovirus*, which causes common colds, also belongs to genus *Enterovirus* of Picornaviridae family. However, common colds caused by rhinovirus infection should not be confused with "flu" caused by influenza virus. The symptoms of the common cold are not as severe as "flu"; nonetheless it still represents a serious disease burden. In addition, *enterovirus 71* and *coxsackie virus*, which cause hand-foot and mouth disease (HFMD) in children, also belongs to genus *Enterovirus*. Finally, *EMCV* (encephalomyocarditis virus) and *hepatitis A virus* (HAV) also belongs to Picornaviridae family.

Animal picornaviruses cause significant veterinary disease in livestock as well. Foot-and-mouth disease virus (FMDV) causes a fatal disease in cows, pigs, sheep, and goats and FMDV epidemic draws significant concern. The FMDV outbreak occurred in 2010 in South Korea resulted in the loss of 3 million livestock, costing 3 billion USD.

Vaccine: Vaccines for poliovirus were developed in the 1950s by Jonas Salk and Albert Sabin, independently (see Box 25.2). Thanks to the poliovirus eradication campaign initiated by the WHO, poliovirus is now on the verge of eradication (see Fig. 25.2).

11.2 THE VIRION AND GENOME STRUCTURE OF POLIOVIRUS

Virus Particles: It is a naked virus having a diameter of only 30 nm (Fig. 11.1). The capsid is composed of 60 subunits (capsomers), each of which comprises three virion proteins (ie, VP1, VP2, and VP3), representing $T = 3$ symmetry. Inside the capsid, the viral RNA genome is enclosed.

---

1. **Picornavirus** The term "picornavirus" is derived from Latin word for "small"-*pico* + "*RNA*."
2. **Poliomyelitis** The term "poliomyelitis" is derived from Greek word for "gray"-*polio* + for "marrow"-*myelos* + "inflammation"-*titis*. The term implies paralytic poliomyelitis, resulted from destruction of motor neurons within the spinal cord in the CNS.
3. ***Enterovirus*** A genus of picornavirus family that infects gastrointestinal tracts.

Molecular Virology of Human Pathogenic Viruses. DOI: http://dx.doi.org/10.1016/B978-0-12-800838-6.00011-4

TABLE 11.1 The Defining Features of Picornavirus

Genome	Particle Structure	Replication Mechanism
Positive-strand RNA	Nonenveloped	RNA-directed RNA synthesis
VPg-linked genome	Icosahedral symmetry	VPg-primed RNA synthesis
Life cycle	**Host effect**	**Disease**
Cytoplasmic	Translation suppression	Poliomyelitis
IRES	Host shutoff	Common cold

TABLE 11.2 Classification of Picornavirus

Genus	Virus	Host	Tissue	Disease
Aphthovirus	Foot-and-mouth disease virus (FMDV)	Artiodactyla (cow, swine)	Epithelial cell	Foot-and-mouth disease
Cardiovirus	Encephalomyocarditis virus (EMCV)	Human	Heart, CNS	Encephalomyocarditis
Enterovirus	Poliovirus	Human	Gut	Poliomyelitis
	Coxsackie virus	Human	Gut	Hand-foot-mouth disease
	Coxsackie virus A24	Human	Conjunctiva	Conjunctivitis
	Enterovirus 71	Human	Gut	Hand-foot-mouth disease
	Enterovirus 70	Human	Conjunctiva	Conjunctivitis
	Rhinovirus	Human	Upper, lower airway tract	Colds, respiratory diseases
Hepatovirus	Hepatitis A virus (HAV)	Human	Liver	Hepatitis A
Senecavirus	Seneca valley virus	Swine	–	–

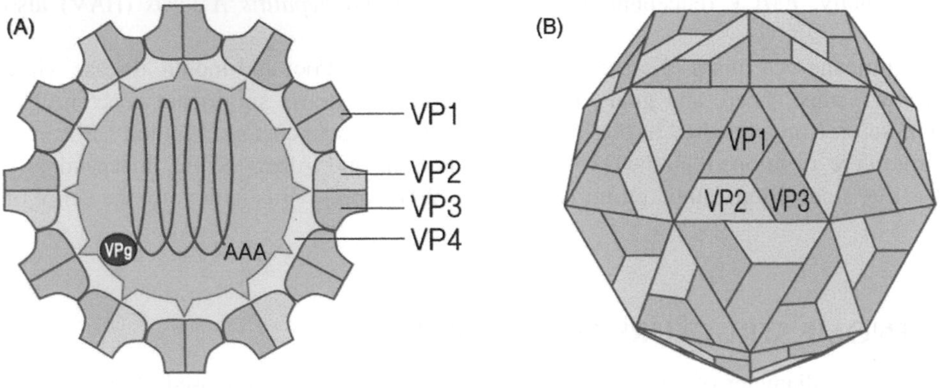

FIGURE 11.1 A diagram showing poliovirus particle. (A) Cross-section view of poliovirus capsid. The viral RNA genome having VPg linked at the 5′ end is packaged inside of the capsid. (B) Schematic diagram of icosahedral capsid structure of poliovirus capsid. VP1, VP2, and VP3, which constitute the assembly subunit, are denoted. VP4 does not contribute to the capsid symmetry.

Genome Structure: The poliovirus has a 7.5 kb single-strand RNA with positive polarity (Fig. 11.2). Notably, a viral protein, dubbed *VPg* (virion protein genome linked), is linked to the 5′ end of the RNA genome. The VPg is linked to the 5′ nucleotide residue (pU) via a tyrosine. On the other hand, the 3′ end of the RNA genome is polyadenylated. The virion RNA (vRNA) itself is "*infectious*," as it can lead to the progeny virus production, when transfected into cells (Box 11.1). Note that "infectivity" of the viral genomic RNA is a hallmark of positive-strand RNA viruses. VPg is

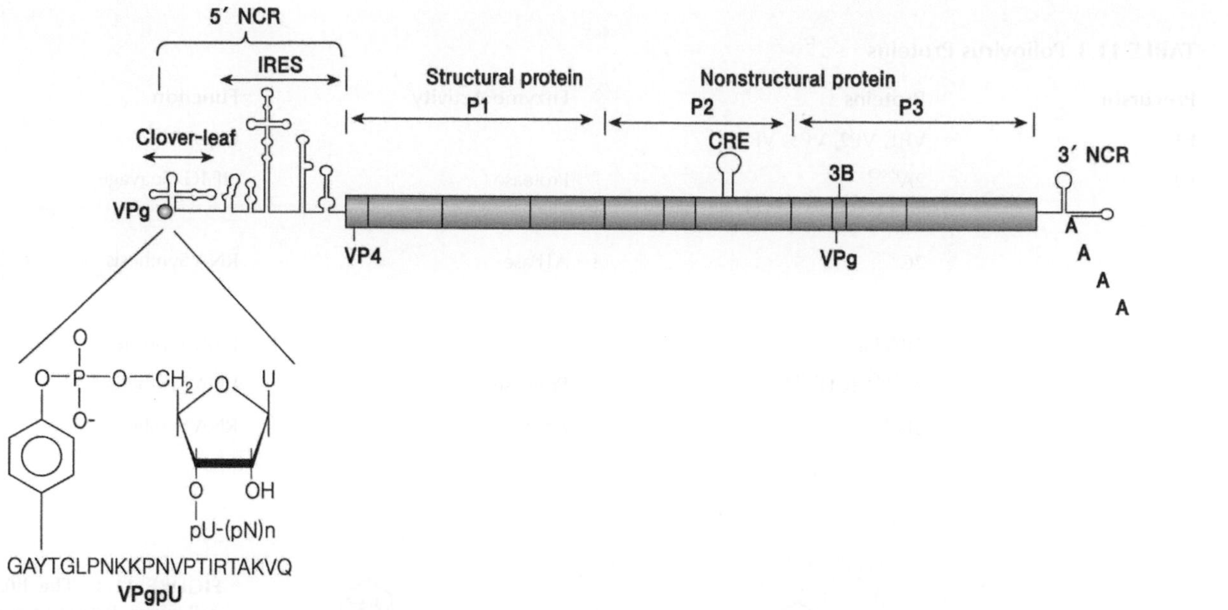

FIGURE 11.2 RNA genome structure of poliovirus. The viral RNA genome encodes one large ORF for the polyprotein. The 5′ end of the RNA genome is covalently linked to the viral protein (ie, VPg) and is expanded below for the detail. The amino acid compositions of the VPg peptide are shown. The 3′ end of the RNA genome has poly (A) tail. The 5′ NCR contains two *cis*-acting elements such as a clover-leaf structure and IRES. NCR, noncoding region; CRE, *cis*-acting replication element.

BOX 11.1 Molecular Tools for Studying Picornavirus

How can the RNA genome be manipulated for molecular studies? Advances in molecular biology allowed us to handle RNA as we handled DNA. First, the RNA genome is converted into cDNA, and then it is inserted into an appropriate plasmid vector for transcription. In doing so, one could generate a "replicon" construct, which can produce the progeny virus, when transfected into cells. In principle, the replicon contains all genetic information essential for viral genome replication. In case of picornavirus, the cDNA of the full-length viral RNA genome is inserted downstream of CMV/T7 promoter to construct a replicon. Transfection of animal cells with the replicon will induce transcription of viral genomic RNA from CMV promoter, which leads to production of the progeny virus. Alternatively, the viral genomic RNA can be synthesized in vitro using T7 RNA polymerase. In vitro synthesized RNAs can be used for transfection. Altogether, these molecular approaches that enable us to handle viral genome by molecular technique are termed "Molecular Virology."

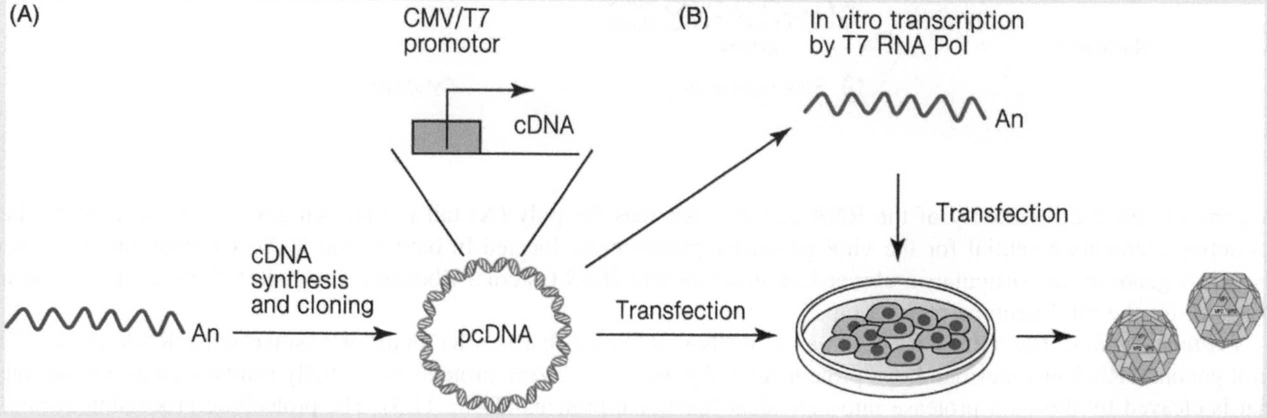

A diagram illustrating molecular approaches in studying picornavirus genome replication. (A) cDNA synthesis: the cDNA is made by reverse transcription of viral RNA genome. The full-length cDNA is then inserted into an appropriate plasmid vector. (B) Transfection: Either transfection of the replicon plasmid itself or in vitro transcribed RNA could lead to production the progeny virus.

TABLE 11.3 Poliovirus Proteins

Precursor	Proteins	Enzyme Activity	Function
P1	VP1, VP2, VP3, VP4	–	Capsid
P2	$2A^{PRO}$	Protease	eIF4G cleavage
	2B	–	–
	2C	ATPase	RNA Synthesis
P3	3A	–	–
	3B/VPg	–	Protein-primer
	$3C^{PRO}(3CD^{PRO})$	Protease	RNA Synthesis
	$3D^{POL}$	RdRp	RNA Synthesis

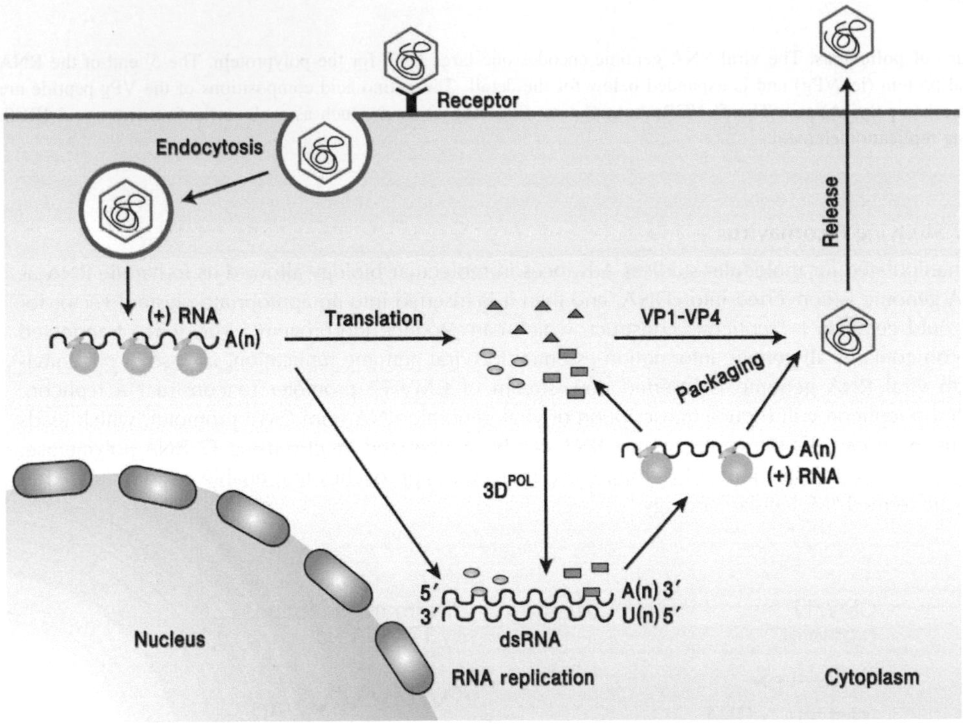

FIGURE 11.3 The life cycle of poliovirus. Poliovirus capsid enters the cell via receptor-mediated endocytosis. The viral genomic RNA itself is utilized as mRNA for translation upon entry. The negative-strand RNA is transiently made during viral genome replication, resulting in double-strand RNA intermediates. The positive-strand RNA is selectively packaged into nascent assembled viral capsid. Note that the life cycle of poliovirus is restricted to the cytoplasm.

dispensable for the infectivity of the RNA genome, whereas the poly (A) tail is indispensable for the infectivity. Few *cis*-acting elements essential for the viral genome replication are located in both 5′ and 3′ *NCR* (noncoding region) of the RNA genome. In particular, a clover-leaf structure and *IRES* (internal ribosome entry site) element at 5′ NCR are essential for the viral genome replication.

Protein Coding: The vRNA itself serves as mRNA, a feature that is a hallmark of positive-sense RNA viruses. The viral genomic RNA encodes one large protein, termed *polyprotein*.[4] Viral proteins are initially synthesized as a polyprotein that is cleaved by the viral protease into individual functional proteins (Table 11.3). The proteolytic processing occurs in two steps: first into three precursors, termed P1, P2, and P3 precursor proteins, and then into individual proteins.

4. **Polyprotein** It refers to a large protein that is later processed to multiple functional proteins.

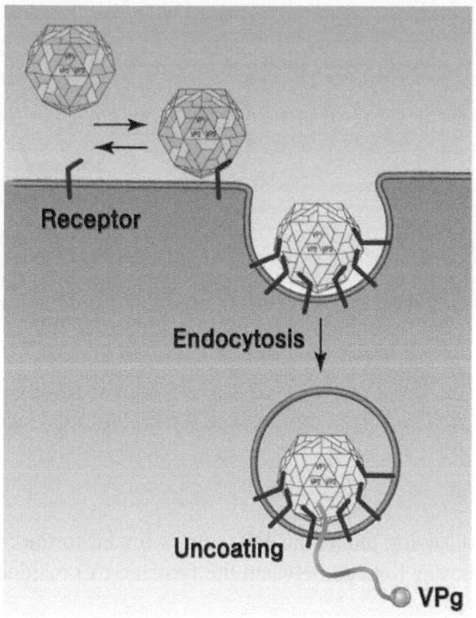

FIGURE 11.4 Steps involved in the entry of poliovirus. Engagement of the receptor (ie, CD155/Pvr) to the viral capsid triggers endocytosis. The native virion undergoes a receptor-mediated conformational transition, as highlighted by distinct color. In the endosome, the acidic pH causes the conformational changes of the viral capsid protein, in particular VP1, resulting in a channel formation such that the RNA genome can pass through. Via the uncoating process, the viral RNA genome penetrates to the cytoplasm. The viral genomic RNA is denoted in blue, while the VPg is denoted in orange.

TABLE 11.4 Cellular Receptors for Picornaviruses

Viruses	Genotype	Receptor	Protein Family
Poliovirus		Pvr or CD155	IgG superfamily
Human Rhinovirus	91	ICAM-1	IgG superfamily
Human Rhinovirus	10	LDL receptor	–
Coxsackie virus	3	ICAM	IgG superfamily
Coxsackie virus	B	CAR	IgG superfamily
EMCV	1	VCAM-1	IgG superfamily
Enterovirus 71	–	SCARB2, PSGL1	–

Cell Tropism: Poliovirus transmits via the fecal–oral route. One ingests the virus and viral replication occurs in the alimentary tract. Poliovirus targets epithelial cells in the intestines. The presence of CD155, which is an entry receptor of poliovirus, defines the animals and tissues that can be infected by poliovirus. CD155 is found only on the cells of humans, higher primates, and Old World monkeys. Poliovirus is however strictly a human pathogen, and does not naturally infect any other species.

11.3 THE LIFE CYCLE OF POLIOVIRUS

The life cycle of poliovirus is confined to the cytoplasm (Fig. 11.3).

Entry: The poliovirus enters the cells via receptor-mediated endocytosis using *CD155* (Pvr) as an entry receptor (Fig. 11.4). Many cellular receptors for the picornavirus family belong to immunoglobulin superfamily (Table 11.4). For instance, *ICAM-1* (intracellular adhesion molecule) molecule is a receptor for human rhinovirus genotype 91, while *LDL receptor* is a receptor for human rhinovirus genotype 10. Notably, coxsackie virus genotype B utilizes *CAR* (coxsackie virus adenovirus receptor) as a receptor, which is also used as a receptor for adenovirus (see Fig. 8.5). Note that cellular membrane proteins having an intrinsic function are co-opted for the viral entry. Upon the engagement on the receptor, the capsid penetrates into the cytoplasm, and becomes located inside endosomes. Upon an acidic pH (pH 5.5) of the endosomes, the capsid undergoes structural changes involving the exposure of the hydrophobic region of the

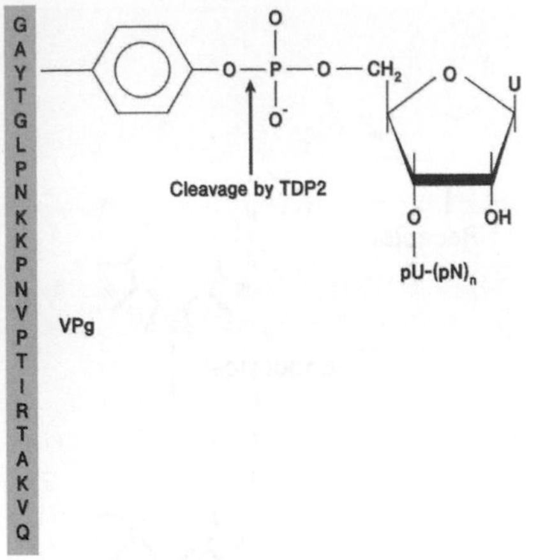

FIGURE 11.5 **Cleavage of VPg-RNA linkage.** Following entry, the VPg that is linked to the 5′ terminus of the polioviral RNA is removed by the cleavage catalyzed by cellular TDP2. The phosphodiester linkage between the tyrosine (Y) residue of VPg and the first U residue of the viral RNA is cleaved.

capsid protein, which is poised for interaction with the endosomal membrane (see Fig. 11.4). Consequently, a channel is formed in the capsid, from which the vRNA is released. Intriguingly, the vRNA exits from the capsid from its 5′ end. Note that the uncoating mechanism of the poliovirus is different from endosome lysis that is used by adenovirus or membrane fusion, a process that occurs for almost all enveloped viruses (see Fig. 3.3).

Following the release from the capsid, the VPg of the RNA genome is removed by so-called VPg unlinkage activity. For over three decades, the identity of this cellular activity responsible for the cleavage of the VPg-RNA linkage remained elusive. Recently, the VPg unlinkage activity was identified to be 5′-tyrosyl-DNA phosphodiesterase 2 (*TDP2*)[5] (Fig. 11.5). Intriguingly, the DNA repair enzyme is involved in the cleavage of the tyrosine-RNA linkage. Currently, the biological importance of the removal of VPg by TDP2 during a picornavirus infection is unclear. The removal of VPg by TDP2 may be required to stimulate efficient viral RNA replication by inhibiting premature vRNA packaging, because only VPg-linked RNA is encapsidated.

Following the removal of VPg from the 5′ end, the genomic RNA itself serves as mRNA. The clover-leaf structure at 5′ NCR is necessary and sufficient to protect the RNA from nucleases following the removal of VPg.

Translation: Translation in eukaryotic cells proceeds in a cap-dependent manner (Box 11.2). However, the RNA genome of picornavirus does not have a cap structure at the 5′ end. So a question arises, how is translation initiation of the picornavirus RNA accomplished? A short answer to this question is that ribosome recognizes the IRES element on the viral RNA and enters the RNA internally, as opposed to the 5′ end, for translation. This mode of translation initiation is termed "cap-independent translation." As the polyprotein is being translated, it is processed by viral $2A^{PRO}$ to yield P1 and P2 polypeptides, and then by $3C^{PRO}$ to yield multiple individual proteins (Fig. 11.6).

Genome Replication: The RNA genome replication of poliovirus follows a strategy of the positive-strand RNA viruses (see Fig. Part III-1). The negative-strand RNA is synthesized by using the vRNA as a template, and then the nascent negative-strand RNA serves as a template for the positive-strand RNA synthesis. The genomic RNA is a template for RNA genome replication as well as for translation. The mechanism that streamlines these apparently competitive processes has been described (Box 11.3).

Negative-Strand RNA Synthesis: One salient feature of picornavirus RNA genome replication is that the RNA synthesis is initiated by *protein-priming*[6] (Fig. 11.7). In other words, a protein (ie, VPg), rather than nucleotide, acts as a primer for RNA synthesis by $3D^{POL}$. Specifically, the hydroxyl group of a tyrosine residue of VPg peptide is linked to the first nucleotide UTP, a process dubbed *VPg-uridylylation*. As a result, VPg-linked oligonucleotide (ie, VPg-pUpUOH) is synthesized. Following the cleavage of the 3AB precursor to VPg by $3C^{PRO}$, the RNA synthesis continues to yield a full-length negative-strand RNA. It is notable that protein-priming occurs in association with the vesicular membrane, where the 3AB precursor is attached. The membrane-associated RNA genome replication occurs for almost all positive-strand RNA viruses (see Fig. 12.7).

5. **TDP2 (5′-tyrosyl-DNA phosphodiesterase 2)** An enzyme that is known to cleave a tyrosine-DNA phosphodiester linkage found in topoisomerase II-DNA adducts.

6. **Protein-priming** Refers to RNA synthesis that is initiated by protein-primer.

BOX 11.2 IRES

The 5′ end of eukaryotic mRNA harbors the cap structure, which plays a pivotal role in translation. In fact, ribosome engages the mRNA via recognition of the cap structure. Specifically, eIF4F (eIF4A + eIF4G + eIF4E), an eukaryotic translation initiation factor, is associated with mRNA via the interaction of the cap with eIF4E, a cap-binding factor. Then, the 40S ribosomal subunit, which is associated with eIF3, engages mRNA via eIF3-eIIF4G interaction, and begins translation. This process represents a typical translation mechanism, so-called "cap-dependent translation."

On the other hand, the 5′ end of picornavirus RNA genome lacks the cap structure. Instead, the picornavirus RNA harbors an IRES element at 5′ NCR. It was shown that the ribosome could enter mRNA directly via interaction with IRES without the cap structure. This kind of translation mechanism is termed *"cap-independent translation."* In cells that are infected by poliovirus, the N-terminal domain of eIF4G is cleaved by viral $2A^{PRO}$ protease so that cap-dependent translation of cellular mRNAs is suppressed. This suppression is the underlying mechanism for the *"host shutoff"* function induced by picornavirus. By contrast, a ribosome could enter the viral RNA because the N-terminal domain of eIF4G is dispensable for binding to IRES, thereby following the cap-independent translation.

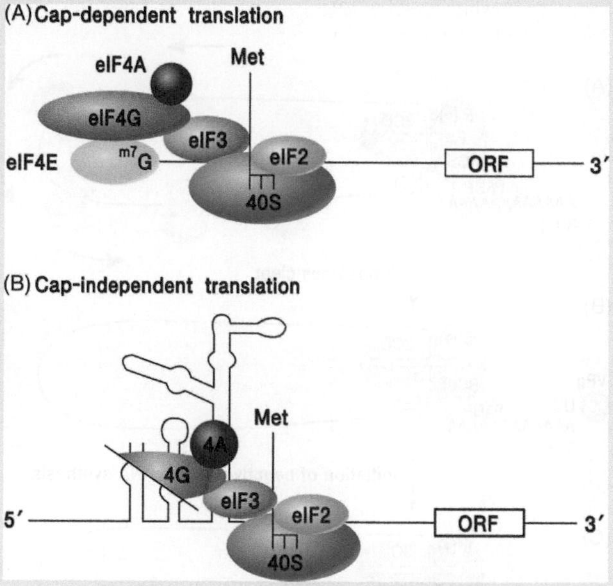

(A) Cap-dependent translation. The eIF3, bound to mRNA via eIF4F (eIF4A + eIF4G + eIF4E), recruits 40S ribosome subunit that is associated with eIF2-methionine (Met) charged tRNA (fork shape) for translation initiation. (B) Cap-independent translation. The eIF4G binds directly to the IRES, thereby recruiting eIF3-40S ribosome. Following the cleavage of eIF4G by $2A^{PRO}$ (as denoted by *solid line*), the remaining eIF4G domain is sufficient to recruit eIF3/40S ribosome complex.

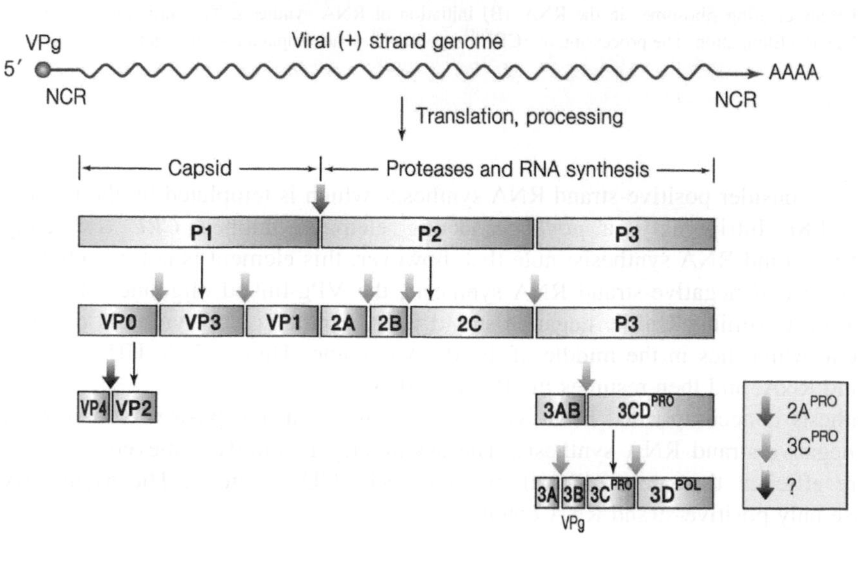

FIGURE 11.6 Proteolytic processing of the poliovirus polyprotein. The polyprotein is proteolytically processed into three precursors (P1, P2, and P3), and then further processed to generate multiple individual proteins. The cleavage sites for $2A^{PRO}$ and $3C^{PRO}$ are denoted. The protease responsible for the cleavage of VP0 to VP4 and VP2 is unknown. Superscript denotes the biochemical activities (eg, $3CD^{PRO}$ = protease and $3D^{POL}$ = polymerase).

BOX 11.3 Switch from Translation to Genome Replication

The positive-strand RNA of picornaviruses serves a dual role in that it is mRNA for translation, but also it is a template for RNA genome replication. These two processes could not occur simultaneously in a given RNA, since ribosomes move from 5′ to 3′ direction, while RNA polymerase moves from 3′ to 5′ direction. Otherwise, they could collide in the middle of the RNA genome, which would be futile. How can these two otherwise competitive processes be streamlined? A clue to this puzzle was hinted at in a recent finding, supporting the circularization of the positive-strand RNA. Intriguingly, the sequence element at the 5′ end contributes to the event occurring at the 3′ end of the RNA via circularization. Two observations supported this notion. First, $3CD^{PRO}$ binds not only to 3′ NCR but also 5′ NCR. Second, *PCBP* [poly (rC) binding protein], which binds to the clover-leaf structure at 5′ NCR, interacts with *PABP* [poly (rA) binding protein] at 3′ NCR. According to this scenario, the circularization of the RNA, which is mediated by the cross-talk between the 5′ and 3′ ends of the RNA via protein–protein interaction, is instrumental for the switch from translation to RNA genome replication. Intriguingly, the circularization of RNA genome was also reported in other positive-strand RNA viruses such as flaviviruses (eg, dengue virus, hepatitis C virus, and Japanese encephalitis virus). Moreover, the circular form of flavivirus RNA genome (eg, dengue virus) was visualized by atomic force microscopy, providing physical evidence for the circularized RNA. After all, the circularization of RNA genome became a rule, rather than an exception among positive-strand RNA viruses.

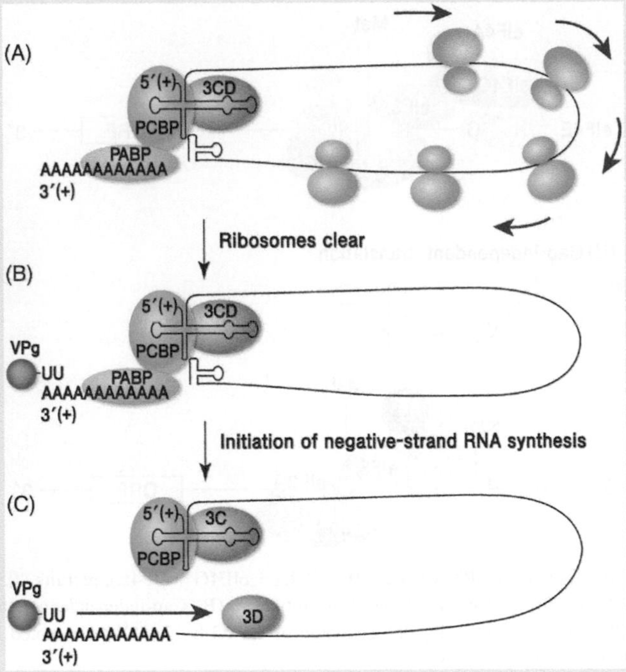

Circularization of the RNA genome is instrumental for the switch. (A) Blockade of ribosome entry. Circularization via a cross-talk between the 5′ and 3′ ends prevents the entry of ribosome, thereby clearing ribosomes in the RNA. (B) Initiation of RNA synthesis. The initiation by VPg-uridylylation starts using 3′ poly (A) tail as a template. (C) Elongation. The processing of $3CD^{PRO}$ into $3D^{POL}$ is accompanied by elongation.

Positive-Strand RNA Synthesis: Next, we consider positive-strand RNA synthesis, which is templated by the nascent full-length negative-strand RNA (Fig. 11.8). Intriguingly, a novel *cis*-acting element, dubbed *CRE* (*cis*-acting replication element), is essential for positive-strand RNA synthesis; note that, however, this element is not essential for negative-strand RNA synthesis. Similar to that of negative-strand RNA synthesis, the VPg-linked oligonucleotide (ie, VPg-pUpUOH) is first synthesized via protein-priming. Unlike negative-strand RNA synthesis, the synthesis of VPg-UU primer is templated by a CRE element, which lies in the middle of the RNA genome. Then, a VPg-UU primer is translocated to the 3′ end of negative-strand RNA, and then resumes the RNA synthesis.

One notable feature is that RNA synthesis is accomplished in an asymmetric manner in that positive-strand RNA synthesis occurs more abundantly than negative-strand RNA synthesis. The asymmetry is simply achieved by CRE-templated VPg-UU synthesis being more efficient than poly (A) tail-templated VPg-UU synthesis. The asymmetry makes sense in a viral perspective, because only positive-strand RNA counts.

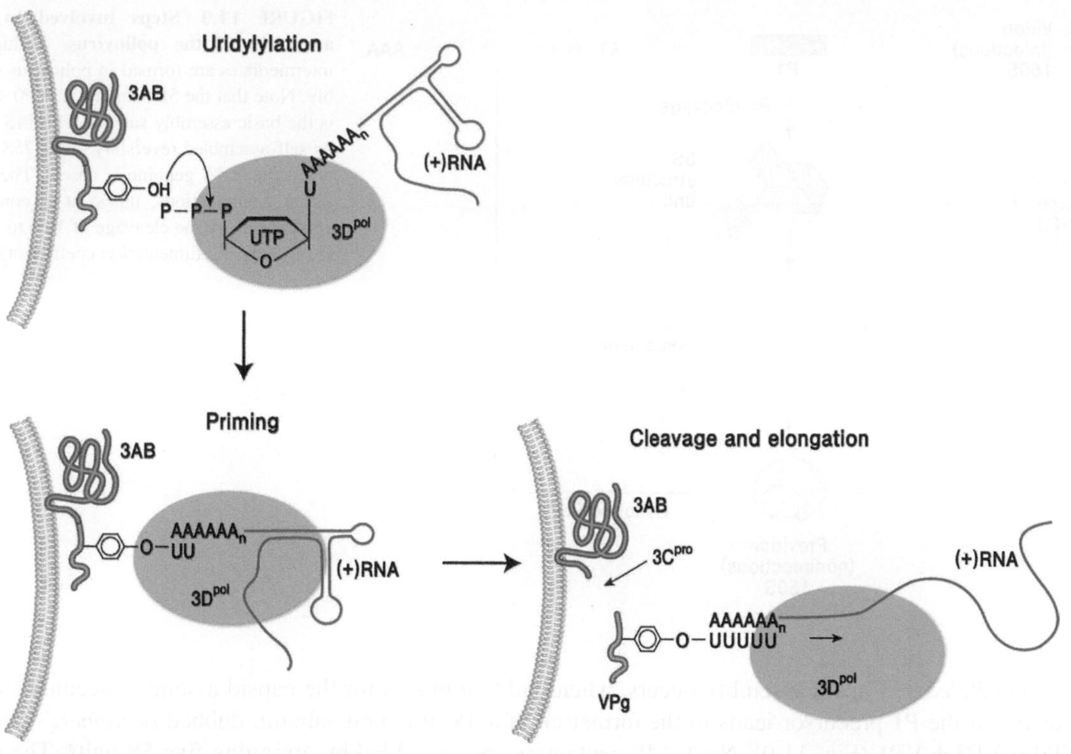

FIGURE 11.7 Protein-priming by poliovirus 3D^{POL}. The initiation of (–) RNA synthesis by 3D^{POL} is primed by 3AB precursor, which is associated with the vesicular membrane (eg, endoplastic reticulum). The nucleophilic attack by a hydroxyl group of a tyrosine residue covalently links the 3AB precursor to the first U residue. VPg-linked RNA is released from the membrane following the cleavage by 3C^{PRO}.

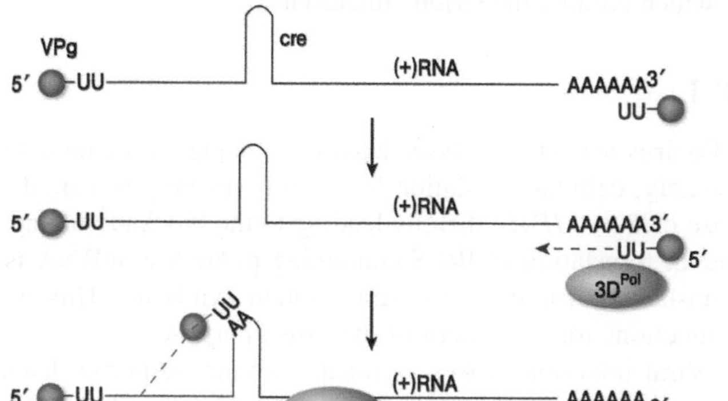

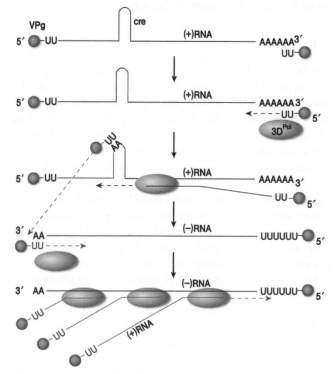

FIGURE 11.8 The RNA genome replication of the poliovirus. The initiation of (–) RNA synthesis is templated by 3′ poly (A) tail, whereas the initiation of (+) RNA synthesis is templated by CRE element. The translocation of VPg-UU primer from CRE to the 3′ end of (–) RNA is denoted by dashed lines. CRE element, which holds a hairpin structure, encodes "AA" residues on the loop region of the hairpin. Nascent (–) RNA is colored by red, while nascent (+) RNA is colored by blue.

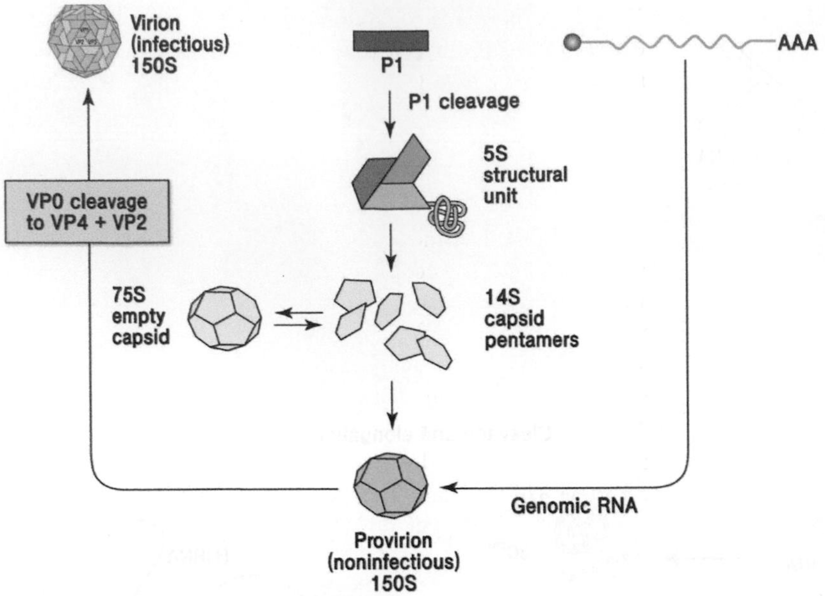

FIGURE 11.9 Steps involved in the capsid assembly of the poliovirus. Multiple assembly intermediates are formed in poliovirus capsid assembly. Note that the 5S subunit (ie, VP0 + VP3 + VP1) is the basic assembly subunit. The 14S subunit could be self-assembled reversibly to the 75S empty capsid when the RNA genome is absent. The 150S provirion is noninfectious, unless it is converted to the 150S virion by the cleavage of VP0 to VP4 and VP2 (S, Svedberg sedimentation coefficient).

Assembly and Release: Capsid assembly occurs, when building blocks for the capsid assembly accumulate. The proteolytic cleavage of the P1 precursor leads to the formation of a *5S* structural subunit, dubbed *protomer*,[7] which is composed of VP0 + VP3 + VP1 (Fig. 11.9). Next, 14S pentamers are assembled by arranging five 5S units. The recognition of the genomic RNA by 14S pentamers drives not only the RNA packaging, but also the 150S provirion assembly. The provirion, although it contains the RNA genome, is noninfectious. The assembled poliovirus particles (ie, provirion) are released through cell lysis. Finally, the proteolytic cleavage of VP0 to VP4 + VP2 occurs after the viral release. This process is called "*maturation*," which confers the virion "infectivity."

11.4 EFFECTS ON HOST

The productive infection by poliovirus leads to cell lysis, leaving a plaque on the monolayer cells in culture.

Translation Suppression: Notably, cellular translation function is severely impaired in poliovirus infected cells. For instance, the viral $2A^{PRO}$ protease cleaves *eIF4G*, thereby leading to the blockade of cap-dependent translation (see Box 11.2). In contrast, cap-independent translation of IRES-containing poliovirus mRNA is still to occur. In other words, poliovirus subverts the host translation machinery for viral protein synthesis. This is one example of showing how viruses cunningly explore host functions for the benefit of their own purpose.

Suppression of IFN Action: Viral infection evokes an innate immune response, leading to interferon (IFN) production. IFN induces and activates *PKR*,[8] which phosphorylates *eIF2α* and thereby leads to the inhibition of translation (see Fig. 5.8). In principle, the antiviral effect of IFN would equally affect the translation of both viral mRNA and cellular mRNA. Nonetheless, the host cells are less vulnerable, since host proteins have already accumulated to some extent. Nonetheless, poliovirus successfully manages to lead to the productive infection under these circumstances. Then, one wonders how poliovirus avoids the antiviral action of IFN. Recently, it was shown that picornaviruses avoid the antiviral effect of IFN by switching from earlier eIF2α-dependent translation to *eIF2α-independent translation* in the late phase of infection. After all, picornaviruses co-opt host translation machinery to complete their life cycle.

Pathogenesis: Poliovirus transmits via the fecal–oral route and viral replication occurs in the alimentary tract. In 95% of cases, only a primary and transient presence of viremia (virus in the blood) occurs, and the poliovirus infection is asymptomatic. In about 5% of cases, the virus spreads and replicates in other sites such as muscle. The sustained viral replication causes secondary viremia and leads to the development of minor symptoms such as fever, headache, and sore throat. Paralytic poliomyelitis occurs in less than 1% of poliovirus infections. Paralytic disease

7. **Protomer** It refers to a structural subunit of poliovirus capsid. It is also called 5S structural unit, based on its sedimentation coefficient (ie, Svedberg sedimentation coefficient).

8. **PKR (protein kinase R)** A serine/threonine protein kinase that is activated by double-strand RNA.

occurs when the virus enters the central nervous system (CNS) and replicates in motor neurons within the spinal cord, brain stem, or motor cortex, resulting in the selective destruction of motor neurons leading to temporary or permanent paralysis. However, the underlying mechanism by which poliovirus invades the CNS and causes poliomyelitis remains unknown.

11.5 PERSPECTIVES

Since its discovery in 1908, poliovirus has been intensively studied to better understand and control this formidable pathogen. Poliovirus vaccines developed in the 1950s have been paramount in controlling the pathogen (see chapter: Vaccines). The history of poliovirus research is not, however, limited to the fight against the disease. Poliovirus replication studies also have played importantly roles in the development of modern virology. Poliovirus was, for example, the first animal RNA virus to have its complete genome sequence determined; the first animal RNA virus of which an infectious clone was constructed; and along with the related rhinovirus, the first human virus for which its three-dimensional structure was solved by X-ray crystallography. In particular, the studies on poliovirus have contributed greatly to our current understanding of eukaryotic translation. For instance, the IRES element first discovered in the poliovirus was instrumental for our current understanding on cap-independent translation, one that is distinct from conventional cap-dependent translation. Importantly, enteroviruses other than poliovirus have emerged as serious threats to public health. These include *enterovirus 71*, responsible in infants and young children for hand, foot, and mouth disease with the potential for severe nervous system complications, and *enterovirus D68*, detected in children hospitalized with severe lower respiratory symptoms and asthma. In addition, many other members of the picornavirus family present serious health concerns. For instance, HAV is responsible for epidemic hepatitis in the Western Hemisphere (see chapter: Hepatitis Viruses). It is hoped that knowledge learned from poliovirus as the prototype should be instrumental in better understanding other related picornaviruses that represent unmet medical needs.

11.6 SUMMARY

- *Picornavirus*: Picornaviruses are found in diseases of many organs including gastrointestinal tracts, respiratory tracts, neuronal tissue, and muscles. Poliovirus is a prototype of picornavirus family possessing a single-strand RNA genome.
- *Virion structure*: It is a nonenveloped capsid with $T = 3$ symmetry. The capsid is composed of 60 subunits, each of which constitutes three virion proteins (ie, VP1, VP2, and VP3).
- *Viral genome*: The virion RNA (vRNA) is a positive-strand RNA about 7.5 kb in length. VPg is linked to the 5′ end of the vRNA. The IRES element located at 5′ NCR allows "cap-independent translation." The RNA genome encodes one large polyprotein, which is subsequently processed into individual viral proteins.
- *RNA genome replication*: The RNA genome replication is initiated via protein-priming by using VPg as a primer.
- *Host effect*: Poliovirus infection causes cell lysis, resulting in plaque formation. The cleavage of eIF4G by $2A^{PRO}$ blocks cap-dependent translation.

STUDY QUESTIONS

11.1 A novel RNA virus was discovered. Intriguingly, the 5′ terminus of the RNA genome is covalently linked to a viral protein. (1) Please hypothesize the role of the 5′-linked viral protein. (2) How would you test your hypothesis?

11.2 Poliovirus progeny particles are produced if the viral genomic RNA is transfected into appropriate cells. Please state whether the progeny virus is produced when the respective viral RNA is transfected. (1) VPg-free RNA, (2) Δ clover-leaf structure-RNA, (3) Δ IRES-RNA, (4) Δ CRE-RNA, (5) poly (A) tail-free RNA.

11.3 Compare and contrast the differences between (−) RNA synthesis and (+) RNA synthesis of picornaviruses.

SUGGESTED READING

Feng, Z., Hensley, L., McKnight, K.L., Hu, F., Madden, V., Ping, L., et al., 2013. A pathogenic picornavirus acquires an envelope by hijacking cellular membranes. Nature. 496 (7445), 367–371.

Leveque, N., Semler, B.L., 2015. A 21st century perspective of poliovirus replication. PLoS Pathog. 11 (6), e1004825.

Nagy, P.D., Pogany, J., 2012. The dependence of viral RNA replication on co-opted host factors. Nat. Rev. Microbiol. 10 (2), 137–149.

Virgen-Slane, R., Rozovics, J.M., Fitzgerald, K.D., Ngo, T., Chou, W., van der Heden van Noort, G.J., et al., 2012. An RNA virus hijacks an incognito function of a DNA repair enzyme. Proc. Natl Acad. Sci. USA. 109 (36), 14634–14639.

Walsh, D., Mohr, I., 2011. Viral subversion of the host protein synthesis machinery. Nat. Rev. Microbiol. 9 (12), 860–875.

JOURNAL CLUB

- Feng, Z., Hensley, L., McKnight, K.L., Hu, F., Madden, V., Ping, L., et al., 2013. A pathogenic picornavirus acquires an envelope by hijacking cellular membranes. Nature 496 (7445), 367–371.

  Highlight: Picornavirus has a naked capsid lacking an envelope component. Intriguingly, this article showed that hepatitis A virus (HAV), a picornavirus causing acute hepatitis, is cloaked in host-derived membranes, thereby protecting the virion from antibody-mediated neutralization. These enveloped viruses ("eHAV") resemble exosomes and are fully infectious and circulate in the blood of infected humans. Thus, membrane hijacking by HAV blurs the classical distinction between "enveloped" and "nonenveloped" viruses and has broad implications for host immune response.

Chapter 12

Flaviviruses

Chapter Outline

Flavivirus[1] infection leads to diverse diseases in both human and animals. Many vector-borne diseases including yellow fever, dengue fever, and Japanese encephalitis are caused by flavivirus. In fact, yellow fever virus (YFV) was the first human virus discovered in 1902 (see Fig. 1.3). Since then, YFV has served as a prototype of the flavivirus family. Besides these mosquito-borne flaviviruses, hepatitis C virus (HCV), which was discovered in 1989 as an etiological agent for non-A, non-B viral hepatitis (see chapter: Hepatitis Viruses), belongs to the flavivirus family. Due to its clinical importance in the Western hemisphere, HCV has been extensively studied in recent decades. Hence, this chapter will focus on HCV, in particular the molecular aspect of the viral life cycle. The clinical aspects of HCV infection including epidemiology, pathology, and treatments will be described in chapter "Hepatitis Viruses."

12.1 CLASSIFICATION OF FLAVIVIRUSES

The flavivirus family can be subdivided into three genera: *Flavivirus*, *Pestivirus*, and *Hepacivirus* (Table 12.1). Almost all viruses belonging to the genus *Flavivirus* are insect-borne and *zoonotic viruses*.[2] For instance, YFV, Dengue virus, and Japanese encephalitis virus (JEV) are transmitted via mosquitoes. Over 100 million people annually are infected by dengue virus (DENV), and 500,000 people suffer from dengue fever. In fact, flavivirus is the most prevalent virus worldwide infecting more people than any other virus (Box 12.1). In addition, animal viruses belonging to the *Pestivirus* genus include bovine viral diarrhea virus and classical swine fever virus (see Table 12.1). HCV differs from other animal flaviviruses with respect to the mode of transmission and the infection pathology. As a result, HCV is now classified in its own genus, *Hepacivirus*.

12.2 HEPATITIS C VIRUS

The genome structure and the viral life cycle of flaviviruses, including HCV, are fundamentally similar to those of picornaviruses (Table 12.2). HCV infection causes chronic *hepatitis*[3] as well as acute hepatitis.

Epidemiology: Worldwide, over 180 million people (~2.8%) are chronically infected with HCV. The viral prevalence even approaches 2% in countries in the Western Hemisphere such as the United States, Europe, and Japan. HCV is primarily transmitted via blood or body fluids; however, transmission via blood transfusion is rare nowadays due to blood screening. The hallmark of HCV infection is the high rate of chronicity. Over 70% of newly infected individuals become chronically infected after a period of acute infection. These individuals suffer a lifelong infection and face a

1. **Flavivirus** The term "flavivirus" is derived from the Latin word for "yellow"-*flavus.*

2. **Zoonotic virus** A virus that is transmitted between species (sometimes by a vector) from animals to human.

3. **Hepatitis** It refers to a medical condition defined by the inflammation of the liver and is characterized by the presence of inflammatory cells in the tissue of the organ.

Molecular Virology of Human Pathogenic Viruses. DOI: http://dx.doi.org/10.1016/B978-0-12-800838-6.00012-6

TABLE 12.1 Taxonomy of Flavivirus

Genus	Viruses	Vector	Host	Diseases
Flavivirus	Yellow fever virus (YFV)	Mosquito	Human, monkey	Hemorrhagic fever
	Dengue virus (DENV)	Mosquito	Human, monkey	Hemorrhagic fever
	Japanese encephalitis virus (JEV)	Mosquito	Human	Encephalitis
	West Nile virus (WNV)	Mosquito	Human, bird	Fever
	Tick-borne encephalitis virus (TBEV)	Tick	Human	Encephalitis
Pestivirus	Bovine viral diarrhea virus	–	Cow	Diarrhea
	Classical swine fever virus	–	Swine	Hog cholera
Hepacivirus	Hepatitis C virus (HCV)	–	Human	Hepatitis, liver cancer

BOX 12.1 Zoonotic Flaviviruses

The vast majority of flaviviruses are *vector-borne*; for example YFV and DENV are mosquito-borne, while the rest including tick-borne encephalitis virus (TBEV) are tick-borne (see Table 12.1). These vector-borne flaviviruses are zoonotic viruses, which infect both animals and humans (see Fig. 21.5). In other words, zoonotic flaviviruses spread via mosquito bites. Importantly, infections with flaviviruses are the most prevalent viral infections worldwide, however their impact on public health, are less appreciated than other viral diseases, partly because the most affected are developing countries in Africa and Southeast Asia. An estimated 390 million infections occur annually, of which 96 million have clinical manifestations, leading to approximately 25,000 deaths. In addition, YFV infections are responsible for 200,000 infections and 30,000 deaths every year. Although these vector-borne flaviviruses are responsible for significant morbidity and mortality throughout tropical and subtropical regions of the world, the preventive vaccines are not available yet. Recent clinical trials for Dengue virus vaccine were promising, and give some hope that preventing Dengue become reality. Importantly, Zika virus, that causes recent outbreak in Brazil, is also a mosquito-borne flavivirus.

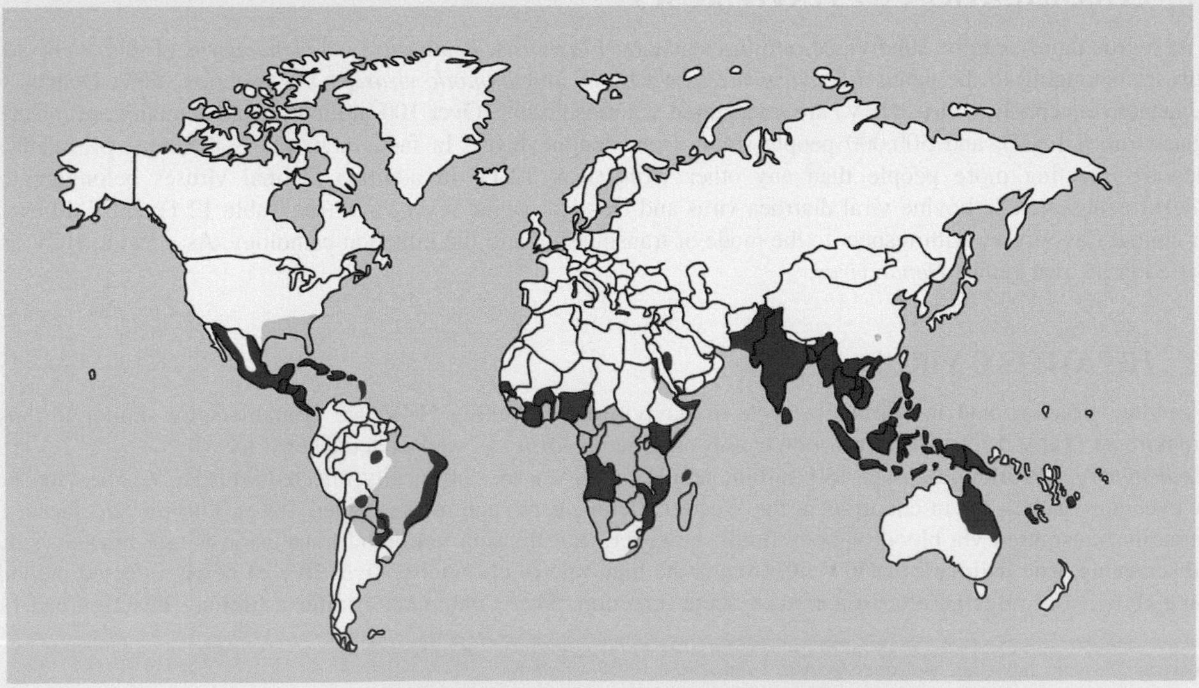

Geographic regions affected by dengue virus infection in year 2006. The regions affected by the dengue epidemic are colored in pink. The regions harboring mosquitoes (*Aedes aegypti*) without epidemics are colored in blue.

TABLE 12.2 The Defining Features of HCV

Genome	Particle Structure	Replication Mechanism
Positive-strand RNA	Enveloped	RNA-directed RNA synthesis
No cap and no poly (A) tail	Icosahedral capsid	IRES
Life Cycle	**Host Effect**	**Disease**
Cytoplasmic	Immune evasion	Liver diseases (hepatic steatosis, HCC)
Membrane-associated	Cell transformation	Liver cancer (HCC)

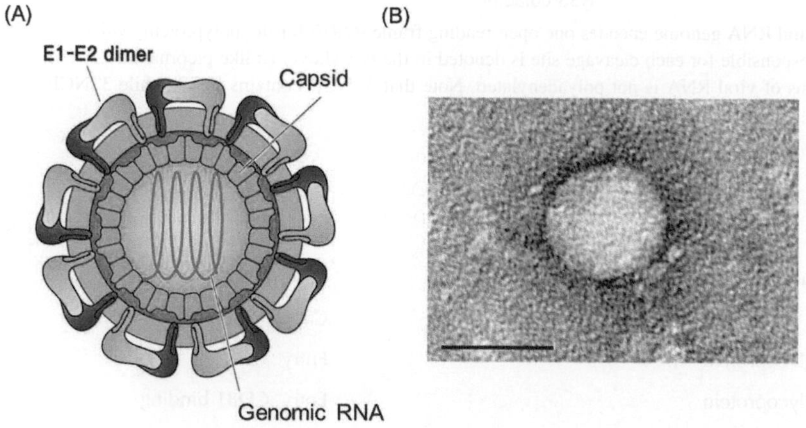

FIGURE 12.1 A diagram illustrating HCV particle. (A) A cross-section view of HCV virion. E1-E2 envelope glycoprotein dimers on the lipid envelope, and the capsid inside of the lipid envelope are denoted. The viral genomic RNA is packaged inside the capsid. (B) Electron micrographs of HCV purified from cell culture. Scale bar is 50 nanometers.

high risk of developing liver diseases, such as liver cirrhosis and hepatocellular carcinoma (HCC). Recently, several effective antiviral drugs have been developed for the treatment of chronic infection: boceprevir, telaprevir, sofosbuvir, and ledipasvir (see Fig. 26.10). The combination of these antiviral drugs is fairly effective in controlling chronic HCV infection (see Fig. 23.12). Although strenuous effort has been put toward HCV prevention, a preventive vaccine is not yet available.

Genotypes and Subtypes: The nucleotide sequence heterogeneity is high, ranging from 30% to 35%. In light of sequence divergence, HCV is subdivided into seven genotypes, genotype 1 to genotype 7. Additionally, the virus is further subdivided into subtypes (a, b, c, etc.) which differ up to 25% in the RNA sequences.

12.3 THE VIRION AND GENOME STRUCTURE OF HCV

Virion Structure: The HCV virion is an enveloped particle with a diameter of 40–60 nm (Fig. 12.1). The respective homodimers of E1 and E2 envelope glycoproteins are embedded into the lipid bilayer; in particular, E2 glycoprotein is responsible for the recognition of CD81 receptor on hepatocytes. The icosahedral nucleocapsid encompassing the viral genomic RNA is enclosed inside the viral envelope.

Genome Structure: The HCV genome is a 9.6 kb positive-stranded RNA (Fig. 12.2). It encodes one large polyprotein composed of 3010 amino acid residues. Being a positive-strand RNA virus, the viral RNA itself serves as mRNA upon entry into the cells. HCV genomic RNA does not have a cap structure at the 5′ terminus and lacks poly (A) tail at the 3′ terminus. Like picornavirus, it contains an internal ribosome entry site (IRES) element in the 5′ noncoding region (NCR), which facilitates cap-independent translation. The X region in the 3′ NCR is highly conserved and is crucial for efficient viral RNA replication.

Protein Coding: The viral RNA encodes one large polyprotein, which is composed of 3010 amino acid residues. The polyprotein is subsequently processed into 10 individual functional proteins: three structural proteins (ie, core, E1, and E2) and seven nonstructural proteins (ie, NS2 to N5B) (Table 12.3). Many of the nonstructural proteins exhibit enzymatic activities essential for viral genome replication. In particular, NS2 has a *cysteine protease* activity, while

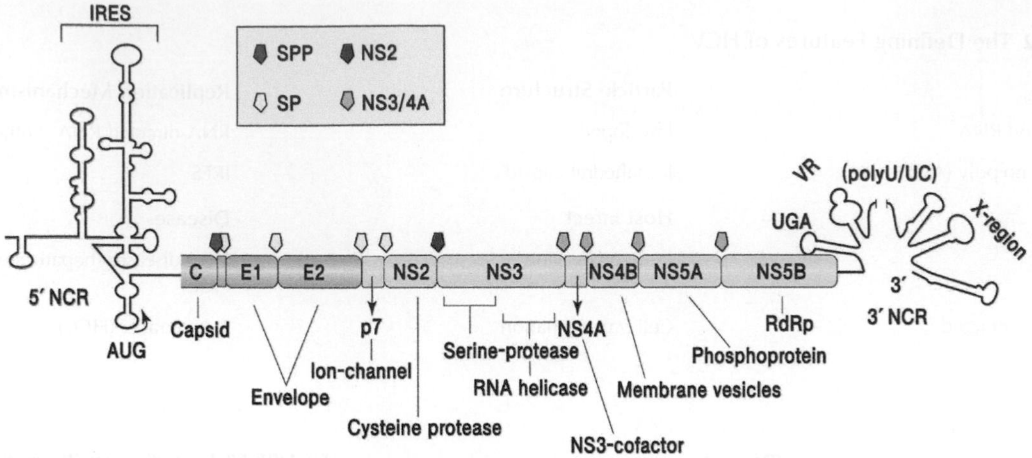

FIGURE 12.2 RNA genome structure of HCV. The viral RNA genome encodes one open reading frame (ORF) for the polyprotein, which is subsequently cleaved to 10 individual proteins; the proteases responsible for each cleavage site is denoted in the *box* above. Unlike picornavirus, 5′ terminus of viral RNA does not have cap structure, and 3′ terminus of viral RNA is not polyadenylated. Note that 5′ NCR contains IRES, while 3′ NCR contains a highly conserved X region.

TABLE 12.3 Viral Proteins of HCV

Protein	Biochemical Activity	Function
Core	Capsid	Capsid assembly
E1	Enveloped glycoprotein	Entry
E2	Enveloped glycoprotein	Entry, CD81 binding
P7	Ion channel	Virion release
NS2	Cysteine protease	Polyprotein processing
NS3	Serine protease/RNA helicase	NS3/4A serine protease
NS4A	NS3 binding	NS3/4A serine cofactor
NS4B	Membrane vesicle	Membrane alteration
NS5A	Phosphoprotein, RNA binding	RNA replication, capsid assembly
NS5B	RNA-dependent RNA polymerase	RNA replication

NS3 has a *serine protease* activity. NS4A serves as a cofactor for NS3 serine protease, constituting a heterodimer NS3/NS4A. Besides a serine protease, the C-terminus of NS3 exhibits an RNA helicase activity as well. Finally, NS5B is the *RNA-dependent RNA polymerase* (RdRp) for HCV RNA genome replication.

Cell Tropism: HCV exhibits a narrow host range and tissue tropism, as it infects the hepatocytes of only human and chimpanzee. Recently, a mouse model that supports HCV infection and replication was established (see Journal Club).

12.4 THE LIFE CYCLE OF HCV

Similar to other positive-strand RNA viruses, the HCV life cycle is confined to the cytoplasm (Fig. 12.3).

Entry: HCV virion enters the cell via receptor-mediated endocytosis (Fig. 12.4). HCV entry is a highly coordinated process, requiring a plethora of host cell factors, some of which contribute to the hepatotropism. HCV virion is characteristically associated with a *low density lipoprotein*[4] (LDL) particle, and therefore, attaches to hepatocytes via the LDL

4. **Low density lipoprotein (LDL)** It refers to a nanoparticle of 22 nm diameter composed of phospholipid, cholesterol, and apolipoproteins.

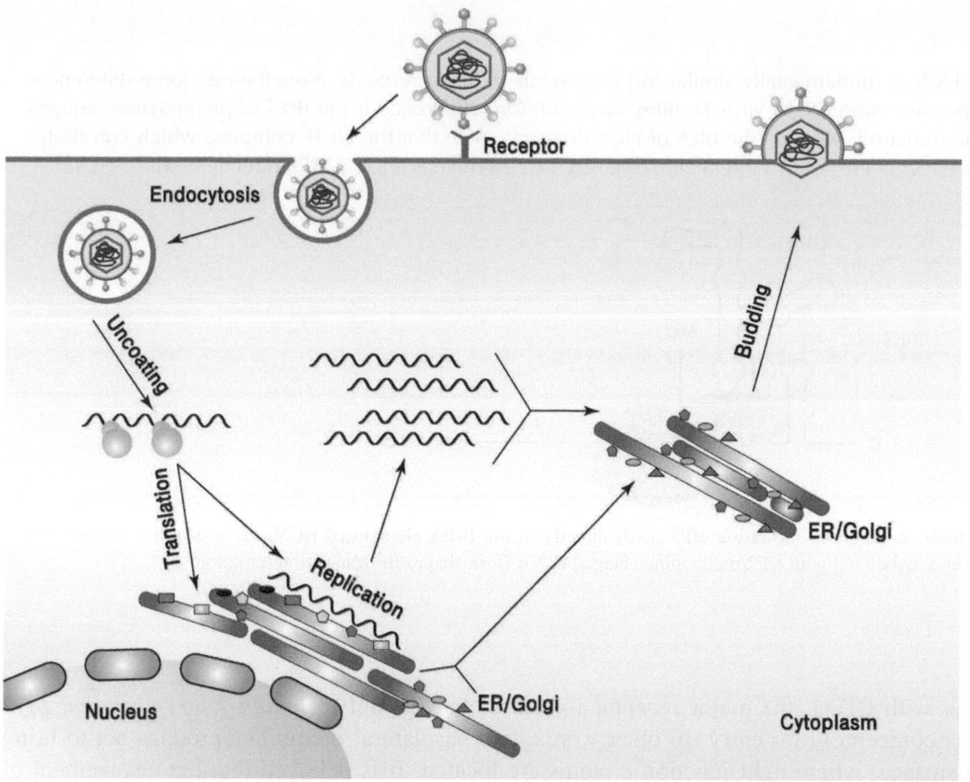

FIGURE 12.3 The life cycle of flavivirus. The life cycle of flavivirus is restricted to the cytoplasm. The virus particle enters the cell via receptor-mediated endocytosis. Upon entry to cytoplasm, the viral genomic RNA itself serves as mRNA for translation. The viral RNA genome replication takes place in association with vesicular membrane structure, where the positive-strand RNA is selectively packaged by nascent capsid particles. The capsid is enveloped by viral envelope proteins in the ER, and is then released through the secretory pathway.

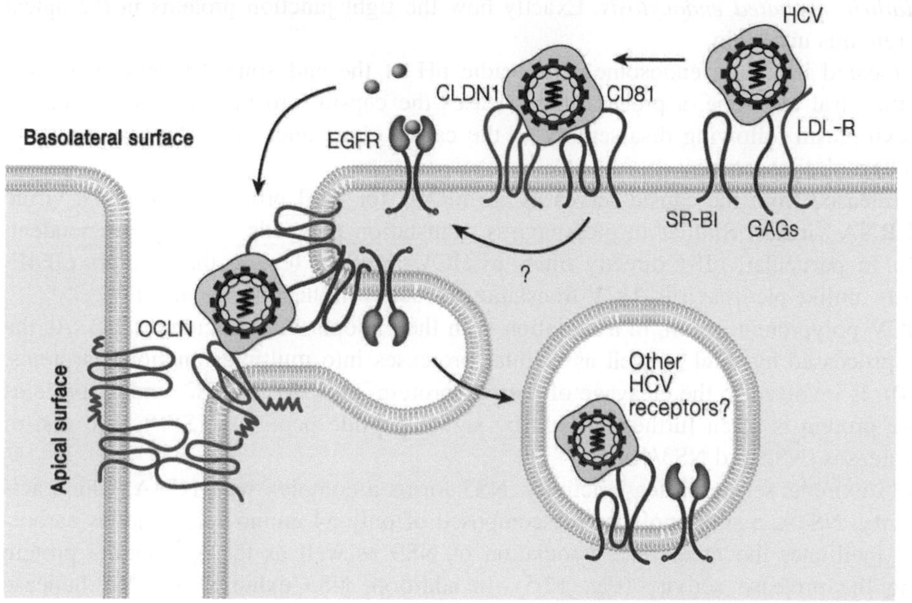

FIGURE 12.4 Entry of HCV particle in polarized hepatocytes. At least five cellular membrane proteins are sequentially involved in steps from attachment to endocytosis, starting from the LDL receptor for attachment, two basolateral membrane proteins (ie, SRB1 and CD81) for early steps of entry, and two tight junction proteins (ie, claudin-1 and occludin) for the late steps of entry. Although EGFR does not directly interact with the HCV particle, EGFR signaling promotes the formation of CD81-claudin-1 complex that is essential for HCV entry. Three surfaces of hepatocyte are denoted: basal and lateral membranes are collectively called "basolateral membrane." Tight junction is located in the apical surface, where claudin-1 and occludin are expressed.

receptor. Perhaps, the interaction between the LDL receptor and LDL apolipoproteins mediates the attachment. Following attachment, four membrane proteins contribute to HCV entry. First, two basolateral membrane proteins (ie, *SRB1*[5] and *CD81*[6]) serve as receptors for HCV entry. E2 glycoprotein of HCV envelope first encounters SRB1, which

5. **SRB1 (scavenger receptor class B member 1)** A protein highly expressed in the liver which can bind high-density lipoproteins (HDLs) and very-low-density lipoproteins (VLDLs), mediating the uptake of cholesterol from these lipoproteins by hepatocytes.
6. **CD81** A ubiquitously expressed protein and a member of the tetraspanin superfamily.

BOX 12.2 HCV IRES

As a positive-strand RNA virus, HCV is fundamentally similar to picornavirus in many respects. Nonetheless, some differences between the two prototypes of positive-strand RNA virus families stand out. One difference is the IRES of picornavirus requires the eIF4F complex for translation initiation, whereas the IRES of HCV does not. Note that the eIF4F complex, which constitutes eIF4A, eIF4G, and eIF4E recruits eIF3 via eIF4G-eIF3 interaction (see Box 11.2). Whereas, HCV's IRES directly binds to eIF3.

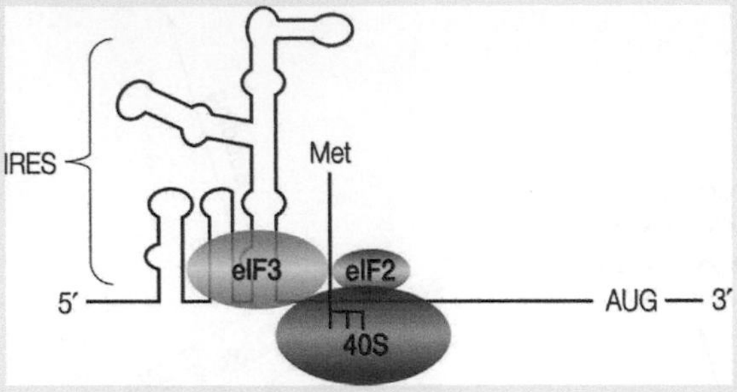

Direct binding of eIF3 to HCV IRES. A diagram illustrating eIF3 binds directly to the IRES element of HCV. eIF3, bound to mRNA via IRES, recruits the 40S ribosome subunit associated with eIF2/methionine charged tRNA (fork shape) for translation initiation.

in turn facilitates the interaction with CD81, the major receptor for entry. Interestingly, *claudin-1* and *occludin, tight junction*[7] proteins, participate as coreceptors for entry. In other words, two basolateral membrane proteins act to bring the HCV particle to the apical surface, where tight junction proteins are located. It is believed that the engagement of claudin-1 by the virion triggers actin-dependent relocalization of claudin-1 to the apical surface. The HCV particle penetrates into the cytoplasm via *clathrin-mediated endocytosis*. Exactly how the tight junction proteins in the apical membrane facilitate the endocytosis remains uncertain.

Upon endocytosis, the virion is located inside an endosome. The acidic pH of the endosome triggers membrane fusion which induces uncoating of the viral envelope, a process that releases the capsid into the cytoplasm. Finally, the viral RNA is released into the cytoplasm following disassembly of the capsid. It is uncertain whether uncoating of the viral envelope is coupled with capsid disassembly.

Translation: The genomic RNA released from the capsid serves as an mRNA for viral protein synthesis, a virtue that is shared by all positive-strand RNA viruses. Similar to picornavirus, translation proceeds in a cap-independent, IRES-dependent manner (Box 12.2). In particular, eIF3 directly binds to HCV IRES, a feature that confers eIF4F-independent translation. In other words, unlike picornavirus, HCV translation proceeds in the absence of eIF4F.

Importantly, the translation of HCV polyprotein occurs in association with the endoplasmic reticulum (ER). As the polyprotein is being translated, it is processed by viral as well as cellular proteases into multiple functional proteins. A host protease, *signal peptidase* (SP), is involved in the cleavage of the polyprotein into core, E1, E2, and p7 proteins (see Fig. 12.2). In addition, the core protein is often further cleaved by signal peptide peptidase (SPP). The rest of cleavage is catalyzed by two viral proteases (NS2 and NS3/4A).

The N-terminal subdomain of NS3 exhibits serine protease activity. NS3 forms a complex with NS4A, which acts as a cofactor for serine protease activity. NS4A, a small polypeptide composed of only 54 amino acid residues harboring a transmembrane (TM) domain, facilitates the membrane association of NS3 as well as the appropriate protein folding of NS3, thereby augmenting the protease activity (Fig. 12.5). In addition, NS3 exhibits an RNA helicase activity. The C-terminal domain of NS3 belongs to the DEAD-box RNA helicase family. The RNA helicase is likely to facilitate unwinding of a duplex structure present during RNA replication.

RNA Genome Replication: As viral proteins accumulate, the viral genome replication follows, in which negative-strand RNA is newly synthesized using the genomic RNA as a template. The nascent negative-strand RNA serves, in turn, as a template for positive-strand RNA synthesis (see Fig. 12.3). No primer is required for HCV RNA synthesis, a feature dubbed "*de novo synthesis*." Several HCV proteins are involved in RNA genome replication (see Table 12.3)

7. **Tight junction** A kind of junction in epithelial cells which confers polarity by blocking lateral diffusion. Hepatocytes are epithelial cells (Box 23.1).

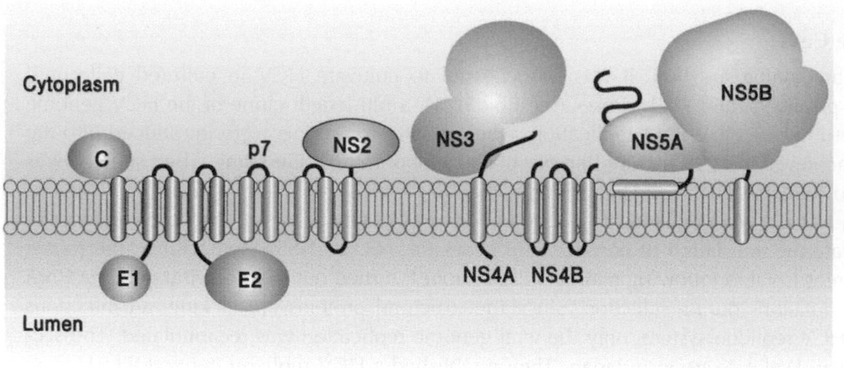

FIGURE 12.5 Membrane-associated domain structure of HCV proteins. All HCV proteins are characteristically membrane-associated. The viral proteins are schematically drawn with TM domain, globular domain, and unstructured loop domains. NS3 is membrane bound via interaction with NS4A. NS5A is also associated with NS5B via protein–protein interaction.

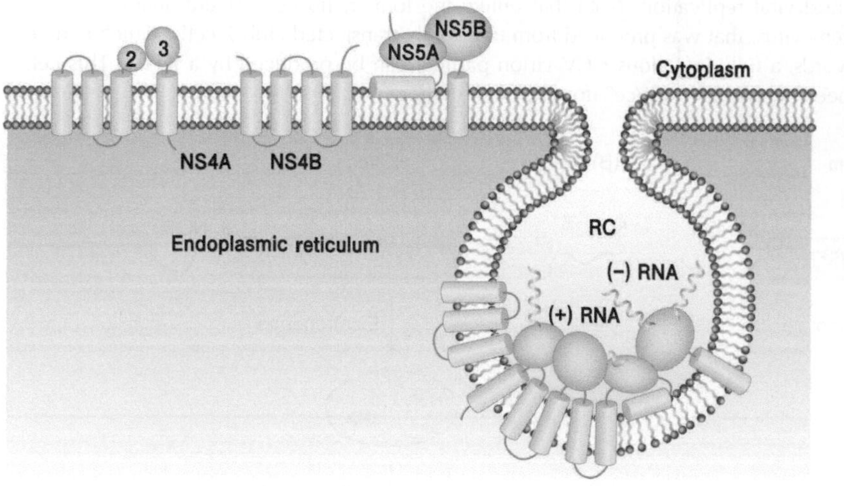

FIGURE 12.6 HCV RNA genome replication occurring in association with membrane structures. The HCV genome replication occurs in association with the vesicular membrane structures termed a membraneous web. The membraneous web is primarily derived from ER. The formation of invagination that is triggered by NS4B or NS5A is the site for viral RNA replication. This peculiar membrane structure is considered as a viral strategy to sequester viral RNAs (ie, double-strand RNA) from being sensed by pattern recognition receptors.

First, *NS5B* (RdRp) provides the catalytic activity for the RNA genome replication. NS5B is associated with the membrane via its TM domain at its C-terminus (see Fig. 12.5). In addition, *NS5A* (phosphoprotein with RNA-binding ability) contributes to the RNA genome replication by modulating the catalytic activity of NS5B.

One peculiar feature of HCV-infected cells are that they contain multiple vesicles forming an intracellular membrane structure, termed as a *membranous web (MW)*. The MW is derived primarily from the ER and is formed as invaginations of the ER membrane. Importantly, MW, which is induced by NS4B, is the site for the viral RNA replication, where HCV *replication complex*[8] *(RC)* resides (Fig. 12.6).

As such, what then constitutes the HCV RC? Besides the viral RNA genome, multiple nonstructural proteins including NS5A, NS5B, NS3/4A, and NS4B are involved. First, NS5A, a phosphoprotein with RNA-binding ability, induces a *double membrane vesicle*[9] *(DMV)*, a dominant constituent of the MW structure. A recent report demonstrating DMV represents the site for viral RNA replication has drawn attention. It is likely that NS5A converts the NS4B-induced MW to a DMV. On the other hand, the contribution of the RNA helicase activity of NS3 to viral genome replication is unclear. Perhaps, the helicase activity may facilitate the initiation of viral RNA replication by melting out the RNA stem-loop structures present in both 5′ and 3′ NCR. In addition, it might facilitate the elongation driven by NS5B (RdRp) via removing secondary structures of the RNA template, thereby enhancing the "processivity."

To investigate the viral RNA genome replication, it is crucial to get "infectious clone" or "*replicon*,"[10] from which autonomous replicating RNA is expressed. Recently, an HCV replicon and "infectious clone" have been established (Box 12.3).

8. **Replication complex (RC)** It refers to a subcellular site, where viral RNA replication is confined. It is also referred to as a "replication factory."

9. **Double membrane vesicle (DMV)** It refers to double membrane structures formed as protrusions from the ER membrane into the cytosol, frequently connected to the ER membrane via a neck-like structure.

10. **Replicon** A replicon in genetics is a region of DNA or RNA that replicates from a single origin of replication. It refers to an autonomous replicating RNA in this context.

BOX 12.3 Propagation of HCV in Culture Cells

Despite strenuous efforts since its molecular cloning in 1989, it was unsuccessful to cultivate HCV in cultured cells until recently. In other words, unlike most of the positive-strand RNA viruses (see Box 11.1), a full-length clone of the HCV genome in an appropriate expression plasmid failed to induce HCV RNA replication, when in vitro transcripts were introduced into the hepatoma cells. The first breakthrough was achieved in 1999 by Ralf Bartenschlager's laboratory in Germany. Their strategy was to insert the neomycin selection marker into the HCV genome. The rationale is to select HCV replicating cells, if any. To this end, firstly, the structural genes (ie, core, E1, and E2) were replaced by the neomycin marker and secondly, encephalomyocarditis virus (EMCV) IRES was inserted to facilitate the translation of nonstructural genes (ie, NS3 to NS5). Remarkably, the robust HCV RNA replication was observed in colonies isolated following neomycin selection. It turned out that the viral genome RNA acquired quite a few cell culture-adaptive mutations during selection. Since the structural proteins are lacking, an infectious HCV virion could not be produced. In this HCV replicon system, only the viral genome replication was recapitulated. The second breakthrough was made in 2005 by Takaji Wakita's group in Japan. They established a HCV replicon (termed JFH-1) from an HCV isolate from a fulminant hepatitis patient. Surprisingly, transfection of in vitro transcripts made from the full-length JFH-1 clone into Huh-7 cells led to the robust viral replication. Note that unlike the former, no cell culture adaptation was required. Remarkably, for the first time progeny virus, that was produced from transiently transfected Huh-7 cells, which can be used to infect naïve Huh-7 cells. In other words, a fully infectious HCV virion particle can be produced by a JFH-1. This cell culture derived infectious viral particle has been dubbed as "*HCVcc*" (for HCV cell culture).

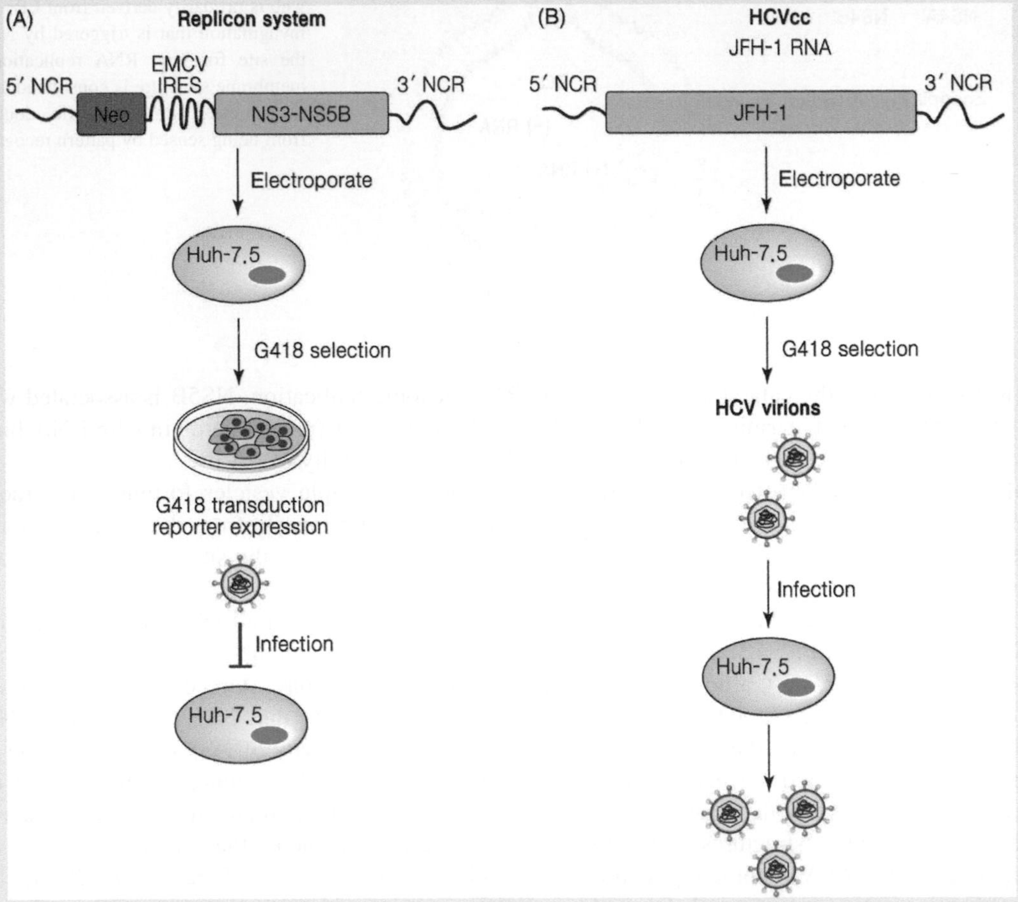

HCV replication in cultured cells. (A) A strategy to design the HCV replicon with neomycin selection marker. To accommodate the inserts (the neomycin selection marker gene and EMCV IRES element), the minimal HCV genome essential for the viral RNA replication were retained, including 5′ NCR, 3′ NCR and the nonstructural proteins (ie, NS3 to NS5). EMCV IRES was inserted for the translation of NS3-NS5 proteins. Huh-7.5 human hepatoma cell line. (B) A schematic diagram showing the procedure involved in HCVcc production. The full-length HCV clone derived from a Japanese fulminant hepatitis patient (JFH-1) is drawn on the top. HCV progeny produced from the transfected cell could reinfect naïve Huh-7 cells.

Assembly and Release: As described above, HCV RNA genome replication occurs in close proximity with intracellular membrane structures. What about the capsid assembly, a step following RNA genome replication? Peculiarly, *lipid droplet*[11] *(LD)* serves as the site of capsid assembly. In fact, the viral core protein associated with LD recruits viral RNA associated with NS5A, a component of viral RC (see Fig. 12.6). Then, the RNA becomes enclosed by core proteins. The core-NS5A interaction not only triggers the RNA packaging but also leads to ensuing envelopment by E1 and E2 envelope proteins. Finally, the assembled capsids bud into lumen of ER and are released extracellularly (see Fig. 12.3). One could envision that capsid assembly is coupled with the budding process. In line with this notion, p7 (see Fig. 12.2), an ion channel protein, is critical not only for envelopment, but also for the capsid assembly.

12.5 EFFECTS ON HOST

Chronic HCV infection is associated with liver cirrhosis and HCC. HCV propagation in cultured cells, however, does not induce any significant cytopathic effect. Therefore, HCV associated liver pathology is most likely a consequence of the immune response. The underlying mechanism for HCV-induced pathogenesis is now only beginning to be uncovered.

Blockade of Innate Immunity: The vast majority of HCV infection ($>80\%$) progresses to chronic infection. The high chronicity of HCV infection reflects the effective suppression of the host innate immune response. In fact, HCV proteins counteract host innate immune response by the suppression of both IFN induction and IFN action.

Suppression of IFN Induction: HCV blocks two major signaling pathways involved in the induction of innate immunity. The blockade is achieved by the cleavage of two adapter proteins by *NS3/4A serine protease* (see Fig. 5.18). First, the invading HCV is sensed by TLR3, which relays the signal downstream to the *TRIF* adapter. The TRIF adapter is specifically cleaved by NS3/4A protease, thereby blocking the TLR3 signaling. Second, *MAVS*, an adapter for cytoplasmic sensor RIG-I, is likewise cleaved by NS3/4A protease. Overall, HCV blocks two pathways of innate immunity effectively via specific inactivation of adapter molecules by the viral protease.

Suppression of Interferon Action: Some HCV proteins block interferon action as well. First, *NS5A* acts as antagonist of interferon action. NS5A also induces IL-8 expression, which inhibits interferon action. In addition, NS5A binds to PKR, thereby inhibiting its antiviral activity. Besides, *E2* glycoprotein also inhibits PKR via direct binding.

Pathogenesis: In addition, chronic HCV infection is often associated with *hepatic steatosis*, also known as a fatty liver disease. Accumulation of LDs in HCV-infected hepatocytes, which serves as the platform for virion assembly, is related to the pathogenesis.

Tumors: Chronic HCV infection often leads to the development of liver cancer. Still, it remains uncertain how chronic HCV infection induces tumor formation. Multiple viral proteins including core, NS3, NS5A, and NS5B are reportedly involved in the regulation of tumor suppressor genes via modulation of a signaling pathway, which is reminiscent of DNA tumor viruses. For instance, NS5B, in association with *E6AP*,[12] acts to target *Rb*[13] protein for degradation via an ubiquitin-proteasome pathway (Fig. 12.7). E6AP is a cellular protein that was identified as an HPV E6 binding protein (see Fig. 7.10). Remarkably, the proteins encoded by two unrelated viruses (ie, HCV NS5B and HPV E6) co-opt E6AP for Rb degradation. It is of great interest to see whether NS5B and E6 share structural similarity with respect to E6AP binding.

Besides the direct mechanism by HCV proteins, it was suggested that prolonged inflammation during persistent viral infection indirectly contributes to cancer development. It is worth noting HCV is the only tumor virus that does not have DNA intermediates in its viral life cycle. Note that all tumor viruses are either DNA tumor viruses (ie, polyomavirus, papillomavirus, adenovirus, and herpesvirus) or retroviruses that have DNA intermediates (see Table 24.2).

12.6 PERSPECTIVES

Since its discovery in 1989, HCV has been extensively investigated due to its clinical importance. The molecular details of the HCV life cycle have been uncovered to a great extent, parallel to that of HIV. For instance, HCV replicons have been established which support HCV replication in cultured cells. Remarkably, the five cellular receptors responsible

11. **Lipid droplet (LD)** Lipid-rich cellular organelles which regulate the storage and hydrolysis of neutral lipids. They are found largely in adipose tissue. They also serve as reservoirs of cholesterol and acyl-glycerols for membrane formation and maintenance.

12. **E6AP (E6-associated protein)** A host factor that was first identified as HPV E6 binding protein. It is now classified as a member of HECT (homologous to E6AP carboxyl-terminus)-type ubiquitin E3 ligase family.

13. **Rb (retinoblastoma)** A tumor suppressor gene that was first identified as a defective gene in retinoblastoma.

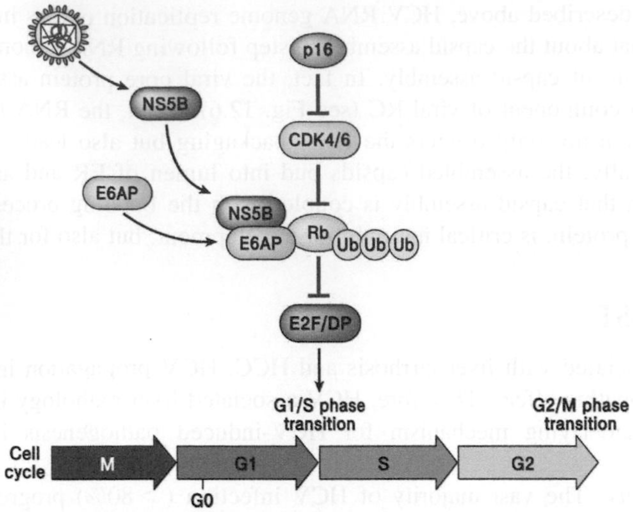

FIGURE 12.7 Regulation of Rb by NS5B. Rb protein, a gatekeeper of cell cycle regulation, blocks cell cycle progression from G1/S transition via blocking E2F transcription factor. In HCV-infected cells, NS5B, in association with E6AP, targets Rb protein for ubiquitin-mediated degradation, thereby allowing progression to the S phase of the cell cycle.

for HCV entry have all been discovered. The crystal structures of almost all the HCV proteins have been resolved including NS3/4A serine protease and NS5B RdRp. Despite the remarkable progress in our understanding of HCV biology, there is still much to be learned about the pathogenic mechanisms as well as the mechanistic details of HCV replication, particularly, how the virus weaves the functions of host cell activities into complex web of intracellular events in its life cycle. Importantly, several direct acting antiviral (DAA) drugs—boceprevir, telaprevir, and sofosbuvir—have been developed and have received FDA approval. Still, efforts to develop a preventive vaccine have been unsuccessful. A recently generated humanized mouse model supporting the complete HCV life cycle will facilitate studies on infection pathology and vaccine development (see Journal Club). In contrast to the remarkable progress made in HCV field, classical mosquito-borne flaviviruses have received much less attention over the years, although over 50 million people are infected every year (see Box 12.1). In the meantime, West Nile virus (WNV), another mosquito-borne flavivirus, has arrived in North America (see Fig. 21.4). The WNV outbreak in North America demonstrated the epidemiological importance of mosquito-borne flaviviruses. It is hoped that knowledge learned from HCV can be applied to these and other vector-borne flaviviruses.

12.7 SUMMARY

- *Flavivirus*: The flavivirus family includes many mosquito-borne viruses including yellow fever virus and dengue virus. HCV is classified in its own genus *Hepacivirus*.
- *HCV*: HCV is a small, enveloped, icosahedral virus which possesses a single positive-strand RNA of 9.6 kb.
- *HCV genome replication*: The RNA genome replication of HCV takes place in association with intracellular membrane structures, which is referred to as a membranous web.
- *Host effect of HCV*: Chronic HCV infection leads to cirrhosis and liver cancer. It remains uncertain how chronic HCV infection induces liver cancer.
- *Antiviral drugs*: In addition to broad spectrum antivirals such as IFN and ribavirin, three direct acting antiviral (DAA) drugs—boceprevir, telaprevir, and sofosbuvir—have been developed for the treatment of chronic HCV infection.
- *Preventive vaccine*: Preventive vaccine for HCV infection is not yet available.

STUDY QUESTIONS

12.1. Describe two *cis*-acting elements essential for HCV RNA genome replication and state their functions.

12.2. One salient feature of HCV infection is the high chronicity, demonstrating HCV effectively suppresses the innate immune response. Describe what is known about the immune evasion mechanisms of HCV infection.

12.3. HCV can infect human hepatocytes (eg, Huh-7.5 cells) but cannot infect rodent hepatocytes (eg, AML12 cells). It was then hypothesized that a host restriction factor in rodent cells may limit HCV infection. How would you test this hypothesis?

SUGGESTED READING

Dorner, M., Horwitz, J.A., Donovan, B.M., Labitt, R.N., Budell, W.C., Friling, T., et al., 2013. Completion of the entire hepatitis C virus life cycle in genetically humanized mice. Nature. 501 (7466), 237–241.

Dreux, M., Garaigorta, U., Boyd, B., Decembre, E., Chung, J., Whitten-Bauer, C., et al., 2012. Short-range exosomal transfer of viral RNA from infected cells to plasmacytoid dendritic cells triggers innate immunity. Cell Host Microbe. 12 (4), 558–570.

Lemon, S.M., McGivern, D.R., 2012. Is hepatitis C virus carcinogenic? Gastroenterology. 142 (6), 1274–1278.

Lindenbach, B.D., Rice, C.M., 2013. The ins and outs of hepatitis C virus entry and assembly. Nat. Rev. Microbiol. 11 (10), 688–700.

Paul, D., Madan, V., Bartenschlager, R., 2014. Hepatitis C virus RNA replication and assembly: living on the fat of the land. Cell Host Microbe. 16 (5), 569–579.

JOURNAL CLUB

- Dorner, M., Horwitz, J.A., Donovan, B.M., et al., 2013. Completion of the entire hepatitis C virus life cycle in genetically humanized mice. Nature 501 (7466): 237–241.

  Highlight: This seminal article reports the establishment of a mouse model for HCV infection. Building on the observation that CD81 and occludin comprise the minimal set of human factors required to render mouse cells permissive to HCV entry, this work demonstrated transgenic mice stably expressing human CD81 and occludin can support HCV entry.

Chapter 13

Other Positive-Strand RNA Viruses

Chapter Outline

In the two preceding chapters, picornaviruses and flaviviruses were covered. Besides these two positive-strand RNA viruses, caliciviruses, togaviruses, and coronaviruses are also clinically important human pathogens. Here, these positive-strand RNA viruses will be covered only briefly with an emphasis on classification and the genome features.

13.1 CALICIVIVIRUS

Caliciviruses[1] (family Caliciviridae) are similar to picornaviruses in many respects. Caliciviruses are small (35–40 nm), nonenveloped, and icosahedral viruses that possess a positive-strand RNA genome of 7–8 kb. Caliciviruses are important human and veterinary pathogens which are associated with a broad spectrum of diseases in their hosts. Here, norovirus, a prototype of calicivirus and the causative agent of nonbacterial gastroenteritis in humans, will be covered.

Classification: Caliciviruses comprise four genera (Table 13.1). Norwalk virus, a member of the genus *Norovirus*, causes epidemic gastroenteritis. Sapporovirus, a member of the genus *Sapovirus*, also causes epidemic gastroenteritis. Besides human caliciviruses, two veterinary caliciviruses are found. Rabbit hemorrhagic disease virus (RHDV), a member of the genus *Lagovirus*, is associated with a fatal liver disease in rabbits. Feline calicivirus (FCV), a member of the genus *Vesivirus*, causes respiratory disease in cats.

13.1.1 Norovirus

Norovirus[2] represents a prototype of calicivirus. Noroviruses are the causative agents of nonbacterial gastroenteritis in humans and are responsible for almost all viral gastroenteritis outbreaks worldwide.

Epidemiology: Noroviruses are transmitted via the oral-fecal route due to contaminated water and food, or directly from person to person. They are extremely contagious. Transmission can be aerosolized when those stricken with the illness vomit, or when a toilet flushes with vomit or diarrhea in it; infection can occur by eating food or breathing air near an episode of vomiting, even if cleaned up. The viruses continue to be shed after symptoms have subsided and shedding can still be detected many weeks after infection.

Noroviruses are responsible for over 50% of gastroenteritis worldwide (Fig. 13.1). Each year, human noroviruses cause at least around 267 million episodes and over 200,000 deaths in developing nations as well as approximately 900,000 cases of pediatric gastroenteritis in industrialized nations. Even in the United States, they cause over 20,000 episodes and 300 deaths. Shellfishes (oysters in particular), which are often consumed uncooked, are frequently the

1. **Calicivirus** The term is derived from Greek word for "cup"-*calyx.*

2. **Norovirus** The norovirus was originally named the "Norwalk agent" after Norwalk, Ohio, in the United States, where an outbreak of acute gastroenteritis occurred among children at the elementary school in 1968. The virus was given the name "Norwalk virus."

Molecular Virology of Human Pathogenic Viruses. DOI: http://dx.doi.org/10.1016/B978-0-12-800838-6.00013-8

TABLE 13.1 Classification of Calicivirus

Genus/Species	Acronym	Host	Transmission	Disease
Norovirus/Norwalk virus	NoV	Humans	Fecal-oral	Epidemic gastroenteritis
Sapovirus/Sapporovirus	SV	Humans, pigs	Fecal-oral	Epidemic gastroenteritis
Vesivirus/Feline calicivirus	FCV	Cats	Contact	Respiratory disease
Lagovirus/Rabbit hemorrhagic disease virus	RHDV	Rabbits	Fecal-oral	Hemorrhages

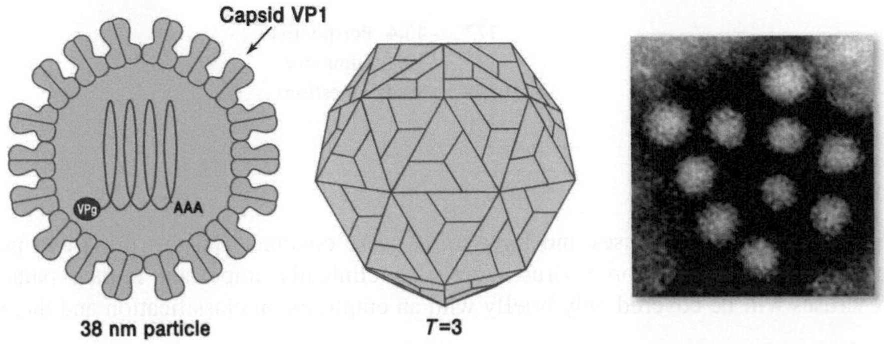

FIGURE 13.1 Norovius particle. (Left) A cross section of norovirus capsid particle of 38 nm in diameter. VPg is linked to 5′ terminus of the positive-strand RNA genome. The capsid is an icosahedral particles having $T = 3$ symmetry. (Right) The electron microscopic image of norovirus particles.

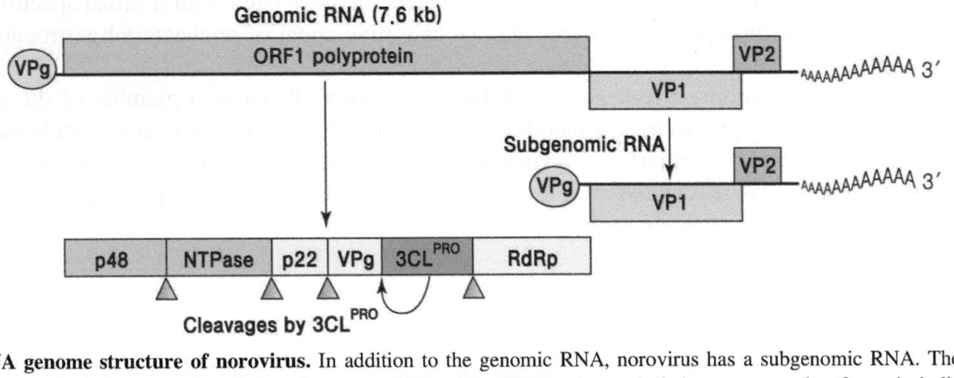

FIGURE 13.2 RNA genome structure of norovirus. In addition to the genomic RNA, norovirus has a subgenomic RNA. The VPg is covalently linked to the 5′ termini of both the genomic and the subgenomic RNAs. The ORF1 polyprotein is later processed to form six individual proteins by a virus-encoded $3CL^{PRO}$ protease. RdRp, RNA-dependent RNA polymerase.

source of infection. Oysters act as a filter to concentrate the virus particles present in sea water. Symptoms include vomiting, diarrhea, and dehydration. No vaccine or antiviral drugs are yet available.

Genome Structure: The RNA genome of norovirus is fundamentally similar to that of picornaviruses, except that a *subgenomic RNA*[3] is expressed (Fig. 13.2). *VPg* (virion protein genome-linked) is covalently linked to both the genomic and the subgenomic RNAs. The ORF1 polyprotein encoded by the genomic RNA encodes nonstructural proteins such as $3CL^{PRO}$ and RdRp that are essential for viral RNA replication. On the other hand, the subgenomic RNA encodes the structural proteins such as VP1 and VP2. Unlike picornaviruses, internal ribosome entry site (IRES) element is not

3. **Coronavirus** The name "coronavirus" is derived from the Latin *corona*, meaning crown or halo.

BOX 13.1 VPg of Norovirus in Translation

VPg is linked to the 5′ terminus of norovirus RNAs, which is a reminiscent of picornavirus. However, unlike picornavirus, the IRES element, which is essential for the cap-independent translation, is not present in the 5′ NCR of norovirus RNAs. An immediate question is then how norovirus genome RNA is translated. Surprisingly, it was found that VPg has the ability to bind eIF4E and, in doing so, it facilitates the translation initiation of viral RNAs in eIF4E-dependent manner. In other words, VPg substitutes the role of the cap structure of eukaryotic mRNAs in recruiting eIF4E, a cap-binding protein. Overall, norovirus co-opts the cellular eIF4E-dependent translation mechanism via a novel VPg-eIF4E interaction.

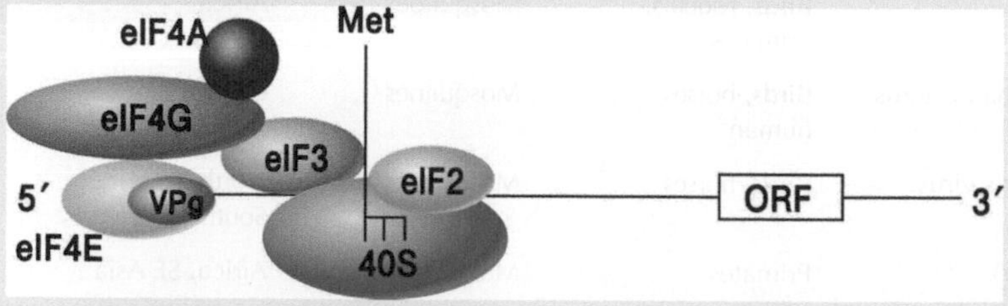

Translation initiation factors associated with norovirus RNA. eIF4F = eIF4A + eIF4G + eIF4E.

found in the 5′ NCR. Instead, the VPg recruits eIF4E, a cap-binding translation initiation factor, to initiate translation (Box 13.1). In other words, the VPg substitutes for the cap structure in recruiting host translation factors.

Cultivation in Cell Culture: The investigation on human norovirus has been hampered by the lack of cell lines that support the virus infection. Since human norovirus has remained uncultivable, the studies on viral genome replication could be performed only by transfection of viral genomic RNA into appropriate cells. Recently, major progress in norovirus research was made by a successful demonstration of a human norovirus infection in a cell culture (see Journal Club). Successful establishment of an in vitro cultivation system of human norovirus will facilitate the development of prophylatic vaccine and antivirals.

Animal Model: Animal model for human norovirus infection is not yet available. Murine norovirus was recently discovered and will be explored as an animal model for human norovirus.

13.2 TOGAVIRUS

Togaviruses (family Togaviridae) are enveloped and icosahedral viruses that possess a positive-sense single-strand RNA genome of 11 kb. Members of this family are frequently referred to as alphaviruses, a genus of this family.

Sindbis virus and Semliki Forest virus (SFV) have been extensively studied as the prototype of togaviruses, since these viruses are only weakly pathogenic to human.

Classification: Family Togaviridae is constituted by two genera: genus *Alphavirus* and genus *Rubivirus* (Table 13.2). Chikungunya virus, an emerging virus, is an important human pathogen belonging to the genus *Alphavirus*. Togaviruses are important veterinary pathogens, and are transmitted via mosquitoes. Some of them are zoonotic viruses, including Venezuelan equine encephalitis virus (VEEV). Rubella virus is the only member of togavirus family that causes significant disease in human—German measles.

Epidemiology: Togaviruses are largely transmitted via mosquitoes. Chikungunya virus, an emerging virus, caused a massive outbreak in some countries in Africa including Kenya and Madagascar in 2004–2006. VEEV, a zoonotic virus, could infect birds, horses, and human. VEEV infection causes rash, fever, and encephalitis. Rubella virus, that causes German measles in human, is unique among togaviruses in that it is not transmitted via mosquitoes.

Virion Structure: Togavirus virions are small (70–80 nm), enveloped, icosahedral capsid inside (Fig. 13.3). Togavirus virion structure stands out in that not only the capsid but also the envelope has an icosahedral symmetric structure. Indeed, 240 molecules of E1/E2 dimer form an icosahedral structure having $T = 4$ symmetry. Likewise, 240 molecules of capsid proteins form an icosahedral structure having $T = 4$ symmetry. Togavirus is unprecedented in that the envelope as well as the capsid has a symmetric structure (see Fig. 2.9).

Genome Structure: Togavirus genome is a positive-strand RNA of 11 kb (Fig. 13.4). In addition to the genomic RNA that encodes a polyprotein for nonstructural proteins, a subgenomic RNA is expressed that encodes a polyprotein for structural proteins. The 5′ terminus of RNA is capped, whereas the 3′ terminus is polyadenylated. Note that the

TABLE 13.2 Classification of Togavirus

Genus/Species	Vertebrate Hosts	Invertebrate Host	Geographic Distribution	Human Disease
Alphavirus				
Sindbis virus (SINV)	Birds	Mosquitoes	Africa, Australia, Middle East	Fever, arthritis, rash
Semliki Forest virus (SFV)	Birds, rodents, primates	Mosquitoes	Africa	Fever, encephalitis
Venezuelan equine encephalitis virus (VEEV)	Birds, horses, human	Mosquitoes	North America, South America	Fever, encephalitis
Western equine encephalitis virus (WEEV)	Birds, horses	Mosquitoes	North America, South America	Fever, encephalitis
Chikungunya virus (CHIKV)	Primates	Mosquitoes	Africa, SE Asia	Fever, arthritis, rash
Rubivirus				
Rubella virus	Human		German measles	

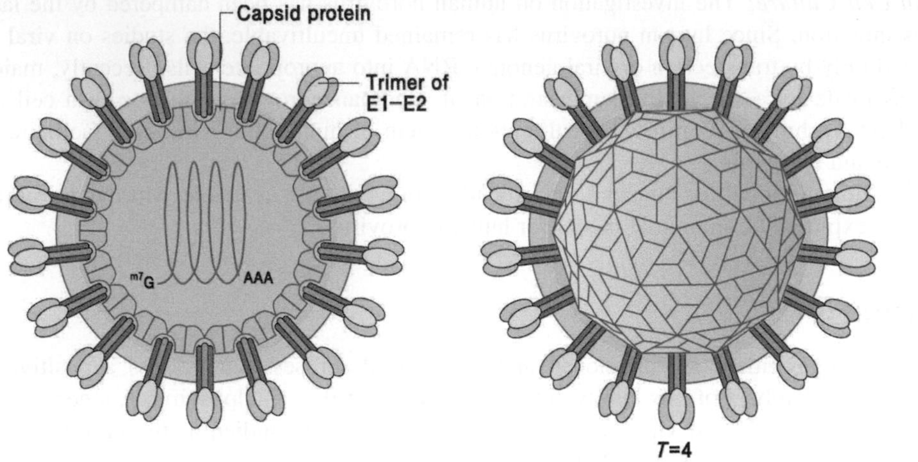

FIGURE 13.3 Sindbis virus particle. Cross-section view of sindbis virus particle. Note that trimers of the E1/E2 dimer envelope glycoproteins are tightly associated with the capsid proteins, revealing a parallel symmetric arrangement of envelope glycoproteins.

capping at the 5′ terminus is facilitated by the viral capping enzyme (ie, nsP1) (see Box 16.3). The polyproteins are processed into individual proteins either by viral protease (ie, nsP2 protease) or by host proteases such as signal peptidase and furin.

13.3 CORONAVIRUS

Coronaviruses[4] (family Coronaviridae) are enveloped, spherical, and about 120 nm in diameter and possess a single-strand RNA genome of approximately 30 kb.

Mouse hepatitis virus (MHV) is a prototype of coronavirus. MHV outbreak in animal laboratory facilities represents a serious concern. In some cases, the facility has to be closed for many years until reuse. On the other hand, human coronavirus has been considered clinically unimportant, until *SARS* (severe acute respiratory syndrome) coronavirus

4. **Subgenomic RNA** It refers to a viral RNA that is smaller than the full-length viral RNA.

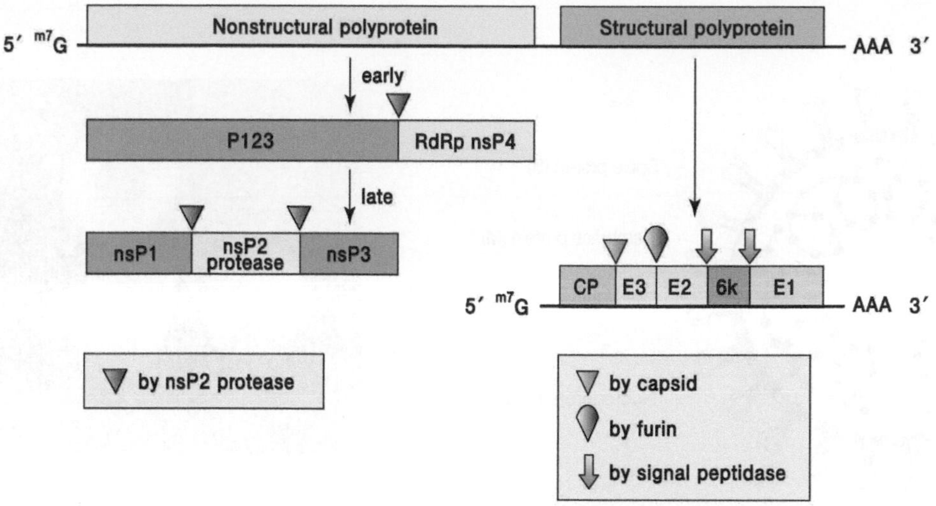

FIGURE 13.4 RNA genome structure of sindbis virus. In addition to the genomic RNA, sindbis virus has a subgenomic RNA. The nonstructural proteins are translated from the genomic RNA, while the structural proteins are translated from the subgenomic RNA. The cleavage sites of polyproteins are denoted by symbols of respective proteases. RdRp, RNA-dependent RNA polymerase.

TABLE 13.3 Classification of Coronavirus

Genus	Virus Species	Host	Disease
Alpha Coronavirus	Human coronavirus-229E	Human	Colds, pneumonia
	Human coronavirus-NL63	Human	Colds, pneumonia
	Transmissible gastroenteritis virus (TGEV)	Swine	Gastroenteritis
Beta Coronavirus	Mouse hepatitis virus (MHV)	Rodents	Hepatitis
	Human coronavirus-HKU1	Human	Pneumonia
	SARS-coronavirus (SARS-CoV)	Human	Pneumonia, Gastroenteritis
	MERS-coronavirus (MERS-CoV)	Human, Camel	Respiratory infection
Gamma Coronavirus	Avian infectious bronchitis virus (AIBV)	Avian	Kidney infection

was discovered in 2003, as a newly emerging virus. More recently, the *MERS* (Middle-East respiratory syndrome) outbreak in the Middle East has drawn attention to human coronaviruses again (see Fig. 21.11).

Classification: Family Coronaviridae is subdivided into three genera: alpha, beta, and gamma coronaviruses (Table 13.3). SARS-coronavirus (SARS-CoV) and MERS-coronavirus (MERS-CoV) are classified as *Beta coronavirus*.

Epidemiology: Human coronaviruses were known to cause only mild respiratory infections, until the SARS outbreak. The SARS-CoV outbreak that occurred in 2003 has drawn attention, since it was fatal, causing SARS (see below). Moreover, a novel human coronavirus, MERS-CoV, is responsible for a new emerging respiratory infection that occurred in 2012 in Middle East countries, including Saudi Arabia (see Fig. 21.11).

Virion Structure: Coronaviruses are enveloped and contain a large helical nucleocapsid inside (Fig. 13.5). The viral envelope is studded with spike glycoprotein trimer (S), hemagglutinin-esterase dimer (HE), membrane protein (M), and envelope protein (E). In particular, the protruding spike protein (S) characterizes the crown shaped virion, which it was named after. "Corona" (crown) refers to the characteristic appearance of virions under electron microscopy with a fringe of large, bulbous surface projections creating an image reminiscent of the solar corona.

Genome Structure: The genome of coronavirus represents a large single-strand RNA of 27–32 kb (Fig. 13.6). It is the largest RNA genome among animal RNA viruses. The ORF1a and ORF1ab are translated as polyproteins, which are subsequently processed to 16 nonstructural proteins. In addition to the large genomic RNA, coronaviruses have

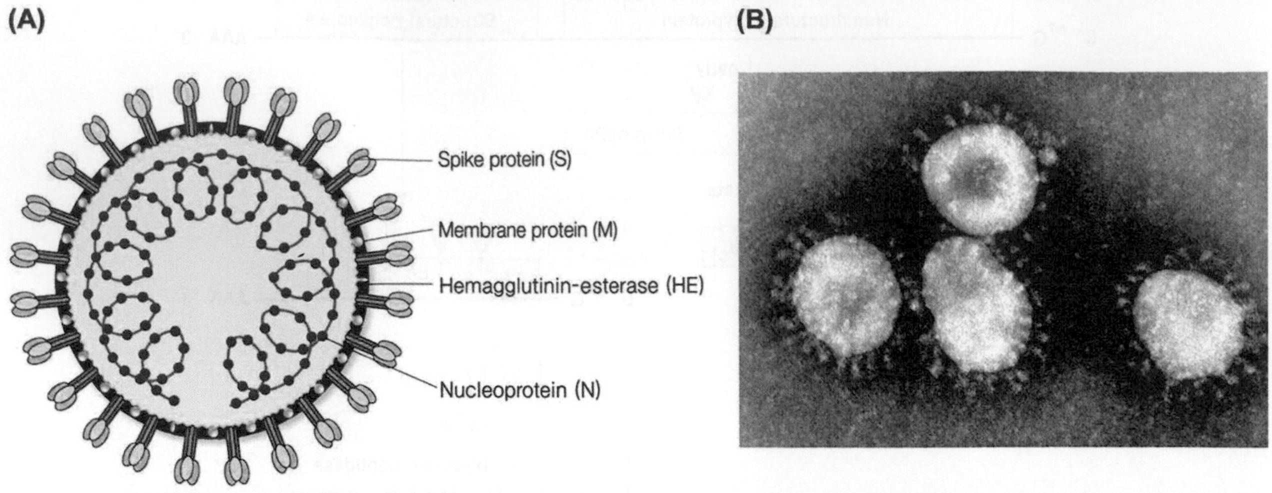

FIGURE 13.5 Coronavirus particle. (A) Envelope proteins are denoted: spike glycoprotein (S), hemagglutinin-esterase (HE), and membrane protein (M), enveloped small membrane pentamer (E), and nucleocapsid protein (N). A large helical nucleocapsid (N) inside encircles the viral RNA genome. (B) Electron microscopy of SARS-CoV. Coronaviruses are a group of viruses that have a halo, or crown-like (corona) appearance when viewed under an electron microscope.

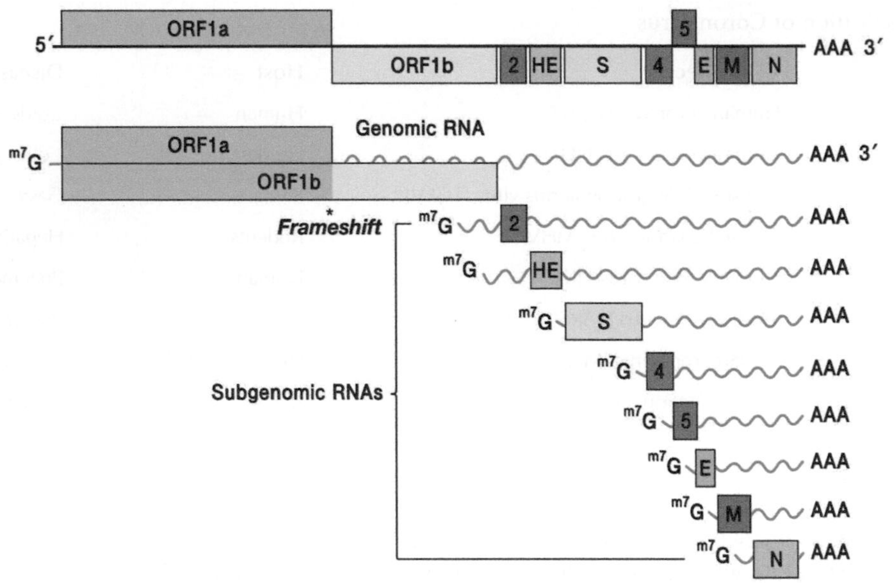

FIGURE 13.6 RNA genome structure of coronavirus. In addition to the large genomic RNA, coronavirus has eight subgenomic RNAs. The genomic RNA encodes two ORFs: ORF1a and ORF1b. The ORF1b is translated as an ORF1a-ORF1b fusion protein following a frame-shift (denoted by an *asterisk*). Each subgenomic RNA encodes one protein, as denoted by a *box*: spike glycoprotein (S), hemagglutinin-esterase (HE), membrane protein (M), enveloped small membrane pentamer (E), and nucleocapsid protein (N).

eight subgenomic RNAs, each of them encoding one structural protein. In particular, the S (spike) envelope glycoprotein binds to the host cell receptor and determines tissue tropism and host range.

SARS-CoV: SARS outbreak first occurred in Southern China (including Hong Kong), and spread to South East Asia and Northern America within a few weeks (see Fig. 21.10). The initial casualty and media hype made the public paranoid, canceling overseas travel and international meetings. SARS outbreaks resulted in over 8000 infections and 700 deaths in 20 countries. It turned out that bats were the reservoir for SARS-CoV. It is believed that bat coronavirus had acquired the ability to infect human, extending the host range, by having a few mutations in the spike protein. Recently, *ACE* (angiotensin-converting enzyme 2) was identified as the cellular receptor for virus entry of SARS-CoV to human infection.

MERS-CoV: MERS outbreak first occurred in the Middle East (mainly Saudi Arabia) in 2012, and spread to European countries in a limited manner (see Fig. 21.11). As of June 2015, MERS-CoV caused 1266 cases and 470 deaths reported in multiple countries. MERS-CoV cases have been reported in 23 countries, including Saudi Arabia, Malaysia, Jordan, Qatar, Egypt, the United States, and South Korea. The fatality of MERS-CoV is considerably higher than that of SARS-CoV, approaching 30%. It is speculated that the virus spreads from bats to human via dromedary camel (see Fig. 21.11). The risk of sustained person-to-person transmission appears to be very low. Recently, dipeptidyl peptidase 4 (DPP4 or also known as CD26) was identified as the cellular receptor for virus entry of MERS-CoV. Further, a mouse model for MERS-CoV infection was established that expresses the DPP4. It is hoped that the established mouse model will facilitate the development of a vaccine and antiviral drugs.

13.4 PERSPECTIVES

In this chapter, three miscellaneous positive-strand DNA viruses are described. Noroviruses, a prototype of caliciviruses, are responsible for almost all viral gastroenteritis outbreaks worldwide. Nonetheless, no vaccine and antiviral drugs are available to control norovirus infection. The lack of susceptible cell lines and animal model for norovirus infection have imposed barriers to basic research until recently. A recent successful demonstration of human norovirus infection using B lymphocytes (see Journal Club) deserves more attention in this regard. Such progress in norovirus infection system will greatly advance our understanding on the infection pathology of human norovirus and at the same time facilitate antiviral drug discovery and preventive vaccine development. Sindbis virus and SFV have been extensively studied as the prototype of togaviruses. Surveillance on zoonotic togaviruses, such as VEEV, has become more important. Finally, the emergence of deadly human coronaviruses—SARS-CoV and MERS-CoV—have bolstered research in these viral and often zoonotic pathogens. Accordingly, great advances, such as identification of host cell receptor, the establishment of reverse genetics, and small animal model for infection, have been made in the past decade.

13.5 SUMMARY

- *Calicivirus*: Noroviruses, the prototype of caliciviruses, are responsible for almost all viral gastroenteritis outbreaks worldwide.
- *Togavirus*: Sindbis virus and Semliki Forest virus (SFV) have been extensively studied as the prototype of togaviruses. Chikungunya virus, an emerging virus that caused an outbreak in Africa during 2004–2006, belongs to the togavirus family.
- *Coronavirus*: Coronaviruses possess the larger RNA genome of approximately 30 kb. SARS-coronavirus was discovered as a newly emerging virus that caused the 2003 SARS outbreak and 2012 MERS outbreak.

STUDY QUESTIONS

13.1 Consider that a novel positive-strand RNA virus was discovered. The RNA genome structure is organized similar to that of picornaviruses, encoding one large polyprotein. Moreover, a viral protein is covalently linked to the 5′ terminus of the RNA genome. (1) Please hypothesize the role of the 5′ terminus-linked viral protein. (2) How would you test your hypothesis?

13.2 Sindbis virus has subgenomic RNA as well as genomic RNA. (1) State your hypothesis on the mechanism by which the subgenomic RNA is transcribed. (2) How would you test your hypothesis?

SUGGESTED READING

Jones, M.K., Watanabe, M., Zhu, S., Graves, C.L., Keyes, L.R., Grau, K.R., et al., 2014. Enteric bacteria promote human and mouse norovirus infection of B cells. Science. 346 (6210), 755–759.

Karst, S.M., Wobus, C.E., 2015. A working model of how noroviruses infect the intestine. PLoS Pathog. 11 (2), e1004626.

Li, W., Moore, M.J., Vasilieva, N., Sui, J., Wong, S.K., Berne, M.A., et al., 2003. Angiotensin-converting enzyme 2 is a functional receptor for the SARS coronavirus. Nature. 426 (6965), 450–454.

Raj, V.S., Mou, H., Smits, S.L., Dekkers, D.H., Muller, M.A., Dijkman, R., et al., 2013. Dipeptidyl peptidase 4 is a functional receptor for the emerging human coronavirus-EMC. Nature. 495 (7440), 251–254.

Zust, R., Cervantes-Barragan, L., Habjan, M., Maier, R., Neuman, B.W., Ziebuhr, J., et al., 2011. Ribose 2′-O-methylation provides a molecular signature for the distinction of self and non-self mRNA dependent on the RNA sensor Mda5. Nat. Immunol. 12 (2), 137–143.

JOURNAL CLUB

- Jones, M.K., Watanabe, M., Zhu, S., et al., 2014. Enteric bacteria promote human and mouse norovirus infection of B cells. Science 346 (6210), 755–759.

  Highlight: The biggest hurdle in norovirus research has been that norovirus is not cultivatable in a cell culture. It has been speculated that noroviruses primarily target intestinal epithelial cells, which line the intestine and protect it from pathogens. Surprisingly, the authors here demonstrated that human norovirus can be propagated in B cells, with the help of enteric bacteria. It is an unprecedented finding that bacteria facilitates the virus infection.

Part III-2

Negative-Strand RNA Viruses

RNA viruses can be divided into two classes with respect to their genome polarity: positive-strand RNA viruses (PSVs) and negative-strand RNA viruses (NSVs). The PSVs were covered in the previous three chapters. The NSVs will be covered in the next three chapters. Rhabdovirus and influenza virus will be covered in chapter "Rhabdovirus" and chapter "Influenza Viruses," respectively. Other miscellaneous NSVs, including paramyxovirus, filovirus, and bunyavirus will be briefly described in chapter "Other Negative-Strand RNA Viruses." In addition, reovirus will be covered at the end of chapter "Other Negative-Strand RNA Viruses," even though it is a double-strand RNA virus, since the replication strategy is similar to that of negative-strand RNA viruses, where only one strand (ie, negative-strand) is used as a template.

NSVs are taxonomically divided into two orders, depending on whether their genomes are segmented or not. The NSVs possessing a single RNA genome are called **mononegavirales**, while the NSVs possessing multiple RNA genomes are called **multinegavirales** (Table PIII-2). Frequently, the term "**split genome**" is used to refer to multiple genomes of the multinegavirales.

The genome replication strategy of NSVs is a bit more complicated than that of PSVs, as they synthesize two distinct kinds of positive-strand RNAs. In other words, the full-length positive-strand RNA as well as the subgenomic mRNAs are transcribed from the negative-strand RNA (Fig. PIII-2). In the case of mononegavirales, in contrast to multinegavirales, a novel mechanism, which allows the synthesis of multiple mRNAs from a single RNA template, should be invoked.

A few characteristic features are shared by all NSVs (Table PIII-3). First, the RNA genome molecule of NSVs does not possess "**infectivity**." Unlike PSVs, transfection of the viral RNA genome into cells does not lead to viral genome replication or production of progeny virus. This is a defining distinction between PSVs and NSVs. Second, RNA polymerase (RdRp) molecules are packaged into virus particles in NSVs; by contrast, RdRp is not packaged into virions in PSVs. The virion-incorporated RdRp transcribes viral mRNAs from the viral genome immediately following infection. In fact, the incorporation of RdRp into the virion particle is the hallmark of NSVs. Third, one outstanding feature of NSVs is that their nucleocapsids have a helical symmetry (linear symmetry), in which the viral RNA is a string in the center of the nucleocapsid ribbon. Fourth, the nucleocapsid of NSVs, instead of the naked genome, serves as the template for viral RNA synthesis. Importantly, the RNA genome is always encapsidated by the nucleocapsid protein in the NSVs. During both transcription and replication, the RNA template is transiently released from the nucleocapsid by RdRp, and is utilized as a template for the RNA synthesis.

TABLE PIII-2 Classification of Negative-Strand RNA Viruses

Order	Family	Genus
Mononegavirales	Rhabdovirus	VSV (vesicular stomatitis virus)
		Rabies virus
	Paramyxovirus	Measles virus
		Sendai virus
	Filovirus	Ebola virus
	Bornavirus	Borna disease virus
Multinegavirales	Orthomyxovirus	Influenza virus
	Bunyavirus	Hantavirus

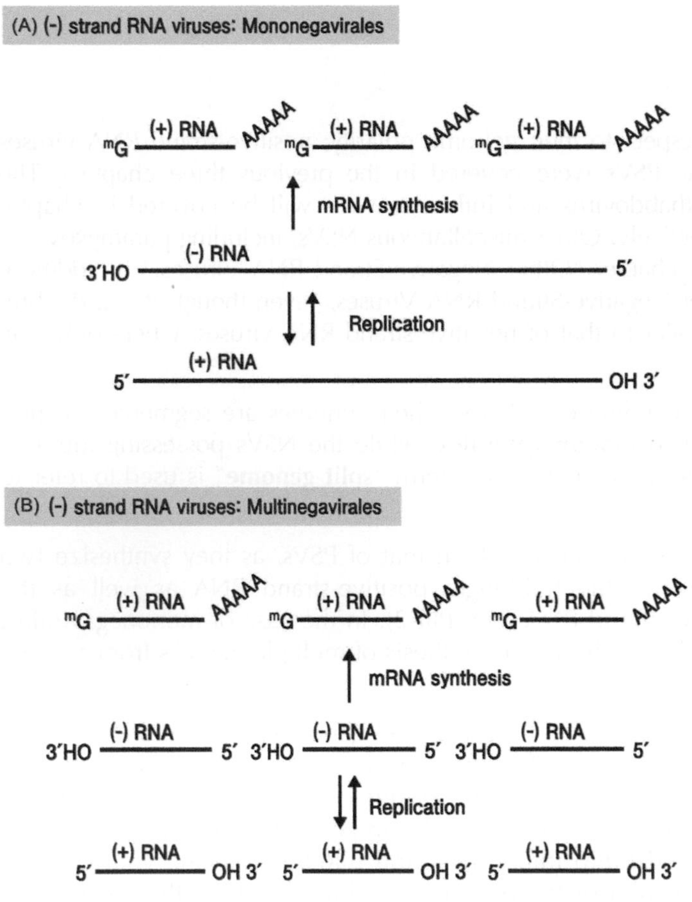

FIGURE PIII-2 The RNA genome replication strategies of negative-strand RNA viruses. A hypothetical RNA virus that has three protein coding genes (ie, mRNA) is shown for sake of brevity. (A) Mononegavirales. Two positive-strand RNAs are synthesized from a single negative-strand RNA genome (vRNA): multiple subgenomic mRNAs and a single complementary genome-length RNA (cRNA). (B) Multinegavirales. Two positive-strand RNAs are synthesized from a single negative-strand RNA genome (vRNA): one mRNA and one complementary RNA (cRNA) per each negative-strand RNA genome (vRNA). Note that the 5′ end of the negative-strand RNA genome is positioned to the right. The viral mRNAs are capped at the 5′ end and polyadenylated at the 3′ end. (+), positive-strand; (−), negative-strand.

TABLE PIII-3 Major Features of Negative-Strand RNA Viruses

Characteristics	Rhabdovirus	Paramyxovirus	Orthomyxovirus
Prototype	VSV	Sendai virus	Influenza virus
Genome	Nonsegmented	Nonsegmented	Segmented
Reassortment	N/A	N/A	Yes
Capsid symmetry	Helical	Helical	Helical
Site for RNA replication	Cytoplasm	Cytoplasm	Nucleus
Surface antigen	G protein	HA + NA	HA + NA

Chapter 14

Rhabdovirus

Chapter Outline

Rhabdovirus is a prototype of *Mononegavirales*, which has a single negative-strand RNA as a genome (Table 14.1). *VSV*[1] and *rabies virus*[2] belong to Rhabdoviridae.[3] This chapter will focus on vesicular stomatitis virus (VSV) that infects cattle, as VSV has been extensively studied among animal viruses that belong to the rhabdovirus family.

14.1 CLASSIFICATION

Classification: Rhabdoviruses are found in invertebrates as well as in vertebrates. Hence, its host range is wide, ranging from mosquitoes to cattle. It is composed of two genera, in which VSV is the prototype of vesiculovirus genus, and rabies virus is the prototype of lyssavirus genus (Table 14.2).

Epidemiology: VSV is transmitted via arthropods such as mosquitoes. VSV causes *vesicular stomatitis*,[4] which is a significant, but not fatal, disease in cattle and pigs. Humans are rarely infected by VSV via laboratory exposure; VSV infection causes flu-like symptom to humans.

Rabies virus causes fatal encephalomyelitis, resulting in more than 55,000 deaths annually. In fact, rabies virus, that causes fatal rabies, is the only kind of human rhabdovirus. Rabies virus is a neurotropic virus that invades the central nervous system via the peripheral nerve. The incubation period between infection and the first flu-like symptoms is typically 2 to 12 weeks. Symptoms may soon expand to slight or partial paralysis, anxiety, and insomnia. The patient may suffer hydrophobia. Significantly, its fatality is nearly 100% after the onset of the symptom. The only treatment option available to the infected individual is the postexposure immunization that was developed by Louis Pasteur a century ago (see Box 25.1).

14.2 THE VIRION AND GENOME STRUCTURE OF VSV

Virus Particles: VSV virions are bullet-shaped enveloped particles (Fig. 14.1). G protein, an envelope glycoprotein, is embedded into the lipid bilayer, while M protein, a matrix protein, coats the inner leaflet of the lipid bilayer. Inside the envelope, the nucleocapsid with helical symmetry is found, in which the RNA genome is encapsidated by numerous N proteins, nucleocapsid protein. L/P protein complex is bound to one tip of the nucleocapsid. The L protein is the viral RNA-dependent RNA polymerase (*RdRp*), and the associated P protein acts as a cofactor of RdRp (Table 14.3). Incorporation of RdRp into the viral particle is a hallmark of negative-strand RNA virus (NSV). The virion-associated RdRp carries out the viral mRNA transcription immediately after infection.

1. **VSV (vesicular stomatitis virus)** VSV can infect insects, cattle, horses, and pigs. It leads to a flu-like illness in humans.
2. **Rabies virus** The rabies virus causes rabies in humans and animals.
3. **Rhabdoviridae** The term *rhabdo* is a Greek word for "rod."
4. **Vesicular stomatitis** It is a disease in cattle, appearing as mucosal vesicles and ulcers in the mouth.

Molecular Virology of Human Pathogenic Viruses. DOI: http://dx.doi.org/10.1016/B978-0-12-800838-6.00014-X

TABLE 14.1 The Defining Features of Rhabdovirus		
Genome	**Particle Structure**	**Replication Mechanism**
Negative-strand RNA	Helical nucleocapsid	RNA-directed RNA synthesis
Nonsegmented RNA	RdRp encapsidated	Antitermination
Life Cycle	**Host Effect**	**Disease**
Cytoplasmic	Translation suppression	Rabies
–	Host shutoff	Stomatitis

TABLE 14.2 Classification of Rhabdoviruses

Genus	Virus	Vector	Host	Disease
Vesiculovirus	VSV-New Jersey	Mosquito	Mammal	Flu-like symptom
	VSV-Indiana	Mosquito	Mammal	Flu-like symptom
Lyssavirus	Rabies virus	Canine, bats	Human	Fatal encephalomyelitis

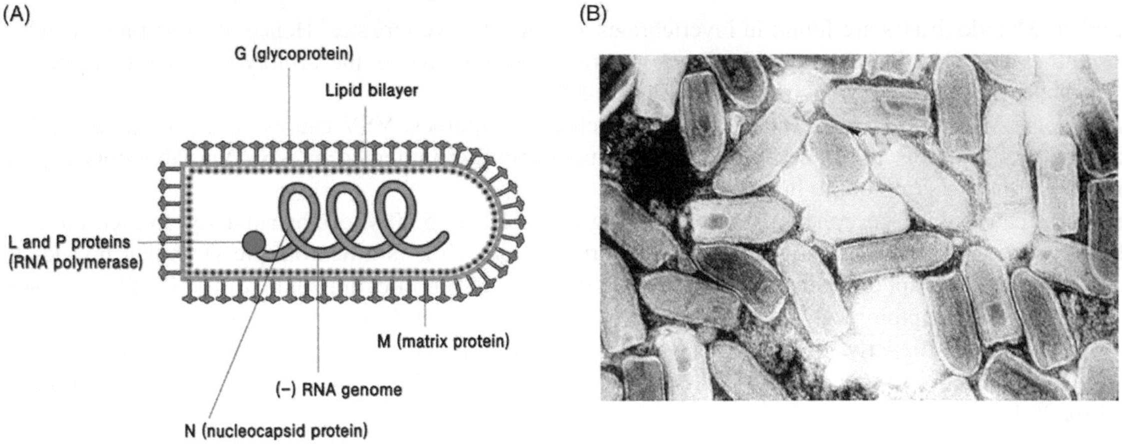

FIGURE 14.1 VSV particles. (A) A diagram showing a VSV particle. The VSV virion is a bullet-shaped particle 70 nm in diameter and 175 nm in length. The nucleocapsid found within the lipid bilayer contains the viral RNA genome. G protein (envelope glycoprotein), M protein (matrix), and N (nucleocapsid) are denoted. L/P protein complex (RdRp) is bound to the end of the nucleocapsid. (B) Transmission electron micrograph of VSV particles, which shows the characteristic bullet-shaped virion. Magnification = 40,000 ×.

TABLE 14.3 VSV Proteins and Their Functions

Protein	Function
N	Nucleocapsid protein
P	Phosphoprotein, RdRp cofactor
M	Matrix protein
G	Envelope glycoprotein
L	RdRp, capping, methyl transferase

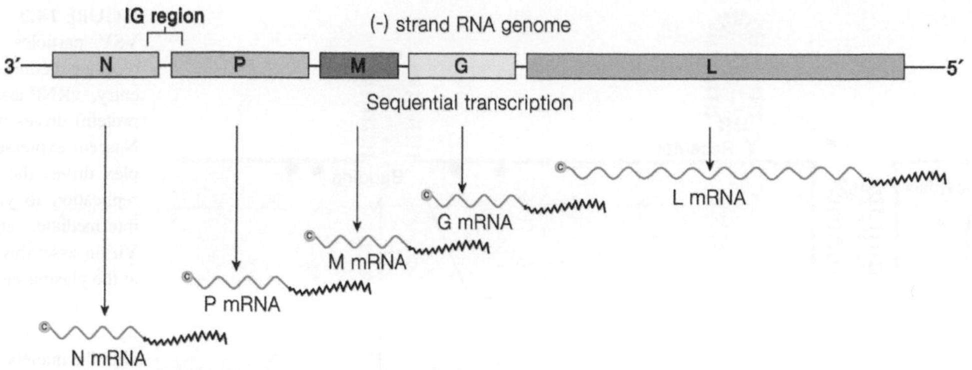

FIGURE 14.2 The RNA genome of VSV. The structure of VSV genome RNAs is drawn on the top. Five VSV genes (ORFs) are distinctly colored. Note that the 5′ end of negative-strand RNA genome is positioned to the right. IGs (ie, noncoding region) positioned between coding regions are denoted by brackets. The viral mRNAs are capped at the 5′ end and are polyadenylated at the 3′ end. (+) and (−) denotes the polarity of the RNA.

Genome Structure: VSV has a single-strand RNA genome of 11 kb in length (Fig. 14.2). Its 5′ terminus has the triphosphate group, which reads (p)ppACG−, while its 3′ terminus has no phosphate group, which reads −CGU-OH. As it has a negative polarity, it does not possess a poly (A) tail. Transfection of the viral RNA into cells does not lead to the productive infection; its RNA genome is "not infectious." Five viral genes are positioned in the order of 5′-L-G-M-P-N-3′. Each gene expresses one mRNA and one protein. In addition to five mRNAs, two small noncoding RNAs are transcribed. A noncoding RNA, termed leader RNA (l), is transcribed from near the 3′ end of the genome, while another noncoding RNA, termed trailer RNA (tr), is transcribed from near the 5′ end of the genome.

Protein Coding: Five mRNAs are transcribed from the VSV RNA genome. One viral protein per mRNA is expressed (see Table 14.3).

14.3 THE LIFE CYCLE OF VSV

The life cycle of VSV is confined to the cytoplasm (Fig. 14.3).

Entry: VSV particles enter the cell via recognition of *phosphatidyl serine*[5] on the plasma membrane. It is notable that a phospholipid molecule, instead of glycoprotein, acts as an entry receptor. Upon entry, the virus particle is located inside the endosome. Uncoating of the virus particles occurs in a manner coupled with cytoplasmic trafficking (see Fig. 3.6). The viral nucleocapsid penetrates to the cytoplasm via membrane fusion between the endosome and the viral envelope. The membrane fusion is triggered by the lower pH of endosome, which induces the exposure of hydrophobic region of *G protein*.[6]

RNA Synthesis: L/P protein complex associated with the nucleocapsid serves as *RdRp* for the viral RNA synthesis. Two positive-strand RNAs are transcribed from the negative-strand RNA genome: the subgenomic mRNAs and the genome-length RNA (Fig. 14.4). The mRNA synthesis is referred to as "transcription," while the genome-length RNA synthesis is referred to as "RNA replication." Importantly, the nucleocapsid or ribonucleoprotein complex (*RNP*[7]), in which the RNA molecule is completely encircled by numerous N proteins, serves as the template. Note that the usage of the RNP complex as a template, instead of a naked RNA, is the hallmark of all NSVs.

Transcription: Two short noncoding RNAs as well as the viral mRNAs are transcribed from the incoming negative-strand RNP viral ribonucleoprotein (vRNP) as the template (see Fig. 14.4). Both the leader RNA and trailer RNA are neither capped or polyadenylated, while the five mRNAs are capped and polyadenylated. RdRp begins to transcribe from 3′ end of the template, starting from the leader RNA, then N-, P-, M-, G-, L mRNA, and then the trailer RNA. In other words, RdRp executes multiple RNA synthesis without being dissociated from the template, a process termed "*single-site initiation*." The viral mRNAs are capped by capping activity associated with the L protein and the G residue of the cap structure is methylated by the methyl transferase activity associated with L protein (Table 14.3; see Table 16.3). For host mRNAs, the posttranscriptional modifications take place in the nucleus. Since its life cycle is confined to the cytoplasm, VSV has acquired the ability of posttranscriptional modifications during evolution.

5. **Phosphatidyl serine** A kind of phospholipid, which is a major constituent of membrane lipid.
6. **G protein** Envelope glycoprotein of VSV.
7. **RNP (ribonucleoprotein) complex** It refers to a RNA/protein complex. vRNP refers to the viral nucleocapsid.

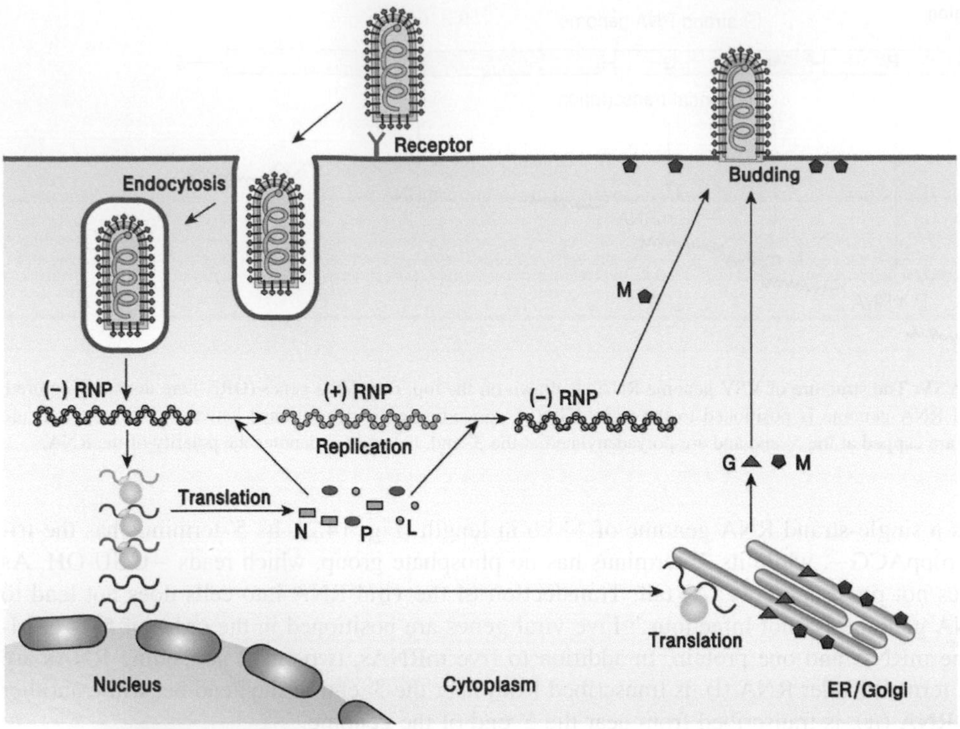

FIGURE 14.3 Life cycle of VSV. VSV particles enter the cell via receptor-mediated endocytosis. Upon entry, vRNP-associated RdRp (L/P protein) drives mRNA transcription. Nascent expressed L/P protein complex drives the viral RNA genome replication to yield (+) RNP as an intermediate, and then (–) RNP. Virion assembly and budding occur at the plasma membrane. (–) RNP or (+) RNP is used to refer to the polarity of the viral genome associated with the nucleocapsid.

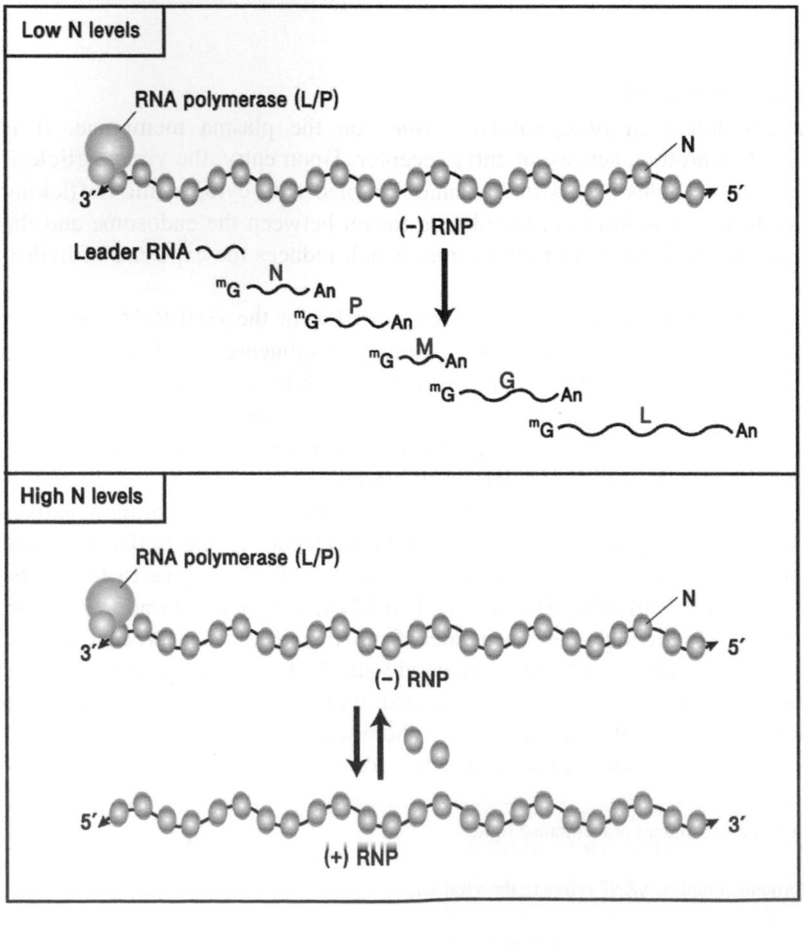

FIGURE 14.4 Regulation of VSV RNA synthesis. (–) RNP serves as the template for both transcription and genome replication. It is known that N protein level determines whether transcription or genome replication occurs. In the early phase of infection, when the N protein level is low, the viral mRNA synthesis by transcription mainly occurs. Thus, five mRNAs are made. In the later phase of infection, as N protein accumulates, excess N protein binds to the nascent transcribed RNA, and acts as a transcription terminator. Consequently, the full-length (+) RNA, instead of mRNA, is made.

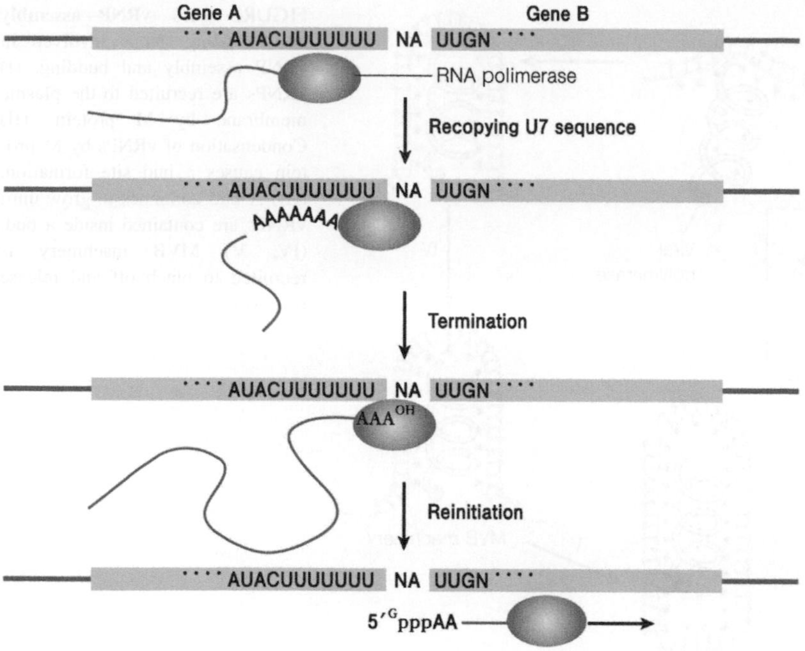

FIGURE 14.5 Polyadenylation and reinitiation of VSV mRNA transcription. The poly (A) tail of mRNA is synthesized by recopying a short run of U sequences located at the end of the coding region. The polyadenylation continues until over 200 A nucleotides are added by stuttering. It is believed that the longer poly (A) tail triggers not only transcription termination but also transcription reinitiation of the downstream gene. IGs (ie, NA dinucleotides, which is highlighted by bold face) are located between the short runs of U and the first nucleotide of the downstream gene.

The single-site initiation by RdRp allows transcription of seven RNAs from the negative-strand RNA genome. Mechanistically, one wonders how RdRp continues to reinitiate mRNA transcription from the downstream genes without being released from the template. Notably, the intergenic regions (*IGs*) between the genes constitute only two nucleotides (eg, NA dinucleotides) (Fig. 14.5). IGs direct the transcriptional termination and the transcriptional reinitiation of the downstream genes. As a matter of fact, following transcription of the upstream genes, RdRp pauses at IGs and only a subset of RdRp resumes transcription of the downstream genes. As a consequence, mRNA level gradually decreases from 3′ to 5′: leader $>$N$>$P$>$M$>$G$>$L$>$ trailer, a phenomena termed *attenuation*.

On the other hand, the poly (A) tail at the 3′ end of viral mRNA is added via a template-dependent manner. A short run of U residues (ie, U_{6-7}) at the end of the coding region are repeatedly copied to yield the poly (A) tail, a process termed "*stuttering*" (see Fig. 14.5). Unlike cellular mRNA, the poly (A) tail is copied from the template. Following the addition of the poly (A) tail to the transcript, RdRp continues to initiate transcription from downstream genes.

Genome Replication: The N protein level is a critical factor for transit from transcription to genome replication (see Fig. 14.4). Specifically, the N proteins in excess bind to and encircle the nascent transcribed positive-strand RNA and act as *antiterminators*, resulting in the synthesis of the full-length RNA or (+) RNPs. Such synthesized (+) RNPs are in turn used as a template for the synthesis of (−) RNPs.

Translation: Five mRNAs are translated to five viral proteins in the cytoplasm. The nascent N proteins as well as P/L proteins become associated with the RNP complex involved in the viral RNA synthesis (see Fig. 14.3). M (matrix) proteins are targeted to the plasma membrane, attached to the cytoplasmic side of the membrane. On the other hand, G envelope glycoprotein is translated via endoplasmic reticulum, and located in the plasma membrane.

Assembly and Release: Capsid assembly and envelopment occur in a sequential manner, in that vRNP assembly occurs first in the cytoplasm and then the assembled vRNP complex is recruited to the plasma membrane for budding (see Fig. 3.9). As described above, vRNPs are assembled during the RNA genome replication, as they are being synthesized (see Fig. 14.4). vRNPs are recruited to the cell surface by M proteins located at the cytoplasmic side of the plasma membrane (Fig. 14.6). The budding, which involves pinching (or membrane fusion), is triggered by the M protein. Specifically, the "*L domain*" of the M protein recruits cellular machinery termed the *multivesicular body (MVB) pathway*[8] to carry out the budding process (see Box 3.3).

Defective Interfering (DI) Particles: In addition to infectious virions, subviral particles are frequently produced during cultivation of VSV at high multiplicity of infection (MOI; Box 14.1).

8. **Multivesicular body (MVB)** It refers to an intracellular structure that is generated by the inward vesiculation in late endosomes.

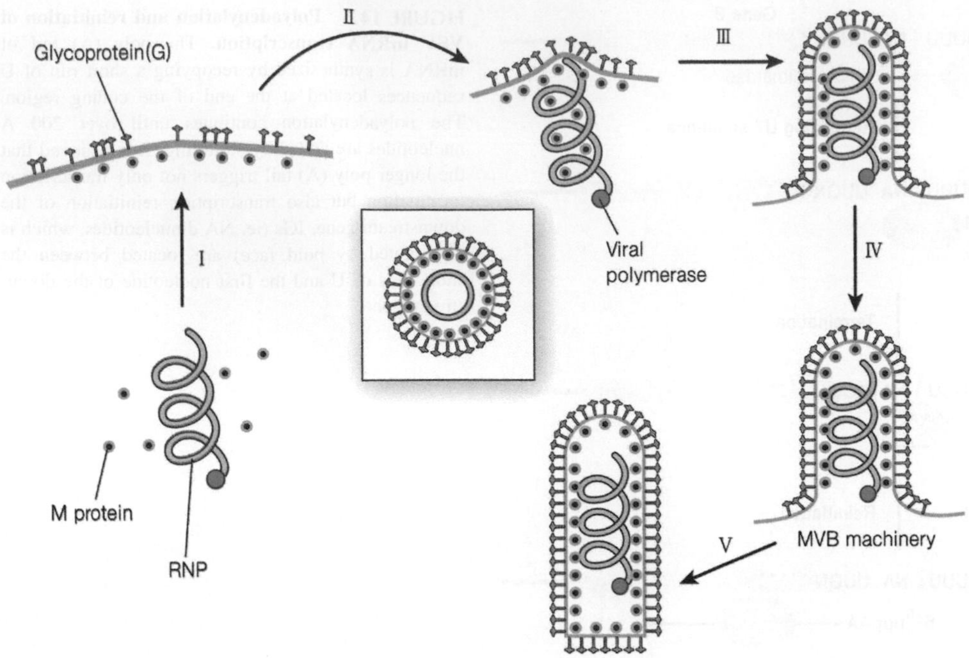

FIGURE 14.6 vRNP assembly and budding. Steps involved in vRNP assembly and budding. (I) vRNPs are recruited to the plasma membrane by M protein. (II) Condensation of vRNPs by M protein causes a bud site formation. (III) A bud continues to grow until vRNPs are contained inside a bud. (IV, V) MVB machinery is recruited to pinch off and release the virion.

BOX 14.1 Defective Interfering (DI) Particles

When growing RNA viruses at high MOI, characteristic mutant virus particles are frequently generated. They are viral variants having a substantial deletion in their genome and they do coexist with wild-type virus. These subviral particles are *DI particles*. What is the biological importance of DI particles? In the early days, virologists passaged virus culture medium without dilutions from cells to cells to generate DI particles. It was soon uncovered that DI particles keep all sequence elements essential for the viral genome replication such as promoters, origin of replication, and packaging signals. One interpretation is that DI particles are the consequence of viral evolution in keeping the *cis*-acting sequence element essential for the viral genome replication, while the protein-coding regions, which can be provided by a coinfecting wild-type virus, are dispensed.

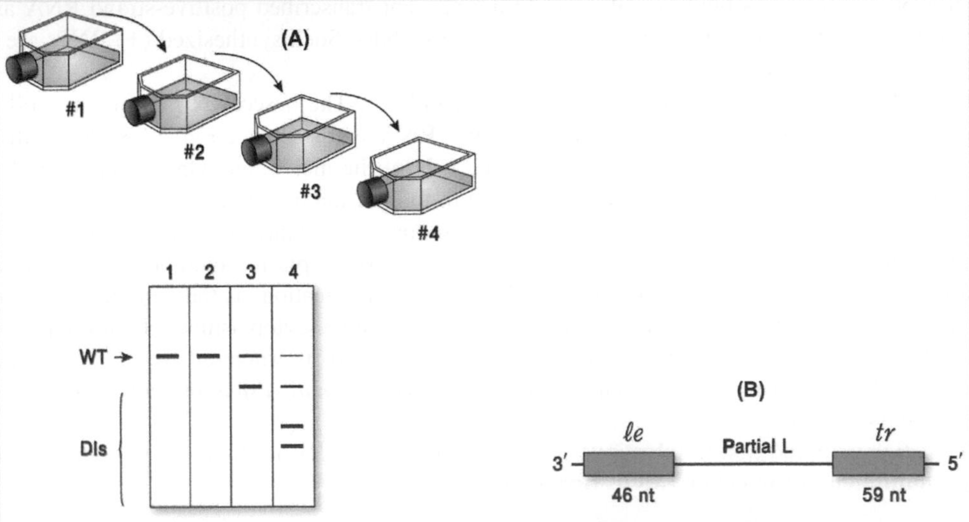

Generation of DI particles. (A) A diagram showing serial passage of virus stock (culture medium) without dilution. (Bottom) Gel electrophoresis of the viral genome extracted from each passage. As the passage number increases, DI particles having a deleted genome become more prominent. (B) The RNA structure of VSV DI particle. Essentially all protein-coding regions are deleted in the DI particle, except a subset of L gene. Note that the leader (le) and trailer (tr) genes located at the terminus of VSV genome remain intact, as they represent the minimal *cis*-acting elements required for the VSV genome replication.

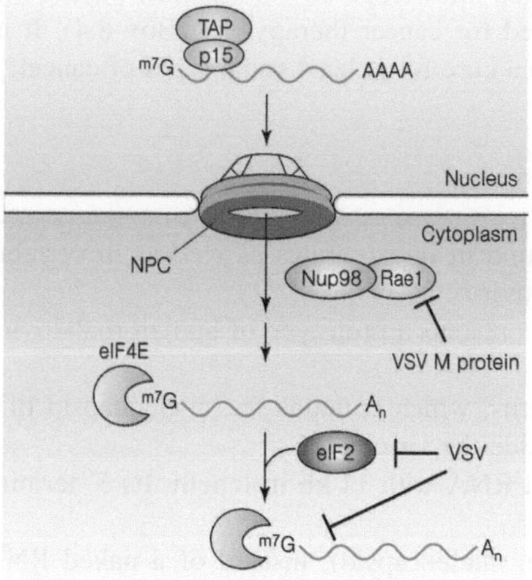

FIGURE 14.7 Host shutoff functions by VSV M protein. Cellular pre-mRNAs are exported to cytoplasm via nuclear pore. TAP/p15 complex serves as a transport receptor for pre-mRNAs, and recruits pre-mRNAs to NPC (nuclear pore complex). VSV M protein blocks the nuclear export of pre-mRNAs via its interaction with Rae1 molecule, a nuclear export factor. Note that Rae1 acts as a nuclear export factor via its interaction with Nup98, a component of NPC. In addition, VSV inhibits cellular translation via two mechanisms: (1) by enhancing eIF2 phosphorylation and (2) by suppressing eIF4E phosphorylation.

These subviral particles are aptly termed *DI particle*, because they are defective and they interfere the wild-type virus. In fact, the coexisting wild-type virus serves a helper virus in providing functions (ie, proteins) missing in DI particles.

14.4 EFFECTS ON HOST

CPE: VSV-infected cells are lysed and as a result, plaques are formed. How does VSV infection lead to cell lysis? VSV infection blocks host cell protein synthesis, while viral protein synthesis remains unaffected. Two mechanisms for "host shutoff" function have been revealed, as shown below.

RNA Processing: First, VSV infection blocks the nuclear export of cellular pre-mRNA. Specifically, it was shown that viral *M protein* binds to *Rae1*,[9] a nuclear export factor, thereby blocking the nuclear export of cellular pre-mRNA (Fig. 14.7). It should be noted that this inhibitory mechanism does not affect the viral mRNA synthesis, as the viral mRNA synthesis is limited to the cytoplasm. Hence, this mechanism is also dubbed "host shutoff."

Translation: In addition, VSV infection blocks cellular translation. First, the phosphorylation of *eIF4E*[10] is reduced in VSV-infected cells, thereby reducing *eIF4F*[11] complex formation. On the other hand, the phosphorylation of eIF2 is increased in VSV-infected cells, thereby decreasing translation. Perhaps, the phosphorylation of *eIF2*[12] by IFN-activated *PKR* greatly contributes to the translation suppression (see Fig. 5.8). An intriguing point is that VSV protein translation remains unaffected and the viral proteins accumulated, although host translation function is substantially impaired. The question of how selectively VSV mRNA translation is unaffected remains unknown.

14.5 PERSPECTIVES

Rabies virus, a human rhabdovirus, had been the focus of attention during the 19th century in Europe, as it causes a fatal transmissible disease. In fact, Louis Pasteur (1822–1895) studied rabies virus, even before the official discovery of "virus" as a submicrobial agent. He eventually developed a postexposure vaccine for the treatment of rabies infected individuals (see Box 25.1). On the other hand, VSV has been an extensively studied rhabdovirus, although VSV is not a

9. **Rae1** A nuclear export factor involved in cellular pre-mRNA export.
10. **eIF4E** Eukaryotic translation initiation factor 4E.
11. **eIF4F** A trimeric complex constituted by eIF4A, eIF4G, and eIF4E.
12. **eIF2** Eukaryotic translation initiation factor 2.

significant human pathogen. Accumulated knowledge on VSV is now paying off by the recent discovery of the oncolytic activity of VSV that can be explored for cancer therapy (see Box 8.4). It is hoped that further exploitation of the oncolytic activity of VSV could lead to a cure for at least some types of cancer.

14.6 SUMMARY

- *Rhabdovirus*: Rhabdoviruses are found in invertebrates as well as in vertebrate, ranging from mosquitoes to cattle. VSV is a prototype of animal rhabdovirus.
- *Human rhabdovirus*: Rabies virus represents a prototype of human rhabdovirus. Rabies virus causes fatal encephalomyelitis in human.
- *VSV particle*: It is an enveloped virus, which contains the nucleocapsid of a helical symmetry inside. L/P protein complex (RdRp) is encapsidated inside the virion.
- *VSV genome*: It is a negative-strand RNA with 11 kb in length. Its 5′ terminus is not capped, and its 3′ terminus is not polyadenylated.
- *VSV genome replication*: vRNP (ie, nucleocapsid), instead of a naked RNA, serves as the template for the RNA synthesis. The N protein level regulates the transit from transcription to the genome replication.
- *Host effect of VSV*: VSV infection leads to host shut-off and cell lysis.

STUDY QUESTIONS

14.1 VSV particles contain the nucleocapsid having a RNA genome with negative polarity [(−) NC]. In addition, L/P RdRp complex is associated with the (−) NC. (1) State whether VSV progeny particles are produced, when (−) NC isolated from VSV particle is introduced to cells. (2) State whether VSV progeny particles are produced, when (+) NC is introduced to cells. (3) What about if (−) NC or (+) NC is introduced to the N protein expressing cells.

14.2 A few temperature-sensitive mutants (*ts*) of VSV were isolated. Compare the phenotype of the mutants with that of wild-type. Note that *ts* mutants produce progeny virus as wild-type at permissive temperature (ie, 33°C), but not at nonpermissive temperature (ie, 39°C). (1) *ts* mutant of N protein gene, (2) *ts* mutant of M protein gene, and (3) *ts* mutant of G protein gene.

SUGGESTED READING

Chai, Q., He, W.Q., Zhou, M., Lu, H., Fu, Z.F., 2014. Enhancement of blood–brain barrier permeability and reduction of tight junction protein expression are modulated by chemokines/cytokines induced by rabies virus infection. J. Virol. 88 (9), 4698–4710.

Green, T.J., Zhang, X., Wertz, G.W., Luo, M., 2006. Structure of the vesicular stomatitis virus nucleoprotein-RNA complex. Science. 313 (5785), 357–360.

Green, T.J., Cox, R., Tsao, J., Rowse, M., Qiu, S., Luo, M., 2014. Common mechanism for RNA encapsidation by negative-strand RNA viruses. J. Virol. 88 (7), 3766–3775.

Ruigrok, R.W., Crepin, T., Kolakofsky, D., 2011. Nucleoproteins and nucleocapsids of negative-strand RNA viruses. Curr. Opin. Microbiol. 14 (4), 504–510.

JOURNAL CLUB

- Chai, Q., He, WQ., Zhou, M., Lu, H., Fu, Z.F., 2014. Enhancement of blood–brain barrier permeability and reduction of tight junction protein expression are modulated by chemokines/cytokines induced by rabies virus infection. J. Virol. 88, 4698–4710.

  Highlight: Rabies virus is a neurotropic virus that invades the central nervous system (CNS) via the peripheral nerve. The blood–brain barrier (BBB) is a highly selective permeability barrier that separates the circulating blood from the brain extracellular fluid in the CNS. This article demonstrated that the enhancement of BBB permeability by rabies virus is critical for host survival, and the enhancement is caused by the reduction of tight junction protein expression, which allows cytokines to enter CNS, and clear the rabies virus from CNS.

Chapter 15

Influenza Viruses

Chapter Outline

Influenza virus possess multiple negative-strand RNA genomes. This chapter covers influenza virus, which is a prototype of Orthomyxoviridae,[1] Some of us have been infected by influenza virus in almost every year. In other words, the lack of immunity from past infection episodes represents an outstanding feature of influenza virus infection. We will learn this unique attribute of influenza virus with focuses on molecular aspects of the virus life cycle.

15.1 INFLUENZA VIRUS AND FLU

Influenza and common colds: *Influenza*,[2] commonly known as "*flu*," is an acute respiratory disease caused by the influenza virus. Symptoms can be mild to severe. Its symptoms include: high fever, runny nose, sore throat, muscle ache, headache, coughing, and feeling tired. Flu differs from the "common cold," which is associated with mild respiratory symptoms (Table 15.1). Common colds are caused by either *rhinovirus*[3] or adenovirus infection. Flu is a more severe form of respiratory infection. Flu may not be a severe disease to most adults, but it can be severe or even fatal to children and elderly people, whose immunity are compromised. Worldwide, about 300,000 to 500,000 people per year die of seasonal flu epidemics and the mortality rate is estimated to be 0.2%. In the United States, about 30,000 to 40,000 peoples succumb to flu every year. Influenza virus infects and damages the *mucous membrane*[4] of the upper respiratory tract, which otherwise acts to eliminate invading microbes. As a result, secondary bacterial infection to the lower respiratory tract can be incurred. A fatal outcome is caused by secondary bacterial pneumonia infection.

15.1.1 Vaccines and Therapy

The most effective way to control influenza infection is vaccination. Influenza vaccines have been traditionally prepared from embryonated eggs. Such prepared so-called "*split vaccines*"[5] have been used for years for influenza vaccines. The split vaccines are composed of two viral antigens—hemagglutinin (HA) and neuraminidase (NA)—fractionated from virus particles propagated in embryonated eggs. More recently, a live attenuated vaccine has been

1. **Orthomyxoviridae** The term is a combined word *ortho* + *myxo*, which is a Greek word for "standard" and "mucus," respectively.
2. **Influenza** The term is derived from an Italian word for "influence"—*Influenza*. During 18th century, it was believed that the influenza epidemics were influenced by the constellation of stars.
3. **Rhinovirus** A member of family Picornaviridae, which is the major cause of common colds.
4. **Mucous membrane** The mucous membrane (or mucosa) is the lining covered in epithelium, which is involved in absorption and secretion. It lines cavities that are exposed to the external environment and internal organs. The sticky and thick fluid secreted by the mucous membranes termed "mucus" plays a critical role in eliminating invading microbes.
5. **Split vaccine** It refers to the vaccine manufacturing process, which involves fractionation (split) of immunogen (ie, virus particles).

Molecular Virology of Human Pathogenic Viruses. DOI: http://dx.doi.org/10.1016/B978-0-12-800838-6.00015-1

TABLE 15.1 The Defining Features of Influenza Virus

Genome	Virion Structure	Genome Replication
Negative-strand RNA Segmented genome	Enveloped, helical nucleocapsid RdRp packaged	RNA-directed RNA polymerase Cap-snatching
Life Cycle	**Interaction with Host**	**Diseases**
Nucleus Genome reassortment	Evasion of innate immunity Blockade of RNA processing	Respiratory infection –

BOX 15.1 Antiviral Drugs for Influenza Virus

The neuraminidase activity of NA protein plays a major role in the release of nascent viral particles assembled in infected cells (see Fig. 15.2). It facilitates the release of viral particles from cells by the cleavage of a sialic acid residue of glycan moiety linked to cellular glycoproteins in the plasma membrane. A small molecule that inhibits the neuraminidase activity will block the release of newly assembled particles and thereby progeny virus will not spread to neighboring cells. Two drugs with such activity were developed upon the approval of the FDA. Oseltamivir and zanamivir are analogs of sialic acid residue, which is a substrate for neuraminidase. In particular, osetamivir was aptly used during the 2009 H1N1 pandemic.

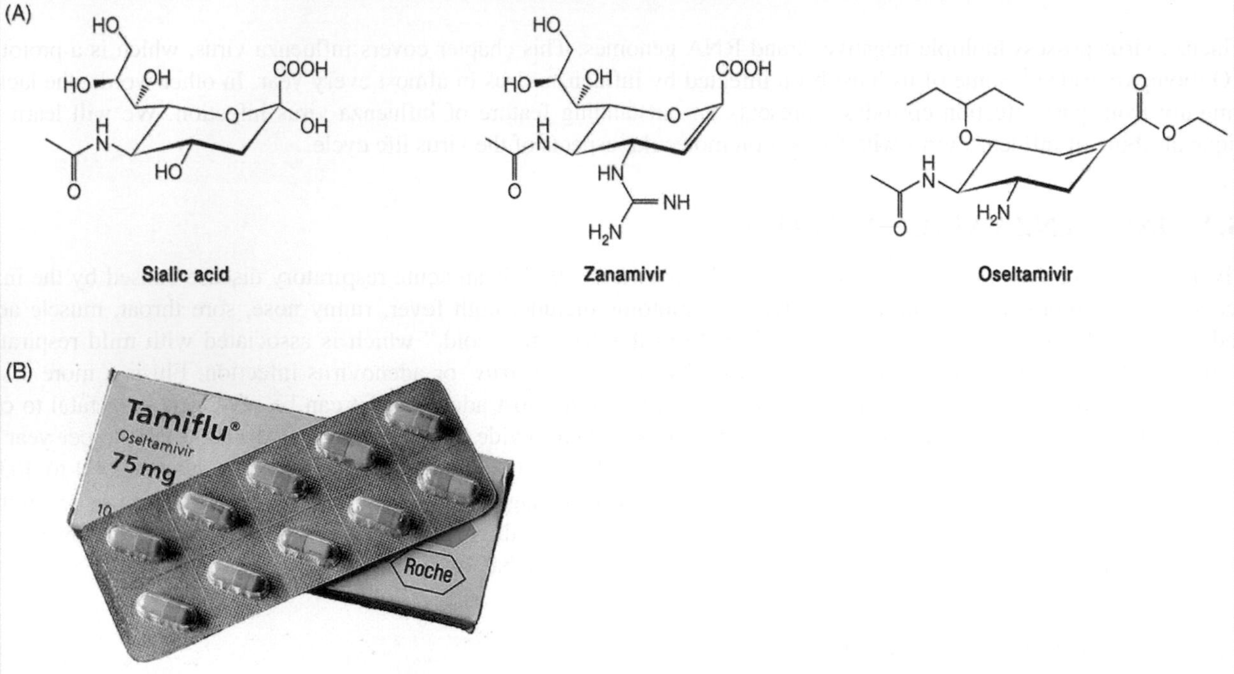

NA inhibitors and sialic acid. NA inhibitors are chemical analogs of a sialic acid. Oseltamivir is also called by trade name, Tamiflu.

developed (see Box 25.2). Flu epidemics each year are caused by novel strains (ie, subtypes) of influenza viruses. Thus, it is believed that antibodies elicited by past infection or vaccination cannot provide protective immunity against seasonal flu strains. Therefore, unlike other viral vaccines, influenza vaccine needs to be administrated every year. The neuraminidase inhibitors (ie, oseltamivir and zanamivir) are used as antiviral drugs (Box 15.1). These drugs are analogs of sialic acid, which is a substrate of neuraminidase.

15.2 CLASSIFICATION

Classification: Influenza virus comprises three genera (ie, A, B, and C), based on the antigenicity of viral NP and M proteins (Table 15.2). Type B and C virus infect only humans, while type A can infect not only humans but also birds,

TABLE 15.2 The Classification of Influenza Virus

Genus	Host	Subtypes
Type A	Human, birds, pigs, horses	16 HA subtype + 9 NA subtypes
Type B	Human	
Type C	Human	

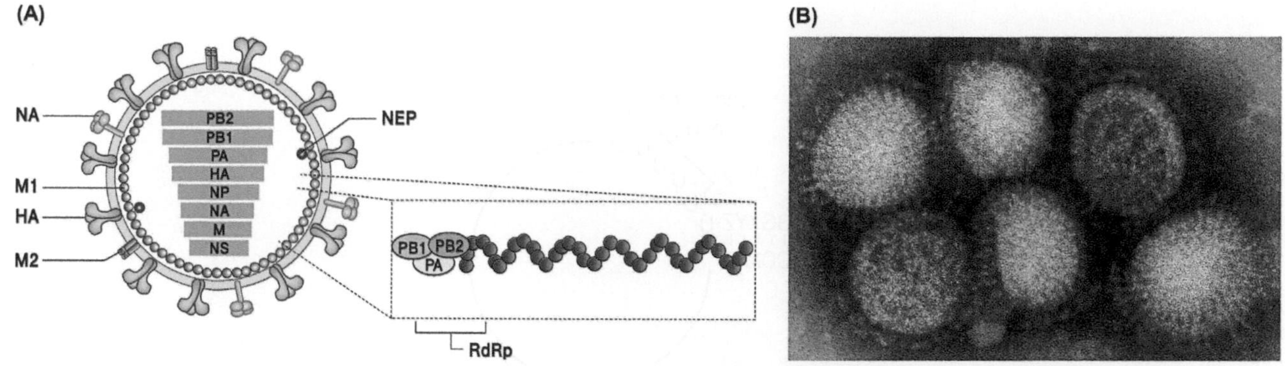

FIGURE 15.1 A diagram showing influenza virus particle. (A) Two envelope glycoproteins embedded in the envelope are denoted: HA and NA. M2 protein is also found in the envelope. Inside the envelope, eight RNA segments are denoted with their coding proteins. In addition, a few molecules of NEP/NS2 protein are attached to M1 matrix proteins, which line the inner leaf of the viral envelope. [Inset] One RNA segment is expanded in an insert, which shows a helical nucleocapsid structure composed of an RNA genome (ie, vRNA) being encompassed by numerous NP. Note that the viral RdRp (a trimeric complex of PB1, PB2, and PA protein) is attached to the end of the nucleocapsids. (B) Electron microscopic image of influenza virus particles.

pigs, and horses. Thus, here type A is referred to, unless otherwise indicated. In particular, type A could be further divided into many *subtypes*,[6] depending on the antigenicity of the HA and NA antigens. Currently, 16 HA subtypes and 9 NA subtypes have been reported. Therefore, a certain strain can be described as the combination of one HA and one NA subtype: for example, H1N1 or H5N1 strain. In theory, up to 144 subtypes can be generated by combining 16 HA and 9 NA subtypes. However, so far only H1N1 and H3N2 subtypes along with influenza B virus are known to cause seasonal flu epidemics; for instance, H1N1 and H3N2 subtypes and influenza B virus are responsible for the 2013 flu epidemics.

Nomenclature: In addition to the multiplicity of subtypes, the country (city) and year of isolation are explicitly indicated in the nomenclature of influenza virus, just like the naming of car models. For instance, A/Puerto Rico/8/1934 (H1N1) refers to the virus isolated in Puerto Rico in 1934 that is classified as type A and an H1N1 subtype. In short, it can be abbreviated as either A/PR/8/1934 (H1N1) or A/PR/8.

15.3 THE VIRION AND GENOME STRUCTURE

Virus Particles: Influenza viruses are enveloped particles with a cylinder shape having a diameter of 80 to 120 nm (Fig. 15.1). Two viral envelope glycoproteins (ie, HA and NA) are studded in the envelope. HA has a propensity to bind and aggregate red blood cells, as its name implies. This property of HA is exploited for the detection of influenza virus (see Fig. 4.11). Importantly, HA, which is a trimer, is the viral protein that recognizes the cellular receptor for the entry. In addition to HA and NA, M2 protein, which acts as an ion channel, is also embedded into the viral envelope. M1 protein, which is a matrix protein, is attached to the inner leaflet of the lipid bilayer. Inside the envelope, eight nucleocapsids are enclosed, in which each viral RNA segment (ie, vRNA) is coated by NP. The nucleocapsids represent a helical structure. The viral *RdRp* is bound to the tip of the nucleocapsids (see Fig. 15.1).

6. **Subtype** It refers to a rank that below "species" in taxonomic rank.

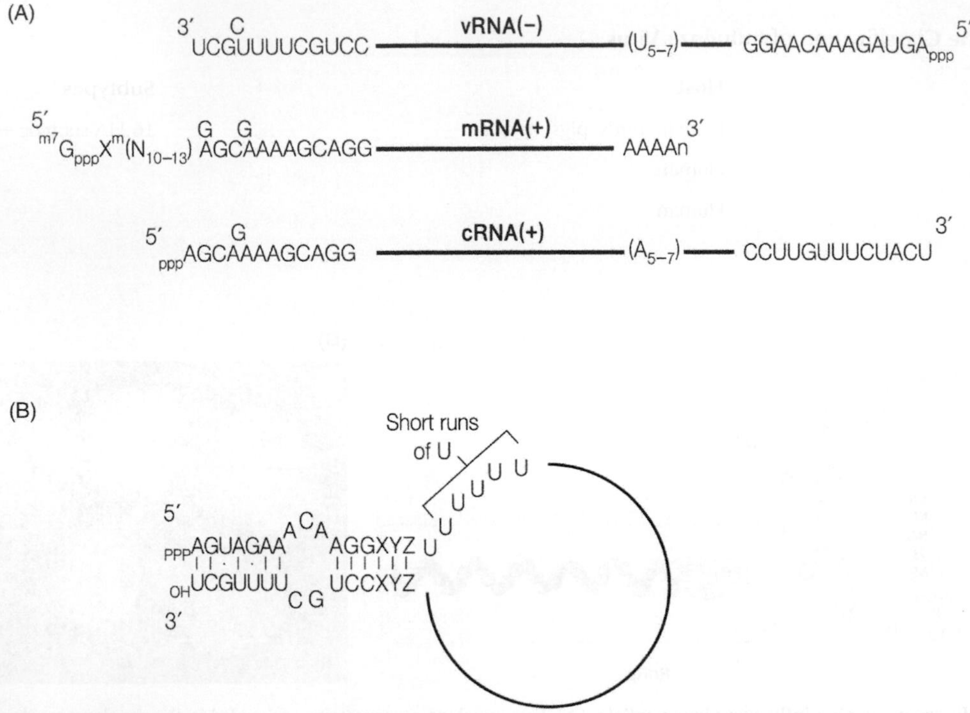

FIGURE 15.2 Influenza viral RNAs. (A) Three viral RNAs. Three viral RNAs (ie, vRNA, mRNA, and cRNA) derived from one typical RNA segment is shown. The noncoding regions are expanded out of scale for clarity. vRNA: the negative-strand RNA found in virions. mRNA: the viral mRNA that is capped at the 5′ end and is polyadenylated at the 3′ end. cRNA: the positive-strand RNA that is precisely complementary to vRNA. (+)(−) denotes the polarity of the RNA. Note that the 5′ end of negative-strand RNA is positioned to the right. (B) A panhandle structure of a viral vRNA. vRNA folds into a typical secondary structure, called a panhandle, due to a partial complementarity between 5′ and 3′ end sequences of vRNAs.

Genome Structure: Influenza virus contains eight RNA segments. It is also called a "segmented genome" or "split genome," because the genome is split into eight segments. The multiplicity of the genome is attributable for the emergence of viral variants through genetic recombination. Each RNA segment is a single-strand RNA 0.9−2.3 kb in length (Fig. 15.2A). It is also called "*vRNA*."[7] Being a negative-strand RNA, it is neither capped at the 5′ end or polyadenylated at the 3′ end. Instead, it has a triphosphate group at the 5′ end (ie, pppAp---) and a hydroxyl group at the 3′ end (ie, CUUUUGCU-OH-3′). In particular, the 12−13 nt *NCR*[8] regions at either end are complementary to each other, folding into a panhandle structure (Fig. 15.2B). Importantly, the integrity of this second structure is critical in transcription and RNA genome replication.

Protein Coding: Each RNA segment encodes one protein, except for the M1 and NS1 segments, which encode two proteins via alternative splicing (Fig. 15.3). The M1 segment expresses M2 as well as M1 protein, while the NS1 segment expresses *NEP/NS2*[9] and NS1. Moreover, a novel viral protein, called *PB1-F2*,[10] is newly described (see Box 15.4). PB1-F2 has drawn attention due to its association with enhanced virulence.

15.4 THE LIFE CYCLE

Unlike other RNA viruses, influenza viral genome replication occurs in the nucleus (Fig. 15.4).

Entry: Influenza virus enters the cell via recognition of the cellular receptor, a *sialic acid*[11] (Fig. 15.5). The engagement of the HA timer to a sialic acid moiety of *glycan*[12] on the cell membrane triggers endocytosis. The sialic acids are

7. **vRNA (virion RNA)** It refers to an RNA encapsidated inside virion.
8. **NCR (noncoding region)** It refers to the region of the genome in which no protein is encoded.
9. **NEP/NS2** It was originally called NS2, but was renamed as NEP (nuclear export protein), as its nuclear export function is revealed.
10. **PB1-F2** It refers to a novel protein derived from PB1 segment in +1 reading frame (frame 2).
11. **Sialic acid** It refers to a kind of amino sugar derived from mannose.
12. **Glycan** It refers to the carbohydrate moiety of glycoproteins.

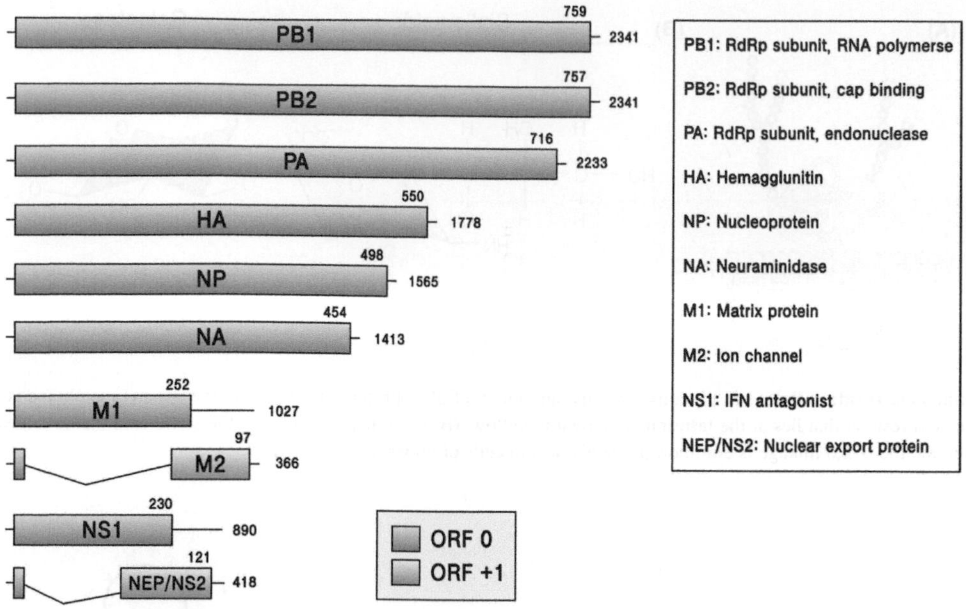

FIGURE 15.3 Viral proteins coded by influenza virus. Eleven viral proteins encoded by eight RNA segments are indicated by boxes in the corresponding genome. Proteins encoded in +1 reading frame are shaded. The spliced RNAs are indicated by lines. To the right, the function of each viral protein is denoted.

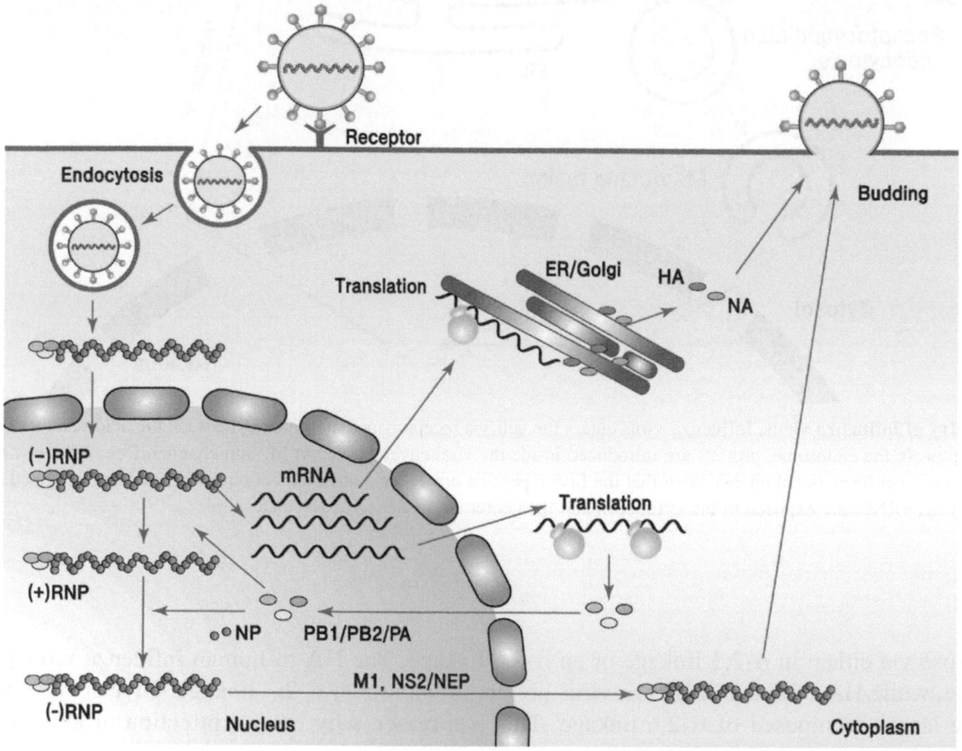

FIGURE 15.4 Life cycle of influenza virus. Unlike most other RNA viruses, the influenza viral genome replication takes place in the nucleus. Virions enter the cell via receptor-mediated endocytosis. Upon entry, vRNP, the nucleocapsid, enters the nucleus via a nuclear pore. vRNP-associated RdRp drives mRNA transcription, while nascent RdRp drives the viral genome replication. Newly synthesized (−) vRNPs are exported out to the cytoplasm. Virion assembly and budding occur at the plasma membrane. Only one RNA segment is drawn for clarity.

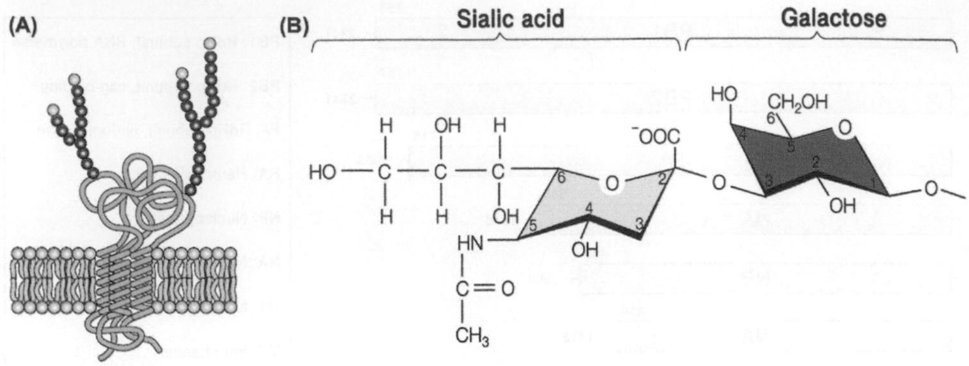

FIGURE 15.5 Sialic acid residues linked to glycans. (A) Glycan moiety of glycoprotein embedded in the plasma membrane is shown in orange; in particular, the sialic acid residue that lies at the terminus is shown in yellow. (B) Chemical structure of the sialic acid that is linked to galactose via an α-2,3 linkage. In human, the α-2,6 linkage is common in the glycans in cells of upper respiratory tract.

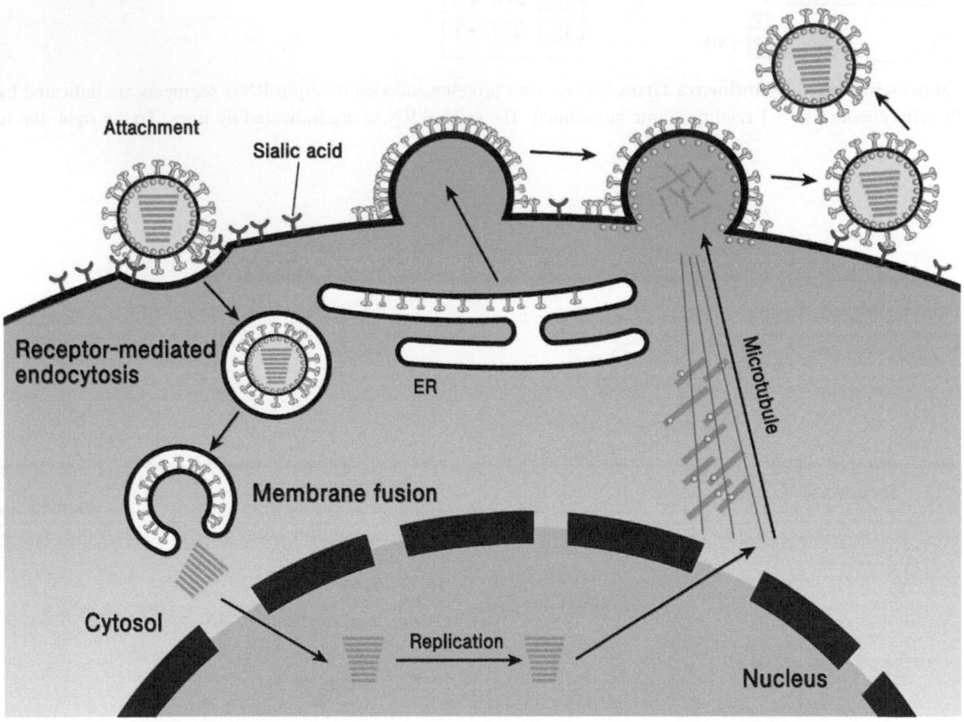

FIGURE 15.6 Entry of influenza virus. Influenza virus enters the cell via receptor-mediated endocytosis. Sialic acid residues on plasma membrane act as an entry receptor. In the endosome, protons are introduced inside the viral envelope via an M2 ion channel. Due to the low pH in the endosome, the HA trimer undergoes conformational change such that the fusion peptide otherwise embedded becomes exposed and activated to trigger membrane fusion. Upon uncoating, vRNPs are exposed to the cytoplasm and then enter the nucleus via a nuclear pore.

linked to galactose via either an α-2,3 linkage or an α-2,6 linkage. The HA of human influenza virus prefers to bind to an α-2,6 linkage, while HA of avian influenza virus prefers to bind to an α-2,3 linkage. Glycans in human upper respiratory tracts are largely composed of α-2,6 linkage. This is a reason why human infection of avian influenza virus is restricted.

Upon endocytosis, the virus particles are located inside the endosome. The acidic pH in the endosome triggers membrane fusion between the two membranes (Fig. 15.6). Via the M2 ion channels, protons are imported inside of the viral envelope and the resulting lower pH induces conformational change of HA trimer such that an embedded *fusion*

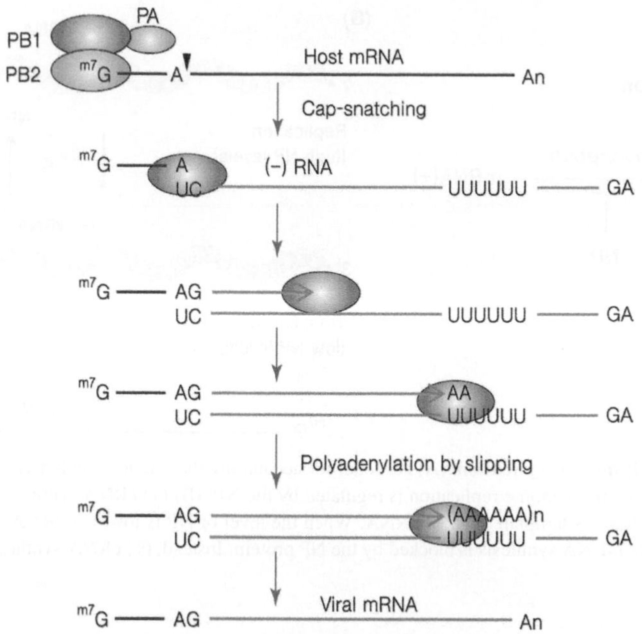

FIGURE 15.7 Viral mRNA transcription via cap-snatching mechanism. Viral RdRp composed of 3 subunits (ie, PB1, PB2, and PA subunit) catalyzes viral mRNA synthesis. The mRNA synthesis is primed by a capped RNA fragment that is snatched from cellular mRNA, a process called cap-snatching. The cleavage site by a PA subunit, which exhibits endonuclease activity, is denoted by an *arrow*. For clarity, only a PB1 subunit is shown following the cap-snatching step. Poly (A) tail is made by repeatedly copying the short runs of U residue, a process called "shuttering." As a result, the viral mRNAs are capped at the 5′ end and polyadenylated at the 3′ end, just like host mRNAs. Cellular mRNA is denoted by a *red line*.

peptide[13] domain becomes unfolded and activated (see Box 3.2). This activated fusion peptide triggers membrane fusion between two membranes and as a result, the nucleocapsids inside the envelope are released to cytoplasm. The nucleocapsids are constituted by vRNA, NP, and RdRp (ie, PB1 + PB2 + PA). This *RNP complex*[14] now enters the nucleus via the nuclear pore (see Fig. 15.4).

Transcription: Most RNA polymerases do not need an RNA primer for transcription initiation (ie, *de novo* initiation). However, influenza viral RdRp needs a primer for transcription initiation (Fig. 15.7). Moreover, influenza virus RdRp steals a capped RNA fragment from the cellular mRNAs and utilizes them as primers. Viral RdRp are composed of 3 subunits: PB1 acts as a RNA polymerase, PB2 exhibits cap-binding ability, and PA exhibits endonuclease activity. In particular, the endonuclease activity of PA cleaves a capped RNA fragment from the 5′ end of cellular mRNAs, which is then used as a primer for viral mRNA transcription. This process is referred to as "*cap-snatching*."[15]

Unlike cellular mRNA, the poly (A) tail is copied from the template during transcription (see Fig. 15.7). A short run of U residues (ie, U_{6-7}) at the 5′ end of vRNA are repeatedly copied to make the tail (see Fig. 15.2B). In other words, poly (A) tail at the 3′ end of viral mRNA is added via template-dependent manner. The mechanism is called "*stuttering*." Viral mRNA sequences are not complementary to the 5′ end of vRNA (see Fig. 15.2), because the viral transcription is terminated after polyadenylation. Resulting viral mRNAs are exported to cytoplasm and used as mRNA for the viral protein synthesis.

Genome Replication: Influenza virus genome replication takes place in the nucleus. Hence, the viral proteins essential for viral genome replication needs to be imported into the nucleus (see Fig. 15.4). For instance, NP (nucleocapsid protein), RdRp (PB1 + PB2 + PA subunit), NS1, and NEP are imported to the nucleus (see Fig. 15.10). Upon the entry of *vRNPs*[16] into the nucleus, viral genome replication as well as viral mRNA synthesis occurs. Note that two distinct positive-sense RNAs (ie, mRNA and cRNA) are synthesized by using the negative-sense RNA as a template (ie, vRNA) (Fig. 15.8A).

13. **Fusion peptide** It refers to a hydrophobic peptide domain that triggers membrane fusion (see Box 3.2).
14. **RNP (ribonucleoprotein) complex** It refers to a molecular complex that is composed of RNA and protein.
15. **Cap-snatching** It refers to a step in transcription process, which involves the cleavage of RNA fragments from the 5′ end of cellular mRNAs.
16. **vRNP** It refers to a ribonucleoprotein (RNP) complex, in which vRNA is encapsidated.

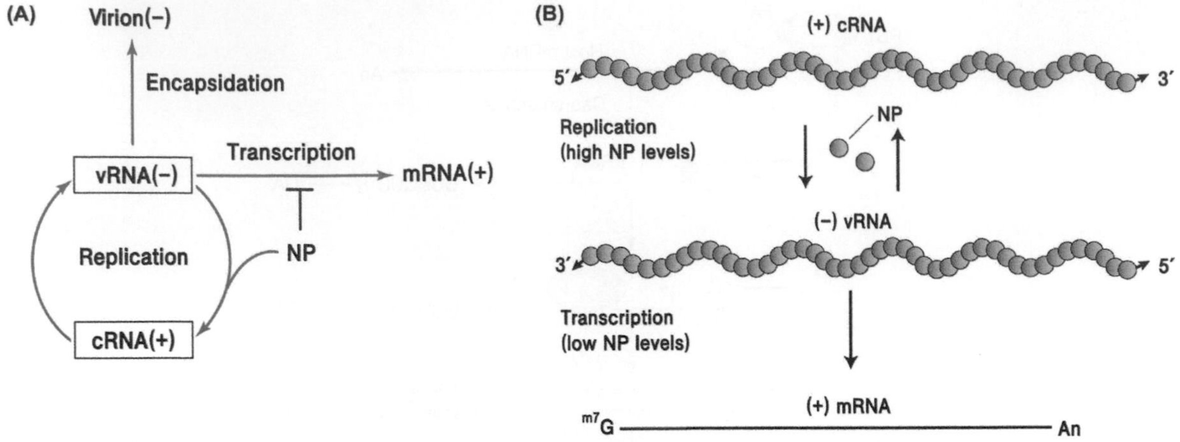

FIGURE 15.8 Switch from transcription to replication. (A) A model for accounting the relationship between vRNA, mRNA, and cRNA syntheses. The switch from the transcription to the viral genome replication is regulated by the NP. (B) (+) cRNA synthesis versus (+) mRNA synthesis. Note that both cRNA synthesis and mRNA synthesis is templated by (−) vRNA. When the level of NP is low, (−) vRNA is mainly utilized for mRNA synthesis. However, when the level of NP is high, mRNA synthesis is blocked by the NP protein. Instead, (+) cRNA synthesis begins.

How then are these two exclusive processes regulated? A short answer to this question is that the abundance of NP is the determinant for the regulation (Fig. 15.8B). For instance, viral mRNA synthesis predominates in the early phase of infection, when the NP level is low. As the NP accumulates, mRNA synthesis is suppressed and instead, cRNA synthesis is concomitantly induced. Note that cRNA synthesis does not require a primer. Thus, the 5′ end of cRNA is precisely complementary to the vRNA (see Fig. 15.2A). Likewise, the 3′ end of cRNA is precisely complementary to the vRNA. Conversely, vRNA is synthesized by using cRNA as a template. Nascent vRNA is encapsidated by the NP, resulting in vRNP formation.

Translation: Viral mRNAs are exported to cytoplasm, and then serve as mRNA for the viral protein synthesis (see Fig. 15.4). In particular, HA and NA envelope proteins are translated in association with the ER and Golgi body through a secretary pathway (Fig. 15.9). HA and NA play an important role in entry and exit, respectively. HA protein is processed by a cellular protease furin (or trypsin-like protease) to two disulfide-linked subunits (ie, HA1 and HA2). The fusion peptide domain lies at the N-terminus of the HA2 subunit following the cleavage. Thus, the cleavage by furin is critical for the infectivity of nascent virus particles.

Nuclear Export: Nascent vRNPs are exported to the cytoplasm via nuclear pore. M1 protein as well as NEP facilitates the nuclear export of vRNPs (Fig. 15.10). Specifically, M1 binds to vRNP, and then NEP binds indirectly to vRNP via its interaction to M1. NEP, which harbors *NES*,[17] is exported via *Crm-1*-mediated pathway. In other words, vRNPs are exported to the cytoplasm via the protein export pathway.

Assembly and Release: The vRNAs exported from the nucleus traffic to the plasma membrane, a process called "cytoplasmic transport." In this process, vRNAs are associated with the microtubule and move toward the cell surface in a microtubule-dependent manner (Fig. 15.11). Intriguingly, microtubule-driven transport moves the nucleocapsid in the opposite direction to the entry (see Fig. 3.7). Upon the arrival near the plasma membrane, a set of vRNPs grouped together interacts with HA and NA that are already embedded into the plasma membrane. Importantly, all eight RNA genome segments (ie, vRNP) need to be packaged during assembly to yield an infectious progeny particle. However, it is unclear yet how the selective packaging of eight vRNPs is achieved. Overall, viral particle assembly and budding proceed in a linked manner.

The *neuraminidase*[18] activity of NA facilitates the release of nascent particles from the infected cells (see Fig. 15.11). It cleaves the sialic acid moiety, as nascent virus particles are attached to the cell surface via its HA protein. Oseltamivir, a neuraminidase inhibitor, blocks the viral spread by inhibiting the release of nascent virus particles (see Box 15.1).

---

17. **NES (nuclear export signal)** It refers to a peptide motif encoded by nuclear export proteins that is essential for nuclear export.
18. **Neuraminidase** An enzyme that cleaves a sialic acid of glycan.

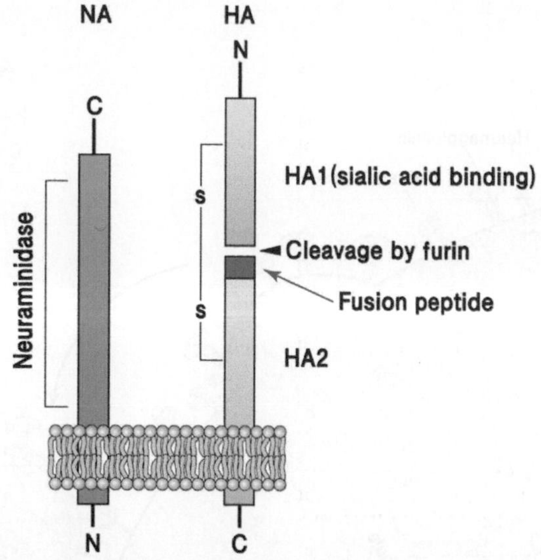

FIGURE 15.9 Domain structure of NA and HA protein on envelope. The ectodomain of the NA protein exhibits the neuraminidase activity. Two subunits of the HA protein are linked by a disulfide bond. The HA1 domain binds to sialic acid, while the HA2 domain harbors the fusion peptide at the N-terminus.

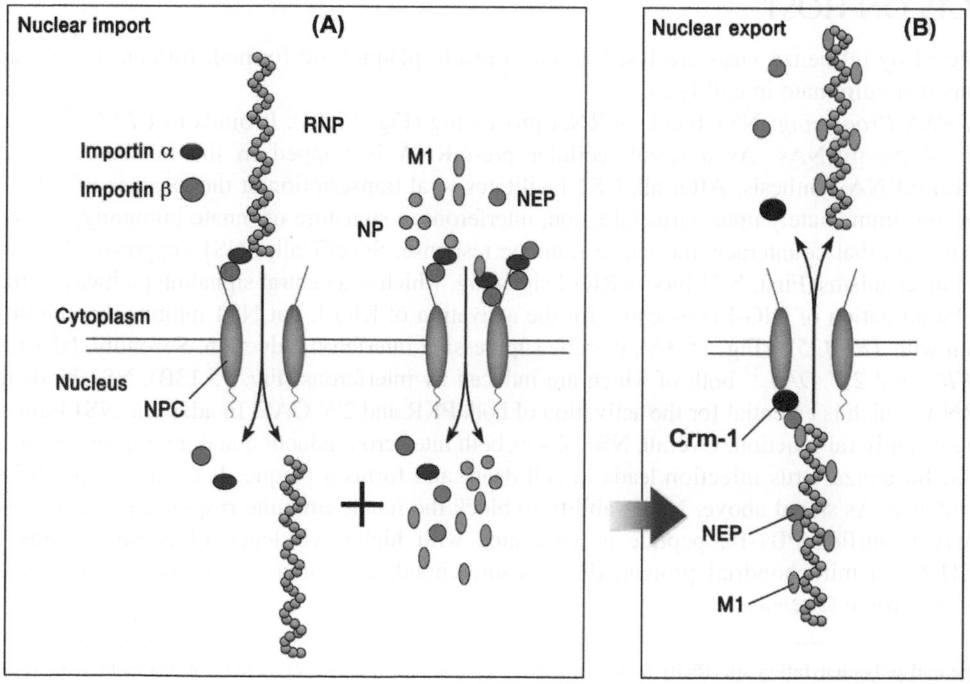

FIGURE 15.10 Nuclear import and nuclear export of vRNPs. (A) Nuclear import. After penetration into the cytoplasm, vRNPs enter the nucleus, where viral genome replication takes place. In this process, importin α or -β, a receptor of the nuclear import, binds to vRNPs via its association with viral NP. Other viral proteins, such as NP, M1, and NEP, are also imported via importin α or -β. (B) Nuclear export. Nascent vRNPs in the nucleus are exported out to the cytoplasm via a Crm-1 dependent mechanism. vRNPs are recognized by Crm-1 via its associated NEP protein. NEP is attached to vRNP indirectly via its association with the M1 protein.

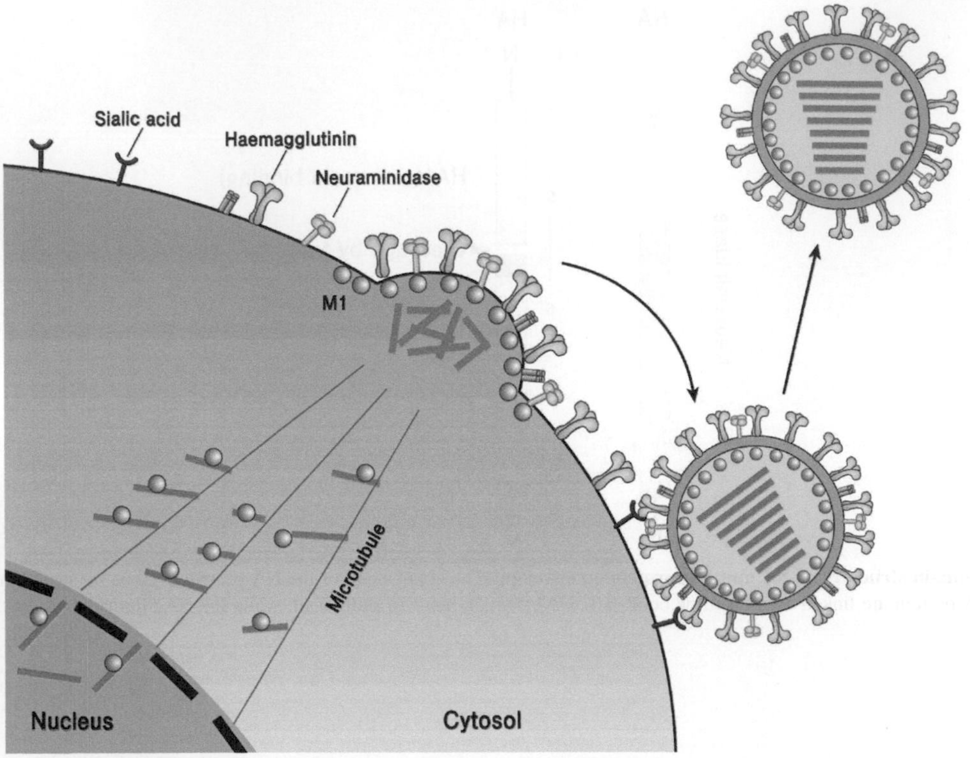

FIGURE 15.11 **Virion assembly and release.** M1 matrix protein plays a central role in the assembly and budding process. An interaction between M1 and envelope glycoprotein (HA and NA) triggers a bud formation. M1 protein recruits a set of eight vRNPs, which are packaged into a virion particle. HA in the nascent virions attached to the plasma membrane via a sialic acid is cleaved by neuraminidase activity of NA.

15.5 EFFECTS ON HOST

CPE: Cells infected by influenza virus are lysed and as a result, plaques are formed. Influenza virus shuts off multiple cellular functions that culminate in cell lysis.

Blockade of RNA Processing: NS1 blocks mRNA processing (Fig. 15.12). It binds to *CPSF*,[19] thereby blocking the polyadenylation of pre-mRNAs. As a result, cellular pre-mRNA is trapped in the nucleus and subjected to cap-snatching for viral mRNA synthesis. After all, NS1 facilitates viral transcription at the expense of cellular mRNAs.

Immune Evasion: Immediately upon virus infection, interferons, a signature of innate immunity, are induced. NS1 protein is the viral protein that counteracts the innate immune response. Specifically, NS1 suppresses host immune response via three distinct mechanisms. First, NS1 blocks RIG-I signaling, which is a central signaling pathway of the innate immune response. The ubiquitination of RIG-I is essential for the activation of RIG-I, but NS1 inhibits the ubiquitination of RIG-I via its interaction with *TRIM25*[20] (Fig. 15.3A), thereby suppressing interferon induction. Secondly, NS1 inhibits the antiviral action of *PKR*[21] and *2′5′ OAS*,[22] both of which are induced by interferons (Fig. 15.13B). NS1 binds to and sequesters double-strand RNA, which is essential for the activation of both PKR and 2′5′ OAS. In addition, NS1 binds to PKR directly, thereby blocking its antiviral function. Overall, NS1 blocks both interferon induction and interferon action.

Pathogenesis: Influenza virus infection leads to cell death and forms a plaque. It is known that NS1 is the determinant of viral virulence. As stated above, NS1's ability to block the innate immune response is related to its virulence. In addition, a newly identified PB1-F2 peptide is associated with higher virulence of certain strains (see Box 15.4). Intriguingly, PB1-F2, a mitochondrial protein, disrupts mitochondrial membrane potential and induces apoptosis via mitochondrial cytochrome C release.

---

19. **CPSF (cleavage and polyadenylation specificity factor)** A cellular protein that acts on polyadenylation via its binding to AAUAAA element of pre-mRNA.
20. **TRIM (tripartite motif) family** It refers to a protein family that harbors tripartite motif (ie, RING domain, B-box domain, and coiled-coil domain).
21. **PKR (protein kinase R)** A serine/threonine protein kinase that is activated by dsRNA.
22. **2′5′ OAS** Oligoadenylate synthetase.

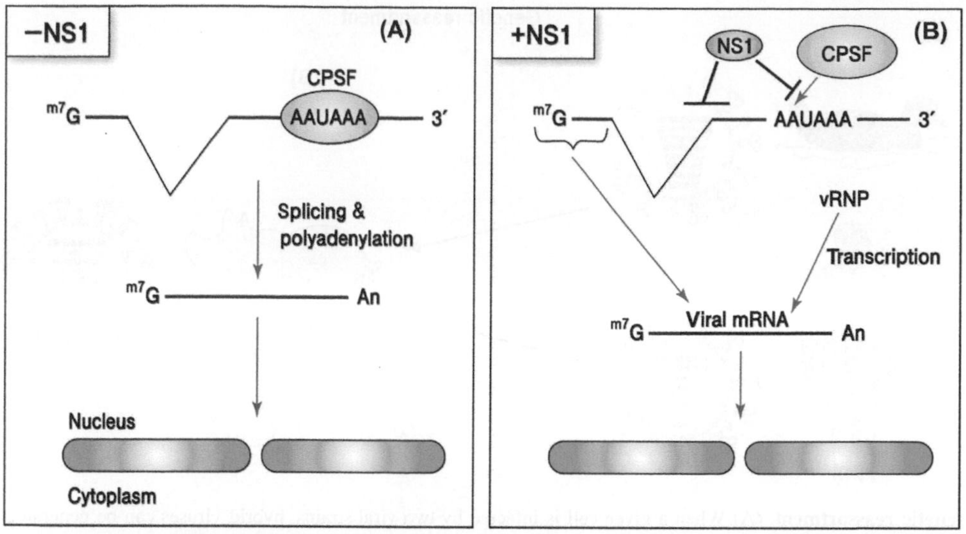

FIGURE 15.12 NS1-mediated blockade of RNA processing. (A) In normal cells, CPSF recognizes the polyadenylation signal (AAUAAA) at the 3′ end of pre-mRNA and facilitates 3′ polyadenylation of pre-mRNA in the nucleus. Nuclear export of mRNA is preceded by polyadenylation. (B) In infected cells, NS1 binds to CPSF, thereby blocking the polyadenylation and nuclear export of pre-mRNA. Consequently, pre-mRNAs are trapped in the nucleus and are subjected to cap-snatching.

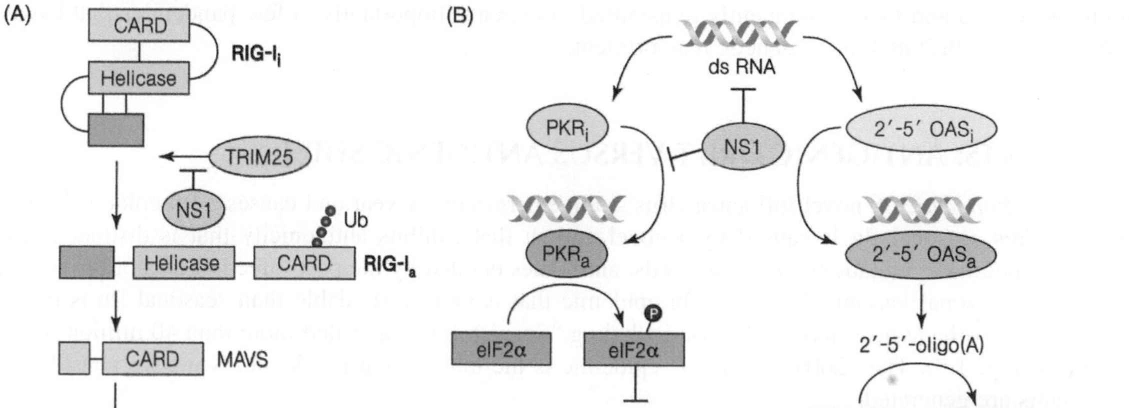

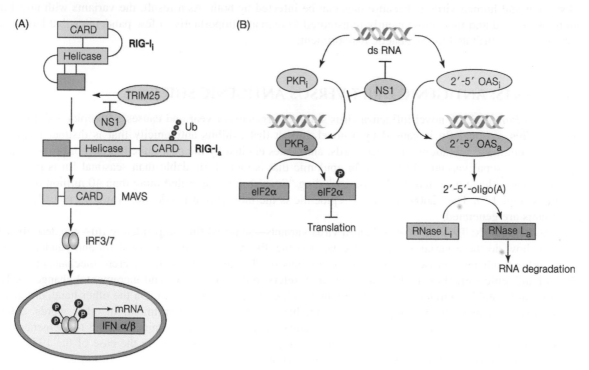

FIGURE 15.13 Blockade of interferon induction and interferon action by NS1. (A) NS1 blockade of interferon induction. RIG-I are expressed as an inactive form, unless activated by ubiquitination by TRIM25. The activated RIG-I then interacts with mitochondrial antiviral signaling (MAVS) to trigger IRF signaling, which leads to interferon induction. NS1 blocks the RIG-I ubiquitination via its interaction with TRIM25. (B) NS1 blockade of interferon action. NS1 inhibits the protein kinase activity of PKR via direct binding. In addition, NS1 prevents the activation of OAS by the sequestration of double-strand RNA.

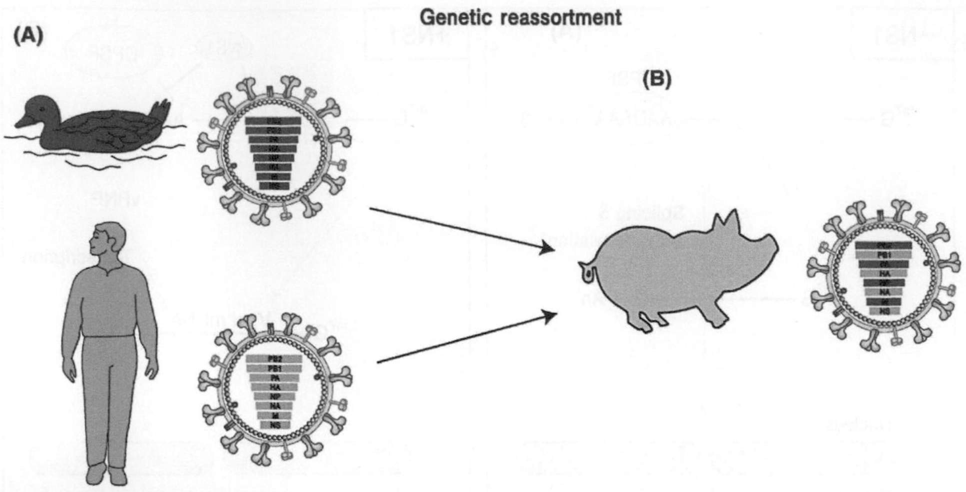

FIGURE 15.14 Genetic reassortment. (A) When a given cell is infected by two viral strains, hybrid viruses can be generated by exchanging their segmented genome, a mechanism called "genetic reassortment." (B) Pigs can be infected by both avian and human influenza viruses, because the glycans in the upper respiratory tract express sialic acids with not only an α-2,3 linkage but also an α-2,6 linkage. When avian and human influenza viruses coinfect pigs, hybrid viruses could be generated.

15.6 GENETIC REASSORTMENT

When cells are infected by two strains of influenza viruses with distinct subtypes, a recombinant virus can be generated during the assembly. This recombination is accomplished by co-packaging RNA segments derived from two subtypes, a process called *genetic reassortment*[23] (Fig. 15.14). Genetic reassortment is a kind of genetic recombination, which pertains to a virus with a segmented genome. In particular, genetic reassortment could occur when pigs are simultaneous infected by avian and human viruses, because pigs can be infected by both. As a result, the variants with novel antigenicity can be generated and then subsequently transmitted to humans. Importantly, a few pandemics that have occurred in the past have been attributed to the genetic reassortment.

15.7 VARIANTS: ANTIGENIC DRIFT VERSUS ANTIGENIC SHIFT

Seasonal Flu and Pandemic: A novel influenza virus strain emerges every year and causes a flu epidemic, the so-called *seasonal flu*.[24] This seasonal flu is caused by a novel variant that exhibits antigenicity that is distinct from the past strains due to accumulated mutations. In other words, antibodies elicited by the past infection cannot provide protective immunity against seasonal variants. In fact, a flu epidemic that is more formidable than seasonal flu is pandemic flu. A few flu *pandemics*[25] have occurred in the past, including "Spanish flu" that killed more than 40 million people during 1918–1919 (see Fig. 1.7). The "2009 H1N1 flu" epidemic is the most recent pandemic. Now, we consider how these influenza variants are generated.

Emergence of Variants: Two kinds of influenza virus variants—seasonal flu and pandemic flu—are described above. The question is how do these variants emerge and what is the difference between seasonal flu and pandemic flu? The viral variants refer to changes in two viral envelope proteins (ie, HA and NA). Two different mechanisms exist: antigenic drift and antigenic shift (Fig. 15.15). *Antigenic drift* refers to the progressive and incremental changes of HA and NA antigenicity attributable to mutations. Antigenic drift is the cause of seasonal flu. On the other hand, *antigenic shift* refers to the drastic changes in the antigenicity attributable to the genetic reassortment. In other words, a block of genetic information contained in a RNA segment is exchanged between different strains, leading to the generation of a hybrid virus. Antigenic shift is the cause of pandemic flu (Fig. 15.16). For instance, in the case of the 1957 Asian flu, three RNA segments of avian influenza virus were transmitted to human (Box 15.2).

23. **Genetic reassortment** It refers to a kind of genetic recombination occurring during assembly of segmented genome.
24. **Seasonal flu** A flu epidemic that comes regularly during winter.
25. **Pandemic** An epidemiology term that refers to an epidemic that affects multiple continents.

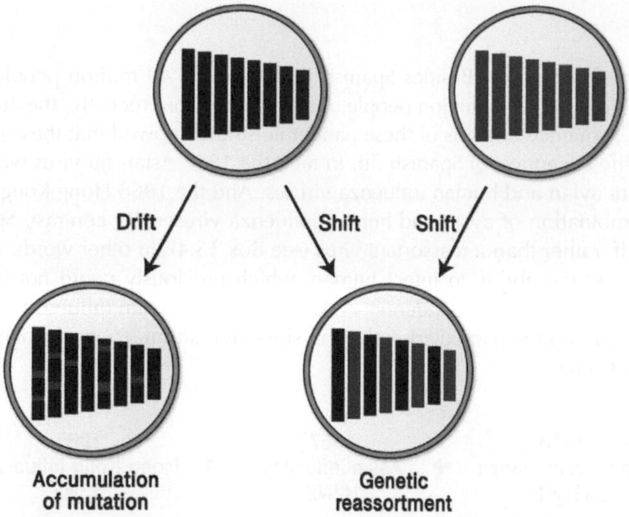

FIGURE 15.15 Antigenic drift and antigenic shift. Antigenic drift: mutations in the RNA genome are highlighted in pink. Antigenic shift: antigenic differences between two strains are denoted in different colors.

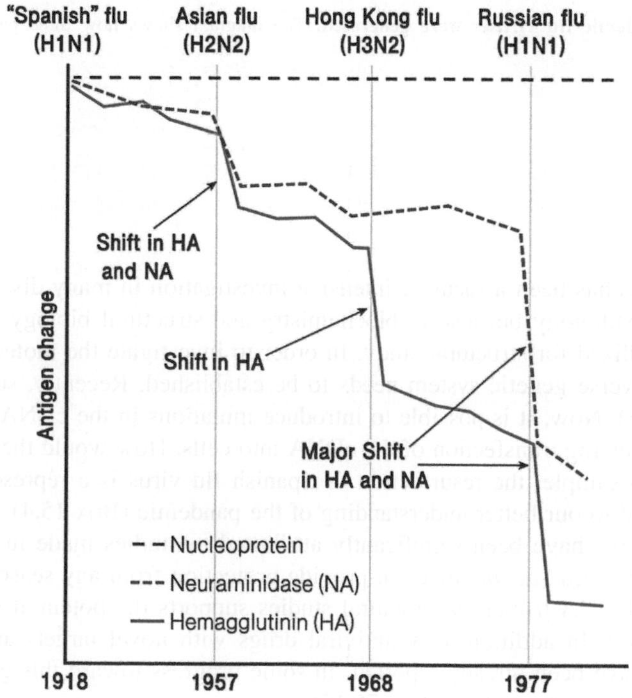

FIGURE 15.16 Antigenic changes in viral proteins of Flu epidemic. The extent of antigenic changes in three viral proteins (HA, NA, and NP) are compared. The major shifts in antigenicity of HA and NA proteins occurred during three pandemics between 1930 and 1970. In contrast, the antigenicity of NP remained unchanged.

BOX 15.2 Flu Pandemic

Four flu pandemics have occurred until now. Besides Spanish flu that killed 40 million people, 1957's "Asian flu" and1968's "Hong Kong flu" affected over 700,000 and 1 million people, respectively. More recently, the 2009 H1N1 flu pandemic affected over 16,000 people worldwide. Sequence analysis of these pandemic strains showed that these pandemic flu viruses emerged as a result of antigenic shift, with the exception of Spanish flu. In fact, the 1957 Asian flu virus was found to be an H2N2 subtype resulting from a recombination of avian and human influenza viruses. And the 1968 Hong Kong virus was found to be an H3N2 subtype resulting from the recombination of avian and human influenza viruses. By contrast, Spanish flu virus was found to be H1N1 avian influenza virus itself, rather than a reassortant virus (see Box 15.4). In other words, Spanish flu virus was, in fact, an avian influenza virus that acquired the ability to infect human, which previously could not infect human. In addition to the genetic shift, the host change contributed to the pandemic. A question of how avian influenza virus acquired the ability to infect human remains enigmatic. This question is particularly relevant, since human infection of avian influenza virus has occurred in 2003 in Vietnam and in 2012 in China.

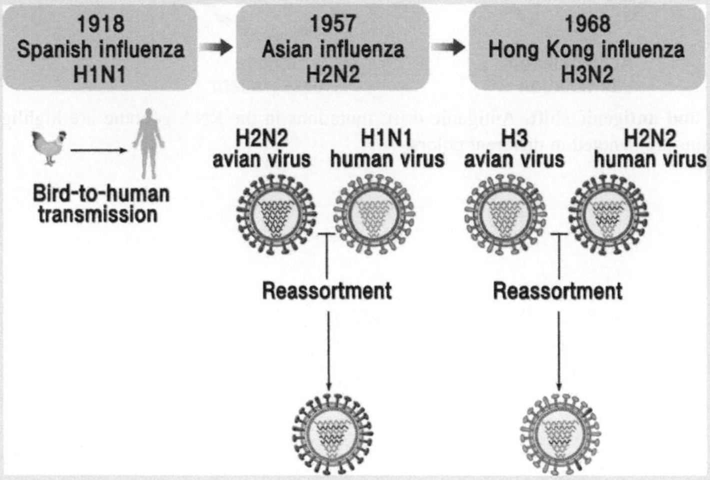

A diagram illustrating how pandemic flu viruses were generated. The diagram shows how three pandemics occurred in 20th century are generated.

15.8 PERSPECTIVES

Needless to say, influenza virus has been a focus of intensive investigation in many disciplines of modern biology: not only in virology and immunopathology but also in biochemistry and structural biology. For instance, HA protein was one of the first proteins crystallized for structural study. In order to investigate the molecular genetics of RNA viruses, including influenza virus, a reverse genetic system needs to be established. Recently, such a *reverse genetic*[26] system has been established (Box 15.3). Now, it is possible to introduce mutations in the cDNA of the influenza viral genome and examine its phenotype following transfection of the cDNA into cells. How would the reverse genetic system benefit influenza virus research? For example, the resurrection of Spanish flu virus is a representative example in which the reverse genetics has contributed to our better understanding of the pandemic (Box 15.4). In particular, advances of vaccine technology or "vaccinology" have been significantly attributed to studies made in influenza vaccines. An unmet medical need is a *universal influenza vaccine* that can provide protection from any seasonal flu strain. Identification of highly conserved epitopes in the HA trimer by structural studies supports the potential feasibility of a universal influenza vaccine (see Journal Club). In addition, new antiviral drugs with novel targets are urgently needed, given that oseltamivir-resistant variants have been already reported. In some progress toward this goal, an antiviral drug targeting RdRp is reportedly on the verge of approval (see Fig. 26.12).

26. **Reverse genetic** A genetic method to examine phenotypic changes caused by experimental mutagenesis. This methodology is a reverse of classical genetic, which is to examine genetic changes associated with phenotype.

BOX 15.3 Reverse Genetics

To investigate the molecular genetics of RNA viruses, it is essential to have the cDNA version of the full-length RNA genome. By inserting cDNA in the appropriate expression plasmids, an infectious RNA can be synthesized in vitro and then, transfected into cells to produce a progeny virus particle. Alternatively, transfection of the cDNA-inserted plasmid into cells could also lead to viral replication and progeny production. These kinds of experimental methods are called reverse genetics. Although reverse genetics was established and used in the positive-strand RNA viruses, such as picornaviruses, it was not immediately utilized for negative-strand RNA viruses. In fact, it has been difficult to synthesize the full-length RNA of the negative-strand RNA viruses, since their 5′ and 3′ ends are distinct from cellular mRNAs. Furthermore, for influenza virus, it was proven more difficult to establish the reverse genetic system due to the multiple nature of the segmented RNA genome. Recently, Yoshihiro Kawaoka and his colleagues at the University of Wisconsin have successfully established a reverse genetic system for influenza virus. Intriguingly, both RNA Pol II and RNA Pol I promoters were inserted at either end of the viral cDNA but in opposite directions so that both mRNA and vRNA are synthesized from one plasmid. Unlike RNA Pol II transcripts, RNA Pol I transcripts are neither capped at the 5′ end nor polyadenylated at the 3′ end. Thus, RNAs, whose 5′ and 3′ ends are precisely identical to vRNA are synthesized. Such transcribed mRNAs are utilized for protein synthesis, while vRNAs are utilized as a template for viral genome replication in transfected cells. When eight plasmids corresponding to the eight RNA segments are transfected into cells, up to 10^8 virus particles/mL are produced. Now, it is possible to introduce any kind of mutations to the influenza viral genome to study their impact on the viral life cycle [Curr. Top Microbiol. Immunol. 283:43–60 (2004)].

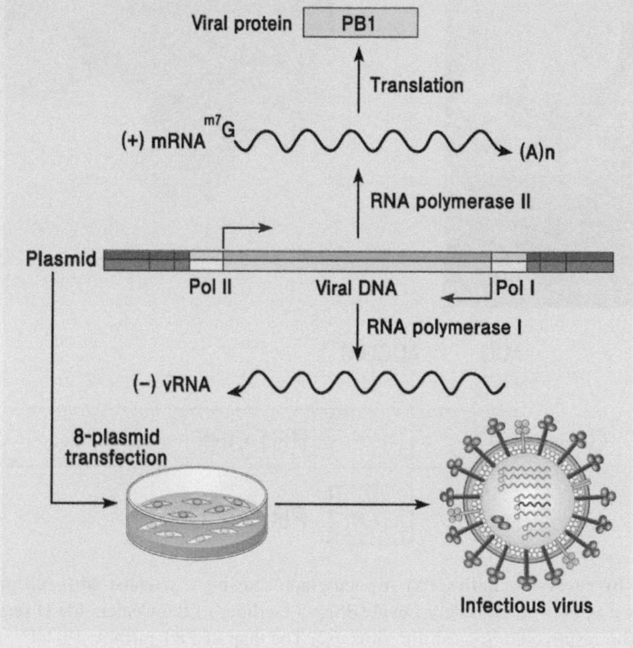

A diagram illustrating the strategy of reverse genetics. The diagram shows only the PB1 RNA segment as an example. The direction of RNA Pol I and RNA Pol II transcriptions is indicated by *arrows*. The polarity of mRNA and vRNA is denoted by (+) and (−), respectively. Eight plasmids, in which each corresponds to eight segmented RNA genomes, are required to produce an infectious influenza virus particle.

BOX 15.4 Resurrection of Spanish Flu Virus

During 1918–1919, when World War I was not yet over, 40 million people were killed by a pandemic flu, called Spanish flu. The outbreak started in the summer of 1918 among American soldiers staying in France and spread to all continents. The fatality of Spanish flu (2%) was higher than those of seasonal flu (0.1–0.2%). The question was why the fatality of Spanish flu was conspicuously higher? To answer this question, a team led by Jeffery Taubenberger at Armed Forces Institute of Pathology collected the frozen remains of Spanish flu victims in Alaska, where the viral RNAs were more likely to be intact. They cloned out the residual viral RNA fragments and stitched them to make an intact RNA segments. Remarkably, they succeeded in cloning the full-length viral genome of the Spanish flu. In collaboration with Peter Palese at Mt. Sinai Medical School, he was able to

(Continued)

BOX 15.4 (Continued)

resurrect Spanish flu virus. It was an occasion where the imagination of Jurassic Park became real. Importantly, further characterization of the resurrected Spanish flu virus revealed that it was, in fact, avian influenza virus with an H5N1 subtype. It was not a reassortant virus, as was predicted. Furthermore, it was revealed that avian influenza virus changed its host range and acquired the ability to infect human.

Importantly, the resurrected Spanish flu virus provided the clue for its higher virulence. First, the HA gene of Spanish flu virus was similar to ones found in the highly pathogenic avian influenza virus (HPAI) strain. Second, the Spanish flu virus was found to encode a novel open reading frame (ORF) in the PB1 genome, dubbed PB1-F2. Interestingly, PB1-F2 is related to its higher virulence. PB1-F2 encodes a 90 amino acid polypeptide in a reading frame that is +1 shifted from PB1 ORF. It was discovered as a novel reading frame that encodes a CTL epitope associated with higher virulence of the Spanish flu virus. PB1-F2 is a mitochondrial inner protein that disrupts mitochondrial membrane potential and induces the release of cytochrome C from mitochondria, thereby leading to apoptosis. Furthermore, in the mouse model, it was shown that PB1-F2 is associated with a higher incidence of secondary pneumonia infection, which causes death [Science 310:77–80 (2005)].

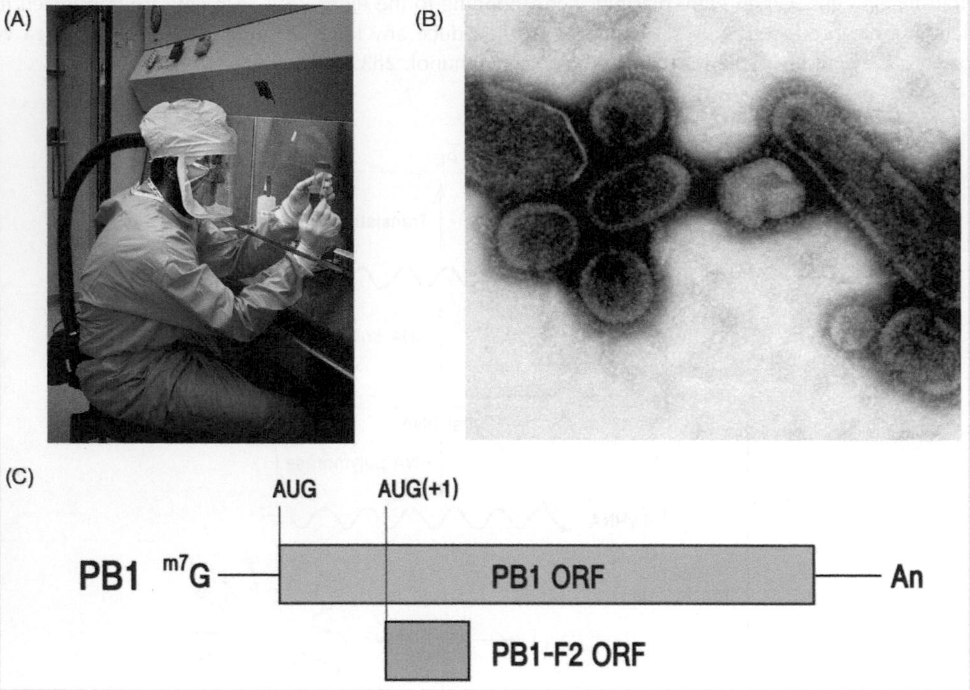

Resurrection of Spanish flu virus by reverse genetics. (A) A photograph showing a scientist with full protection gadget examining a reconstructed Spanish flu virus. The work was done in Biosafety Level (BSL)-3 facilities of the Centers for Disease Control (CDC) in Atlanta, United States. (B) Electron micrograph of the resurrected Spanish flu virus. (C) The map of PB1 mRNA. PB1-F2, as its name implies, is derived from the second reading frame (+1) of PB1 mRNA.

15.9 SUMMARY

- *Virus particles*: HA and NA are embedded into the envelope. Each particle contains eight segmented genomes (ie, vRNPs), each of which is encompassed by NP. Inside of the particles, RdRp (PB1 + PB2 + PA subunit) is associated to vRNPs.
- *Viral genome*: The viral genome constitutes eight segmented genomes, which are negative-strand RNA. vRNA folds into a panhandle structure.
- *Genome replication*: The viral genome replication occurs in the nucleus. NP regulates the switch from transcription to genome replication. The viral mRNA transcription is primed by cap-snatching mechanism.
- *Host effect*: NS1 is the virulence factor. The NS1 suppresses not only host mRNA processing, but also suppresses host innate immune response.

- *Variants*: Antigenic drift is caused by high mutation rate. Antigenic shift occurs as a consequence of genetic reassortment between two distinct strains.
- *Antivirals and vaccine*: Oseltamivir, an inhibitor of neuraminidase activity of NA, is available as an effective antiviral drug for influenza virus infection. Vaccines are available, but yearly vaccination is required for effective protection.

STUDY QUESTIONS

15.1 Three kinds of viral RNAs are found in cells infected by influenza virus. (1) Please depict three viral RNA species with characteristic noncoding region and explain the difference between them. (2) Please indicate the nonviral sequence, and explain how they are made.

15.2 A few temperature-sensitive mutants (*ts*) of influenza virus were isolated. Compare the phenotype of the mutants with that of wild-type. Note that *ts* mutants produce progeny virus as wild-type does at a permissive temperature (ie, 33°C), but not at a nonpermissive temperature (ie, 39°C). (1) *ts* mutant of NS1 gene and (2) *ts* mutant of NA gene.

15.3 Unlike other vaccines, influenza vaccine needs to be administered every year for effective prevention. Explain why?

SUGGESTED READING

Eisfeld, A.J., Neumann, G., Kawaoka, Y., 2015. At the centre: influenza A virus ribonucleoproteins. Nat. Rev. Microbiol. 13 (1), 28–41.

Everitt, A.R., Clare, S., Pertel, T., John, S.P., Wash, R.S., Smith, S.E., et al., 2012. IFITM3 restricts the morbidity and mortality associated with influenza. Nature. 484 (7395), 519–523.

Jagger, B.W., Wise, H.M., Kash, J.C., Walters, K.A., Wills, N.M., Xiao, Y.L., et al., 2012. An overlapping protein-coding region in influenza A virus segment 3 modulates the host response. Science. 337 (6091), 199–204.

Medina, R.A., Garcia-Sastre, A., 2011. Influenza A viruses: new research developments. Nat. Rev. Microbiol. 9 (8), 590–603.

Rehwinkel, J., Tan, C.P., Goubau, D., Schulz, O., Pichlmair, A., Bier, K., et al., 2010. RIG-I detects viral genomic RNA during negative-strand RNA virus infection. Cell. 140 (3), 397–408.

JOURNAL CLUB

- Impagliazzo, A.F., et al., 2015. A stable trimeric influenza hemagglutinin stem as a broadly protective immunogen. Science, 349, 1301–1306.

  Highlight: Influenza vaccine needs to be administered annually. The protective antibodies are directed against invariant molecular components of antigen to elicit broadly neutralizing antibodies. This notion inspires the concept of a *universal influenza vaccine* that can provide protection from any seasonal flu strain. This seminal paper demonstrated that antibodies elicited by a stable HA stem antigen (mini-HA) neutralizes H5N1 viruses. This result provides proof-of-concept for design of HA stem mimics that elicit broadly neutralizing antibodies.

Chapter 16

Other Negative-Strand RNA Viruses

Chapter Outline

In the two preceding chapters, two negative-strand RNA viruses—rhabdovirus and influenza virus—were described. In addition, paramyxoviruses, filoviruses, and bunyaviruses are also important human pathogens possessing negative-strand RNA genome. Here, these negative-strand RNA viruses will be briefly covered with an emphasis on the genome structures. Reovirus is included in this chapter for the sake of breadth, although it possesses a double-strand RNA (dsRNA) genome.

16.1 PARAMYXOVIRUS

Paramyxoviruses[1] (family Paramyxoviridae) are enveloped viruses, and contain helical nucleocapsid particles that possess a negative-strand RNA of 15–18 kb. As the name implies, paramyxoviruses share the tropism with orthomyxoviruses in that they infect the host via the mucous layer. However, their genome organizations are distinct from those of orthomyxoviruses. In fact, their genome structure is rather similar to that of rhabdoviruses (see chapter: Rhabdovirus). Here, Sendai virus will be mainly described.

Classification: Paramyxoviruses are composed of five genera (Table 16.1). Sendai virus is a prototype of paramyxoviruses. Sendai virus was originally isolated from a mouse and causes only mild disease in human. Measles virus represents an important human pathogen in this family. Measles is considered as a fever that everyone gets in their infancy. Being transmitted via aerosol, it is highly contagious. It is believed that the majority of North and South American Indians succumbed to the measles that was brought by European invaders during the 16th century. Live vaccines, first developed in 1961, effectively prevent youngsters from infection. Although the fear of measles has disappeared, nearly 10 million children each year are infected by measles virus in Africa and South America, resulting in over 120,000 deaths.

Besides measles virus, *respiratory syncytial virus (RSV)* is also an important human pathogen. RSV causes bronchiolitis to children and is the main pathogen for viral pneumonia in children. *Syncytium*[2] is the major cytopathic effect, as the name implies. For adults, however, RSV causes only mild symptoms, often indistinguishable from common colds and minor illnesses. In fact, RSV infection is one of the most common respiratory infections in infants and most children get RSV infection at least once within 2 years after birth. RSV infection can be fatal, as it costs the lives of over 160,000 children yearly. Despite its clinical importance, no vaccine is yet available. In addition, Nipah virus, a newly emerging virus, is a paramyxovirus (see Table 16.1). The first outbreak of Nipah virus occurred in Malaysia in 1999 (see Table 21.3). Hendra virus, another newly emerging virus, is also a paramyxovirus (see Table 21.3). Hendra virus is now

1. **Paramyxovirus** The term is derived from the Greek word for "alternative"-*para* and for "mucus"-*myxo*.

2. **Syncytium** A syncytium (*pl.*, syncytia) refers to a multinucleated cell that can result from multiple cell fusions of uninuclear cells. The term is derived from the Greek word for "together"-*syn* and for "box (cell)"-*kytos*.

Molecular Virology of Human Pathogenic Viruses. DOI: http://dx.doi.org/10.1016/B978-0-12-800838-6.00016-3

TABLE 16.1 Classification of Paramyxoviruses

Genus	Species	Disease
Respirovirus	Sendai virus (SeV)	Respiratory infection (mild)
	Human Parainfluenza virus (hPIV)	Respiratory infection
Rubulavirus	Mumps virus	Mumps
Morbillivirus	Measles virus	Measles
Pneumovirus	Respiratory syncytial virus (RSV)	Bronchiolitis and pneumonia to children
Henipavirus	Hendra virus	Hemorrhage in lung
	Nipah virus (Niv)	Encephalitis

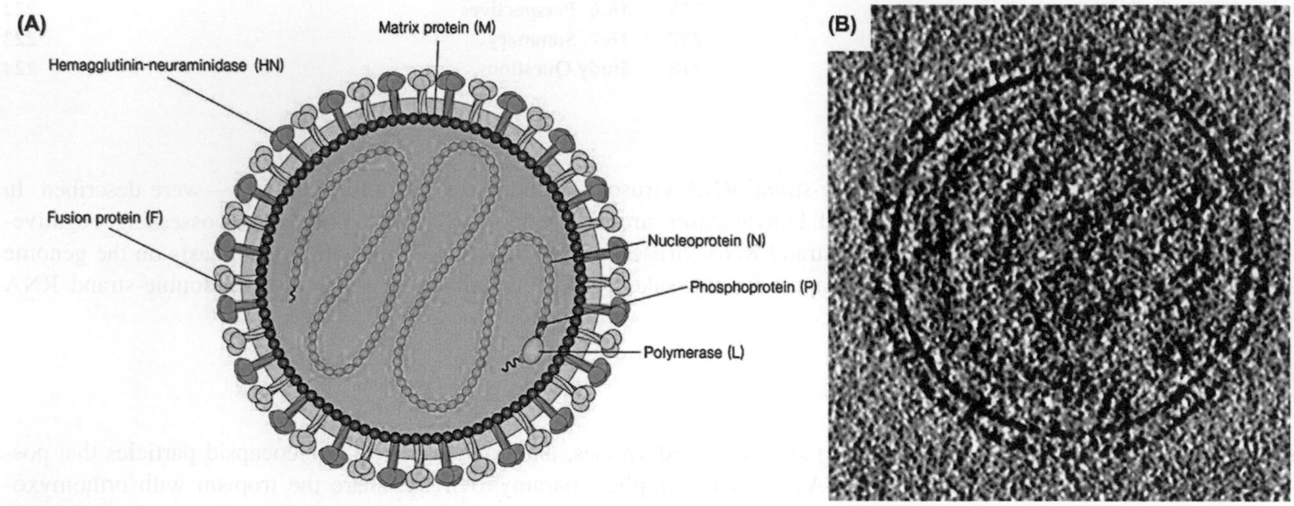

FIGURE 16.1 Virion structure of paramyxovirus. (A) A diagram shows the virion structure of Sendai virus. The envelope is studded with two envelope glycoproteins, HN and F (fusion). A helical nucleocapsid, in which the RNA genome is encapsidated by N (nucleocapsid) proteins, is associated with L (polymerase) and P (phosphoprotein). (B) Transmission electron micrograph (TEM) of Sendai virus particle.

classified as a newly established genus *Henipavirus*, together with the distantly related Nipah virus (see Table 16.1). Both Nipah virus and Hendra virus are *zoonotic viruses* in that they infect animals (ie, pigs and horses, respectively) as well as humans.

16.1.1 Sendai Virus

Discovery: Sendai virus was originally isolated from a mouse in the city of Sendai in Japan. Sendai virus causes severe respiratory infection to a mouse, but only mild disease in human, which makes it suitable for research.

Virion Structure: Sendai virus virions are round with a pleomorphic enveloped particle, ranging from 150 to 350 nm in size. It has a helical nucleocapsid inside (Fig. 16.1). An envelope glycoprotein, termed *hemagglutinin-neuraminidase (HN)*, corresponds to a fused version of the HA and NA of influenza virus.

Genome Structure: The RNA genome of Sendai virus is a negative-strand RNA of 15–18 kb (Fig. 16.2). The genome organization is similar to that of rhabdovirus in three aspects. First, it has "leader" and "trailer" RNA genes at the 5′ and 3′ termini, respectively. Second, multiple genes are encoded in a single RNA molecule; moreover, their order from the 3′ terminus is similar to that of rhabdovirus, starting from N to L gene (see Fig. 14.2). Third, "intergenic regions" lie between genes (see Fig. 16.2).

Protein Coding: Sendai virus encodes six genes (Table 16.2). A similarity between Sendai virus and rhabdovirus is also noted in protein coding. Sendai virus has a P/C/V gene, instead of the P gene of rhabdovirus; otherwise protein

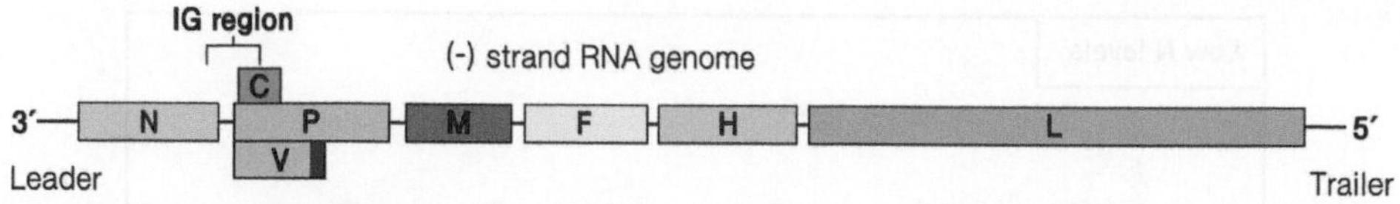

FIGURE 16.2 The RNA genome of paramyxovirus. The RNA genome of Sendai virus is shown. Note that the 5′ end of negative-strand RNA genome is positioned to the right. Two noncoding RNA genes, "leader" and "trailer," are denoted at the termini. Intergenic (IG) regions (ie, noncoding region) positioned between coding regions are denoted by a bracket.

TABLE 16.2 Viral Genes of Paramyxovirus

Gene	Protein	Function
N	N (nucleocapsid)	Nucleocapsid protein
P/C/V	P (phosphoprotein)	RdRp cofactor
	C	Unknown
	V	IFN signaling inhibitor
M	M (matrix protein)	Matrix protein
F	F (fusion protein)	Fusion protein
HN	HN	Envelope glycoprotein
L	L (large)	RdRp

coding is identical. The coding strategy of the P/C/V gene is extraordinarily complex, as three distinct proteins are synthesized from one mRNA. It involves mRNA editing and an alternative translation initiation codon (ie, ACG codon instead of AUG).

Genome Replication: Again, the genome replication strategy is similar to that of rhabdovirus in that the genomic RNA can serve as the template for RNA genome replication as well as RNA transcription (Fig. 16.3). Similar to that of rhabdovirus, the N protein level is believed to regulate the transit from RNA transcription to RNA replication. The genomic RNA serves as the template for the viral RNA transcription until the N protein is accumulated soon after infection. As the N protein accumulates, the transition from transcription to the RNA genome replication takes place.

16.2 FILOVIRUS

Filovirus (family Filoviridae) is a family of nonsegmented, negative-strand RNA viruses. The virions are enveloped but filamentous (not spherical or round) and contain a helical nucleocapsid encompassing 19 kb RNA genome.

Classification: Ebola virus, first discovered in 1976, is now classified in the family Filoviridae together with Marburg virus, discovered earlier in the city of Marburg in Germany (Table 16.3). The first Ebola virus outbreak occurred in Zaire and Sudan in 1976; the virus was named after the Ebola river in the region of Zaire. It is one of the most horrifying pathogens, as its fatality approaches 90%.

Ebola virus represents the most horrifying pathogen that leads to fatal consequences (~90% fatality). One might wonder where this dangerous culprit came in the first place? The movie "Outbreak," starring Dustin Hoffman, which is based on the episode of Ebola virus outbreak, featured monkeys as the natural reservoir. Unlike the movie, bats were identified as the natural reservoir for Ebola virus. Earlier, primates such as monkeys were suspected as being a reservoir, but primates turned out to be victims rather than a reservoir of Ebola virus.

16.2.1 Ebola Virus

Ebola virus, a prototype of *Filoviruses*,[3] is similar to rhabdovirus in some aspects, having a large nonsegmented, negative-strand RNA genome.

3. **Filovirus** The name is derived from the Latin noun *filum* (alluding to the filamentous morphology).

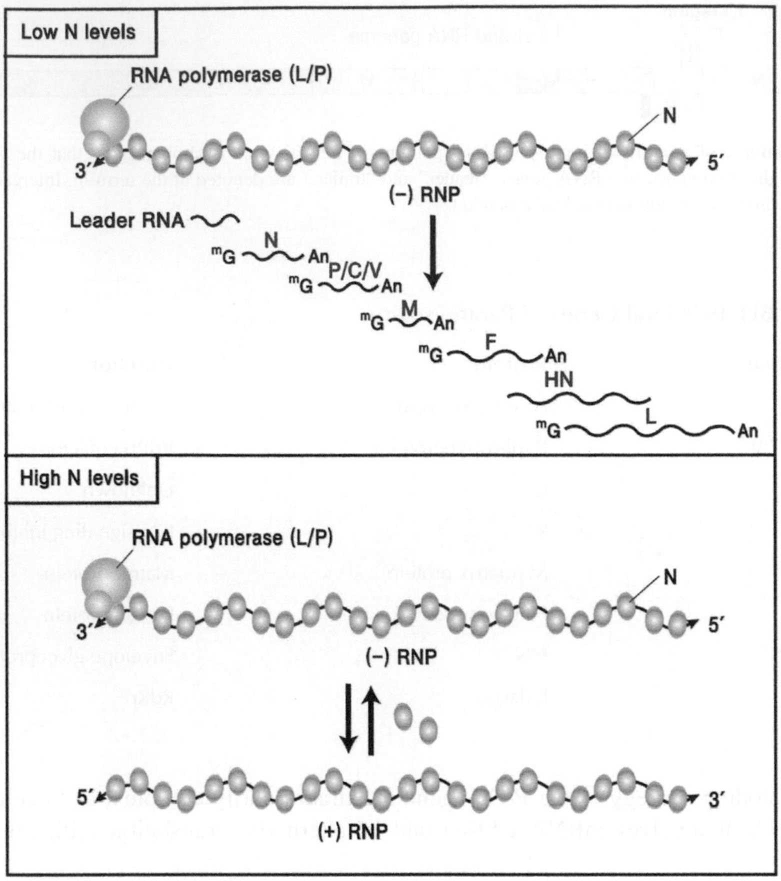

FIGURE 16.3 **Regulation of RNA replication and transcription of paramyxovirus.** The genomic RNA serves as the template for both transcription and RNA genome replication. At early infection, when the N protein level is low, mainly viral mRNA synthesis occurs. When the N protein level is high at late infection, the full-length (+) RNA, instead of mRNA, is synthesized.

TABLE 16.3 Classification of Filoviruses

Genus/Species	Host	Reservoir	Outbreak
Ebolavirus			
Sudan ebolavirus	Primates, human	Bat	Africa
Zaire ebolavirus	Primates, human	Bat	Africa
Ivory coast ebolavirus	Primates, human	Bat	Africa
Reston ebolavirus	Primates, swine	Bat	Africa, Philippine
Marburgvirus			
Lake Victoria marburgvirus	Primates, human	Bat	Africa

Virion Structure: The viral filament characteristically appears in various shapes including a coil or branched resulting in pleomorphic particles in images (Fig. 16.4). The filaments are reported to be between 60 and 80 nm in diameter, but the length of a filament is extremely variable—usually 1000 nm but up to 14,000 nm in length has been reported. The helical nucleocapsid is encapsidated by NP protein and associated with RdRp (L + VP35) at the tip.

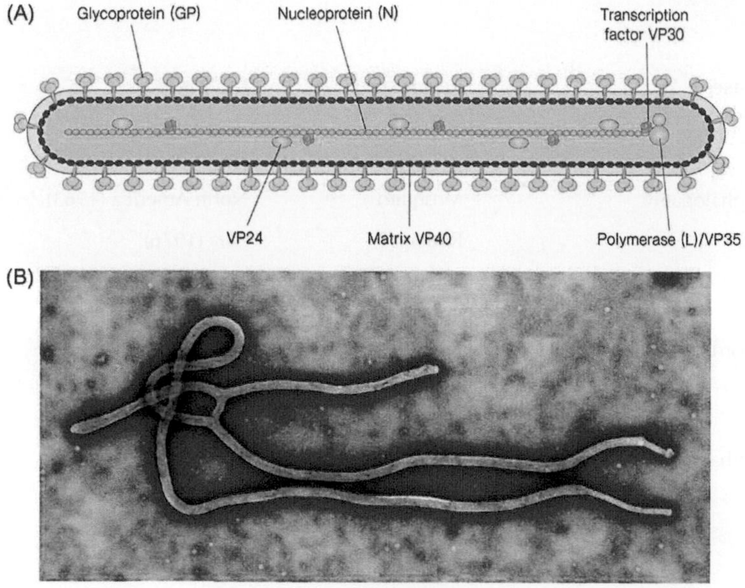

FIGURE 16.4 Virion structure of Ebola viruses. (A) A diagram illustrating the filamentous shape of the Ebola virus particle. Embedded in the enveloped are trimeric glycoprotein spikes. Inside the envelope, the nucleocapsid, in which the negative-strand RNA is encapsidated by NP proteins, is enclosed. Note that the RdRp complex (ie, L/VP35 complex) is attached to the end of the nucleocapsid. (B) An electron micrograph of an Ebola virus particle revealing the characteristic filamentous structure.

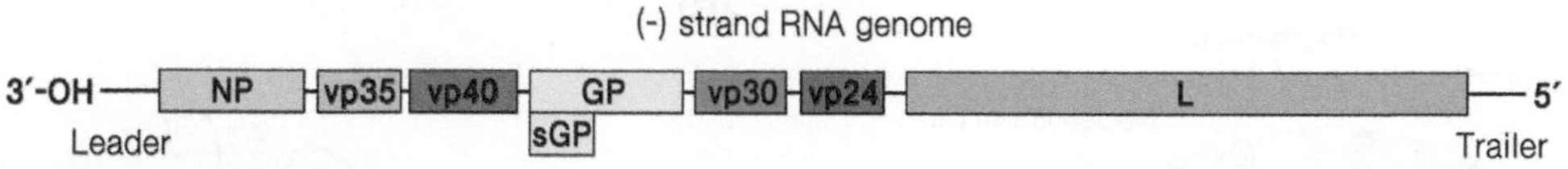

FIGURE 16.5 The RNA genome of Ebola virus. A diagram illustrating the negative-strand RNA genome of Ebola virus. Note that the 5′ end of the negative-strand RNA genome is positioned to the right. Two noncoding RNA genes, "leader" and "trailer," are denoted at the termini. NP, a major nucleoprotein; VP35, a component of RdRp complex; VP40, matrix protein; GP, glycoprotein; VP30, a minor nuclear protein with RNA-binding ability; VP24, a minor matrix protein; L, RdRp.

Genome Structure: The RNA genome is a negative-strand RNA of 19 kb (Fig. 16.5). The genome organization is similar to that of rhabdovirus and paramyxovirus, having noncoding RNA genes at the both termini and "intergenic regions" (see Figs. 14.2 and 16.2).

Epidemiology: Ebola virus represents an important zoonotic virus in infecting both primates and humans. Within a few days, most infected individuals succumb to the virus with hemorrhagic fever, with fatality approaching near 90%. Fortunately, secondary infection to comforting families or medical personnel is rare, implicating that transmission via the airborne route is insignificant. The Zaire ebolavirus is the most dangerous of the six species of Ebola viruses, causing an extremely severe hemorrhagic fever in humans and other primates. Ebola virus outbreak remains a threat not only to humans but also to endangered primates, such as gorilla. Bats have been identified as the natural reservoir of Ebola virus. Ebola virus is a select agent, which is classified as WHO Risk Group 4 Pathogen, requiring Biosafety Level 4-equivalent containment for handling (see Fig. 4.14).

Treatment: Broad spectrum antiviral agents, such as interferon and ribavirin, are ineffective against Ebola virus infection. No treatment is available.

16.3 BUNYAVIRUS

Bunyavirus (family Bunyaviridae[4]) is a family of negative-strand RNA viruses, possessing three segmented RNA genomes.

Classification: Family Bunyaviridae is composed of diverse members (more than 100), which can be divided into four genera (Table 16.4). Bunyaviruses are vector-borne viruses and zoonotic viruses infecting both animals and human. Transmission occurs via an arthropod vector (mosquitoes or tick). Bunyamwera virus was the first member of this family, which was isolated from an outbreak in Bunyamwera, a town in Uganda. Hantaviruses are transmitted through contact with mice feces. Hantaan virus, a member of the genus *Hantavirus*, was isolated from a mouse near Hantaan river in Korea. Sin Nombre virus was isolated from an outbreak that occurred in the Four Corners region of the western

4. **Bunyavirus** It is named for Bunyamwera, a town in Uganda, where the virus was isolated.

TABLE 16.4 Classification of Bunyaviruses

Genus	Species	Disease	Reservoir (Vector)	Outbreak
Orthobunyavirus	Bunyamwera virus	–	–	Africa
	La Crosse virus	Encephalopathy	Mosquito	North America (1963)
Hantavirus	Hantaan virus	HFRS	Rodents	Asia (1976)
	Sin Nombre hantavirus	Hantavirus pulmonary syndrome	Deer mouse	North America (1993)
Nairovirus	Crimean Congo hemorrhagic	Hemorrhagic fever	Tick	East Europe, Africa (1944)
	Fever virus			
Phlebovirus	Rift Valley Fever virus	Encephalopathy	Mosquito	Africa (1931)
	SFTS virus	SFTS	Tick	China (2009)

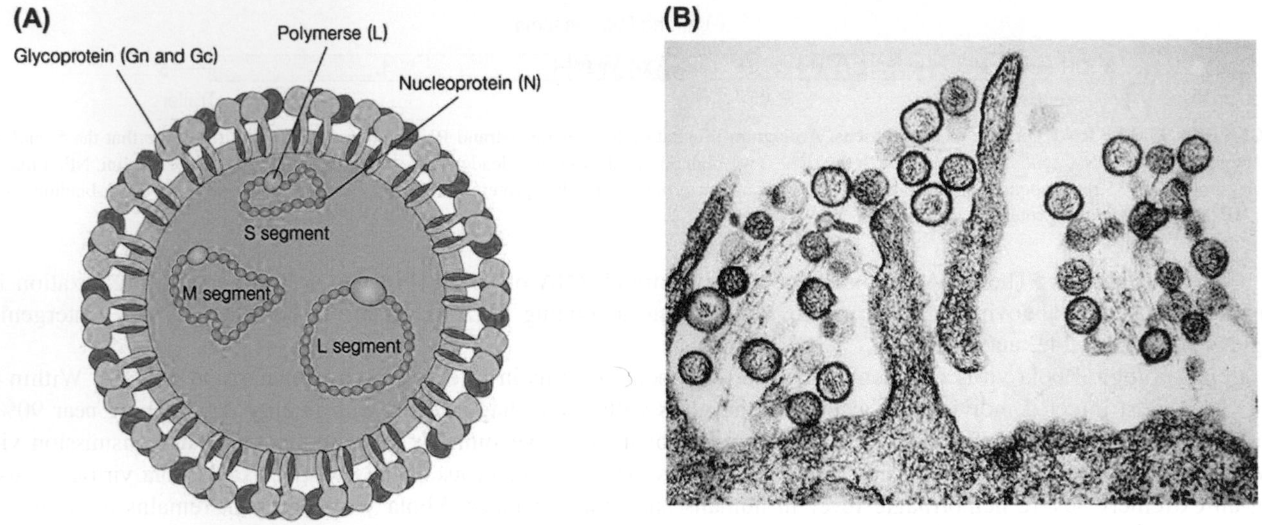

FIGURE 16.6 Virion structure of hantaviruses. (A) A diagram illustrating Hantavirus virion. Three helical nucleocapsids are shown: L, M, and S segment. The nucleocapsids are believed to be in a circular configuration by base-pairing of both termini. The negative-strand RNA is encapsidated by N (nucleocapsid) protein. L protein (RdRp) attached to the end of the nucleocapsid is drawn expanded. (B) TEM of the Sin Nombre Hantavirus. The enveloped virus particles released from infected cells are shown.

United States (see Fig. 21.8). SFTS virus is a new member of the Bunyavirus family; it was recently isolated in China as an emerging virus that causes severe fever with thrombocytopenia syndrome (SFTS). It is now classified as a new member of genus *Phlebovirus*. An SFTS outbreak occurred later in Japan and Korea. SFTS virus is a tick-borne virus and its fatality is high (nearly 20%).

Here, *Hantavirus*, a genus of Bunyavirus family, will be mainly described.

16.3.1 Hantavirus

Hantavirus is the prototype of bunyavirus. In particular, Sin Nombre virus has been extensively studied.

Virion Structure: Hantavirus is enveloped (100 nm diameter) and has two envelope glycoproteins, Gn and Gc (Fig. 16.6). Three RNA segments are presented as helical nucleocapsids, which are encapsidated by N protein. L protein (RdRp) is also associated with the nucleocapsids.

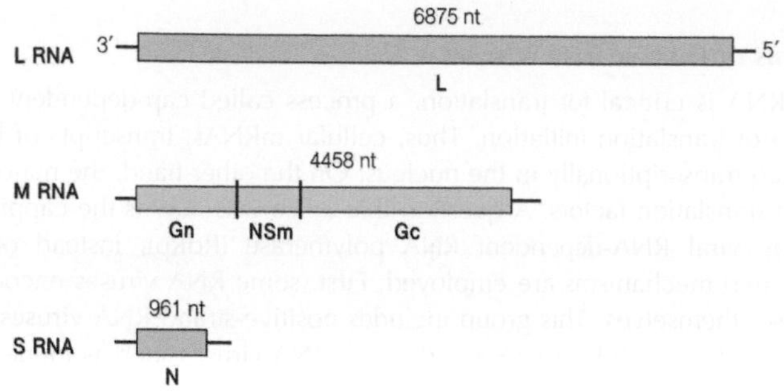

FIGURE 16.7 The RNA segments of hantaviruses. Three RNA segments (L, M, and S) are shown. Note that the 5′ end of negative-strand RNA genome is positioned to the right. The viral proteins encoded by each RNA segments are denoted below. L, RdRp; Gn, envelope glycoprotein; Gc, envelope glycoprotein, N, nucleocapsid protein.

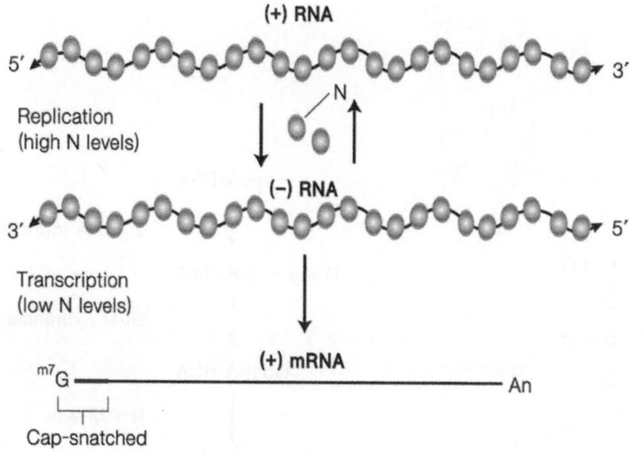

FIGURE 16.8 Regulation of RNA replication and transcription of hantavirus. The genomic RNA serves as the template for both transcription and RNA genome replication. In early in infection, when the N protein level is low, the viral mRNA synthesis by transcription occurs mainly. As the N protein accumulates, the full-length (+) RNA, instead of mRNA, is made. The cap snatched from cellular mRNA is denoted by a red line.

Genome Structure: Hantavirus has three negative-strand RNA genomes of 6.8, 4.5, and 0.9 kb, termed L, M, and S segments (Fig. 16.7). Three mRNAs (ie, L, M, and S) are transcribed from three RNA segments. L protein (RdRp) is encoded by L mRNA; three proteins (ie, Gc, Gn, and NSm) are encoded by M mRNA; two proteins (N, and Ns) are encoded by S mRNA.

Genome Replication: The RNA genome replication strategy of hantavirus is similar to those of other negative-strand RNA viruses including rhabdovirus and influenza virus. First of all, N protein level determines the transition from RNA transcription to RNA genome replication (Fig. 16.8). Earlier in infection, when the N protein level is low, the RNA transcription proceeds. As N protein accumulates later in infection, the N proteins, as a trimeric form, begin to bind to the viral RNA, resulting in formation of a (+) ribonucleoprotein (RNP) structure, which then serves as a template for viral genome replication. Second, intriguingly, the RNA primer utilized for the viral transcription is snatched from cellular mRNA (Box 16.1). In other words, the RNA primer, which is the capped 10–18 nt RNA fragment, is obtained by the cap-snatching mechanism, which is reminiscent of influenza virus (see Fig. 15.7).

16.4 ARENAVIRUS

Arenaviruses (family Arenaviridae[5]) are enveloped, spherical particles with a diameter from ~120 nm. Arenavirus contains segmented genomes, like orthomyxovirus and bunyavirus. Arenavirus infects rodents and occasionally humans.

5. **Arenavirus** The name *Arena* comes from the Latin root meaning *sand* (see Fig. 16.9).

BOX 16.1 Capping Mechanisms of Diverse RNA Viruses

The cap structure of cellular mRNA is critical for translation, a process called cap-dependent translation. The engagement of eIF4E on the cap is the first step of translation initiation. Thus, cellular mRNAs, transcripts of host RNA polymerase II, are all capped and the capping occurs co-transcriptionally in the nucleus. On the other hand, the majority of viral mRNAs have a cap, as their translation relies on host translation factors. A question that arose was how is the capping achieved by RNA viruses, in which RNAs are transcribed by viral RNA-dependent RNA polymerase (RdRp), instead of cellular RNA polymerase II? Depending on the RNA viruses, two mechanisms are employed. First, some RNA viruses encode capping enzymes (ie, guanyl transferase and methyl transferase) themselves. This group includes positive-strand RNA viruses such as flavivirus, coronavirus, and negative-strand RNA viruses such as rhabdovirus. Second, some RNA viruses such as influenza virus and bunyavirus snatch the cap structure from cellular mRNAs, a mechanism dubbed cap-snatching. In this case, the virus encodes endonuclease required for the cleavage of the capped RNA (eg, PA endonuclease of influenza virus).

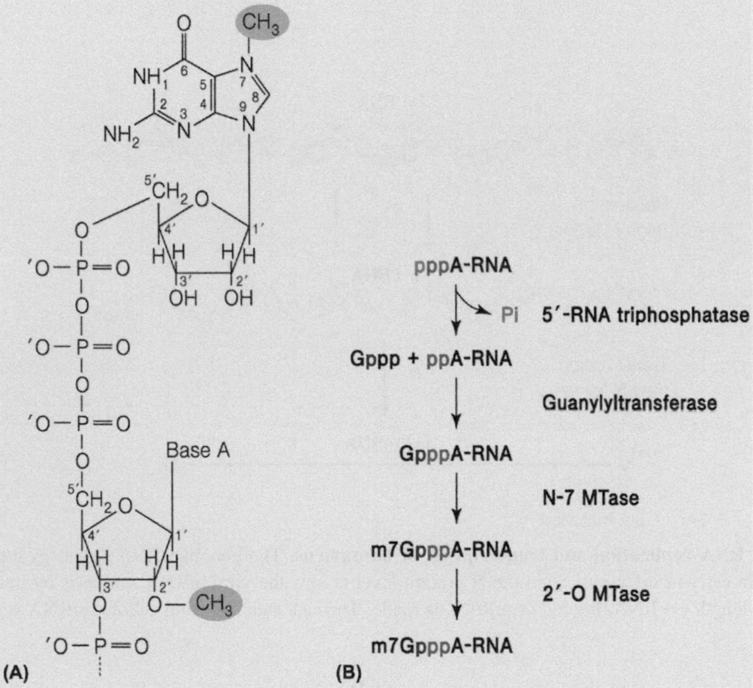

(A) Chemical characteristics of cap structure. The cap is a special kind of nucleotide, represented as "7-methyl-GpppA," in which the G nucleotide is linked via 5′-5′ linkage to the transcript. In addition, the N7 position of the G residue is methylated and the first nucleotide is also 2′-O-methylated. (B) Capping processes and the enzymes involved. Four enzymatic reactions are required for the capping: (1) 5′-RNA triphosphatase, (2) guanyl transferase, (3) N7-methyl transferase, and (4) ribose 2′-O-methyl transferase.

Although *arenaviruses* are classified as negative-strand RNA viruses, strictly speaking, arenavirus genomes are "*ambisense*" in that both strands encode viral proteins. Otherwise, arenaviruses are very similar to bunyaviruses in many respects: (1) having segmented genomes of negative-strand RNA, (2) epidemiological association with rodents (zoonotic virus), and (3) causing hemorrhagic fever in humans.

Discovery and Classification: Lassa virus was first isolated in an outbreak occurred in a region called Lassa in Nigeria in 1969. The virus is zoonotic or animal-borne and can be transmitted to humans (Table 16.5). Lymphotropic choriomeningitis virus (LCMV), a rodent-borne virus, is also a member of the arenaviruses. LCMV is naturally spread by the common house mouse. LCMV infection manifests itself in a wide range of clinical symptoms, and may even be asymptomatic for immunocompetent individuals.

Aseptic meningitis, a severe human disease that causes inflammation covering the brain and spinal cord, can arise from the LCMV infection.

TABLE 16.5 Classification of Arenaviruses

Genus	Species	Disease	Reservoir	Outbreak
Arenavirus	Lymphotropic choriomeningitis virus (LCMV)	Encephalitis	Rodents	–
	Lassa virus (LASV)	Lassa fever/ Hemorrhagic fever	Rodents	Nigeria in 1969, annually ~500,000 cases, 5000 deaths
	Tacaribe virus (TCRV)	Hemorrhagic fever	Rodents	–

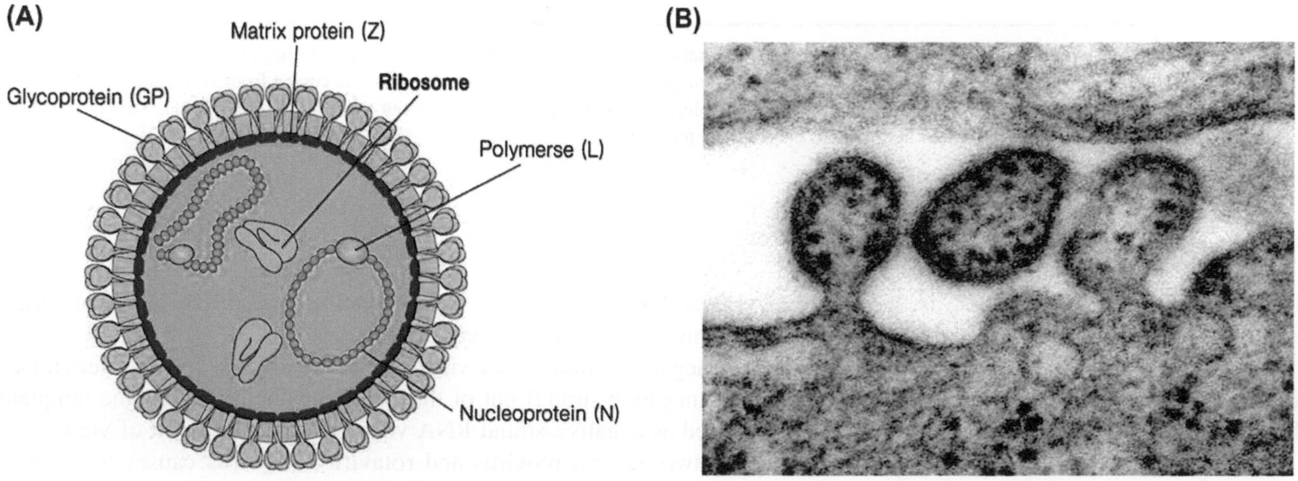

FIGURE 16.9 Virion structure of arenaviruses. (A) A diagram illustrating Lassa virus virion. Viewed in cross section, it shows grainy particles that are ribosomes acquired from their host cells. It is from this characteristic that they acquired the name *Arena* which comes from the Latin root meaning *sand*. These are round, pleomorphic, and enveloped with a diameter of 60 to 300 nm. Two RNA segments are contained. Ribosomes are denoted. (B) This highly magnified TEM depicts some of the ultrastructural details of a number of Lassa virus virions adjacent to some cell debris.

Here, Lassa virus will be mainly described. Note that LCMV has been extensively studied by "viral immunologists" as a model for chronic viral infection.

16.4.1 Lassa Virus

Lassa virus is the most significant human virus in the arenaviruses family.

Epidemiology: *Lassa fever*, primarily caused by Lassa virus infection, is a significant cause of morbidity and mortality: annually 300,000–500,000 infection cases, resulting in 5000 deaths. While Lassa fever is mild or has no observable symptoms in about 80% of people infected with the virus, the remaining 20% suffer from a severe disease. The symptoms include a flu-like illness characterized by fever, general weakness, cough, sore throat, headache, and gastrointestinal manifestations. Approximately 15–20% of patients hospitalized for Lassa fever die from the illness. Overall about 1% of infections with Lassa virus result in death.

Virion Structure: Lassa virus virions are round, pleomorphic, and enveloped with a diameter of 120 nm. The virus contains two negative-stranded RNA segments (Fig. 16.9). The nucleocapsid consists of a nucleic acid enclosed in a protein coat. Peculiarly, ribosomes are encapsidated inside arenavirus particles, although their significance remains uncertain.

Genome Structure: Arenaviruses have two RNA segments: L and S segments, 7.5 and 3.5 kb, respectively (Fig. 16.10). The genome replication strategy is expected to be largely similar to that of bunyaviruses, except that the positive-strand RNAs as well as the negative-strand RNAs code for the viral proteins.

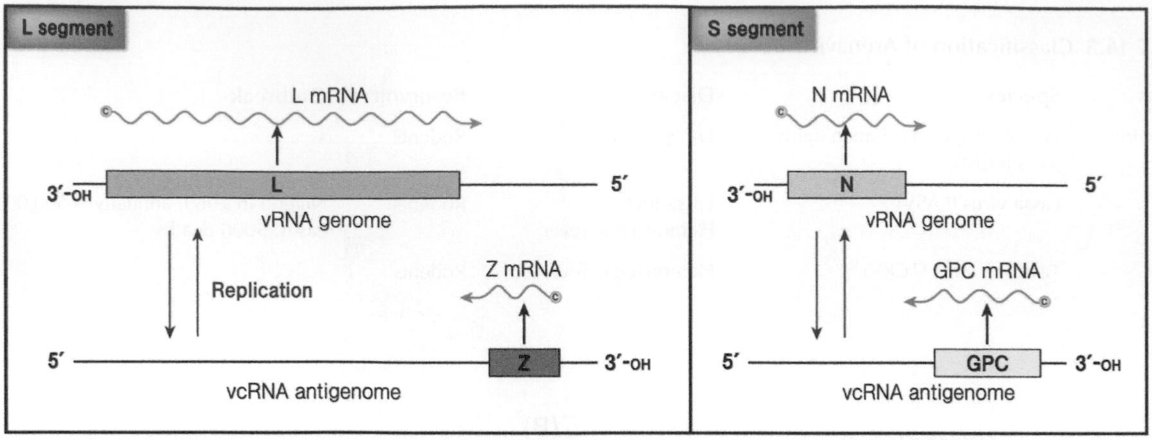

FIGURE 16.10 RNA genomes of arenaviruses. Arenaviruses possess two segments of RNA genome, which are negative-strand RNA genomes of 7.5 and 3.5 kb, respectively. (A) L segment. L mRNA is copied from genomic vRNA (−), while Z mRNA is copied from antigenomic vcRNA (+). (B) S segment. N mRNA is copied from the genomic vRNA (−), while GPC mRNA is copied from the antigenomic vcRNA (+). Strictly speaking, arenavirus genomes are "ambisense" in that both strands encode viral proteins.

16.5 REOVIRUS

Reoviruses (family Reoviridae[6]) is the family of viruses that can affect the gastrointestinal system (such as Rotavirus) and respiratory tract. Reoviruses have genomes consisting of 8−11 segmented, dsRNA.

In fact, reoviruses are dsRNA viruses rather than negative-strand RNA viruses. For the sake of brevity, reoviruses are included in this chapter. Since only one strand (ie, negative-strand) out of the two stands is utilized as the template for the viral RNA replication, reoviruses can be regarded as negative-strand RNA virus in molecular point of view.

Classification: Family Reoviridae is composed of two genera: reovirus and rotavirus. Reovirus causes respiratory infection to children but its associated symptom is mild or subclinical.

Epidemiology: In contrast to reovirus, rotavirus is the main cause of gastroenteritis in the winter. Rotavirus infection leads to diarrhea and vomiting, resulting in dehydration. Rotavirus is estimated to cause about 40% of all hospital admissions due to diarrhea among children under 5 years of age worldwide—leading to some 100 million episodes of acute diarrhea each year that result in 350,000 to 600,000 child deaths. The infection episode can be life-threatening, unless properly treated. Rotavirus vaccine is available. However, no therapeutic antiviral drug is available for the treatment.

Virion Structure: Rotavirus virions are naked, nucleocapsid particles, 70−90 nm in diameter, containing 11 segments of dsRNA genome (Fig. 16.11). One peculiarity is that the nucleocapsid is double-shelled so that inside the outer shell is another layer of shell, the inner shell. Twelve spikes project from the inner layer at each of the 12 vertices.

Genome Structure: Rotavirus possesses 11 RNA segments, which are dsRNA (Fig. 16.12). The replication strategy is similar to that of negative-strand RNA virus, where only one strand (ie, negative-strand) is copied during replication (see Fig. Part III-2). Each RNA segment encodes one protein (open reading frame, ORF), except that two segments (Segments 9 and 11) express an additional related protein by using the second AUG codon.

16.6 PERSPECTIVES

In this chapter, five negative-strand RNA viruses were described with an emphasis on their genome structure. The molecular studies on negative-strand RNA viruses have begun only recently, because it has been difficult to establish an infectious clone. Now, the infectious clones (ie, reverse genetic tool) are established for the majority of negative-strand RNA viruses, including Ebola virus, Sendai virus, hantavirus, and reovirus, we expect to learn a great detail on the virus life cycle via molecular approaches in the near future. Importantly, many newly emerging viruses belong to the negative-strand RNA viruses, including Nipah virus, Ebola virus, Sin Nombre virus, and so on. Importantly, the recent 2013−14 Ebola outbreak in Western Africa has drawn a lot of attention because of its record death toll.

6. **Reovirus** The name is derived from three initials of *respiratory enteric orphan*. The term "orphan" means that a virus is not associated with any known disease.

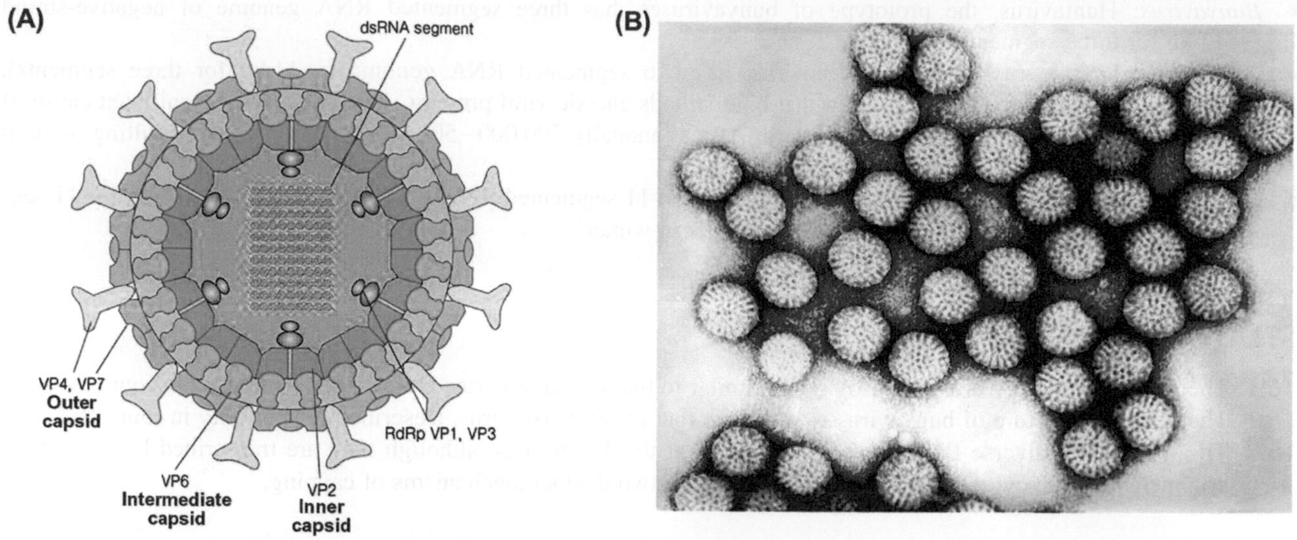

FIGURE 16.11 Virion structure of rotaviruses. (A) A diagram illustrating the double shell structure of a rotavirus particle. Spikes (yellow) project from the vertices of the outer shell. The outer shell (green) is composed of two layers of capsids (ie, outer capsid and intermediate capsid). Inside the inner shell (blue), 11 RNA segments are enclosed. The outer shell has $T=13$ symmetrical structure, while the inner shell has $T=2$ symmetry. (B) TEM of rotavirus particles.

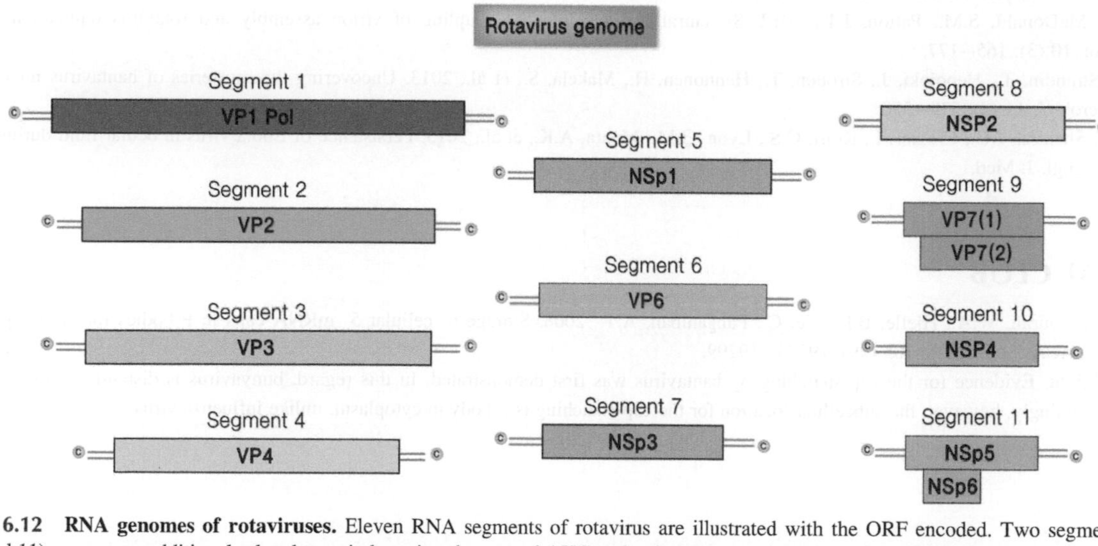

FIGURE 16.12 RNA genomes of rotaviruses. Eleven RNA segments of rotavirus are illustrated with the ORF encoded. Two segments (ie, segments 9 and 11) express an additional related protein by using the second AUG codon in a process dubbed "leaky scanning."

According to the World Health Organization (WHO), the cost for combating the epidemic was set to be about a minimum of $1 billion. The preparedness for future Ebola outbreak including preventive vaccine development (see chapter: New Emerging Viruses) has become an important global agenda.

16.7 SUMMARY

- *Paramyxovirus*: Sendai virus, the prototype of paramyxoviruses, has one large negative-strand RNA genome (15–18 kb). Its genome replication strategy is strikingly similar to that of rhabdovirus. Respiratory syncytial virus (RSV) is a clinically significant human paramyxovirus, which causes bronchiolitis in children and is the main pathogen for viral pneumonia in children.
- *Filovirus*: Ebola virus, the prototype of filoviruses, has one large negative-strand RNA genome (~19 kb). Ebola virus represents the most horrifying pathogen that leads to fatal consequences (~90% fatality).

- *Bunyavirus*: Hantavirus, the prototype of bunyaviruses, has three segmented RNA genome of negative-strand (~12 kb for three segments).
- *Arenavirus*: Lassa virus, a human arenavirus, has two segmented RNA genome (~11 kb for three segments). Arenavirus genomes are "ambisense" in that both strands encode viral proteins. Lassa fever is a significant cause of morbidity and mortality in an endemic area in Africa: annually 300,000–500,000 infection cases, resulting in 5000 deaths.
- *Reovirus*: Rotavirus, the prototype of reoviruses, has an 11 segmented RNA genome of dsRNA (~18 kb for 11 segments). Rotavirus is the main cause of gastroenteritis in winter.

STUDY QUESTIONS

16.1 The genome structure of paramyxovirus is similar to that of rhabdovirus. Describe the similarity in four respects.
16.2 The genome structure of bunyavirus is similar to that of influenza virus. Describe the similarity in four respects.
16.3 The mRNAs of diverse RNA viruses are capped at the 5′ terminus, although they are transcribed by viral RNA polymerase in the cytoplasm. Describe and compare two distinct mechanisms of capping.

SUGGESTED READING

Elliott, R.M., 2014. Orthobunyaviruses: recent genetic and structural insights. Nat. Rev. Microbiol. 12 (10), 673–685.

Marzi, A., Halfmann, P., Hill-Batorski, L., Feldmann, F., Shupert, W.L., Neumann, G., et al., 2015. Vaccines. An Ebola whole-virus vaccine is protective in nonhuman primates. Science. 348 (6233), 439–442.

Trask, S.D., McDonald, S.M., Patton, J.T., 2012. Structural insights into the coupling of virion assembly and rotavirus replication. Nat. Rev. Microbiol. 10 (3), 165–177.

Vaheri, A., Strandin, T., Hepojoki, J., Sironen, T., Henttonen, H., Makela, S., et al., 2013. Uncovering the mysteries of hantavirus infections. Nat. Rev. Microbiol. 11 (8), 539–550.

Varkey, J.B., Shantha, J.G., Crozier, I., Kraft, C.S., Lyon, G.M., Mehta, A.K., et al., 2015. Persistence of Ebola virus in ocular fluid during convalescence. N Engl. J. Med.

JOURNAL CLUB

- Mir, M.A., Duran, W.A., Hjelle, B.L., Ye, C., Panganiban, A.T., 2008. Storage of cellular 5' mRNA caps in P bodies for viral cap-snatching. Proc. Natl Acad. Sci. U.S.A. 105 (49), 19294–19299.

  Highlight: Evidence for the cap-snatching by hantavirus was first demonstrated. In this regard, bunyavirus is distantly related to influenza virus. Intriguingly, however, the subcelluar location for the cap-snatching is P body in cytoplasm, unlike influenza virus.

Part IV

RT Viruses

Animal viruses can be divided into DNA and RNA viruses, depending on the nature of their genomes. DNA and RNA viruses are described in Part II and Part III, respectively. In Part IV, two virus families, which replicate their genome via reverse transcription (RT), will be described (Fig. PIV.1). Although these viruses have either an RNA or DNA genome, they are classified as reverse transcribing viruses or RT viruses. **Retroviruses**[1] have an RNA genome, while hepadnaviruses have a DNA genome.

Retroviruses, which have an RNA genome, are divided into simple retroviruses and complex retroviruses. MLV (murine leukemia virus) is the prototype of the simple retrovirus, while HIV is the prototype of the complex retrovirus. In addition, hepatitis B virus, which is a DNA-containing RT virus, will be described in chapter "Hepadnaviruses."

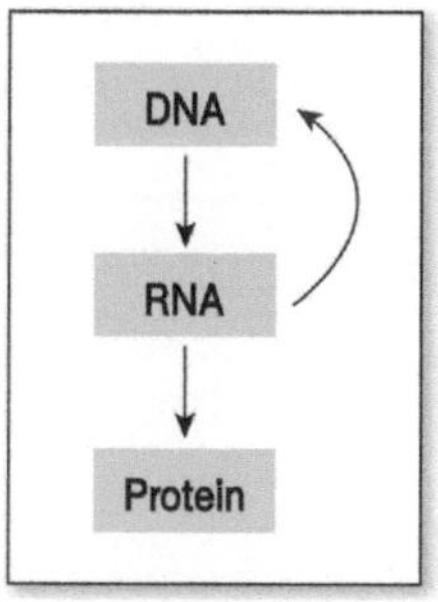

FIGURE PIV.1 Reverse transcription and central dogma. A central dogma states that the information flows from DNA to RNA, then to protein. The central dogma was challenged by Howard Temin and David Baltimore, who codiscovered reverse transcriptase. Howard Temin and David Baltimore were awarded the 1975 Nobel Prize for their discovery of reverse transcriptase (see Box 17.1).

1. **Retrovirus** The term "retro" is derived from Latin word for "backward"-*retro*.

Chapter 17

Retroviruses

Chapter Outline

Retroviruses have played a significant role in the advance, not only of modern virology, but also of molecular biology and molecular oncology (Box 17.1). Moreover, retroviruses have been at the center of all biomedical research since the discovery of HIV. In this chapter, the principles of retroviruses will be described using MLV (murine leukemia virus) as a prototype of simple retroviruses and HIV as a prototype of complex retroviruses (Table 17.1). In particular, HIV will be described with emphasis on six accessory proteins: Tat, Rev, Nef, Vif, Vpu, and Vpr.

17.1 DISCOVERY AND CLASSIFICATION

Discovery: The first retrovirus discovered was an avian virus that causes a tumor in chickens. It was discovered in 1911 and the virus, Rous sarcoma virus (RSV), was named after the discoverer Peyton Rous (see Box 24.1). Since the discovery of RSV, retroviruses were abundantly discovered in many other animals, such as mice, rats, and primates. The first human retrovirus, human T-lymphotropic virus (**HTLV-1**), was discovered in 1977 by Robert Gallo. Only a few years after HTLV-1, HIV was discovered as an etiologic agent of acquired immune deficiency syndrome (**AIDS**).

Classification: Retroviruses can be largely classified in two groups: **simple retroviruses** and **complex retroviruses** (Table 17.2). The former encodes three polyproteins, termed Gag, Pol, and Env, while the latter encodes six accessory proteins in addition to the three polyproteins.

17.2 SIMPLE RETROVIRUS: MLV

Murine leukemia virus (MLV) is the prototype of simple retroviruses. Thus, MLV will be mainly described, unless otherwise stated.

Virion Structure: Retrovirus virion is an enveloped particle with a diameter of 100 nm (Fig. 17.1). Two envelope glycoproteins, termed **SU**[1] and **TM**, are present in the viral envelope. Inner leaflet of the viral envelope is encircled by matrix proteins, termed **MA**. Capsid proteins, termed **CA**, form the icosahedral capsids. Two copies, instead of one, of RNA genomes are found inside the capsid. The RNA genomes are coated by nucleocapsid proteins, termed **NC**. Further, tRNA molecules are attached to the 5′ end of the RNA genome. Three other viral proteins are packaged inside capsids, including **RT** (reverse transcriptase), **PR** (protease), and **IN** (integrase).

Genome Structure: As stated above, two copies of RNA genomes are packaged inside the capsid. Having two identical copies of the genome (ie, diploid genome) is unprecedented; however, its biological significance remains unclear. Intriguingly, two RNA genomes are attached to each other near the 5′ end.

1. **SU (surface protein)** Retroviral proteins are often referred by two-lettered abbreviations.

Molecular Virology of Human Pathogenic Viruses. DOI: http://dx.doi.org/10.1016/B978-0-12-800838-6.00017-5

BOX 17.1 Provirus Hypothesis

During his studies on RSV (Rous sarcoma virus), an RNA tumor virus, Howard Temin proposed the so-called DNA provirus hypothesis in 1964, which stated that the viral RNA is converted to a genetically stable DNA in infected cells. While he studied both the replication of RSV and RSV-induced cell transformation, he noted that the replication of RSV was fundamentally different from that of other RNA viruses. He found that cells transformed by RSV stably maintain transformed phenotype, even in the absence of virus replication. On the basis of these findings, he hypothesized that a more stable form of the viral genome (ie, DNA) is attributable to the stable transformed phenotype. Evidence that the provirus is DNA was then derived from experiments with metabolic inhibitors. First, actinomycin D, which inhibits DNA-directed RNA synthesis, was found to inhibit virus production by RSV-infected cells. Second, inhibitors of DNA synthesis were found to inhibit early stages of cell infection by RSV. It thus appeared that DNA synthesis was required early in infection and that DNA-directed RNA synthesis was subsequently needed for the production of progeny virus. These findings led to the proposal that the provirus was a DNA copy of the viral RNA genome.

A photo of Howard M. Temin (1934–1994).

It was a radical proposal that contradicts with the widely accepted "central dogma" of molecular biology: that there is a unidirectional flow of genetic information from DNA to RNA to protein. Temin's hypothesis was not accepted by the scientific community. It was the discovery of the viral enzyme capable of carrying out the synthesis of DNA from an RNA template that finally led to widespread acceptance of the DNA provirus hypothesis in 1970. He shared the Nobel Prize with David Baltimore in 1975 for their discovery of reverse transcriptase.

TABLE 17.1 The Defining Features of Retroviruses

Genome	Virion Structure	Genome Replication
ssRNA (7–10 kb)	Enveloped	Reverse transcription
Dimeric RNA	Icosahedral symmetry	
Life Cycle	**Interaction with Host**	**Diseases**
DNA synthesis during entry	Immune evasion	Tumor
Provirus	Viral oncogenesis	AIDS

The retroviral genome RNA itself is a typical RNA polymerase II transcript, having a cap structure at the 5′ end and poly (A) tail at the 3′ end (Fig. 17.2). The ORFs of three polyproteins, termed Gag, Pol, and Env, lie in the middle of the RNA genome. In contrast, *cis*-acting elements essential for the viral reverse transcription are all located in the noncoding region. The sequence elements, termed **R** (repeat), are positioned at both ends of the RNA genome. Next to the

TABLE 17.2 The Classification of Retroviruses

Class	Genus	Prototype	Host
Simple retroviruses	Alpha retroviruses	Avian leucosis virus (ALV)	Chicken
		Rous sarcoma virus (RSV)	Chicken
	Beta retroviruses	Mouse mammary tumor virus (MMTV)	Mouse
	Gamma retroviruses	Murine leukemia virus (MLV)	Mouse
Complex retrovirus	Lentivirus	Human immunodeficiency virus	Human
		Human T-cell leukemia virus (HTLV)	Human
		Simian immunodeficiency virus (SIV)	Monkey

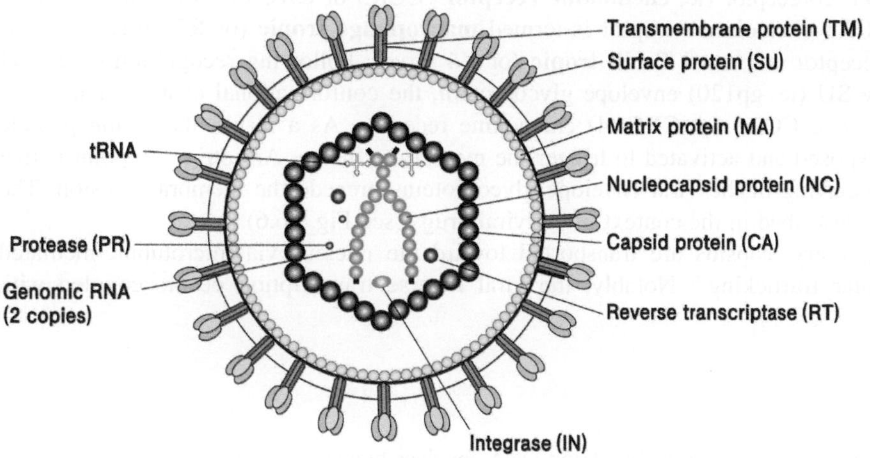

FIGURE 17.1 The virion structure of retrovirus. Two envelope glycoproteins (SU and TM) are found in the viral envelope, in which SU is attached to TM by a disulfide bond. An icosahedral capsid (CA) is found inside of the viral envelope. Two RNA genomes encapsidated by nucleocapsid protein (NC) are found inside the capsid. Two tRNA molecules attached to the viral RNAs are indicated. Three viral enzymes (ie, RT, PR, and IN) are denoted. Two-letter nomenclatures of viral proteins are shown in parenthesis.

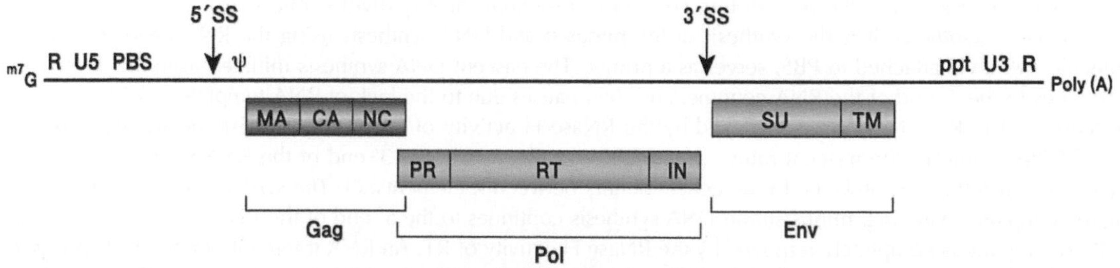

FIGURE 17.2 The RNA genome structure of retrovirus. Retroviral genome is a single-strand RNA of 8–10 kb in length. The ORFs for Gag, Pol, and Env are denoted by constituent viral proteins. Three *cis*-acting elements in the 5′ noncoding region (ie, R, U5, and PBS), and three *cis*-acting elements in the 3′ noncoding region (ie, PPT, U3, and R) are denoted. The 5′ and 3′ splice sites (SS) are denoted, respectively. ψ for packaging signal.

R element, the sequence elements, termed **U5** and **U3**, are present in the 3′ and 5′ side of the R element, respectively. In addition, the primer binding site (**PBS**), which is the binding site for tRNA, is located downstream of the U5 element. In fact, tRNA molecule is utilized as an RNA primer for viral reverse transcription. The isoforms of tRNA used as primers are different among retroviruses: a proline-tRNA is used for MLV, whereas a lysine-tRNA is used for HIV.

Lastly, a **PPT**[2] element located in the 3′ side is utilized as an RNA primer for the second strand DNA synthesis during viral reverse transcription (Box 17.2).

Protein Coding: The viral mRNA is essentially identical to the genome RNA. However, unlike the positive-strand RNA viruses, the genomic RNA in the virion does not serve as mRNA. This feature distinguishes retroviruses from other positive-strand RNA viruses. Retroviral proteins are initially made as part of a polyprotein, which is then cleaved by the viral protease into individual functional proteins. The genomic RNA encodes three polyproteins—**Gag**,[3] **Pol**, and **Env**. Gag polyprotein is processed to MA, CA, and NC; Pol polyprotein is processed to PR, RT, and IN; Env polyprotein is processed to SU and TM (Table 17.3).

17.3 THE LIFE CYCLE OF RETROVIRUS

The life cycle of a retrovirus is carried out in the nucleus as well as in the cytoplasm (Fig. 17.3).

Entry: Simple retroviruses, such as MLV, enter the cell via endocytosis following the attachment to the viral receptor. In contrast, complex retroviruses, such as HIV, enter the cell via endocytosis or direct fusion (Fig. 17.4). Furthermore, the complex retroviruses require a co-receptor in addition to the main entry receptor. For example, in addition to the main entry receptor CD4, coreceptor (ie, **chemokine receptor**[4] CCR5 or CXCR4) is required for HIV entry. Specifically, HIV that uses a CCR5 chemokine receptor is termed **macrophage tropic** (or R5 virus), whereas HIV that uses a CXCR4 chemokine receptor is termed **T-cell tropic** (or X4 virus). Following recognition of a CD4 molecule on a helper T lymphocyte by SU (ie, gp120) envelope glycoprotein, the conformational change induced by CD4 interaction allows gp120 to bind to a CCR5 (or CXCR4) chemokine receptor. As a result, the fusion peptide embedded in TM (ie, gp41) becomes exposed and activated to trigger the membrane fusion. An intriguing point is that the multiple conformational changes occurring in the viral envelope glycoproteins precede the membrane fusion. The molecular details of the interactions are described in the context of antiviral drugs (see Fig. 26.6).

Intracellular Trafficking: Following entry, capsids are transported toward the nucleus via microtubule-mediated transport, a process termed "intracellular trafficking." Notably, the viral reverse transcription occurs coupled with

BOX 17.2 Reverse Transcription

Reverse transcription, which converts single-strand RNA to double-strand DNA, involves two steps of DNA synthesis. The first step represents "RNA-directed DNA synthesis," while the second step is "DNA-directed DNA synthesis." The former is also called the first-strand DNA synthesis or minus-strand DNA synthesis, the latter is called the second-strand DNA synthesis or plus-strand DNA synthesis. Here, the reverse transcription can be conveniently divided into nine reactions for explanation. (1) *The first-strand DNA synthesis*: It is the synthesis of the minus-strand DNA synthesis using the RNA genome as a template. A tRNA molecule, which is attached to PBS, serves as a primer. The nascent DNA synthesis initiated from the 3′ end of the tRNA primer continues to the 5′ end of the RNA genome, and then pauses due to the lack of RNA template. (2) *RNase H digestion of the RNA genome*: The RNA template is removed by the RNase H activity of RT. (3) *The first-strand transfer*: To continue the minus-strand DNA synthesis, the nascent minus-strand DNA translocates to the 3′ end of the RNA template. This minus-strand transfer (template switching) is mediated by sequence identity between R elements. (4) *The synthesis of full-length genome copy*: Following the template switching, minus-strand DNA synthesis continues to the 5′ end of the RNA template. (5) *RNase H digestion*: The RNA template is completely removed by the RNase H activity of RT. An RNA fragment, termed PPT (polypurine tract), is resistant to the digestion and is utilized as an RNA primer for the second strand DNA synthesis. (6) *The second-strand DNA synthesis*: The second, plus-strand DNA synthesis initiated from the PPT continues to the end. (7) *RNase H digestion*: PPT as well as tRNA is removed by the RNase H activity of RT. (8) *The second-strand transfer*: To continue the plus-strand DNA synthesis, the nascent plus-strand DNA translocates to the 3′ end of the DNA template. This plus-strand transfer (template switching) is mediated by sequence identity between PBS elements. (9) *Extension*: The extension of both DNA strands results in duplex DNA having LTR (long terminal repeat) at the 5′ and 3′ ends.

(Continued)

2. **PPT** Polypurine tract, which is composed of multiple purine residues (A or G).
3. **Gag** The retroviral structural polyprotein composed of three domains: MA, CA, and NC. It was named the "group-specific antigen" after its antigenic property.
4. **Chemokine receptor** It refers to cytokine receptors found on the surface of certain cells that interact with a type of cytokine called a chemokine. It belongs to the family of G-protein coupled receptor (GPCR) that has a 7 trans-membrane (7TM) domain.

BOX 17.2 (Continued)

As detailed above, the biochemical reactions involved in viral reverse transcription appears complicated. The salient features of two strand DNA synthesis are as follows. First, two template switching events are involved during viral reverse transcription: once for the minus-strand DNA synthesis, and once for the plus-strand DNA synthesis. At a glance, the template switching (strand transfer) appears mechanistically difficult to carry out. In fact, the template switching needs to occur in order for the DNA synthesis to continue due to the lack of template. Second, the sequence identity is involved in template switching. The first template switching is mediated by sequence identity between R elements, and the second template switching is mediated by sequence identity between PBS elements. Third, compared to the RNA genome, the final product of reverse transcription has an additional U3 element at 5′ UTR and an additional U5 element at 3′ UTR. As a result, both ends of the dsDNA product have LTR composed of U3-R-U5.

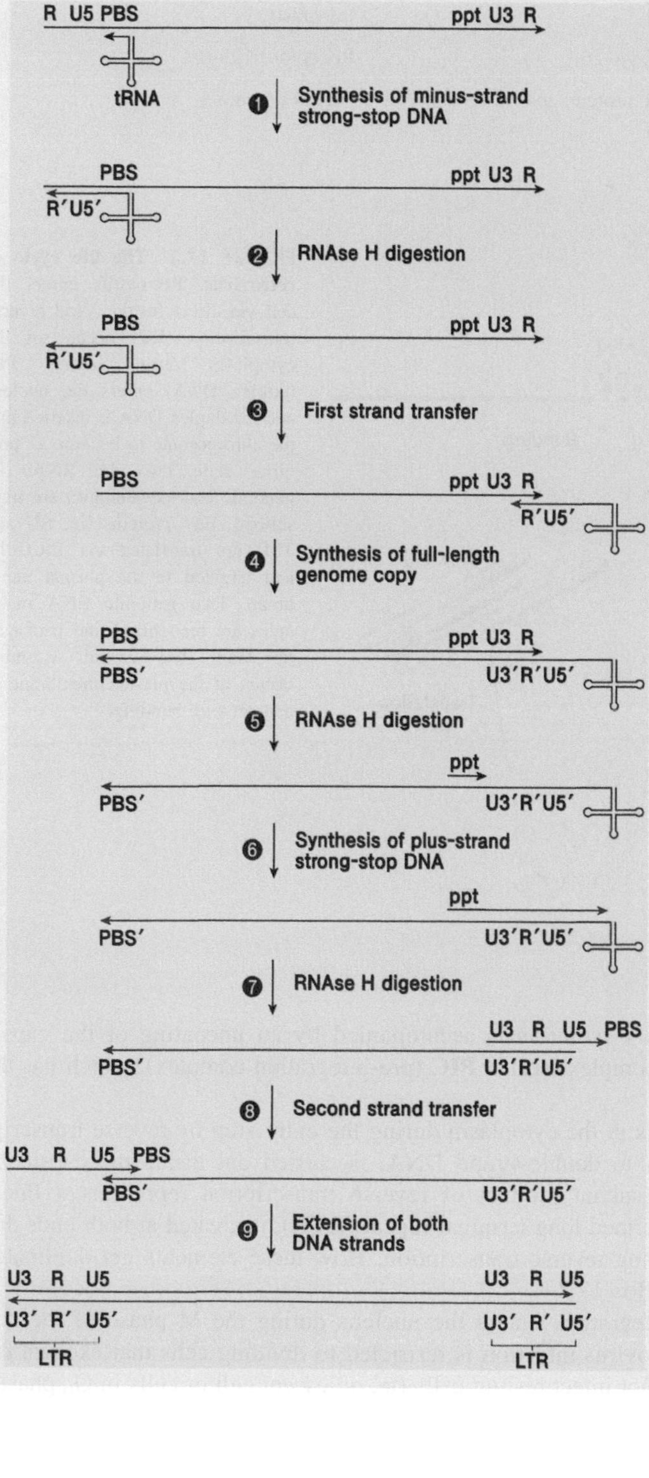

The steps involved in the viral reverse transcription. The newly synthesized DNA is denoted by red lines.

TABLE 17.3 Retroviral Proteins and Their Functions

Polyprotein	Abb	Protein	Functions/Activities
Gag	MA	Matrix	Matrix
	CA	Capsid, p24	Capsid
	NC	Nucleocapsid	RNA binding
Pol	PR	Aspartate protease	Protease
	RT	Reverse transcriptase	Reverse transcriptase, RNase H
	IN	Integrase	Integrase
Env	SU	Surface protein, gp120	Receptor binding
	TM	Transmembrane protein, gp41	Fusion domain

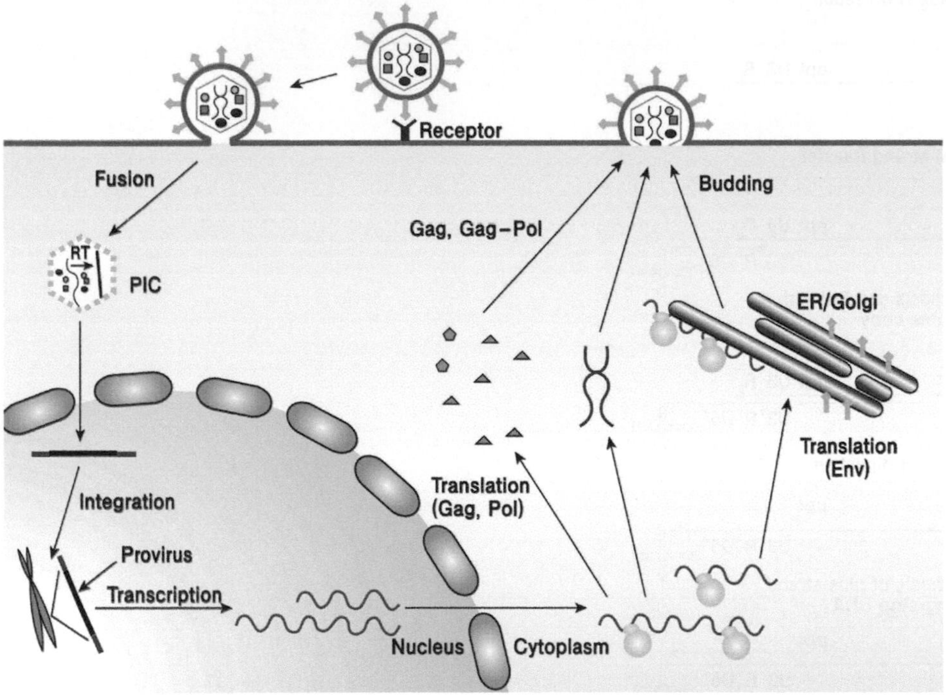

FIGURE 17.3 The life cycle of retrovirus. Retrovirus enters the cell via direct fusion. Viral reverse transcription (RT) occurs in the cytoplasm during entry. PIC (duplex DNA) enters the nucleus and the duplex DNA is inserted into the chromosome to become a "provirus" state. Two viral RNAs (ie, genomic and subgenomic) are transcribed. Env proteins (ie, SU and TM) are translated via ER/Golgi and targeted to the plasma membrane. Two genomic RNA molecules are recognized and packaged by Gag. The capsid assembly occurs at the plasma membrane in concert with budding.

intracellular trafficking. During the transit, the viral DNA synthesis is accompanied by an uncoating of the capsid (Fig. 17.4). As a result, the capsid is now converted to a complex termed **PIC** (pre-integration complex), which has the double-strand DNA.

Reverse Transcription: The viral DNA synthesis occurs in the cytoplasm during the entry step by reverse transcription. The reverse transcription, which converts the RNA to double-strand DNA, is carried out inside the capsid by **RT** that possesses reverse transcriptase activity. The resultant product of reverse transcription represents a linear double-strand DNA (Fig. 17.5). A novel repeat element, termed long terminal repeat (**LTR**), is created at both ends due to the addition of U3 at the 5′ and U5 at the 3′ end during reverse transcription. How these elements get duplicated during viral reverse transcription is explained in detail in Box 17.2.

Nuclear Entry: PIC, a precursor for chromosomal integration, enters the nucleus during the M phase of the cell cycle, when the nuclear membrane is broken. Hence, retrovirus infection is restricted to dividing cells that execute the M phase of the cell cycle. In fact, simple retroviruses cannot infect resting cells (ie, quiescent cell or cells in G_0 phase).

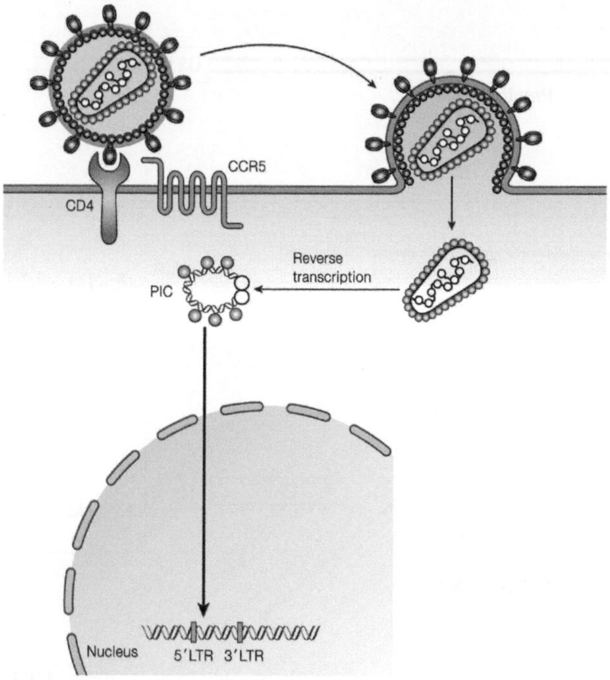

FIGURE 17.4 Model for HIV entry via direct fusion. The gp120 envelope glycoprotein of HIV particles binds to the CD4 molecule expressed on the plasma membrane of $CD4^+$ T lymphocyte. The gp120-CD4 binding induces the conformational changes in gp120, which allows coreceptor (CCR5 or CXCR4) binding of gp120. As a result of the interaction between gp120 and the coreceptor, the "fusion peptide" embedded in gp41(TM) becomes exposed, and is inserted into the plasma membrane, thereby triggering the membrane fusion. Consequently, the viral capsid penetrates into cytoplasm via direct fusion.

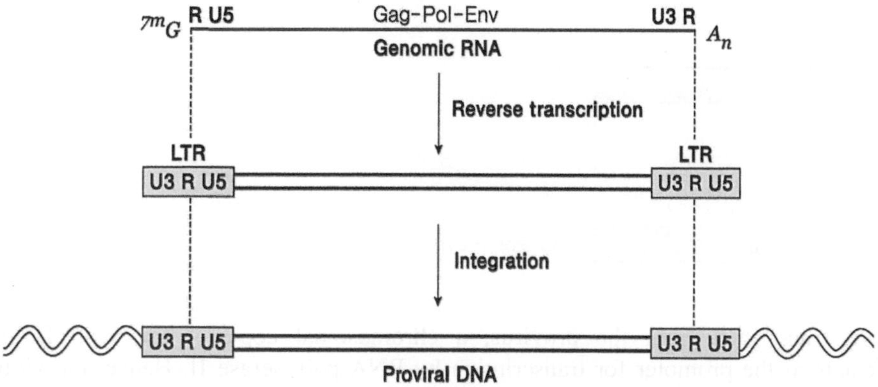

FIGURE 17.5 The structure of DNA genome. (A) The structure of retrovirus RNA genome. In addition to three polyprotein ORFs (ie, Gag–Pol–Env), R (repeat) and U3 element at the 5′ noncoding region, and U3 and R element at the 3′ noncoding region are denoted. (B) The structure of linear duplex DNA, a product of viral reverse transcription. Note that the U3 element is located at the 5′ LTR, while the U5 element is located at the 3′ LTR. *LTR*, long terminal repeat.

In contrast to simple retroviruses, complex retroviruses, such as HIV, can enter the nucleus via a nuclear pore, and thus could infect nondividing cells as well. This feature divides the lentivirus vector from the MLV-based retrovirus vector with respect to versatility as a gene therapy vector (see chapter: Virus Vectors).

Integration: PIC, a precursor for chromosomal integration, enters the nucleus and the DNA becomes integrated into the chromosome. The integration occurs randomly with respect to the chromosomal site, but precisely with respect to the viral sequence (Fig. 17.5). In other words, the nucleotide sequence at the very end of LTR is preserved during integration. The inserted viral DNA is termed "**provirus**,"[5] which implies the precursor of virus.

5. **Provirus** It refers to the retroviral DNA integrated into the chromosomal DNA.

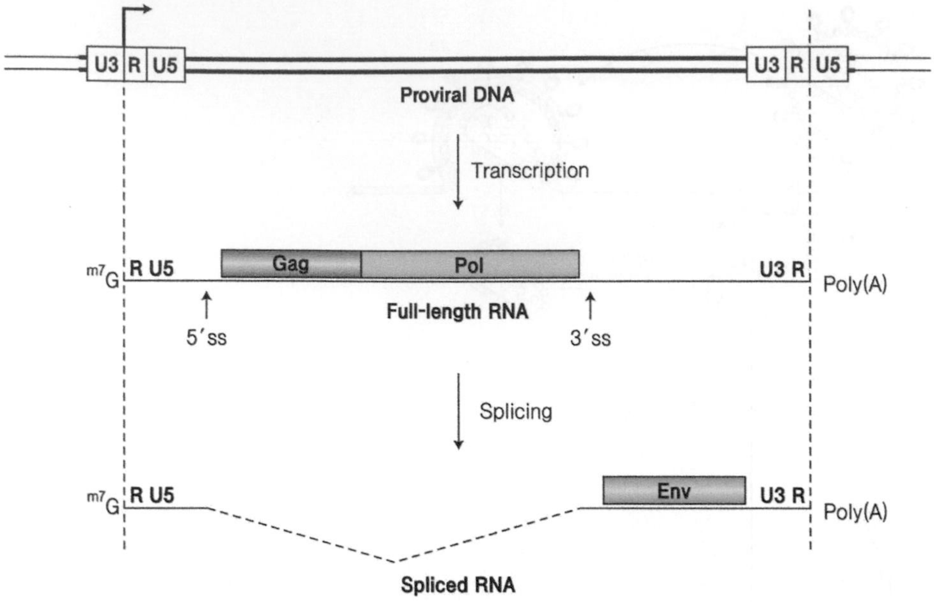

FIGURE 17.6 The genomic RNA versus subgenomic RNA. The provirus DNA (red), which is integrated into chromosome, is shown on the top. The viral RNA transcribed from the provirus has a cap structure at the 5′ end and poly (A) tail at the 3′ end. The full-length genomic RNA encodes the Gag–Pol ORF. The full-length genomic RNA is alternatively spliced into a subgenomic RNA, which encodes the Env ORF. SS is an abbreviation for splice site.

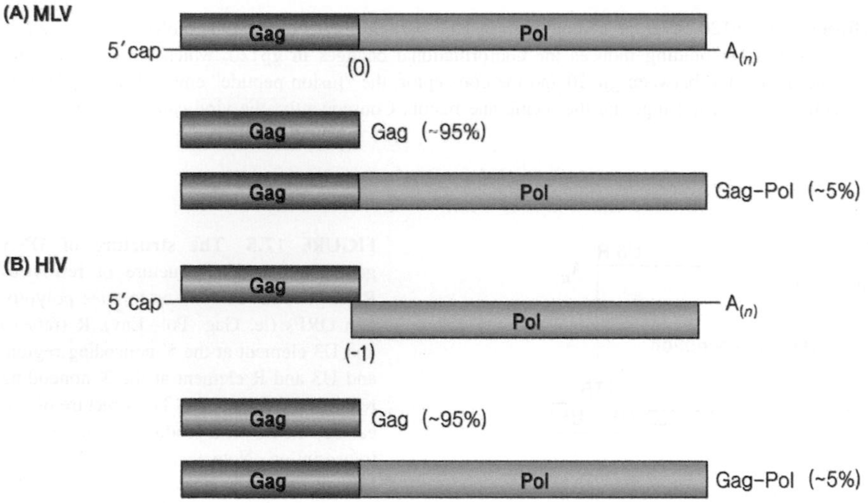

FIGURE 17.7 Translation of Gag–Pol fusion proteins. (A) In case of MLV, Gag and Pol ORFs are in the same reading frame (0). The stoichiometry of two polyproteins synthesized are denoted in parenthesis. (B) In case of HIV, Pol ORF is located in a different reading frame relative to that of Gag ORF (−1).

Transcription: Retroviral RNAs are transcribed from the provirus, a chromosomal copy of the viral DNA (Fig. 17.6). The U3 element at 5′ LTR acts as the promoter for transcription by RNA polymerase II. Hence, the viral RNAs have the 5′ cap and a poly (A) tail at the 3′ end. The viral RNAs are alternatively spliced to a subgenomic RNA.

The full-length genomic RNA serves a dual role. It serves as an mRNA for translation of Gag–Pol polyprotein, and also as an RNA template for viral reverse transcription. On the other hand, the spliced subgenomic RNA serves as mRNA for translation of Env polyprotein. Via alternative splicing, retrovirus expands its protein coding capacity, encoding two ORFs from a single promoter without overriding the "one mRNA–one gene rule."

Translation: All retroviruses basically encode three polyproteins (ie, Gag, Pol, and Env). In fact, Gag–Pol as well as Gag polyprotein is translated from the genomic RNA. The Pol ORF, which is the second ORF, is made as a Gag–Pol fusion protein, which is later processed into Gag and Pol polyprotein.

In eukaryotic cells, the first AUG codon from the 5′ end is recognized as a translation initiation codon of mRNA, a mechanism termed "**the first AUG rule**."[6] Then, how is the 2nd ORF (Pol) translated? In fact, two distinct mechanisms are adopted, depending on the arrangement of two ORFs (Fig. 17.7). In the case of MLV, in which two ORFs are in the

6. **The first AUG rule** The first AUG located at the 5′ end of mRNA is utilized as the translation initiation codon.

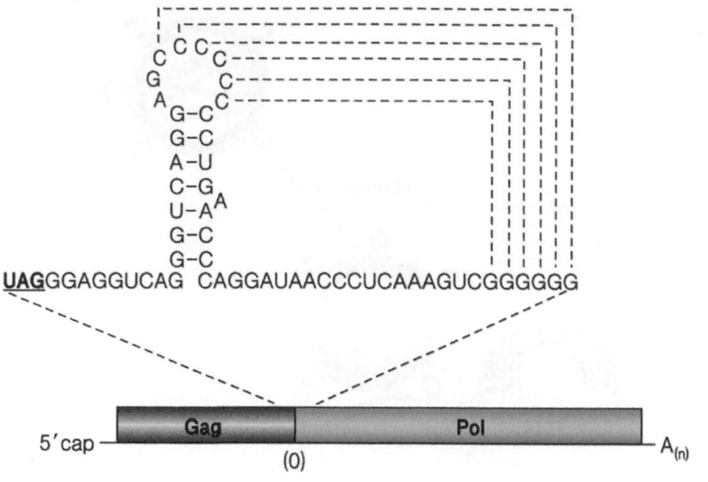

FIGURE 17.8 A pseudoknot structure involved in the suppression of translation termination. The pseudoknot structure in the MLV genome is highlighted with base-pairs, which are denoted by dotted lines. The UAG codon of Gag ORF is underlined.

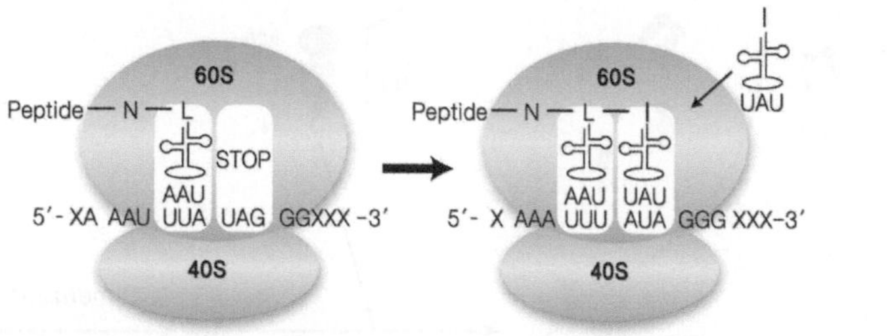

FIGURE 17.9 Ribosomal frame shift for the translation of Gag–Pol fusion protein. The entry sites for peptidyl tRNA (P site) and aminoacyl tRNAs (A site) are drawn as white boxes in the middle of the 60S ribosomal subunit. (A) Before the shift. The UAG termination codon in mRNA is normally positioned in the A site of the ribosome, leading to translation termination. (B) After the shift. The AUA codon, instead of UAG codon, is positioned in the A site due to the ribosomal frameshift, leading to read through translation of the Gag–Pol polyprotein.

same reading frame, the **suppression of translation termination** is the mechanism that translates the Pol ORF. The UAG termination codon lies at the end of the Gag ORF. When the translation of the Gag ORF is terminated at the termination codon, Gag is made. However, when translation termination is occasionally suppressed, Gag–Pol fusion protein is made. In fact, the read-through of the termination codon occurs at the efficiency of 5%. In other words, the UAG codon is misread as CAG by glutamine-tRNA, resulting in the synthesis of Gag–Pol fusion protein. A **pseudoknot structure**,[7] a second RNA structure, lying downstream of the termination codon is essential for the read-through (Fig. 17.8).

On the other hand, in case of HIV, where two ORFs are arranged in different reading-frame, a mechanism called "ribosomal frame-shift" is adopted to translate the 2nd ORF. Gag polyprotein is translated when the termination codon of the Gag ORF is normally recognized. At some efficiency (ie, ~5%), the UAG termination codon of the Gag ORF is not recognized due to ribosomal frame-shift (Fig. 17.9). As a result, the Gag–Pol fusion protein is translated, instead of Gag. Altogether, notably, without violating "the first AUG rule," retroviruses have acquired novel mechanisms to translate two proteins from a single mRNA during evolution.

What is the advantage of translating two proteins from one mRNA? First, the two viral proteins are synthesized in a fixed ratio: high Gag (structural protein) to Pol (enzyme) ratio (ie, 20:1) in this case. More importantly, Pol protein can be packaged via protein–protein interaction between Gag and Gag–Pol during the capsid assembly (see below).

Packaging and Assembly: Retroviral assembly begins in the cytoplasm, where the Gag binds to the genomic RNA and initiates assembly. Gag molecules (or Gag/RNA complex) are recruited to the plasma membrane, leading to the capsid assembly (Fig. 17.10). In other words, retroviral capsid assembly is coupled with the packaging of viral RNA genome. The RNA genome packaging is triggered by recognition of the packaging signal (**Psi or ψ**[8]) by the NC domain

7. **Pseudoknot structure** A secondary structure of RNA, in which unpaired loop region of hairpin structure is base-paired with other region within the same RNA molecule.

8. **Psi (ψ)** It refers to a RNA sequence element that is necessary and sufficient for packaging of retroviral RNA genome.

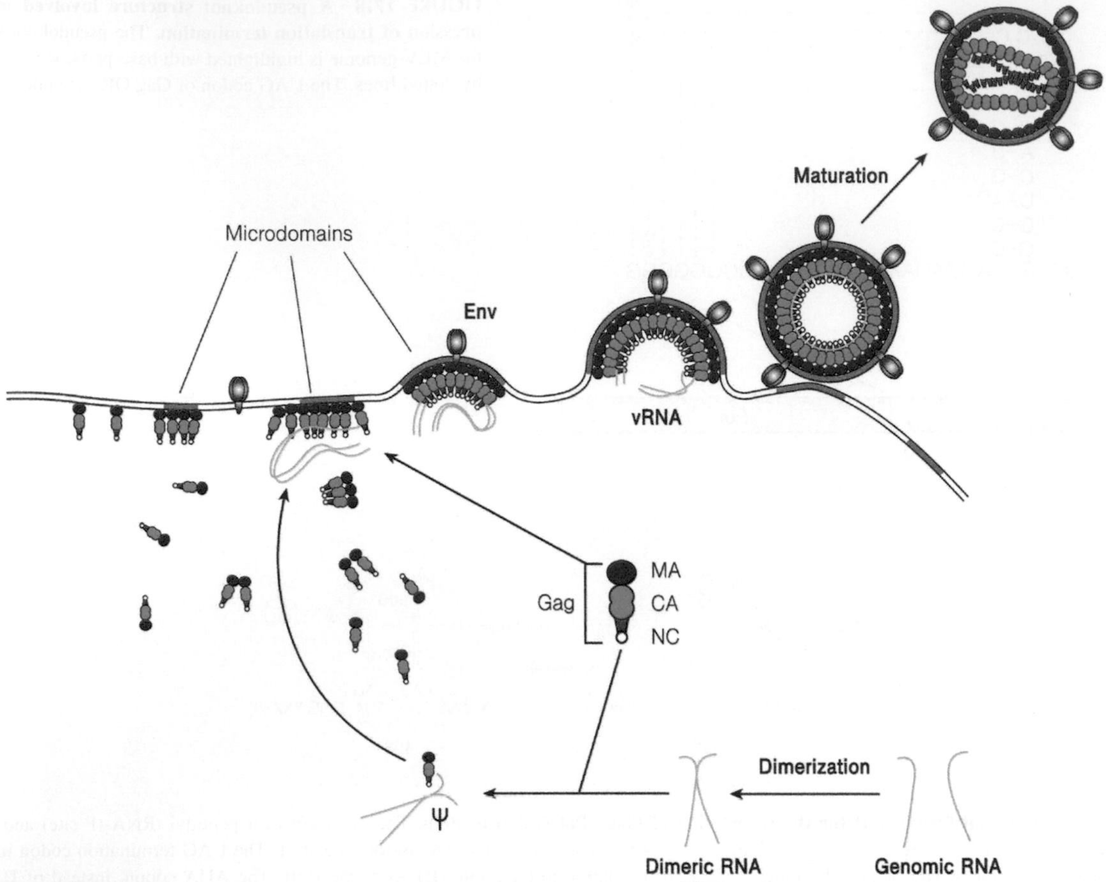

FIGURE 17.10 The capsid assembly and release. The dimerization of the RNA genome, which is mediated via a DIS (dimerization initiation site) element, precedes RNA packaging. The RNA genome packaging is initiated by the recognition of the packaging signal (ψ) in a dimeric RNA by Gag polyprotein (NC domain). Gag–Pol as well as Gag polyprotein is targeted to the plasma membrane, where the capsid assembly occurs. Envelopment of the capsids occurs coupled with the capsid assembly. Maturation of the virion, which is initiated by the activation of PR protease, takes place after the virion release. Morphological changes such as capsid condensation accompany it.

of Gag. Since the packaging signal, which is located just downstream of the 5′ splice site in the MLV genome, is spliced out, the subgenomic RNA is excluded from packaging (see Fig. 17.2). Importantly, the dimeric RNA genome, rather than the monomer, is recognized and packaged by Gag. In addition to Gag, Gag–Pol polyproteins are also copackaged into the capsids by the interaction between Gag and Gag–Pol polyprotein.

Release and Maturation: Env polyprotein is glycosylated and processed into SU and TM proteins, which are then located to the plasma membrane. Gag proteins, which are modified with a **myristate**,[9] are recruited to the plasma membrane. The envelopment of the capsid is facilitated by an interaction between TM and Gag at the plasma membrane. In other words, the capsid assembly is linked with the envelopment (see Fig. 3.9). Assembly of Gag polyproteins (ie, immature core) at the plasma membrane induces membrane curvature, ultimately resulting in a virus particle that buds from the cell. In this process, the **L domain** encoded in Gag (ie, MA domain) plays a key role in recruiting **ESCRT** machinery involving in multivesicular bodies (MVB) pathway (see Box 3.3). The nucleocapsids of released virus particles are composed of Gag (or Gag–Pol). Following the release to extracellular space, the viral protease encoded by the PR domain of Gag–Pol protein becomes activated and cleaves Gag polyprotein to liberate the constituent proteins (ie, MA, CA, and NC). Gag–Pol is also cleaved to the individual proteins (MA, CA, NC, PR, RT, and IN). The polyprotein processing is characteristically accompanied by morphological changes such as capsid condensation. This final step of the virion assembly is termed "**maturation**," as it occurs outside the cells following release. Importantly, the maturation is essential for the infectivity of the progeny virions.

9. **Myristate** It is a kind of saturated fatty acid with 14 carbons.

BOX 17.3 Discovery of HIV

AIDS was first reported in 1981 as acquired immune deficiency disease among homosexual males. The mysterious disease was named AIDS, reflecting the pathological characteristic of patients, who manifested an abrupt drop in $CD4^+$ T-lymphocyte count in peripheral blood. Epidemiologic studies indicated the transmission of diseases among high-risk groups, such as homosexual males and blood transfusion recipients, implicating an infectious pathogen (ie, virus or microorganism). Only a few years after the first report of the disease, the culprit was identified as a human retrovirus, which we now called HIV. To uncover the etiologic pathogen, a group led by Luc Montagnier and Francoise Barre-Sinoussi of the Pasteur Institute in Paris examined the lymph node of a patient with lymphadenopathy, at an early phase of disease, who still had some $CD4^+$ T lymphocytes harboring the pathogen. By cultivating T cells in vitro, they hoped that the pathogen could be propagated enough for further characterization. They succeeded in cultivating the T cells by supplementing naïve T cells from healthy donors. To their surprise, reverse transcriptase activity was detectable in their preparation, a compelling evidence for the presence of retroviruses. Consistently, retrovirus-like particles were revealed under an electron microscope. The discovery of a novel human retrovirus in lymphadenopathy patients provided a keystone to make the causal link between HIV and AIDS. The discovery of HIV, only a few years after the first clinical report, led to the development of diagnostic tools for screening bloods for transfusion, thereby preventing transmission. Importantly, current antiviral drug therapy is capable of preventing disease progression to AIDS. In retrospect, the impact of the discovery of HIV is truly enormous. As would be expected, the two French virologists were awarded the Nobel Prize in 2008 for their discovery of HIV.

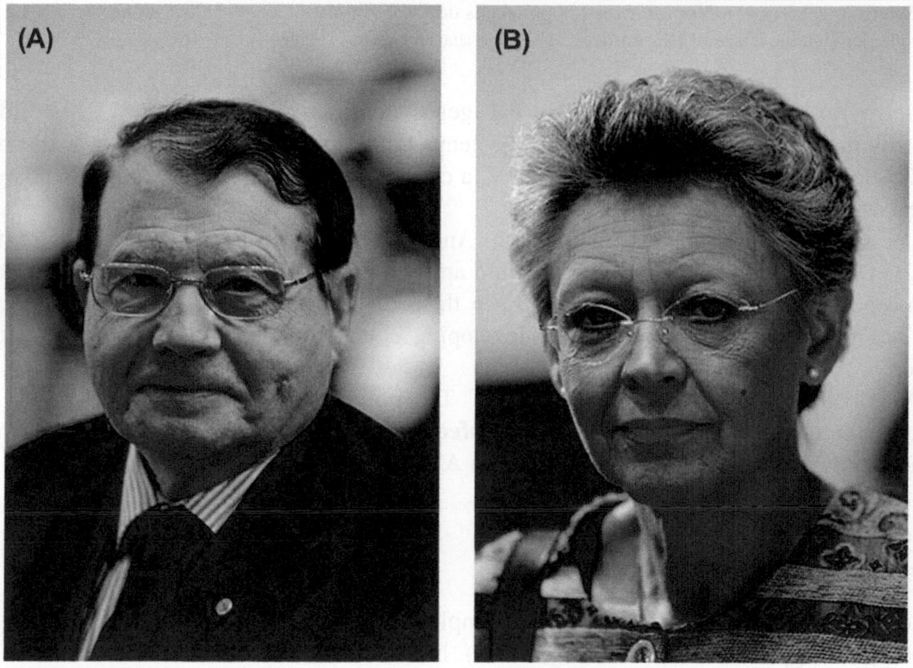

(A) Luc Montagnier (1932–) and (B) Francoise Barre-Sinoussi (1947–).

17.4 LENTIVIRUS: HIV

As stated above, retroviruses are divided into two genera: simple retroviruses and complex retroviruses. The complex retroviruses encode six accessory proteins, in addition to Gag, Pol, and Env polyproteins. The accessory proteins are mainly involved in the regulation of viral propagation and pathogenesis. Here, the functions of six accessory proteins of HIV, the prototype of complex retroviruses, are described. In addition, clinical aspects of HIV infection and AIDS will be described in chapter "HIV and AIDS."

17.4.1 Discovery and Classification

Discovery: Immune deficiency disease was first reported in 1981 among male homosexuals living in Los Angeles. The epidemiological characteristic and the disease drew media attention. The disease was named AIDS, reflecting the immune deficiency among adults. Only 2 years later from the first clinical report, the culprit was identified as a retrovirus by a group led by two French virologists, Luc Montagnier and Francoise Barre-Sinoussi (Box 17.3).

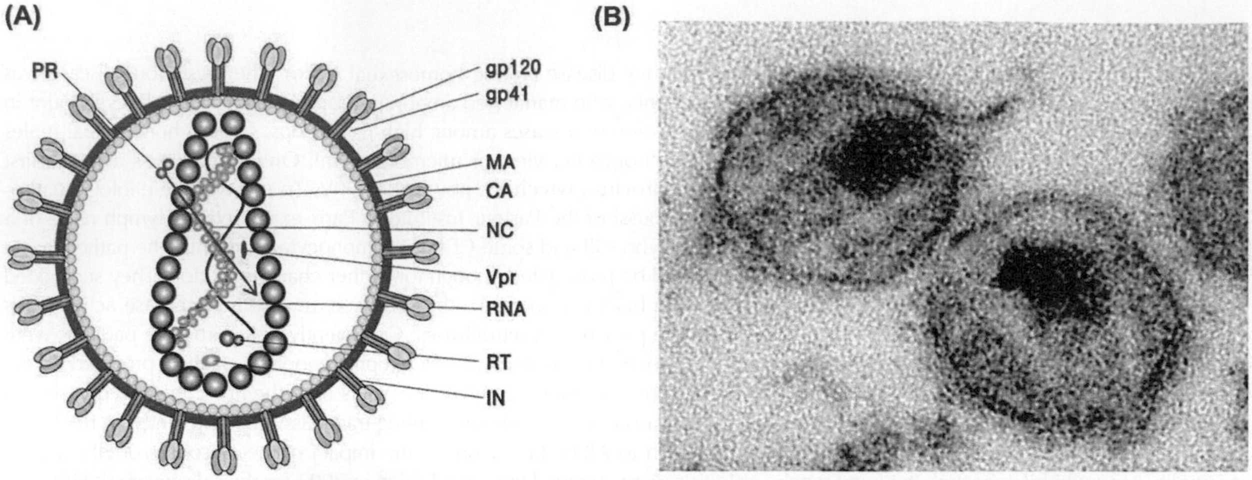

FIGURE 17.11 **The virion structure of HIV.** (A) Schematic diagram of HIV virion. The characteristic cone-shaped capsid contains two RNA genomes. Each viral protein is denoted by a two letter abbreviation. Vpr, an accessory protein, is denoted. For clarity, only one out of the two RNA genomes coated with NC is drawn. (B) Electron micrographic image of HIV particles. The cone-shaped capsid is evident.

Classification: The complex retroviruses are composed of only one genus *Lentivirus*[10] (see Table 17.2). Besides HIV, HTLV-1, the first human retrovirus discovered, also belongs to the genus Lentivirus. In fact, HTLV-1 was discovered a few years earlier than HIV. Soon after the discovery of HIV, SIV, a cousin of HIV, was discovered in monkeys. Since its discovery, SIV has served as an animal model for HIV.

Epidemiology: HIV epidemic is a significant threat to public health. An estimated 37 million people are currently infected worldwide (see Fig. 22.2). During the 1980s, the HIV epidemic appeared more serious in Western countries, such as the United States and Europe, than the rest of the world. Since then, the epidemiologic landscape has been changed and the HIV epidemic has become more serious in the developing countries such as sub-Saharan Africa, Southeast Asia, and South America.

Pathology: A hallmark of HIV infection is the reduced level of $CD4^+$ T lymphocyte in peripheral blood, resulting in immunodeficiency. As a result of immune deficiency, **opportunistic infection**[11] such as pneumonia and viral herpes occurs. In particular, KS herpesvirus (HHV-8) causes Kaposi's sarcoma in AIDS patients.

17.4.2 Virion and Genome Structure

Virion structure: The virion structure of HIV is parallels that of the simple retroviruses. The viral envelope has the spike, which consists of trimer of gp120–gp41 heterodimers (Fig. 17.11). Nonetheless, a few differences are notable. First, unlike the simple retrovirus, the HIV capsid is cone-shaped. It is an elongated deviation of a spherical icosahedral capsid. The cone-shaped capsid of HIV is unique among animal viruses. Second, some accessory proteins, such as Vpr, are enclosed inside of the nucleocapsid.

Genome structure: The genomic RNA of HIV is 9 kb in size, which is a bit longer than that of the simple retroviruses. In addition to Gag, Pol, and Env, HIV encodes six accessory proteins such as Tat, Rev, Nef, Vif, Vpr, and Vpu (Fig. 17.12). The genomic RNAs are alternatively spliced into two classes of subgenomic RNAs: 4 kb singly spliced RNA and 2 kb doubly spliced RNA. The singly spliced RNA encodes Env, Vpr, Vif, and Vpu, while the doubly spliced RNA encodes Tat, Rev, and Nef. All six accessory proteins are encoded by the subgenomic RNAs.

Genotypes: HIV is grouped into two genotypes: HIV-1 and HIV-2 (see Fig. 22.5). In fact, HIV-1 is largely responsible for the current global HIV epidemic, whereas HIV-2 causes AIDS in regions of West Africa. Overall, the genome structure of HIV-2 is almost identical with that of HIV-1, except that Vpu is not found in HIV-2; instead, Vpx is found in HIV-2 (see Fig. 22.6). Here, we are focusing on HIV-1.

10. **Lentivirus** The term "Lenti" is derived from Latin word for "slow."

11. **Opportunistic infection** An opportunistic infection is an infection caused by pathogens, particularly "opportunistic pathogens" that usually do not cause disease in a healthy host.

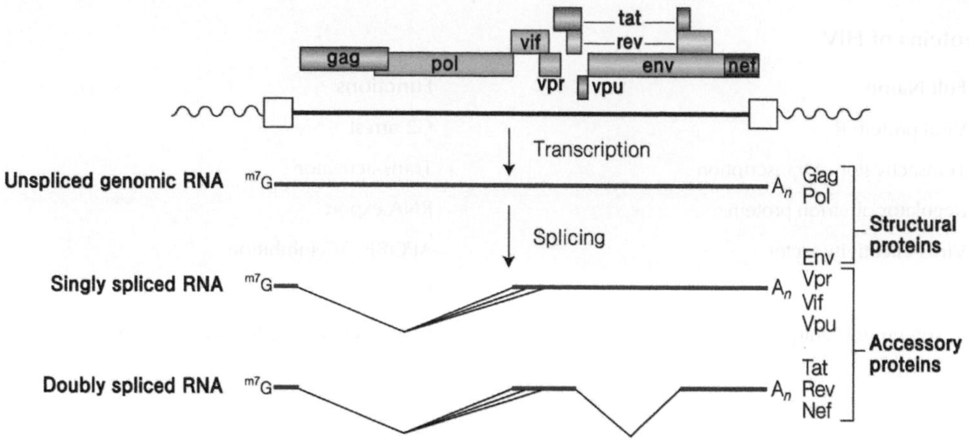

FIGURE 17.12 The RNA structure of HIV. The ORFs of six accessory proteins are denoted on the provirus genome. Three HIV RNAs are shown below proviral DNA: full-length genomic RNA, singly spliced RNA, and doubly spliced RNA. The multiple 3′ spliced sites of the first intron, as indicated by three lines, are employed to encode three accessory proteins via distinct reading frame. Viral proteins encoded by each HIV RNAs are shown to the right of the RNAs.

TABLE 17.4 The Simple Retrovirus Versus the Complex Retrovirus

Features	Simple Retrovirus	Complex Retrovirus
Prototype	MLV	HIV
Genus	Retrovirus (alpha-)	Lentivirus
Accessory proteins	None	Tat, Rev, Nef, Vif, Vpu, Vpr
Entry	Endocytosis	Direct fusion
Nuclear entry	Via nuclear membrane breakdown during mitosis	Via nuclear pore
Susceptible cells	Dividing cells	Nondividing cells
CPE	No	Yes (cell lysis)

17.4.3 The Life Cycle of HIV

The life cycle of HIV largely parallels that of the simple retroviruses. Nonetheless, a few significant differences were noted, which will be highlighted below (Table 17.4).

Entry: As stated above, unlike the simple retroviruses, which enter the cells via receptor-mediated endocytosis, the HIV enters the cells mainly via direct fusion (see Fig. 17.4). In the latter, the viral capsid has to overcome actin barrier that restricts the penetration of the viral capsids to cytoplasm. Intriguingly, the signaling triggered by chemokine coreceptor (CCR5 or CXCR4) induces actin depolymerization, thereby obviating the actin barrier. Further, the resultant actin dynamics facilitates postentry processes, such as cytoplasmic trafficking. Note that CCR5 is expressed in macrophage, while CXCR4 is expressed in $CD4^+$ T lymphocyte.

Nuclear entry: One notable difference from simple retroviruses is that HIV can infect nondividing cells as well as dividing cells (see Table 17.4). As stated above, the viral capsids of the simplex retroviruses enter the nucleus through nuclear membrane breakdown occurring during the M phase of dividing cells (proliferating cells). By contrast, the HIV capsid has the ability to enter the nucleus via a nuclear pore in nondividing cells (ie, resting cells or cells in G_0 phase). Although the underlying mechanism for nuclear entry of HIV capsids remains unclear, this became a critical point in using the lentivirus vector for gene delivery (see chapter: Virus Vectors), because the majority of human cells are nondividing cells (eg, neuron, hepatocytes, and so on).

CPE: Unlike the simple retroviruses, which do not manifest a cytopathic effect (CPE), HIV infection induces cell death. The HIV infected cells are lysed via **syncytium**[12] formation, which is attributable to the membrane fusion activity of the Env protein (gp120/gp41).

12. **Syncytium** A syncytium (*pl.*, syncytia) refers to a multinucleated cell that can result from multiple cell fusions of uninuclear cells.

TABLE 17.5 The Accessory Proteins of HIV

Abb	Full Name	Functions
Vpr	Viral protein R	G2 arrest
Tat	Transactivator of transcription	Trans-activator
Rev	Regulator of virion protein	RNA export
Vif	Virus infectivity factor	APOBEC3G inhibition
Nef	Negative factor	CD4 downregulation
Vpu	Virion protein unique	CD4 degradation, tetherin inhibition

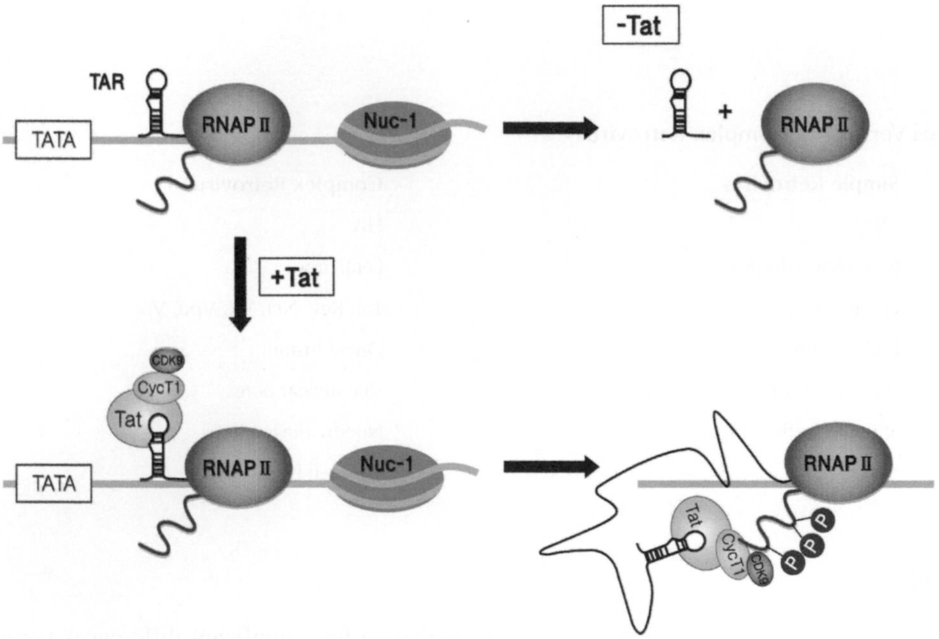

FIGURE 17.13 Transcriptional trans-activation by Tat. In the absence of Tat: Following the transcription of TAR region, viral transcripts are released from the RNA Pol II complex, a process dubbed "premature termination." Thus, no full-length viral transcripts are synthesized. In the presence of Tat: Tat binds to TAR region, forming P-TEFb complex (positive transcription elongation factor b) by recruiting CDK9 and cyclin T1. Such recruited CDK9 phosphorylates the CTD of RNA Pol II, thereby enhancing the processivity of RNA Pol II. *CTD*, the C-terminal domain of RNA Pol II.

17.4.4 Accessory Proteins

In addition to the Gag, Pol, and Env polyproteins, HIV encodes six small ORFs in the 3′ portion of the genome, termed "accessory proteins" (see Fig. 22.6). Six accessory proteins are Tat, Rev, Nef, Vif, Vpu, and Vpr (Table 17.5). All accessory proteins are relatively small, ranging from 10 to 25 kDa and they are mainly involved in regulation and viral pathogenesis. Two of these proteins regulate transcription and mRNA nuclear export (Tat and Rev, respectively) and are indispensable both in vitro (cell culture) and in vivo (animal). On the other hand, the remaining four accessory proteins (ie, Nef, Vif, Vpu, and Vpr) are dispensable in vitro but indispensable in vivo. These six accessory proteins provide a fascinating insight not only into HIV biology but also into cell biology and immunology.

Tat: **Tat** (**T**rans-**a**ctivator of **t**ranscription) is a transcriptional transactivator essential for the establishment of HIV infection (Fig. 17.13). It is an RNA-binding protein, which specifically recognizes an RNA stem-loop structure of HIV, termed TAR (**Tat-R**esponsive element). Notably, unlike most other transcriptional factors, Tat binds to the RNA. In the absence of Tat, RNA polymerase II (Pol II) falls off the template immediately after transcribing TAR RNA. Consequently, viral transcription prematurely terminates. In the presence of Tat, Tat binds to TAR, and then recruits a **P-TEFb complex** (positive transcription elongation factor), which is composed of cyclin T1 and cyclin-dependent kinase 9 (**Cdk9**). The phosphorylation of the C-terminal domain (CTD) of RNA polymerase II by Cdk9 enhances the **processivity**[13] of RNA polymerase II, thereby allowing transcription of the full-length viral RNA. Overall, Tat/TAR acts as an antiterminator of the viral transcription.

13. **Processivity** It refers to the extent of DNA or RNA synthesis per engagement of RNA/DNA polymerase to template.

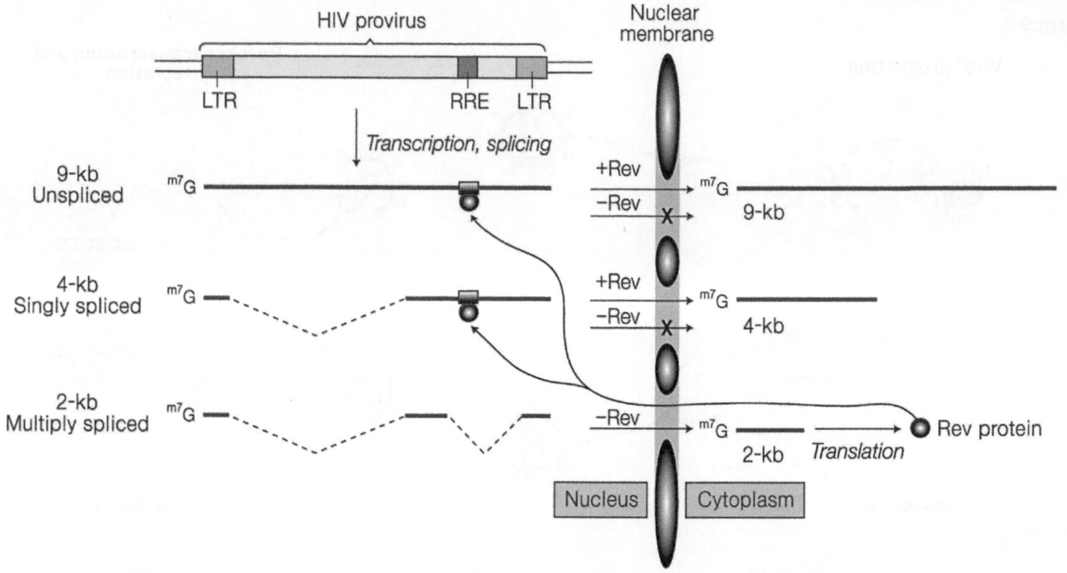

FIGURE 17.14 Rev-mediated RNA export. In the absence of Rev, only the doubly spliced RNAs are exported out to the cytoplasm, which encodes Tat, Rev, and Nef. As Rev accumulates, Rev binds to both the unspliced and the singly spliced RNA, facilitating their nuclear export to the cytoplasm.

Rev: Pre-mRNAs of eukaryotic cells are exported out to the cytoplasm following the excision of introns through RNA splicing. Moreover, the complete removal of introns is a prerequisite for nuclear export of pre-mRNAs. In the case of retroviruses, not only the subgenomic RNAs (both singly and doubly spliced RNAs) but also the genomic RNA containing introns needs to be exported out to the cytoplasm. Hence, retroviruses need a specific mechanism to obviate the complete removal of introns. HIV has adopted an intriguing mechanism for the nuclear export of the intron-containing RNAs via acquisition of a **Rev** (**R**egulator of **e**xpression of **v**irion protein) during evolution (Fig. 17.14). Rev is an RNA-binding protein that binds to an RNA sequence element termed **RRE** (**R**ev **R**esponse **E**lement). It is a shuttling protein that crosses the nuclear membrane via a nuclear pore in a **Crm1**-dependent manner. Rev binds to an RRE element, which lies in the middle of the second intron, and carries the intron-containing RNAs to the cytoplasm via a protein export mechanism. After all, the nuclear export of the unspliced RNAs is achieved by co-opting the protein export mechanism of Rev. In fact, the nuclear export signal (**NES**),[14] which is composed of a leucine-rich motif, was first discovered in Rev.

Vif: Soon after its discovery, **Vif** (**V**irus **i**nfectivity **f**actor) was believed to be important in determining infectivity, as its name implies. In fact, Vif is dispensable for the infection of stable cell lines, such as HEK293 cell and HeLa cells (nonpermissive), but indispensable for the infection of primary T cells (permissive). A hypothesis was that **a host restriction factor**[15] present in nonpermissive cells restricts HIV infection. In the HIV infected cells, Vif counteracts the host restriction factor, thus permitting HIV infection. Surprisingly, the host restriction factor was revealed as **APOBEC3G**,[16] an RNA editing enzyme.

Then, how does the host restriction factor APOBEC3G restrict HIV infection in nonpermissive cells and how does Vif counteract the host restriction factor APOBEC3G (Fig. 17.15). In the presence of Vif (ie, wild-type HIV), Vif targets the APOBEC3G for the degradation. Hence, APOBEC3G is not packaged into the HIV virion particle. Therefore, the virus infection proceeds. In contrast, in the absence of Vif (ie, ΔVif mutant), APOBEC3G is packaged into the HIV particles. When ΔVif HIV infects the target cell, the APOBEC3G in the capsid causes the deamination of C to U during entry, resulting in either hypermutation or degradation of viral DNA. Thus, the virus infection is blocked. Then, how does Vif target APOBEC3G for degradation? Intriguingly, Vif constitutes a cullin-based ubiquitin E3 ligase for the degradation of APOBEC3G (Box 17.4). After all, HIV co-opts a cullin-based ubiquitin E3 ligase to counteract a host restriction factor that limits its infection.

14. **NES (nuclear export signal)** It refers to a peptide motif encoded by nuclear export proteins that is essential for nuclear export.
15. **Host restriction factor** A host factor that restricts the virus infection (Box 17.5).
16. **APOBEC3G (apolipoprotein B editing enzyme)** An RNA editing enzyme with cytidine deaminase activity.

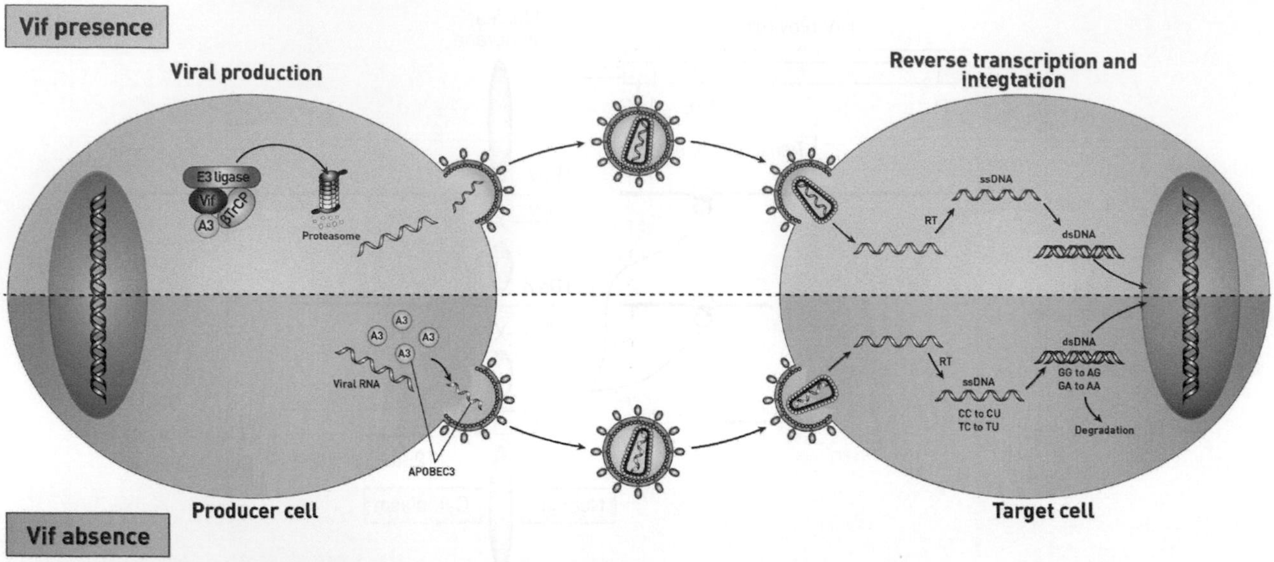

FIGURE 17.15 **Vif counteracts APOBEC3G, a host restriction factor.** In the presence of Vif (ie, wild-type), APOBEC3G is subjected to Vif-mediated ubiquitination and subsequent ubiquitin-proteasome degradation in the produce cell. Then, APOBEC3G (A3) is not packaged into the matured HIV virion.

In the absence of Vif (ie, ΔVif mutant), APOBEC3G is packaged into the capsid in the producer cells. In subsequent infection to target cells, APOBEC3G causes the hyper-mutation or degradation of the viral genome.

Nef: **Nef** (**Ne**gative **f**actor) is a myristoylated protein of about 27 kDa associated with cytoplasmic membranes. Nef is critical for the maintenance of high virus loads and accelerates disease progression in HIV-infected individuals. Nef manifests multiple activities, which culminate in immune evasion and viral propagation. First, Nef downregulates multiple plasma membrane proteins, including CD4, MHC class I, and CXCR4 receptors, by recruiting these membrane proteins via interactions with adaptor protein 2 (AP2) to endocytic machinery or by rerouting them to lysosomes for degradation (Fig. 17.16). Consequently, the receptor downregulation impairs immune function by: (i) preventing virus-infected T cells from CTL lysis via the down-regulation of MHC class I molecule and (ii) impairing helper T cell function via the downregulation of CD4 and CXCR4. In addition, downregulation of CD4, the main receptor for HIV entry, enables HIV to avoid gp120/CD4-mediated retention of virions at the cell surface during exit, promoting the virion release. After all, Nef is a misnomer, because Nef is a positive factor for HIV infection.

Vpu: **Vpu** (**V**iral **p**rotein **u**nique) is an integral membrane protein of 16 kDa. Vpu interacts with two host factors, and in doing so, it facilitates the virion assembly and release. First, it interacts with newly synthesized CD4 in the endoplasmic reticulum and recruits the SCF ubiquitin ligase complex to mediate polyubiquitination and proteasomal degradation (Fig. 17.16). In other words, Vpu serves as an adaptor for recruiting CD4 for degradation. CD4 is the primary receptor for HIV, and thus its degradation may facilitate the viral release by preventing the formation of gp120/CD4 complexes in the endoplasmic reticulum. Second, Vpu is required for efficient viral particle release by another reason. In the absence of Vpu, the virion remains attached to the plasma membrane, as **tetherin**[17] blocks the viral release. Intriguingly, tetherin is thought to form physical cross-links between cell and virion by virtue of their dual membrane anchors. Vpu blocks the virion incorporation of tetherin either by surface downregulation or sequestration of tetherin in the Golgi (Fig. 17.17).

Vpr: **Vpr** (**V**irion **p**rotein **R**) is a virion-associated factor of 14 kDa. Unlike other accessory proteins, Vpr is found packaged inside the capsid (about 100–700 copies per capsid) (see Fig. 17.11). Vpr is not required for HIV propagation in dividing cells, but is required for nondividing cells, such as macrophages. Multiple activities of Vpr have been reported including cell cycle arrest at G2/M, and induction of apoptosis. Intriguingly, G2/M arrest involves the interaction of Vpr with the cullin 4A-**DDB1**[18]-**DCAF1**[19] (see Box 17.4). In other words, Vpr constitutes a cullin-based

17. **Tetherin** It is an interferon-induced membrane protein that inhibits the release of enveloped virus particles from infected cells.

18. **DDB1 (DNA damage binding protein)** A large subunit of DNA damage binding protein, which is a heterodimer composed of large (DDB1) and small subunit (DDB2). This protein functions in nucleotide excision repair.

19. **DCAF (DDB1-Cul4A-associated WD40 domain protein)** It serves as a substrate receptor, linking DDB1 to Vpr. It is also called VprBP (Vpr-binding protein).

BOX 17.4 Ubiquitin E3 Ligase of HIV

Being intracellular parasites, viruses co-opt the diverse cellular functions to make their life in the host environment. It is increasingly clear that numerous host restriction factors, which limit viral infection, exist in host cells. In particular, host restriction factors for lentiviral infection, such as APOBEC3G and tetherin, have received much attention. Lentiviruses encode six accessory genes that have evolved to counteract the effects of host restriction factors: Vif, Vpu, and Vpr. One strategy particularly favored by viruses to achieve these goals is to subvert the host ubiquitin machinery to induce the proteasomal degradation of specific host restriction factors. In particular, cullin-RING ubiquitin ligases (CRLs) are the ones exploited by primate lentiviruses. The CRLs, the major class of ubiquitin E3 ligases, are multisubunit complexes composed of a catalytic core containing the invariant ROC/RBX RING finger protein. Also, it is built with a substrate recognition module comprising various adaptor proteins, nucleated around a cullin scaffold protein, such as Elon B/C, Skp1, and DDB1 (see also Box 8.2). Importantly, the specificity of target molecules for the degradation is determined by substrate receptors, such as βTrCP and DCAF1. In particular, HIV-1 explores three viral accessory proteins to target antiviral proteins (restriction factors) for the degradation: (i) Vif for APOBEC3G, (ii) Vpu for CD4 or tetherin, and (iii) Vpr for UNG2. In other words, via interaction with substrate receptors, the viral accessory proteins confer the target specificity for the CRLs.

Table Box 17.4 Viral Ubiquitin E3 Ligases and Their Functions

Virus Family	Viral Proteins	Ub E3 Ligase	Host Targets	Functions
HIV-1	Vpu	Cul1–Skp1–βTrCp1	CD4, tetherin	Viral release↑
	Vif	Cul5–Elongin B-C	APOBEC3G/F	Host restriction↓
	Vpr	Cul4–DDB1–DCAF1	UNG2	G2 arrest↑
HIV-2	Vpx	Cul4–DDB1–DCAF1	SAMHD1	Reverse transcription↑

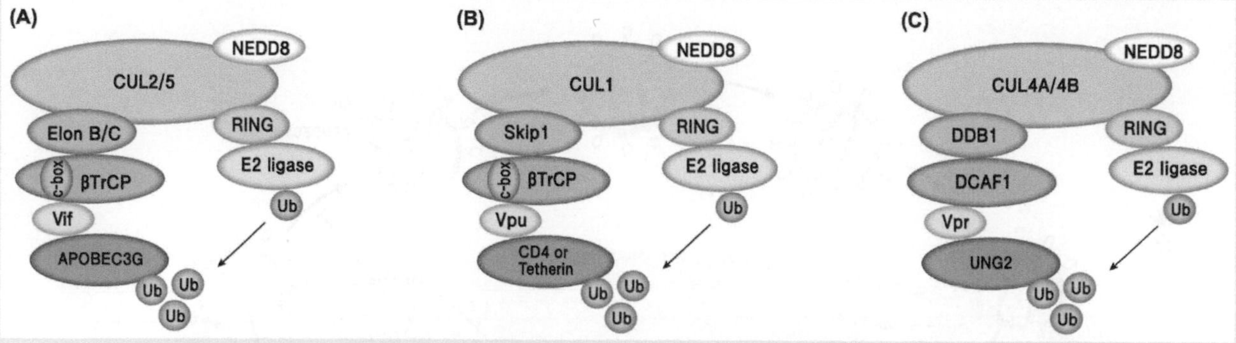

Schematic models illustrating Ubiquitin E3 ligases of HIV. (A) Vif-mediated cullin-RING ubiquitin ligase. (B) Vpu-mediated cullin-RING ubiquitin ligase. (C) Vpr-mediated cullin-RING ubiquitin ligase. DCAF (DDB1-cullin-associated factor). Three accessory proteins of HIV-1 (ie. Vif, Vpu, and Vpr) are in yellow.

ubiquitin E3 ligase, via an interaction with DCAF1. The ubiquitination/degradation of a yet-to-be-known target substrate induces the activation of **ATR**,[20] thereby leading to G2 arrest (see Box 6.2). Another known substrate for Vpr-mediated ubiquitination is **UNG uracil-DNA glycosylase**, which is involved in DNA repair (see Box 17.4).

17.5 HOST RESTRICTION FACTORS

As alluded to above, a few host factors that restrict HIV infection are revealed during the investigation of accessory proteins such as Vif, Vpu, Nef, and Vpr. Unlike Tat and Rev regulatory proteins, these mysterious accessory proteins are required for the virus to replicate in some, but not all, cell types. As the roles of these accessory proteins in virus replication have been unraveled, the theme that has repeatedly emerged is that they serve as a means to counteract host antiviral defense mechanisms. It is believed that HIV has acquired these "accessory" genes that antagonize antiviral host restriction factors, thereby making cells permissive to the viral replication. As described above, Vif counteracts

20. **ATR (ataxia telangiectasia and RAD3-related)** ATM-related gene. ATR is a serine/threonine kinase that is involved in sensing DNA damage and activating DNA damage checkpoint, leading to cell cycle arrest.

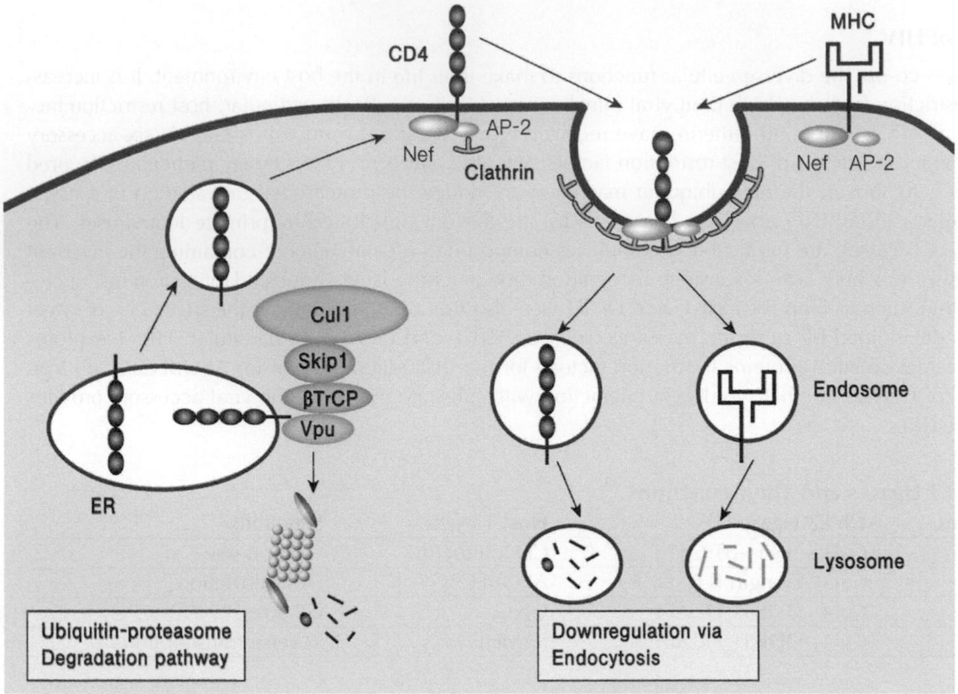

FIGURE 17.16 Nef- and Vpu-mediated downregulation of CD4. (A) Surface downregulation of CD4 by Nef. Nef moves to the plasma membrane via its interaction with AP2 (adaptor protein), then recruits CD4 as well as other receptors such as MHC class I and CXCR4 receptor to the endocytic route for degradation. (B) Vpu-mediated degradation of CD4. A newly synthesized CD4 in ER membrane is recruited by Vpu for proteasomal degradation. Vpu acts as an adaptor for cullin-based ubiquitin E3 ligase.

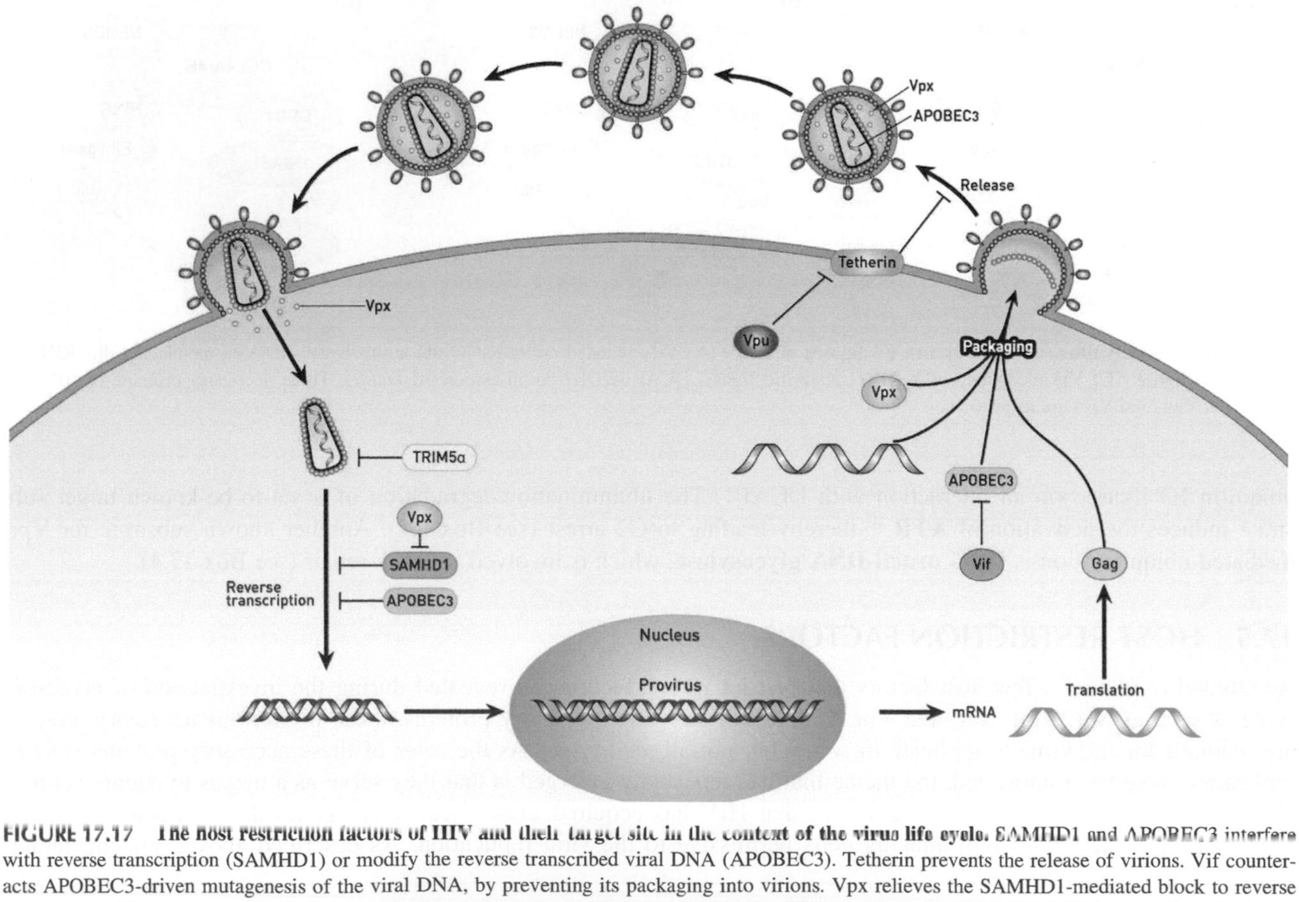

FIGURE 17.17 The host restriction factors of HIV and their target site in the context of the virus life cycle. SAMHD1 and APOBEC3 interfere with reverse transcription (SAMHD1) or modify the reverse transcribed viral DNA (APOBEC3). Tetherin prevents the release of virions. Vif counteracts APOBEC3-driven mutagenesis of the viral DNA, by preventing its packaging into virions. Vpx relieves the SAMHD1-mediated block to reverse transcription. Vpu prevents tetherin from holding on to the virus at the plasma membrane. TRIM5α blocks the lentivirus infection at entry by promoting premature uncoating, contributing to host tropism.

APOBEC3-driven mutagenesis of the viral DNA by preventing its packaging into virions (see Fig. 17.17). Vpu prevents tetherin from holding on to the virus at the plasma membrane. Tetherin, one of the interferon-stimulated genes (ISGs), prevents the release of virions.

Two additional host restriction factors were discovered: TRIM5α and SAMHD1. **TRIM5α**[21] was identified as a host restriction factor that is responsible for determining the host range of the primate lentivirus. As a matter of fact, HIV cannot infect monkeys, whereas primate lentiviruses, such as SIV (simian immunodeficiency virus), cannot infect human. One question is what is the identity of the host factor that determines the host tropism of primate lentiviruses? It turned out that TRIM5α of monkey cells restricts HIV infection of monkey cells, and conversely TRIM5α of human cells restricts SIV infection of human cells. Intriguingly, the blockade by TRIM5α occurs at the entry step of HIV infection by triggering premature disassembly of HIV capsids via interaction with the viral capsids (see Fig. 17.17). On the other hand, **SAMHD1**[22] interferes with reverse transcription (SAMHD1) (see Fig. 17.17). SAMHD1 depletes the pool of NTPs available to a reverse transcriptase for viral cDNA synthesis and thus prevents viral replication. SAMHD1 has also shown nuclease activity and a ribonuclease activity has been recently described to be required for HIV-1 restriction. Importantly, SAMHD1 is responsible for blocking replication of HIV in dendritic cells, macrophages, and monocytes. Vpx relieves the SAMHD1-mediated block to reverse transcription by downregulation of SAMHD1 via the formation of Cul4/DDB1/Vpx ubiquitin E3 ligase (see Fig. 17.17 and Box 17.4). Note that Vpx is encoded only by HIV-2 (see Fig. 22.6).

Overall, two types of host restriction factors are found. The first kind is the host restriction factors, including APOBEC3, tetherin, and SAMHD1, that confer "intrinsic immunity" to the HIV infection and the host restrictions are counteracted by accessory proteins of HIV. Another kind is the host restriction factor, including TRIM5α, that determines the host range of primate lentiviruses. It will be not surprising if host restriction factors that are responsible for the viral tropism are soon discovered in other animal viruses, besides lentiviruses.

17.6 PERSPECTIVES

Retroviruses have been at the center of all biomedical research since the discovery of HIV. Accordingly, the studies on HIV for over three decades have greatly advanced our knowledge not only on HIV biology but also on immunopathology, as reflected by the huge number of publications in HIV and AIDS-related fields. Not surprisingly, numerous findings made in HIV have opened new insights in the field of virology. For instance, the discovery of host restriction factors (ie, APOBEC3G, tetherin, and TRIM5α) that limit HIV infection hints that other animal viruses could be restricted by such cellular factors as well. Although the fundamentals of the HIV life cycle are understood in detail, some aspects of the viral life cycle are incompletely understood. A few unanswered questions are as follows. Unlike simple retrovirus, HIV can infect nondividing cells, because the HIV capsid enters the nucleus via a nuclear pore. However, the mechanism by which the HIV capsid enters the nucleus remains unknown. Second, functions of some accessory proteins are less than clear. For instance, the function of Vpr in the viral life cycle and disease progression remains largely uncertain. Finally, there is still much to be learned about the immunopathology of HIV, particular as to how HIV evades the host immune response. On the other hand, HIV is now being exploited as a gene delivery tool (ie, lentivirus vector) and extensively used in many laboratories. The principle of lentiviral vectors will be described in detail in chapter "Virus Vectors." On the other hand, oncogenic retroviruses, such as Rous sarcoma virus (RSV), were extensively studied as a model for oncogenesis during the 1970s and 1980s. Oncogenic retroviruses and the discovery of the src oncogene will be covered in chapter "Tumor Viruses."

17.7 SUMMARY

- *Classification*: Retroviruses are classified into two groups: simple retrovirus and complex retrovirus (lentivirus). The former encodes three polyproteins termed Gag, Pol, and Env, while the latter encodes six other accessory proteins.
- *Virion structure*: An enveloped virion with a capsid that contains a dimeric RNA genome. Three viral enzymes, such as RT, IN, and PR, are packaged inside the capsid.

---

21. **TRIM (tripartite motif)** Tripartite motif represents three structural motifs composed of RING finger domain, B-box domain, and coiled-coil domain. TRIM family is a large protein family that constitutes over 50 members.

22. **SAMHD1 (SAM domain-and HD domain-containing protein 1)** It is an enzyme that exhibits phosphohydrolase activity, converting nucleotide triphosphates to triphosphate and a nucleoside (ie, nucleotides without a phosphate group).

- *Viral genome*: It is an 8–10 kb RNA having a cap structure at the 5′ end and poly (A) tail at the 3′ end. It encodes three ORFs: Gag, Pol, and Env polyproteins.
- *Genome replication*: The viral RNA genome is converted into DNA via reverse transcription during entry, and the DNA is inserted into chromosome as a provirus.
- *Lentivirus*: HIV, the prototype of lentivirus, encodes six accessory proteins: Tat, Rev, Nef, Vif, Vpu, and Vpr. All accessory proteins are relatively small, ranging from 10 to 25 kDa, and they are involved in regulation and viral pathogenesis.
- *Host restriction factors*: A few host factors that restrict HIV infection have been discovered, including TRIM5α, APOBEC3, tetherin, and SAMHD1. Accessory proteins of HIV counteract these restriction factors to establish HIV infection: APOBEC3 by Vif, tetherin by Vpu, and SAMHD1 by Vpx.

STUDY QUESTIONS

17.1 Describe and compare the RNA genome of the simple retrovirus with the provirus genome. Indicate the difference in the repeat element between the RNA genome and the provirus genome. Explain how this difference in the repeat element is generated.

17.2 Describe what would happen if the following proviral DNA of HIV mutants lacking a corresponding gene are transfected into cells. (1) Δ*Tat*, (2) Δ*Rev*, (3) Δ*Nef*.

17.3 Consider that a certain mutant of HIV (eg, HIV-ΔVif) cannot infect primary T cells (nonpermissive), but can infect stable cell lines such as HEK293 cells (permissive). (1) Propose a hypothesis to explain the difference between permissive and nonpermissive cells. (2) How would you test your hypothesis?

SUGGESTED READING

Campbell, E.M., Hope, T.J., 2015. HIV-1 capsid: the multifaceted key player in HIV-1 infection. Nat. Rev. Microbiol. 13 (8), 471–483.

Carter, C.C., Onafuwa-Nuga, A., McNamara, L.A., Riddell, J.T., Bixby, D., Savona, M.R., et al., 2010. HIV-1 infects multipotent progenitor cells causing cell death and establishing latent cellular reservoirs. Nat. Med. 16 (4), 446–451.

Freed, E.O., 2015. HIV-1 assembly, release and maturation. Nat. Rev. Microbiol. 13 (8), 484–496.

Martin-Serrano, J., Neil, S.J., 2011. Host factors involved in retroviral budding and release. Nat. Rev. Microbiol. 9 (7), 519–531.

Simon, V., Bloch, N., Landau, N.R., 2015. Intrinsic host restrictions to HIV-1 and mechanisms of viral escape. Nat. Immunol. 16 (6), 546–553.

JOURNAL CLUB

- Carter, C.C., et al., 2010. HIV-1 infects multipotent progenitor cells causing cell death and establishing latent cellular reservoirs. Nat. Med. 16 (4), 446–451.

  Highlight: Despite drugs that inhibit viral spread, HIV infection has been difficult to cure because of uncharacterized reservoirs of infected cells that are resistant to highly active antiretroviral therapy (HAART) and the immune response. This paper revealed the presence of distinct populations of active and latently infected hematopoietic progenitor cells (HPCs). Some populations of HPCs are defined as a novel reservoir for persistent HIV infection.

BOOK CLUB

- Cooper, G.M., et al., 1995. The DNA Provirus: Howard Temin's Scientific Legacy. ASM Press.

  Highlight: A proceedings book from the Howard Temin Memorial Symposium, which was held in 1994. Each of his papers is prefaced by introductions that review the scientific context and impact of his work.

Chapter 18

Hepadnaviruses

Chapter Outline

In addition to retroviruses, hepadnaviruses are reverse transcribing viruses or RT viruses (Table 18.1). In contrast to the retroviruses, hepadnavirus possess a DNA genome. Here, we will focus on the genome organization and the molecular aspects of the viral life cycle of hepatitis B virus (HBV) as a prototype of hepadnaviruses, while the clinical aspects of HBV infection are described in chapter "Hepatitis Viruses." Besides HBV, five other 'hepatitis viruses' have been discovered from HAV to HGV, but they are not related to HBV and have been named by alphabetical order of discovery (see Table 23.1).

18.1 DISCOVERY

Discovery and Classification: HBV was discovered by Baruch Blumberg in 1968. Subsequently, animal viruses similar to HBV were discovered in woodchucks (woodchuck hepatitis virus or WHV) and ducks (duck hepatitis B virus or DHBV), making up the **hepadnavirus**[1] family (Table 18.2). In particular, WHV, being a mammalian virus, has a considerable similarity with HBV, while avian DHBV is distantly related (Fig. 18.1). For instance, X ORF is not found in DHBV but found in WHV. Importantly, chronic infection by WHV causes liver cancer in animals. Therefore, woodchucks serve as an important animal model to investigate the HBV-associated liver cancer. Subsequently, hepadnaviruses have been discovered in primates such as chimpanzees and woolly monkeys.

Epidemiology: HBV infection is a major global public health problem, with over 350 million chronically infected patients worldwide. Chronic HBV infection carries a great risk of developing severe liver diseases, including cirrhosis and liver cancer, which result in a million deaths annually. HBV infection is endemic in Northeast Asia, for example, China and Korea, where more than 5% of people are chronically infected and is a major cause of liver disease in this region. HBV is blood-borne and typically transmitted via blood transfusion, while neonatal transmission is common in the endemic region. HBV vaccine is safe and effective for prevention (see Fig. 25.11). A few nucleoside analogs (NUC), such as **lamivudine**, **entecavir**, and **tenofovir**, are available for antiviral therapy. Although these nucleoside analogs are quite effective in suppressing viral genome replication, long-term administration (>5 years) is essential for a clinic benefit (see Box 23.2).

18.2 THE VIRION AND GENOME STRUCTURE

Virion Structure: Three kinds of virus-related particles are present in the serum derived from HBV-infected individuals (Fig. 18.2). The most abundant particles are subviral particles that are entirely composed of HBV envelope proteins (**HBsAg** or hepatitis B virus surface antigen). The subviral particles are either spherical or filamentous in shapes. These subviral particles, aptly called "empty" particles, do not contain viral genetic material and cores or capsids and hence

1. **Hepadnavirus** The term "hepa" is derived from Greek word for "liver". "*dna*" implies that it has DNA genome.

Molecular Virology of Human Pathogenic Viruses. DOI: http://dx.doi.org/10.1016/B978-0-12-800838-6.00018-7

TABLE 18.1 The Defining Features of Hepadnaviruses

Genome	Virion Structure	Genome Replication
Partial duplex DNA (3.2 kb) Protein-linked	Enveloped Icosahedral symmetry	Reverse transcription Protein-priming
Life Cycle	**Interaction with Host**	**Diseases**
cccDNA as episome	HBx-mediated signaling	Chronic hepatitis Liver cancer

TABLE 18.2 Members of the Family Hepadnaviridae

Virus	Host	X ORF	Homology
Hepatitis B virus	Human	Yes	–
Woodchuck hepatitis virus (WHV)	Woodchuck	Yes	70%
Duck hepatitis B virus (DHBV)	Duck	No	30%

FIGURE 18.1 **Animal models of hepadaviruses.** Hepadnaviruses are found not only in humans, but also in primates (chimpanzee, woolly monkey), rodents (woodchuck), and birds (duck). A salient feature of HBV infection is its narrow host range, as it infects only chimpanzee, beside human. Thus, woodchuck and duck hepatitis B viruses and their natural hosts have been utilized as indispensable models for HBV.

are not infectious. Less abundantly detectable particles with a size of approximately 42 nm diameter contain viral capsids and the DNA genome and represent an infectious HBV virion: it is also called **Dane particle** named after the investigator who first reported it. An infectious HBV virion is enveloped with a capsid that has $T=3$ or 4 icosahedral symmetry.

Genome Structure: The size of the viral DNA genome is only 3.2 kb. In fact, it is the smallest genome among animal viruses. Moreover, HBV genome is unique in several aspects. First, the viral genomic DNA has the unusual property of being partly double-stranded, with a single-stranded region of variable length (about 1/3 of the genome) (see Fig. 18.2). Furthermore, unlike other circular DNA viral genomes, neither DNA strand forms a covalently closed circle.

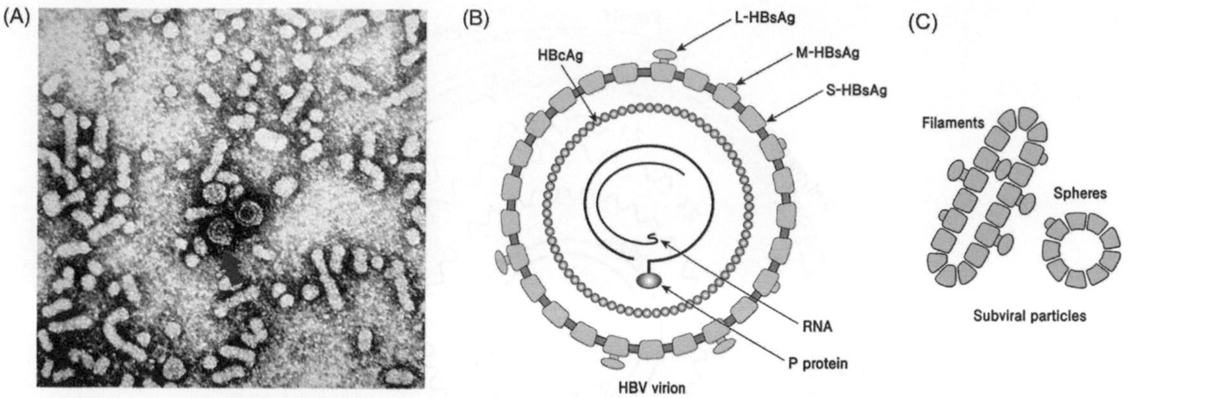

FIGURE 18.2 HBV virions and subviral particles. (A) Electron micrograph of HBV particles from a chronically infected patient. Dane particles (red arrow) or infectious particles, with a diameter of 42 nm, are visible. Subviral particles (spherical or filamentous, with a diameter of 22 nm) that lack the capsid are abundantly detected. (B) Diagram of HBV particles. An infectious HBV virion particle is depicted. Three kinds of HBsAg are embedded in the viral envelope: L-, M-, and S-HBsAg. Inside of the envelope, an icosahedral capsid with $T = 3$ or 4 symmetry is found. Inside of the capsid, a partially duplex double-stranded DNA genome is found. The viral P protein covalently linked to the 5′ terminus of the minus-strand DNA is denoted. The RNA fragment linked to the 5′ terminus of the plus-strand DNA is denoted. (C) Diagram of HBV subviral particles. Subviral particles are either filamentous or spherical shape. These subviral particles lack the viral capsid.

Therefore, the genome is called **circular partial duplex DNA**. Second, the 5′ terminus of two strands is linked to either protein or RNA. In detail, HBV P protein (or Pol) is linked to the 5′ terminus of the **minus-strand DNA**,[2] while an RNA fragment is linked to the 5′ terminus of the **plus-strand DNA**. These two molecules linked to the 5′ termini serve as primers for the corresponding strands during viral reverse transcription.

Protein Coding: HBV encodes four ORFs: C for core, P for polymerase, S for surface antigen, and X for HBx protein (Fig. 18.3). Four ORFs are considerably overlapped, reflecting the compact nature of the viral genome organization. Notably, every nucleotide is used at least once or twice for protein coding, a feature that is unprecedented among animal viruses. In the case of the S ORF, three cocarboxyl terminal HBsAg are synthesized (ie, large, middle, and small or L-, M-, and S-HBsAg, respectively) depending on the start AUG codon used for translation initiation (Table 18.3).

Cell Culture: HBV is characterized by a strict tropism of human hepatocytes. Human hepatoma cell lines such as HepG2 cell and Huh7 cell are not susceptible to HBV infection. Primary human hepatocyte (PHH) and certain PHH-derived cells such as HepaRG cell are susceptible to HBV infection.

18.3 THE LIFE CYCLE OF HBV

Entry: HBV virion attaches to hepatocytes via **HSPG**[3] and enters the cell via receptor-mediated endocytosis (Fig. 18.4). The pre-S1 domain of the large HBsAg version (L-HBsAg) is a viral ligand interacting with the receptor. Recently, **NTCP**[4] was identified as a cellular receptor for entry. Upon penetration into cytoplasm, viral capsids traffic to the nucleus via cytoplasmic trafficking. The viral capsid enters the nucleus via the nuclear pore and the viral DNA is released.

cccDNA Formation: Upon the nuclear entry, the virion relaxed circular (RC) DNA is converted to covalently closed circular DNA (**cccDNA**), which then serves as a template for viral transcription. In other words, cccDNA formation is a prerequisite for establishing a viral infection. Importantly, cccDNA is the molecular basis of viral persistence, as it is stably maintained as an **episome**[5] during chronic infection. Despite its critical importance, little is known about host factors involved in the cccDNA conversion (Fig. 18.5).

In principle, a number of biochemical reactions need to be carried out for the conversion of RC DNA to cccDNA. For instance, the HBV P protein linked to the 5′ terminus of minus-strand DNA and the RNA oligomer linked to the 5′

---

2. **Plus- and minus-strand DNA** The strand that has the same polarity as mRNA is called the plus-strand, while the opposite strand is called the minus-strand.

3. **HSPG (heparan sulfate proteoglycan)** It refers to a kind of proteoglycans present abundantly on cell surface, in which heparin sulfate represents the glycan moiety.

4. **NTCP (sodium taurocholate co-transporting polypeptide)** A cellular receptor for bile acids transport.

5. **Episome** It refers to autonomously replicating extrachromosomal DNA. The term "epi" is derived from Greek word for "above."

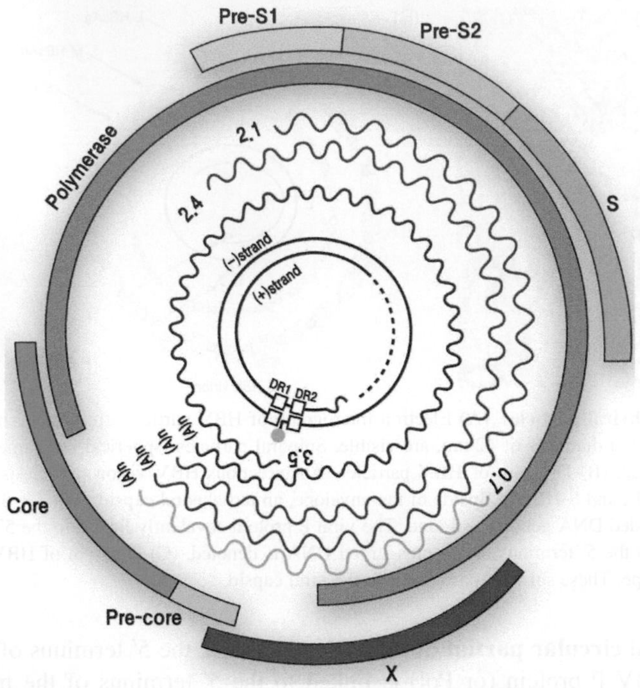

FIGURE 18.3 **HBV genome and viral proteins.** A partial duplex circular DNA genome is shown in the middle, and four viral transcripts (3.5, 2.4, 2.1, and 0.7 kb) are drawn by wavy lines. Four ORFs (ie, Core, Polymerase, S, and X ORFs) are drawn in parallel. The viral genome shows the polarity of two strands, indicated by (+) or (−). The gap in the plus-strand is denoted by dashed lines. The P protein (oval orange) is linked to the 5′ terminus of the (−) stand and the RNA oligomer (pink wavy line) is linked to the 5′ terminus of the (+) strand. Two direct repeat (DR) elements (ie, DR1 and DR2) are highlighted by boxes.

TABLE 18.3 HBV Proteins and Their Functions

ORFs	Antigen	Functions (M.W.)	Copy #/virion
Core (C)	HBcAg	Capsid, p21	240 or 180
	HBeAg	HBeAg, p14–p17	0
Polymerase (P)	–	Reverse transcriptase, p98	1
Surface (S)	L-HBsAg	Envelope glycoprotein, gp42	50
	M-HBsAg	Envelope glycoprotein, gp36	50
	S-HBsAg	Envelope glycoprotein, gp27	200
X	–	HBx, p17	0

terminus of plus-strand DNA have to be removed (see Fig. 18.5). In addition, the gap region present in the plus-strand DNA of the RC DNA needs to be filled. Since the cccDNA conversion entails DNA repair, it has been speculated that cellular DNA repair machinery carries out some of these DNA repair steps. However, the identity of the cellular factors involved in this process remains enigmatic.

Viral RNAs: Four viral transcripts are transcribed by cellular RNA Pol II by using cccDNA as a template: one genomic length RNA and three subgenomic RNAs (Fig. 18.6). As four viral transcripts are driven from four distinct promoters, their 5′ termini are different but their 3′ termini are identical, as they share the polyadenylation signal. In fact, the genomic length RNA is longer (ie, 3.5 kb) than the DNA genome (ie, 3.2 kb) due to the terminal redundancy (denoted by **R** element in Fig. 18.6). Consequently, DR1 and epsilon element present twice in the 3.5 kb RNA. The 3.5 kb RNA is termed "**pregenomic RNA**" (pgRNA), because it serves as an RNA precursor for the viral genome synthesis.

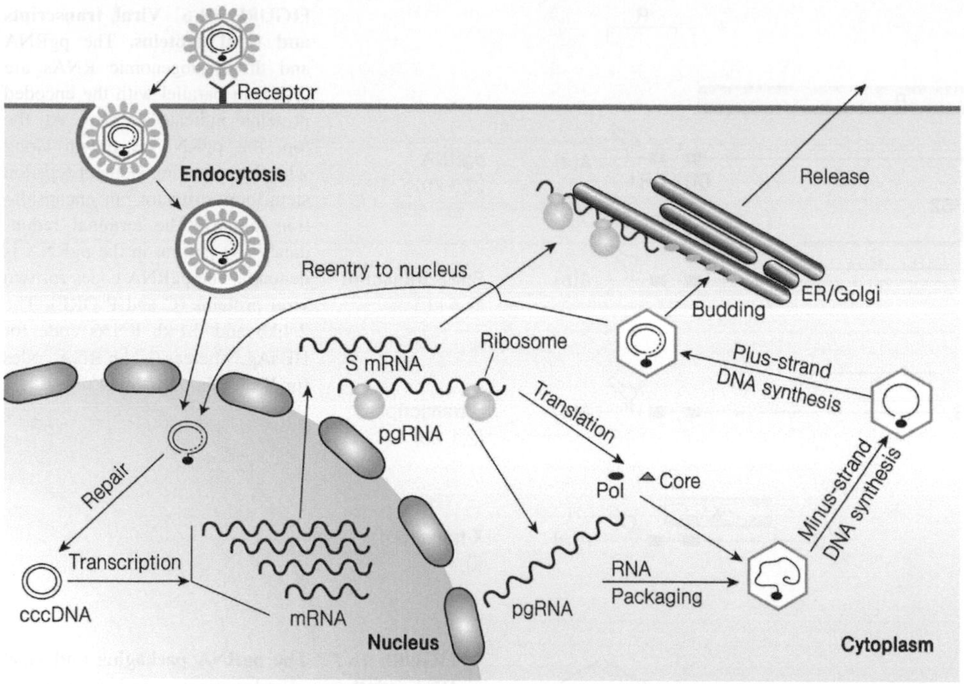

FIGURE 18.4 The life cycle of HBV. HBV enters the cell via receptor-mediated endocytosis, leading to the delivery of the viral genome to the nucleus. The virion RC DNA is repaired to become cccDNA, the template for viral transcription. Viral reverse transcription occurs following the packaging of the viral pregenomic RNA. Viral reverse transcription can be divided into two steps: minus-strand DNA synthesis and plus-strand DNA synthesis. Only matured nucleocapsids are enveloped via budding at the ER and released via the secretory pathway.

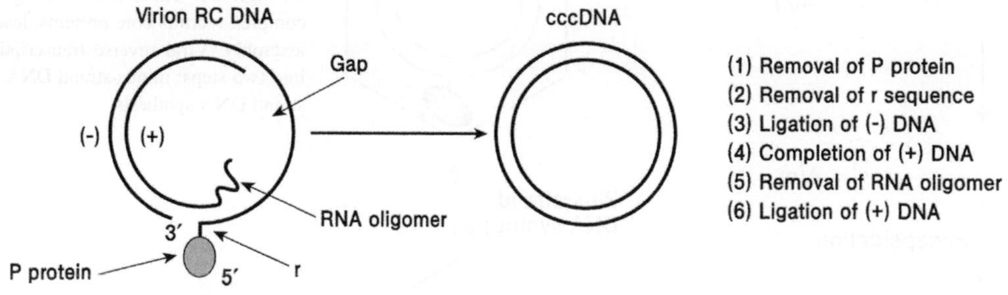

FIGURE 18.5 The conversion of RC DNA to cccDNA. The relaxed-circular (RC) DNA present in the virion shows the two non-DNA molecules which are linked to the termini: HBV P protein inked to the 5′ terminus of the minus-strand DNA, and the RNA oligomer linked to the 5′ terminus of the plus-strand DNA. A short repeat element (r) present at the end of the minus-strand DNA is denoted. The gap region (~1/3 of the genome) present in the plus-strand DNA is indicated by a dotted line. Six biochemical processes are articulated to the right.

Translation: The pgRNA serves two roles: (1) it serves as mRNA for the translation of C and P protein and (2) it serves as an RNA template for viral reverse transcription. Three subgenomic RNAs serve solely as mRNA: 2.4 kb RNA for L-HBsAg, 2.1 kb RNA for M- and S-HBsAg, and 0.7 kb RNA for X protein (HBx). In addition to these four viral proteins, one additional protein, termed "e" antigen or "**HBeAg**," is detected in the serum of infected persons (see Table 18.3). HBeAg is essentially derived from C ORF but exhibits distinct antigenicity from that of HBcAg (ie, core) (see Box 23.3). HbeAg is regarded as an accessory protein, because it is not required for the viral genome replication. Rather, it serves as an important serological marker, indicative of ongoing viral genome replication in the liver tissues (see Box 23.3).

Packaging: Among the four viral transcripts, the pgRNA is exclusively packaged into the capsid and serves as a template for viral reverse transcription. The stem-loop structure, called **epsilon** (ε),[6] located at the 5′ terminus of the pgRNA is recognized by HBV P protein as an encapsidation signal. In fact, the stem-loop structure is present twice in the pgRNA due to terminal redundancy: one at the 5′ and one at the 3′ terminus (see Fig. 18.6). Intriguingly, only the 5′ epsilon is recognized as an encapsidation signal. A question arose as to why only the 5′ epsilon is recognized as

6. **Epsilon** (ε) It refers to the stem-loop structure that serves as an encapsidation signal.

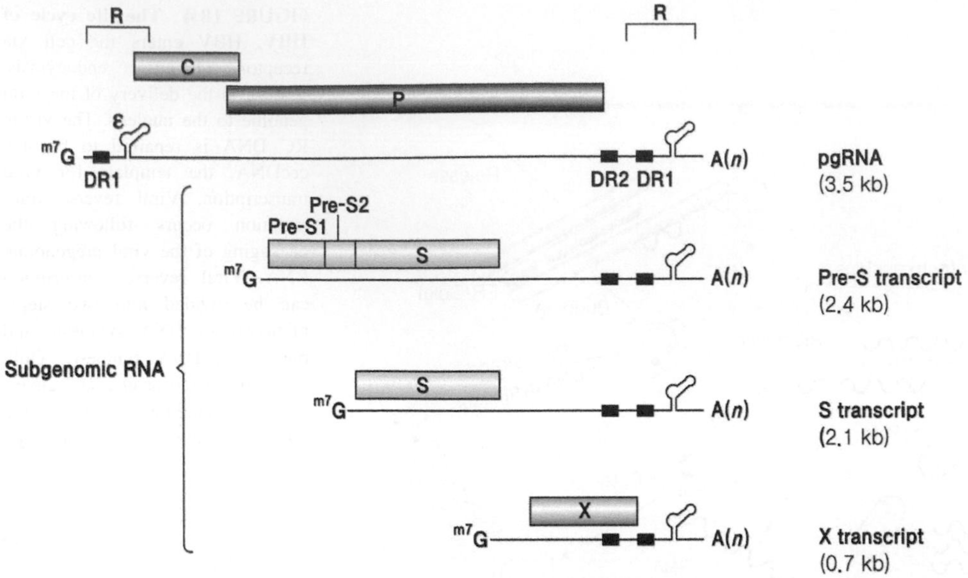

FIGURE 18.6 Viral transcripts and viral proteins. The pgRNA and three subgenomic RNAs are drawn in parallel with the encoded proteins indicated above. At the top, the pgRNA is drawn along with the *cis*-acting element (epsilon stem-loop structure, an encapsidation signal). The terminal redundancy (R) present in the pgRNA is denoted. The pgRNA codes for two viral proteins (C and P ORF). The 2.4 kb and 2.1 kb RNAs code for HBsAg, whereas 0.7 kb RNA codes for X protein. *DR*, direct repeat.

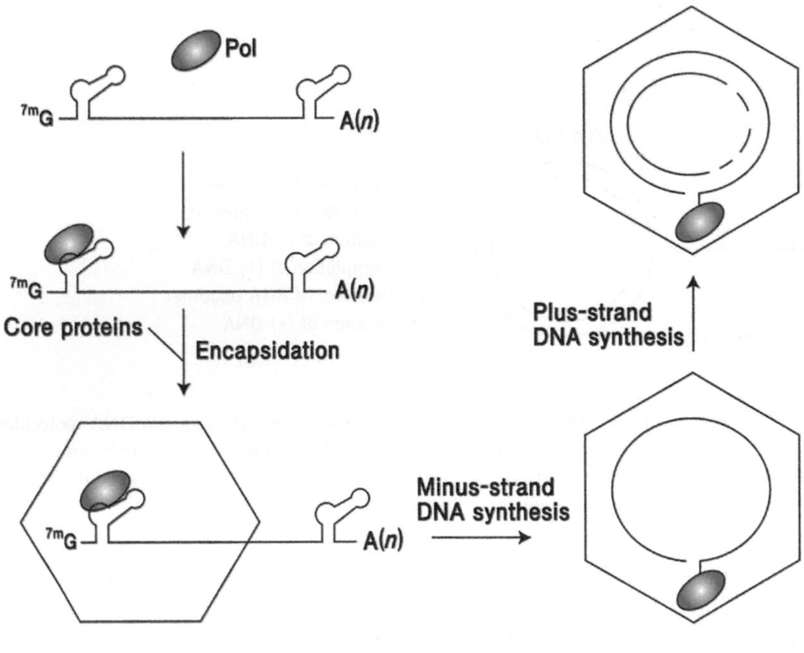

FIGURE 18.7 The pgRNA packaging and viral DNA synthesis. Viral reverse transcription occurs during/after encapsidation of the pgRNA. HBV P protein recognizes the 5′ epsilon stem-loop structure of the pgRNA. Then, the resulting P protein–pgRNA complex recruits core proteins, leading to viral capsid assembly. Viral reverse transcription can be divided into two steps: minus-strand DNA synthesis and plus-strand DNA synthesis.

an encapsidation signal. It was shown later that the viral P protein recognizes not only the epsilon structure, but also the cap structure at the 5′ end for the encapsidation (Fig. 18.7). Altogether, a bipartite signal (ie, 5′ epsilon and 5′ cap) are essential for encapsidation.

Genome Replication: Viral reverse transcription reactions are catalyzed by the HBV P protein. The HBV P protein is composed of four subdomains: in particular, the RT (reverse transcriptase) and RNase H domains exhibit a limited similarity to retroviral RT (Fig. 18.8). Not only the reverse transcriptase activity but also the RNase H activity of P protein is essential for viral DNA synthesis. In addition, two subdomains at the N-terminal extension are found in the P protein. The TP domain harbors an invariant tyrosine residue (Y63) that serves as a primer for viral reverse transcription (ie, **protein-priming**[7]). On the other hand, the spacer domain (SP), that links the TP domain to the RT domain, can be substantially deleted without affecting the activity.

7. **Protein-priming** It refers to the initiation of DNA synthesis, where protein (i.e., tyrosine or serine) is utilized as a primer.

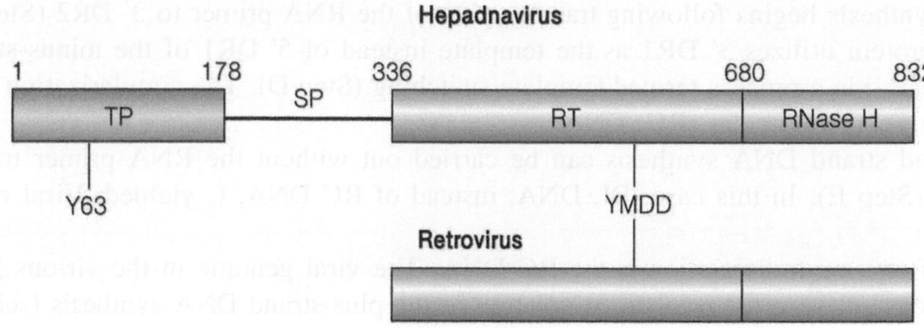

FIGURE 18.8 HBV P protein and its relatedness to retroviral RT. The subdomains of HBV P protein and retroviral RT are shown side by side: TP (terminal protein), SP (spacer), RT (reverse transcriptase), and RNase H domain. The highlighted YMDD motif is critical for the enzymatic activity. The Y63 residue in the TP domain is essential for the protein-priming step to allow minus-strand DNA synthesis. The hydroxyl group of the tyrosine residue is covalently linked to the first nucleotide being synthesized by reverse transcription.

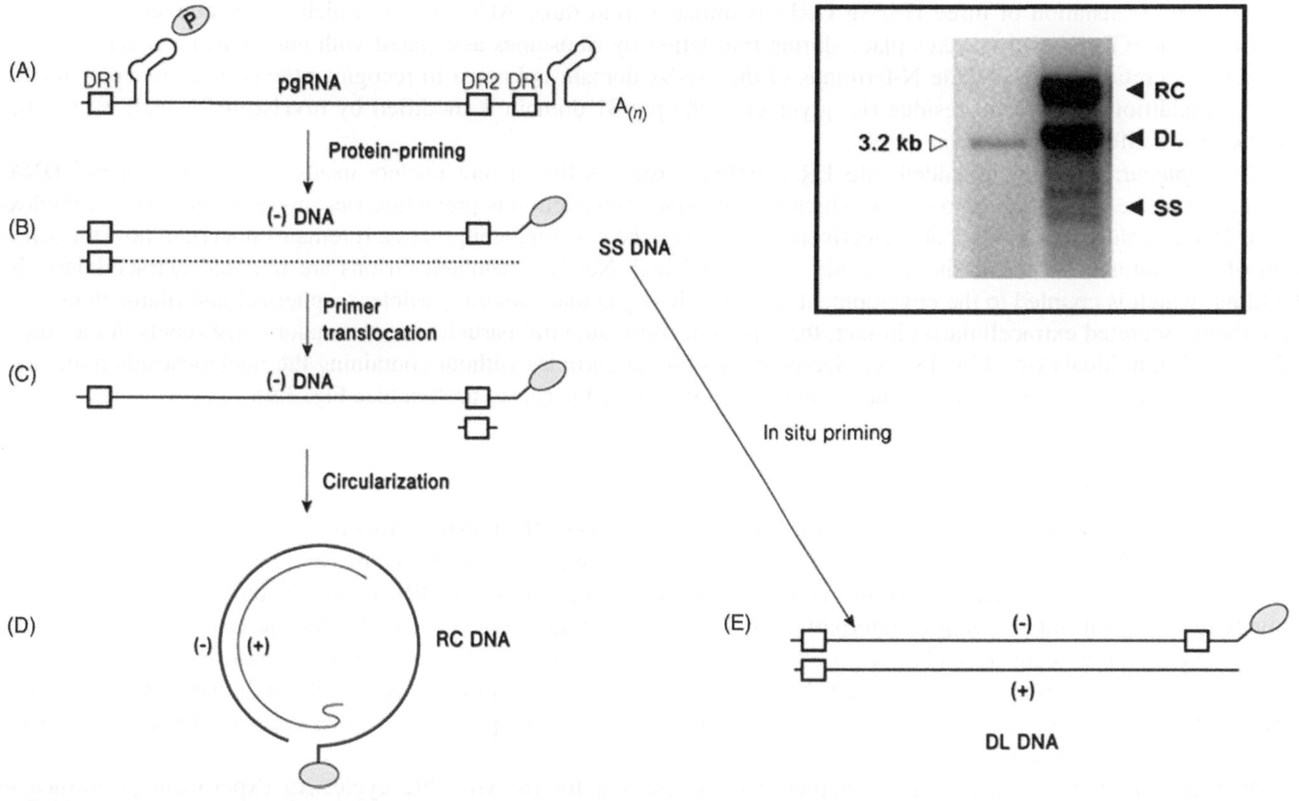

FIGURE 18.9 Viral reverse transcription. Steps involved in the viral reverse transcription are schematically drawn. (A) The pregenomic RNA is drawn with *cis*-acting elements: epsilon, DR1, and DR2. Note that DR1 and epsilon elements are presented twice due to terminal redundancy. (B) The map of minus-strand DNA is shown, in which the viral P protein is linked to the 5′ end. The residual RNA fragment that is resistant to degradation by the RNase H activity of the viral P protein is left behind. (C) The residual RNA fragment undergoes translocation to 3′ DR2. Following translocation, the RNA fragment is used as a RNA primer for the plus-strand DNA synthesis. (D) The map of RC DNA. (E) The map of DL DNA. (Inset) Southern blot analysis of viral replication intermediates extracted from cytoplasmic nucleocapsids. Besides RC DNA, a duplex linear (DL) DNA is found inside of capsids. RC (relaxed circular) DNA, DL (duplex-linear) DNA, and SS (single-strand) DNA are denoted. The DNA size marker (3.2 kb) is denoted.

Viral reverse transcription can be divided into two steps: 1st strand DNA synthesis and 2nd strand DNA synthesis (see Fig. 18.7). The former involves RNA-dependent DNA synthesis, leading to the synthesis of minus-strand DNA, whereas the latter involves DNA-dependent DNA synthesis, leading to the synthesis of plus-strand DNA (Fig. 18.9). Notably, the minus-strand DNA synthesis is initiated by protein-priming, where the P protein itself serves as a primer. Consequently, the P protein is linked to the 5′ terminus of the minus-strand DNA (Step B). Following the completion of minus-strand synthesis, the residual RNA fragment is left behind from the cleavage by the RNase H activity of the HBV P protein. This residual RNA fragment is utilized as an RNA primer for the 2nd strand DNA synthesis.

The plus-strand DNA synthesis begins following translocation of the RNA primer to 3′ DR2 (Step C). To continue the DNA synthesis, the P protein utilizes 3′ DR1 as the template instead of 5′ DR1 of the minus-strand DNA during the plus-strand DNA synthesis, via a process termed template switching (Step D). The circularization during the plus-strand DNA synthesis results in relaxed circular (RC) DNA.

Alternatively, the 2nd strand DNA synthesis can be carried out without the RNA primer translocation, a process called **in situ priming** (Step E). In this case, DL DNA, instead of RC DNA, is yielded. Viral reverse transcription is detailed in Box 18.1.

The final product of reverse transcription is the RC DNA. The viral genome in the virions has a gap in the plus-strand DNA, as the virions get secreted prior to completion of the plus-strand DNA synthesis (see Fig. 18.4). The plus-strand DNA synthesis is only completed in the next cycle of infection.

Unlike retroviruses, HBV has a DNA genome, although they share reverse transcription. What makes the difference? The difference is the step at which viral reverse transcription takes place during the viral life cycle. Viral reverse transcription takes place shortly after entry of retroviruses, while viral reverse transcription takes place during exit phase in HBV (Box 18.2).

Assembly and Release: The viral envelope glycoproteins of HBV are frequently called "surface antigens" or HBsAg. In fact, three viral envelope glycoproteins are found at the surface of virions: L-HBsAg, M-HBsAg, and S-HBsAg (Fig. 18.10). Translation of three HBsAg ORFs is initiated from three AUG codons which are positioned in the same reading frame. Glycosylation takes place during translation by ribosomes associated with endoplasmic reticulum (ER) through a secretary pathway. The N-terminus of the pre-S1 domain is known to recognize the cellular receptor for the entry. In addition, the second residue (ie, glycine) of the pre-S1 domain is modified by **myristate**,[8] which is essential for the virion infectivity.

Envelopment: HBsAg embedded into ER membrane recruits the mature nucleocapsids, which contain RC DNA. Interestingly, the viral nucleocapsids, in which viral reverse transcription is premature (ie, single-strand DNA or duplex-linear DNA-containing capsids), are selectively excluded by the envelopment process. It remains uncertain how the selection of the mature DNA-containing capsids is accomplished. Newly assembled virions are released extracellularly by budding, which is coupled to the envelopment. One peculiarity is that subviral particles—spherical and filamentous—are abundantly secreted extracellularly; in fact, these noninfectious subviral particles are the major constituents in the serum of infected individuals (see Fig. 18.2A). Secretion of subviral particles without containing the nucleocapsids implicates that the budding process is driven by the envelope protein, not by the capsid protein (see Fig. 3.9).

18.4 X PROTEIN

X open reading frame (ORF) encodes a small regulatory protein (ie, HBx), that is composed of only 154 amino acids. It was named "X" because the protein did not exhibit a tangible sequence similarity to other viral or cellular proteins. HBx was presumed to be involved in the viral carcinogenesis, because the X ORF is present only in tumor-associated hepadnaviruses, but not in avian counterparts (see Table 18.2). A question is then whether the HBx contributes to the HBV-associated liver cancer? Unlike most other viral oncogenes, however, X gene transgenic mouse strains do not normally develop liver tumors but do develop liver tumors when treated with other chemical carcinogens. It appears that HBx itself cannot cause tumors but can promote tumor formation in the presence of other carcinogens. It is fair to say that HBx is a moderate oncogene.

Nuclear Functions: A question is whether HBx is essential for the viral life cycle. An experiment performed in woodchucks showed that injection of wild-type viral DNA or viral DNA with a nonfunctional X gene resulted in different outcomes. The X-gene negative viral DNA failed to establish a viral infection, whereas the wild-type viral DNA did, implicating that the X gene is essential for the establishment of a viral infection in animals. Nonetheless, the role of HBx remained uncertain. Recently, it was shown that HBx is essential for viral transcription from the cccDNA template, presumably by inducing epigenetic changes (eg, histone acetylation) (Fig. 18.11). It remains to be seen how HBx induces the epigenetic modification of the nuclear cccDNA template.

Cytoplasmic Functions: HBx activates multiple cytoplasmic signaling pathways (see Fig. 18.11). HBx is predominantly localized in mitochondria, where HBx triggers the release of **ROS**.[9] The resulting ROS induces **NF-kB signaling**.[10] In addition, mitochondrial HBx mobilizes calcium from mitochondria and elevates the cytoplasmic calcium level.

8. **Myristate** A kind of saturated fatty acid with 14 carbons.

9. **ROS (reactive oxygen species)** It refers to chemically reactive molecules containing oxygen. Examples include oxygen ions and peroxides. ROS form as a natural by-product of the normal metabolism of oxygen and have important roles in cell signaling and homeostasis.

10. **NF-kB signaling** A signal transduction pathway that plays a key role in regulating the immune response and cell survival.

BOX 18.1 Viral Reverse Transcription

Reverse transcription represents a biochemical reaction that converts RNA to double-strand DNA. It constitutes two steps of DNA synthesis: 1st strand DNA synthesis, which produces the minus-strand by using the RNA template, and the 2nd strand DNA synthesis, which uses the newly generated minus-strand DNA as the template to generate the plus-strand DNA. The final product of the diverse steps of reverse transcription and DNA synthesis is a double-stranded DNA molecule.

First Strand DNA Synthesis: The HBV minus-strand DNA synthesis, the 1st step of the viral reverse transcription process, is initiated by protein-priming. HBV P protein itself serves as a primer, resulting in the P protein being covalently linked to the 5′ terminus of the minus-strand DNA. The priming reaction is initiated at the epsilon stem-loop structure located near the 5′ end of the pgRNA. Following the synthesis of only 3 or 4 nts, the HBV P protein undergoes translocation to 3′ DR1, and resumes the DNA synthesis [A to B]. This translocation is called template switching or strand-transfer, which is also found in retroviral reverse transcription. The template switching is facilitated by the presence of a short nucleotide sequence (5′-UUCA) present in the epsilon structure and in the DR1 region. The complementarity of the newly synthesized 4 nts linked to the P protein with a sequence in the DR1 region is required for template switching. The minus-strand DNA synthesis continues to the 5′ end of the pgRNA [B to D]. The pgRNA is degraded by the RNase H activity of HBV P protein.

Second Strand DNA synthesis: The minus-strand DNA serves as a template for synthesizing the plus-strand DNA. A residual RNA fragment at the very 5′-end of the RNA template is not digested by the RNaseH, and serves as a RNA primer for the 2nd strand (plus-strand) DNA synthesis. The plus-strand DNA synthesis begins following translocation of the RNA primer to 3′ DR2 near the 5′ end of the minus-strand DNA [D to E]. Again, the 2nd template switching is mediated by the sequence identity of DR1 and DR2, providing a complement sequence of the RNA primer. Immediately after the initiation of the plus-strand DNA synthesis, the 3rd template switching occurs to circularize the template minus-strand DNA [E to F]. The circularization involves the template switch from 5′ DR1 to 3′ DR1 of the minus-strand. The final product of reverse transcription is the relaxed circular (RC) DNA.

Besides RC DNA, a duplex linear (DL) DNA is found inside of capsids. In fact, the 2nd strand DNA synthesis is carried out without the RNA primer translocation, a process called **in situ priming** [D]. In this case, DL DNA, instead of RC DNA, is yielded. DL DNA comprises 20% of the total DNA replication intermediates.

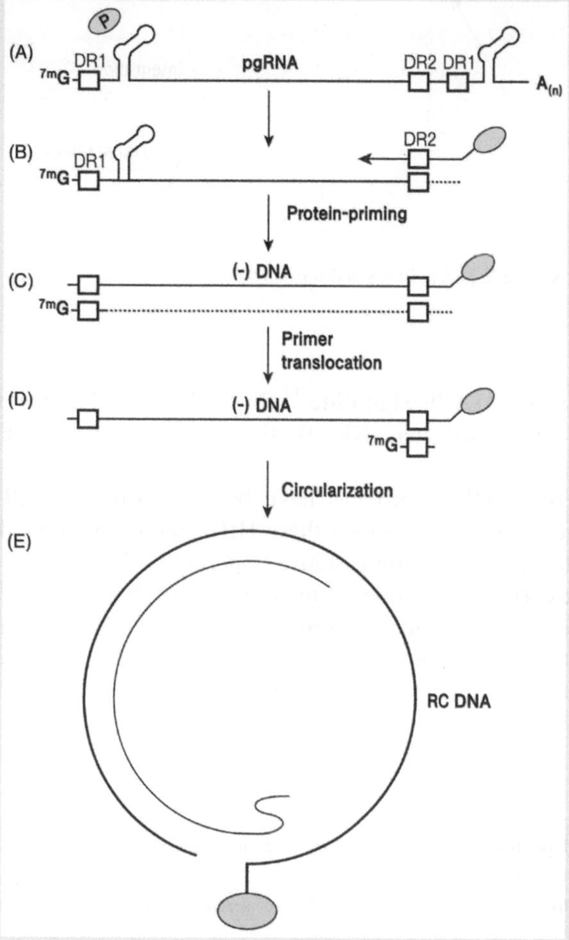

Steps in viral reverse transcription mechanism.

BOX 18.2 HBV Versus Retrovirus

Both retrovirus and HBV replicate via reverse transcription. Despite their fundamental similarities, a retrovirus has an RNA molecule inside the virion particles, whereas HBV has a DNA molecule. In fact, both retrovirus and HBV package the viral genomic RNAs into the viral capsids. However, the step during the virus life cycle at which the reverse transcription step occurs is different. The retroviral capsids exit the cells without reverse transcription. By contrast, the HBV capsids exit the cell following viral reverse transcription. Thus, retroviral virions contain an RNA genome, while HBV virions contain a DNA genome. The retroviral reverse transcription is carried out during the viral entry.

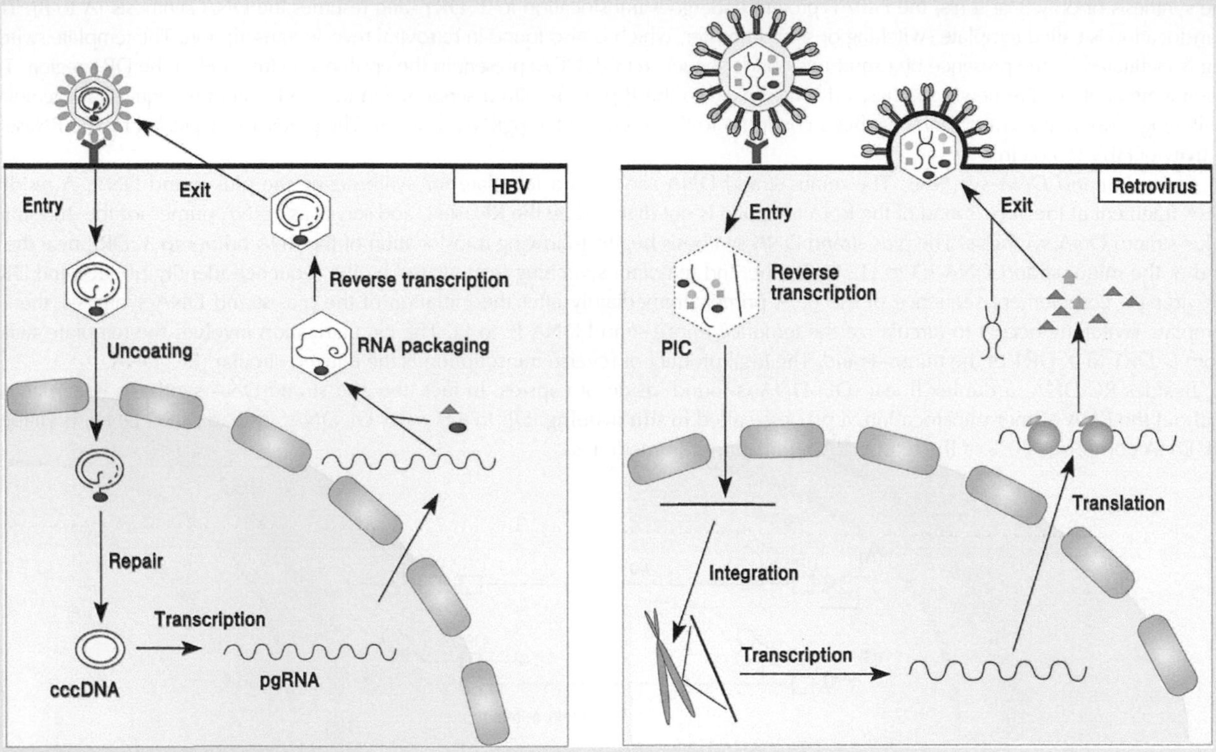

Comparison of the reverse transcription step of hepadanviruses and retroviruses.

The elevated calcium, in turn, activates **MAPK signaling.**[11] Since HBx could induce not only cell survival but also cell proliferation, cross-talk between HBx-activated NF-kB signaling and the MAPK pathway could lead to cell transformation.

HBx-Binding Proteins: Numerous cellular proteins have been identified as HBx-binding proteins; however, the physiological roles remained largely uncertain. Among these HBx-binding proteins, **DDB1**[12] draws much attention, as HBx-DDB1 interaction was validated by a structural study. Importantly, HBx forms a novel Cul4-DDB1 ubiquitin E3 ligase (Fig. 18.12). In this complex, HBx was proposed to act as a "substrate adaptor" via HBx-DDB1 interaction. It is hypothesized that HBx degrades cellular proteins by recruiting them to the HBx-mediated ubiquitin E3 ligase complex. It is tempting to speculate that HBx exhibits numerous functions by targeting multiple cellular proteins to the complex. The cellular proteins that are recruited to the E3 ligase by HBx are yet-to-be identified.

11. **MAPK signaling (mitogen-activated protein kinase)** A signal transduction pathway that plays a role in cell proliferation. Also known as the Ras-Raf-MEK-ERK pathway.

12. **DDB1 (DNA damage binding protein)** A cellular protein that was discovered to bind to the UV-induced DNA damage. It forms a heterodimer with DDB2, which functions in nucleotide excision repair.

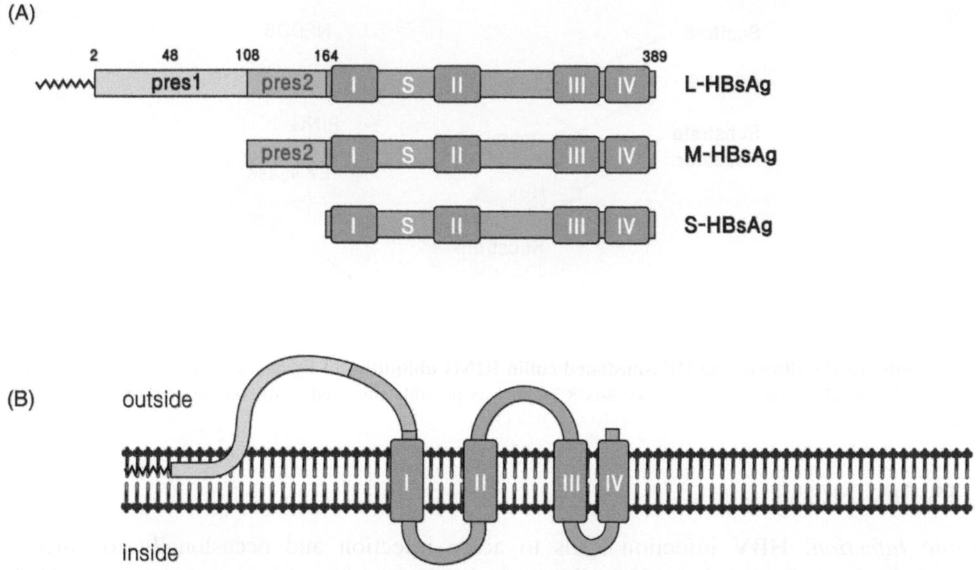

FIGURE 18.10 Subdomains of HBsAg and their membrane topologies. (A) HBV virions contain three envelope (surface) proteins which share the S-domain: L-HBsAg, M-HBsAg, and S-HBsAg. S-HBsAg is composed of the S-domain only, M-HBsAg contains the S-domain and an N-terminal extension, the pre-S2 domain. L-HBsAg is composed of the S-domain, preS2, and an additional N-terminal extension, the pre-S1 domain. (B) Membrane topology of L-HBsAg protein. All three HBsAg proteins share the four TM domains, which are located in the S-domain. The pre-S1 domain interacts with the cellular receptor for the viral entry. The N-terminus of the pre-S1 domain contains a myristoylation site with the essential glycine residue shown at position 2 (wavy line). The leading methionine residue is cleaved prior to myristoylation.

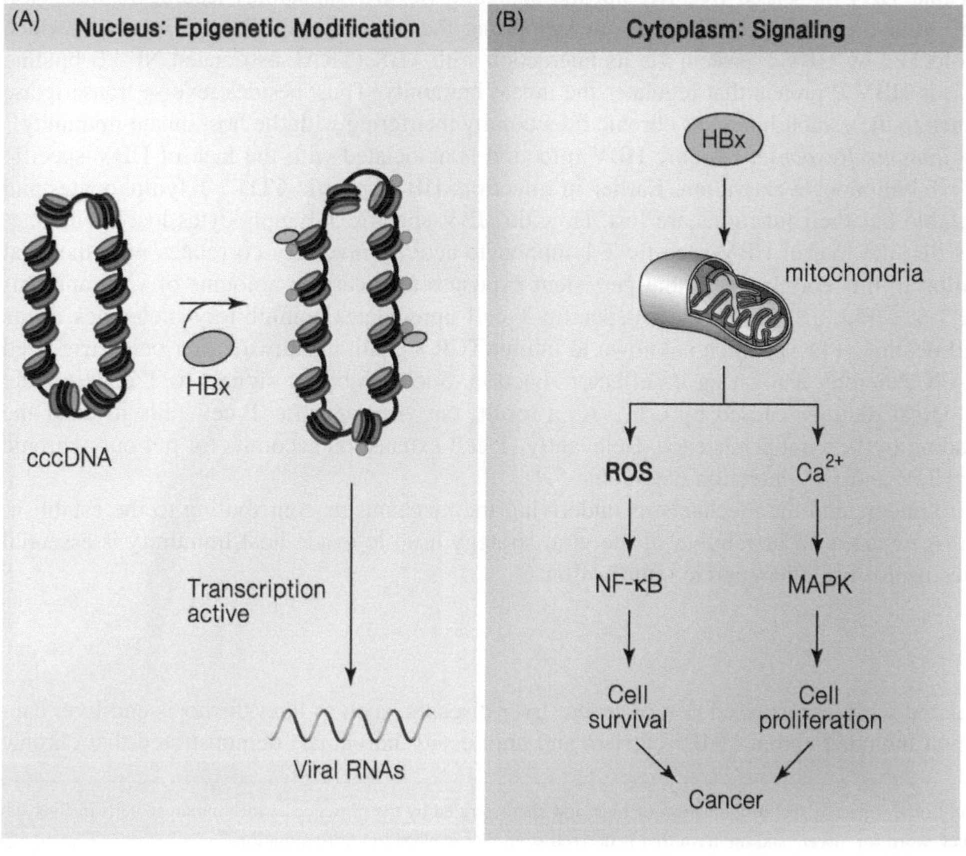

FIGURE 18.11 Distinct roles of HBx in the nucleus and in the cytoplasm. (A) In the nucleus, HBx induces epigenetic changes in the chromatin structure of the cccDNA so that the transcriptionally inactive template becomes transcriptionally active. (B) In the cytoplasm, HBx is associated with the outer mitochondrial membrane, changes the membrane potential and induces the release of ROS, which then activates NF-kB signaling. HBx also induces calcium release from mitochondria, which then activates MAPK signaling.

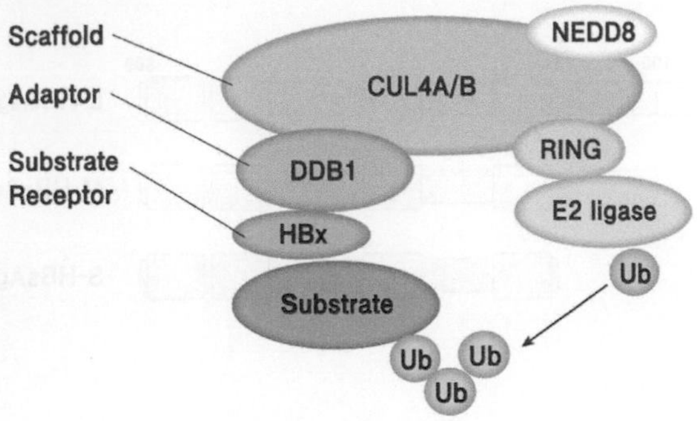

FIGURE 18.12 A schematic model illustrating HBx-mediated cullin RING ubiquitin E3 ligase. A linker such as DDB1 binds to cullin scaffold to form CRL (cullin-based RING ubiquitin E3 ligase) (see Box 8.2). HBx is possibly involved in the recruitment of substrates for ubiquitination via its interaction with DDB1.

18.5 EFFECTS ON HOST

Acute and Chronic Infection: HBV infection leads to acute infection and occasionally to chronic infection (see Fig. 23.7). **Hepatitis,**[13] the inflammation of the liver, results from a host immune response to HBV-infected cells. Despite the host immune response, HBV manages to maintain lifelong chronic or persistent infection. How HBV evades the host immune response represents a major question in this field. A short answer to this question is that HBV suppresses not only the innate immune response but also the adaptive immune response. HBV infection does not cause a cytopathic effect in infected cells.

Suppression of the Innate Immune Response: Unexpectedly, interferon induction early in infection is not observed in HBV-infected chimpanzee. Two hypotheses were proposed to account for the lack of an innate immunity upon HBV infection. The first hypothesis states that HBV is not sensed by the pattern recognition receptors (PRR), such as TLR and RIG-I. The second hypothesis states that HBV is sensed by PRR but the signaling (ie, IRF signaling) leading to interferon induction is potently blocked by viral proteins. Recently, a result supporting the latter hypothesis was reported, which showed that IRF signaling was blocked by HBV P protein via its interaction with TBK (TRAF-associated NF-kB binding kinase 1) (Fig. 18.13). Notably, it is HBV P protein that regulates the innate immunity. Thus, besides reverse transcriptase activity, HBV P protein contributes to the establishment of chronic infection by interfering with the host innate immunity.

Suppression of the Adaptive Immune Response: Chronic HBV infection is associated with the lack of HBV-specific $CD4^+$ T lymphocyte and $CD8^+$ T-lymphocyte activation. Earlier in infection, HBV-specific $CD4^+$ T lymphocytes and $CD8^+$ T lymphocytes are detectable but their functions are lost. How do HBV-specific T lymphocytes lose their functions? A clue to this question is that the loss of HBV-specific T lymphocyte activity inversely correlates with the viral titer in the serum. One explanation to this correlation is that persistent exposure to excessive amounts of viral antigens may lead to **T-cell exhaustion**[14] (see Fig. 18.13). The virus-specific T cell upregulates coinhibitory molecules on its surface such as programmed cell death 1 (PD-1), which is known to inhibit TCR signaling pathway after being triggered by their ligands expressed on APC, thereby abrogating its effector function. Such inhibitory signals to T cells protect the tissues from immunopathological damage caused by CTL. As a result, the virus-specific T cell fails to clear the HBV-infected virus, thereby leading to the viral persistence. Relevantly, T cell exhaustion accounts for not only chronic HBV infection but also chronic HCV and HIV infection (see Table 5.4).

In fact, we have just begun to understand the mechanisms underlying the mechanisms contributing to the establishment of chronic HBV infection. A detailed understanding of the viral strategy used to evade host immunity is essential for the identification of strategies useful for a therapeutic intervention.

18.6 LIVER CANCER

Chronic HBV infection is associated with an increased risk of severe liver diseases, such as liver cirrhosis and liver cancer. An epidemiology study which included chronic HBV carriers and uninfected individuals demonstrated that chronic

13. **Hepatitis** It refers to a medical condition defined by the inflammation of liver and characterized by the presence of inflammatory cells in liver tissue. The term *hepa* is derived from Greek word for "liver" and the term *titis* is derived from Greek word for "inflammation."

14. **T cell exhaustion** a phenomena that leads to dysfunction of activated T lymphocytes.

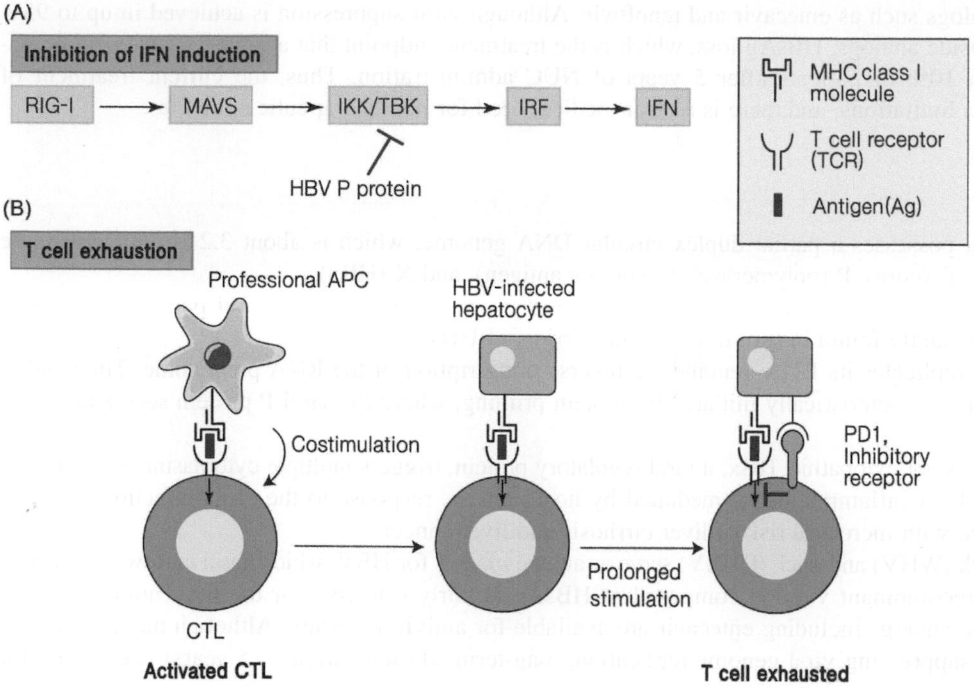

FIGURE 18.13 Viral evasion of host immune response. (A) Blockade of IRF signaling. HBV P protein blocks IRF signaling via its interaction with TBK, thereby suppressing IFN induction. (B) T cell exhaustion. A HBV-specific T lymphocyte that has been activated by an antigen-presenting cell (APC) at an early stage after infection lost its function upon persistent stimulation, a phenomena called T cell exhaustion. T cell exhaustion is accompanied by the expression of PD-1, an inhibitory receptor.

hepatitis B increased the risk for developing liver cancer almost 20-fold. Despite a high correlation between a chronic HBV infection and liver cancer, little is known about the underlying mechanisms for the carcinogenesis.

Three mechanisms have been considered for HBV-associated carcinogenesis. First, HBx contributes to cell transformation via activation of multiple signaling pathways. HBx activates NF-kB signaling and MAPK signaling, which could lead to cell proliferation (see Fig. 18.11). In addition, HBx is shown to activate **Wnt signaling,**[15] which has been implicated in colon cancer. Secondly, chronic inflammation caused by HBV infection is believed to contribute to carcinogenesis (see Fig. 23.8). Chronic HBV infection is accompanied by persistent damage of the liver tissue due to cell-mediated immunity. Excessive cell division of otherwise nondividing quiescent cells is induced to compensate for the cell loss. Consequently, mutations accumulate due to excessive cell division which then leads to cell transformation. Thirdly, viral DNA can integrate into the host chromosomes. The integration can cause genome instability, thereby facilitating the formation of cancer cells. In fact, HBV DNAs are frequently found as integrated forms in the chromosomes of HCC tissues. However, unlike retroviruses (see Fig. 24.16), no evidence is found that the activation of a proto-oncogene is involved in HBV-associated carcinogenesis.

18.7 PERSPECTIVES

A salient feature of HBV infection is its narrow host range, in that it infects only human and chimpanzee, and its strict tissue tropism in that it infects only primary human hepatocytes (PHH). The lack of a susceptible cell line has significantly hampered the progress of HBV research since its discovery. Nonetheless, much progress has been made in our understanding of the virus life cycle, in particular, the viral reverse transcription. However, significant gaps in our knowledge are notable in the area of virus entry and the cccDNA formation in the virus life cycle. The recent discovery of the HBV entry receptor, sodium taurocholate cotransporting polypeptide (NTCP), enabled us to study the complete viral life cycle by establishing an HBV-susceptible HepG2-NTCP cell line (see Journal Club). The HBV susceptible cell line will be instrumental in the examination of the viral entry and the cccDNA formation. Undoubtedly, the establishment of an NTCP-transgenic mouse in the near future will fuel new efforts to study viral pathogenesis as well as the viral life cycle. In a clinical perspective, a preventive HBsAg vaccine had been developed earlier by recombinant DNA technology, which has been proven to be a fairly effective vaccine. On the other hand, antivirals for treatment of chronic HBV patients represent significant unmet medical needs. Specifically, current treatments for chronic hepatitis B

15. **Wnt signaling** The term "Wnt" is a combined word between "*Wingless*" and "*int*," as it was independently discovered as a gene associated with wingless phenotype of *Drosophila* and as a gene activated by the integration of a mouse retrovirus (Wingless/Int).

largely rely on nucleoside analogs such as entecavir and tenofovir. Although viral suppression is achieved in up to 95% of patients treated with nucleoside analogs, HBsAg loss, which is the treatment endpoint that allows for treatment cessation, is achieved in less than 10% of patients after 5 years of NUC administration. Thus, the current treatment of chronic HBV infections has its limitations, and there is a clear medical need for new therapeutic strategies.

18.8 SUMMARY

- *Viral genome*: HBV virion possesses a partial duplex circular DNA genome, which is about 3.2 kb. HBV genome encodes four viral proteins: C (core), P (polymerase), S (surface antigen), and X (HBx).
- *Viral particles*: An infectious virion contains an icosahedral capsid inside. In addition, subviral particles composed of surface antigens are abundantly found in serum of chronic carriers of HBV.
- *Genome replication*: HBV replicates its DNA genome via reverse transcription of the RNA pregenome. The synthesis of the 1st strand DNA is characteristically initiated by protein priming, where the viral P protein serves as a protein primer.
- *Host effect*: HBV infection is not cytopathic. HBx, a viral regulatory protein, triggers multiple cytoplasmic signaling.
- *Pathogenesis*: Hepatitis, a liver inflammation, is mediated by host immune response to the virus infection. Chronic HBV infection is associated with increased risk of liver cirrhosis and liver cancer.
- *Animal models*: Woodchuck (WHV) and duck (DHBV) serve as animal models for HBV, which has a narrow host range.
- *Vaccines and therapy*: A recombinant vaccine composed of HBsAg is fairly effective for the preventions of HBV infection. A few nucleoside analogs including entecavir are available for antiviral therapy. Although nucleoside analogs are quite effective in suppressing viral genome replication, long-term administration (>5 years) is essential for a clinical benefit.

STUDY QUESTIONS

18.1 List three viral DNAs found in cytoplasmic capsids and compare them regarding their maturity as replication intermediates DNA. Describe the viral DNA found in the nucleus and its function in viral replication.

18.2 HBV virion contains a partial duplex circular DNA genome. The 5′ termini of two DNA strands of the genome are linked to non-DNA molecules. Describe the nature of these molecules and their functions during viral reverse transcription.

18.3 HBV replicates its DNA genome via reverse transcription by using the pgRNA as a template. (1) Describe two roles of the pgRNA. (2) Propose a hypothesis of how these two roles are regulated. (3) How would you test the hypothesis?

SUGGESTED READING

Lucifora, J., Xia, Y., Reisinger, F., Zhang, K., Stadler, D., Cheng, X., et al., 2014. Specific and nonhepatotoxic degradation of nuclear hepatitis B virus cccDNA. Science. 343 (6176), 1221–1228.

Neuveut, C., Wei, Y., Buendia, M.A., 2010. Mechanisms of HBV-related hepatocarcinogenesis. J. Hepatol. 52 (4), 594–604.

Seeger, C., Mason, W.S., 2015. Molecular biology of hepatitis B virus infection. Virology. 479-480C, 672–686.

Wang, H., Ryu, W.S., 2010. Hepatitis B virus polymerase blocks pattern recognition receptor signaling via interaction with DDX3: implications for immune evasion. PLoS Pathog. 6 (7), e1000986.

Yan, H., Zhong, G., Xu, G., He, W., Jing, Z., Gao, Z., et al., 2012. Sodium taurocholate cotransporting polypeptide is a functional receptor for human hepatitis B and D virus. Elife. 1, e00049.

JOURNAL CLUB

- Yan, H., et al. NTCP is a functional receptor for hepatitis B virus. eLIFE 2112:e00049. doi:10.7554/eLife.00049.

  Highlight: It is a seminal article that defines NTCP (sodium taurocholate co-transporting polypeptide) as a functional receptor for HBV entry. HepG2 cells became susceptible to HBV infection if NTCP was stably expressed. Identification of the NTCP receptor for HBV entry allows the HBV to be cultivable.

Part V

Other Viruses

From Part II to Part IV, 20 animal virus families are covered with an emphasis on the molecular aspect of the virus life cycle. Ten chapters covered the 10 major animal virus families individually and 3 chapters covered 10 miscellaneous viruses briefly. All of these animal viruses are classified into either DNA viruses, RNA viruses, or RT viruses by the so-called Baltimore classification. Various other related viruses will be covered in Part V, including *Subviral agents, New emerging viruses*, and *Viral vectors*. Subviral agents (see chapter: Virus Vectors) refer to the "virus-like transmissible agents" that do not fit into classical definition of "virus." New emerging viruses (see chapter: Subviral Agents and Prions) refer to viruses that have never been described previously. Viral vectors (see chapter: New Emerging Viruses) refer to the recombinant viruses that are experimentally modified for the purpose of gene delivery.

Chapter 19

Virus Vectors

Chapter Outline

One salient feature of viruses that is distinguished from other organisms is that the virus is an organism that can deliver its genome to target cells. Being an obligate intracellular parasite, the virus has acquired the ability to deliver its genome efficiently during evolution. This unique feature has been exploited to create *gene expression vectors*[1] or *gene therapy vectors*. The former refers to ones for research purposes, while the latter refers to ones for clinical treatment purposes.

19.1 GENE EXPRESSION VECTORS

In order to ectopically express a gene of interest (ie, *transgene*[2]) in target cells, the gene expression vectors, a plasmid form DNA, are utilized. The critical component in plasmid vectors is the so-called *expression cassette*, which is composed of promoters and polyadenylation signal (Fig. 19.1). Some of these elements are often derived from viruses. Diverse kinds of expression vectors are available, but the following two kinds are most frequently used: (1) CMV (cytomegalovirus) plasmid and (2) EBV (Epstein-Barr virus)-based vector.

19.1.1 CMV Expression Plasmid

The so-called pcDNA3, a representative CMV plasmid is widely used for the ectopic gene expression in animal cells (Fig. 19.2). The name was derived from a promoter employed in the plasmid, which is referred to as a CMV promoter, a strong promoter derived from the immediate early gene of CMV. The *multiple cloning site* (MCS) positioned at the immediate downstream of the promoter contains multiple restriction sites, each of which is unique in the plasmid. The unique restriction sites in MCS are conveniently utilized for the insertion of transgenes. Gene expression can be readily examined by Western blot analysis at 2–3 days following *transfection*[3] of cells. Such transfected DNA cannot be maintained for longer than a few days and are only utilized for *transient expression* (Table 19.1). In addition, pcDNA3 can

1. **Vector** It refers to vehicles for the gene delivery. It could refer to either a plasmid DNA for the gene expression (ie, plasmid vector) or a recombinant virus.

2. **Transgene** It refers to the genes of interest that are to be transferred or to be expressed in the target cell.

3. **Transfection** It refers to the process of deliberately introducing nucleic acids into cells. The term is often used for nonviral methods in eukaryotic cells. The word *transfection* is a blend of *trans-* and *infection*.

Molecular Virology of Human Pathogenic Viruses. DOI: http://dx.doi.org/10.1016/B978-0-12-800838-6.00019-9

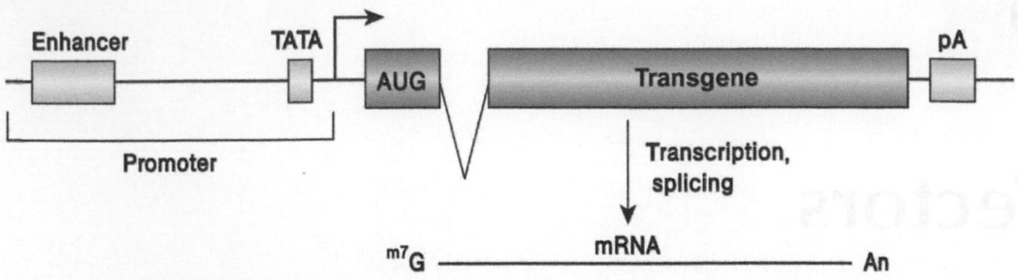

FIGURE 19.1 A schematic diagram illustrating an expression cassette. The open reading frame (ORF) for the transgene is positioned downstream of a typical RNA polymerase II promoter (TATA box and upstream elements) and upstream of the polyadenylation signal. The ORF is defined by starting from the AUG translation initiation codon to the termination codons (ie, UAG or UGA or UAA). The transcription initiation site is denoted by the rightward arrow. The mRNA transcribed from the expression cassette has the cap structure at the 5′ end and a poly (A) tail at the 3′ end.

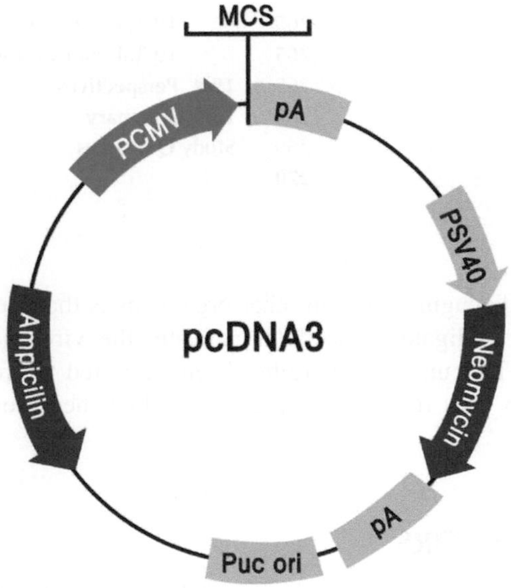

FIGURE 19.2 The plasmid map of pcDNA3. The MCS positioned immediately downstream of the CMV promoter (ie, pCMV) is the site for the insertion of transgenes. The transcript starting from the CMV promoter will terminate by the polyadenylation signal positioned downstream of the MCS. The neomycin gene serves as a selection marker in animal cells. The ampicillin gene is used for selection following transformation in *Escherichia coli*, while pUC ori, an origin of replication, is utilized for the propagation in *E. coli*.

TABLE 19.1 Two Types of Gene Expression

	Selection/Marker	Expression Duration	Chromosome Integration	Vector
Transient expression	No	2–3 days	No	pcDNA3
Stable expression	Yes, neomycin	>1 year	Yes	pcDNA3

also be utilized for long-term *stable expression*, since pcDNA3 contains the selection marker, neomycin gene. Selection of a transfected cell by treatment of *G418*[4] (ie, neomycin marker) for 2 weeks leads to the establishment of stable cell lines that could express the transgene. In stable cell lines, the transfected DNA is stably integrated into the chromosome during selection, and is stably maintained during cell division.

4. **G418** A kind of aminoglycoside antibiotics that is used for selection of the cell transfected with a plasmid expressing Neomycin gene. The neomycin marker confers the resistance to G418.

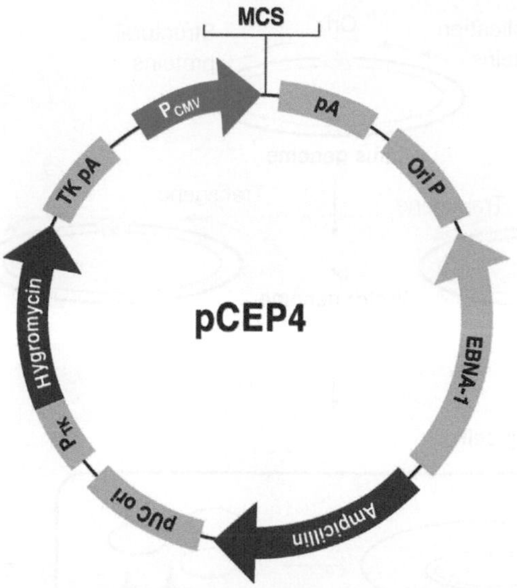

FIGURE 19.3 The plasmid map of pCEP4. Two elements derived from EBV are denoted: Ori P and EBNA-1. The Ori P represents a *cis*-acting element required for the plasmid maintenance of EBV DNA. EBNA (EBV nuclear antigen) is the viral protein that engages on the Ori P, thereby supporting the viral DNA replication. Other symbols are the same as described in Fig. 19.2.

19.1.2 EBV Expression Plasmid

Besides pcDNA3, another plasmid vector derived from EBV, a herpesvirus, is frequently used for stable expression of genes. The EBV vector explores the salient feature of EBV in that EBV maintains its viral DNA as an *episome* during latent infection. For instance, pCEP4 is one kind of a commercially available EBV vector plasmid (Fig. 19.3). The EBV vector contains two elements derived from EBV: (1) the Ori P, the sequence element required for plasmid maintenance and (2) a gene for EBNA-1, an Ori P binding protein. These two viral elements are the minimal requirement for the plasmid maintenance of EBV plasmid. Treatment of antibiotic for the selection marker (ie, hygromycin) allows the selection of cells in that the transfected plasmid is maintained as an episome without the chromosomal integration; the EBV vector is also dubbed "episome vector." Note that the EBV vector is suitable for lower expression of transgene, as the copy number is maintained at a lower level, approximately 20–100 copies per cell.

19.2 RECOMBINANT VIRUS VECTORS

In theory, all viruses can exploited for the transgene expression. Perhaps due to practical reasons, retrovirus, lentivirus, and adenovirus are commonly used for transgene expression in research laboratories. Further, these viral vectors are ones used for gene therapy as well. As a matter of fact, the principle behind both transgene expression vectors and gene therapy vectors are the same. Here, the principle of recombinant viral vectors will be described. Then, the design of retrovirus vectors and lentivirus vectors will be described. Adenovirus vector will be detailed in the section on gene therapy vectors.

19.2.1 Principles

The viral factors essential for viral genome replication are of two kinds: *cis-acting elements* and *trans-acting factors*.[5] The principle underlying the design of recombinant virus (or gene therapy vectors) is to separate these two factors into two disparate plasmids: (1) the vector plasmid and (2) the helper plasmid. Specifically, the transgene is inserted into the vector plasmid that contains all *cis*-acting elements required for the viral genome replication. This vector plasmid is often referred to as the "*gene transfer plasmid.*" On the other hand, the viral proteins that are deleted in the vector plasmid can be provided by the *helper plasmids*. The helper plasmids should not contain the packaging signal so that it is not packaged. Alternatively, a *packaging cell* line can be used to provide the missing viral proteins, instead of the helper plasmids.

When cells are transfected with two plasmids, the helper plasmid will provide the viral proteins essential for the viral genome replication and assembly, while the vector plasmid encoding the transgene will be replicated and packaged

5. ***cis*-acting elements and *trans*-acting factors** The former refers to factors that are not diffusible (ie, the sequence element present in the plasmid), while the latter refers to factors that can be diffusible (ie, proteins).

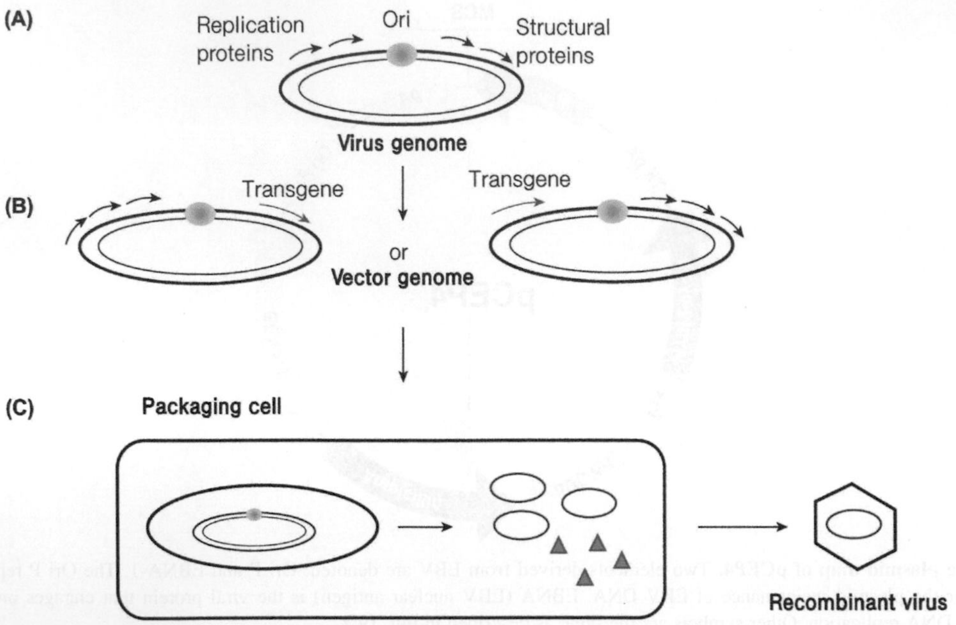

FIGURE 19.4 A diagram illustrating the underlying principle of recombinant viral vectors. (A) The map of virus genome. *Trans*-acting proteins and *cis*-acting elements essential for viral replication are denoted. Ori, the origin of viral genome replication. (B) The map of virus vector DNA. The recombinant viral vector is made by substituting a set of viral genes (either genes for the replication proteins or the structural proteins) with a transgene. The viral vector should contain all *cis*-acting elements essential for the viral genome replication (ie, Ori and packaging signal). (C) Recombinant virus production. The recombinant virus containing the transgene is produced when the recombinant vector is transfected into the packaging cell line that complements the viral genes deficient in the recombinant viral vector.

by the viral proteins derived from the helper plasmid (Fig. 19.4). In other words, *complementation* between two plasmids (ie, one harboring a *cis*-acting element and another expressing a *trans*-acting factor) is the underlying principle for the generation of the recombinant virus vectors.

19.2.2 Retrovirus Vectors

Retrovirus has been considered to be an ideal viral vector for gene therapy, since the viral genome becomes integrated into the chromosome and maintained stably upon cell division. In fact, retroviral vectors are widely used for transgene expression in many laboratories. Two kinds of retroviral vectors are available: (1) murine leukemia virus (MLV)-derived vector and (2) HIV-derived lentivirus vector. MLV is a prototype of a simple retrovirus, whereas HIV is a prototype of a complex retrovirus or *lentivirus*. MLV, a simple retrovirus, cannot infect a resting cell (ie, nondividing cells), while lentivirus can infect a resting cell. Because of the capability of lentivirus to infect nondividing cells as well as dividing cells, the lentiviral vector is now widely used. For the sake of simplicity, MLV vector will be used to describe the principle of retroviral vectors, and some salient features of lentivirus vector will be subsequently described.

The genome organization of a simple retrovirus is suitable for vector design. Specifically, all *cis*-acting elements and *trans*-acting factors are readily separable from each other on the genome, because all the *cis*-acting elements are positioned within two long terminal repeats (LTRs) (ie, 5′ LTR and 3′ LTR), while all viral genes (ie, *trans*-acting) are positioned between two LTRs (see Fig. 17.3). The strategy for the retrovirus vector is to separate *cis*-acting elements and *trans*-acting factors into two different plasmids (see Fig. 19.4). The gene transfer vector is made by replacing the viral *gag*, *pol*, and *env* genes with a transgene. Because retrovirus has a RNA genome, the proviral DNA genome is utilized as a vector DNA for the insertion of a transgene (Fig. 19.5). To generate the recombinant virus, the transfer plasmid is transfected into packaging cells, where Gag, Pol, and Env proteins are constitutively expressed. In packaging cells, the viral RNA encoding the transgene is transcribed from 5′ LTR promoter, and the viral RNA is then packaged into the retroviral vector via the recognition of the packaging signal (Ψ). Such generated recombinant retrovirus retains the ability to infect target cells and to express the transgene. Upon infection of target cells, the viral RNA will be reverse transcribed and integrated into the chromosome as a proviral DNA (Fig. 19.6). In target cells, the viral RNA encoding the transgene will be transcribed, and the transgene will be expressed.

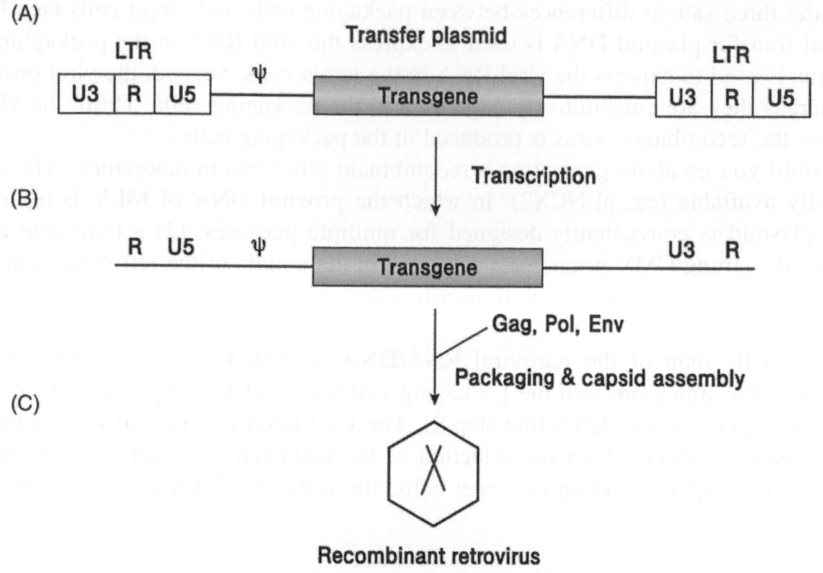

FIGURE 19.5 Steps involved in the generation of recombinant retrovirus. (A) The map of retroviral transfer plasmid is drawn in the context of the proviral DNA. Transgene is inserted between LTRs. (B) The map of retroviral RNA. The retroviral RNA is packaged by three retroviral proteins (Gag, Pol, and Env), which are provided either by the helper plasmid or by packaging cell line. (C) Recombinant retrovirus. Two RNA genomes are packaged inside the recombinant retroviral particles. LTR, U3 + R + U5; Ψ, packaging signal.

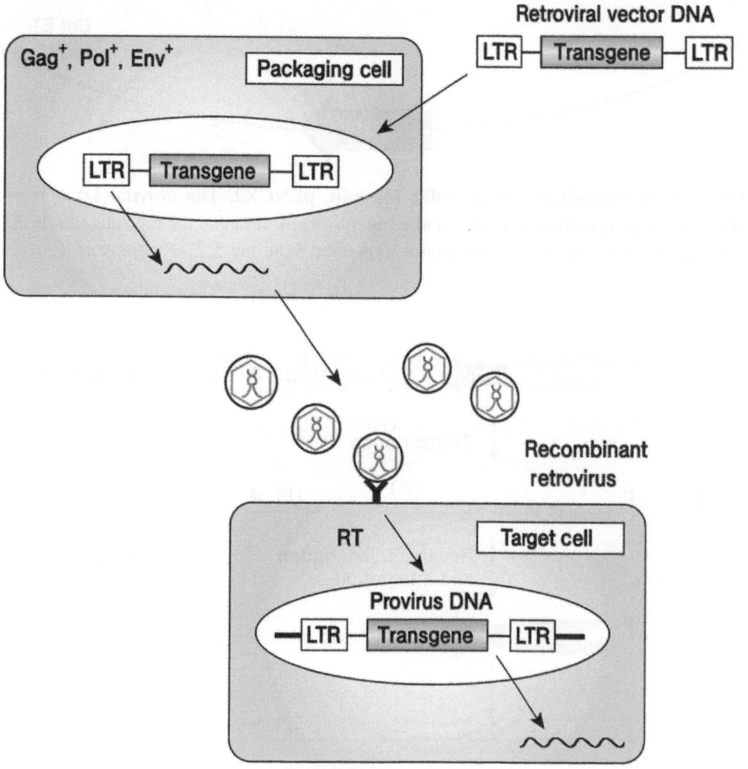

FIGURE 19.6 Recombinant retrovirus: packaging cell versus target cell. In a packaging cell, which constitutively expresses Gag, Pol, and Env proteins, the viral RNA encoding the transgene is packaged into the viral particles and the recombinant retrovirus particles are released. Upon entry into a target cell, the viral RNA is reverse transcribed to DNA in the cytoplasm. The viral DNA is then translocated into the nucleus, where it becomes integrated into the chromosome. In target cells, the transgene is expressed.

It is worth noting the three salient differences between packaging cells and target cells (see Fig. 19.6). First, transfection of the retroviral transfer plasmid DNA is used to express the viral RNA in the packaging cells, while infection of the recombinant virus is used to express the viral RNA in the target cells. Second, the viral proteins are not expressed in the target cells, whereas they are constitutively expressed in the packaging cells. Third, the virus is not produced in the target cells, whereas the recombinant virus is produced in the packaging cells.

Practically, how would you go about generating a recombinant retrovirus in laboratory? The retroviral gene transfer plasmid is commercially available (eg, pLNCX2), in which the proviral DNA of MLV is inserted into pUC plasmid (Fig. 19.7). pLNCX2 plasmid is conveniently designed for multiple purposes: (1) a transgene inserted into MCS site can be transcribed from the strong CMV promoter, positioned in the middle of the retroviral genome, instead of 5′ LTR promoter and (2) the neomycin-resistant marker transcribed from 5′ LTR promoter can be used for the selection of transfected cells.

Let's consider the genetic map of the retroviral RNA/DNA at each step (Fig. 19.8). Transfection of pLNCX2 plasmid (step 1) encoding the transgene into the packaging cell will lead to the production of recombinant retroviral particles that carry the transgene coding RNA (the step 2). The viral RNA is transcribed from the 5′ LTR promoter. In this process, cells are treated with G418 for the selection of the neomycin-resistant cells so that only cells that have been transfected can survive. Upon infection of target cells, the retroviral RNA is reverse transcribed and integrated

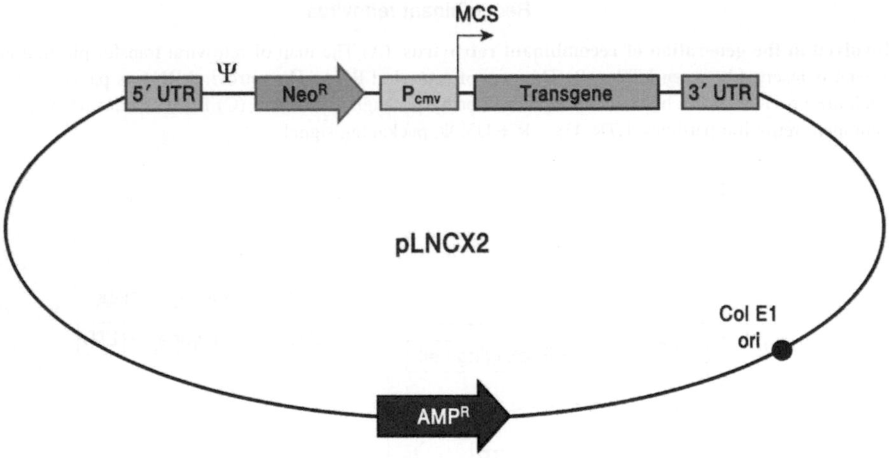

FIGURE 19.7 The map of MLV-based retroviral gene transfer plasmid, pLNCX2. The proviral DNA region drawn as a line on the top is inserted in pUC vector (Col E1 Ori for the propagation in *E. coli* and ampicillin-resistant gene for the selection in *E. coli*). The transgene is expressed from the CMV promoter, whereas a gene for the neomycin selection is expressed from the 5′ LTR promoter. P_{CMV}, cytomegalovirus promoter; MCS, multiple cloning site; Ψ, packaging signal.

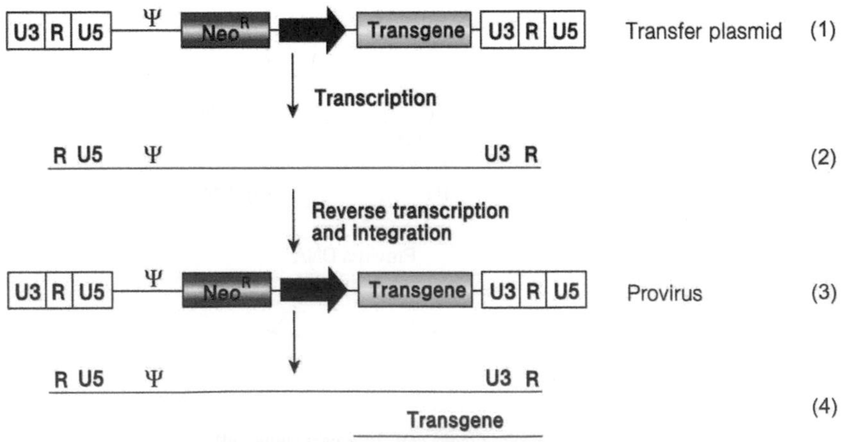

FIGURE 19.8 The map of retroviral DNA in retroviral gene transfer. The map of retroviral transfer plasmid and the proviral DNA in target cell. The ORFs for the neomycin marker (ie, Neo^R) and the transgene are denoted by boxes. The filled arrowed box denotes the CMV promoter. The viral RNA is denoted by red lines.

into the chromosome as the proviral DNA (step 3). In target cells, two RNAs are made: one from the 5′ LTR promoter and another from the internal CMV promoter (step 4). The transgene is expressed.

A few modifications were introduced to improve the gene delivery efficiency of the MLV-based retroviral vector deserve attention. First, the strong CMV promoter, instead of the weak 5′ LTR promoter, was introduced in pLNCX2 vector to express the transgene. In this case, two viral RNAs are expressed: one from the 5′ LTR promoter and another from the internal CMV promoter. It was found that when two promoters are arranged in tandem, the downstream promoter is attenuated, a phenomena dubbed "*promoter occlusion*."

One major drawback of the MLV-based retrovirus vector is that it infects only dividing cells. Recall that the simple retroviral capsid (ie, *PIC*[6]) enters the nucleus via the disruption of the nuclear membrane during mitosis (see chapter: Retroviruses). Since most cells in adults are nondividing cells, it cannot be used in vivo. Thus, the gene therapy is limited to cells that can be explanted and treated ex vivo. On the other hand, lentivirus can infect nondividing cells, as the lentiviral capsid (ie, PIC) enters the nucleus via a nuclear pore, independent of mitosis. Thus, a lentiviral vector is widely used in research laboratories.

19.2.3 Lentivirus Vectors

As stated above, one advantage of the lentiviral vector is that it can infect not only dividing cells but also nondividing cells. In fact, the lentiviral vector is a synonym of the HIV vector, because HIV, a prototype of lentivirus, is the only lentivirus that has been fully exploited for gene delivery. To improve the gene delivery efficiency, the retrovirus particles are frequently pseudotyped by *VSV-G protein* (Fig. 19.9). The VSV-G protein *pseudotyping*[7] of the lentivirus broadens its breadth of tissue tropism, being capable of infecting all kinds of tissue via the G protein. Note that VSV-G protein could mediate the entry of the pseudotyped particles via its receptor, a phosphatidyl serine (see Table 3.1).

In addition, a few additional *cis*-acting elements and *trans*-acting factors are required for the lentiviral vector. One additional viral protein (ie, Rev) is required for efficient HIV infection (Fig. 19.10); accordingly, RRE (ie, Rev-response element) is required. A set of two helper plasmids is often conveniently used to provide all proteins necessary to produce the pseudotyped lentiviral vector: one for Gag, Pol, and Rev, and another for the VSV-G protein.

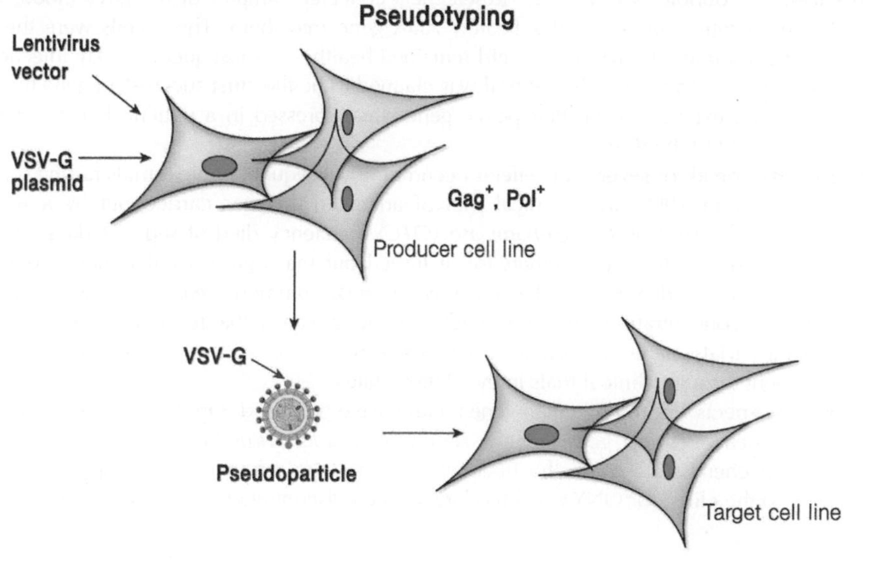

FIGURE 19.9 Pseudotyping of the lentivirus by VSV-G protein. The lentiviral vector plasmid is cotransfected with VSV-G protein expressing plasmid into a packaging cell line that constitutively expressed HIV Gag and Pol proteins. Then, the pseudoparticles that possess the envelope studded with VSV-G protein, are produced from the packaging cell. The VSV-G protein pseudotyped retrovirus particles can infect almost all kinds of cell by using VSV-G protein as ligand.

6. **PIC** Preintegration complex.
7. **Pseudotyping** Pseudotyping is the process of producing viruses or viral vectors in combination with foreign viral envelope proteins. The result is a pseudotyped virus particle. With this method, the foreign viral envelope proteins can be used to alter host tropism or an increased/decreased stability of the virus particles.

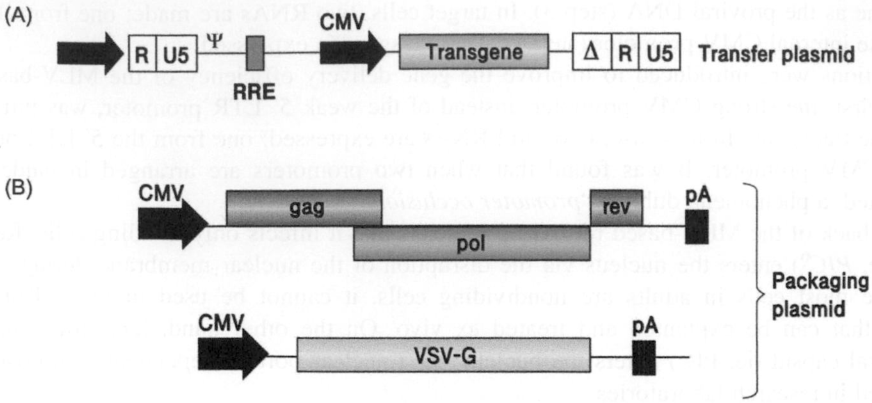

FIGURE 19.10 The map of lentiviral transfer plasmid and the helper plasmids. (A) Transfer plasmid. CMV, cytomegalovirus promoter; RRE, Rev-response element; Ψ, packaging signal. (B) Packaging plasmids. Three HIV proteins (ie, Gag, Pol, and Rev) are expressed from a helper plasmid, while VSV-G protein is conveniently provided by another plasmid. Δ denotes the U3 deletion.

BOX 19.1 Gene Therapy: Up and Down

It is fair to say that most human diseases are caused by the lack of a few specific genes related to the diseases. In theory, complementation of the missing genes by the wild-type could cure the diseases, if the gene is delivered to the diseased tissue effectively via appropriate vehicles. This kind of genetic therapy termed "gene therapy," as opposed to conventional "drug therapy," attracted a great deal of attention in the 1990s with the hope that it could cure many untreatable genetic diseases. However, multiple failures of clinical trials in patients have made investigators or regulatory authorities more cautious on the clinical implementation of gene therapy.

The first clinical trial of human gene therapy was carried out in 1990 by a group led by Dr French Anderson in National Institute of Health (United States). A 4-year-old girl suffering *severe combined immunodeficiency (SCID)* was treated with retrovirus vector harboring a gene for *adenosine deaminase (ADA)* that was deficient in the patient. ADA deficiency is one cause of SCID, which is also known as "bubble boy disease." Researchers collected samples of the girl's blood, isolated some of her white blood cells, and used a retrovirus to insert a healthy ADA gene into them. These cells were then injected back into her body and began to express a normal enzyme. The girl remained healthy by subsequent weekly injection of PEG-ADA therapy, a type of naked DNA therapy. This early clinical trial was claimed to be the "first success" of gene therapy. More precisely, the trial was a "success" to the extent that the therapeutic gene was expressed in a patient, but it was far from a "success" because the patient was not cured of the disease.

Despite the early "success," death or severe side effects occurred in subsequent clinical trials raising serious concerns on the safety of human gene therapy. In 1999, in a clinical trial of adenovirus vector carried out by a group in University of Pennsylvania, a young man with *ornithine transcarbamylase (OTC)* deficiency died of sepsis 3 days after treatment. It was a pity, because the patient was relatively healthy otherwise. It turned out that rigorous inflammation due to the administered adenovirus particle was the cause of death. OTC deficiency is the most common urea cycle disorder in humans. In severely affected individuals, ammonia concentrations increase rapidly causing ataxia, lethargy, and death without rapid intervention. The casualties of gene therapy trials made the regulatory authorities more cautious over clinical trials of gene therapy and this slowed down the advancement through clinical trials in the United States.

Since then, pessimistic prospects regarding human gene therapy have prevailed. On the other hand, a twist that reversed the pessimistic view was in the area of cancer gene therapy. An *oncolytic adenovirus* ONYX-15 was approved in 2005 in China for use in combination with chemotherapy for the treatment of late-stage refractory nasopharyngeal cancer (see Box 8.4). Outside of China, the push to the clinic for ONYX-015 has largely been discontinued for financial reasons.

19.3 GENE THERAPY VECTORS

Upon the completion of human genome sequencing, gene therapy was once considered to be a paradigm of future medicine, since it was then thought that all diseases could be treated or cured by delivery of the disease-related therapeutic genes. Since the first trials of gene therapy were carried out by researchers at National Institute of Health (NIH) in the United States in 1990, more than a hundred clinical trials were carried out until a decade ago (Box 19.1). Although some successful anecdotes were occasionally reported, the most of these clinical trials were soon discontinued due to unacceptable side effects. Gene therapy is not yet approved for clinical treatment.

TABLE 19.2 Characteristics of Gene Therapy Vectors

Vector	Packaging Size	Advantages	Disadvantages	Target Disease
Retrovirus	7–8 kb	Stable expression	Chromosomal integration	Genetic disorders
Adenovirus	10 kb	Higher expression	Transient expression, inflammation	Cancers
Adeno-associated virus (AAV)	4 kb	Stable expression	Transient expression, small packaging capacity	Genetic disorders
Nonviral vector	>10 kb	Packaging limit	Transient expression	–

The major concerns of gene therapy are "safety" and "efficacy." In short, no gene therapy protocols in the past clinical trials could meet both safety and efficacy criteria. Certainly, the pros and cons of the gene therapy vectors are different (Table 19.2). Importantly, the choice of gene therapy vector depends on the diseases.

The central issue of gene therapy is the ability of a vector to efficiently deliver genes to the target cell. Viral vectors are used for the vast majority of clinical trials (~80%), while nonviral vectors (ie, naked DNA or liposomes) are used for the rest (~20%). The viral vectors include retrovirus, adenovirus, and adeno-associated virus (AAV). Here, three viral vectors will be described. Then, a nonviral vector will be briefly described.

19.3.1 Retrovirus Vectors

Retrovirus has been considered to be an ideal viral vector for gene therapy, because the viral genome is stably integrated into the chromosome. In fact, retroviral vectors have been the most extensively used gene therapy vectors in the early stages of clinical trials. Currently, the use of retroviral vectors is limited to research purposes only. Nonetheless, the retroviral vectors are widely used in laboratories for the purpose of gene expression. The underlying principle of the retroviral vector is the same, whether it is for gene therapy or for basic research.

Because the principle for the retroviral vector has been described above, only issues pertaining to clinical application will be considered here. Two kinds of safety issue have been raised. First, retroviral vectors are designed to be replication-defective, but the replication-competent recombinant virus could be generated during the manufacturing process. In other words, recombination could occur between the transfer plasmid and the helper plasmid. Although only a trace amount of the replication-competent recombinant retrovirus is produced, it can be propagated in the target tissue, thereby causing side effects. Second, the consequence of random integration could end up causing other unwanted diseases such as cancer. For instance, the retroviral integration near cellular oncogenes could lead to the activation of the proto-oncogene (see Fig. 24.15). As a matter of fact, such an incidence did occur during a clinical trial of retroviral vector, a case that has darkened optimistic prospects in the future.

19.3.2 Adenovirus Vectors

Adenovirus is a DNA virus, having a 36 kb DNA genome possessing over 50 genes (see Fig. 8.3). Due to its large genome, only the subset of genes needs to be replaced by a therapeutic gene. Thus, unlike retrovirus vector, adenovirus vector still keeps the bulk of the virus genes. For instance, in an adenovirus vector, an essential E1 gene is replaced by a transgene (Fig. 19.11). This adenovirus vector should be produced in cells (ie, *HEK293 cell*) that stably express E1 protein (see Box 8.4). A recombinant adenovirus produced will express the therapeutic gene following entry to target cells. However, in the target cell, the adenovirus will not replicate due to the lack of the E1 gene.

The above-described adenovirus vector lacking the E1 gene is often called the "*first-generation adenovirus vector*." One flaw of the first-generation vector is that replication-competent recombinant adenovirus, although low level, is produced. To obviate this drawback, subsequently, the so-called *second-generation adenovirus vector* was created, which lacks not only the E1 gene but also the E2 and E4 genes. Eventually, to preclude the generation of the replication-competent recombinant viruses, the so-called *third-generation adenovirus vector* that keeps only the inverted terminal repeats (*ITR*) at both ends of the viral genome but lacks the entire coding region was generated.

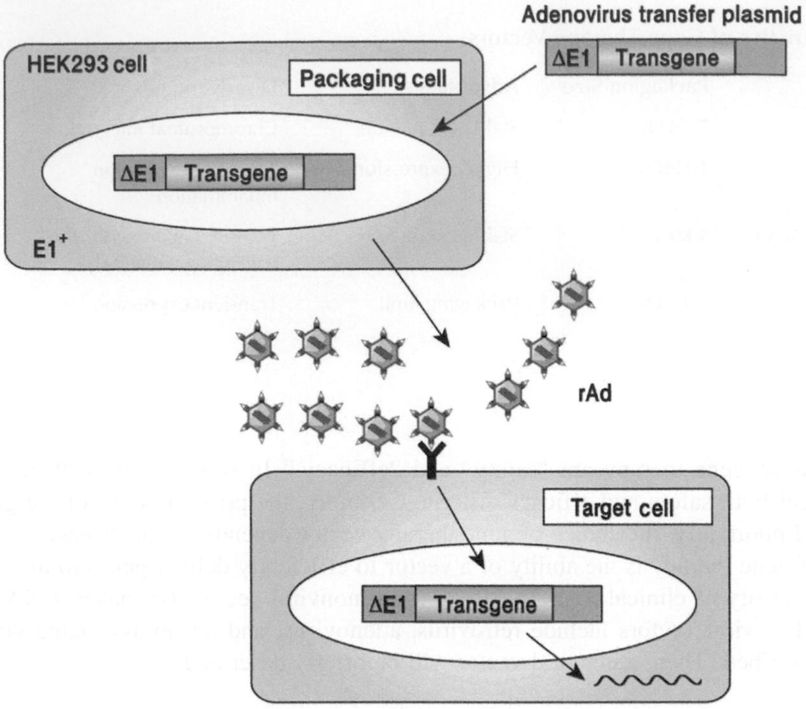

FIGURE 19.11 **Recombinant adenovirus vector.** Adenovirus transfer plasmid (ΔE1) is transfected to the packaging cell (ie, HEK293 cell) that expresses E1 protein. Then, the recombinant adenovirus (ie, rAd) containing the transgene is produced. Upon entry to target cell, the adenoviral DNA is translocated in the nucleus and expressed the transgene.

In this case, the adenoviral proteins are provided by a helper plasmid. Nevertheless, the generation of the replication-competent recombinant needs to be carefully monitored during the manufacturing process.

On the other hand, adenovirus vector has several shortcomings that make it inappropriate for gene therapy. First, the adenovirus particle itself elicits significant inflammatory response to an extent that is unacceptable for the treatment of genetic diseases. Conversely, because rigorous immune response to the adenovirus can be of merit for cancer therapy, it can be exploited for the treatment for cancer. Hence, currently, adenovirus vector is being developed for cancer gene therapy (see Box 8.4). Second, the duration of gene expression from adenovirus vector is only transient, as the viral DNA is not stably maintained in target cells. As a result, multiple administrations are essential to achieve the desirable therapeutic effect. Third, adenovirus vector needs to be injected directly to the target organs (eg, heart or lung), if these nonliver tissues are the target, because the vast majority of intravenously administered adenovirus vector goes to the liver. To obviate this limitation, attempts are made to redirect adenovirus vector to nonliver tissues. Specifically, a ligand protein is expressed as a fusion protein of the fiber of adenovirus particle. Such a modified adenovirus vector can target cells that express the receptor for that specific ligand. This innovative approach is termed "*retargeting*," as adenovirus vector is redirected to otherwise unsusceptible tissues.

19.3.3 AAV Vectors

AAV has drawn attention as a gene therapy vector, because AAV presents some merits that complement the shortcomings of retrovirus and adenovirus vectors. As a matter of fact, AAV vector is considered suitable for gene therapy of genetic diseases such as hemophilia and cystic fibrosis.[8] AAV possesses a small single-strand DNA genome (see Fig. 10.3). AAV is a kind of satellite virus of adenovirus, a helper virus. Hence, in nature, it coexists with a host

8. **Cystic fibrosis** Cystic fibrosis is a genetic disorder that affects the lungs. Affected individuals suffer difficulty breathing and coughing up sputum as a result of frequent lung infections. It is caused by the presence of mutations in both copies of the gene for the protein cystic fibrosis transmembrane conductance regulator.

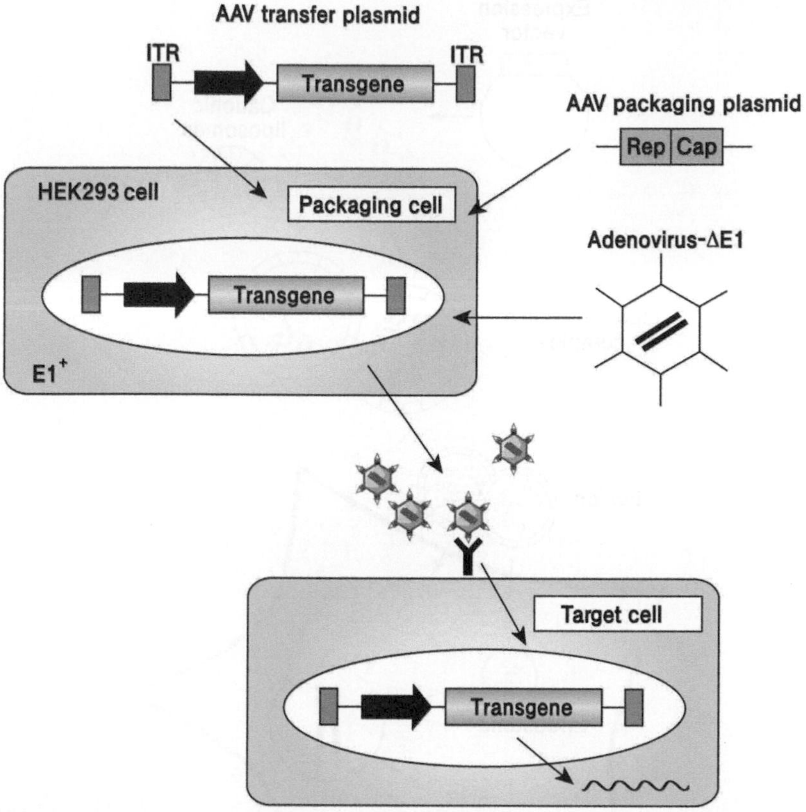

FIGURE 19.12 **Recombinant AAV vector.** Three components are transfected and infected into the packaging cell (ie, HEK293 cell): (1) AAV transfer plasmid, (2) AAV packaging plasmid, and (3) adenovirus lacking El gene, as a helper virus. ITR, inverted terminal repeat. Then, the rAAV particles containing the transgene are produced. In target cell, the rAAV vector DNA integrated into the chromosome and stably expresses the transgene.

virus (ie, helper virus), adenovirus. In the absence of a host virus, AAV causes latent infection, and its genome gets inserted into the chromosome at a specific site (ie, human chromosome 19) (see Fig. 10.4).

AAV genome is organized such that it can be readily used for gene therapy vector (see Fig. 10.3). In other words, *ITR*, an essential *cis*-acting element for the viral genome replication, is located at both termini of the AAV genome. Hence, the transgene can be conveniently inserted between two ITRs (Fig. 19.12). The viral *trans*-acting factors (ie, Rep and Cap proteins) can be complemented via a helper plasmid (ie, packaging plasmid). In addition to AAV proteins, a helper adenovirus is required for AAV replication. To produce AAA vector, these three components are transfected or infected into an HEK293 cell that is used for the production of adenovirus vector. Such produced recombinant AAV vector (ie, rAAV) contains the transgene. In a target cell, the rAAV will express the transgene stably following the chromosomal integration. In contrast to wild-type AAV, the rAAV DNA will integrate randomly with respect to the chromosome. The random integration of rAAV DNA is attributed to the lack of AAV Rep proteins in the target cells that are required for the site-specific integration.

A few merits of AAV vector stand out that complement the shortcomings of established retrovirus and adenovirus vectors. First, one merit of AAV vector is safety. AAV, a satellite virus, cannot replicate in the absence of a helper virus. Thus, the safety concerns raised by retrovirus and adenovirus vectors due to the generation of replication-competent recombinant virus is precluded. Another merit is the broadness of target tissue. AAV can infect nondividing cells as well as dividing cells, a feature that expands the target tissue range. The third merit of AAV vector is the stability of gene expression following integration into the chromosome. Last, unlike adenovirus, AAV does not induce significant inflammatory response. On the other hand, AAV has a few limitations. The packaging limit of AAV is relatively small (ie, about 4.5 kb) due to its small genome. However, since most therapeutic genes are smaller than 4.5 kb, the small packaging limit of AAV vector does not present a significant obstacle in most cases. A more significant concern is that AAV vector DNA integrates into the chromosome randomly. Subsequently, random integration could lead to insertional mutagenesis and chromosomal rearrangement in an unpredictable manner.

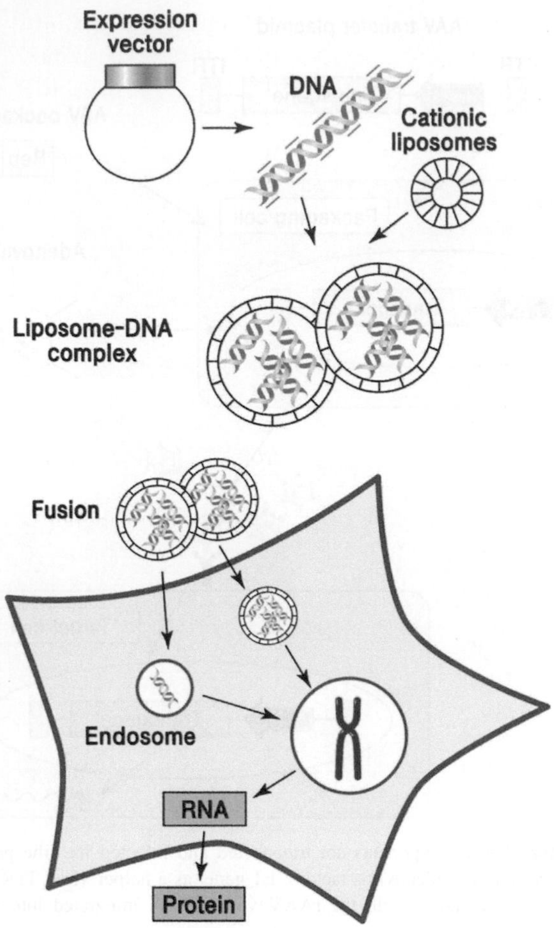

FIGURE 19.13 **Liposome-mediated gene delivery.** Polycationic reagents bind to DNA, encapsulated the DNA into liposomes. Liposomes, positively charged particles, bind to the anionic cell surface and are brought into the cytosol via endocytosis. Once the particle goes inside the cell, the protonation of the amines results in an influx of counter ions and a decrease of the osmotic potential. Osmotic swelling results and bursts the vesicle releasing the polymer-DNA complex (polyplex) into the cytoplasm. Then, the DNA is free to diffuse into the nucleus.

19.3.4 Nonviral Vectors

In addition to the viral vectors, nonviral vectors such as liposomes or naked DNA can be used for gene therapy. Cationic polymers such as *polyethylenimine* (PEI) are used to prepare liposomes. Cationic polymers can readily mix with anionic DNA, facilitating the introduction of DNA into cells (Fig. 19.13). The PEI/DNA complex enters the cell via liposome-mediated endocytosis. The advantages of liposome-mediated delivery include the lack of immune response and the unlimited packaging limit (see Table 19.2). The drawback of liposome is the intrinsic lower efficiency of gene delivery. Since such delivered DNA is not stably maintained in the nucleus, multiple treatments are essential to attain therapeutic effect.

Besides the liposomes, intramuscular injection of plasmid DNA itself is used for gene delivery. This approach is termed "*naked DNA*," as no vehicle is used. In practice, a tool called a "gene gun" is used to inject DNA. This bold idea was conceived by an accidental discovery of immune response (T-cell immunity) to the antigen encoded in the plasmid injected as a negative control in animal experiment. Presumably, the lower level of the protein expression from injected DNA is sufficient to elicit cellular immunity. The simple nature of technology has drawn some attention. However, the efficacy is limited in that the induced T-cell immunity, although detectable, is not sufficient to attain therapeutic benefit. In addition, the naked DNA technology is being explored as a DNA vaccine for the prevention of certain virus infections, such as HIV, where conventional vaccines have been proven to be ineffective.

19.4 PERSPECTIVES

Virus is unique in that it can efficiently deliver its genome to target cells. This feature can be aptly explored for gene delivery for the treatment of diseases. Thus, virus has been a favorite model for the development of gene therapy vector. Although gene therapy is still in the early stages of development, viral vectors are extensively used in research laboratories. In particular, lentivirus vector is conveniently used in many laboratories for experimental gene expression in mammalian cells. Moreover, "*oncolytic viruses*" have drawn attention. These include adenovirus, herpesvirus, poxvirus, reovirus, and measles virus. The aim is to exploit viruses that are capable of killing cancerous cells, but not normal cells. In particular, the adenovirus, dubbed "*ONYX-015*," is undergoing clinical trials for cancer gene therapy (see Box 8.4). Cancer gene therapy by using oncolytic virus is becoming a hope for patients whose cancer cells are hopelessly resistant to chemotherapy. It can be said that modern technology has allowed us to subvert an "enemy" to defeat an "enemy."

19.5 SUMMARY

- *Gene therapy*: Animal viruses have been exploited to deliver transgenes to a target cell. Retrovirus, adenovirus, and AAV are explored for gene delivery purposes.
- *Retroviral vector*: The genome organization of simple retrovirus is suitable for the vector design, as *cis*-acting elements and *trans*-acting factors are readily separable from each other on the genome. MLV-based vector is a prototype of retrovirus vector.
- *Lentiviral vector*: Lentivirus vector is extensively utilized in research laboratories for the purpose of ectopic gene expression. VSV-G protein psuedotyped lentivirus vector can infect a broad range of mammalian cells.
- *Adenovirus vector*: Adenovirus is considered suitable for cancer gene therapy due to its ability to elicit rigorous inflammatory response. An oncolytic adenovirus has been approved in China for use in combination with chemotherapy.
- *AAV vector*: AAV vector is considered suitable for genetic diseases due to its intrinsic ability of stable gene expression.
- *Nonviral vectors*: Nonviral vectors such as liposomes or naked DNA are used for the gene therapy purpose.

STUDY QUESTIONS

19.1 In the retroviral gene transfer, in addition to the transgene RNA, the full-length retroviral genomic RNA is expressed, as depicted in Fig. 19.8. How would you modify the retroviral transfer vector in order to avoid the expression of the full-length retroviral genomic RNA?

19.2 Explain the differences between the retroviral DNA present in a packaging cell and the retroviral DNA present in a target cell. What is the merit of a pseudotyped retrovirus vector?

19.3 Adenovirus vector is more suitable for cancer gene therapy. Explain why. Explain why HEK293 cell is used for the preparation of AAV as well as adenovirus vector.

SUGGESTED READING

High, K.A., 2012. The gene therapy journey for hemophilia: are we there yet? Blood. 120 (23), 4482–4487.

Kirn, D.H., Thorne, S.H., 2009. Targeted and armed oncolytic poxviruses: a novel multi-mechanistic therapeutic class for cancer. Nat. Rev. Cancer. 9 (1), 64–71.

McCormick, F., 2003. Cancer-specific viruses and the development of ONYX-015. Cancer Biol. Ther. 2 (4 Suppl 1), S157–160.

Parato, K.A., Senger, D., Forsyth, P.A., Bell, J.C., 2005. Recent progress in the battle between oncolytic viruses and tumours. Nat. Rev. Cancer. 5 (12), 965–976.

Russell, S.J., Peng, K.W., Bell, J.C., 2012. Oncolytic virotherapy. Nat. Biotechnol. 30 (7), 658–670.

JOURNAL CLUB

- High, K.A., 2012. The gene therapy journey for hemophilia: are we there yet? Blood 120 (23), 4482–4487.

  Highlight: Hemophilia is a genetic disease that is caused by a gene for factor VIII or IX deficiency. A recent clinical trial of hemophilia patients who underwent intravenous infusion of an AAV vector expressing factor IX demonstrated successful conversion of severe hemophilia to mild disease in six adult patients. This review article discusses the obstacles and safety issues that need to be addressed before more wide spread used of this AAV vector gene therapy.

Chapter 20

Subviral Agents and Prions

Chapter Outline

A number of virus-like transmissible agents have been discovered, which do not comply with the classical definition of "virus". These virus-like agents cannot be classified as viruses in the strict sense; nonetheless, they were first considered as "viruses" until their nonviral features were unfolded later. Now, these virus-like transmissible agents are collectively termed "subviral agents." (see Table 1.4).

20.1 SUBVIRAL AGENTS

Subviral agents are composed of three kinds: satellite viruses, viroids, and prions (Table 20.1). These transmissible agents are classified as subviral agents as they are less than a virus in some respects. The first subviral agent is the "satellite virus," which is morphologically indistinguishable from ordinary virus particles, but it depends on another virus, a host or helper virus, for propagation. Therefore, a satellite virus is often called "a parasite of a parasite," as it relies on another parasite, a virus. The second subviral agent are "viroids" found in plants. Viroids are comprised of "RNA only," and are devoid of any protein component. Remarkably, it is a circular RNA molecule itself about only 0.3 kb in length. Nonetheless, the viroid RNA is transmissible and causes pathogenic lesions in plants. The third subviral agent are "prions," which are etiological agents of neurodegenerative diseases, such as mad cow disease. In contrast to the viroids, prions are transmissible agents that are comprised of "protein only," and are devoid of any nucleic acid components.

20.2 SATELLITE VIRUSES

Satellite viruses[1] are mainly found in plants. The replication of a satellite virus depends on a helper virus, while the replication of the helper virus does not depend on the satellite virus. Importantly, no sequence homology is found between host virus and satellite virus, implicating no genetic relatedness of satellite viruses to their hosts.

Satellite viruses found in plants typically possess their own capsid (Table 20.2). For instance, tobacco mosaic virus (TMV) is a helper virus of satellite tobacco mosaic virus (STMV). Unlike TMV, which has a long helical capsid, STMV has a spherical capsid. Satellite tobacco necrosis virus (STNV), the first plant satellite virus, was discovered as a virus-like particle abundantly present in cultured medium of tobacco necrosis virus (TNV), a host virus. Its identity as a satellite virus was revealed by its dependence on TNV for propagation. STNV encodes only one protein, a capsid protein. Its host dependence is specific for TNV. In other words, host dependence of satellite viruses is generally specific. STNV does not encode its own RNA-dependent RNA polymerase (RdRp), a helper function that is presumably provided by TNV.

1. **Satellite virus** It refers to a virus, which depends on the host virus as a helper.

Molecular Virology of Human Pathogenic Viruses. DOI: http://dx.doi.org/10.1016/B978-0-12-800838-6.00020-5

TABLE 20.1 The Major Features of Subviral Agents

Features	Satellite Virus	Viroids	Prions
Nucleic acid	RNA or DNA	RNA	No
Protein-coding	Yes (capsid)	No	Yes (host)
Protein in particles	Yes (capsid)	No	Yes (PrP)
Helper-dependency	Yes	No	No
Infectivity	Yes	Yes	Yes
Disease	Yes	Yes	Yes

TABLE 20.2 Satellite Viruses in Plants

Plant Satellite Virus	Genome Size (nt)	Protein
Satellite tobacco necrosis virus (STNV)	1239	Capsid (20 kDa)
Satellite tobacco mosaic virus (STMV)	1059	Capsid (18 kDa)
Satellite panicummosaic virus (SPMV)	826	Capsid (18 kDa)
Satellite maize whited line tobacco mosaic virus (SMLMV)	1168	Capsid (24 kDa)

In contrast to the abundance of satellite viruses in plant viruses, only a few satellite viruses are found in animal viruses. Two representative satellite viruses among human viruses are adeno-associated virus (AAV) and hepatitis delta virus (HDV). AAV is a satellite virus of adenovirus (Fig. 20.1), while HDV is a satellite virus of hepatitis B virus (HBV) (see Fig. 23.13). AAV depends on adenovirus for its genome replication (see Fig. 10.4), while HDV relies on HBV for its envelope glycoprotein (ie, HBsAg).

20.3 VIROIDS

Viroids are the smallest infectious pathogens known, comprised solely of a short circular RNA without protein coats. Viroids are plant pathogens with economic importance. Viroid genomes are extremely small in size, only about 300 nucleotides. Viroids have been found in agricultural products, such as potatoes, tomatoes, apples, and coconuts. Several viroids-induced diseases are of considerable economic importance (Table 20.3). For example, yield losses can be high in potatoes infected with potato spindle tuber viroid (PSTVd), citrus infected with Citrus exocortis viroid, coconut palms infected with Coconut cadang-cadang viroid (CCCVd), and avocado infected with Avocado sunblotch viroid (ASBVd) (Fig. 20.2). Exclusion or eradication of infected material is the most effective means of controlling viroid diseases. Viroid infections are transmitted by cross-contamination following mechanical damage to plants as a result of horticultural or agricultural practices. Some are transmitted by aphids and they can also be transferred from plant to plant by leaf contact.

Viroids, the first known representatives of a new domain of "subviral pathogens," stand out in many respects. First, viroids are the only infectious agents that lack protein components such as capsids. In general, the role of the viral capsid is the protection of the viral genome from degradation. Second, the viroid has a circular RNA genome, unlike most RNA viruses. Moreover, the viroid is not included in the Baltimore classification (see Fig. 1.9). The circular configuration confers resistance to exonucleases on the viroid RNA, a feature that makes the protein coat dispensable. Notably, human HDV also has a circular RNA genome (see chapter: Hepatitis Viruses); hence, HDV is often called

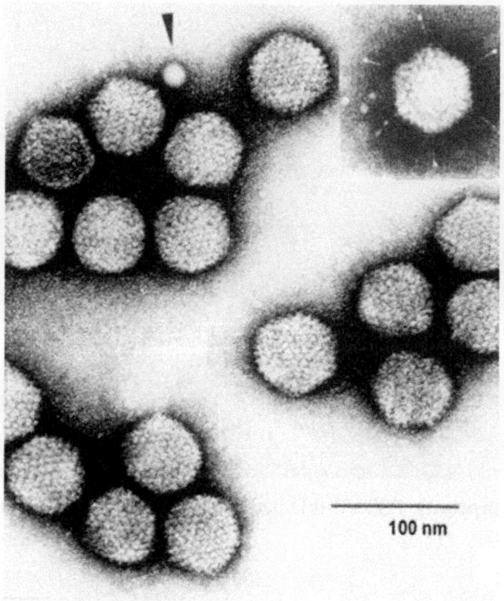

FIGURE 20.1 Electron micrograph of adenovirus and its satellite virus, adeno-associated virus (AAV). Electron microscopic image of AAV particles. AAV particle (arrow head) is seen in the midst of adenovirus particles, a helper virus. Note that the morphology of AAV, a satellite virus, is distinct from that of adenovirus particle, a host virus.

TABLE 20.3 The Main Features of Plant Viroids

Viroid	Abbreviation	Size (nt)
Potato spindle tuber viroid	PSTVd	356–360
Coconut cadang-cadang viroid	CCCVd	246, 247
Tomato apical stunt viroid	TASVd	360
Apple scar skin viroid	ASSVd	360
Avocado sunblotch viroid	ASBVd	246–250

"viroid" in animals. Intriguingly, both viroids and HDV possess a rod-shaped RNA genome, in which about 70% of bases are base-paired (Fig. 20.3). Third, viroids do not code for any proteins. The genome size is indeed too small to encode any functional protein. Fourth, despite the small genome, a helper virus is not required for viroids. In contrast, HDV encodes one viral protein (ie, delta antigen) and relies on the envelope protein of HBV for the virion assembly (see Fig. 23.13). Hence, viroids are not satellite viruses. Overall, the viroid does not fit into the classical definition of "virus," although it is a "submicroscopic infectious agent" having a nucleic acid genome.

Then, how does a viroid replicate its RNA genome? The so-called rolling-circle mechanism has been proposed to account for the RNA genome replication of the viroids (Fig. 20.4). Note that the rolling-circle mechanism has been proposed for herpesvirus and bacteriophage lambda (see Box 10.2). According to the mechanism, by using the infected circular RNA (+), the multimeric linear RNA (−) is synthesized. Then, the multimeric linear RNA is converted to a circular RNA (−) following cleavage to a genome-length. Such yielded circular RNA (−) is then used as a template for the synthesis of the (+) strand RNA.

At least three enzymatic activities are required to fulfill the rolling-circle mechanism for the viroid replication. First, RdRp is needed to carry out the RNA synthesis. Since hosts do not code for RdRp and the viroids do not code for any protein, the identity of the RdRp seems enigmatic. Intriguingly, the viroid RNA replication is α-*amanitin*-sensitive, which potently inhibits cellular RNA polymerase II. An interpretation is that RNA polymerase II, which normally utilizes a DNA template, uses the viroid's RNA as a template. In other words, the viroid subverts the host DNA-dependent RNA polymerase to replicate its RNA genome. It is believed that RNA polymerase II recognizes the

FIGURE 20.2 Potatoes with spindle tuber symptoms. PSTVd-infected tubers may be small, elongated, from which the disease derives its name, misshapen, and cracked.

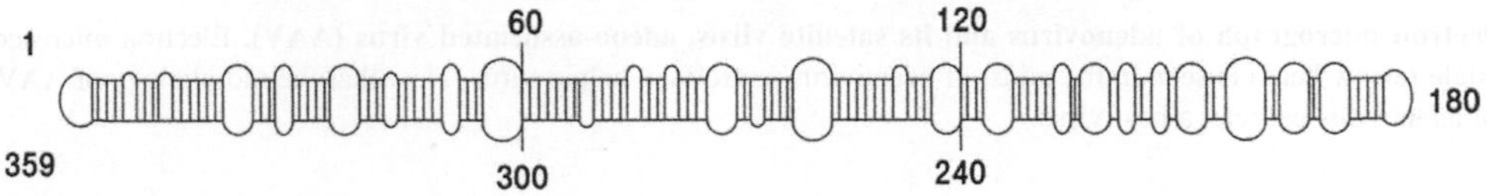

FIGURE 20.3 A secondary structure of plant viroids. A rod-shaped secondary structure of a circular RNA genome of PSTV is drawn. The numbers indicate the nucleotide number of the PSTV genome. The base-pair is denoted by lines.

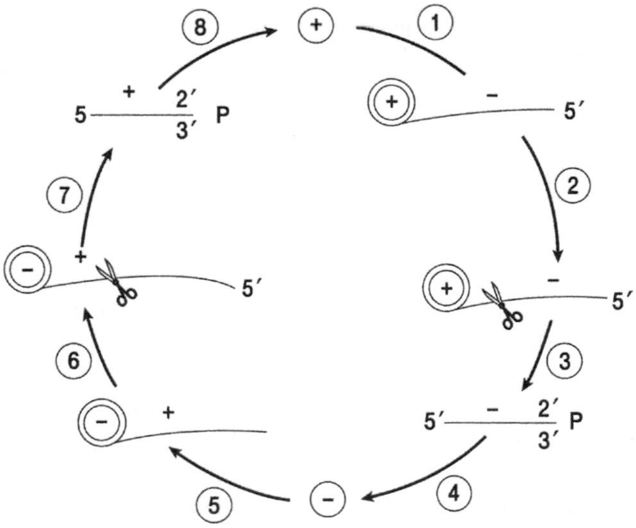

FIGURE 20.4 A rolling-circle mechanism of viroid replication. (Steps 1–2) Multimeric negative-strand RNA is synthesized by using infected viroid RNA (+) as template. Here, the infected viroid RNA is defined as positive-strand (+). (Step 3) The multimeric RNA is cleaved. The cleavage site is denoted by a scissor. P denotes the 2′, 3′ phosphate group at the cleavage site. (Step 4) The unit length RNA is then circularized to an antigenomic RNA (−). (Steps 5–8) The synthesis of positive-strand RNA by using antigenomic RNA is carried out, as steps 1–4.

rod-shaped secondary structure of the viroid RNA, mimicking the double-strand DNA, as a template. It is intriguing that the viroids explore cellular DNA-dependent RNA polymerase for their RNA replication, although the notion has not been formally proven.

Another enzyme required for the viroid RNA replication is an endonuclease that cleaves the multimeric linear RNA to a genome-length and also an RNA ligase that circularizes the cleaved genome-length RNA. Surprisingly, these two enzyme activities are attributed to the viroid RNA itself. In fact, the *ribozyme* (ie, RNA enzyme) encoded in the viroid RNA catalyzes these two enzymatic reactions. Specifically, the secondary structure of viroid RNA, dubbed "hammer head," carries out the sequence-specific endonuclease reaction (Fig. 20.5). Following the cleavage, the rod-shaped

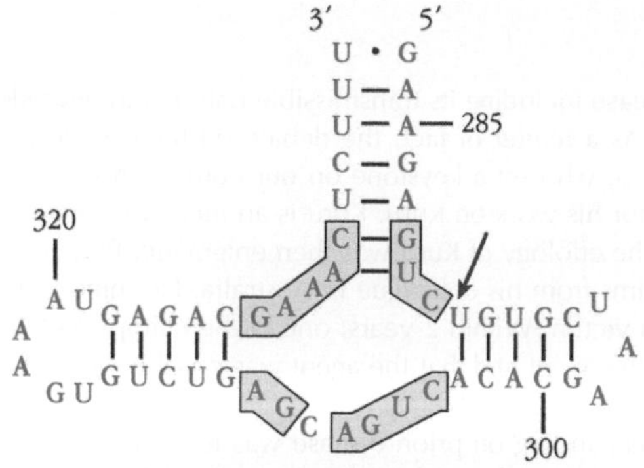

FIGURE 20.5 Predicted secondary structure of hammerhead ribozyme of plant viroids. The nucleotides conserved in all hammerhead ribozymes are boxed and shaded, the site of cleavage is marked by an arrow. Note that the ribozyme acts only once, because the segment harboring the cleavage site constitutes the catalytic unit of the ribozyme. The 5′ and 3′ termini of the ribozyme segment are denoted.

secondary structure of the viroid RNA promotes the ligation reaction to form a circle. Nonetheless, many fundamental questions regarding the viroids remain unresolved. For instance, how could viroid RNA infect plants without having a protein component? What is the underlying mechanism for the pathogenesis? What are the origins of viroids?

20.4 PRIONS AND MAD COW DISEASE

Prion[2] is an etiologic agent of transmissible neurodegenerative diseases. In contrast to viroids, prions are constituted of "protein only" and have no nucleic acid components. The prion disease best known to the public is the so-called mad cow disease. Here, the features of prion diseases are described with an emphasis on mad cow disease.

Discovery of Prions: Prion is a transmissible agent that was first isolated from a sheep afflicted by a disease termed "*scrapie*,"[3] which is a transmissible spongiform encephalopathy (TSE) of sheep. Scrapie was described as early as the 18th century in England, France, and Germany, but it only became a subject of scientific investigation after the 1990s. Scrapie was first known as an infectious disease in farm animals. The characterization of the scrapie agent as a "protein only" agent by Stanley Prusiner brought debates on the biological nature of the agent. He named it a "proteinaceous infectious agent" or "prion" to emphasize its lack of a nucleic acid component. The prion hypothesis was a "revolutionary thought," and it is still not widely accepted (Box 20.1).

Prions Diseases: Fatal neurodegenerative diseases similar to scrapie were found in human as well as in animals (Table 20.4). Histological examination of diseased brain tissues revealed the presence of vacuoles (ie, microscopic "holes" in the gray matter), which is seen as a sponge-like appearance (Fig. 20.6A). Reflecting the histological feature, these neurodegenerative diseases are collectively termed "TSE." Moreover, *amyloid*[4] fibrils were observed by electron microscopic examination of samples isolated from the tissues (Fig. 20.6B). It is speculated that amyloids are responsible for the pathological lesions found in TSE. On the other hand, all experimental evidence supports the lack of nucleic acids (DNA or RNA) in the agents, which leads to the "prion hypothesis."

Some pathological features are notable. First, spongiform encephalopathy, which is the loss of neuronal cells, represents the disease pathology. Amyloid plaque observed in the afflicted brain tissue is thought to be responsible for the cell death. Second, unlike most other infectious diseases, the lesions are not associated with inflammation. Third, the latency, a period from the primary infection to the onset of disease, is characteristically long (eg, many years). This pattern of infection is termed "slow infection," as exemplified in HIV infection (see Box 9.1). Fourth, the disease outcome is always fatal.

Importantly, prion diseases were found in human as well (see Table 20.4). The first human prion disease reported was "*Kuru*" that was described in the 1950s. Kuru was transmitted among members of the Fore tribe of Papua New Guinea via funerary cannibalism (Fig. 20.7). The etiological nature of the fatal disease endemic to tribal regions of

---

2. **Prion** The word *prion*, is derived from the words *protein* and *infection*, in reference to a prion's ability to self-propagate and transmit its conformation to other prions.

3. **Scrapie** It refers to a fatal, degenerative disease that affects the nervous systems of sheep and goats. The name "scrapie" is derived from one of the clinical signs of the condition, wherein affected animals will compulsively scrape off their fleeces against rocks, trees, or fences.

4. **Amyloid** It refers to the insoluble fibrous "protein aggregates" sharing specific structural traits.

BOX 20.1 Prion Hypothesis

A few outstanding features of prion disease including its transmissible nature and neurodegenerative disease manifestation made scientists interested in prion diseases. As a matter of fact, the debate on their etiological nature is still ongoing. Nobel Prizes were awarded to two visionary scientists, who set a keystone on our current knowledge of prion disease. First, Peter Gajdusek was awarded the Nobel Prize in 1976 for his work on Kuru. Kuru is an incurable degenerative neurological disorder endemic to tribal regions of Papua New Guinea. The etiology of Kuru was then enigmatic. Peter Gajdusek became interested in the disease and obtained brain tissues of Kuru victims from his colleague in Australia. He injected two chimpanzees with the materials isolated from brain tissue samples of Kuru victim. Within 2 years, one of the chimps had developed Kuru, demonstrating that Kuru is transmissible through inoculation of the agent and that the agent was capable of crossing the species barrier to other primates. This work defines the etiology of Kuru as a transmissible agent.

Another major advance in our understanding on prion disease was made by Stanley Prusiner. Based on early experimental evidence revealing the lack of nucleic acids in scrapie sample, he postulated that "proteineous infectious" materials are etiologic entities or the so-called the protein-only hypothesis.

In this work, he coined the term *prion*, which comes from the words "proteinaceous" and "infectious," in 1982 to refer to a previously undescribed form of infection due to protein misfolding. Stanley Prusiner received the Nobel Prize in Physiology or Medicine in 1997 for his work in proposing an explanation for the cause of BSE (mad cow disease) and its human equivalent, Creutzfeldt-Jakob disease.

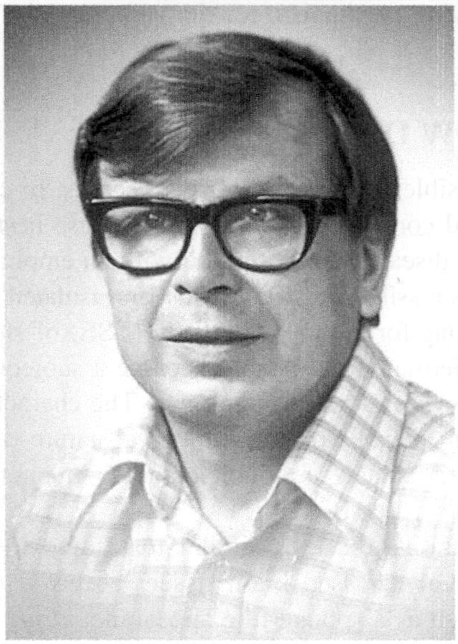

The photos of two pioneers in prion research. (Left) Peter Gajdusek and (right) Stanley Prusiner.

Papua New Guinea was then enigmatic. It was Peter Gajdusek, who first demonstrated Kuru in chimpanzee by injecting infected materials, revealing the transmissible nature of the disease (see Box 20.1).

Although Kuru was the first human prion disease described, *Creutzfeldt-Jacob disease (CJD)* and *Gerstmann-Straussler syndrome (GSS)* are more prevalent prion diseases in human (see Table 20.4). CJD (sporadic CJD or sCJD) is the most common prion disease in human, nonetheless, only one in million per year is afflicted. CJD can be familial (ie, fCJD). Most cases of CJD (ie, 85%) are sporadic, while a fraction (ie, 10−15%) are familial. In the familial prion disease, a mutation occurs in the gene for PrP (ie, PRNP gene). GSS is much rarer (one in 10−100 million per year) and it is familial and associated with a certain mutation in the gene for PrP (ie, P102L mutation or proline to leucine substitution in amino acid number 102). In addition to these human prion diseases, a novel CJD was more recently reported, in which the disease pathology is somewhat distinct from CJD. It was later termed a *variant CJD* (vCJD).

Mad Cow Disease: Until the 1970s, scrapie was the representative prion disease in animals. Prion diseases in other animals, such as transmissible mink encephalopathy (TME) in minks and chronic wasting disease (CWD) in deer and elk, did not get much attention, because the cases were extremely rare. However, prion disease in cow, termed "mad cow disease," first reported in 1985 in the United Kingdom drew a great deal of public attention and became the

TABLE 20.4 The Important Features of Human and Animal Prion Diseases

Human Prion Diseases	Features (Cause, Epidemiology)	Symptoms
Creutzfeldt-Jacob disease (CJD)		
Sporadic CJD	Sporadic, aging	Dementia
Familial CJD	PrP mutation	Dementia
Variant CJD (vCJD)	Human mad cow disease	Progressive dementia
Gerstmann-Straussler syndrome (GSS)	PrP mutation	Fatal neurodegenerative disease
Kuru	Cannibalism (Papua New Guinea)	Fatal neurodegenerative disease
Fatal familial insomnia	PrP mutation	Insomnia
Animal prion diseases	**Host**	**Symptoms**
Scrapie	Sheep, goat	Fatal neurodegenerative disease
Bovine spongiform encephalopathy (BSE)	Cattle	Fatal neurodegenerative disease
Transmissible mink encephalopathy (TME)	Mink	Fatal neurodegenerative disease
Chronic wasting disease (CWD)	Deer and Elk	Fatal neurodegenerative disease

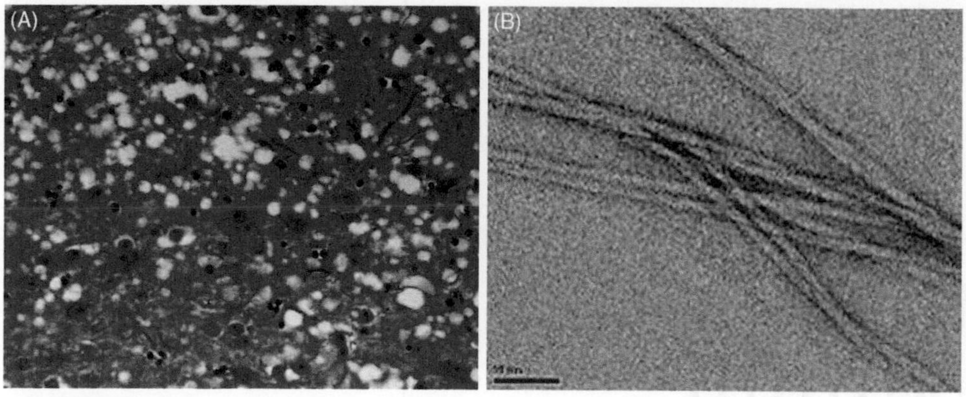

FIGURE 20.6 Pathological lesions in prion diseases and amyloid fibril. (A) This micrograph of brain tissue reveals the histologic lesions seen in BSE. The presence of vacuoles (ie, microscopic "holes" in the gray matter) gives the brain of BSE-affected cows a sponge-like appearance. (B) Electron micrograph image of the amyloid fibrils reconstituted in vitro by the prion peptide.

representative prion disease in animals. It is formally termed *bovine spongiform encephalopathy* (*BSE*).[5] In the United Kingdom, the country that was most affected, more than 180,000 cattle were infected and 4.4 million slaughtered during the eradication program.

In the past, the slaughtered meats of the afflicted animals were recycled as animal feeds after inactivation treatment (heat and chemical treatment) that would completely kill viruses, because scrapie was then regarded as a virus. The epidemic was believed to be caused by the recycling of slaughtered animals for animal feeds prior to recognition of etiological nature of prion diseases. Since 1988, ruminant protein feed was banned to stop the epidemic. Nevertheless, a new form of neurodegenerative disease, which is similar to sporadic CJD, began to occur in 1992 (Fig. 20.8); it was termed as a vCJD But it is commonly known as "human mad cow disease." By June 2014 it had killed 177 people in

5. **Bovine spongiform encephalopathy (BSE)** It is a fatal neurodegenerative disease (encephalopathy) in cattle that causes a spongy degeneration in brain and spinal cord.

FIGURE 20.7 A group of Kuru patients in 1957 at New Guinea. Kuru patients can only stand with the aid of the sticks. All died within 1 year of the photograph being taken. The term "kuru" derives from the Fore word "kuria/guria" (to shake), a reference to the body tremors that are a classic symptom of the disease.

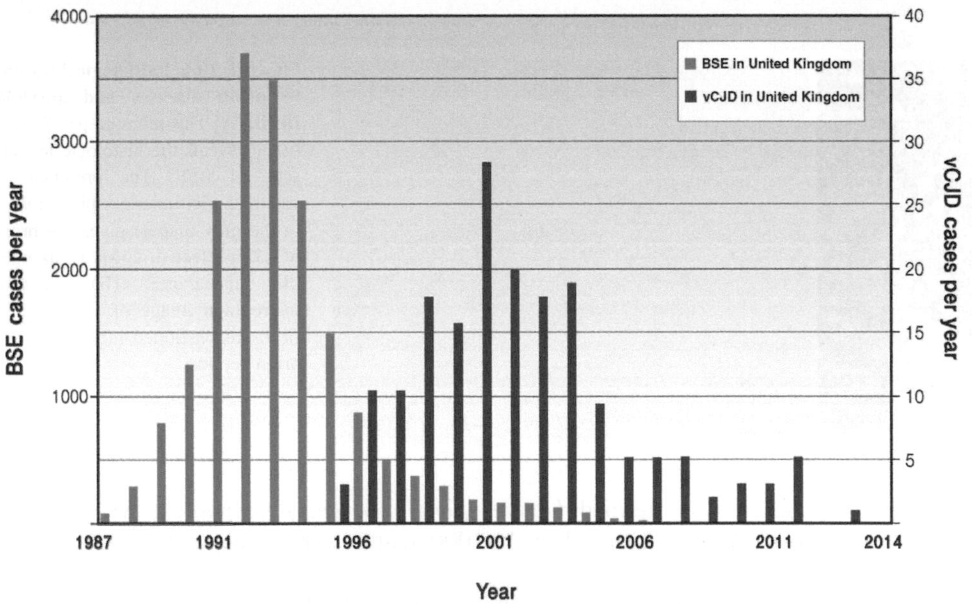

FIGURE 20.8 The relationship between BSE and vCJD. The epidemiologic relationship between BSE and vCJD is evident, as the bell shaped curve of BSE cases is shadowed by the vCJD cases with a gap of about 20 years in between. The gap might represent the latent period for the onset of disease after primary infection.

the United Kingdom, and 52 elsewhere. A causal link between vCJD and BSE was speculated in the early 1990s. The controversy over the cause of vCJD was concluded by the UK government's acknowledgment on the causal relationship between BSE and vCJD in 1996. Thanks to the implementation of stringent monitoring of slaughtered meat products, vCJD as well as "mad cow disease" is practically eradicated.

Prion Hypothesis: The "prion hypothesis" or "a protein-only hypothesis" states that a protein is the only etiologic component of the pathogen that causes the disease. What is the evidence for the lack of nucleic acids in prion agents? Four kinds of biochemical evidences were obtained. First, the infectivity of prions is not inactivated by heating to 90°C for 30 min (or even at 360°C for 1 h), a condition that would completely inactivate any nucleic acid. Second, the infectivity of scrapie agents is resistant to the inactivation by UV radiation and ionizing radiation. Because UV radiation and

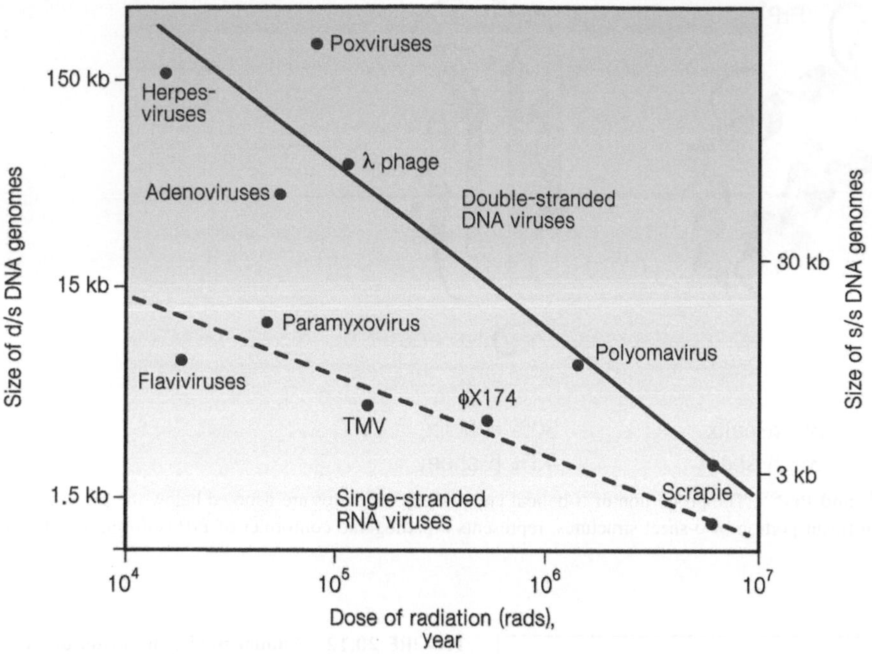

FIGURE 20.9 Correlation between the virus genome size and susceptibility to UV radiation. Inactivation radiation doses of various DNA (solid line) and RNA viruses (dotted line) are plotted. Larger genomes present a larger target and therefore are more sensitive to the inactivation of infectivity by UV than smaller genome. The estimated genome size of the scrapie agent is less than 1 kb, if any. Note log scale on vertical axis.

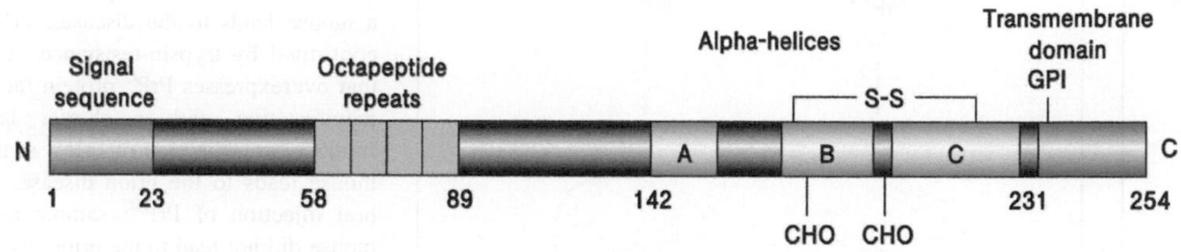

FIGURE 20.10 Protein domains of PrP. The N-terminal signal sequence targets it to the surface of cells. Three α-helical structures (A, B, and C) are denoted. The glycosylation sites are denoted by CHO. A glycosylphosphatidylinositol (GPI) membrane anchor at the COOH-terminal tethers PrP to cell membranes.

ionizing radiation inactivates infectious organisms by causing damage to their nucleic acid genome, there is an inverse relationship between the size of nucleic acid genome and the dose of UV radiation needed for the inactivation (Fig. 20.9). The scrapie agent was found to be highly resistant to both UV light and ionizing radiation, indicating that any nucleic acid present must be extremely small, probably less than 1 kb nucleotides. Third, scrapie agent was resistant to DNase and RNase treatment but sensitive to proteinase treatment, indicating that the infectivity is conferred by proteins. Fourth, scrapie was sensitive to any protein-denaturing agents such as urea, sodium dodecyl sulfate, phenol, and other chaotropic agents.

What is, then, the identity of the protein components of prion agents? Stanley Prusiner found a novel protein with a size of 27–30 kDa in a fraction with higher infectivity. It was termed *PrP* (prion protein) (Fig. 20.10). Surprisingly, the gene for this prion-associated protein is coded by the host chromosome. It was totally unexpected that a gene for the infectious agent is a host gene. In retrospect, the lack of inflammation in prion diseases is consistent with the causative role of host encoded protein. No immune response is invoked, because the infectious agents are self rather than nonself.

It was noted that PrP in scrapie animals is distinct from the one in normal animals in some biochemical properties such as protease-sensitivity. It turned out that the secondary structure of PrP in scrapie animals is distinct from that in normal animals (Fig. 20.11). The PrP in a normal cell has a largely α-helical structure, whereas the PrP in scrapie animals are abundant in a β-sheet structure. Importantly, the conformation change of PrP protein is associated with the disease. The endogenous and properly folded form is denoted PrP^{C} (for *Cellular*), whereas the disease-linked and misfolded form is denoted PrP^{Sc} (for *Scrapie*). PrP^{Sc} forms amyloid fibrils, leading to amyloid plaque formation. The cell death induced by amyloid plaque, eventually results in spongiform encephalitis. Prion disease is often referred to as "protein folding disease," emphasizing the unprecedented etiology involving the conformation change of the disease protein.

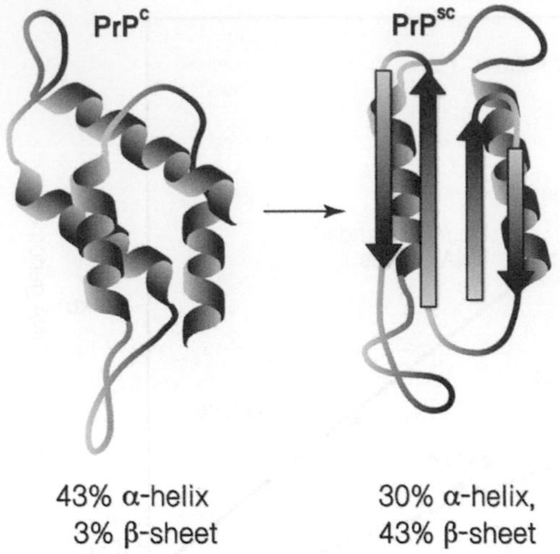

FIGURE 20.11 The protein structure of PrP^C and PrP^{SC}. The proportion of α-helical and β-sheet structures are denoted below the protein structures of PrP^C and PrP^{SC}. PrP^{SC}, which has a significant portion of β-sheet structures, represents a pathogenic conformer of PrP isoform, and it is an aggregation-prone isoform of PrP^C.

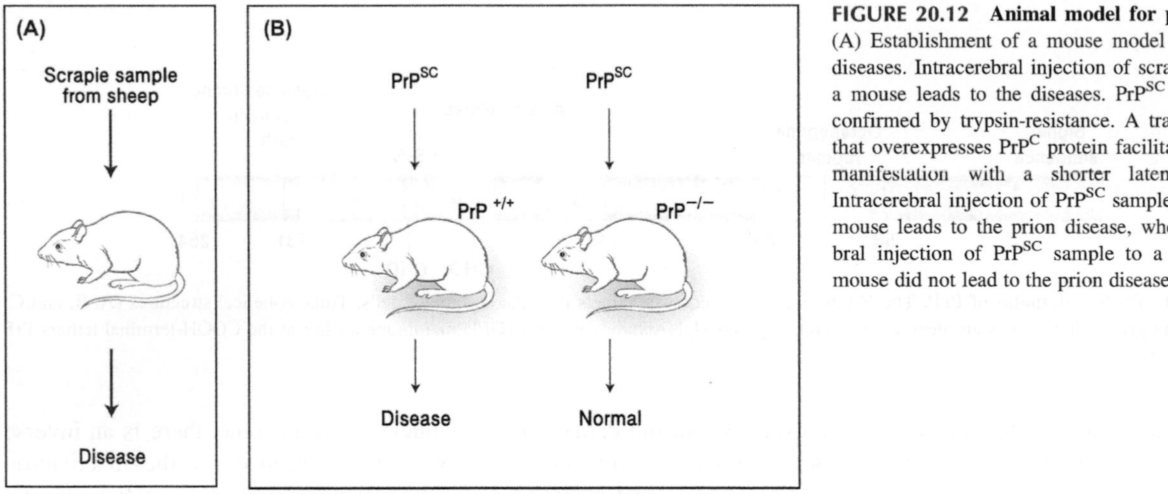

FIGURE 20.12 Animal model for prion diseases. (A) Establishment of a mouse model to study prion diseases. Intracerebral injection of scrapie agents into a mouse leads to the diseases. PrP^{SC} formation was confirmed by trypsin-resistance. A transgenic mouse that overexpresses PrP^C protein facilitates the disease manifestation with a shorter latent period. (B) Intracerebral injection of PrP^{SC} sample to a wild-type mouse leads to the prion disease, whereas intracerebral injection of PrP^{SC} sample to a PrP^C-knockout mouse did not lead to the prion disease.

To prove a protein-only hypothesis, one needs to demonstrate the infectivity of PrP in in vivo circumstances. An animal model for the prion disease has been established in mouse by intracerebral injection of scrapie agents isolated from a scrapie infected sheep (Fig. 20.12A). Remarkably, pathological lesions including spongiform encephalopathy were observed in the brain of mouse about 3 months after the intracerebral administration. Recapitulation of prion diseases in the mouse model enabled the investigation of the pathogenesis of prion disease in a laboratory. An experiment performed using PrP gene knockout mouse further corroborated the prion hypothesis (Fig. 20.12B). An intracerebral injection of PrP^{Sc} sample to a PrP knockout mouse did not cause the disease, revealing that a PrP knockout mouse is resistant to the disease. Moreover, this finding supports the notion that the protein–protein interaction between PrP^C and PrP^{SC} is critical for the disease manifestation.

After all, the prion hypothesis was largely substantiated by the finding described above. In summary, (1) an etiological agent of prion disease is a structural isoform (ie, PrP^{Sc}) of a host-coded PrP^C protein, (2) PrP^{Sc} is essential for the conversion of PrP^C to PrP^{Sc}, and (3) PrP^C exists in monomer, while PrP^{Sc} tends to aggregate to form a multimeric form. A question is, then, how can the amount of infectious agent (ie, PrP^{Sc}) be amplified? The amplification of PrP^{Sc} can be attained by the conversion of PrP^C to PrP^{Sc} via a protein–protein interaction in that PrP^{Sc} is a template for the conversion of PrP^C to PrP^{Sc} (Fig. 20.13). In doing so, the amount of PrP^{Sc} increases.

The fact that a Nobel Prize was awarded to Stanley Prusiner did not necessarily vindicate the protein-only hypothesis. Others still insisted that PrP^{Sc} represents only a subset of infectious materials, arguing the presence of other

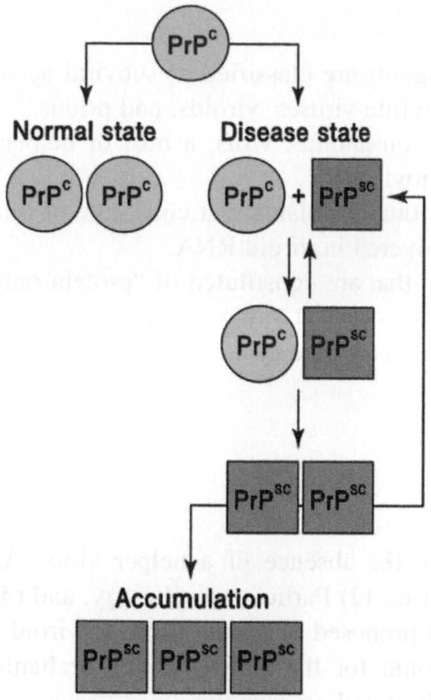

FIGURE 20.13 A model that accounts for the propagation of prion agents. A model states that the conformation change induced by the protein–protein interaction between PrP^C and PrP^{SC} underlies the propagation of prion agents. Specifically, PrP^C is converted into PrP^{SC} upon an interaction with PrP^{SC} monomer, forming a dimer or multimer. Accumulation of PrP^{SC} leads to the formation of amyloid fibrils. Breakage of PrP^{SC} nucleus recycles to amplify PrP^{SC}.

cofactors. The final proof would be a successful demonstration of the disease pathology in the mouse model following intracerebral injection of highly purified recombinant PrP^{Sc} or synthetic PrP^{Sc} made in vitro. Recently, recombinant PrP^{Sc} was successfully generated in vitro from bacterially expressed recombinant PrP^{Sc} protein (see Journal Club). This seminal work proved that the infectivity of the prion disease results from an altered conformation of PrP^C. Although the molecular attributes of prion diseases became clear, the pathogenesis of prion disease remains uncertain. It still remains unclear how prions damage neuronal cells.

20.5 PERSPECTIVES

Two subviral agents, including viroids and prions, deserve a great deal of attention, because they appear to represent the most simplistic infectious agents that one could imagine. They are distinct from other conventional infectious agents with respect to their biochemical constituents, replication mechanism, and host interaction. Viroids are "RNA only" agents, while prions are "protein-only" agents. Regarding viroids, many fundamental questions have not been answered. For instance, it remains uncertain whether cellular RNA polymerase II is entirely responsible for RNA synthesis, and if so, how cellular RNA polymerase II, which normally does not utilize RNA as a template, can utilize viroid RNA as a template. Another important question is how the RNA itself can transmit, and how RNA can induce pathologic lesions in plants. Regarding prions, a great advance has been made in our understanding on the peculiar nature of the disease etiology. Nonetheless, some salient questions remain unanswered. For instance, it is unclear how prions enter the body and reach the central nervous systems (CNS). This raises an intriguing question of how a mere protein aggregate can trespass mucosal barriers, circumvent innate and adaptive immunity, and travel across the *blood–brain barrier*[6] to eventually cause brain disease. Another unanswered question is what is the physiological function of PrP^C besides serving as a substrate for the generation of PrP^{Sc}? Lastly, how does PrP^C misfolding cause neurological disease? These intriguing questions provide challenges for young ambitious scientists.

6. **Blood–brain barrier** It refers to a highly selective permeability barrier that separates the circulating blood from the brain extracellular fluid in the CNS. It allows the passage of water, some gases, and lipid soluble molecules by passive diffusion, as well as selective transport of molecules such as glucose and amino acids that are crucial to neural function.

20.6 SUMMARY

- *Subviral agents*: Some transmissible agents are classified as subviral agents as they do not fit into the conventional definition of "virus." These include satellite viruses, viroids, and prions.
- *Satellite virus*: Satellite virus depends on another virus, a host or helper virus, for propagation. Adeno-associated virus (AAV) is a satellite virus of adenovirus.
- *Viroids*: Viroids are infectious agents found in plants that constitute of only a small circular RNA molecule without a protein coat. The ribozyme was discovered in viroid RNA.
- *Prions*: Prions are transmissible agents that are constituted of "protein only," and are devoid of nucleic acid components. Conversion of cellular prion protein (PrP^C) to a pathogenic isoform (PrP^{Sc}) via protein–protein interaction is the underlying mechanism for prion diseases. The best-known example of prion disease for the general public is so-called mad cow disease in cattle.

STUDY QUESTIONS

20.1 Satellite viruses cannot replicate in the absence of a helper virus. Answer the following question on satellite viruses. (1) Relatedness to host viruses. (2) Particle morphology, and (3) Impact on the growth of helper virus.

20.2 Rolling-circle mechanism has been proposed to account for the viroid RNA replication. (1) List three enzymatic activities that are required to account for the rolling-circle mechanism. (2) Explain how viroids attain these enzymatic activities despite the lack of coding capacity.

20.3 State three experimental evidences that are in favor of a protein-only hypothesis.

SUGGESTED READING

Aguzzi, A., Zhu, C., 2012. Five questions on prion diseases. PLoS Pathog. 8, e1002651.

Chiesa, R., 2015. The elusive role of the prion protein and the mechanism of toxicity in prion disease. PLoS Pathog. 11, e1004745.

Legname, G., Baskakov, I.V., Nguyen, H.O., Riesner, D., Cohen, F.E., DeArmond, S.J., et al., 2004. Synthetic mammalian prions. Science. 305, 673–676.

Prusiner, S.B., 2012. Cell biology. A unifying role for prions in neurodegenerative diseases. Science. 336, 1511–1513.

Wang, F., Wang, X., Yuan, C.G., Ma, J., 2010. Generating a prion with bacterially expressed recombinant prion protein. Science. 327, 1132–1135.

JOURNAL CLUB

- Wang, F., Wang, X., Yuan, C.G., Ma, J., 2010. Generating a prion with bacterially expressed recombinant prion protein. Science 327, 1132–1135.

  Highlight: This seminal paper described the generation of a highly infectious prion in vitro by using PrP purified from *Escherichia coli*. A recombinant prion has the attributes of the pathogenic PrP isoform: aggregated, protease-resistance, and self-perpetuating. After intracerebral injection of the recombinant prion, remarkably, wild-type mice developed neurological diseases in 150 days and succumbed to prion disease. This work proved that the infectivity in prion disease results from an altered conformation of PrP.

Chapter 21

New Emerging Viruses

Chapter Outline

Newly emerging viruses such as the Ebola virus, severe acute respiratory syndrome (SARS)-, Middle East respiratory syndrome (MERS)-coronavirus, and the avian influenza virus are serious threats to public health and have become a global concern. In this chapter, we will learn the basic terminology of the epidemiology of infection and the outbreaks caused by newly emerging viruses.

21.1 EPIDEMIOLOGY

To define the causes and the patterns of an infection, epidemiologic studies are conducted. *Epidemiology*[1] is the study of patterns, causes, and effects of disease conditions in defined populations. Viral infections can be classified depending on their epidemiologic features (Table 21.1). *Epidemic infection* (or *epidemics*[2]) is a type of infection in which the infection is extensive and not limited to a certain region but only lasts transiently (Fig. 21.1). The seasonal flu is an example of an epidemic infection. In the case of the seasonal flu, although the affected region is extensive, the epidemics in different regions are linked because the same strain of influenza virus is often responsible for a season. *Endemic infection* is a type of infection in which the infection persists for a longer period of time in a certain region but the infection is typically limited to only a subset of the population. Hepatitis B viral infection in China and in Korea is an example of endemic infection. A subset of the population (~4–8%) is chronically infected. *Sporadic infection* is a type of infection in which the infection is limited to the affected persons and regions. It lasts transiently. In this case, there are no links between *outbreaks*[3] occurring in different regions. *Pandemics* refer to epidemics occurring across multiple continents.

Surveillance to monitor the emergence of new viruses and their transmission is an important goal of public health authorities. In particular, the explosive expansion of international trade and travel has made it difficult to control viral transmission between countries and continents. *Quarantine*[4] at airports and seaports has become increasingly important for surveillance (Fig. 21.2).

---

1. **Epidemiology** It is derived from Greek *epi*, meaning "upon," *demos*, meaning "people," and *logos*, meaning "study."

2. **Epidemics** It refers to the rapid spread of infectious disease to a large number of people in a given population within a short period of time. The period is usually 2 weeks or less.

3. **Outbreak** In epidemiology, an outbreak is an occurrence of disease greater than expected at a particular time and place. Outbreaks may also refer to epidemics which affect a particular region in a country or a group of countries.

4. **Quarantine** Quarantine is used to separate and restrict the movement of people who may have been exposed to a communicable disease in order to monitor their health. The word comes from the Italian (17th-century Venetian) word "quaranta," meaning forty, which is the number of days ships were required to be isolated (quarantined) before coming ashore during the black death.

Molecular Virology of Human Pathogenic Viruses. DOI: http://dx.doi.org/10.1016/B978-0-12-800838-6.00021-7

TABLE 21.1 Epidemiologic Patterns of Viral Transmission

Infection Pattern	Scale in Population	Region	Linked	Period	Persistency
Endemic	Small to large	Local	Yes	Long (>yr)	Yes
Epidemic	Small to large	Local	Yes	Short (<yr)	No
Sporadic	Small	Multilocal	No	Short (<yr)	No
Pandemic	Large	Multicontinents	Yes	Short (<yr)	No

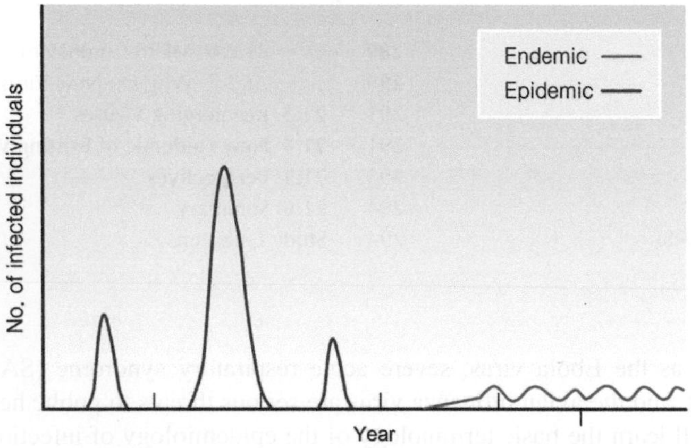

FIGURE 21.1 Epidemic versus endemic infection. Infection period and the infection scale are depicted to contrast epidemic versus endemic infection. A virus infection pattern can be changed from epidemic to endemic. For instance, WNV infection is now considered to be endemic in the United States (see Fig. 21.8).

FIGURE 21.2 Quarantine at the airport. A quarantine official (left) checks people arriving at South Korea's Incheon International Airport for signs of fever during MERS outbreak in 2015.

21.2 NEWLY EMERGING VIRUSES

Since the 1970s, newly emerging viruses of unknown origins have been continuously discovered (Fig. 21.3). The first outbreak of the Ebola virus occurred in 1976 in Zaire and Sudan. AIDS was first described as acquired immunodeficiency syndrome among homosexual males in 1981 and its culprit was soon identified in 1983 as HIV (see Box 17.3). SARS, a respiratory disease caused by SARS-CoV, was first reported in Hong Kong in 2003. More recently, a novel coronavirus, Middle East respiratory syndrome-coronavirus (MERS-CoV), was discovered in 2012 in Saudi Arabia. Newly emerging

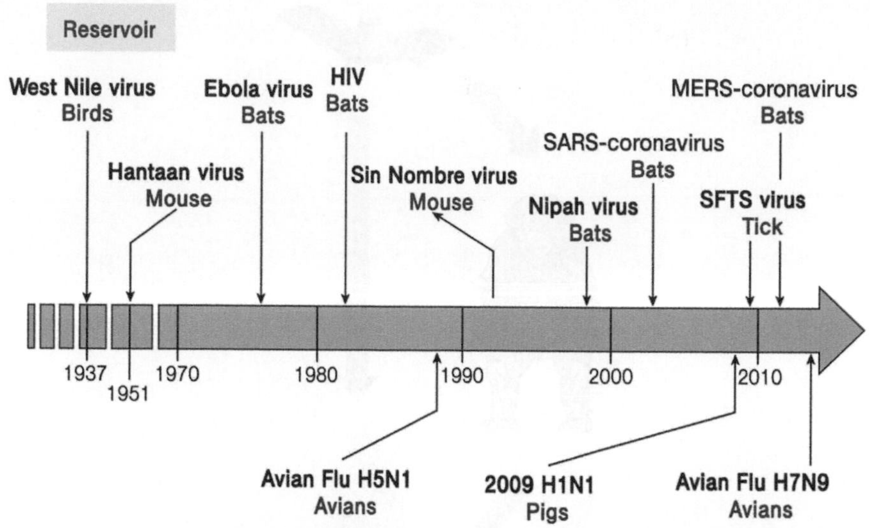

FIGURE 21.3 **Time line of newly emerging viruses.** The outbreaks of newly emerging viruses since 1930 are listed. Influenza outbreaks are shown below the line. Natural reservoirs for each virus are annotated in red.

viruses, such as the Ebola virus, the WNV, Nipah virus, and SARS-CoV, will be described in this chapter. HIV will be covered in a separate chapter due to its huge impact on the global community (see chapter: HIV and AIDS).

21.2.1 Ebola Virus

Since the first Ebola outbreak in 1976, Ebola outbreaks have continued to be reported. The fatality rate of Ebola fever is extremely high (~90%). Fortunately, Ebola outbreaks have been mainly restricted to Africa. The potential for Ebola to become a widespread epidemic is low due to the high case-fatality rate and the rapid demise of patients. The Ebola virus came up in the Hollywood movie, "Outbreak," starring Dustin Hoffman. While a monkey served as the reservoir of the virus in the movie, in reality, bats are the reservoir (Fig. 21.4). Bats and primates are sold in markets and consumed as bush meat in some areas of Africa. In a viral perspective, the host change from bats to humans is a bit challenging as it imposes a selective pressure for mutations. It is thought a variant highly virulent in humans is generated during the adaptation of the virus in a new host. The Ebola virus is a good example of a new emerging virus which originates from the rainforest in Africa.

The recent 2013–14 Ebola outbreak in Western Africa set a record death toll (ie, 11,295 deaths on July 31, 2015) (Fig. 21.5). It has been the most widespread Ebola epidemic in history. It began in Guinea in December 2013 and then spread to Liberia and Sierra Leone. International organizations from around the world have responded to help stop the ongoing Ebola epidemic in West Africa. In July 2014, the World Health Organization (WHO) declared the outbreak as an international public health emergency. In September 2014, the United Nations Security Council declared the Ebola outbreak in the West African region "a threat to international peace and security" and unanimously adopted a resolution urging its member states to provide more resources to fight the outbreak. As of August 2015, the outbreak is almost contained. The reason why the 2013–14 Ebola outbreak was widespread remains debatable. There is currently no treatment for the Ebola virus. More recently, the WHO announced "an extremely promising development" in the search for an effective vaccine for Ebola disease.

21.2.2 West Nile Virus

The WNV[5] epidemic in North America is a good example of the spread of a mosquito-borne zoonotic virus due to climate change. WNV is a member of the family Flaviviridae and is transmitted via mosquito (Fig. 21.6). Thus, it is also known as an *arbovirus*.[6] It was first identified in the West Nile subregion in the East African nation of Uganda in 1937.

5. **West Nile Virus (WNV)** A mosquito-borne zoonotic arbovirus belonging to the genus *Flavivirus* in the family Flaviviridae.

6. **Arbovirus** Arbovirus is a descriptive term applied to hundreds of predominantly RNA viruses that are transmitted by arthropods, notably mosquitoes and ticks. The word *arbovirus* is an acronym (Arthropod-Borne virus).

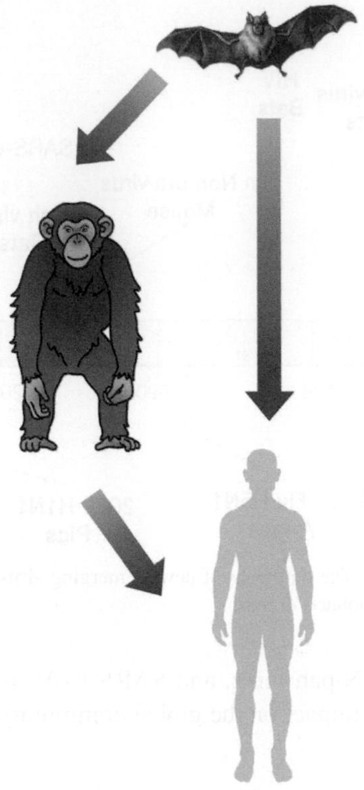

FIGURE 21.4 Ebola virus transmission to humans. The Ebola virus is transmitted by bodily fluids (blood, secretions, etc.) of infected animals such as chimpanzees, gorillas, and fruit bats can be also transmitted from human to human.

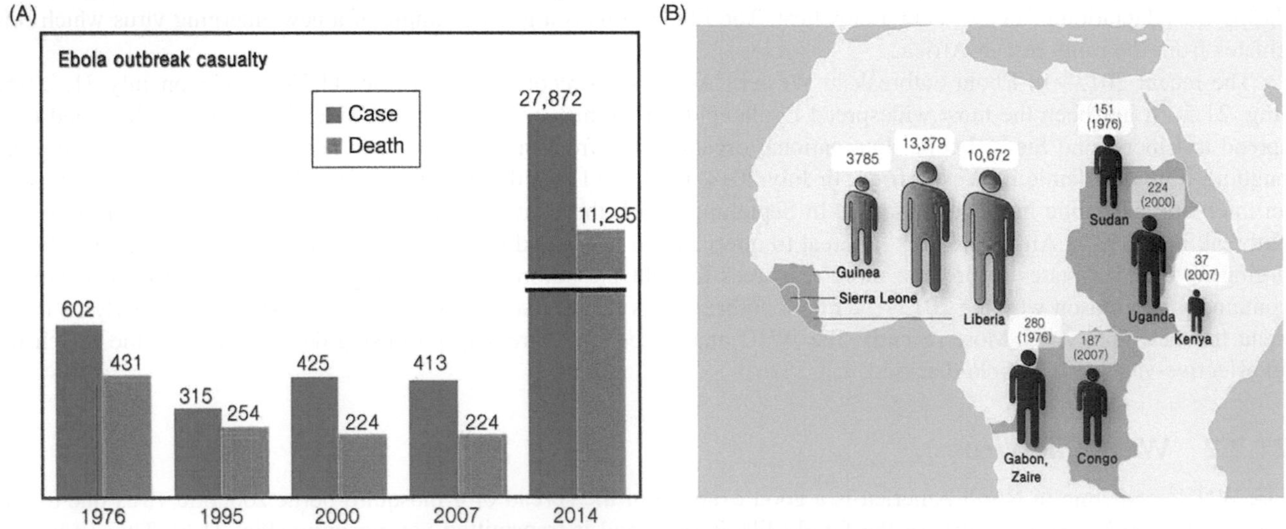

FIGURE 21.5 Ebola outbreaks in Africa. (A) Ebola outbreak casualty. The damage by the past Ebola outbreak was limited to less than 500 casualties per outbreak. By contrast, the 2013–14 Western Africa Ebola outbreak caused 27,872 cases and 11,295 deaths, as of July 2015. The magnitude of casualty was at an unprecedented level. The mortality rate was about 40%. (B) Regions affected by the Ebola outbreak. Since the first Ebola outbreak occurred in 1976 in Zaire, Ebola virus outbreaks were confined to Central Africa including Uganda, Zaire, and Congo. By contrast, the 2013–14 Ebola outbreak differs with respect to the region affected. The outbreak occurred in Western Africa including Guinea, Liberia, and Sierra Leone.

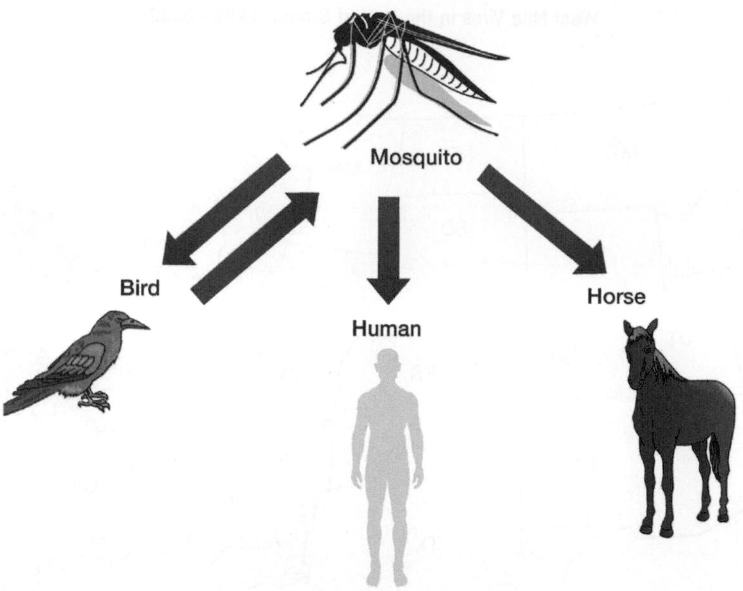

FIGURE 21.6 **Transmission routes of WNV to human.** WNV transmits from birds to humans or to horses via mosquitoes. During the epidemic in the United States, the death of crows in a particular town was considered a sign of WNV arrival.

Approximately 80% of WNV infections in humans are subclinical, meaning that no symptoms are associated with the infection. For the rest, symptoms include fever, headache, fatigue, and rash. Less than 1% of the cases are severe and result in neurological disease.

The WNV has typically been confined to temperate and tropical regions. Prior to the mid-1990s, WNV disease occurred only sporadically and was considered a minor risk for humans. The WNV has now spread globally. The first case in the Northern Hemisphere was identified in New York City in 1999. Over the next 5 years the virus spread across the continental United States and north into Canada (Fig. 21.7). The arrival of an outsider in New York City in 1999 drew the media's attention. Heavy media coverage on the epidemic caused panic among residents. The WNV is now considered to be an endemic pathogen in the United States. In 2012 the United States experienced one of the worst epidemics, in which WNV killed 286 people.

The WNV epidemic in the United States warned of the risk of mosquito-borne *zoonotic virus*.[7] It is believed that global warming is attributed to the spread of tropical mosquito-borne zoonotic viruses to the Northern Hemisphere (Table 21.2).

21.2.3 Sin Nombre Virus

A mystery respiratory illness causing deaths among young Navajo was first reported in 1993 in the "Four Corners" region of the western United States (Fig. 21.8). Clinical resemblance to hantavirus infection, which occurred during the Korean war (1950–53), led investigators to look for the culprit in rodents. Until then no known cases of hantavirus infection had ever been reported in the United States. The culprit was a deer mouse discovered near the home of one of the initial patients. Not surprisingly, newly discovered Sin Nombre virus (SNV) is related to the hantavirus. Hantavirus was discovered by Ho-Wang Lee in Korea in 1776 as the etiological agent responsible for the Korean hemorrhagic fever outbreak which occurred among American and Korean soldiers during the Korean War. Patients with *hantavirus pulmonary syndrome* (HPS) had mild flu-like symptoms such as malaise, headache, cough, fever, with a sudden onset of pulmonary edema, and finally death. In this epidemic, 24 cases were reported with a 50% mortality rate. SNV[8] belongs to the *Hantavirus* genus of family Bunyaviridae (see Table 16.4). SNV infection occurs wherever its reservoir rodent carrier (the deer mouse) is found. This includes the entire populated area of North America except for the far southeastern region. SNV can be contracted through the inhalation of virus-contaminated deer mouse excretion.

7. **Zoonotic viruses** A virus transmitted between species (sometimes by a vector) from animals to humans.

8. **Sin Nombre virus (in Spanish, "the nameless virus") (SNV)** Its original name was "Four Corners virus." The name was changed after local residents raised objections.

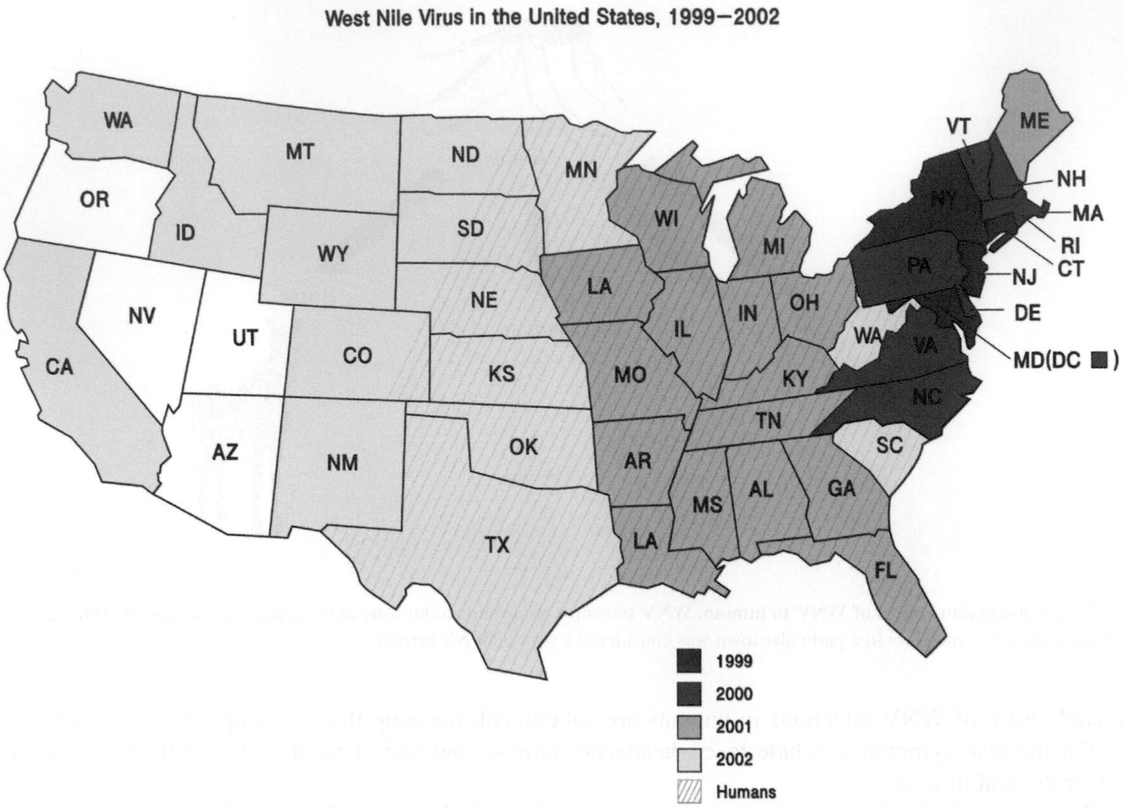

FIGURE 21.7 Geographic distribution of the WNV in the United States. WNV epidemic, which started in New York in 1999, rapidly spread to the West coast in a few years. The states with human cases are stippled.

The fatality rate of SNV-induced HPS in the United States was reported to be about 66.7%. However, since that time the fatality rate has steadily declined as more mild cases have come to be recognized.

21.2.4 Nipah Virus

The *Nipah*[9] virus was first identified in April 1999 on a pig farm in peninsular Malaysia when it caused an outbreak of neurological and respiratory disease. The outbreak resulted in 257 human cases, 105 human deaths, and the culling of 1 million pigs (Fig. 21.9). Symptoms of infection from the Malaysian outbreak were primarily encephalitic in humans and respiratory in pigs. Respiratory illness in humans has been seen in later outbreaks, increasing the likelihood of human-to-human transmission and indicating the existence of more dangerous strains of the virus. Based on seroprevalence data and the data from viral isolations, the primary reservoir for Nipah virus was identified as Pteropid fruit bats. The transmission of Nipah virus from bats to pigs is thought to be due to an increasing overlap between bat habitats and piggeries in peninsular Malaysia.

A related Hendra virus was discovered in September 1994 when it caused the deaths of 13 horses and a trainer at a training complex in Hendra, a suburb of Brisbane in Queensland Australia. Nipah virus, along with Hendra virus, belongs to the genus Henipavirus belonging in the family Paramyxoviridae (see Table 16.1).

21.2.5 SARS-Coronavirus (SARS)

SARS is a viral respiratory disease of zoonotic origin caused by the SARS-CoV. Between November 2002 and July 2003, an outbreak of SARS in southern China led to 8273 cases and 775 deaths in multiple countries. The majority of cases were in Hong Kong (9.6% fatality rate) according to the WHO. Within weeks, SARS spread from Hong Kong to

9. The name "Nipah" refers to the place, Kampung Baru Sungai Nipah in Negeri Sembilan State, Malaysia, the source of the human case from which Nipah virus was first isolated.

TABLE 21.2 The List of Major Zoonotic Viruses

Family	Zoonotic Virus	Outbreak (Reported Year)	Frequency	Source of Human Infection	Reservoir Host	Disease	Human Cases (Fatality), Region
Influenza virus	Avian Influenza H5N1	Vietnam (2003–)	Rare	Chicken	Wild birds	Respiratory disease	~566 (~60%) in SE Asia
	Avian Influenza H7N9	China (2013)	Rare	Chicken	Wild birds	Respiratory disease	~127 (~20%) in China
Filovirus	Ebola virus	Africa (1976–)	Rare	Primates	Bat	Hemorrhagic fever	>800 (~90%) in Africa
Flavivirus	Dengue virus	Asia, Africa, S. America (1953–)	Endemic	Mosquito	Monkeys	Hemorrhagic fever	50–100 million/yr (1–5%) in Southeast Asia
	Japanese encephalitis virus (JEV)	Asia (1935–)	Endemic	Mosquito	Birds, bats	Encephalitis	30,000–50,000/yr in Asia (0.3–60%); ~10,000 death/yr
	West Nile virus	Uganda (1937–)	Endemic	Mosquito	Birds	Encephalitis	>10,000 (~0.7%) in the United States, since 1999
Bunyavirus	Hantaan virus	Korea (1951–)	Common	Mouse	Wild mouse	Hemorrhagic fever	~200,000/yr, (~15%) in Asia
	Sin Nombre virus	United States (1993)	Rare	Mouse	Wild mouse	Hantavirus pulmonary syndrome (HPS)	~12 (~60%) in the four corners region of the United States
Coronavirus	SARS-CoV*	China (2002)	Very rare	Bats	Bats	Severe acute respiratory syndrome (SARS)	8273 (775 death) (9.6%) in China, Hong Kong, etc.
	MERS-CoV*	Saudi Arabia (2012–)	Very rare	Bats	Camel	SARS-like	1084 cases and 439 death (40%) in Middle East and others
Paramyxovirus	Hendra virus	Australia (1994–)	Very rare	Horses	Bats	hemorrhage in lung	~76 horse died (~60%), since 1994
	Nipah virus	Malaysia (1999–)	Rare	Pigs	Bats	Encephalitis	~257 (~40%) in 1999
Rhabdovirus	Rabies virus	Europe (18th)	Endemic	Animals (dog)	Wild animals (bats, racoon, etc.)	Paralysis and hydrophobia	~55,000 (>90%) death/yr, worldwide (Asia and Africa)

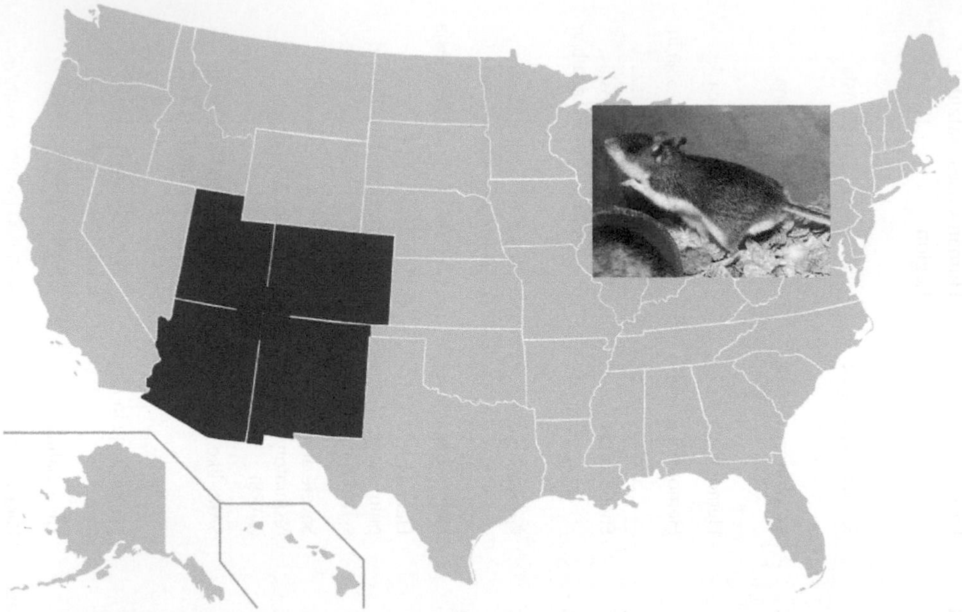

FIGURE 21.8 **The region affected by outbreak in the United States.** SNV A Southwest region of the United States called "Four corners" is where the SNV outbreak occurred. A deer mouse, the natural reservoir of SNV, is shown in insert.

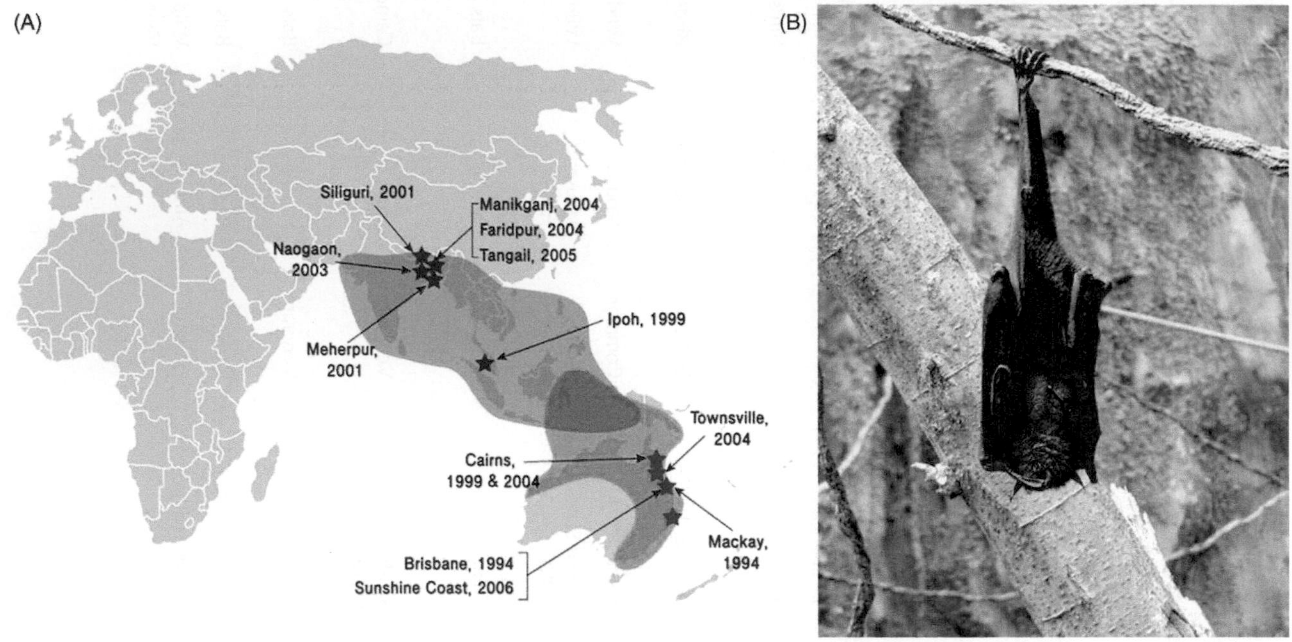

FIGURE 21.9 **Henipah virus outbreaks.** (A) Geographic distribution of Henipah virus outbreaks. The regions affected by Nipah virus (violet) and Hendra virus (red) are denoted. (B) Bats (large flying fox), the natural reservoirs of Nipah virus.

infect individuals in 37 countries in early 2003 (Fig. 21.10). It was eradicated by January of the following year. Phylogenetic analysis of these viruses indicated a high probability SARS-CoV originated in bats and spread to humans either directly or indirectly through animals held in Chinese markets.

21.2.6 MERS-Coronavirus

Another novel human coronavirus, MERS-CoV was isolated from a patient with acute pneumonia in 2012 in Saudi Arabia. As of March 2015, MERS-CoV infection has led to 1084 cases and 439 deaths reported in multiple countries.

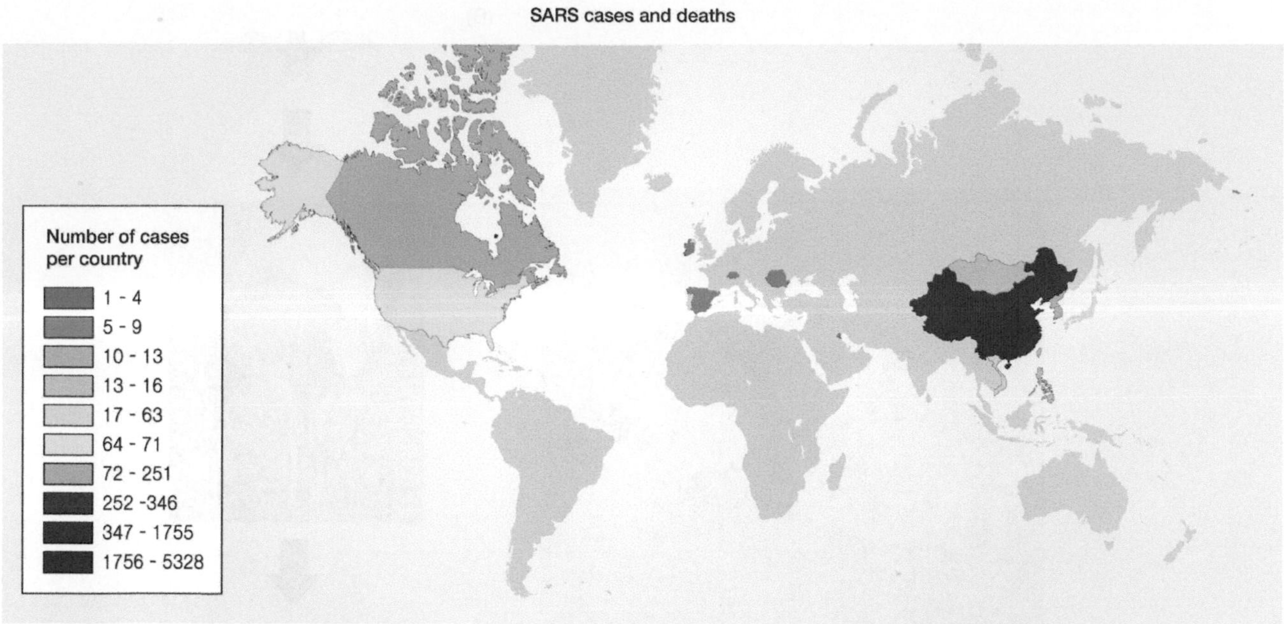

FIGURE 21.10 **SARS-CoV outbreaks.** The map illustrates the areas around the world that were affected by the SARS outbreak of 2002–03.

The fatality of MERS-CoV is considerably higher than SARS-CoV, approaching 30%. MERS-CoV may be originating from bats due to its high sequence homology to the bat virus. It is speculated the virus spreads from bats to human via dromedary camel (Fig. 21.11). Almost all cases have been linked to Saudi Arabia. Although human-to-human transmission seems to be inefficient, it has been shown to spread between people who are in close contact. As of June 2015, MERS-CoV cases have been reported in 23 countries including Saudi Arabia, Malaysia, Jordan, Qatar, Egypt, the United States, South Korea, and China. Most importantly, MERS-CoV transmission is not yet under control. A recent MERS-CoV outbreak in South Korea started in May 2015 from a single patient who visited Arab countries. It has been the largest outbreak outside of the Middle Eastern region leading to 182 reported cases including 33 deaths (case-fatality rate 16%) as of June 2, 2015.

21.2.7 Why Do New Viruses Emerge?

Why are new human pathogenic viruses continually emerging? In most cases, outbreaks have been known to occur in tropical regions in which there were no human inhabitants. An increase in contact with wild animals, due to the expansion of the human habitat, is believed to be the main cause for the emergence of new viruses. Changes in the environment such as rainforest developments have led to an increase in contact between wild animals and humans. As a result, viruses which only existed in rainforests are able to be transmitted to a new human host. Climate changes such as global warming are another cause for the emergence of new viruses. This is demonstrated by the WNV outbreaks in the Northern Hemisphere. Many of these newly emerging viruses are zoonotic viruses. In particular, bats serve as reservoirs to many newly emerging viruses. Bats are natural reservoirs for the Ebola virus, SARS-CoV, and the Nipah virus. What makes bats so special? Bats are unique because they are mammals that can fly. As mammals, they are a closer relation to humans than to birds. In addition, it is speculated that the immunity of bats is conspicuously tolerable to viruses. Many types of viruses found in bats (ie, *virome*[10]) are currently being analyzed with the implementation of *next generation sequencing* technology. The bat virome is expected to provide insight on newly emerging viruses that have yet been discovered.

21.3 REEMERGING VIRUSES

Besides newly emerging viruses, the variants of existing viruses also cause serious epidemics. These viruses have infected humans in the past. However, they continue to appear in drug-resistant forms or reappear after apparent control

10. **Virome** It refers to the collection of all the viruses (or viral genome) in a given animal.

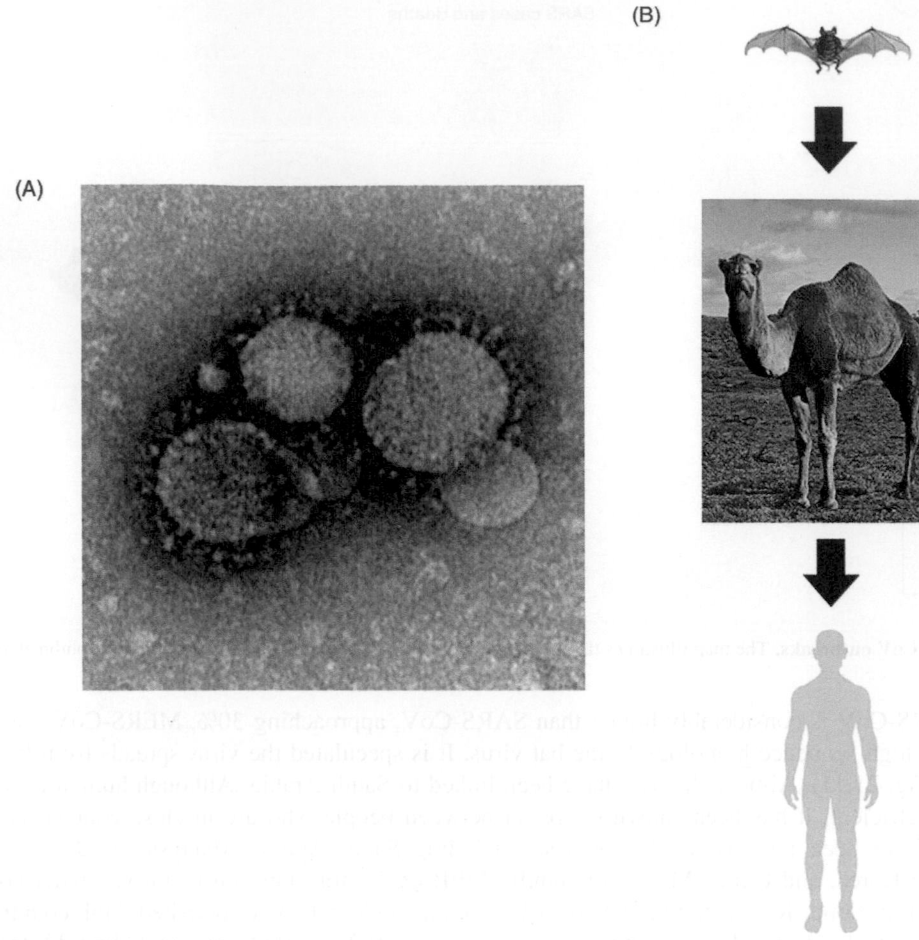

FIGURE 21.11 **MERS-CoV.** (A) Electron micrograph of MERS-CoV. MERS-CoV particles as seen by negative stain using electron microscopy. Virions contain characteristic club-like projections emanating from the viral membrane. (B) The transmission route of MERS-CoV. It is transmitted from cave bats to humans via dromedary camel.

or elimination. These variant viruses are termed "*reemerging viruses.*" Influenza virus represents the best example. Three distinct kinds of influenza virus variants are responsible for flu epidemics: the *seasonal flu*, the *pandemic flu*, and the *avian flu*. Since the mechanisms underlying seasonal and pandemic flu emergence have been described in chapter "Influenza Viruses," these two kinds of variants will be described briefly. On the other hand, the avian influenza virus will be described with an emphasis on the mechanism of zoonotic infection.

Seasonal Flu: The seasonal flu, as its name implies, is a flu epidemic that occurs yearly. Why are influenza viruses able to infect people who have previously been infected in the past? Multiple reinfection capabilities represent a salient feature of influenza virus infection. The reason for this is because variants with distinct antigenicity emerge each year. This is known as "antigenic drift" (see Fig. 15.15). The antibodies circulating in the person previously infected are unable to neutralize a new seasonal strain.

Pandemic Flu: The pandemic flu which occurred in the 20th century has already been described (see Box 15.2). Here, specific features of the 2009 H1N1 influenza pandemic are described. The epidemic started in Mexico in April of 2009. It rapidly spread globally, resulting in up to 300 million infections and over 16,000 deaths. During the pandemic, many international meetings were canceled because many people were afraid to travel abroad without proper protection. The culprit was quickly identified to be a swine influenza virus which first infected a boy living near a pig farm in Mexico. The lack of an appropriate flu vaccine during the pandemic frightened the public. An antiviral drug, Tamiflu was the only protective means until a vaccine became available (see Box 15.1). The 2009 H1N1 pandemic was notable. First, it was the first swine influenza virus that caused a pandemic. Second, it was not the typical double reassortment. It was an unprecedented triple reassortment (Fig. 21.12).

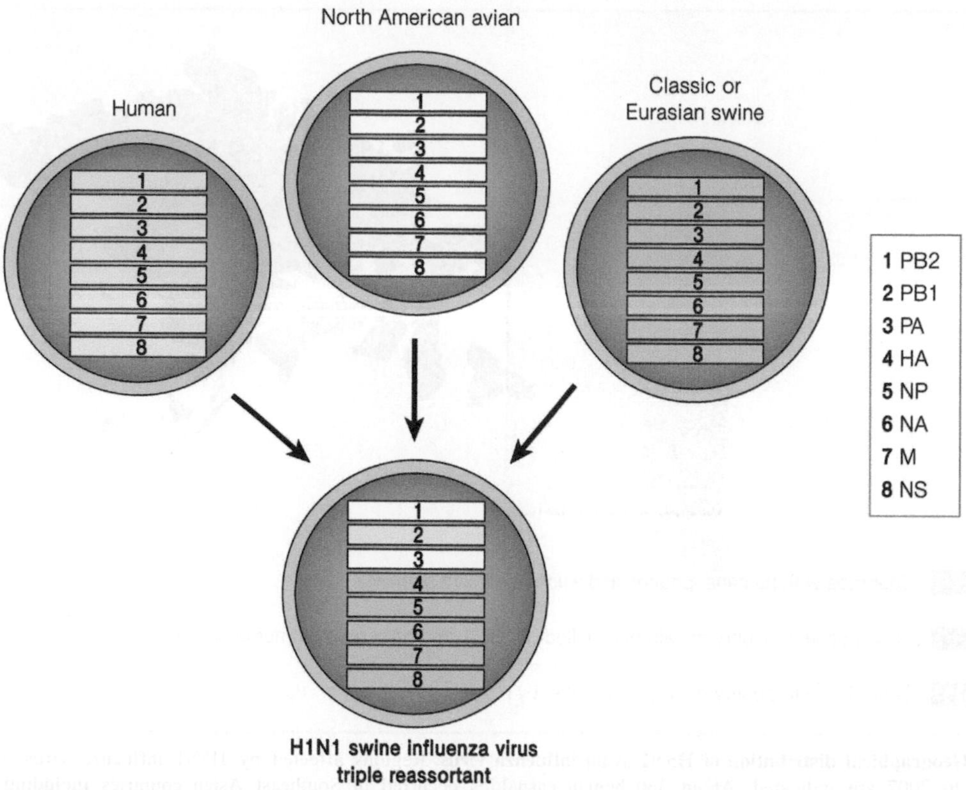

FIGURE 21.12 Gene segments of 2009 H1N1 influenza virus. The 2009 H1N1 swine virus is a triple reassortant. It includes segments from swine, avian, and human influenza viruses. Gene segments, host, and the year of introduction are denoted. PA and PB2 segments are derived from the avian virus. The PB1 segment is derived from a human virus that entered swine in 1979. The other five segments (ie, HA, NP, NS1, NA, and M) are derived from the avian virus which entered either classical swine in 1918 or Eurasian swine in 1979.

H5N1 Avian Flu: It was established that avian flu virus does not normally infect human. However, transmission of the avian virus to humans has occurred, although rarely, in the past decade. Thus, the avian influenza virus has become a global concern as a potential pandemic threat. Since the first H5N1 outbreak occurred in 1987, the highly pathogenic influenza subtype has killed millions of poultry in many countries throughout Asia, Europe, and Africa (Fig. 21.13). Because a significant species barrier exists between birds and humans, the virus does not easily cross species. Since 2003 human cases of H5N1 have been reported in 15 countries: 359 people have died from H5N1 in 12 countries (including China, Cambodia, and Vietnam) as of August 10, 2012. The mortality rate for humans with H5N1 is 60%. One of the growing concerns is that it could be mutated further and spread via human-to-human infection.

Why is avian H5N1 often fatal? The short answer is it has to do with different sites of infection of the seasonal H1N1 virus versus the avian H5N1 virus (Fig. 21.14). HA of human influenza virus prefers to bind the sialic acid linked to glycan via α-2,6 linkage typically found in glycans distributed in the upper respiratory tract. In contrast, HA of the avian virus prefers to bind the sialic acid linked to glycans via α-2,3 linkage typically found in glycans distributed in the lower respiratory tract (see Fig. 15.5). The lack of α-2,3 linked glycans in the upper respiratory track serves as a barrier to prevent avian flu virus transmission to humans.

H7N9 Avian Flu: Influenza virus subtype H7N9 is a novel avian influenza virus first reported to have infected humans in 2013 in China. Most of the reported cases of human infection have resulted in severe respiratory illness. As of the end of June 2013, 133 cases had been reported and 43 deaths (32% mortality). The number of cases detected after April 2013 fell abruptly. The decrease in the number of new human H7N9 cases may have resulted from the containment measures taken by Chinese authorities (closing live bird markets), a change in seasons, or possibly a combination of both factors.

21.4 NEW EPIDEMIC OF EXISTING VIRUSES

Besides new emerging viruses, new epidemics caused by existing viruses represent serious public health concerns. The measles virus is a good example of this in the Western Hemisphere.

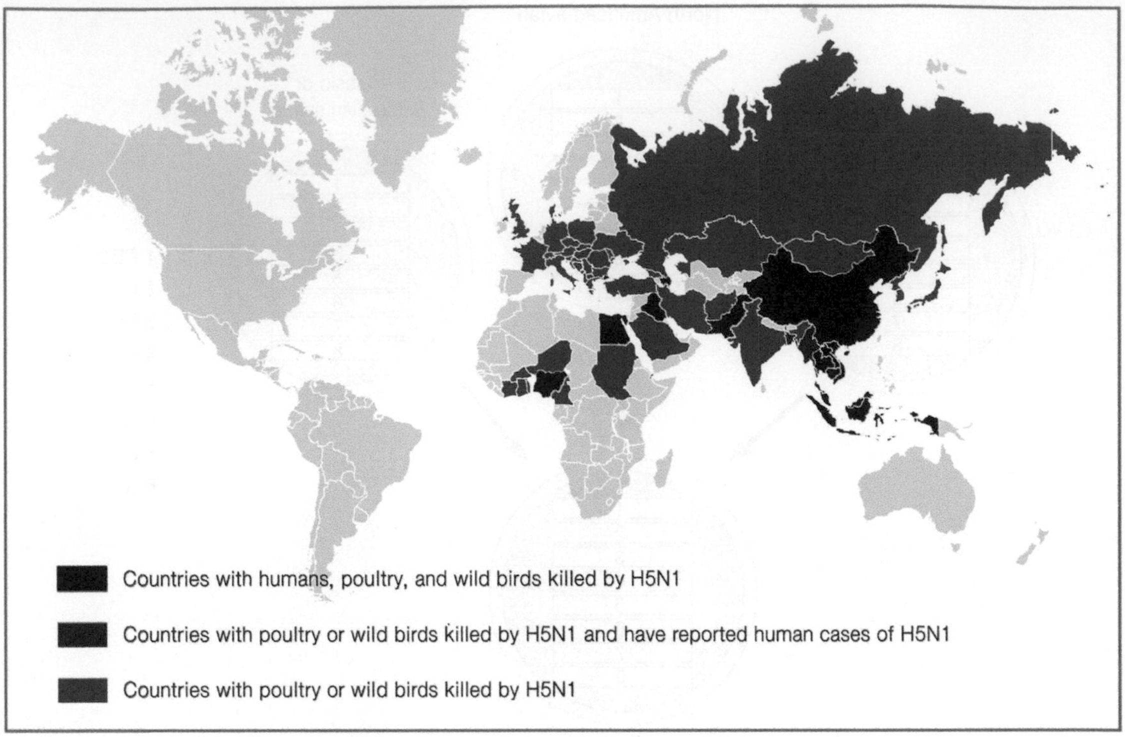

FIGURE 21.13 **Geographical distribution of H5N1 avian influenza virus.** Regions affected by H5N1 influenza virus in poultry and wild birds from 2003 to 2007 are indicated. About 360 human casualties occurred in Southeast Asian countries including China, Cambodia, and Vietnam.

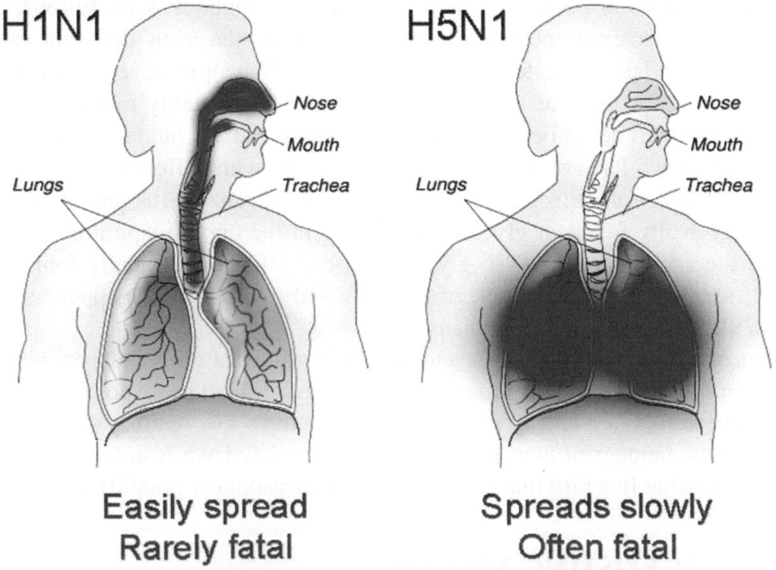

FIGURE 21.14 **Infection pathology of human influenza virus versus avian influenza virus.** Seasonal H1N1 influenza virus infects the upper respiratory tract and is easily spread but, rarely fatal. In contrast, avian H5N1 influenza virus infects the lower respiratory tract, spreads slowly, and is often fatal. The affected area is colored in red.

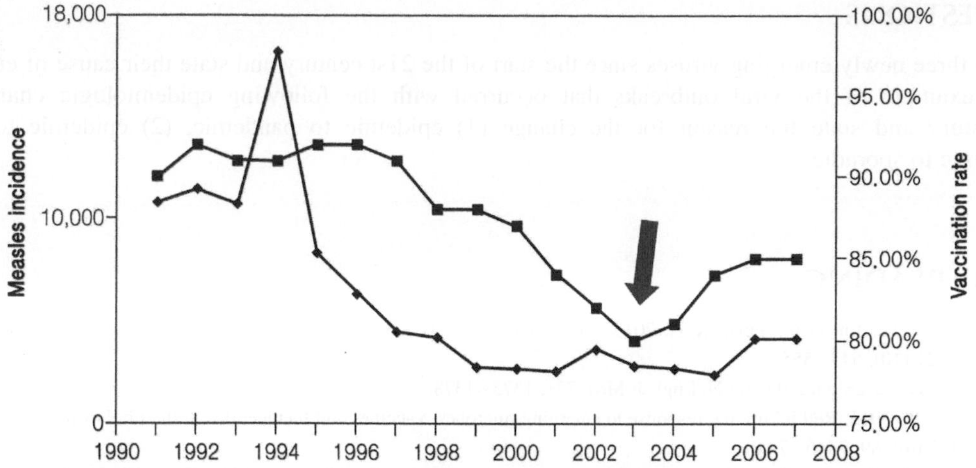

FIGURE 21.15 **Measles incidence and vaccination rates in England and Wales in 1991–2007.** Measles incidence (square) has begun to increase from 2003 (red arrow), a decade following the sudden drop of vaccination rates (diamond) in England.

Measles Virus: Measles is caused by measles virus, which belongs to the family Paramyxoviridaee (see chapter: Other Negative-Strand RNA Viruses). Live vaccines, first developed in 1961, effectively prevent youngsters from infection. Measles outbreaks continue to occur in the 21st century. Although an effective measles vaccine is available, a controversy regarding the popular *MMR vaccine*[11] formulation in the late 1990s led to a reduction of vaccinations in some countries. This disrupted efforts toward the eradication of measles. After the MMR vaccine controversy began, MMR vaccination compliance dropped sharply in the United Kingdom from 92% in 1996 to 84% in 2002 (Fig. 21.15). After the vaccination rates dropped the incidence of two out of the three diseases increased greatly in the United Kingdom. In 2008, for the first time in 14 years, measles was declared endemic in the United Kingdom. The disease was sustained within the population. This was caused by the preceding decade's low MMR vaccination rates which created a population of susceptible children able to spread the disease.

21.5 PERSPECTIVES

The emergence of new infectious diseases has been recognized in human history well before the discovery of causative infectious agents. Despite extraordinary advances in the development of countermeasures (diagnostics, therapeutics, and vaccines), the ease of world travel and an increase in global interdependence has added layers of complexity to containing these infectious diseases which could affect not only the health but the economic stability of societies. Surveillance such as quarantine at the airport and seaport locations has become critically important. In particular, during an epidemic, special attention should be paid to contain airborne viruses such as the influenza virus and SARS-CoV. These viruses can spread rapidly to other continents via air travel. Countermeasures to control these emerging viruses should be made available in preparation for future pandemics. Recent **Zika virus**[12] outbreak started in May 2015 in Brazil reminded us the importance of mosquito-borne flaviviruses (see Box 12.1).

21.6 SUMMARY

- *Newly emerging viruses*: Increased contact with animals, primarily due to the expansion of the human habitat, is the cause for the emergence of new viruses. Newly emerging viruses include HIV, Ebola virus, SARS-CoV, and MERS-CoV.
- *Reemerging viruses*: Besides newly emerging viruses, the variants of existing viruses also cause serious epidemics. Influenza virus represents the best example.
- *New epidemics of old viruses*: New epidemics caused by old viruses represent serious public health concerns. The measles outbreak in the Western Hemisphere is a good example of this.

11. **MMR vaccine** MMR vaccine is an immunization against measles, mumps, and rubella. It is a mixture of live attenuated viruses of the three diseases, administered via injection.
12. **Zika virus** A mosquito-borne flavivirus that was first discovered in 1947 in Zika forest in Uganda.

STUDY QUESTIONS

21.1 Describe three newly emerging viruses since the start of the 21st century and state their cause of emergence.

21.2 List an example of the viral outbreaks that occurred with the following epidemiologic change during the 20th century and state the reason for the change (1) epidemic to pandemic, (2) epidemic to endemic and (3) endemic to sporadic.

SUGGESTED READING

Chan, J.F., To, K.K., Tse, H., Jin, D.Y., Yuen, K.Y., 2013. Interspecies transmission and emergence of novel viruses: lessons from bats and birds. Trends Microbiol. 21 (10), 544–555.

Feldmann, H., 2014. Ebola—a growing threat? N. Engl. J. Med. 371, 1375–1378.

Knipe, D.M., Whelan, S.P., 2015. Rethinking the response to emerging microbes: vaccines and therapeutics in the Ebola Era—a conference at harvard medical school. J. Virol. 89, 7446–7448.

Mokili, J.L., Rohwer, F., Dutilh, B.E., 2012. Metagenomics and future perspectives in virus discovery. Curr. Opin. Virol. 2 (1), 63–77.

Wang, T.T., Palese, P., 2009. Unraveling the mystery of swine influenza virus. Cell. 137 (6), 983–985.

JOURNAL CLUB

- Xu, R., Ekiert, D.C., Krause, J.C., Hai, R., Crowe Jr., J.E., Wilson, I.A. 2010. Structural basis of preexisting immunity to the 2009 H1N1 pandemic influenza virus. Science 328 (5976), 357–360.

Highlight: The 2009 H1N1 swine flu is the first influenza pandemic in decades. Interestingly, the 2009 pandemic largely spared the elderly. This structure paper reported the hemagglutinin antigenic structure of the 2009 H1N1 flu virus is extremely similar to that of the 1918 H1N1 Spanish flu virus. This revealed conservation of the epitope in both pandemic viruses which are separated by 91 years. Antigenic similarities between two pandemic viruses provide an explanation of the age-related immunity against the 2009 H1N1 flu virus.

Part VI

Viruses and Disease

So far, we have focused on the molecular description of human pathogenic viruses; clinical consequences of virus infection and treatments have been only briefly described. Human viruses we covered in the preceding chapters cause significant disease to human (Box PVI.1). Although human pathogenic viruses cause diverse pathological lesions and clinical symptoms, here we focus on three important clinical consequences of the virus infections: AIDS, viral hepatitis, and cancers. Clinical aspects of HIV infection and AIDS will be covered in chapter "HIV and AIDS," and clinical aspects of viral hepatitis will be covered in chapter "Hepatitis Viruses." Tumor viruses will be covered in chapter "Tumor Viruses." Finally, preventive vaccines and therapeutic antiviral drugs will be covered in chapters "Vaccines" and "Antiviral Therapy," respectively.

BOX PVI.1 Why Viruses Cause Disease to Their Hosts?

It is counterintuitive that viruses cause disease to their host organism. This question could be a philosophical rather than scientific inquiry. A prevailing thought is that "the aim of virus evolution is to maximize its spread, not to cause disease to its host." In other words, the diseases that viruses cause are the unintended outcomes of virus infection. In analogy to a murder case, it is "accidental." Then, how does a virus cause disease or to what extent is the disease relevant to viral propagation in the host? A short answer to this question is that the viral disease is a consequence of host response to the virus infection (see Box 24.2). The majority of viral diseases are attributable to host immune response (eg, inflammation) that is activated to inhibit the viral propagation. For instance, the hepatitis B virus (HBV) itself is noncytopathic to the cells it infects but the host immune cells kill the HBV-infected hepatocytes, resulting in tissue damage.

Chapter 22

HIV and AIDS

Chapter Outline

The AIDS pandemic continues to confront us with unique scientific and public health challenges. The causative pathogenic agent, HIV infects predominantly immune cells, thereby weakening the host immune system. As a consequence, opportunistic infections such as pneumonia occur and are the main cause of death of HIV infected patients. Since the first reports of AIDS (acquired immunodeficiency syndrome) in the early 1980s, HIV, the etiological agent of AIDS, has drawn a great deal of attention from the media as well as public concerns. The AIDS epidemic is often called the "black death" of our time. Consequently, AIDS/HIV has been at the center of biomedical research in terms of research focus and resource allocation. Since the molecular aspects of HIV are covered in chapter "Retroviruses", this chapter will focus on clinical aspects of AIDS/HIV, such as its epidemiology, clinical course, diagnosis, prevention, and therapy.

22.1 EPIDEMIOLOGY

The AIDS pandemic continues to confront us with unique scientific and public health challenges. The causative pathogenic agent, HIV infects predominantly immune cells, thereby weakening the host immune system. As a consequence, opportunistic infections such as pneumonia occur and are the main cause of death of HIV infected patients. Since the first reports of AIDS in the early 1980s, HIV, the etiological agent of AIDS, has drawn a great deal of attention from the media as well as public concerns. The AIDS epidemic is often called the "black death" of our time. Consequently, AIDS/HIV has been at the center of biomedical research in terms of research focus and resource allocation. Since the molecular aspects of HIV are covered in chapter "Retroviruses", this chapter will focus on clinical aspects of AIDS/HIV, such as its epidemiology, clinical course, diagnosis, prevention, and therapy.

AIDS epidemiology has been characteristically associated with certain groups, termed high-risk groups, such as male homosexuals and intravenous drug users. The association with high-risk groups has now become somewhat blurred, since heterosexual transmission has become a common route of HIV transmission.

Discovery: In 1981, patients suffering from an immunodeficiency syndrome without any known causes and associated with male homosexuals were reported in California. The patients were ill with *opportunistic infections*[1] similar to infections that are associated with immunocompromised individuals. Predominantly, young male individuals with man-to-man sexual relationships were affected. In addition, intravenous drug users, hemophilia patients, and individuals who had received a blood transfusion were identified as the high-risk groups implicating an infectious nature of transmission. Because of its unusual epidemiological association and the fatal nature of the disease, a great deal of attention was paid to identifying the cause of the disease. In 1983, only two years after the publication of the first report on

1. **Opportunistic infection** It refers to the infection by pathogens, such as bacteria, viruses, yeast or protozoa, that usually do not cause diseases in a healthy host but cause disease in an immune compromised host.

Molecular Virology of Human Pathogenic Viruses. DOI: http://dx.doi.org/10.1016/B978-0-12-800838-6.00022-9

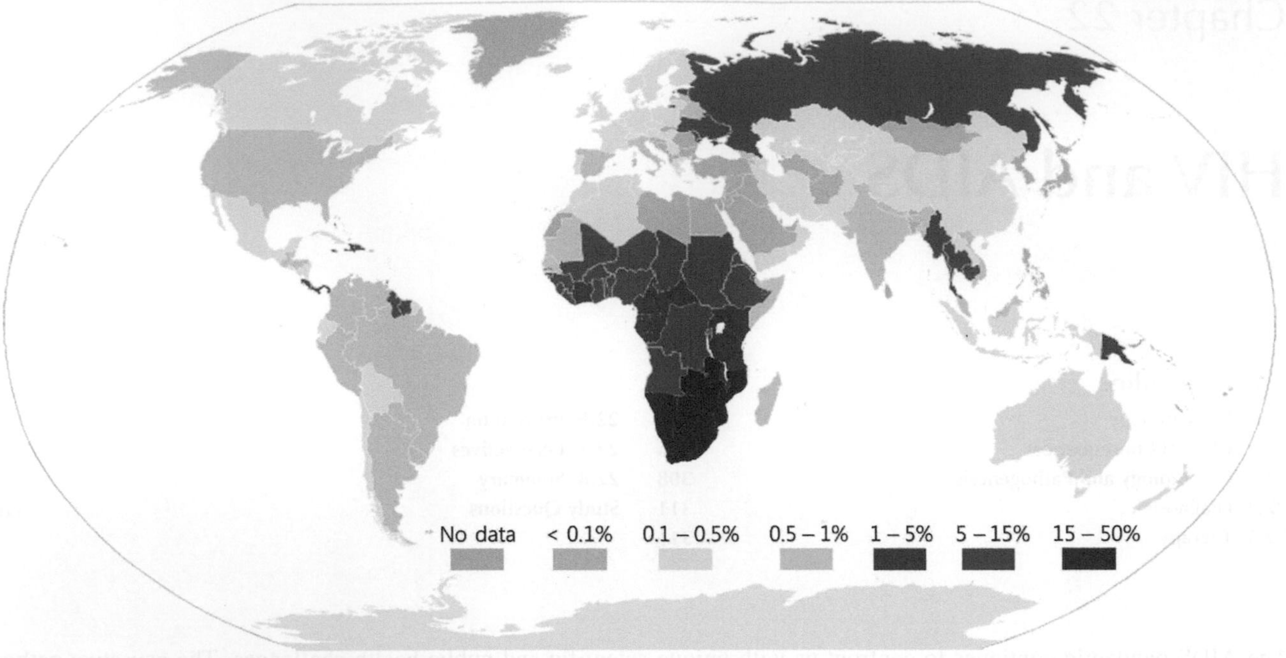

FIGURE 22.1 Geographic distribution of the HIV/AIDS prevalence. An estimated 37 million people are currently infected with HIV, which is about 0.5% of world population. Note the high rate of HIV infection in sub-Saharan Africa.

the emerging disease, two French virologists, Luc Montagnier and Francoise Barre-Sinoussi, found that a retrovirus is the causative pathogen. Indeed, they detected the reverse transcriptase (RT) activity from cultured $CD4^+$ T lymphocytes derived from a patient with lymphadenopathy, which occurs at an early phase of the disease. The detection of RT activity is a compelling evidence for the presence of retroviruses (see Box 17.3).

Epidemiology: It is estimated that over 60 million people have been infected by HIV, and the death toll is currently approaching 25 million. HIV infections are most prevalent in the sub-Saharan Africa region with more than 10% of the population infected (Fig. 22.1). Worldwide, nearly 37 million people are HIV-positive (Fig. 22.2). Moreover, the majority of new infections occur in sub-Saharan Africa. The reduced average life expectancy in this region is compelling evidence for the social impact of the AIDS epidemic (Fig. 22.3). The AIDS epidemic in sub-Saharan Africa represents a major public concern to the international community, such as the United Nations. Educational programs and improved access to antiviral drugs introduced by the United Nations and World Health Organization seem to have improved the situation.

Transmission: HIV spreads via blood or sexual contact. Poor medical practices such as the reuse of syringes have been a route of transmission in some areas. Transmission via blood products can be prevented by the diagnostic testing of all donor bloods. The transmission efficiency varies, depending on the mode of transmission. For instance, in the case of sexual transmission, the male-to-male transmission rate is higher than heterosexual transmission.

22.2 CLINICAL CONSEQUENCES

As stated above, the primary clinical consequence of HIV infection is the "immunodeficiency" that is caused by the depletion of $CD4^+$ T lymphocytes. The resulting immunodeficiency leads to opportunistic infections. The clinical course of HIV infection can be divided into three phases: acute infection, clinical latency, and AIDS (Fig. 22.4).

Acute infection: The initial period after contracting HIV is called "acute infection" or acute HIV. After the virus enters the body, there is a period of rapid viral replication, leading to an abundance of virus in the peripheral blood. During primary infection, the level of HIV may reach several million virus particles per milliliter of blood. This response is accompanied by a marked drop in the number of circulating $CD4^+$ T cells. The acute viremia is almost invariably associated with activation of $CD8^+$ T cells, which kill HIV-infected cells, and subsequently with antibody production, or seroconversion.

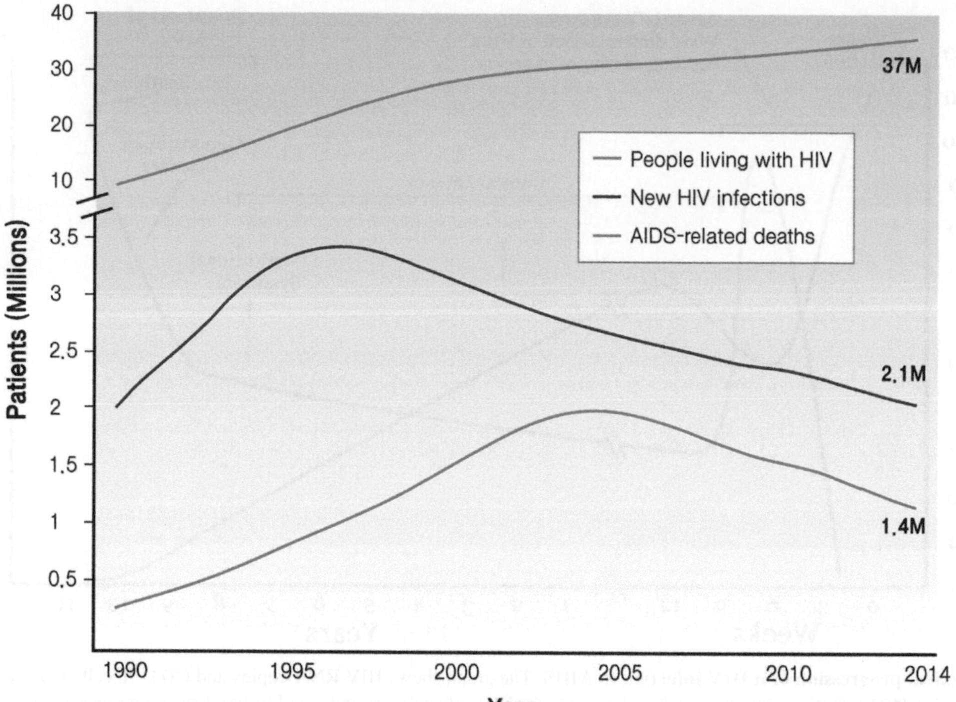

FIGURE 22.2 **HIV statistics.** The number of people living with HIV has continually increased to nearly 37 million (M). New HIV infections have been decreasing since 1996, while AIDS-related deaths are also gradually decreasing after peaking at 2.3 million deaths in 2005. Note that the scale on the *Y*-axis is broken.

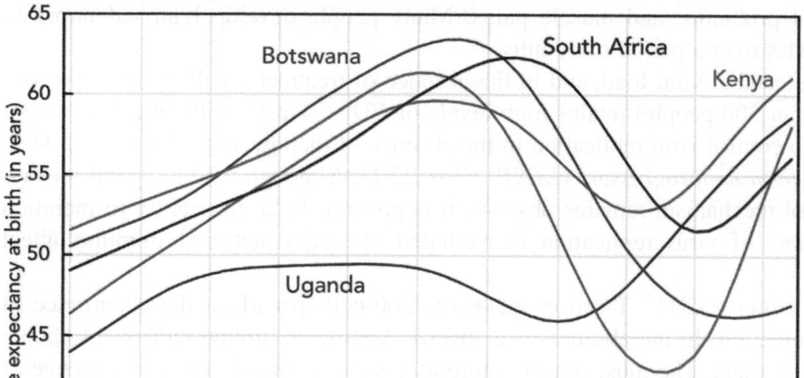

FIGURE 22.3 **Life expectancy of African countries severely affected by the HIV/AIDS epidemic.** Life expectancy in some AIDS-ravaged southern African countries from 1960 to 2012 is shown. The rapid drop in life expectancy that started in the 1990s became reversed after 2005.

Many individuals develop an influenza-like illness 2–4 weeks postexposure, while others have no significant symptoms. Symptoms occur in 40–90% of cases and most commonly include fever, enlarged and tender lymph nodes, throat inflammation, rash, headache, and/or a sore mouth and genitals. The duration of the symptoms varies, but it is usually one or two weeks. HIV-infected individuals characteristically suffer *lymphopenia*, which is the condition of having an abnormally low level of lymphocytes in the blood.

Clinical latency: The initial symptoms are followed by the stage known as "clinical latency," or asymptomatic HIV, or chronic HIV. In this context, the latency refers to the absence of clinical symptoms but not an absence of the viral replication. Without treatment, this second phase of an HIV infection can last from 3 years to over 20 years (on

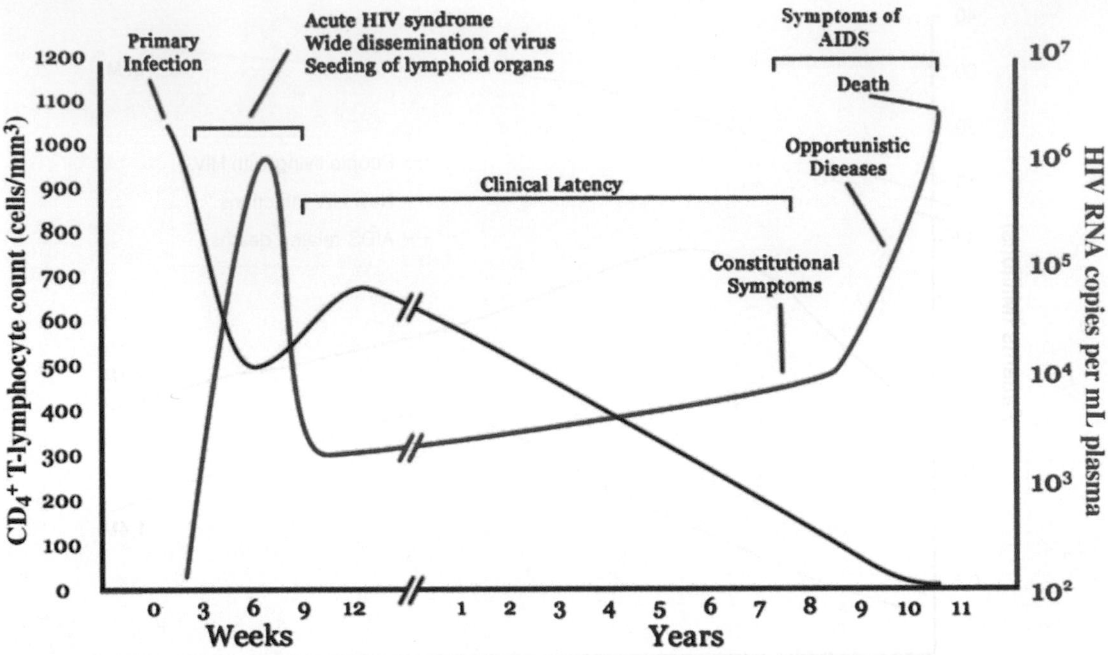

FIGURE 22.4 Clinical progression of a HIV infection to AIDS. The graph shows HIV RNA copies and CD4$^+$ T-cell counts in an infected individual of a treatment-naive HIV infection. The three clinical phases of a HIV infection are indicated in brackets: acute phase, clinical latency, and AIDS phase.

average, about 8 years). While typically there are few or no symptoms at first, near the end of this phase, many people experience fever, weight loss, gastrointestinal problems, and muscle pains. Many people develop lymphadenopathy, characterized by enlargement of the lymph nodes over a period of months.

Most HIV-1 infected individuals have a detectable viral load, and in the absence of treatment, will eventually progress to AIDS. However, a small proportion (1 in 300 people) retains high levels of CD4$^+$ T cells without antiretroviral therapy. These rare individuals who are able to control viral replication in the absence of therapy and achieve long-term asymptomatic infection are classified as *long-term nonprogressors (LTNP)* (Box 22.1). Although the biological mechanism responsible for this so-called elite control mechanism remains obscure, it is presumed that the ability to maintain CD4$^+$ T-cell counts and undetectable amounts of virus replication is mediated by active antiviral immunological control.

AIDS phase: AIDS is defined in terms of either a CD4$^+$ T-cell count below 200 cells per μL or the occurrence of specific disease in association with an HIV infection. In the absence of a specific treatment, around half of all people infected with HIV will develop AIDS within 10 years. The most common initial conditions that alert to the presence of AIDS are *opportunistic infections* such as pneumonia (40%), cachexia in the form of HIV wasting syndrome (20%), and esophageal candidiasis. Another common sign is recurring respiratory tract infections.

People with AIDS have an increased risk of developing various virally induced cancers, including Kaposi's sarcoma and Burkitt's lymphoma. Kaposi's sarcoma is the most common cancer, occurring in 10–20% of people with HIV. The second most common cancer is the Burkitt's lymphoma, which is the cause of death in nearly 16% of people with AIDS and is the initial sign of AIDS in 3–4% of infected individuals. Kaposi's sarcoma is associated with the KS-herpesvirus, while Burkitt's lymphoma is associated with EBV (see Box 9.3).

22.3 HIV BIOLOGY AND PATHOGENESIS

Human and primate lentiviruses: Besides HIV-1, another related human lentivirus, HIV-2, has been isolated. HIV-1 is believed to be a descendant of a simian immunodeficiency virus infecting chimpanzees in Central Africa (SIVcpz) (Fig. 22.5). Many species of African monkeys are natural hosts for simian immunodeficiency virus (SIV) but generally do not develop a disease as a consequence of an infection. In contrast, infection of Asian macaques, which are not

BOX 22.1 Elite Controller

Unlike most HIV-1 infected individuals, a small proportion of individuals (1 in 300) retains a high level of $CD4^+$ T cells without antiretroviral therapy for more than 5 years. These individuals are classified as "elite controllers" or *long-term nonprogressors (LTNP)*. Many of these patients have been HIV-positive for 30 years without progressing to the disease. They have been the subject of a great deal of research, since an understanding of their ability to control HIV infection may lead to the development of immune therapies or a therapeutic vaccine. LTNPs typically have viral loads under 10,000 copies RNA per mL blood, do not take antiretroviral drugs, and have $CD4^+$ counts within the normal range. Although the biological mechanism responsible for this so-called elite control mechanism remains obscure, it is presumed that the ability to maintain $CD4^+$ T-cell counts and undetectable amounts of virus replication is mediated by active antiviral immunological control. Evidence supporting the immunological control mechanisms, especially the action of HIV-specific $CD8^+$ T cells, arises for the fact that these rare people show considerable enrichment for alleles encoding certain HLA class I molecules (ie, MHC class I) or so-called protective HLA alleles. Another genetic trait that may affect progression includes CCR5 coreceptor mutation (ie, Δ32 variant of CCR5). The CCR5 mutation has drawn attention, because CCR5 is the coreceptor for HIV-1 entry. *CCR5-Δ32* is an allele of CCR5. The deleted portion of the CCR5 gene consists of 32 base pairs that correspond to the second extracellular loop of the receptor; the mutated receptor is nonfunctional and does not allow M-tropic HIV-1 virus entry, thus resulting in infection resistance. This allele is found in 5–14% of Europeans (of which 1% are homozygotes), but it is rare in Africans and Asians. It has been speculated that this allele was favored by natural selection during the smallpox epidemics that spread throughout Europe. The allele has a negative effect on T-cell function but appears to protect against smallpox and HIV. One study found that homozygotes for the mutated allele were strongly resistant to HIV-1 infection, and heterozygotes showed some level of resistance. The presence of the identified advantageous mutations, however, is not a unifying theme among LTNPs and is observed in an exceedingly small number of these patients. A fundamental challenge ahead is the development of effective ways of translating insights that emerge from studies of rare elite controllers into innovations for the prevention or treatment or infection with HIV.

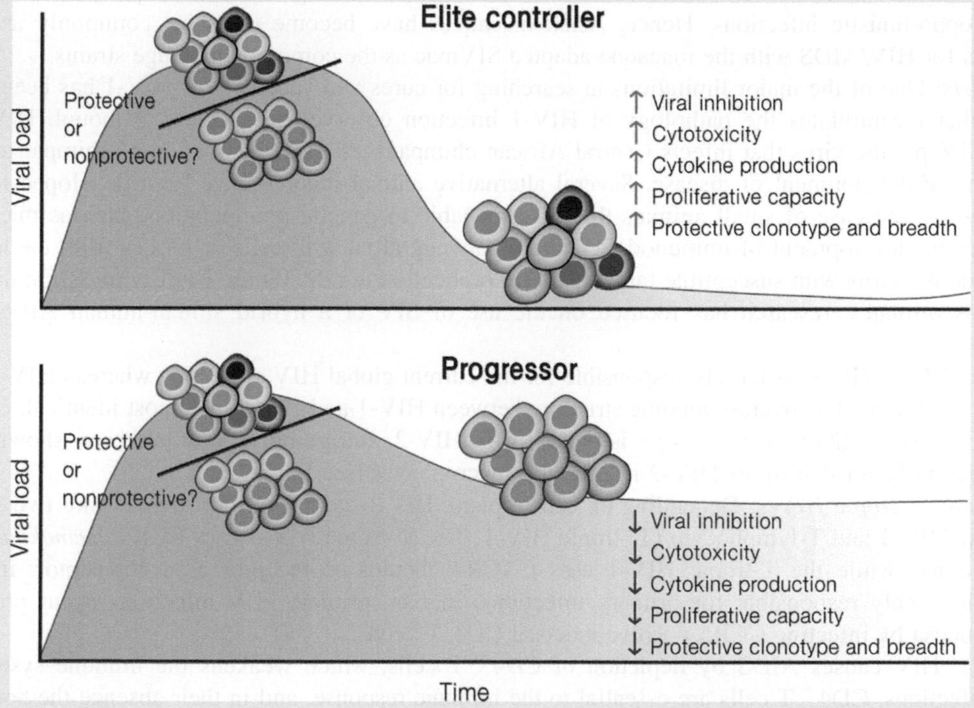

A schematic accounting for the protective $CD8^+$ T-cell in HIV elite controller. HIV specific $CD8^+$ T cells in elite controllers are characterized by greater functionality, greater breadth, and distinct TCR clonotypic profiles relative to those of $CD8^+$ T cells from progressors that recognize the same viral epitope. An important question is whether this difference of HIV specific $CD8^+$ T cells in elite controllers is the cause or the consequence of viral control.

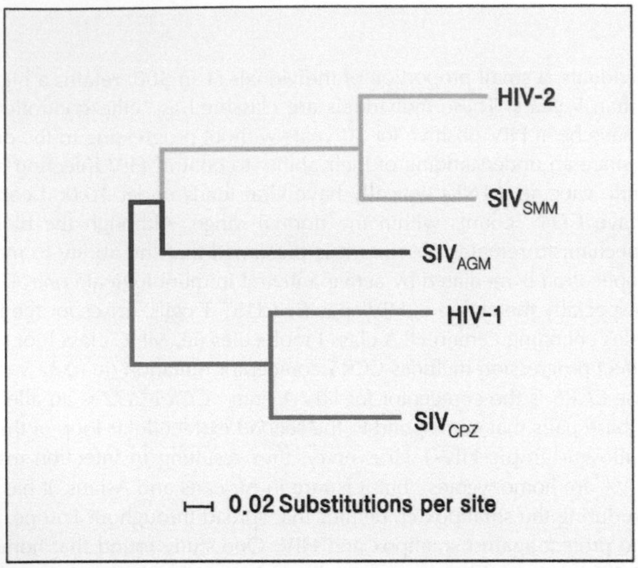

FIGURE 22.5 Phylogenetic relationship of primate lentiviruses. HIV-1 is related to SIV isolate from chimpanzees. HIV-2 represents a distinct lineage that is closely related to SIV of sooty mangabey monkey (SMM). SIV of African green monkeys (AGM) form other lineages that are more closely related to HIV-2 than to HIV-1. Horizontal branch lengths are to the scale shown. SIV_{SMM}: soothy mangabey monkey; SIV_{AGM}: African green monkey; SIV_{CPZ}: chimpanzee.

natural hosts for primate lentiviruses, with certain strains of SIV results in high viral loads, progression of $CD4^+$ T-cell depletion, and opportunistic infections. Hence, Asian macaques have become the most commonly used and widely accepted models for HIV/AIDS with the macaque-adapted SIVmac as the common challenge strains.

Animal models: One of the major limitations in searching for cures and vaccines for HIV-1 has been the lack of an animal model that recapitulates the pathology of HIV-1 infection observed in humans. Although HIV-1 is a direct descendant of SIVcpz, the virus that infects Central African chimpanzees, HIV-1 infection of chimpanzees in captivity rarely results in the development of disease. Several alternative animal models have been developed to study HIV-1 infection in vivo. In the case of small animals that are amenable to genetic manipulation, such as mice, efforts have been focused on the development of immunodeficient animals engrafted with cells or tissues from the human immune system to provide the virus with susceptible target cells for replication (eg, SCID-hu-Thy/Liv mice). In the case of large animals, such as primates, research has focused on the use of SIV or a hybrid simian-human AIDS virus dubbed "SHIV."

HIV-1 versus HIV-2: HIV-1 is largely responsible for the current global HIV epidemic, whereas HIV-2 causes AIDS in regions of West Africa. The overall genome structure between HIV-1 and HIV-2 is almost identical, except that Vpu is absent in HIV-2 (Fig. 22.6). Instead, Vpx is encoded by HIV-2. Intriguingly, Vpx has been shown to counteract SAMHD1 restriction factor that limits HIV-2 infection in macrophages (see Fig. 17.17).

M-tropic versus T-tropic HIV-1: Depending on cell tropism, HIV-1 can be further divided into two groups: macrophage (M)-tropic HIV-1 and T-lymphocyte (T)-tropic HIV-1. The M-tropic HIV-1 uses CCR5 *chemokine receptor*[2] as a coreceptor for entry, while the T-tropic HIV-1 uses CXCR4 chemokine receptor as a coreceptor. Importantly, the M-tropic HIV is largely responsible for primary infections. In fact, primary HIV infections occur predominantly in the intestinal mucosa by infecting CCR5 positive mucosal $CD4^+$ T cells.

Pathogenesis: HIV causes AIDS by depletion of $CD4^+$ T cells, which weakens the immune system and allows opportunistic infections. $CD4^+$ T cells are essential to the immune response, and in their absence the body cannot fight infections. The mechanism of $CD4^+$ T-cell depletion differs in the acute and chronic phases. During the acute phase, HIV-induced cell lysis and killing of infected cells by cytotoxic T cells accounts for $CD4^+$ T-cell depletion, although apoptosis may also be a contributing factor. During the chronic phase, the consequences of generalized immune activation, coupled with the gradual loss of the ability of the immune system to generate new T cells, appears to account for the slow decline in $CD4^+$ T-cell numbers. Although the symptoms of immune deficiency do not appear for years after

2. **Chemokine receptor** It refers to cytokine receptors found on the surface of certain cells that interact with a type of cytokine called a chemokine. It belongs to the family of G-protein coupled receptor (GPCR) that has a 7 transmembrane (7TM) domain.

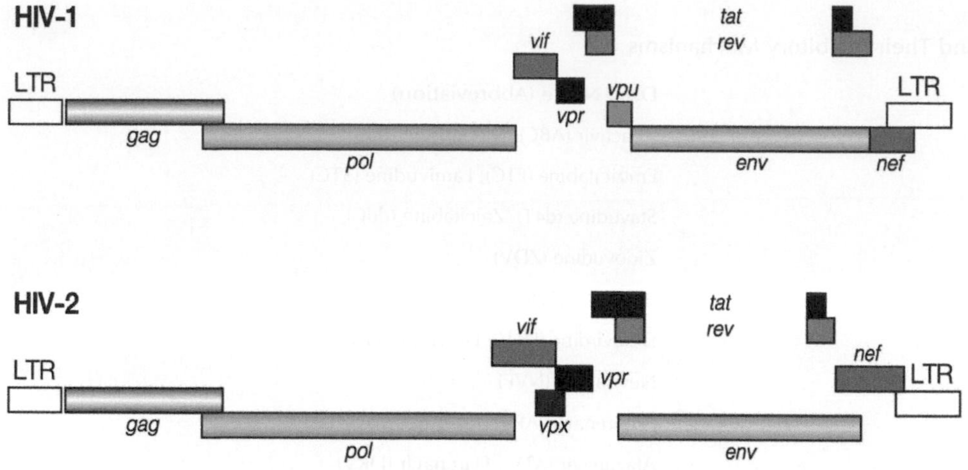

FIGURE 22.6 Genetic map of HIV-1 and HIV-2. The genetic organization of HIV-1 is compared to that of HIV-2. Note that Vpu is encoded only in HIV-1, whereas Vpx is encoded only in HIV-2.

a person is infected, the bulk of the $CD4^+$ T-cell loss occurs during the first weeks of infection (see Fig. 22.4), especially in the intestinal mucosa, which harbors the majority of the lymphocytes found in the body. The reason for the preferential loss of mucosal $CD4^+$ T cells is the expression of CCR5 chemokine receptor by the majority of mucosal $CD4^+$ T cells, whereas only a small fraction of $CD4^+$ T cells in the bloodstream express CCR5 chemokine receptor.

22.4 DIAGNOSIS

HIV/AIDS is diagnosed via laboratory testing and then assessed based on the presence of certain symptoms. HIV screening is recommended for all people 15–65 years of age including all pregnant women. In addition, testing is recommended for all those at high risk, which includes individuals diagnosed with a sexually transmitted illness.

Most people infected with HIV develop specific antibodies (ie, seroconversion) within 3–12 weeks after the initial infection occurred. ELISA is a common diagnostic method for the detection of HIV antibodies. Once an individual is found to be positive by ELISA, immunoblotting are used to confirm the presence of HIV antigens in the blood (see Fig. 4.3). Primary HIV infections (ie, acute infection) before seroconversion are identified by measuring the presence of HIV-RNA or p24 antigen in the blood. Positive results obtained by antibody or PCR testing are confirmed by using a different antibody with diagnostic relevance or by PCR.

22.5 THERAPY

Anti-HIV drugs: More than 25 anti-HIV drugs have been developed over the past three decades. As far as antiviral drugs are concerned, HIV stands out in numbers compared to other viruses (see Fig. 26.1). These drugs are composed of five classes, nucleoside and nonnucleoside drugs targeting the RT, protease and integrase inhibitors, and drugs acting on the viral envelopes (Table 22.1). The first anti-HIV drug approved by the FDA in 1987 was zidovudine (AZT), which was initially developed as a cancer drug. AZT is a nucleoside analog, acting as an inhibitor of HIV RT. Related nucleoside analogs drugs include lamivudine (3TC) and emtricitabine (FTC). These nucleoside RT inhibitors (*NRTIs*) are less toxic than AZT, nonetheless the emergence of drug-resistant variants has been a major drawback. In addition to nucleoside analogs, nonnucleoside RT inhibitor (*NNRTIs*) targeting RT have been developed such as nevirapine (NVP) (see Fig. 27.7). Because these nonnucleoside analogs bind to a region distinct from the active site of HIV RT, they were expected to complement the drawback of the nucleoside analog drugs. Nonetheless, treatment with nonnucleoside analog drugs also generate the emergence of drug-resistant variants.

In addition to the RT activity of HIV, two other viral enzyme activities have been exploited as anti-HIV drug targets: aspartate protease (PR) and integrase (IN). In particular, the resolution of the crystal structures of the viral proteins has promoted drug discovery by *structure-based drug design*. In particular, PR inhibitors have been successfully developed by implementing such a rational drug design (see Fig. 26.8). As an outcome of this an approach, over 10 PR inhibitors have been developed including ritonavir (RTV) and saquinavir (SQV) (see Table 22.1). Likewise, two integrase inhibitors have been developed, raltegravir (RAL) and elvitegravir (EVG) (see Table 22.1).

TABLE 22.1 Anti-HIV Drugs and Their Inhibitory Mechanisms

Mechanism	Drug Name (Abbreviation)
RT inhibitors (nucleoside)	Abacavir (ABC), Didanosine (ddI)
	Emtricitabine (FTC), Lamivudine (3TC)
	Stavudine (d4T), Zalcitabine (ddC)
	Zidovudine (ZDV)
RT inhibitor (nucleotide)	Tenofovir (TDF)
RT inhibitors (nonnucleoside)	Delavirdine (DLV), Etravirine (EFV)
	Nevirapine (NVP)
PR inhibitors	Amprenavir (APV), fos-Amprenavir (fAPV)
	Atazanavir (ATV), Darunavir (DRV)
	Indinavir (IDV), Lopinavir (LPV)
	Nelfinavir (NFV), Saquinavir (SQV)
	Ritonavir (RTV), Tipranavir (TPV)
Integrase inhibitor	Raltegravir (RAL), Elvitegravir (EVG), Dolutegravir (DTG)
Entry inhibitor	Enfuvirtide (ENF), Maraviroc

In addition to targeting enzymes involved in replication, protein processing and integration, other steps in the HIV life cycle are also exploited as antiviral targets. Recently, two entry inhibitors have been developed: enfuvirtide (ENF) and maraviroc. First, enfuvirtide (ENF) targets the gp41 (TM) envelope glycoprotein of HIV, thereby blocking membrane fusion, a step necessary for viral entry. Second, maraviroc essentially acts as an antagonist of the CCR5 coreceptor to prevent viral entry (see Fig. 26.6). It blocks the interaction between CCR5 and gp120, a critical step prior to the membrane fusion.

None of these anti-HIV drugs alone is effective in suppressing HIV propagation or the progression of the disease. Fortunately, the availability of multiple anti-HIV drugs has made the so-called combination therapy possible that has proved fairly effective in suppressing the viral replication and disease progression.

Combination therapy: Current treatment consists of a highly active antiretroviral therapy (*HAART*[3]) which slows the progression of the disease. It also includes preventive and active treatment of opportunistic infections.

Current HAART option is the therapy using a combinations (or "cocktails") consisting of at least three medications belonging to at least two types, or "classes," of antiretroviral agents (Fig. 22.7). Initially, treatment typically consists of a NNRTI plus two nucleoside analog reverse transcriptase inhibitors (NRTIs). Typical NRTIs include: AZT or tenofovir (TDF) and 3TC or FTC. Combinations including protease inhibitors (PIs) are used if the above regimen loses effectiveness.

The World Health Organization recommends anti-HIV drug treatment in all adolescent, adults, and pregnant women with a CD4 count less than 500 per μL with this treatment being especially important in those with counts less than 350 per μL or those with symptoms regardless of CD4 count. For a successful treatment outcome to reduce the viral load, treatment has to be ongoing without any interruptions. The desired outcome of treatment is to obtain plasma HIV-RNA count below 50 copies per mL in the long run. The benefits of treatment include a decreased risk of progression to AIDS and a decreased risk of death. Additional benefits include a decreased risk of transmission of the disease to sexual partners and a decrease in mother-to-child transmission. The effectiveness of the treatment depends on compliance to a large part to a large part. Reasons for nonadherence include poor access to medical care, inadequate social supports, mental illness, and drug abuse.

3. **HAART (highly active antiretroviral therapy)** It refers to antiretroviral therapy in which two or three antiviral drugs are combined and given as a mixture.

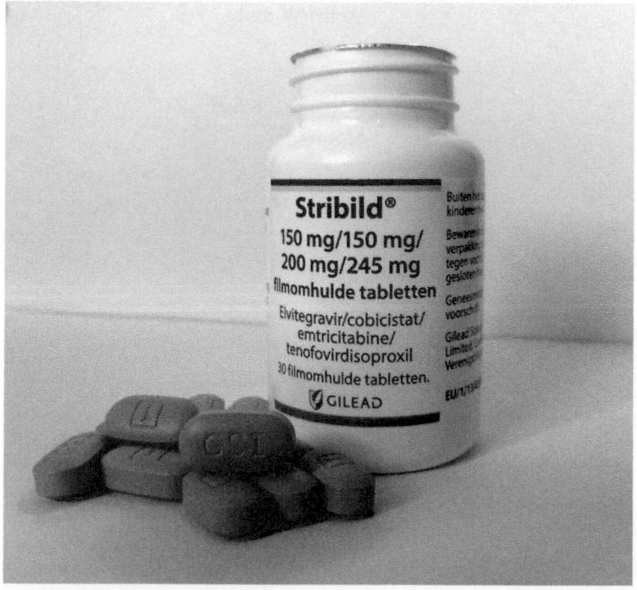

FIGURE 22.7 One tablet medication used for combination therapy. A tablet contains a combination of three anti-HIV drugs: elvitegravir/cobicistat/emtricitabine/tenofovir (brand name Stribild) is a combination drug for the treatment of HIV/AIDS. Elvitegravir, emtricitabine, and tenofovir directly suppress viral reproduction. Cobicistat increases the effectiveness of the combination by inhibiting the liver and gut wall enzymes that metabolize elvitegravir.

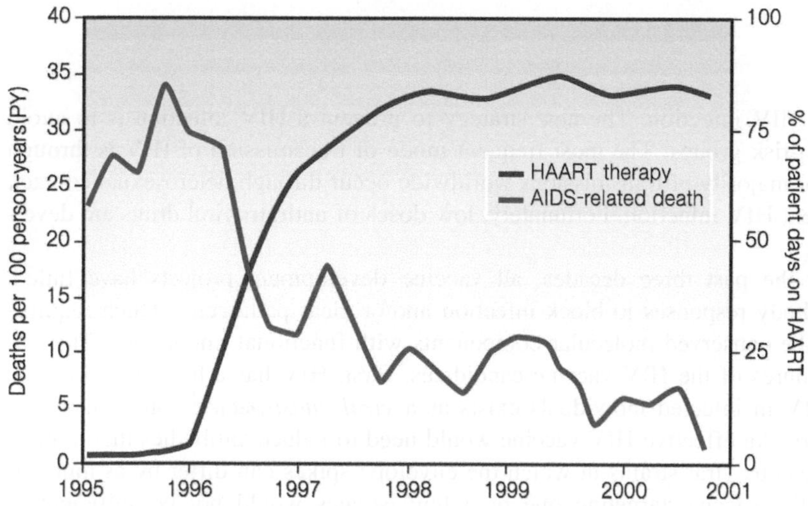

FIGURE 22.8 Combination therapy and its impact on the AIDS-related deaths. Implementation of combination therapy, termed HAART therapy, has dramatically decreased the number of AIDS-related deaths in the United States.

Since the implementation of HAART in 1996, its outcome has been tremendous. The AIDS-related deaths rate in the United States has dramatically decreased (Fig. 22.8). In most treated individuals, HIV in the blood becomes undetectable using standard diagnostic tests, and the $CD4^+$ T-cell count recovers to near normal levels. Although the virus itself is not cleared completely from the body, progression to AIDS can be prevented. The introduction of the HAART approach has changed HIV/AIDS from an "acute fatal disease" to a "chronic manageable disease." HAART is expected to change the trajectory of the epidemic.

HIV-1 reservoirs: A question arises as to why HIV infection is not cured with such an effective antiviral treatment. A strong cocktail of anti-HIV drugs reduces the HIV load in the blood to undetectable levels on standard tests, but these drugs have not cured anyone. A reason is the presence of a small population of long-lived "*reservoir*"[4] cells with a quiescent, replication-competent provirus inserted in their chromosomes (Fig. 22.9). The long-lived population of

4. **Reservoir** It refers to a small population of HIV-infected T cells in resting state, but can be reactivated later.

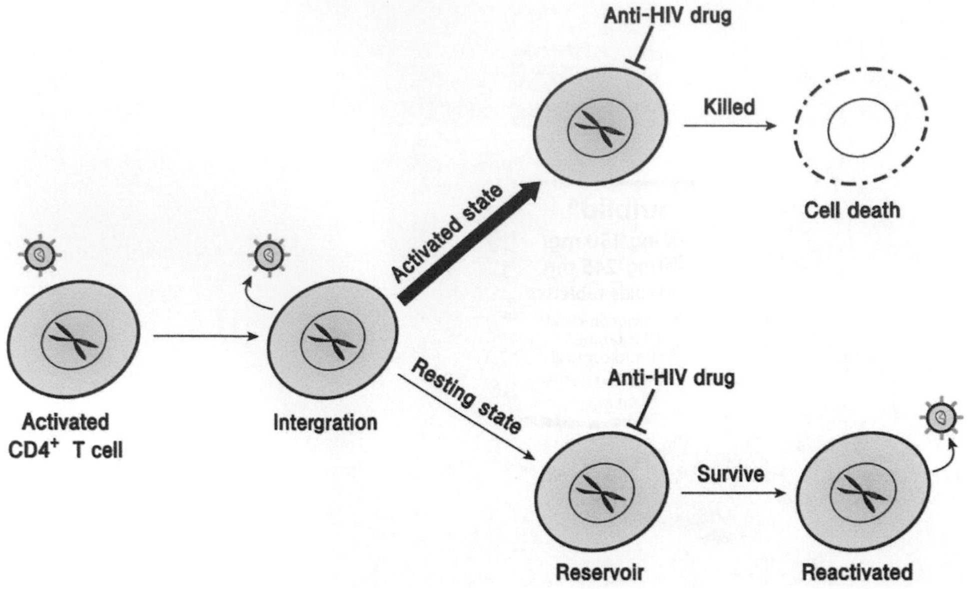

FIGURE 22.9 Viral latency and long-lived reservoir T cells. HIV infection and integration of viral DNA into the host chromosomes of $CD4^+$ T cells leads to active replication but cells can also enter a resting state. These latently infected memory cells can live for years. The dormant proviral DNA is capable of reestablishing new rounds of infection, if therapy is stopped.

memory T cells is capable of reigniting new rounds of infection if the therapy is interrupted. This latent pool of virus is established within days of infection and is impervious to current therapy.

22.6 PREVENTION

Currently, no vaccine is available to prevent HIV infection. The best strategy to prevent a HIV infection is to avoid social behaviors that are characteristic of high-risk groups. The most frequent mode of transmission of HIV is through sexual contact with an infected person, and the majority of transmissions worldwide occur through heterosexual contact. Safe sex is therefore the best strategy to prevent HIV infection. Fortunately, low doses of antiretroviral drugs are developed for preventative applications (Box 22.2).

Vaccines: Despite strenuous efforts over the past three decades, all vaccine development projects have failed. Effective vaccines usually induce potent antibody responses to block infection and/or clear pathogens, which requires that the protective antibodies are directed to the conserved molecular components with functional importance. The following difficulties have contributed to the failures of the HIV vaccine candidates. First, HIV has a high mutation rate (10^{-5} per replication). As a consequence, HIV in infected individuals exists as a *viral quasispecies*[5] rather than as a single viral species with sequence homogeneity. An effective HIV vaccine would need to induce antibodies that target a large number of rapidly evolving contemporaneous viral strains in which the envelope spikes can differ by as much as 35% in their amino acid sequences. Thus, the vaccine targeting one or a few isolates would not be sufficient to neutralize the rapidly mutating viral strains.

Second, the HIV envelope glycoprotein spike has a formidable neutralizing-evasion mechanism, including extensive glycosylation, hypervariability of amino acid sequences, conformational masking, and inaccessibility of the conserved sites (Fig. 22.10). In other words, the *structural features of the envelope spike* enable HIV to hide conserved epitope from neutralizing antibodies. The importance of structural attributes in limiting antibody potency such as "conformational masking" of the receptor binding sites, are well established in rhinovirus and influenza virus. In addition to these attributes, the low density of envelope spikes on the virion, a distinguishing feature of the HIV envelope, also help the HIV to escape neutralizing antibodies by impeding bivalent binding of immunoglobulin G antibodies, normally used to achieve high affinity binding and potent neutralization. The lower avidity of otherwise potent neutralizing antibody allows HIV to evade humoral immunity. Moreover, a novel mode of HIV transmission allows HIV to escape neutralizing antibodies. HIV can spread via cell-to-cell spread without being released from the infected cell by budding (see Box 3.4). All of the above-described mechanisms allow the virus to evade the host's humoral immune response.

5. **Viral quasispecies** It refers to a group of closely related viruses which are distinguished by a high number of mutations.

BOX 22.2 Preexposure Prophylaxis Against HIV Infection

Recently, the Food and Drug Administration (FDA) approved antiretroviral drugs used for the treatment of HIV as preexposure prophylaxis (PrEP) against HIV infection. This approach is the first preventive medicine against HIV infection. *Truvada*, developed by Gilead Sciences, is a fixed-dose combination of two antiretroviral drugs (tenofovir and emtricitabine) used for the treatment of HIV. Yearly 50,000 individuals are newly infected by HIV in the United States. The FDA has contended that "truvada represents a revolutionary means of preventing an HIV epidemic in the US." In studies, tenofovir decreased the risk of contracting HIV by 51%. On the other hand, FDA approval was not unanimously welcomed. Some have warned that long-term use of low doses of antiretroviral drugs would increase the emergence of drug-resistant variants. The verdict on this debate will follow in a few years.

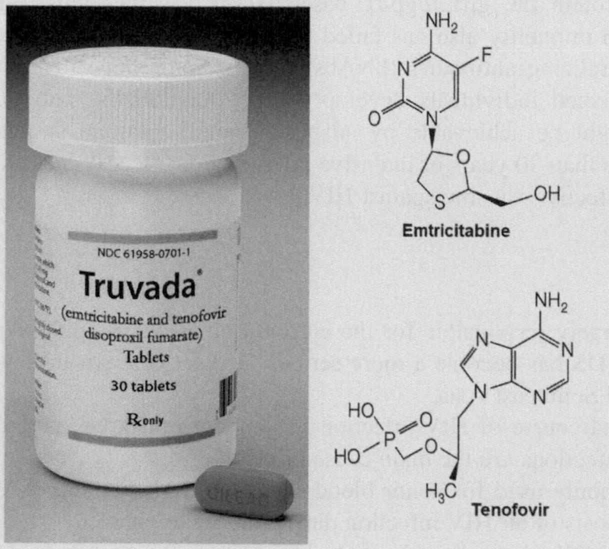

Truvada. (Left) A photo of the "Truvada" bottle. (Right) Chemical structure of two antiretroviral drugs that constitute truvada: tenofovir and emtricitabine.

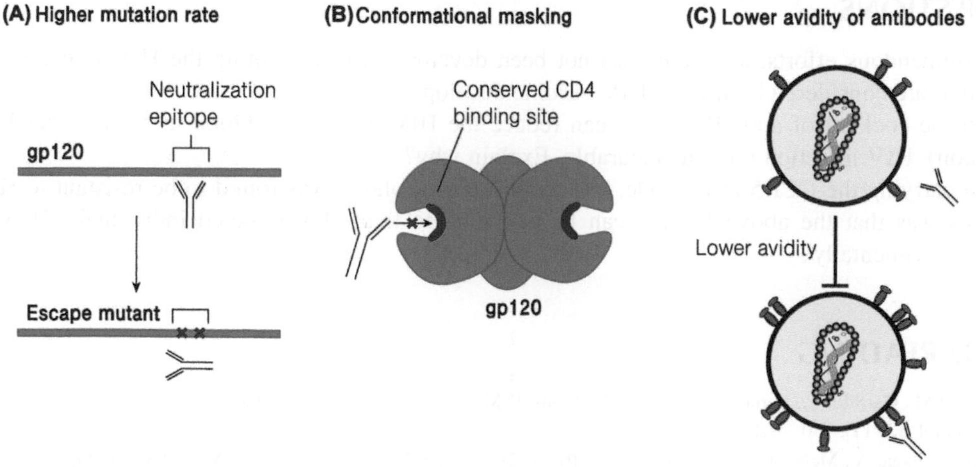

FIGURE 22.10 Attributes related to HIV vaccine failure. (A) High mutation rate. As a result, an infected individual possesses a high number of HIV variants, termed "quasispecies." The presence of quasispecies makes it difficult to neutralize the virus by vaccination, which can target only one or a few variants. (B) Conformational masking. The conserved site on the envelope spike (ie, gp120) is positioned at the base in a "canyon," which is inaccessible to the antibodies. (C) Low numbers of viral glycoproteins present in the viral envelope. Inefficient bivalent binding of antibodies to the envelope spikes due to the low density of the viral envelope glycoproteins results in a low avidity of antibodies.

22.7 PERSPECTIVES

Despite great advances, AIDS/HIV remains a threat to public health worldwide. The "war" against the AIDS pandemic has been the major agenda in developed countries over the past three decades. Unparalleled resources have been allocated to fight the HIV/AIDS pandemic. Since the discovery of HIV in 1983, many efforts have been undertaken to prevent the viral transmission such as vaccine development and antiviral drug discovery. Importantly, the development of an antiretroviral therapy has been a major triumph. Thanks to the anti-HIV drugs, the outlook for disease control is not pessimistic. For instance, AIDS-related deaths and new HIV infections are now gradually decreasing. Nevertheless, AIDS remains a global concern. In particular, serious concerns exist in developing countries, such as sub-Saharan Africa, South America, and Southeast Asia, where antiviral drugs are unaffordable. Efforts to make the drugs affordable clearly represent the unmet medical needs. On the other hand, all the vaccine development has been unsuccessful. For instance, all envelope glycoprotein (ie, gp120/gp41)-based HIV-1 vaccines have failed. The vaccine effort focused exclusively on T cell-mediated immunity also has failed to prevent HIV-1 infection to date. The recent discoveries of naturally arising, broadly neutralizing antibodies (bNAbs) provide some hope that a vaccine is possibly feasible. The finding that a fraction of infected individuals develop broadly neutralizing antibodies supports the idea that new approaches to vaccination might be achievable by adapting natural immune strategies or by using structure-based immunogen design. After more than 30 years of intensive investigation, we still lack the basic scientific knowledge necessary to produce a safe and effective vaccine against HIV-1.

22.8 SUMMARY

- *Epidemiology*: HIV-1 is largely responsible for the current global AIDS epidemic. HIV-2 epidemic is confined to West African countries. AIDS has become a more serious medical concern in developing countries in sub-Saharan Africa, South America, and Southeast Asia.
- *Clinical course*: The clinical course of HIV infection is composed of three stages: acute infection, clinical latency, and AIDS. Opportunistic infections are the main cause of death.
- *Diagnosis*: ELISA is commonly used for donor blood screening to detect anti-HIV antibodies. PCR-based methods are used for the early diagnosis of an HIV infection during the acute phase.
- *Therapy*: More than 25 anti-HIV drugs have been developed over the past three decades. The combination therapy, termed HAART, is based on a cocktail of two or more antiretroviral drugs. The therapy is remarkably effective in suppressing viral propagation and reducing AIDS-related deaths.
- *Vaccines*: No vaccine is available yet, despite tremendous efforts. The high mutation rate and sequence heterogeneity of the viral sequences help explain the vaccine failure.

STUDY QUESTIONS

22.1 Despite tremendous efforts, a vaccine has not been developed for preventing the HIV infection. Describe three reasons that are considered hurdles in HIV vaccine development.

22.2 Although the cocktail of anti-HIV drugs can reduce the HIV titer in the blood to an undetectable level (~99% suppression), HIV infection remains incurable. Explain why?

22.3 A person carrying the CCR5-Δ32 allele, termed "Elite controller," was found to be resistant to HIV infection. A hypothesis was that the above finding can be exploited to treat HIV-infected individuals. How would you go about it experimentally?

SUGGESTED READING

Archin, N.M., Sung, J.M., Garrido, C., Soriano-Sarabia, N., Margolis, D.M., 2014. Eradicating HIV-1 infection: seeking to clear a persistent pathogen. Nat. Rev. Microbiol. 12 (11), 750–764.

Carter, C.C., Onafuwa-Nuga, A., McNamara, L.A., Riddell, Jt, Bixby, D., Savona, M.R., et al., 2010. HIV-1 infects multipotent progenitor cells causing cell death and establishing latent cellular reservoirs. Nat. Med. 16 (4), 446–451.

Cohen, J., 2011. The emerging race to cure HIV infections. Science. 332 (6031), 784–785, 787–789.

Feinberg, M.B., Ahmed, R., 2012. Born this way? Understanding the immunological basis of effective HIV control. Nat. Immunol. 13 (7), 632–634.

Klein, F., Mouquet, H., Dosenovic, P., Scheid, J.F., Scharf, L., Nussenzweig, M.C., 2013. Antibodies in HIV-1 vaccine development and therapy. Science. 341 (6151), 1199–1204.

JOURNAL CLUB

- Tebas, P., et al. 2014. Gene editing of CCR5 in autologous CD4 T cells of persons infected with HIV. N. Engl. J. Med. 370 (10), 901–910.

  Highlight: Elite controller carrying the CCR5-Δ32 allele is resistant to HIV infection. To exploit this observation, CCR5 gene was disrupted by gene editing technology ex vivo. Then, HIV infected people were infused with autologous T cells in which the CCR5 gene was disrupted. Remarkably, the authors found that the cells with disrupted CCR5 remained capable of engraftment, long-term survival, and trafficking to mucosal sites in patients, proving the approach to be safe. This work represents a significant step toward the customized cell therapy for the treatment of HIV infection.

INTERNET RESOURCES

The Joint United Nations Program on HIV and AIDS, or UNAIDS: http://aidsinfo.unaids.org

Chapter 23

Hepatitis Viruses

Chapter Outline

Viruses that cause hepatitis are collectively referred to as *hepatitis*[1] viruses. Certainly, this is based on the similarity of clinical symptoms the viruses can cause but "hepatitis viruses" does not reflect a valid classification, which is based on genome type, genome organization, viral structure, and further genetic and biochemical characteristics. Six viruses were identified which are exclusively associated with hepatitis, and they were named according to their order of discovery: hepatitis A virus (HAV) to hepatitis G virus (HGV). They belong to different virus families, except that hepatitis C virus (HCV) and HGV belong to the family Flaviviridae (Table 23.1). HAV belongs to the family Picornaviridae, hepatitis B virus (HBV) belongs to the family Hepadnaviridae, hepatitis D virus (HDV) is not assigned to a family and classified in the genus "*Deltavirus*," and hepatitis E virus (HEV) is assigned to the family Hepeviridae. Certainly, neither similarity in genome organization nor nucleotide sequence homology is evident among these six hepatitis viruses.

Being discovered in the 1960s, HAV and HBV are well established as etiological agents for A-type and B-type viral hepatitis, respectively. Nonetheless, viral hepatitis that was not attributable to HAV and HBV was clinically recognized in the 1970s and then aptly dubbed as "*non-A, non-B hepatitis*." Thanks to recombinant DNA technology, four hepatitis viruses were identified by molecular technologies and cloned in the 1990s: HCV, HDV, HEV, and HGV (see Table 23.1). It is now believed that the identity of almost all viruses responsible for non-A, non-B hepatitis has been uncovered.

The molecular features of each of these viruses have been described in Part II to Part IV. Although these hepatitis viruses belong to different unrelated virus families by classification, they are often investigated or related to each other in the context of liver diseases such as chronic inflammation (hepatitis), fibrosis (liver cirrhosis), and cancer (hepatocellular carcinoma, HCC). In that regards, it is worth noting similarities and differences among hepatitis viruses. First, all these viruses cause acute hepatitis, an inflammation of liver tissues. In particular, two water-borne viruses (ie, HAV and HEV) cause only acute hepatitis, whereas two blood-borne viruses (ie, HBV and HCV) cause not only acute hepatitis but also chronic hepatitis (Fig. 23.1; see Table 23.1). Moreover, HBV and HCV cause liver cancer, HCC. Out of five human tumor viruses two are hepatitis viruses (see Table 24.2). Second, one salient feature of hepatitis viruses are their liver *tropism*.[2] All these viruses specifically target and infect hepatocytes, a parenchymal cell of liver tissue, but rarely infect nonliver cells. Hepatocyte-specific receptors are believed to confer hepatotropism.

1. **Hepatitis** It refers to a medical condition defined by the inflammation of liver and characterized by the presence of inflammatory cells in the liver tissue. The term *hepa* is derived from Greek word for "liver" and the term *titis* is derived from Greek word for "inflammation."

2. **Tropism** It refers to the cell specificity of viral infection. The term tropism is derived from Greek word for "a turning"—*tropos*, indicating growth or turning movement of a biological organism.

Molecular Virology of Human Pathogenic Viruses. DOI: http://dx.doi.org/10.1016/B978-0-12-800838-6.00023-0

TABLE 23.1 Clinical Features of Human Hepatitis Viruses

Virus	Family	Genome	Transmission	Persistency	Cancer
Hepatitis A virus	Picornaviridae	RNA	Oral-fecal	Acute	–
Hepatitis B virus	Hepadnaviridae	DNA	Blood	Acute and chronic	HCC
Hepatitis C virus	Flaviviridae	RNA	Blood	Acute and chronic	HCC
Hepatitis D virus	Unassigned; only genus "*Deltavirus*"	RNA	Blood	Chronic	
Hepatitis E virus	Hepeviridae	RNA	Oral-fecal	Acute	–
Hepatitis G virus	Flaviviridae	RNA	Oral-fecal	Acute	–

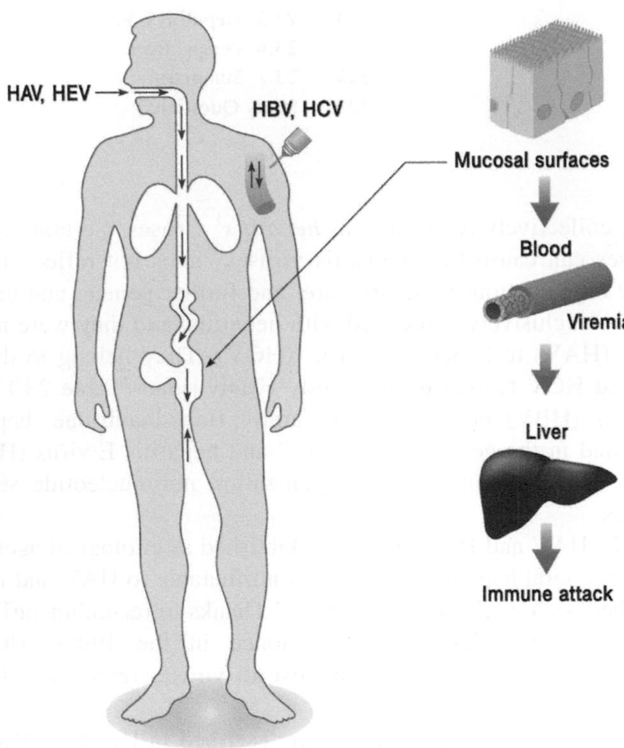

FIGURE 23.1 Transmission routes of hepatitis viruses. Water-borne-HAV and HEV transmit via the oral-fecal route, whereas HBV and HCV transmit via blood or sexual contact. In the latter case, the virus enters the bloodstream via the mucosal layer of reproductive organs.

23.1 HEPATITIS A VIRUS

HAV is a RNA virus, which belongs to the family Picornaviridae (see Table 23.1). HAV is transmitted via the oral-fecal route from contaminated water or foods, and its infection causes acute hepatitis, and was formerly known as *infectious hepatitis*.

Epidemiology: HAV is usually spread by eating or drinking food or water contaminated with infected feces. Shellfish which have not been sufficiently cooked is a relatively common source. It may also spread through close contact with an infectious person. Most of the infected individuals (ie, 90% of children and 25–50% of adults) are asymptomatic. A person is immune for the rest of their life after a single infection. Globally around 1.5 million symptomatic cases occur each year with likely tens of millions of infections in all (Fig. 23.2). It is prevalent in regions with poor sanitation and not enough clean water. In the developing world about 90% of children have been infected by age 10 and thus are immune by adulthood. HAV outbreaks often occur in moderately developed countries where children are not exposed when young and there is not widespread vaccination. In the year 2010, acute hepatitis A (AHA) resulted in 102,000 deaths.

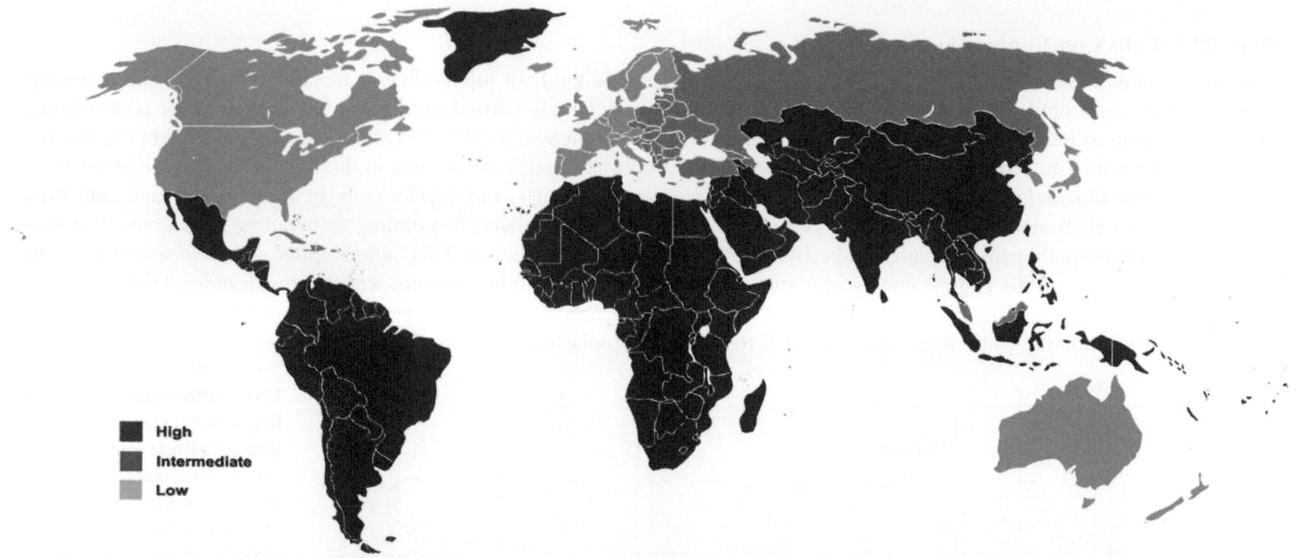

FIGURE 23.2 Geographic distribution of HAV prevalence. The prevalence rates of anti-HAV antibodies are high in most countries in Asia, Africa, and Latin America, but low in Europe, the United States, Australia, South Korea, and Japan. High (red): prevalence >8%; intermediate (orange): 2–7%; low (gray): <2%.

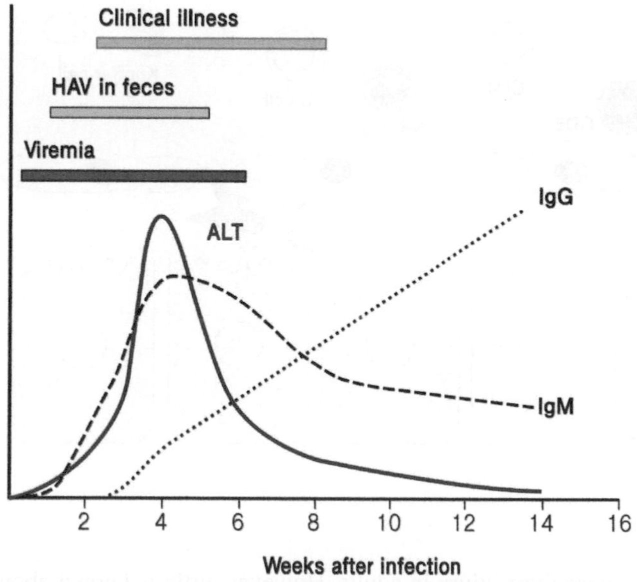

FIGURE 23.3 Clinical course of HAV infection. Antibody response of HAV infection (serum IgG, IgM), and ALT level is plotted. The periods for the viremia (ie, virus in blood), HAV shedding in feces, and clinical illness are denoted by *bold lines*. Note that HAV particles are shed in feces almost 2 weeks earlier than the onset of clinical illness.

Pathogenesis: Many cases show little or no symptoms especially for youngsters. The incubation period is typically 2 to 7 weeks until *serum ALT* (ie, alanine aminotransferase) value is abnormally high (Fig. 23.3). Most symptoms last 8 weeks and may include nausea, vomiting, diarrhea, yellow skin, fever, and abdominal pain. Following ingestion, HAV enters the bloodstream through the epithelium of the oropharynx or intestine. The blood carries the virus to its target organ, the liver, where it multiplies within hepatocytes and *Kupffer cells* (liver-resident macrophages) (Box 23.1). Virions are secreted into the bile and released in stool. HAV is excreted in large quantities approximately 11 days prior to the appearance of symptoms or anti-HAV *IgM* antibodies in the blood (see Fig. 23.3). Thus, the diagnosis relies on the IgM in the blood.

BOX 23.1 Cells Constituting Liver

The liver is often called the "chemical factory" of our body, because the liver fulfills diverse metabolic functions (ie, carbohydrate, proteins, and lipids). The metabolic functions of the liver are largely carried out by the hepatocytes, liver parenchymal cells. In addition to liver parenchymal cells, however, diverse nonparenchymal cells (NPCs) constitute the liver. In fact, hepatocytes constitute only 40% of the cells in the liver, while NPCs constitute 60% of the cells in the liver. The representative NPCs include the *liver sinusoid endothelial cells (LSEC)*, *hepatic stellate cells (HSC)*, and *Kupffer cells*. In addition, immune cells from the bloodstream such as dendritic cells, NK cell, and T cells infiltrate the liver, responding to invading pathogens. The liver clears blood-borne pathogens through uptake by hepatic scavenger cells, such as LSEC and Kupffer cells. Nonetheless, some pathogens (such as hepatitis viruses) can escape immune control and persist in hepatocytes, causing the diseases. Lifelong persistent infections caused by some viruses (HBV and HCV) lead to liver *cirrhosis* and HCC. Liver cirrhosis results from transformation of HSC to myofibroblasts, while liver cancer or HCC results from neoplastic transformation of hepatocytes.

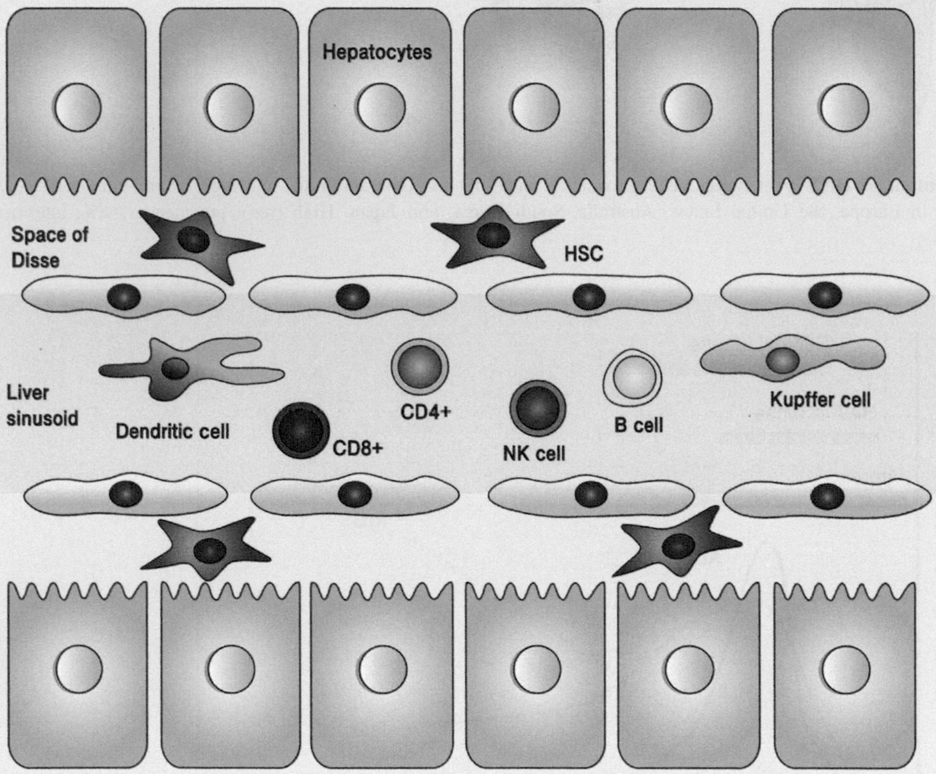

Liver microanatomy showing hepatocytes and NPCs that constitute the liver tissue. Liver sinusoids are lined by a fenestrated monolayer of sinusoidal endothelial cells (LSEC). Sinusoidal cell populations (Kupffer cells, LSEC, and HSC) form a loose physical barrier between hepatocytes and the blood circulating within the sinusoids. The *space of Disse* contains the stellate cells in a loose extracellular matrix.

AHA is accompanied with severe liver injury in adults. However, little is known about the underlying mechanism of liver injury, except that hepatocyte injury is T cell-mediated. A recent study showed that the suppressive activity of peripheral *regulatory T cells*[3] is attenuated during acute hepatitis, implicating that the lack of Treg activity is responsible for liver injury (see Journal Club).

Prevention and Treatment: The hepatitis A vaccine is effective for prevention. Some countries recommend it routinely for children and those at higher risk who have not been previously vaccinated. HAV vaccination appears to be effective for life. Other preventative measures include hand washing and properly cooking food. There is no specific treatment, with rest and medications for nausea or diarrhea recommended. Infections usually resolve completely and without ongoing liver disease. The mortality is less than 0.5%.

3. **Regulatory T cells (Treg)** A subset of T lymphocytes that suppress immune responses of other cells. This is an important "self-check" built in the immune system to prevent excessive reactions. Regulatory T cells come in many forms with the most well-understood being those that express CD4, CD25, and Foxp3.

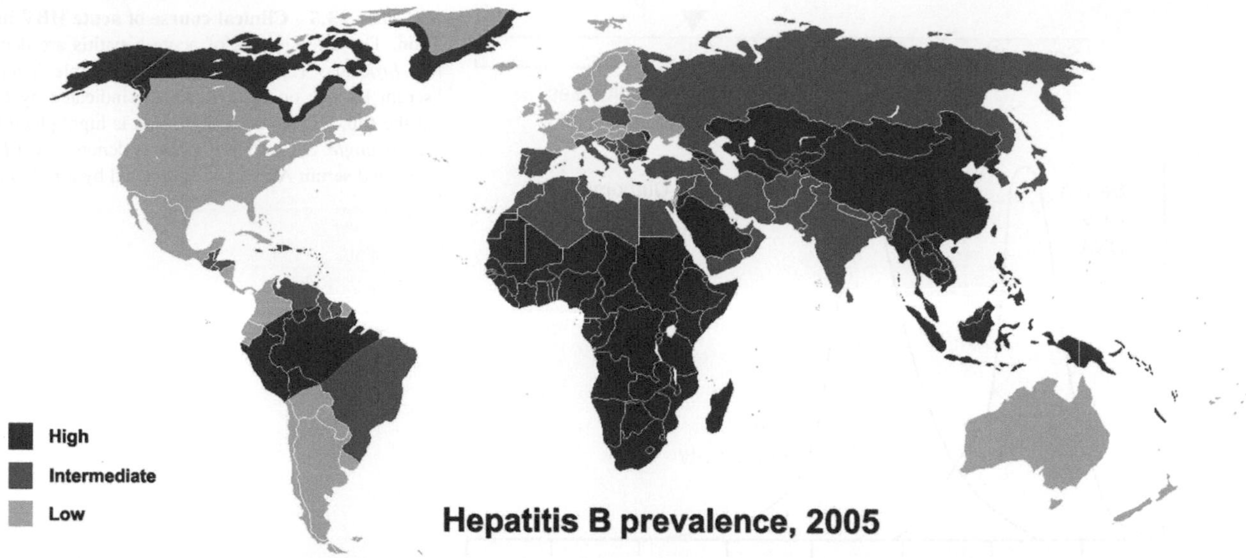

FIGURE 23.4 Geographic distribution of HBV prevalence. The prevalence of HBsAg (hepatitis B surface antigen) is shown. Note the high prevalence in China, Southeast Asian countries, sub-Saharan Africa, and Central Latin America. High (red): >8%; intermediate (orange): 2–7%, low (gray): <2%.

23.2 HEPATITIS B VIRUS

HBV is a DNA virus, belonging to the family Hepadnaviridae (see Table 23.1). HBV infection can lead to an acute illness followed by recovery with viral clearance, but can also lead to a chronic hepatitis and liver cancer. The molecular aspects of HBV have been described in chapter "Hepadnaviruses," clinical features will be described here.

Epidemiology: Worldwide, an estimated 350 million people (~5%) are persistently infected with HBV (Fig. 23.4). Chronic Hepatitis B (CHB) does not necessarily lead to a fatal liver disease, approximately one out of four carriers will suffer from liver cirrhosis or HCC; severe liver disease is assumed to cause worldwide more than 780,000 deaths per year. In particular, HBV infection is endemic in China and some Southeast Asian countries, where over 8% of populations are chronically infected, and HBV infection is the major cause of liver cancer in these regions. HBV transmits via blood and body fluids; however, sexual transmission only rarely occurs. Vertical transmission (ie, mother-to-child transmission) is the major route of transmission in endemic region.

Diagnosis: Early diagnosis for hepatitis B is made on the basis of either clinical symptoms such as jaundice, elevated serum ALT (Fig. 23.5). Hepatitis B serologic testing involves measurement of several HBV-specific antigens and antibodies (Table 23.2). Different serologic "markers" or combinations of markers are used to identify different phases of HBV infection and to determine whether a patient will progress from an acute to a chronic HBV infection. A prolonged presence of HBV DNA and surface (envelope) antigen (HBsAg) in the serum are indicators for developing a chronic disease. Serological markers are also useful to determine the immune status of an individual as a result of prior infection or vaccination. The presence of antibodies specific for the surface antigen is indicative for a protective status.

Prevention and Treatment: The hepatitis B vaccine is effective for prevention (see Fig. 25.11). The WHO recommends hepatitis B immunizations, especially immunizations for infants immediately after birth. The primary immunization should be followed by two or three boost immunizations. The vaccine induces protective antibody levels in more than 90% of infants, and adults. The HBV vaccine has an excellent record of safety and effectiveness. The vaccination is highly recommended for people in high-risk groups (eg, health-care workers).

Interferon α, a broad spectrum antiviral cytokine, has been approved for the treatment of CHB infection, but its use has been limited due to the lack of efficacy and undesirable side effects. Recently, a pegylated form of interferon α (*peg-IFN*[4]), which is an improved version of IFN, has been used for the treatment of CHB. *Lamivudine*, a nucleoside analog, was approved by the *FDA*[5] in 1998 for the treatment of CHB infection. However, the emergence of drug-resistant mutants followed (Box 23.2). The so-called second-generation nucleoside analog antiviral drugs such as

4. **peg-IFN** It is an abbreviation for the pegylated interferon. Covalent attachment of polyethylene glycol (PEG) polymer chains, dubbed "pegylation," prolongs its circulatory time by reducing renal clearance, thereby improving the efficacy of drugs.
5. **FDA (Food and drug administration)** FDA is the United States government authority, which approves new drugs for medical use.

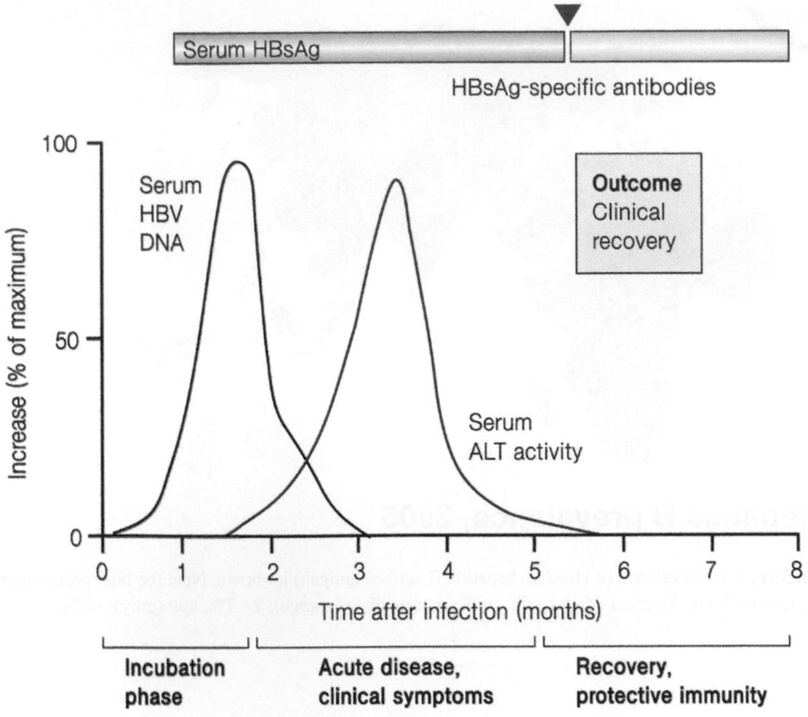

FIGURE 23.5 Clinical course of acute HBV infection. The three phases of acute hepatitis are denoted by *brackets*. The periods with detectable levels of serum HBsAg or anti-HBsAg are indicated by *boxes* at the top; HBsAg seroconversion is highlighted by a *red triangle*. Serum HBV DNA is denoted by a *black line*, and serum ALT level is denoted by a *red line*.

TABLE 23.2 Interpretation of the Results of Serological Test of HBV Infection

Serological Markers	Susceptible to Infection	Immune Due to Subclinical Infection	Immune Due to Vaccination	Acutely Infected	Chronically Infected
HBsAg	Negative	Negative	Negative	Positive	Positive
Anti-HBs	Negative	Positive	Negative	Negative	Negative
Anti-HBc	Negative	Positive	Positive	Positive	Positive
IgM anti-HBc	–	–	–	Positive	Negative

entecavir and *tenofovir* are fairly effective in suppressing the viral replication without the emergence of drug-resistant mutants. The antiviral drugs have become the standard of care (SOC) for CHB infections.

23.2.1 Clinical Course

HBV infections can be transient causing *acute hepatitis* but can also cause lifelong persistent infections causing *chronic hepatitis*. The likelihood that a HBV infection becomes chronic depends on the age at which a person becomes infected. Children less than 6 years of age who are infected with HBV most likely develop chronic infections (eg, 80–90% of infants). In contrast, the majority of adults infected with HBV develop an acute infection, and less than 5% of acutely infected individuals progress to chronic infections. It appears that the outcomes and pathogenesis of HBV infection are largely determined by the vigor of immune response to the viral infection, which is generally related to the age at the time of infection.

Acute Infection: HBV infections are self-resolved in most adult individuals (>90%). The clinical course of an acute infection can be divided into three phases: incubation period, acute disease phase, and recovery (see Fig. 23.5). Serum HBV DNA as well as serum HBsAg are the first detectable serological markers for an

BOX 23.2 Antiviral Drugs for the Treatment of Chronic HBV Infection

Five nucleoside analogs are now available for the treatment of chronic HBV infection. Since the approval of lamivudine in 1998, four additional nucleoside analogs, including adefovir, entecavir, and tenofovir, were developed. Lamivudine was originally developed as an inhibitor of the HIV reverse transcriptase (RT) for the treatment of HIV infection. It was subsequently approved for the treatment of chronic HBV infection, as lamivudine was effective in suppressing HBV replication as well, even though HBV RT is only distantly related to HIV RT. However, the rapid emergence of the drug-resistant mutants during long-term therapy limits its efficacy. The so-called second-generation nucleoside analogs (ie, *entecavir* and *tenofovir*) are associated less frequently with drug-resistant mutants (<1% for 5 years), and became the SOC for the treatment of chronic HBV infection. The second-generation drugs convert chronic HBV infection from a lifelong incurable disease to a manageable disease, just like diabetes or hypertension. Moreover, tenofovir was shown to reverse liver cirrhosis or at least halt the disease progression of liver cirrhosis, giving a hope of recovery. Despite its potent antiviral activity, even long-term therapy failed to clear the virus completely. "HBsAg loss," as a serological indicator for the therapeutic endpoint, is achieved only in less than 5% of patients even after a 5-year long-term therapy.

A question that arose was why the virus could not be completely cleared following long-term antiviral therapy in most of the chronically infected patients. A short answer is that an episomal HBV DNA, which is dubbed *"cccDNA,"* remains stable in the nucleus and serves as the template for the viral transcription (see Fig. 18.5). In fact, the serum HBV DNA became clinically undetectable during the therapy; however, HBsAg persists for a long period and only gradually decreases. Nucleoside analogs, inhibitors of viral RT, cannot eliminate cccDNA, which persists at a copy of 5–50 molecules per cell. Therefore, a new drug targeting cccDNA represents an unmet medical need.

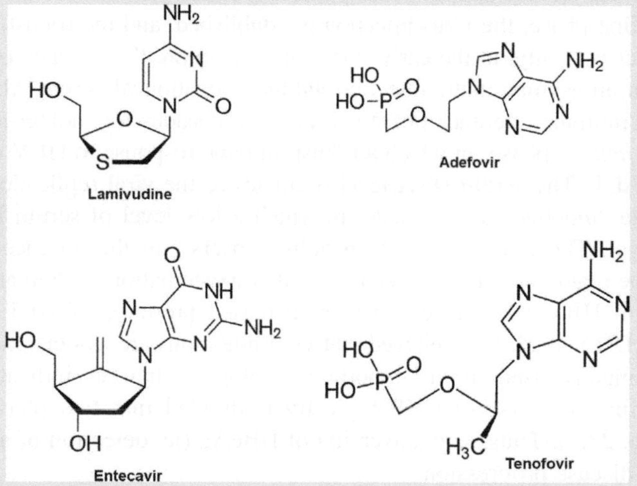

Chemical structures of anti-HBV nucleos(t)ide analog drugs. Lamivudine and entecavir are nucleoside analogs, whereas adefovir and tenofovir are nucleotide analogs having a monophosphate group. Both nucleoside and nucleotide analogs are prodrugs that are converted to the triphosphate forms in vivo.

HBV infection during the incubation period. During the acute disease phase, the serum HBV DNA level is decreased, whereas serum ALT level is elevated. ALT, a liver-specific enzyme participating in amino acid metabolism, is detectable in the serum as the result of inflammatory processes and hepatocyte lysis mediated by HBV-specific cytotoxic T lymphocytes. In the recovery phase, the inflammation declines, the ALT level normalizes and the liver tissue is regenerated. HBV is eliminated, and the diagnostic relevant anti-HBsAg antibodies are detectable, which confer the protective immunity. *Seroconversion*, the transition from the presence of the serum HBsAg antigen to anti-HBsAg antibodies (ie, *neutralizing antibodies*), represents a hallmark of viral clearance.

Chronic Infection: HBV infection often leads to chronic lifelong persistent infection (ie, chronic carriers): the risk of developing CHB is very high in infants (~90%), but low in adults (5–10%) (Fig. 23.6). Following an asymptomatic period, which can last for several decades, approximately 30% of chronically infected individuals will develop a *chronic active hepatitis (CAH)*, a status of actively ongoing inflammation. The liver injury caused by the CAH often leads to liver cirrhosis, liver failure, and liver cancer (ie, HCC) (see Fig. 23.6).

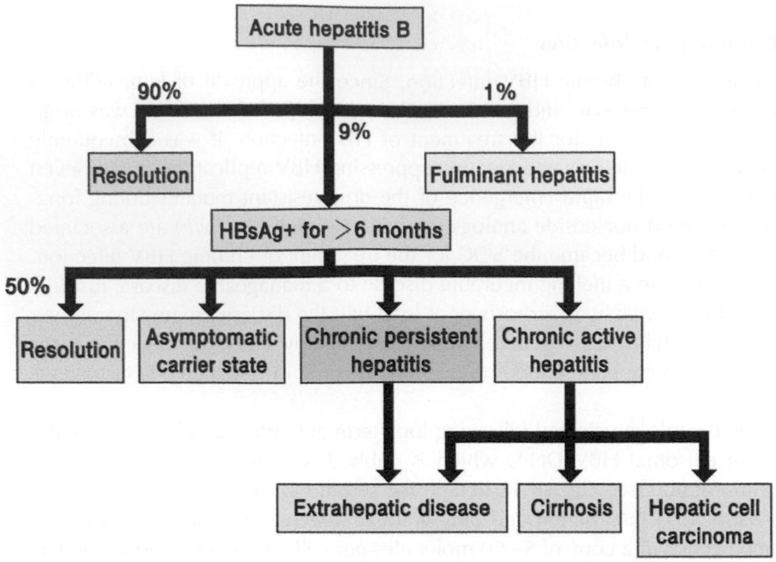

FIGURE 23.6 Liver diseases associated with chronic hepatitis B infection. The likelihood of disease progression toward a clinical pattern is denoted above the lines. CHB can lead to the development of liver cancer in the presence or absence of liver cirrhosis.

The clinical course of CHB can be divided into four phases: immune tolerance, immune active, inactive carrier, and reactivation (Fig. 23.7). In the first phase, the virus infection is established, and the robust viral replication leads to a high serum HBV DNA level. Characteristically, at the early stage after infection, the absence of a prominent immune reaction against HBV antigens results in an asymptomatic infection without elevation of serum ALT. This period is aptly dubbed "*immune tolerance*" phase. An immune tolerance phase can last two decades for mother-to-child (vertical) transmissions. The next phase is the "*immune active*" phase, in which a robust immune response to HBV antigens occurs, leading to liver injury and elevation of serum ALT. The serum DNA level oscillates as the viral replication is suppressed by the immune response. The third phase is the "*inactive carrier*" state, in which a low level of serum HBV DNA is detectable in the presence of a normal ALT level. The individuals are "healthy carriers," in the absence of evident clinical symptoms; the "healthy carrier" state can be maintained for life without further exacerbations. Alternatively, "*reactivation*" can occur, which results in elevated serum HBV DNA levels and an increased presence of ALT. In this phase, the ALT level oscillates again with the serum DNA level. It is believed that viral mutations are associated with the reactivation.

In addition, *HBeAg* is clinically used as an important serological marker indicative of ongoing viral genome replication (Box 23.3). The clinical course of CHB typically is divided into two phases: HBeAg-positive CHB and HBeAg-negative CHB (see Fig. 23.7). Thus, seroconversion of HBeAg (ie, detection of anti-HBeAg antibody) serves as an important landmark for the disease progression.

Liver Cancer: CHB leads to liver cirrhosis, and liver cancer (ie, *HCC*[6]). Epidemiologic studies revealed that CHB increases the incidence of HCC by 20-fold. The correlation is significant, considering that smoking increases the incidence of lung cancer by 10-fold. According to the WHO, worldwide about 50% of HCC is attributable to CHB. Worldwide, 780,000 people die each year from acute or chronic consequences of hepatitis B. Despite the strong association of CHB to HCC, little is known about the underlying mechanism of HBV-induced HCC. HBx, a small regulatory protein of HBV, is believed to be a viral oncoprotein; its oncogenic potency is only modest. Some evidence indicates that HBx contributes to HCC in the context of other oncogenic stimuli. Additional mechanisms contributing to the formation of HCC by HBV have been considered in chapter "Hepadnaviruses" (see Fig. 18.11).

The progression to HCC can be considered as a multiple step process (Fig. 23.8). Hepatocytes in the normal liver remain in the Go phase of the cell cycle for an extensive time period. Liver injury caused by a chronic HBV infection (similar to HCV, excessive alcohol consumption, and aflatoxin B1) leads to a chronic inflammation, which is induced by the infiltration of circulating lymphocytes into the liver tissue. Inflammatory cytokines (ie, TNFα and IFNγ) produced by infiltrating immune cells induces cell deaths and tissue damages. To compensate for the cell loss, cell proliferation is induced. As a result, excessive cell proliferation of otherwise quiescent hepatocytes leads to the accumulation of genetic mutations, leading to oncogenic transformation. Chronic inflammation often results in liver

6. **HCC (hepatocellular carcinoma)** HCC is the most common type of liver cancer, resulting from transformation of hepatocyte.

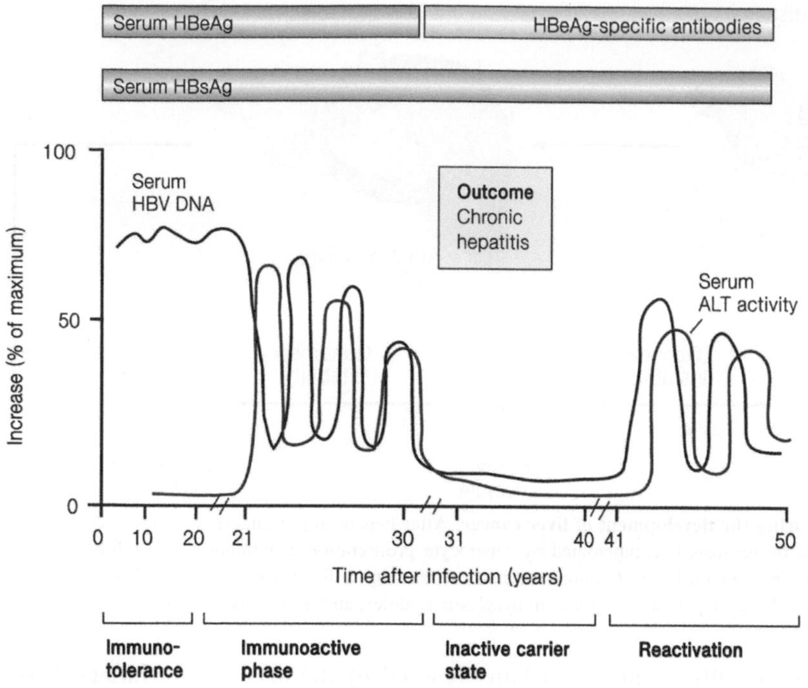

FIGURE 23.7 Clinical course of chronic HBV infection. Four phases of chronic hepatitis are denoted by *brackets*: immune tolerance, immunoactive phase, inactive carrier state, and reactivation phase. Serum HBV DNA and serum ALT levels are denoted by a *black line* and a *red line*, respectively.

BOX 23.3 HBeAg

In addition to core protein (ie, HBcAg), another protein, termed HBeAg, is expressed from the C open reading frame (ORF). HBeAg is regarded as an accessory protein, because it is not required for the viral genome replication. In fact, HBeAg is translated from the pre-C RNA, which is slightly longer than the pgRNA. Due to the presence of the additional upstream AUG initiation codon, which is in frame with C ORF, the pre-C domain is opened, resulting in the synthesis of the pre-C polypeptide. The leader sequence present in the pre-C domain directs the pre-C polypeptide to endoplasmic reticulum. Such made pre-C polypeptide is cotranslationally processed at both N-terminus and C-terminus in endoplasmic reticulum by cellular proteases and secreted extracellularly via the secretory pathway. Although HBeAg shares a significant portion of its constituent amino acid residues with HBcAg, it exhibits characteristically distinct antigenicity. Further, HBeAg is a secreted, nonparticulate form of the viral nucleocapsid. Importantly, HBeAg correlates well with ongoing viral genome replication in liver tissue, although it is not essential for the viral genome replication. Thus, HBeAg serves as an important serological marker, indicative of the viral replication. Nonetheless, the biological function of HBeAg remained uncertain. It has been claimed that HBeAg acts as an "immune tolerogen," suppressing host immune response, thereby contributing to persistent chronic infection. However, the evidence for this hypothesis is circumstantial.

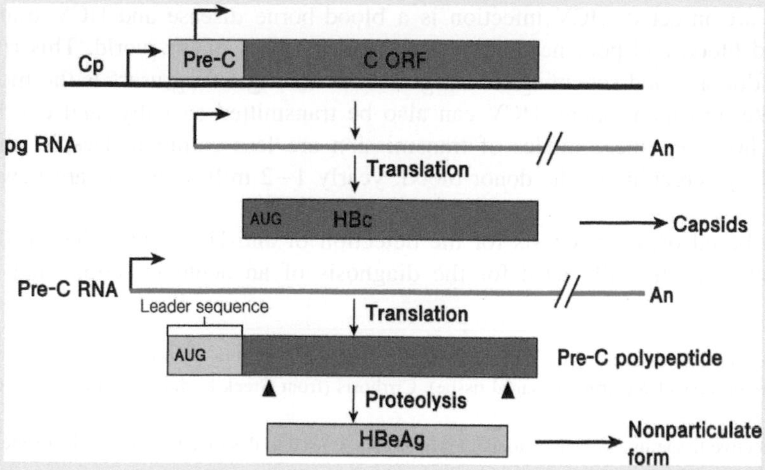

Biogenesis of HBeAg. The C ORF (red) and the pre-C domain (pink) are drawn on the top along with two rightward *arrows* indicating the direction of transcription. The pre-C domain is an N-terminal extension of the C ORF. Two viral genomic RNAs (blue) having slightly different 5′ ends are shown below: pgRNA and pre-C RNA. Two AUG initiation codons are denoted: one for the C ORF and another for the pre-C domain. The cleavage sites by cellular proteases are denoted by *triangles*. Cp, core promoter.

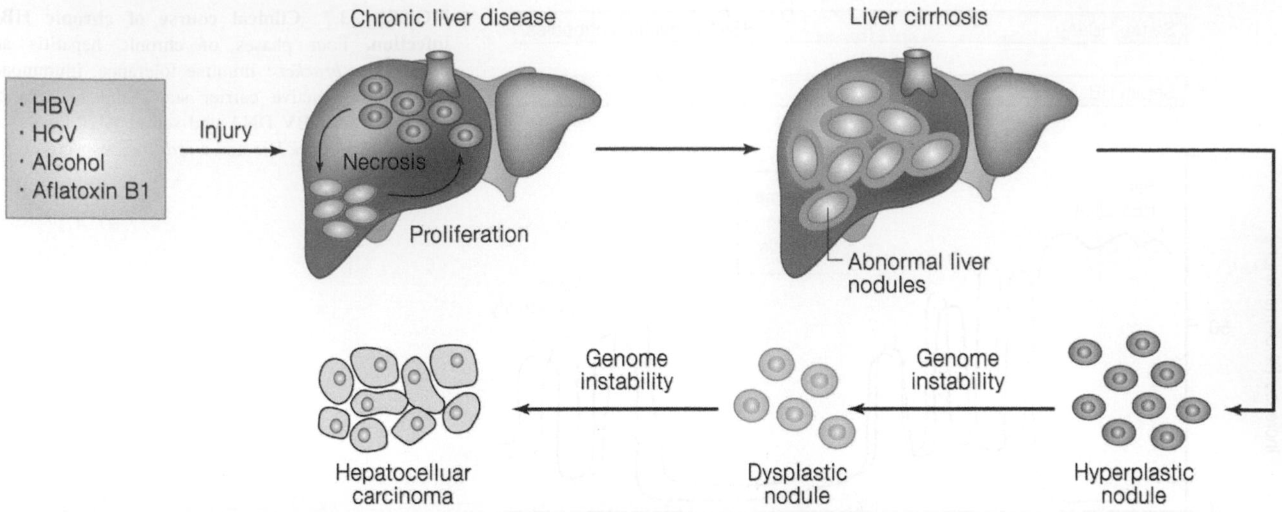

FIGURE 23.8 **Pathological and molecular changes during the development of liver cancer.** After hepatic injury incurred by any one of several factors (HBV, HCV, alcohol, and aflatoxin B1), cell loss by necrosis is compensated by hepatocyte proliferation. Continuous cell proliferation provokes a chronic liver disease condition that culminates in liver cirrhosis. Cirrhosis is characterized by a formation of abnormal liver nodules. Accumulation of mutations triggered by genome instability drives hyperplastic nodules to dysplastic nodules, and ultimately to HCC.

cirrhosis,[7] which is a consequence of *fibrosis*.[8] Finally, genome instability (caused by the loss of p53) further drives preneoplastic cells to malignant tumor, HCC.

Although the etiology of HCC is well known, the molecular events leading to HCC are not completely understood. For instance, cellular genes (ie, oncogenes or tumor suppressor genes) directly linked to each of the stage involved in HCC development are not yet clearly described.

23.3 HEPATITIS C VIRUS

HCV was first recognized as an agent causing liver inflammation following blood transfusion in the 1970s. In 1989, the cause of blood transfusion-related hepatitis was identified as an RNA virus, belonging to the family Flaviviridae (see Table 23.1). HCV infections can result in an acute or chronic hepatitis. Chronic hepatitis C is associated with an increased risk for developing liver cancer.

Epidemiology: Worldwide, an estimated 180 million people ($\sim$2.0%) are persistently infected with HCV (Fig. 23.9). The most affected regions are Central and East Asia and North Africa. Unlike HBV, HCV prevalence is significant ($>$2%) in the Western Hemisphere. For instance, over 4 million people are chronically infected in the United States. Over 350,000 people die of HCV-associated liver diseases. Both acute and chronic HCV infections are largely asymptomatic. As a result, less than 5% of the world's HCV-infected population, and only 50% of the United States' HCV-infected population, are aware that they are infected. HCV infection is a blood-borne disease and HCV transmission occurs through receipt of contaminated blood and poor needle hygiene in many parts of the world. This risk has been reduced in most countries due to the donor blood screening. Sharing needles amongst drug users is the most common route for HCV transmission in the Western hemisphere. HCV can also be transmitted sexually, and can be passed from an infected mother to her baby; however, these modes of transmission are less common. Despite that transmission via blood transfusion is precluded by screening of the donor blood, yearly 1–2 million people are newly infected.

Diagnosis: The standard HCV diagnosis is based on ELISA tests for the detection of anti-HCV antibodies in the serum. Alternatively, quantitative PCR methods are clinically used for the diagnosis of an acute infection, and to

7. **Cirrhosis** Cirrhosis is a result of advanced liver disease. It is characterized by the replacement of liver tissue by fibrous connective tissue (scar tissue) and regenerative nodules (lumps that are formed in the process of repairing damaged tissue). Cirrhosis (from Greek kirrhos "yellowish" and the common Greek suffix -sis meaning "condition").

8. **Fibrosis** Fibrosis is the formation of excess fibrous connective tissue in an organ or tissue. Fibrosis can be used to describe the pathological state of excess deposition of fibrous tissue, as well as the process of connective tissue deposition in healing.

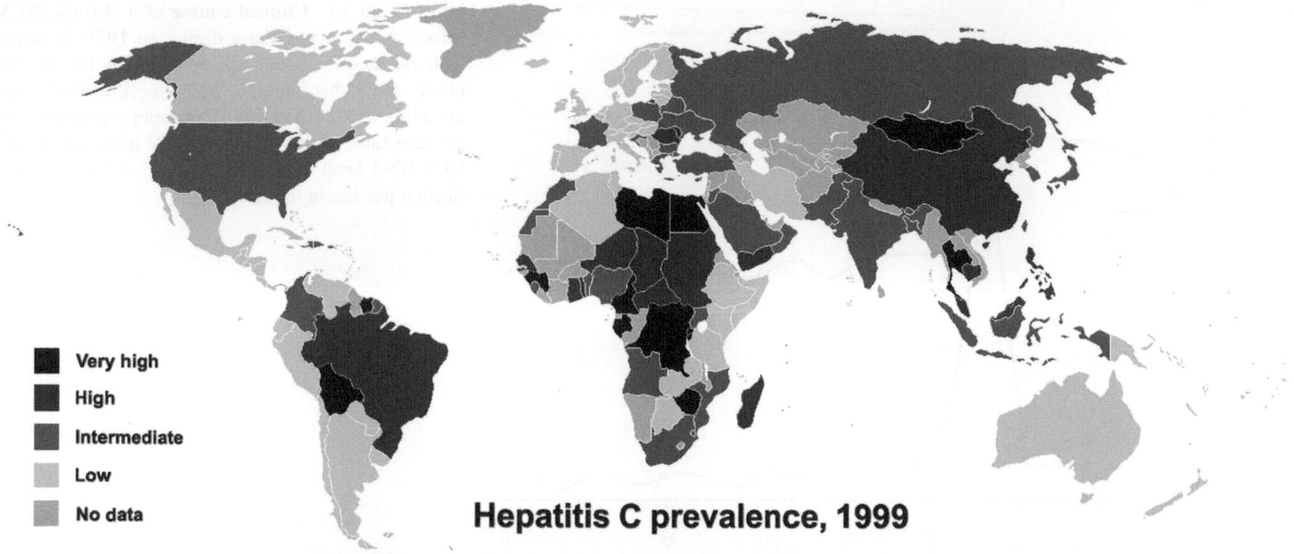

FIGURE 23.9 Geographic distribution of HCV prevalence. HCV prevalence rates are very high in Central and East Asia, North Africa, and the Middle East, they are intermediate in South and Southeast Asia, sub-Saharan Africa, Central and Southern Latin America, and Europe; and they are low in Asia Pacific, Tropical Latin America, and North America. Very high (dark brown): >5%; high (red): 2.5–5%; intermediate (orange): 1–2.5%; low (light orange) <1%; no data (gray).

determine the amount of HCV genomes in the serum during disease progression. Further, the PCR is often employed to monitor the antiviral response during antiviral therapy.

Genotype: The nucleotide sequence heterogeneity of HCV RNA genome is high, showing about 30–35% divergence among HCV isolates. Based on the sequence heterogeneity, HCV can be grouped into six genotypes, from genotype 1 to genotype 6. The six genotypes are further subdivided into 12 subtypes; genotype 1 is subdivided into genotype *1a* and *1b*. Significantly, the response to a certain antiviral therapy (eg, IFNα) varies, depending on the genotype.

Pathogenesis: Unlike HBV infections in adults, the vast majority of HCV infections (~70%) are not self-resolved, and progress to a chronic infection. The high chronicity is the outstanding feature of an HCV infection. HCV replication persists in the liver, leading to severe liver diseases, such as cirrhosis and liver cancer. A cell-mediated immune response is responsible for liver injury.

23.3.1 Clinical Course

HCV causes both transient (ie, *acute hepatitis*) and lifelong persistent infection (ie, *chronic hepatitis*).

Acute and Chronic Infection: An acute HCV infection is usually asymptomatic, and is only very rarely associated with a life-threatening disease. In fact, most people with an HCV infection become aware of their status after testing of blood samples. About 20% of infected persons spontaneously clear the virus within 6 months of infection without any treatment. The remaining ~80% of persons will develop a chronic HCV infection. In this case, HCV replication persists lifelong and serum ALT oscillates (Fig. 23.10). The risk of chronic HCV carriers to develop liver cirrhosis is increased by a factor of 20-fold, and the formation of an HCC within 20 years after infection is increased 5-fold (Fig. 23.11). In addition, chronic hepatitis C is often associated with *hepatic steatosis*, also known as fatty liver disease.

Prevention and Treatment: Despite strenuous efforts, the development of an HCV vaccine has not yet been accomplished. In contrast, a few antiviral drugs have been recently developed. The combination therapy of IFN and ribavirin has long been used for the treatment of a chronic HCV infection. Although both are broad spectrum antiviral drugs, the combination of both leads to a sustained virologic response (*SVR*[9]) in up to 70% of chronic HCV patients (Fig. 23.12). Recently, novel antiviral drugs that directly inhibit viral proteins have been approved by the FDA. These so-called *direct acting antivirals* (DAA) include inhibitors of the NS3/4A serine protease and an inhibitor of the HCV NS5B RdRp. For instance, boceprevir and telaprevir, the viral protease inhibitors, are structural analogs of a substrate peptide (see Fig. 26.11). The NS3/4A serine protease is essential for the processing of the viral polyprotein and for the

9. **SVR (sustained virologic response)** SVR refers to "no detection of HCV RNA" in the serum at 24 weeks after cessation of antiviral therapy.

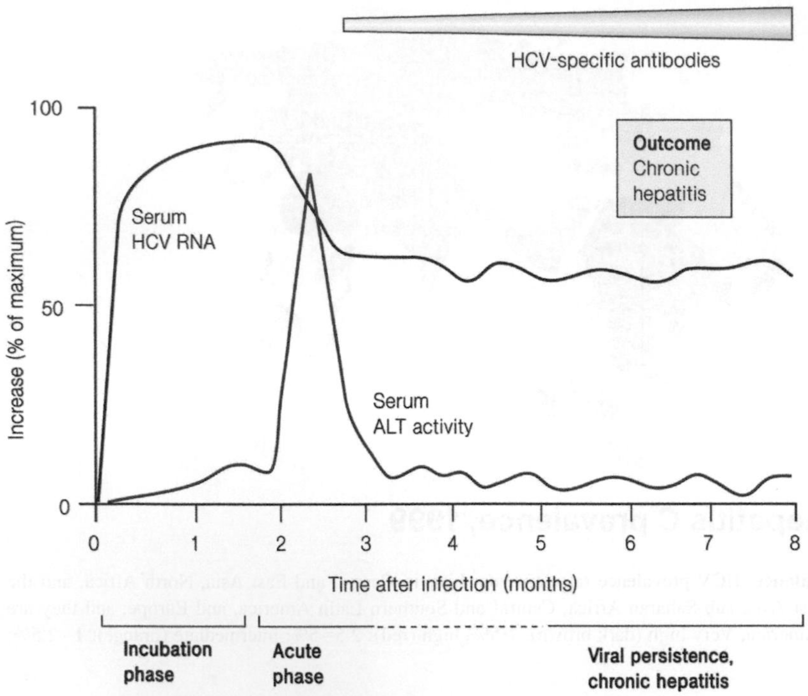

FIGURE 23.10 Clinical course of a chronic HCV infection. An acute phase during an HCV infection with elevated ALT values is followed by the chronic phase. Note that multiple HCV-specific antibodies, including anti-E1 and anti-E2 envelope glycoprotein, are detectable during a chronic HCV infection. Serum HCV RNA level is readily detectable, as the viral replication persists in the liver.

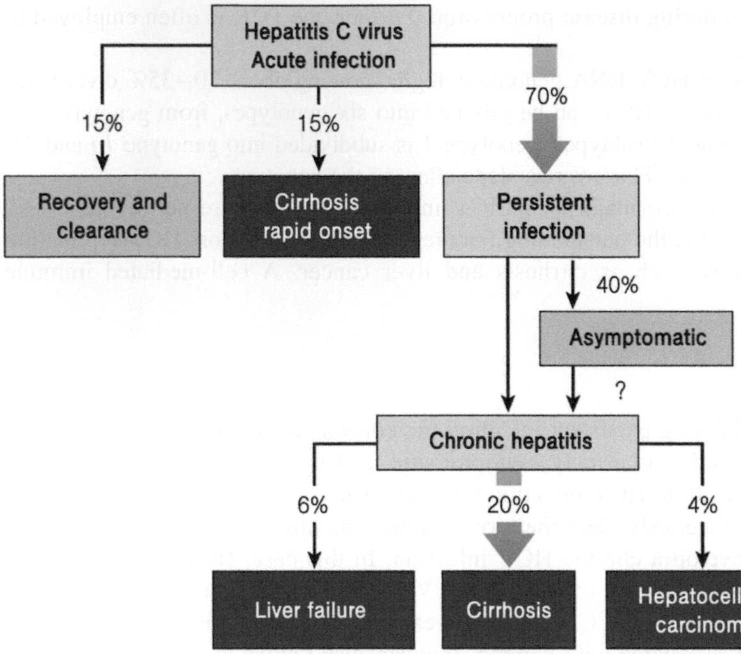

FIGURE 23.11 An HCV infection can progress toward different clinical patterns. The likelihood of developing a specific clinical disease pattern is denoted by percentage. Note the high chronicity of HCV infection (~70%), a hallmark of HCV infection. Manifestation of liver diseases is colored red, whereas asymptomatic outcomes are colored green.

inactivation of TRIF and MAVS adapters that are critical for innate immune response (see Fig. 5.19). Sofosbuvir is an inhibitor of the HCV NS5B RdRp. Remarkably, the result of clinical trials indicated that sofosbuvir monotherapy leads to nearly 100% SVR, demonstrating that sofosbuvir monotherapy is sufficient to control the viral replication completely without the emergence of drug-resistant variants (see Fig. 23.12). A cure for a chronic HCV infection appears in sight, but the cost of these new DAAs is prohibitive for most countries with moderate to high HCV prevalence.

Liver Cancer: Chronic hepatitis C leads to liver cirrhosis and liver cancer (ie, HCC) (see Fig. 23.8). The underlying mechanism for HCV-induced HCC is not completely understood. The dysregulation of tumor suppressor gene products (ie, p53 and Rb) by multiple HCV proteins (eg, core, NS3, NS5A, and NS5B) have been described. In particular, NS5B-mediated Rb degradation has been experimentally demonstrated (see Fig. 12.8). It is worthy of note that HCV is

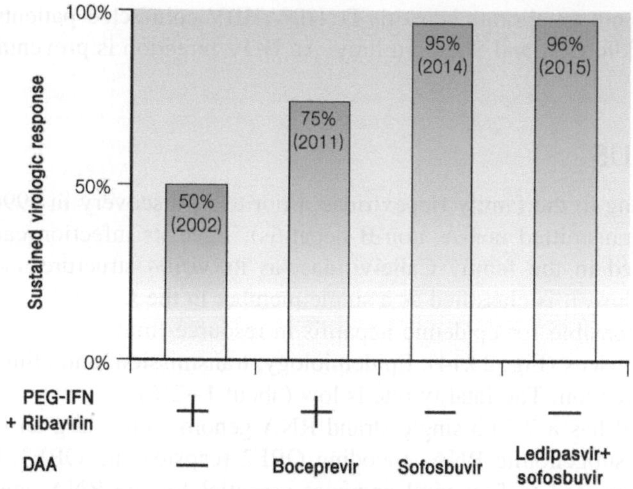

FIGURE 23.12 Advances in anti-HCV therapy. The response to antiviral drug treatment, measured by sustained virologic response (SVR), is shown. Addition of boceprevir to the established peg-IFN + ribavirin regimen improved the SVR to up to 75%. Remarkably, sofosbuvir monotherapy leads SVR to 95%. The year of FDA approval is denoted in *parenthesis*. DAA, direct acting antiviral.

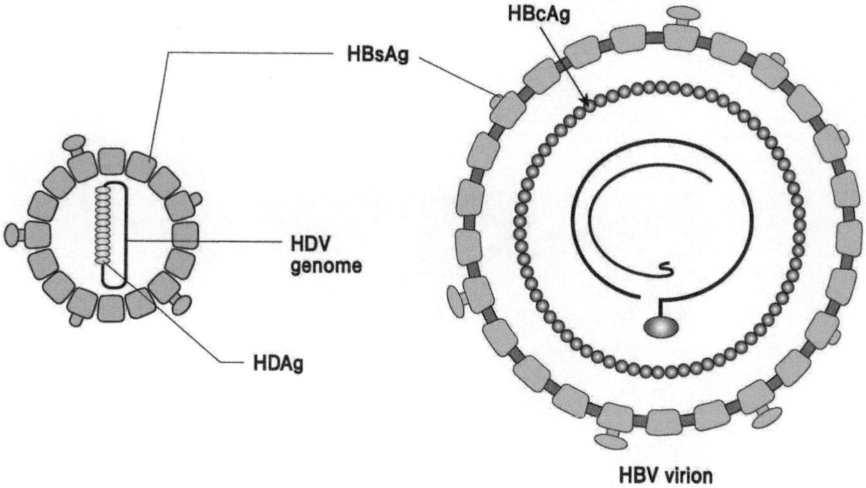

FIGURE 23.13 Virion structure of HDV. HDV shares the envelope glycoproteins of HBV. Inside of the envelope, the single-strand circular RNA genome is associated with the small delta antigen. The large delta antigen, an isoform of small delta antigen, is essential for virion assembly and also packaged into the virion.

the only tumor virus that does not have DNA intermediates during its viral life cycle. In this sense, it is intriguing to see how the virus, whose life cycle is confined to the cytoplasm, could cause cell transformation, which is a nuclear event after all.

23.4 HEPATITIS D VIRUS

HDV is the smallest virus known to infect human. In fact, HDV is a satellite virus of HBV, which provides the envelope proteins that are essential for HDV assembly (Fig. 23.13). HDV recruits the HBV envelope proteins as a required step to form HDV virion particles, and therefore HBV and HDV share the envelope glycoproteins. Another peculiarity is that HDV possesses a circular RNA genome of only 1.7 kb in length. Note that HDV is the only animal virus having a circular RNA. In this regard, HDV is often related to a plant agent, viroids (see Fig. 20.3).

About 15 million people worldwide are infected by HDV among those 240 million infected by its helper HBV. Because of the dependency of HDV on the presence of the HBV envelope proteins, HDV can only establish successful infections by HBV/HDV coinfecting healthy individuals. HDV can superinfect individuals who are already HBV carriers. Viral hepatitis D is considered as one of the most severe forms of human viral hepatitis. HDV infection often exacerbates the liver disease of CHB patients, and often leads to *fulminant hepatitis* or even liver failure. About 5% of CHB carriers are coinfected with HDV. No specific antivirals are currently available to treat HDV infection and antivirals against HBV do not ameliorate hepatitis D. HDV/HBV coinfected patients are more frequently found in certain regions such as the Middle East and Southern Italy. An HDV infection is preventable by the HBV vaccine.

23.5 HEPATITIS E VIRUS

HEV is an RNA virus, belonging to the family Hepeviridae. Prior to its discovery in 1990, the viral hepatitis used to be called *ET-NANB* (enterically transmitted non-A, non-B hepatitis), since its infection causes gastroenteritis. Moreover, HEV was temporarily classified in the family Caliciviridae, as its virion structure and its genome organization are similar to those of norovirus. Now, it is classified as a single member in the family Hepeviridae, hepatitis E-like virus.

Epidemiology: HEV is responsible for epidemic hepatitis in resource-limited countries with limited access to clean water, sanitation, and health services (Fig. 23.14). Epidemiology, transmission, and clinical outcomes of HEV infection are similar to those of HAV infection. The fatality rate is low (about 1–2%).

Genome Organization: HEV has a 7.2 kb single-strand RNA genome, encoding an ORF1 polyprotein (Fig. 23.15). In addition, HEV expresses a subgenomic RNA, encoding ORF2 (capsid) and ORF3. ORF1 encodes a nonstructural protein that is cleaved and processed to four viral enzymes essential for the RNA genome replication: guanyl transferase, methyl transferase, helicase, and RdRp. Two nucleotide modifying enzymes (ie, guanyl transferase and methyl transferase) are involved in cap modification at the 5′ terminus of viral RNAs (see Box 6.2). The subgenomic RNA encodes two viral proteins. The ORF2 encodes capsid protein, while the function of ORF3 protein is unknown.

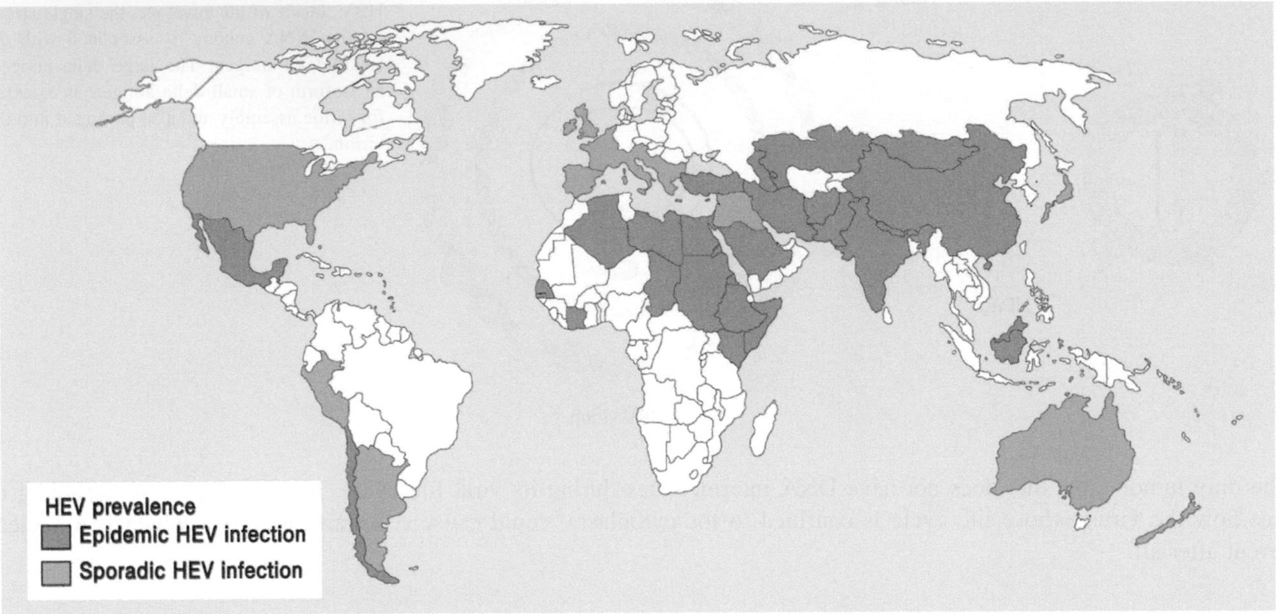

FIGURE 23.14 Geographic distribution of HEV prevalence. HEV is epidemic in Asia, the Middle East, and Northern Africa. Sporadic appearances occur in Western countries including European countries and the United States. An estimated 20 million infections occurs worldwide and 56,600 deaths annually.

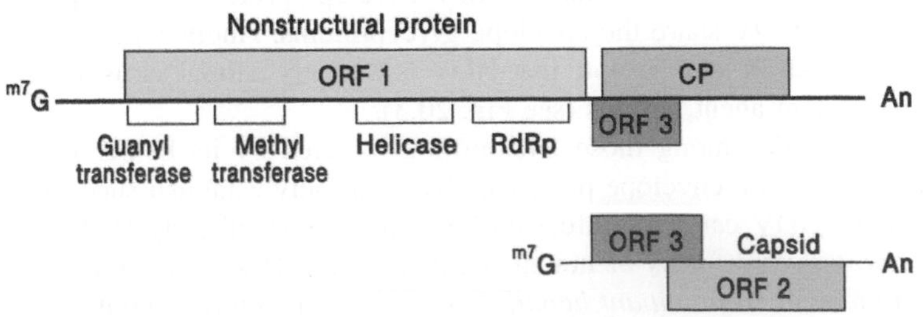

FIGURE 23.15 The genome organization of HEV. The genome organization of HEV is similar to that of norovirus, in that both have a subgenomic RNA as well as the genomic RNA. A notable difference is that HEV has a cap structure at the 5′ terminus, while norovirus has VPg, instead (see Fig. 13.2). Four enzymatic activities of the ORF1 are denoted.

23.6 PERSPECTIVES

HAV and HBV were discovered in the 1960s, while four other hepatitis viruses (ie, HCV, HDV, HEV, and HGV) were discovered in the 1980s and 1990s. Since their discovery, remarkable progress has been made in the field of viral hepatitis including diagnosis, vaccines, and antiviral drugs. Now, vaccines are available and licensed for HAV and HBV. In particular, remarkable advances have been achieved in the development of antiviral drugs against HBV and HCV. The antiviral efficacy of second-generation nucleoside analog anti-HBV drugs including entecavir and tenofovir is remarkably effective in suppressing HBV replication. It should be noted, however, that HBV is not cleared even after long-term antiviral therapy, as the viral episomal DNA (ie, cccDNA) persists in liver cells. A novel drug targeting cccDNA represents an unmet medical need. For HCV, the impact of a recently developed DAA drug able to overcome chronic hepatitis C is huge, in that the cure rate is approaching nearly 100%. According to some experts, HCV infection is considered to be "curable." The fact is that the cost of some of these HCV drugs (ie, ~100,000 USD for 24 week treatment of sofosbuvir) is prohibitively high, making it inaccessible to most patients. Thus, the development of more affordable anti-HCV drugs is needed.

23.7 SUMMARY

- *Viral hepatitis*: Six hepatitis viruses, from HAV to HGV by alphabetical order, have been discovered. They belong to different families, but all cause hepatitis, which is inflammation of liver tissue.
- *HAV*: HAV is water-borne, and an HAV infection causes self-limited acute infection. Safe and effective prophylactic vaccines are available to prevent HAV infection.
- *HBV*: HBV is blood-borne and HBV infection leads to chronic hepatitis, leading to severe liver diseases, such as cirrhosis and cancer. Safe and effective prophylactic vaccines and effective antiviral drugs are available to prevent and treat HBV infection.
- *HCV*: HCV is blood-borne and HCV infection leads to chronic hepatitis, leading to severe liver diseases, such as cirrhosis and cancer. No vaccine is available, but highly effective antiviral drugs are available to control HCV infection.
- *HDV*: HDV is a satellite virus of HBV. HDV infection often exacerbates the liver disease of chronic hepatitis B patients. An HDV infection is preventable by the HBV vaccine, however, no specific antivirals are available to treat HDV infection.
- *HEV*: HEV is water-borne, and HEV infection causes self-limited acute infection.

STUDY QUESTIONS

23.1 List and compare serological markers detectable in serum of chronically infected individuals by the following viruses: (1) HBV and (2) HCV.

23.2 List the representative antiviral drugs for the treatment (standard of care) of chronic HBV and chronic HCV, along with their modes of inhibition.

23.3 The genome organization of HEV is very similar to that of norovirus. Point out the similarity and difference in the genome RNA between HEV and norovirus with respect to the 5′ terminus. Describe how the difference affects the viral gene expression.

SUGGESTED READING

Baker, C.J., 2007. Another success for hepatitis A vaccine. N. Engl. J. Med. 357 (17), 1757–1759.

Choi, Y.S., Lee, J., Lee, H.W., Chang, D.Y., Sung, P.S., Jung, M.K., et al., 2015. Liver injury in acute hepatitis A is associated with decreased frequency of regulatory T cells caused by Fas-mediated apoptosis. Gut. 64 (8), 1303–1313.

McMahon, B.J., 2009. The natural history of chronic hepatitis B virus infection. Hepatology. 49 (5 Suppl.), S45–55.

Protzer, U., Maini, M.K., Knolle, P.A., 2012. Living in the liver: hepatic infections. Nat. Rev. Immunol. 12 (3), 201–213.

Trepo, C., Chan, H.L., Lok, A., 2014. Hepatitis B virus infection. Lancet. 384 (9959), 2053–2063.

JOURNAL CLUB

- Choi, Y.S., et al., 2015. Liver injury in acute hepatitis A is associated with decreased frequency of regulatory T cells caused by Fas-mediated apoptosis. Gut 64(8), 1303–1313.

  Highlight: Acute hepatitis A is accompanied by severe liver injury in adults. However, little is known about the underlying mechanisms causing the liver injury, except that it involves mediated T cells. This work shows that the suppressive activity of peripheral regulatory T cells (Treg) is attenuated during acute hepatitis A, implicating that the lack of Treg activity contributes to the liver injury.

INTERNET RESOURCES

- World Health Organization (WHO) Media Center webpage
  - URL: http://www.who.int/mediacentre/factsheets/fs328/en/

Chapter 24

Tumor Viruses

Chapter Outline

Cancer represents one of the most formidable scourges to mankind. Cancer-causing agents (ie, carcinogens) are remarkably diverse, ranging from chemical carcinogens (eg, smoking and aflatoxin) to radiations (ultraviolet and X-ray irradiation). In addition, viruses can cause cancer, representing an important carcinogen. As a matter of fact, about 15% of human cancer is attributable to viruses. In other words, a significant portion of human cancer can be preventable, if human tumor viruses are eradicated. For instance, cervical carcinoma, which afflicts 500,000 women worldwide each year, is caused by human papillomavirus (HPV) infection. A recently developed HPV vaccine will reduce the number of women suffering HPV-induced cervical carcinoma.

Tumor viruses have served as an excellent experimental model for cancer research since the 1980s. A representative epitome is that the discoveries of the *oncogene*[1] and the *tumor suppressor gene*[2] were achieved by the investigation of tumor viruses. Tumor viruses could induce not only *transformation*[3] of cells in culture but also tumor formation in experimental animals. As cells transformed by tumor viruses manifest characteristics of tumor cells, oncogenesis can be investigated using tumor viruses. Historically, tumor viruses were first discovered in animals (eg, mouse), and subsequently related viruses were discovered in humans.

Tumors[4] can be divided into benign tumors and malignant tumors, depending on its ability of metastasis. A benign tumor is not fatal, because it can be surgically removed. For example, warts are a type of a benign tumor. In contrast to benign tumors, a malignant tumor (ie, *cancer*[5]), being capable of migrating from the primary site to other parts of the body (ie, metastasis), can be fatal.

Tumor cells acquire their ability of uncontrolled growth as a result of multiple genetic mutations. The mutations in oncogenes and tumor suppressor genes drive the conversion of normal cells to tumor cells. This conversion process is termed *tumorigenesis* or *oncogenesis*. As an analogy to a car, oncogenes and tumor suppressor genes act as the accelerator and brake of the *cell cycle*,[6] respectively (Fig. 24.1). Oncogenes, the accelerator, are normally present in an inactive form (ie, *proto-oncogenes*[7]), and the mutations of proto-oncogenes (ie, activation) could drive the uncontrolled

1. **Oncogene** An oncogene is a gene that has the potential to cause cancer. In tumor cells, the oncogenes are often mutated (activated) or expressed at high levels.
2. **Tumor suppressor gene** Tumor suppressor gene, or "anti-oncogene," is a gene that protects a cell to become cancerous.
3. **Transformation** (Malignant) transformation is the process by which cells acquire the properties of cancer. This may occur as a primary process in normal tissue, or secondarily as *malignant degeneration* of a previously existing benign tumor.
4. **Tumor** Tumor, referred to as a neoplasm (from Ancient Greek—*neo*—"new" and *plasma* "formation, creation"), is an abnormal growth of tissue.
5. **Cancer** Cancer is a group of diseases involving abnormal cell growth with the potential to invade or spread to other parts of the body. It is synonym of a malignant tumor. Not all tumors are cancerous; benign tumors do not spread to other parts of the body.
6. **Cell cycle** It refers to the series of events that take place in a cell leading to its division and duplication (replication) that produces two daughter cells.
7. **Proto-oncogene** It refers to a normal gene that can become an oncogene due to mutations or increased expression.

Molecular Virology of Human Pathogenic Viruses. DOI: http://dx.doi.org/10.1016/B978-0-12-800838-6.00024-2

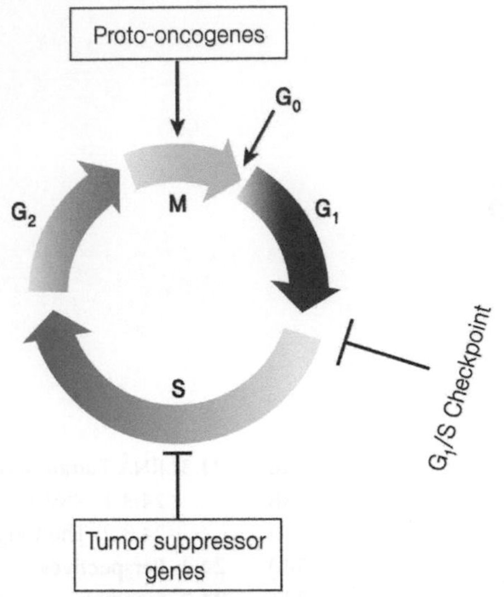

FIGURE 24.1 Oncogenes and tumor suppressor genes in the context of cell cycle. Four phases of the cell cycle (ie, G_1, S, G_2, and M phase) are drawn with an emphasis on the role of proto-oncogenes and tumor suppressor genes. Cell cycle checkpoints are the control mechanisms that ensure the fidelity of cell division in eukaryotic cells. M, mitosis; S, synthesis; G, gap.

TABLE 24.1 Nomenclature of Cancers

Cancer Type	Cell Type	Example
Carcinoma	Epithelial cell	Hepatocellular carcinoma, cervical carcinoma
Leukemia	Leukocyte	T-cell leukemia
Lymphoma	Lymphocyte	Burkitt's lymphoma
Sarcoma	Muscle cell	Kaposi's sarcoma

growth. In contrast, tumor suppressor genes, the brake, are normally expressed in an active form, and the mutations of tumor suppressor genes (ie, inactivation) could drive the uncontrolled growth. Multiple mutations in oncogenes and tumor suppressor genes are required for tumor formation. In the context of tumor viruses, one or more viral oncogenes encoded by tumor viruses contribute to viral tumorigenesis.

Cancer can be divided into four kinds, depending on the cell types they are derived from. Cancer of epithelial cells is termed "carcinoma," which represents over 80% of human tumors. Hepatocellular carcinoma (ie, cancer of hepatocyte, which is the epithelial cells of liver tissues) and cervical carcinoma are examples (Table 24.1). Leukemia is a cancer of leukocytes, and lymphoma is a cancer of lymphocytes, while sarcoma is a cancer of muscle cells.

24.1 DISCOVERY OF TUMOR VIRUSES

The first tumor virus ever discovered was an avian retrovirus. In 1908, Drs Ellerman and Bang reported their observation that virus causes leukemia in chickens. This historic finding had not been appreciated back then, because leukemia was not considered as cancer at the time of discovery. In 1911, Peyton Rous made a similar observation that a virus could induce sarcoma, a tumor of muscle cells, in chickens. The chicken tumor virus was later named *Rous sarcoma virus* (RSV) after the discoverer. Subsequently, similar RNA tumor viruses were found in rodents (Table 24.2). Since these RNA tumor viruses could induce tumor in experimental animals (ie, mouse and chicken), they have been greatly exploited as experimental models for tumorigenesis during earlier days of tumor biology studies. Not surprisingly, the first oncogene ever discovered, *Src* oncogene, was also found in RSV genome (Box 24.1). Another twist in these RNA tumor viruses is that these RNA tumor viruses turned out to be retroviruses, as they synthesize their genome via reverse transcription (see Box 17.1).

TABLE 24.2 Representative of Tumor Viruses

Family	Virus	Tumor
RNA Virus		
Retrovirus	Rous sarcoma virus (RSV)	Mammary carcinoma
	Mouse mammary tumor virus (MMTV)	Leukemia
	Murine leukemia virus (MLV)	Leukemia
	Avian leukosis virus (ALV)	Sarcoma
	Human T-lymphotropic virus (HTLV)	T-cell leukemia[a]
Flavivirus	Hepatitis C virus (HCV)	Hepatocellular carcinoma[a]
DNA Virus		
Adeonovirus	Adeonovirus 5	Various solid tumor
Hepadnavirus	Hepatitis B virus (HBV)	Hepatocellular carcinoma[a]
Herpesvirus	Epstein-Barr virus	Burkitt's lymphoma[a]
		Nasopharyngeal carcinoma[a]
	KS-Herpes virus	Kaposi's sarcoma[a]
Poliovirus	SV40, Polyomavirus	Various solid tumor
Papillomavirus	HPV-16, HPV-18	Cervical carcinoma[a]

[a]*Human cancer.*

BOX 24.1 Rous Sarcoma Virus and Nobel Prize

RSV stands out, in that the scientists who studied it have been awarded the Nobel Prize three times. First, Peyton Rous was awarded the 1966 Nobel Prize for the discovery of RSV (see Fig. 24.11). Second, Howard Temin and David Baltimore shared the 1975 Nobel Prize for their discovery of reverse transcriptase (see Box 17.1). Third, Michael Bishop and Harold Varmus shared the 1989 Nobel Prize for their discovery of Src oncogene. It is remarkable that these seminal discoveries were all made in one RNA tumor virus. In particular, the discovery of the cellular homolog (ie, c-Src) of the viral oncogene (ie, v-Src) was received as a surprising finding, since it implicated that the precursor form of the cancer-causing gene (oncogene) is encoded by the human genome. This discovery laid the keystone for modern tumor biology. In retrospect, the recognition of discoveries made in RSV with three independent Nobel Prizes reflects the breath of the impact of RSV on the disciplines of "Molecular Virology" and "Tumor biology."

Photos of three legendary retrovirologists, who received Nobel Prizes for their work on RSV.

Although tumor-inducing animal retroviruses were abundantly discovered in the earlier half of the 20th century, human counterparts had not been discovered despite strenuous effort until 1978. The first and unique tumor-inducing human retrovirus is the *human T-cell leukemia virus* (HTLV-1 and 2) (see Table 24.2). HTLV-1 is an etiological agent for *adult T-cell leukemia* (ATL). Worldwide 20 million people are chronically infected, and most of the infection appears asymptomatic, but it leads to ATL in about 5–10% of infected individuals. Interestingly, the geographical distribution of HTLV-1 is confined to Central America, Caribbean islands, South America, some African countries, and Southern Japan (Kushu).

Besides retroviruses, which are often called RNA tumor viruses, DNA tumor viruses were also found. The first DNA tumor virus discovered was a papillomavirus isolated from a cotton tail rabbit by Dr Shoppe in 1933. Subsequently, the related papillomavirus was discovered in cattle (ie, bovine papillomavirus). The lack of a cell culture system to propagate papillomavirus prevented it from being exploited as a model to study the viral oncogenesis. The cell culture system to propagate the DNA virus was achieved by the discovery of polyomavirus (ie, SV40) and adenovirus, which cause tumor in rodents. Therefore, studies on SV40 and adenovirus have greatly contributed to our current understanding of the viral oncogenesis.

24.2 DNA TUMOR VIRUSES

Among human DNA viruses, five of them are known to be associated with human tumors (Table 24.3). Among these, three of them (ie, polyomavirus, papillomavirus, and adenovirus) are related in terms of viral oncogenic mechanisms in that the viral oncogenes dysregulate cellular tumor suppressor gene products, such as p53 and Rb. For the sake of brevity, here we focus on these small genome containing DNA tumor viruses. Two cancer-associated herpesviruses (ie, Epstein-Barr virus and KS-herpesvirus) are not detailed here, because they are beyond the scope of this book (see Box 9.3). However, the mechanisms of HBV-induced liver cancer are described in chapter "Hepadnaviruses."

24.2.1 Polyomavirus

SV40, the prototype of polyomavirus, has been extensively studied as a model of DNA tumor virus, since SV40 induces tumor in rodents and it can be readily propagated in cultured cells. SV40 was serendipitously discovered from a monkey cell line that was being used for the production of poliovirus vaccine in the 1960s. SV40 became a highly controversial subject after it was revealed that millions were exposed to the virus after receiving a contaminated polio vaccine produced between 1955 and 1982 (or later). However, no evidence is available supporting its association with human cancer.

SV40 does not induce tumor in monkey cells, a natural host, since monkey cells are permissive to SV40 infection, leading to cell lysis (see Fig. 6.10). By contrast, SV40 induces tumor in a rodent cell, a nonpermissive host. In the nonpermissive cells, only the early genes, but not the late genes, are expressed (Fig. 24.2). An early gene product (ie, *T-antigen*) is attributable for cell transformation or tumor formation. In other words, T-antigen is the viral oncogene of SV40.

TABLE 24.3 Viral Oncogenes of DNA Tumor Viruses

DNA Tumor Viruses	Viral Genes	Functions
Poliomavirus		
SV40	T-antigen	p53 binding, Rb binding
Papillomavirus		
HPV-16, 18	E6	p53 binding
	E7	Rb binding
Adenovirus		
Ad-5	E1A	Rb binding
	E1B-19k	Bcl-2 homolog
	E1B-55k	p53 degradation

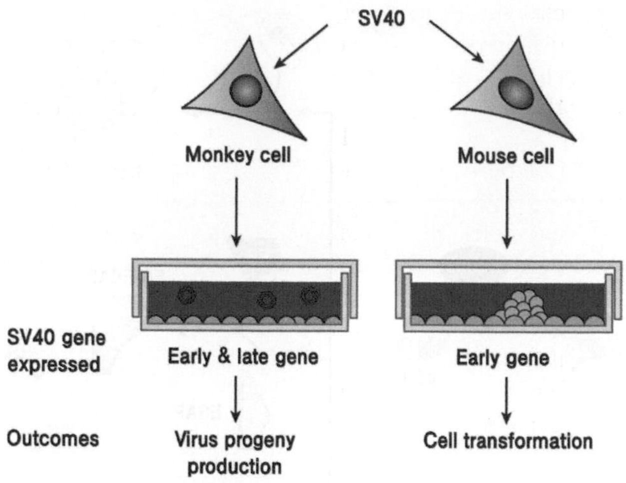

FIGURE 24.2 SV40 genome and the early gene expression. Only the early gene (ie, T-antigen) is expressed in nonpermissive cells. Two isoforms of T-antigens are expressed from the early gene by alternative splicing: large T-antigen and small T-antigen. Large T-antigen is largely responsible for cell transformation.

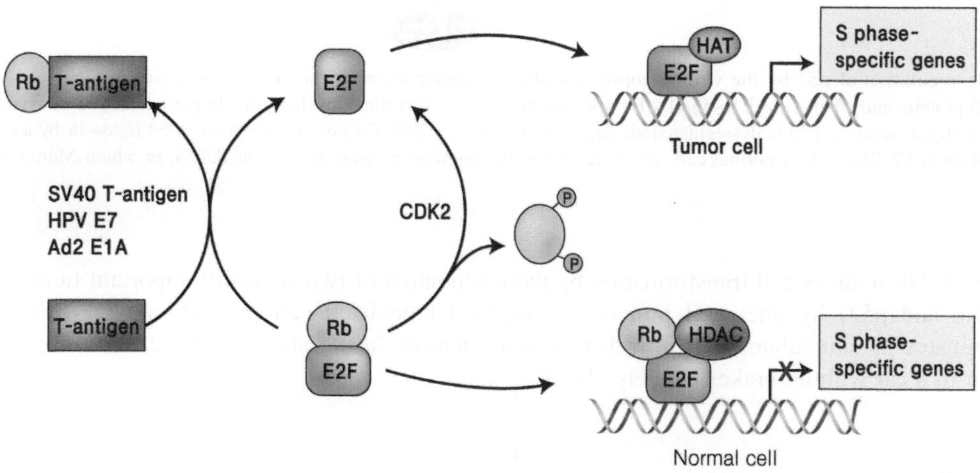

FIGURE 24.3 Dysregulation of Rb function by the viral oncoproteins of DNA tumor viruses. The viral oncoproteins of small genome DNA tumor viruses (ie, SV40 T-antigen, HPV E7 protein, and adenovirus E1A protein) bind and inactivate Rb, which blocks the S-phase entry via an interaction with E2F transcription factor. Rb plays the role of the brake in an analogy to a car. CDK2, cyclin-dependent kinase; HAT, histone acetyl transferase; HDAC, histone deacetylase.

A question then is how does T-antigen induce tumor formation? Notably, SV40 T-antigen dysregulates two tumor suppressor gene products (ie, *Rb*[8] protein and *p53*[9]) via protein–protein interaction (see Table 24.3). First, T-antigen releases *E2F*[10] transcription factor from the transcriptionally inactive Rb/E2F complex via Rb/T-antigen interaction (Fig. 24.3). In a resting cell (ie, G_0 phase of cell cycle), E2F-driven S-phase specific gene transcription is blocked by the engagement of Rb on E2F factor. Upon S-phase entry, the phosphorylation of Rb by cyclin-dependent kinase 2 (CDK2) induces Rb degradation. The free E2F binds to the promoter of the S-phase specific gene and induces transcription. In an SV40 infected cell, T-antigen binding to the Rb triggers E2F release, relieving inhibition and results in S-phase specific gene transcription (ie, cyclin A, cyclin E, etc.). Second, T-antigen inactivates p53 via sequestration (Fig. 24.4).

8. **Rb (retinoblastoma)** A tumor suppressor gene first identified as gene defective in retinoblastoma.
9. **p53** It was discovered as a T-antigen binding protein with a molecular mass of 53 kDa.
10. **E2F** A transcription factor that was discovered as a DNA-binding protein of adenovirus E2 gene promoter. It is a representative S-phase specific transcription factor.

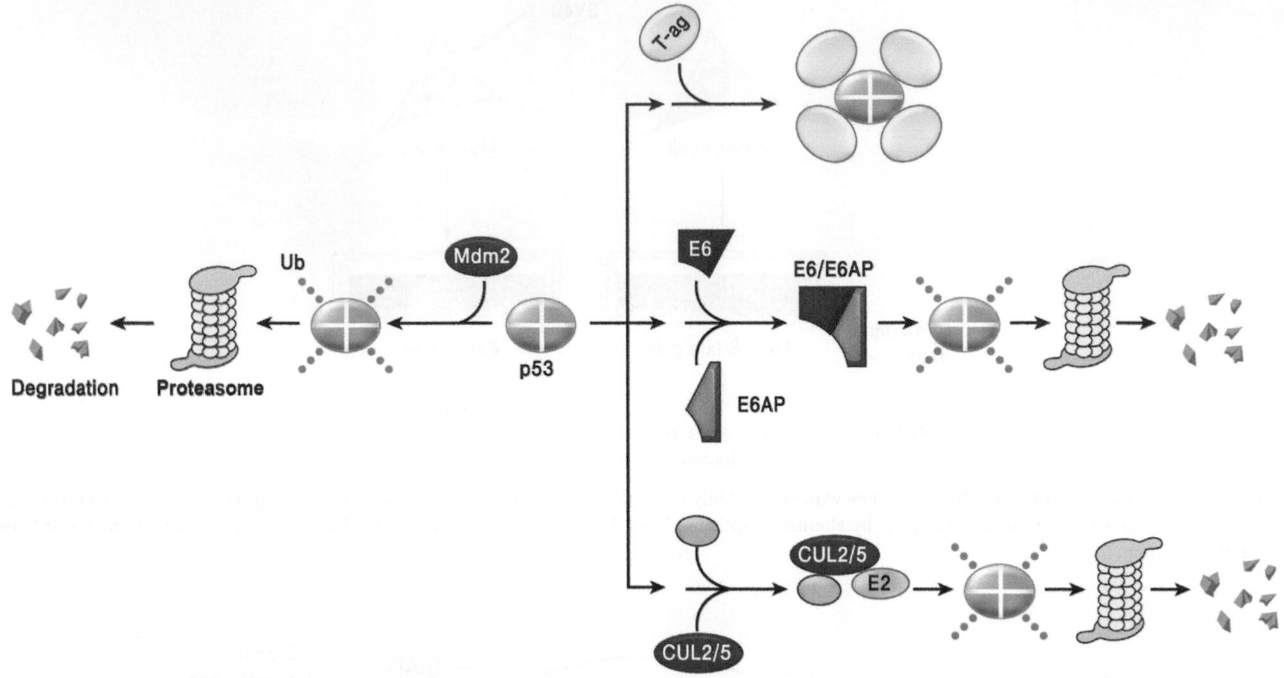

FIGURE 24.4 Dysregulation of p53 by the viral oncoproteins of DNA tumor viruses. The viral oncoproteins of DNA tumor viruses (ie, SV40 T-antigen, HPV E6 protein, and adenovirus E1b-55 kDa/E4 orf6) inactivate p53 via distinct mechanisms. In particular, p53 is dysregulated via sequestration by SV40 T-antigen, whereas p53 is dysregulated via degradation either by HPV E6-associated ubiquitin E3 ligase or by adenovirus E1B-55/E4 orf6-associated ubiquitin E3 ligase. In a normal cell, p53 is regulated by ubiquitin proteasome system (UPS), in which Mdm2 acts as ubiquitin E3 ligase.

In summary, SV40 induces cell transformation by the inactivation of two critically important tumor suppressor gene products (ie, Rb and p53) by one viral protein, T-antigen. Unregulated cell proliferation (cell division) driven by T-antigen culminates in unregulated cell growth or transformation. In the analogy to a car, T-antigen induced tumor cells are similar to a car without brakes (see Fig. 24.1).

24.2.2 Papillomavirus

HPV is associated with cervical carcinoma. Out of over 150 HPV subtypes currently reported, most of them are associated with warts, a benign tumor. In particular, *HPV-16*, and *-18* are known to cause cervical carcinoma (see Box 7.1). Worldwide, approximately half a million women are newly diagnosed in each year with cervical cancer, of which most are associated with HPV (see Box 7.5).

How does HPV cause cancer? HPV DNAs are found chromosomally integrated in cells of cervical cancer. Frequently, more than one copy of HPV DNA is integrated. Note that HPV DNA exists in an episomal state without chromosomal integration in its life cycle (Fig. 24.5). The site of integration is random with respect to the chromosome site. In contrast, the site of integration of HPV DNA is not random but it is inserted via the E1 and E2 regions of HPV DNA. Consequently, E1 and E2 genes are disrupted, but E6 and E7 genes are preserved in cancer cells.

Interestingly, transcription of E6 and E7 genes is suppressed by E2 protein, which acts as a transcriptional suppressor, and is expressed at a low level in normal cells. In contrast, E6 and E7 proteins are overexpressed in cancer cells via two distinct mechanisms. First, E6 and E7 genes are overexpressed, since E2 gene, a transcriptional suppressor of E6 and E7 genes, was disrupted during integration. Secondly, E6 and E7 mRNAs are stabilized in cancer cells. In fact, as the 3′ splice sites of E6 and E7 genes are lost during integration, alternative splicing sites from downstream gene are utilized. Consequently, an AU-rich sequence element, which confers RNA instability, is lost. Overall, overexpression of E6 and E7 genes contributes to the HPV-mediated viral oncogenesis.

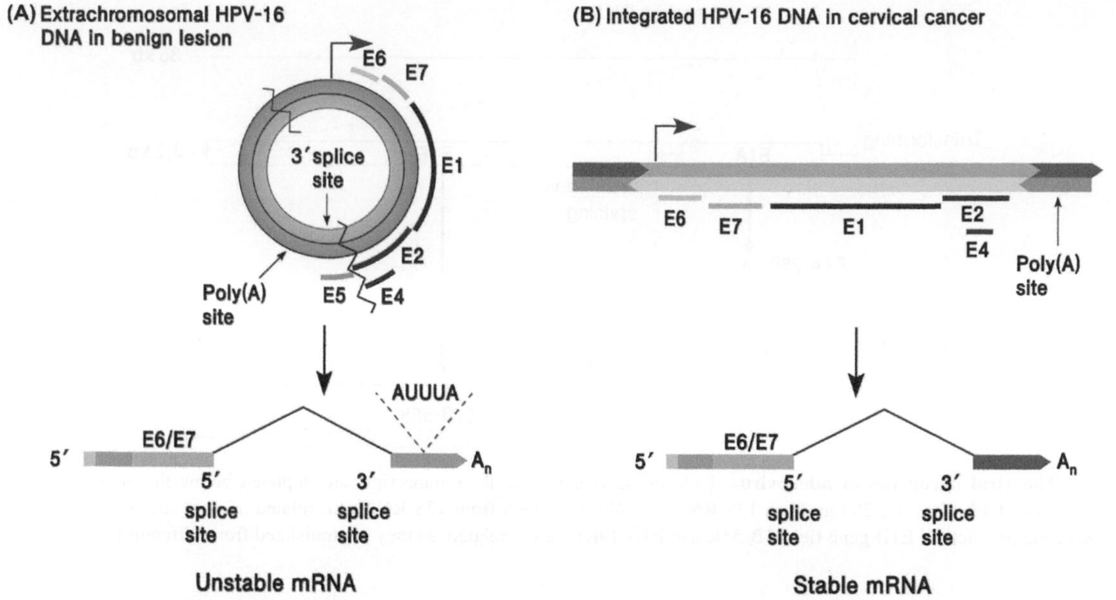

FIGURE 24.5 **The integrated HPV DNA in cervical cancer.** (A) The HPV DNA episome sequence found in normal cells. The sequence element, AUUUA, present in 3′ untranslated region (UTR) of E6 and E7 genes, confers instability to mRNA. (B) The HPV DNA sequence found in cervical cancer. E6 and E7 mRNAs are alternatively spliced into the cellular 3′ splice site, resulting in an enhanced mRNA stability.

How then do E6 and E7 proteins induce tumor? E6 protein targets p53 for degradation (see Fig. 24.4). Specifically, E6 protein forms an ubiquitin E3 ligase, in association with *E6AP*,[11] a cellular factor. As a result, p53 is subjected to ubiquitin-mediated proteolysis. On the other hand, just like SV40 T-antigen, HPV E7 targets Rb, thereby suppressing Rb function in regulating cell cycle progression (see Fig. 24.3). Overall, the viral oncoproteins of HPV dysregulate two tumor suppressor gene products, thereby leading to viral oncogenesis.

24.2.3 Adenovirus

Adenovirus is considered as a DNA tumor virus. Although adenovirus does not cause cancer in human, it induces tumor in rodents (ie, mouse and hamster). In permissive human cells, adenovirus fully executes its life cycle and produces progeny virus via cell lysis. In contrast, in rodent cells, the late phase of life cycle (ie, replication) is not permissive, and only early genes are expressed. In particular, two early genes (ie, *E1A* and *E1B*) are attributable to cell transformation (Fig. 24.6). E1A engages Rb, thereby releasing *E2F* (see Fig. 24.3), a mechanism reminiscent of SV40 T-antigen and HPV E7. On the other hand, E1B gene expresses two gene products—E1B-55 and E1B-19 kDa. E1B-55 kDa targets p53 for ubiquitin-mediated degradation (see Box 8.2), similar to SV40 T-antigen and HPV E6 (see Fig. 24.4). On the other hand, E1B-19 kDa is a member of the *Bcl-2 family*[12] (see Table 24.3). Just like oncogenic Bcl-2, it engages *Bax*,[13] thereby blocking apoptosis. Overall, adenovirus encodes three viral oncoproteins: E1A, E1B-55 kDa, and E1B-19 kDa.

A question arose as to how these two viral genes work together to induce tumor? In other words, what happens if only one viral gene is expressed? Expression of E1A only transforms cells in culture and results in colony formation (Fig. 24.7). However, when the cells isolated from the colony were administered into a mouse, tumor formation was not observed. In contrast, tumor was observed when cells were isolated from the colony formed by transfecting E1A and E1B genes. This experiment highlights the discrepancies between in vitro (cell culture) and in vivo (mouse) experiments. Evidently, cell transformation in vivo is more demanding than that in vitro. To achieve full transformation

11. **E6AP (E6-associated protein)** A host factor that is first identified as HPV E6-binding protein. It is now classified as a member of HECT (homologous to E6AP carboxyl-terminus)-type ubiquitin E3 ligase family.
12. **Bcl-2 (B-cell lymphoma)** It is a founding member of the Bcl-2 family, an apoptosis regulator. It was identified as a gene overexpressed in B-cell lymphoma.
13. **Bax** It is a pro-apoptotic member of the Bcl-2 gene family. Bax promotes apoptosis by binding to and antagonizing Bcl-2 protein.

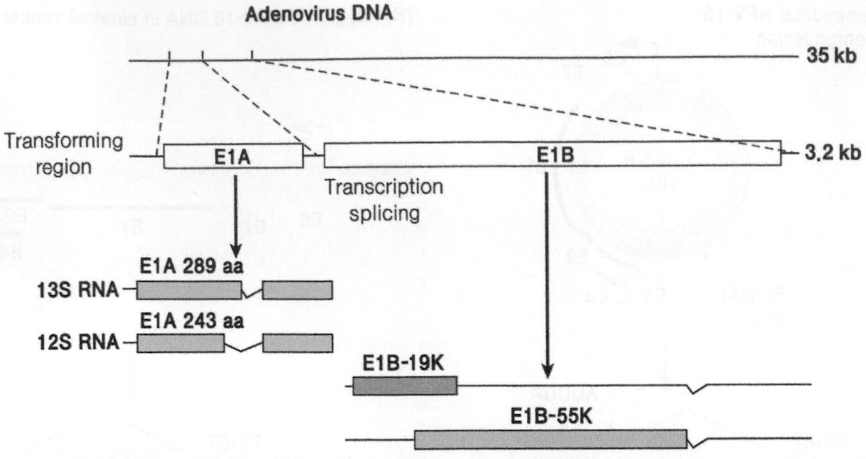

FIGURE 24.6 The viral oncogenes of adenovirus. E1A and E1B gene and their transcripts are depicted below the adenovirus DNA genome. Two gene products of E1A gene (ie, 289 aa from 13S RNA and 243 aa protein from 12S RNA) are related, as they are in the same reading frame. In contrast, two gene products of E1B gene (ie, E1B-55K and E1B-19K) are not related, as they are translated from different reading frames.

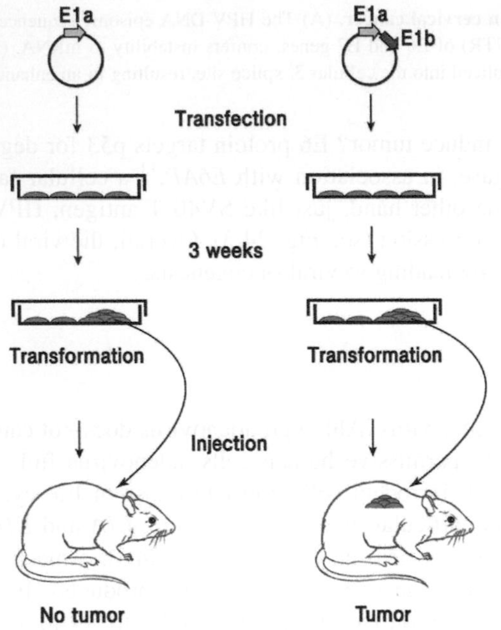

FIGURE 24.7 Collaborative roles of E1A and E1B genes in tumorigenesis. Cultured cells that are stably transfected either by E1A gene or by E1A plus E1B can form colonies that are indistinguishable. However, cells transformed by E1A plus E1B genes form tumors in mouse when cells in a single colony are subcutaneously administered into a mouse, emphasizing the complementing roles of two viral oncogenes in the viral tumorigenesis.

of cells in vivo, both E1A and E1B genes are required, emphasizing the complementing roles of two viral oncogenes in the viral tumorigenesis (see Fig. 8.11).

24.2.4 Attributes of DNA Tumor Viruses

A few attributes are common among DNA tumor viruses. First, DNA tumor viruses cause persistent infection in the host (Box 24.2). In other words, the viral replication continues to occur for decades until tumor is formed. A good example for this lifelong infection is HPV infection. Second, the mechanism for viral oncogenesis does not require the completion of the virus life cycle. In some tumor cells, only a subset of the viral DNA genome (ie, viral oncogenes) is detectable in the absence of the functionally replicating viral genomes. Third, unlike retroviral oncogenes, no cellular homologs of the viral oncogenes of DNA tumor viruses are found, reflecting that they are not derived from host genes

BOX 24.2 DNA Tumor Viruses and Tumors They Cause

Why does a virus evolve to be a tumor-causing virus in the first place? Let's put it in a different way. What is the benefit to the virus by inducing tumors in their host? In short, tumor, a fatal disease to their host, is a dead-end for the virus as well as the host. It is generally believed that tumor is an "unwanted consequence for both viruses and host." Although tumor formation is a very rare event occurring only in one cell per organism; it is fatal to the host, but not to the viruses.

Then, how did viruses acquire the oncogene during evolution? Let's consider SV40, as their oncogenesis mechanisms are uncovered. Tumor viruses including SV40 infect terminally differentiated cells that no longer divide. Since the viral DNA replication relies on the S phase of the cell cycle, SV40 triggers the cells to enter S phase via dysregulation of Rb, a gate keeper. During evolution, DNA tumor viruses acquired the ability to dysregulate tumor suppressor protein in order to undergo the viral propagation.

Then, how is tumor formed in the virus infected cell? Once the viral infection is established in nondividing cells, the viral life cycle is fully executed, and the viral progeny is released. Alternatively, the viral persistence is established in certain infected cells when the virus is not cleared by host immune system. In this circumstance, the viral DNA becomes integrated into the chromosome, which leads to the loss of ability to replicate due to mutations. Cell lysis is precluded due to the inability to execute the viral genome replication, but instead, stable expression of the viral oncogene from the integrated viral genome leads to cell transformation. In short, the loss of the ability to replicate due to mutations occurring during chromosomal integration is a prerequisite for tumor formation.

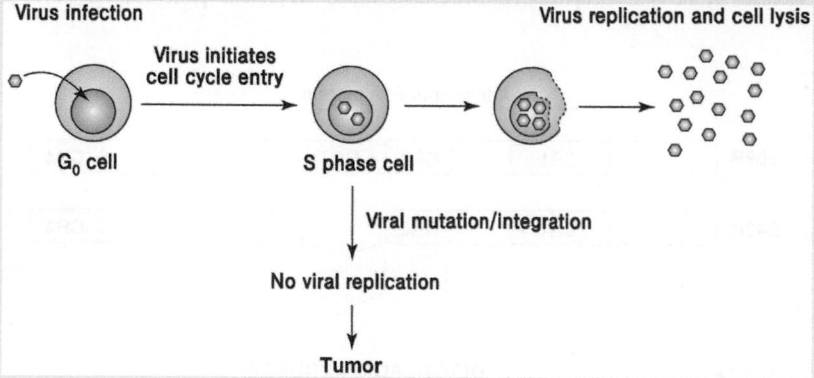

A diagram illustrating steps involved in the viral oncogenesis by DNA tumor viruses. Chromosomal integration is not an obligate step in a viral life cycle, but it occurs not infrequently in persistently infected cells. The loss of the ability to replicate is a prerequisite for tumor formation.

(see Table 24.3). For example, SV40 T-antigen does not have a cellular homolog. Fourth, the viral oncogenes are essential for viral genome replication (eg, SV40 T-antigen).

Notably, three DNA tumor viruses share a common strategy to induce tumor in that the viral oncogenes dysregulate two critically important tumor suppressor proteins (ie, Rb and p53) (Fig. 24.8). Conceivably, during evolution DNA tumor viruses acquired the ability to inhibit both Rb and p53 proteins via protein–protein interaction. Since the nucleotide sequences of these three DNA tumor viruses are not related, the viral oncogenes are not derived from a common ancestor. Conceivably, the virus acquired the viral oncogenes independently during evolution, a process termed "*convergent evolution.*"

24.2.4.1 Dysregulation of Rb Function

In eukaryotic cells, cell cycle checkpoints are the control mechanisms that ensure the fidelity of cell division in eukaryotic cells. In particular, the *G_1/S checkpoint* of the cell cycle, which makes the key decision of whether the cell should divide or delay division, is tightly regulated (see Fig. 24.1). *Mitogenic stimulus* (eg, growth factors) induces cell signaling, leading to the activation of *cyclin-dependent kinase*,[14] Cdk4/6, to phosphorylate Rb. Then, the phosphorylated Rb releases E2F from the Rb/E2F complex, and the released E2F drives S-phase specific gene transcription, such as cyclin D.

14. **Cyclin-dependent kinases (Cdks)** Cdks are a family of protein kinases first discovered for their role in regulating the cell cycle. Cdk binds a regulatory protein called a cyclin for the activation.

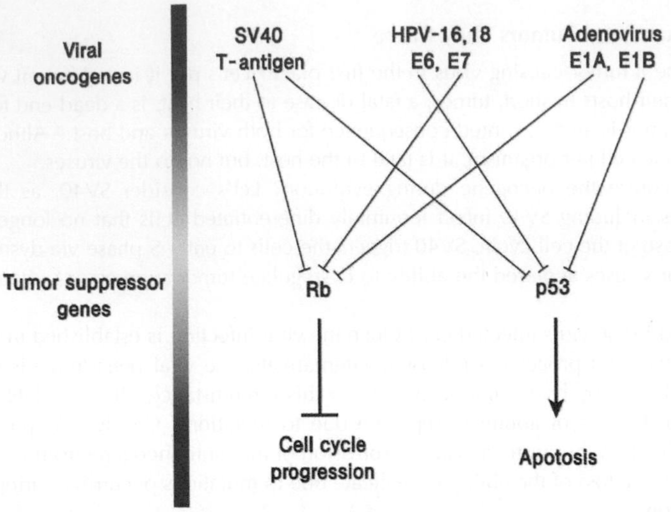

FIGURE 24.8 The protein–protein interactions between the viral oncoproteins of DNA tumor viruses versus tumor suppressor gene products. Rb and p53 functions are antagonized by the viral oncoproteins of DNA tumor viruses via protein–protein interaction, as denoted by T lines.

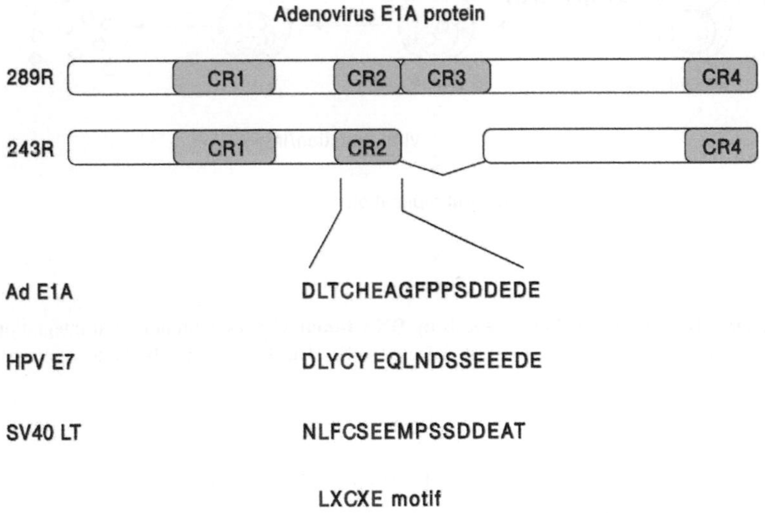

FIGURE 24.9 The conserved amino acid residues in the Rb binding region of three viral oncoproteins. Two conserved regions (CR1 and CR2) are found in the RB binding regions of E1A protein. For comparison, the RB binding region of HPV E7 and SV40 large T-antigen is shown below. The LXCXE motifs found in the viral oncoproteins are highlighted in red. L, leucine; C, cysteine, E, glutamic acid, X, any amino acid residues.

On the other hand, the viral oncoproteins confer cells the capability of overriding the G_1/S checkpoint, leading to an excessive cell proliferation even in the absence of the mitogenic stimulus. For instance, via protein–protein interaction, the viral oncoproteins (ie, SV40 T-antigen, HPV E7, and adenovirus E1A) dysregulate the Rb's role in cell cycle regulation (see Fig. 24.3). In this context, the viral oncoproteins are considered as a *mitogen* that stimulates cell proliferation.

One question that arises is how do the viral oncoproteins of seemingly unrelated DNA tumor viruses commonly dysregulate Rb? An answer to this question was obtained by structural studies of the viral oncoproteins. The structure of the Rb binding region was first resolved by the structural analysis of E1A. Two conserved regions (CR1 and CR2) of E1A protein were identified as Rb binding regions (see Fig. 8.6). Surprisingly, despite no sequence homology at the amino acid level between SV40 T-antigen or HPV E7 protein and adenovirus E1A, a limited but significant similarity was found in the CR2 region of E1A (ie, *LXCXE motif*) (Fig. 24.9). Subsequent structural analysis confirmed that both SV40 T-antigen and HPV E7 interacts with Rb via LXCXE motif. It is worth noting that structural studies could lead to a novel discovery that otherwise would not be readily made.

24.2.4.2 Dysregulation of p53 Function

In addition to the dysregulation of Rb, the inactivation of p53 is invariably accompanied by oncogenic transformation. Ironically, p53 was first discovered as an SV40 T-antigen binding protein. Its importance was realized only after the discovery of the critical importance of p53-T-antigen interaction in cell transformation. p53 has been described as "the guardian of the genome" because of its prominent role in conserving genome stability. Upon DNA damage, the activated ATM/ATR stabilizes p53, which in turn activates p21 transcription, thereby leading to cell cycle arrest (Fig. 24.10). Cells are arrested at a checkpoint of cell cycle until the DNA damage is repaired. In addition to the DNA damage response, p53 plays a critical role in apoptosis. More importantly, the critical significance of p53 in cancer was further validated by the fact that p53 is one of the most frequently mutated genes in cancer.

How is the stability of p53 regulated? In a normal cell, p53 is present only at a low level due to its short half-life (ie, $t_{1/2} = 30$ min). The stability of p53 is regulated by the ubiquitin-mediated proteolysis, where *Mdm2*,[15] ubiquitin E3 ligase, targets p53 for the degradation (Fig. 24.11). The stability of p53 can be augmented in two ways. First, upon

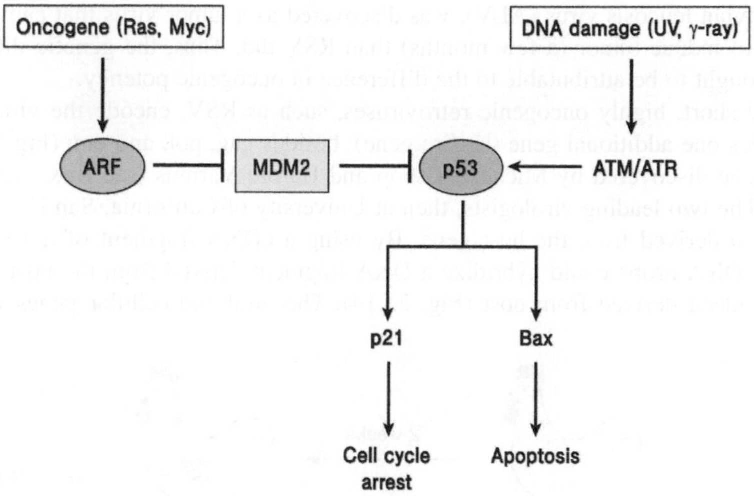

FIGURE 24.10 The effects of p53 on cellular functions. p53 can be activated either by DNA damage response or oncogene signaling. p53, a transcription factor, induces transcription of multiple downstream genes such as p21, and Bax. Depending on the cellular context, the activated p53 could lead to cell cycle arrest, and apoptosis.

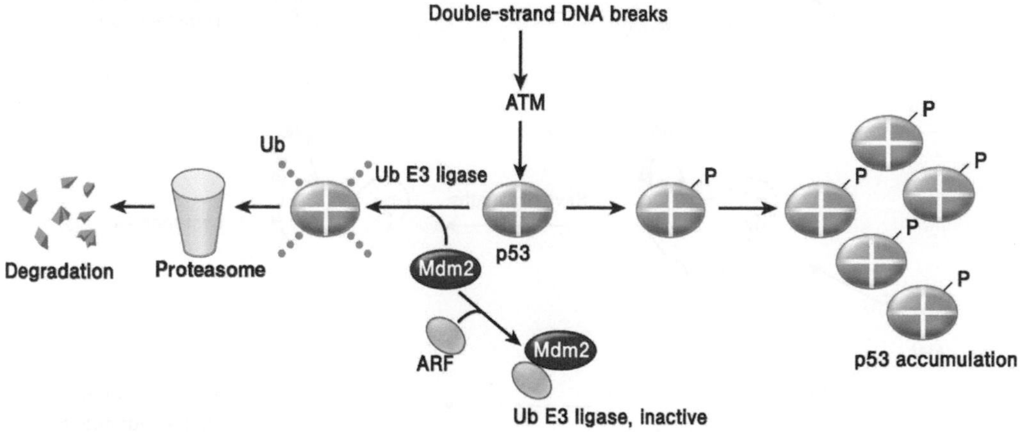

FIGURE 24.11 Regulation of p53 stability. In a normal cell, p53 is regulated by the UPS and has a short half-life. Specifically, p53, a tetramer, is subjected to ubiquitin-mediated proteolysis involving Mdm2 ubiquitin E3 ligase. Upon DNA damage, such as double-strand DNA breaks, p53 is phosphorylated by the activated ATM serine/threonine kinase and becomes stabilized.

---

15. **Mdm2 (murine double minute)** The murine double minute (*mdm2*) oncogene, which codes for the Mdm2 protein, was originally cloned from the transformed mouse cell line 3T3-DM.

DNA damage, the activated ATM kinase phosphorylates p53, thereby leading to p53 accumulation. Second, the binding of *p14ARF*,[16] Mdm2 inhibitor, inactivates Mdm2, leading to p53 accumulation.

24.3 RNA TUMOR VIRUSES

All RNA tumor viruses are retroviruses, with the exception of hepatitis C virus (HCV). However, the underlying mechanism of HCV-induced tumor is unrelated to that of tumor retroviruses (see chapter: Hepatitis Viruses). Historically, the term "RNA tumor virus" has been used as a synonym of "tumor retrovirus."

24.3.1 RSV Versus ALV

Since the discovery of tumor-inducing avian retroviruses, avian retroviruses have become a focus of tumor biology investigation. Moreover, the fact that RSV could cause tumor (ie, sarcoma) within 2 weeks, and that RSV readily transforms cultured cells makes it a favorite experimental model for tumor formation (Fig. 24.12). Subsequently, another related avian retrovirus, avian leucosis virus (ALV), was discovered as a tumor virus that causes lymphoma in chicken, but it took much longer to induce tumor (a few months) than RSV did. Thus, the genetic difference between the two avian retroviruses was thought to be attributable to the difference in oncogenic potency.

To make a long story short, highly oncogenic retroviruses, such as RSV, encode the viral oncogene (Table 24.4). Unlike ALV, RSV encodes one additional gene (ie, Src gene), besides gag, pol, and env (Fig. 24.13). In fact, Src oncogene was the first oncogene discovered by Michael Bishop and Harold Varmus (see Box 24.1), and it exhibits protein tyrosine kinase activity. The two leading virologists, then at University of California, San Francisco, speculated that the Src gene would have been derived from the host gene. By using a cDNA fragment of the viral Src gene as a probe, they discovered that the cDNA probe could hybridize a DNA fragment derived from the host chromosome, implicating that the viral gene was indeed derived from host (Fig. 24.14). The viral and cellular genes were coined as *v-Src* and

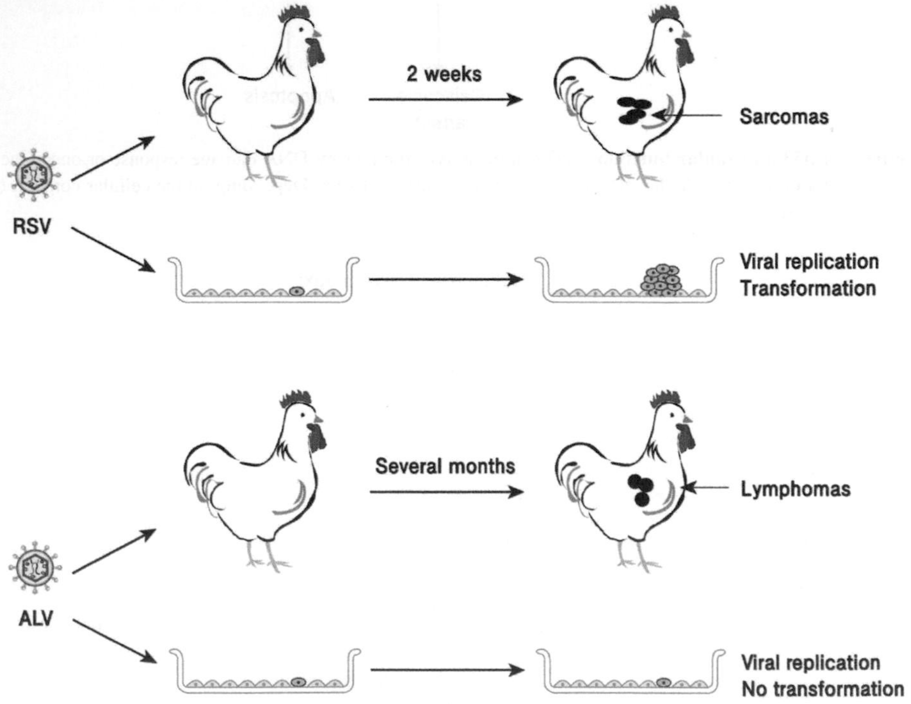

FIGURE 24.12 Comparison of RSV and ALV in tumorigenesis. RSV, a highly oncogenic retrovirus, can cause tumors in 2 weeks in chicken and can transform cultured cells. ALV, a weak oncogenic retrovirus, can cause tumors in few months and cannot transform cultured cells.

16. **p14ARF (alternative reading frame)** It refers to a tumor suppressor gene product, whose gene was discovered as an alternate reading frame (ARF) product of the CDKN2A locus, and is an inhibitor of Mdm2.

TABLE 24.4 Features of Tumor Retroviruses

Features	Highly Oncogenic Retrovirus	Weakly Oncogenic Retrovirus
Prototype	RSV, Ab-MuLV	ALV
Tumor frequency	High (100%)	Low
Latency	A few day	A few month
Replication	No[a]	Yes
Oncogene	Yes (v-src or c-Abl)	No
Tumor mechanism	Viral oncogene	Activation of proto-oncogene
In vitro cell transformation	Yes	No

[a]*RSV is an exception in that it has replication capability.*

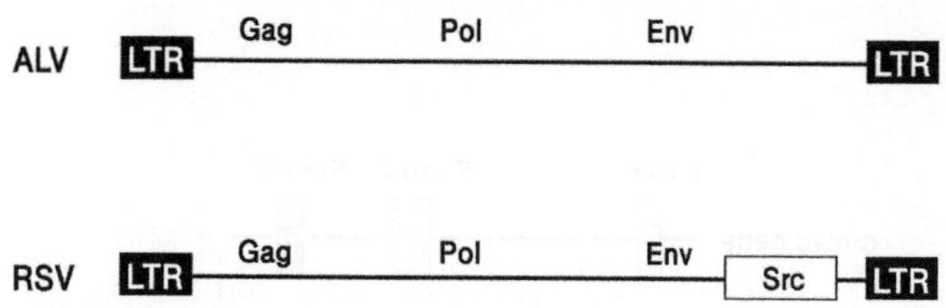

FIGURE 24.13 Comparison of ALV- versus RSV-proviral genome. ALV has a typical retroviral genome organization, having Gag-Pol-Env genes in between two LTRs. In contrast, RSV has an additional gene, v-Src gene, near 3′ LTR.

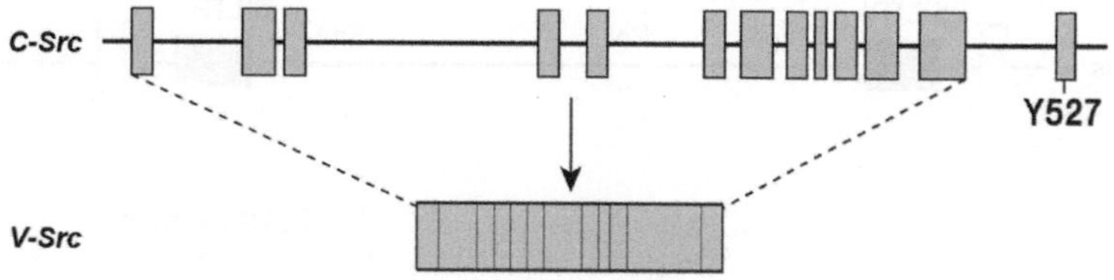

FIGURE 24.14 Comparison of c-Src gene versus v-Src gene. Basically, the v-Src gene is the cDNA version of c-Src gene, lacking introns. The phosphorylation of Y527 residue inactivates the protein tyrosine kinase activity of c-Src protein. The exons are indicated by *boxes*, while introns are indicated by a *solid line*. The Y527 residue (ie, tyrosine) at the C-terminal exon is denoted.

c-Src, respectively. Although they are highly homologous, two important differences should be noted. First, c-Src, the cellular gene, has introns, while v-Src gene lacks introns. Second, v-Src gene lacks the C-terminal exon harboring the Y527 residue (ie, tyrosine), which is subjected to negative regulation via phosphorylation. By contrast, v-Src gene is constitutively activated due to the lack of Y527 residue. The cellular counterparts of viral oncogenes are collectively termed "*proto-oncogene*," implying the precursor of the activated viral oncogenes.

24.3.2 The Origin of RSV

Then, a question is how was RSV generated? It is believed that RSV was originated from ALV during evolution. A hypothesis is that a hybrid RNA transcribed from an ALV provirus integrated near the c-Src gene became packaged into the viral capsid (Fig. 24.15). In other words, the transduction of the cellular c-Src gene into ALV represents a key event for the generation of RSV. The transduction of cellular genes reflects an extraordinary feature of retroviruses, which is that they insert their DNA into chromosomes during their life cycle.

ALV, a weak oncogenic retrovirus, could nonetheless cause tumors, although less potent than RSV, which is a highly oncogenic retrovirus. How can ALV, which lacks a viral oncogene, cause tumor? Following infection, the

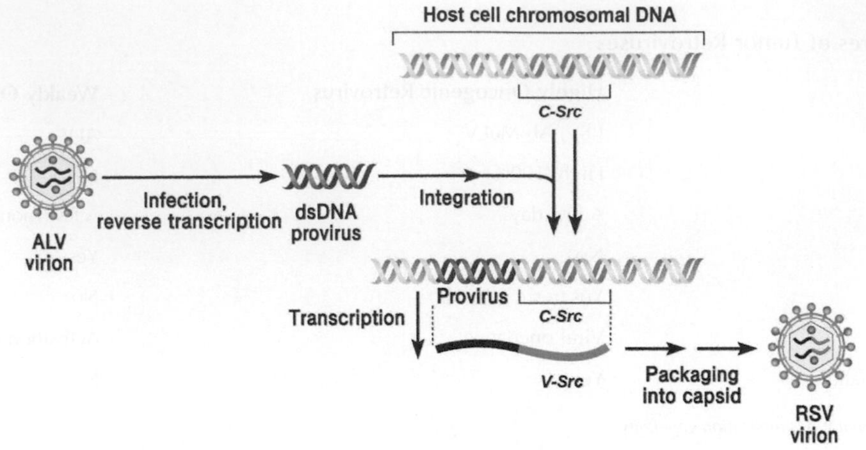

FIGURE 24.15 A model illustrating a hypothesis on the biogenesis of RSV. Following infection of cells by ALV, the integration of the proviral DNA near the c-Src gene could lead to a hybrid viral transcript containing c-Src RNA. Subsequent packaging of the hybrid RNA into the viral capsid could yield RSV. Note that the v-Src genes correspond to the spliced intronless version of c-Src gene.

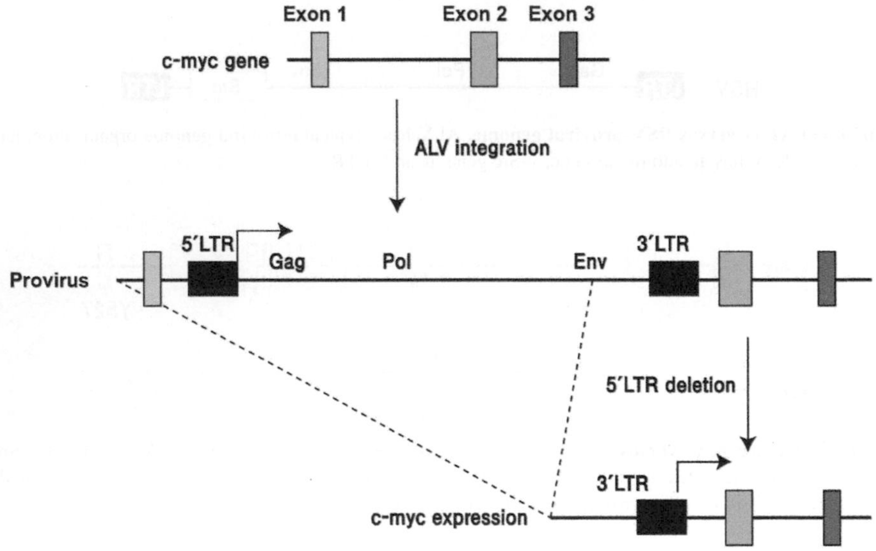

FIGURE 24.16 A schematic diagram showing the steps involved in the activation of c-myc proto-oncogene by ALV-mediated insertional activation. The first step is the integration of ALV provirus into c-myc gene. The second step is the activation of 3′ LTR promoter by the deletion of 5′ LTR; note that transcription from 3′ LTR promoter remains functionally inactive in the presence of 5′ LTR promoter. Three exons of c-myc gene are drawn on the top. The ALV provirus inserted into the c-myc gene is demarcated by two LTRs. The deletion of the proviral DNA is denoted by the *dotted lines*. The transcription of 3′LTR is indicated by the *rightward arrow*. Note that the exon 1 of c-myc gene corresponds to the UTR of c-myc gene.

proviral DNA of ALV becomes integrated into the chromosome. Its chromosomal integration sites are more or less random with respect to chromosome loci. In the case of ALV, the proviral DNA became frequently inserted into c-myc gene, a cellular proto-oncogene; in particular, between exon 1 and exon 2 (Fig. 24.16). Subsequent deletion of 5′ LTR could lead to the c-myc gene overexpression via activation of the otherwise inactive 3′ LTR promoter. This kind of activation of otherwise inactive cellular oncogenes by the retroviral insertion is termed "*insertional activation.*" The insertional activation represents an underlying mechanism for the retroviral oncogenesis of weakly oncogenic retroviruses that lack viral oncogenes. More importantly, the discovery of the insertional activation of cellular proto-oncogenes by oncogenic retroviruses in 1970s shed light on elucidating the "genetic principle of oncogenesis."

24.4 PERSPECTIVES

Cancer represents the most formidable disease in mankind and approximately 15% of human cancer is attributable to viruses. In other words, 15% of human cancer is preventable by vaccination against these human tumor viruses, including HBV, HCV, and HPV. Fortunately, the preventive vaccines for HBV and HPV are now available, and eventually the cancers caused by these two viruses will be decreased as the vaccination campaign progresses. Since the earlier discovery of RSV, RNA tumor viruses (ie, tumor retroviruses) greatly contributed to the progress of not only tumor virology but also tumor biology and oncology. They served as an excellent experimental tool, as they induce tumor readily in animals and transform cells in culture. Not surprisingly, a few seminal discoveries were recognized by Nobel Prizes (see Box 24.1). In addition, a novel human tumor virus, Merkel cell polyomavirus (MCV), was discovered (Box 6.2). MCV was discovered in a malignant form of skin cancer, Merkel cell carcinoma, and its causal association with the tumor was evident. Knowledge acquired from studying animal polyomaviruses, such as SV40, can be directly applicable to the study of MCV, revealing the importance of basic science in the investigation of the disease of clinical importance.

24.5 SUMMARY

- *Tumor viruses*: Up to 15% of human cancers are caused by tumor viruses such as HBV, HCV, and HPV.
- *DNA tumor viruses*: Small genome DNA tumor viruses, including polyomavirus, papillomavirus, and adenovirus, exploit a parallel mechanism for viral oncogenesis via dysregulation of Rb and p53.
- *RNA tumor viruses*: RSV, an avian retrovirus, has been studied extensively as a prototype of RNA tumor viruses. Highly oncogenic retroviruses (ie, RSV) encode the viral oncogene, besides Gag, Pol, and Env genes, while weakly oncogenic retroviruses (ie, ALV) do not encode the viral oncogene. ALV, a prototype of weakly oncogenic retrovirus, causes tumor via insertional activation of proto-oncogene.

STUDY QUESTIONS

24.1 The oncoproteins of small genome DNA tumor viruses, such as SV40, HPV, and adenovirus, cause tumor by dysregulation of tumor suppressor proteins. Compare their mechanism of dysregulating tumor suppressor proteins.

24.2 List two kinds of viral oncoproteins of DNA tumor viruses that exhibit ubiquitin E3 ligase activity and describe their functions related to viral oncogenesis.

24.3 ALV infection leads to lymphoma in chickens. Propose a hypothesis on viral tumorigenesis. How would you test the hypothesis?

SUGGESTED READING

Howley, P.M., Livingston, D.M., 2009. Small DNA tumor viruses: large contributors to biomedical sciences. Virology. 384 (2), 256–259.

Miest, T.S., Cattaneo, R., 2014. New viruses for cancer therapy: meeting clinical needs. Nat. Rev. Microbiol. 12 (1), 23–34.

Moody, C.A., Laimins, L.A., 2010. Human papillomavirus oncoproteins: pathways to transformation. Nat. Rev. Cancer. 10 (8), 550–560.

Moore, P.S., Chang, Y., 2010. Why do viruses cause cancer? Highlights of the first century of human tumour virology. Nat. Rev. Cancer. 10 (12), 878–889.

Weinberg, R.A., 2014. The Biology of Cancer. Garland Science, New York.

JOURNAL CLUB

- Moore, P.S., Chang, Y., 2010. Why do viruses cause cancer? Highlights of the first century of human tumour virology. Nat. Rev. Cancer 10 (12), 878–889.

  Highlight: This review article provides insightful thoughts on the origin of viral oncogenesis. The authors argued that viral oncogenesis is a by-product of the "molecular parasitism" by viruses to promote their own replication. Cells respond to virus infection by activating Rb and p53 to inhibit virus replication as part of the innate immune response. To survive, tumor viruses have evolved the means for inactivating these inhibitors placing the cell at risk for cancerous transformation.

Chapter 25

Vaccines

Chapter Outline

Vaccines are considered as one of the most remarkable discoveries of mankind and have greatly influenced our lives. The principal component of a vaccine is an antigen (ie, protein) that leads to an immune response following its inoculation into a host organism. Vaccines can be grouped into two general types: live attenuated vaccines and killed vaccines. In this chapter, the historical accounts of the most successful cases of vaccination (ie, smallpox and poliomyelitis vaccine) will first be described. Then, we will learn how different kinds of vaccines are manufactured and learn their pros and cons.

25.1 POXVIRUS VACCINE

The effectiveness of vaccines is best exemplified by the eradication of smallpox on earth. Smallpox epidemics have proved one of the most formidable disasters in human history causing over 300 million deaths in 20th century alone. However, smallpox is the only viral disease that has been eradicated. Descriptions of smallpox were found in the historical literatures long before the 20th century. The earliest credible clinical evidence of smallpox is found in medical writings from ancient India (as early as 1500 BC) and China (1122 BC), and smallpox has been detected in the Egyptian mummy of Ramses V, who died more than 3000 years ago (1145 BC). According to the literatures of ancient China, smallpox was described as "a disease that affects an individual only once in a lifetime." According to ancient writings, attempts were made in 11th century China to prevent a smallpox epidemic by inoculating pus from affected individuals.

It was Edward Jenner (1749–1823), who succeeded in conducting the first smallpox vaccination and establishing a major milestone in the vaccination technology. An English practitioner, he noted that milkmaids who had been affected by cowpox were resilient to smallpox. To exploit the observation, he inoculated a healthy boy with pus from a cowpox infection and then challenged him with smallpox 2 weeks later (Fig. 25.1). Remarkably, the boy was unaffected by the smallpox challenge, suggesting that prior exposure to pus via inoculation could prevent the viral infection. Thus, the usefulness of the vaccination was demonstrated for the first time. In retrospect, such human experimentation is unethical today without proper approval.

Despite Jenner's success, the second vaccine, the *rabies*[1] vaccine (Box 25.1), was only developed 100 years later by Louis Pasteur. In fact, the term "vaccine" was coined by Louis Pasteur from "vacca," a Latin word for cow, to honor Edward Jenner's achievement.

25.2 POLIOVIRUS VACCINES

Poliovirus is the etiologic agent of *poliomyelitis*,[2] an acute flaccid paralysis affecting 1–2% of infected individuals and, on rare occasions, causing death (see chapter: Picornavirus). A striking feature of infection is lifelong disabilities that

1. **Rabies** It refers to a viral disease that causes acute encephalitis in animals. Rabies virus belongs to the family Rhabdoviridae (see chapter: Rhabdovirus). Rabies is a Latin word for "madness."

2. **Poliomyelitis** The term poliomyelitis is derived from Greek word for "gray"-*polio* + for "marrow"-*myelos* + "inflammation"-*titis*. The term implies paralytic poliomyelitis, resulted from destruction of motor neurons within the spinal cord in the CNS.

Molecular Virology of Human Pathogenic Viruses. DOI: http://dx.doi.org/10.1016/B978-0-12-800838-6.00025-4

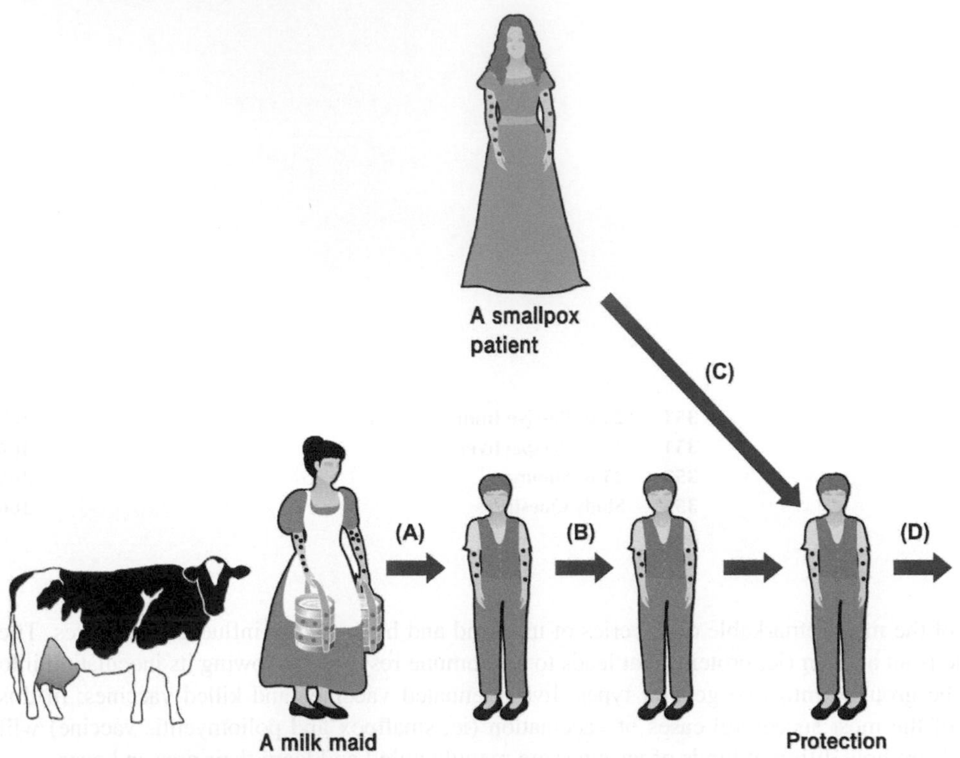

FIGURE 25.1 Steps taken by Edward Jenner to generate smallpox vaccine. Edward Jenner, the father of vaccination, created the first vaccine by inoculating a boy with cowpox, a virus related to smallpox virus. (A) A boy was inoculated with cowpox pus from a milkmaid infected with cowpox. (B) The boy suffered a mild case of cowpox. (C) The boy was then inoculated with scabs collected from a smallpox patient. (D) The boy was unaffected. Protection from smallpox infection was complete.

may affect survivors of the acute disease. Transmitted by the fecal-oral route, the virus was one of the most feared pathogens in industrialized countries during the 20th century affecting hundreds of thousands of children every year.

As we learned from the eradication of smallpox, vaccination is the most effective measure for controlling or containing a viral epidemic. Likewise, poliovirus vaccines emancipated people from their fears by significantly reducing the number of victims of this disease (Box 25.2). Inactivated polio vaccine (IPV) developed by Jonas Salk in 1956 substantially reduced the poliomyelitis cases in the United States. Subsequently, live oral polio vaccine (OPV) developed by Albert Sabin in 1961 further dampened the cases (Fig. 25.2).

With the aim of eradicating poliovirus from our planet, the World Health Organization (WHO) initiated a vaccination campaign. The outcome of this campaign has been remarkable. Although the complete eradication of poliovirus, the original goal of the campaign, was not achieved, no case of poliovirus infection has been reported in the western hemisphere since 1991. As of 2014, fewer than 500 infection cases had consistently been reported in countries such as India and its neighbors, and in some African countries (Fig. 25.3). The reason for the difficulty in completely eradicating poliomyelitis is primarily attributable to the high rate of asymptomatic infection. In fact, only 1 out of 100–200 infected individuals manifests symptoms. Thanks to the WHO-initiated global vaccination campaign, we have practically been freed from the fear of poliomyelitis.

25.3 BASIC ELEMENTS OF VACCINES

As stated above, the early efforts in vaccine development have been miraculously successful, despite our primitive knowledge of immunology and virology. In contrast, the vaccine development for HIV and HCV infections has not been successful despite strenuous efforts. It is clear that innovative vaccine technology that is distinct from the established technology used for smallpox and polio vaccines is needed. We will consider here the three basic elements of effective vaccines: efficacy, safety, and feasibility.

Efficacy: The aim of vaccination is to induce a sufficient level of immunity, which can effectively protect the host from infection. In other words, the induction of so-called protective immunity is the goal. Following vaccination, the vaccine-induced antibody needs to be maintained in the bloodstream as well as in the regions of our body coming into contact with the virus (eg, the upper respiratory tract for respiratory viruses). Importantly, an effective vaccine is expected to induce not only humoral immunity but also cellular immunity. Another important attribute of effective

BOX 25.1 Louis Pasteur and Rabies Vaccine

Rabies has been known since about 2000 BC. The first written record of rabies is from Mesopotamian literature (c. 20th century BC). Rabies spread in the 1880s in Europe as dogs became a popular pet animal. It was Louis Pasteur (1822–95) who conducted the first rabies research. His attempt to cultivate the rabies pathogen, as he did bacteria, was unsuccessful. Then, he realized that the rabies pathogen was not a bacterium, but was "filterable." Instead, he utilized successfully animals such as dogs and rabbits for cultivation. Then, he passaged the inoculum from one rabbit to another rabbit, and obtained an "attenuated" strain. Eventually, he successfully exploited the attenuated strain for the treatment of rabies infection.

The success of Louis Pasteur in treating rabies-infected individuals, who would otherwise have hopelessly succumbed to the disease, is miraculous in many respects. First, he noted that the incubation period for the disease manifestation in human or animals was rather long (about 1–2 months after being bitten by a rabid animal). This characteristically long incubation period led him to consider exploiting the "attenuated pathogen" for therapy. To his surprise, the administration of the attenuated rabies pathogen to dogs who were in incubation period after being bitten by a rabid dog was able to stop the disease manifestations. He then used the attenuated rabies pathogen as a therapeutic vaccine for humans. Since the first successful treatment of a human in 1885, the attenuated rabies virus developed by Louis Pasteur had been the only treatment for rabies virus-infected people, until very recently, when it was replaced by antibody therapy (ie, rabies immunoglobulin) (see Table 26.3). Among Pasteur's numerous accomplishments, rabies therapy is the one that made him well known in Europe.

Louis Pasteur's accomplishment was truly insightful in many respects. First, it should be noted that his work on rabies was done in 1885, even before the official discovery of "viruses" as filterable agents. Recall that the first virus was only discovered a few years later (1892) by Dimitri Ivanowski, a Russian scientist (see Fig. 1.2). As a matter of fact, he studied rabies without any knowledge of viruses. Second, he used animals for his experiment. The use of animals for experimentation was "innovative" at that time, when the cultivation of bacteria was the frontline of microbial technology. His failure to cultivate the pathogen in bacterial media encouraged him to consider using dogs (the natural host) and rabbits instead to cultivate the pathogen. Third, he made the attenuated virus by cultivating the wild-type virus in a different host animal (ie, rabbits). By changing the host from dogs to rabbits, he was able to generate an attenuated strain with a loss of virulence. This insightful concept was explored in the 1950s for the development of the live poliovirus vaccine (Sabin OPV vaccine) (see Box 25.2). Fourth, he is the one who first utilized a vaccine for therapy, as opposed to prophylaxis. More surprisingly, no therapeutic vaccine has been developed as yet except for Pasteur's rabies vaccine.

Louis Pasteur in his laboratory, painting by A. Edelfeldt in 1885.

BOX 25.2 Salk Versus Sabin Vaccine

Both Jonas Salk and Albert Sabin contributed greatly to the control of poliovirus epidemic by developing the so-called killed IPV and live OPV, respectively (see Fig. 25.2). The success of these polio vaccines is miraculous in many respects and has greatly affected the establishment of *modern virology* as a scientific discipline. In addition to their scientific contribution, these legendary virologists left a footmark that remains. The IPV vaccine, the first effective polio vaccine developed in 1956 by Jonas Salk, is based on three virulent strains grown in a type of monkey kidney tissue culture (the Vero cell line) (see Fig. 25.8), which is then inactivated with formalin. For this work, Jonas Salk became an international hero and a household name. Because Jonas Salk was the first to prove that a "killed"-virus could prevent polio, many scientists concurred that Salk should have been awarded the Nobel Prize. Despite his success in the polio vaccine development, he had no interest in deriving personal profit from it. When asked who owned the patent to it, Jonas Salk said, "There is no patent. Could you patent the sun?" In the years after Salk's discovery, many supporters "helped him build his dream of a research complex for the investigation of biological phenomena 'from cell to society.'" In 1960, he founded the Salk Institute for Biological Studies in La Jolla, California, which is today a center for medical and scientific research. Salk's last years were spent searching for a vaccine against HIV.

On the other hand, Albert Sabin developed a live attenuated polio vaccine in 1961. Although the vaccine strains were attenuated by propagating them in different host cells, clinical testing of the Sabin vaccine was considered unacceptable in the United States because of the potential dangers associated with live virus. The Salk vaccine was then already widely used in the United States, so Sabin tested his vaccines in other countries instead. Between 1955 and 1961, Sabin vaccine was tested at least on 100 million people in the USSR, parts of Eastern Europe, Singapore, Mexico, and the Netherlands. Subsequently, Sabin vaccine was tested in the United States in April 1960 on 180,000 Cincinnati school children. Because of its oral administration and lower cost, the Sabin's OPV vaccine has been particularly beneficial in developing countries.

Two legendary virologists who developed poliovirus vaccines. (A) Jonas Salk (1914–1995) and (B) Albert Sabin (1906–1993).

vaccines is that it should be able to induce long-lasting immunity that will persist for long period of time, a property termed "*immune memory.*" Immune memory refers to the development of residual virus-specific lymphocytes (ie, memory B cells or memory T cells) that are maintained even after the termination of virus infection. These immune cells can be reactivated upon subsequent exposure to the same virus. Multiple inoculations are often necessary to induce sufficient protective immunity (Fig. 25.4). For instance, three vaccine inoculations are recommended for hepatitis B virus (HBV) vaccine. Unlike "*killed vaccines*," a single inoculation of "*live vaccine*" is sufficient to induce protective immunity, because the inoculated viruses propagate to some extent. This is an advantage of live vaccine. A critical consideration for live vaccines is that their propagation in vaccine-inoculated individuals is self-limiting. Therefore, virus strains, whose virulence is attenuated, are utilized.

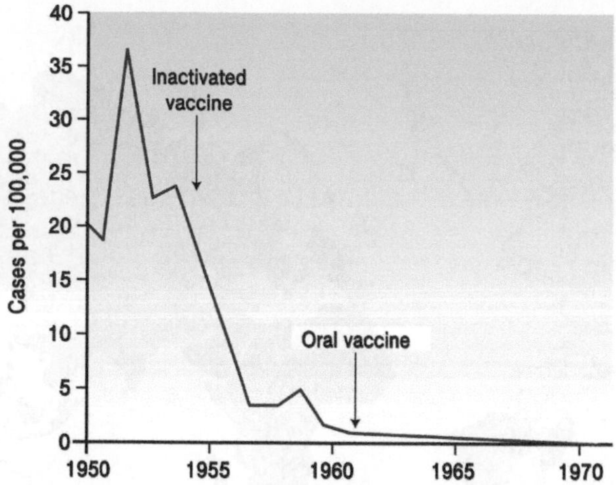

FIGURE 25.2 The impact of poliomyelitis vaccines. The graph eloquently shows the impact of killed vaccine and the subsequently developed OPV on the incidence of poliomyelitis in the United States from 1950 to 1970. The killed vaccine was developed in 1955 by Jonas Salk and the OPV was developed in 1961 by Albert Sabin.

Safety: Safety is the most importance attribute of a vaccine, because vaccines are inoculated to healthy people, including infants and youngsters, for preventive purposes. Any undesirable side effects are intolerable from a public health point of view. In addition, manufacturing processes should preclude any contamination of viruses. Cross-contamination during vaccine injection should be avoided by the use of disposable syringes. It has been claimed that mass vaccination via a "vaccine gun" has facilitated the spread of HCV in some countries (Fig. 25.5). *Anaphylaxis*,[3] which can be a side-effect of vaccination and is potentially fatal, should be minimized.

Feasibility: Feasibility is another important attribute of vaccines. Convenience in administration (ie, oral administration via liquid drop or nasal spray) is useful for mass vaccination, because medical staff are not required for administration. The storage conditions required are also important. For instance, in developing countries, vaccines that can be stored at room temperature, are more feasible than those requiring refrigeration. Affordability is also an important factor for vaccines in developing countries. According to the WHO, the cost for a vaccination should be less than 1.00 USD per dose to vaccinate the vast majority of a population ($>80\%$), which corresponds to the *herd immunity*[4] threshold for well-known infectious diseases such as measles, polio, and rubella.

25.4 TYPES OF VACCINES

The major virus vaccines are listed in Table 25.1 with an emphasis on the type of manufacturing process used to produce them. Basically, two types of vaccines are manufactured: live vaccines and killed vaccines (Fig. 25.6). For example, poliovirus vaccines are available in both live-vaccine and killed-vaccine forms. The concept of live attenuated vaccine was essentially established by Louis Pasteur in 1885 with his accomplishment of the rabies vaccine development (see Box 25.1). *MMR vaccine*[5] (against measles, mumps, and rubella) is a mixture of live attenuated viruses responsible for the three diseases (ie, measles, mumps, rubella), administered via subcutaneous injection. Varicella vaccination is used to prevent zoster (or shingles) in adults as well as chickenpox in youngsters. In addition, thanks to recombinant DNA technology, subunit vaccines or recombinant vaccines have become available. For example, HBV vaccines and human papilloma virus (HPV) vaccines are manufactured by recombinant technology. Important features of the vaccine manufacturing processes will be described below.

Live Attenuated Vaccines: A standard procedure for manufacturing live vaccine is to generate variants (ie, attenuated strains) that can grow well only under restricted conditions. One way to generate attenuated strains is by passage of a virulent strain in a cell line other than the host cell. For instance, cultivation of human viruses in a nonhuman cell line

3. **Anaphylaxis** It is a type of acute severe type I hypersensitivity allergic reaction. The term comes from the Greek words *ana* "against," and *phylaxis* "protection."

4. **Herd immunity** It is a form of indirect protection from infectious disease that occurs when a large percentage of a population has become immune to an infection, thereby providing a measure of protection for individuals who are not immune.

5. **MMR vaccine** MMR vaccine is a vaccine against measles, mumps, and rubella (also called German measles).

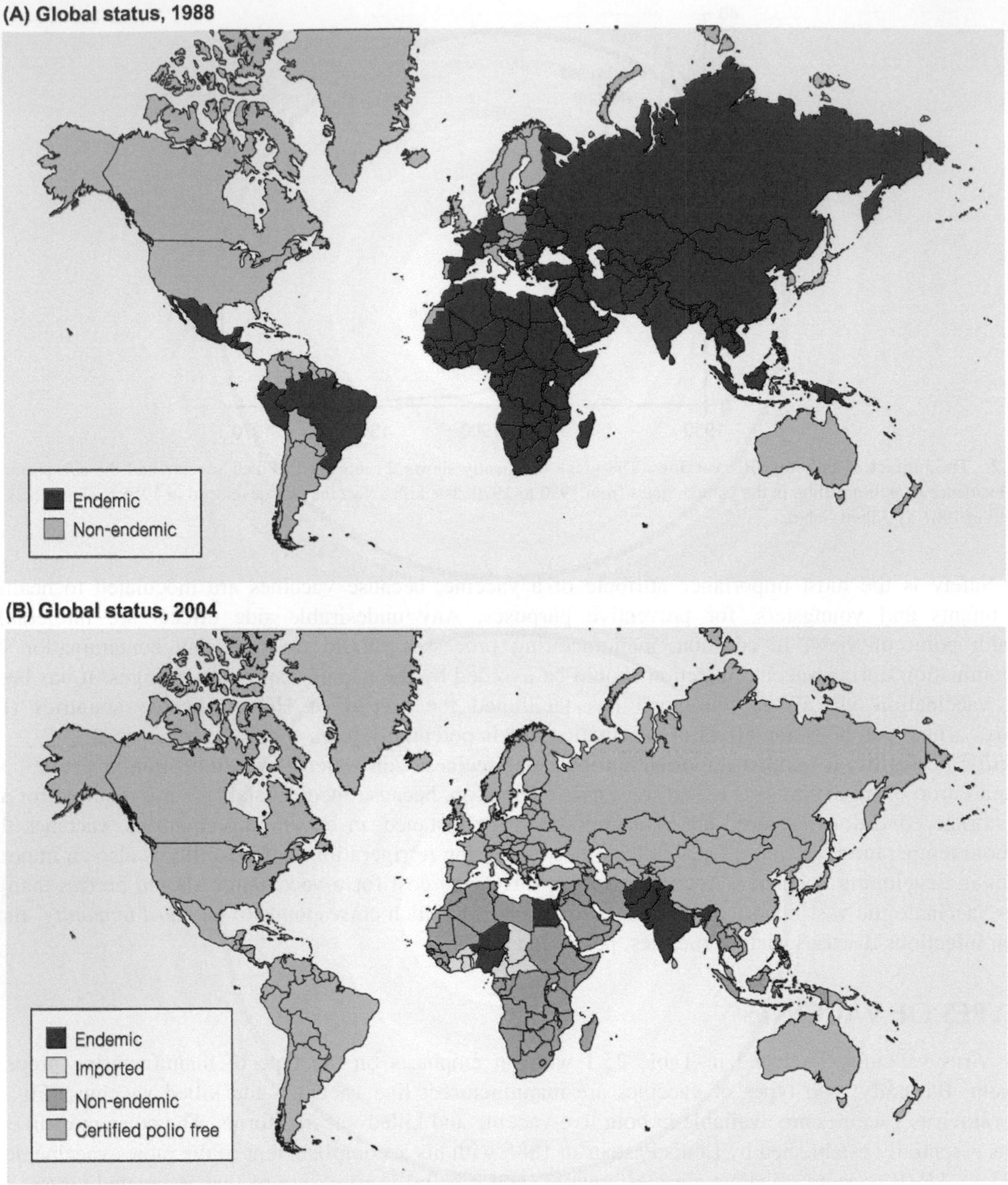

FIGURE 25.3 The progress of the poliovirus eradication program. The poliovirus eradication program, initiated by the WHO has achieved remarkable progress toward its goal since 1988. Recent cases of poliomyelitis infection are confined to only six countries—India, Pakistan, Afghanistan, Egypt, Nigeria, and Niger.

leads to the induction of mutations in the viral genome during adaptation (Fig. 25.7). The accumulation of mutations often leads to the generation of a novel strain that has lost virulence, a process termed "*attenuation*." An attenuated strain that can induce a sufficient immune response to yield a proper level of protection against experimental infection (ie, challenge) is finally selected for the vaccine strain.

The first live attenuated vaccine available was the OPV developed in 1961 by Albert Sabin (see Box 25.2). It was an attenuated strain generated by passage in monkey cells (Fig. 25.8). Production of these vaccine strains was empirical: Viruses were grown using suboptimal conditions and different host cells, and the resulting progeny viruses were tested for virulence, usually using a monkey model. As a matter of fact, Sabin's OPV was mainly used for the poliovirus eradication program initiated by the WHO (see Fig. 25.3).

In addition to OPV, a few other live attenuated vaccines, such as the MMR and varicella vaccines, have been proven to be effective, particularly for children. It is worthy of note that live vaccines have both pros and cons. Three intrinsic

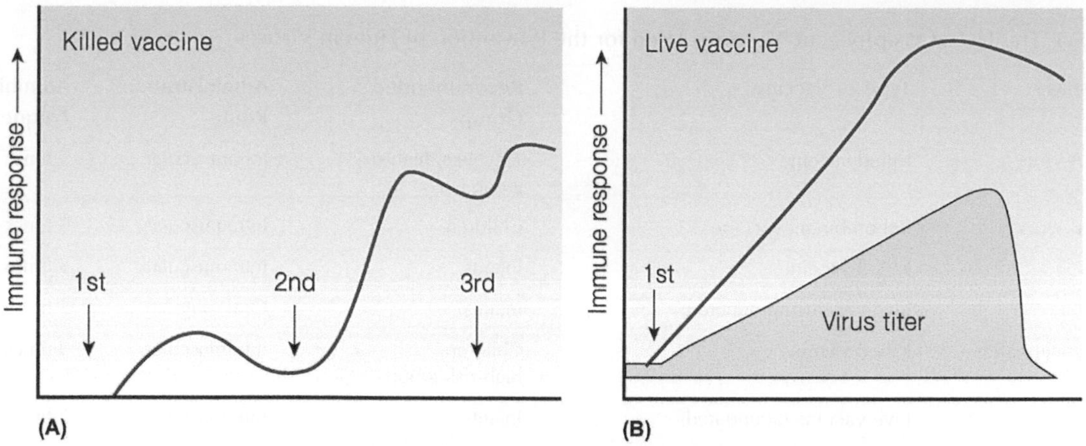

FIGURE 25.4 Killed vaccine versus live vaccine. The immune response induced by vaccination is graphically shown. (A) The magnitude of the immune response is significantly elevated by the 2nd and 3rd booster vaccinations of killed vaccine. (B) A single inoculation of live vaccine is sufficient to induce an adequate immune response because the inoculated virus is amplified through viral replication. Note that the viral propagation is self-limited, since an attenuated strain has been used.

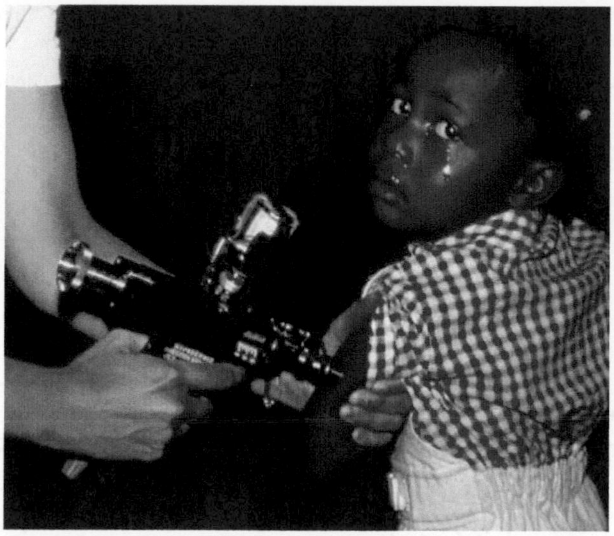

FIGURE 25.5 Vaccine gun. A photo captures the moment of vaccine administration to a girl via a jet injector, a type of vaccine gun. A jet injector is a type of medical injecting syringe that uses a high-pressure narrow jet of the injection liquid instead of a hypodermic needle.

features of live attenuated vaccines are attributable for their effectiveness: (1) The vaccine strains replicate in the host cell only to some extent, but (2) they do not spread to distant tissues, and (3) their infection elicits a sufficient immune response for protection. In particular, live vaccine strains of enteric viruses such as poliovirus are administered via the oral route. Oral administration not only provides convenience in administration but also represents an appropriate route for the induction of IgA, which provides protection in the mucosal layer. Another practical merit is that medical personnel are not required for the administration of Sabin's OPV (Fig. 25.9).

On the other hand, a few concerns should be raised. Being a live virus, the poliovirus in OPV can spread from the vaccinee to nonvaccinated individuals. Indeed, vaccine strains were rarely isolated from nonvaccinated individuals. Cases of *vaccine-associated paralytic poliomyelitis* (VAPP) have been reported in vaccine recipients of OPV and their immediate contacts, and occur at low frequencies. Rarely, primary vaccinees can become paralyzed (1 per 750,000 recipients of the vaccine). VAPP is caused by a reversion of the mutations that attenuate virulence during growth in the vaccinee. Emergence of revertant strains is not terribly surprising for RNA viruses. Caution should be paid by monitoring any vaccine-associated complications, because reversions are unpredictable.

Another concern has to do with the manufacturing process for live attenuated vaccines. Manufacturing biological products requires high purity and sterility during the manufacturing process. Contamination by other animal viruses

TABLE 25.1 The List of Prophylactic Vaccines Used for the Prevention of Human Viruses

Virus/Disease	Type of Vaccine	Recommended Groups	Administration Route	Administration Frequency
Hepatitis A virus	Killed vaccine	Travelers, high-risk group	Intramuscular	3 times
Hepatitis B virus	Recombinant vaccine	Children	Intramuscular	3 times
Poliovirus	Killed vaccine	Infants	Intramuscular	3 times
Poliovirus	Live vaccine (attenuated)	Infants	Oral route	3 times
Japanese encephalitis	Killed vaccine	Children, high-risk group	Intramuscular	3 times
Measles	Live vaccine (attenuated)	Infants	Intramuscular	3 times
MMR or MMRV (measles, mumps, rubella, and varicella)	Live vaccine (attenuated)	Infants	Subcutaneous	Once
VZV/Zoster	Live vaccine (attenuated)	Senior	Intramuscular	Once
Influenza	Killed vaccine or split vaccine (HA + NA)	Senior, high-risk groups	Intramuscular	Once a year
Human papilloma virus	Recombinant vaccine (VLP)	Women (10–25 year)	Intramuscular	3 times

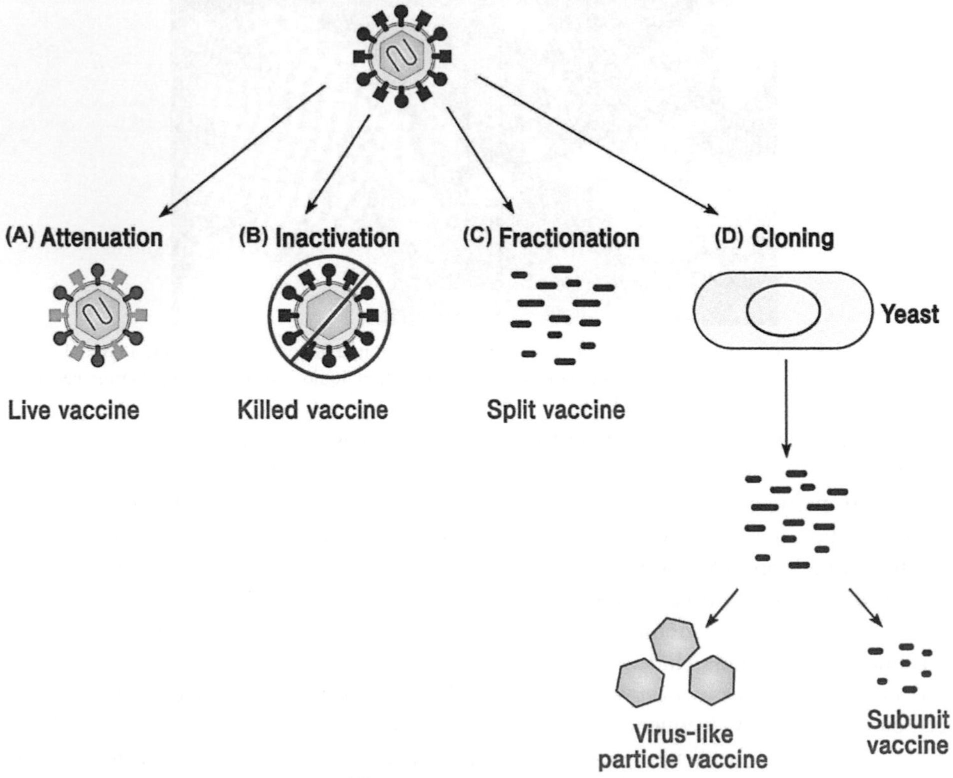

FIGURE 25.6 A schematic illustrating the vaccine manufacturing processes. (A) A live vaccine strain is generated via attenuation through passage in different host cells. (B) A killed vaccine is prepared by inactivation of live virus using prolonged treatment with formalin to remove its infectivity without destroying the antigenicity. (C) A split vaccine is made by biochemical fractionation or purification of viral envelope glycoproteins. (D) Recombinant vaccines are made by heterologous expression of viral antigens in yeasts. VLP formation is often necessary for the induction of protection immunity.

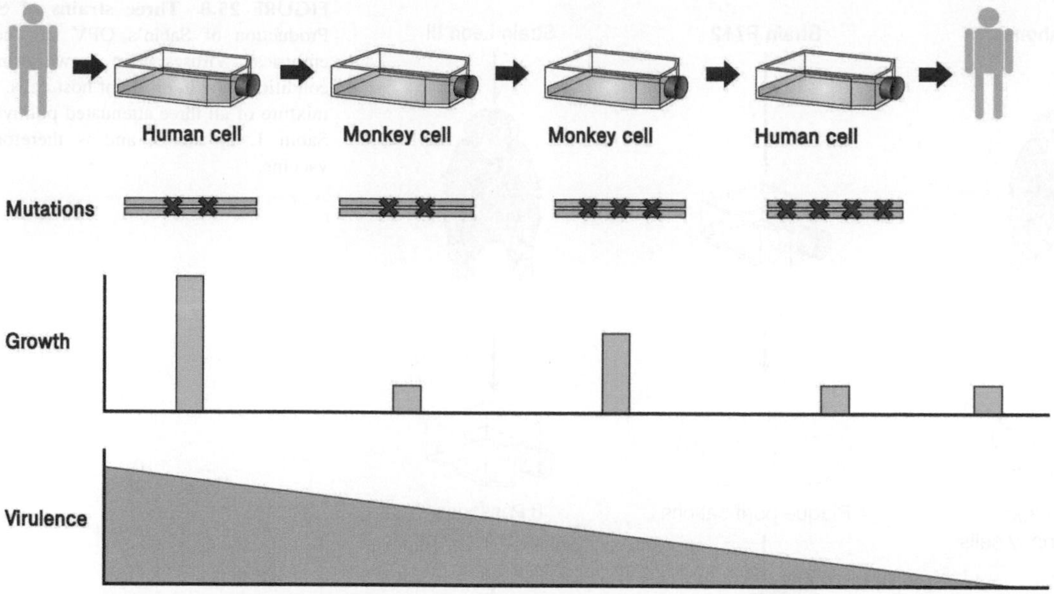

FIGURE 25.7 **Steps involved in the generation of an attenuated virus strain.** Characteristics of the viral variants emerging in each passage are schematically drawn for the purpose of explanation. A virulent strain isolated from a patient is adapted to grow in human cells, then monkey cells. As the mutations accumulate in different host cells, the virulence is accordingly attenuated. Finally, an attenuated strain that has lost its virulence is selected as the vaccine strain.

should be completely eliminated. In fact, OPV manufactured during the 1960s was found to be contaminated by SV40 that had infected the African green monkey cells used for poliovirus propagation. Believe it or not, this is how SV40 was discovered in the first place (see chapter: Polyomaviruses: SV40). Over 20 million doses of SV40-contaminated OPV were administered, and some of them had antibodies against SV40. Fortunately, no side effects caused by SV40 have yet been confirmed.

Another kind of live vaccine widely used clinically is the live influenza vaccine. In particular, reverse genetic technology is being exploited to generate seed strains for the production of this vaccine production (Box 25.3). In this case, attenuated strains are employed as the master donor strain for the generation of seed virus. Typically, six RNA segments are derived from the master donor strains and two RNA segments are derived from the seasonal strains. Each plasmid encodes one gene from each RNA segment (see Box 15.3). A seed virus is generated by transfecting cells with eight such cloned plasmids. Subsequently, a seed virus (ie, 6 + 2 reassortant) is grown in embryonated eggs for vaccine production.

Killed Inactivated Vaccines: Killed vaccines are the most common type of human vaccines. They include IPV, influenza vaccines, and hepatitis A vaccine (see Table 25.1). The inactivation process involves chemical or physical inactivation of viral infectivity without destroying the antigenicity. Disinfectants, such as formaldehyde or β-propiolactone, are used for the inactivation of naked capsids. In the case of enveloped virus particles, nonionic detergents, such as Triton X-100, are used for the inactivation. IPV is a representative killed inactivated vaccine that has been used for a long time.

Influenza vaccine is also a kind of killed vaccine. Fertilized chicken eggs have long been used for the preparation of influenza vaccines. Typically, the virus is grown by injecting it into fertilized chicken eggs (Fig. 25.10). The virus replicates within the allantois of the embryo, which is the equivalent of the placenta in mammals. In fact, the influenza vaccine is frequently referred to as a "*split vaccine*," because the two envelope glycoproteins (ie, HA and NA) of the virus particles are mainly enriched by biochemical fractionation (split) (see Fig. 25.6). The biochemical fractionation involves the purification of the virus particles via centrifugation followed by Triton X-100 treatment to disrupt the envelope. Then, the HA-enriched fraction is treated with formaldehyde to inactivate the virus. One caveat of egg-based vaccines is that manufacturing takes at least 6 months. The longer production process could pose a serious delay to deal with any future pandemic. Alternatively, cell-based vaccines have recently been developed and have become available. The cell-based vaccine technique requires a shorter period (3 months) for production and is expected to be more scalable. It also avoids common problems with eggs, such as the allergic reactions in some individuals and the incompatibility with strains that affect chickens and other birds. The manufacturing process begins following the announcement (typically in

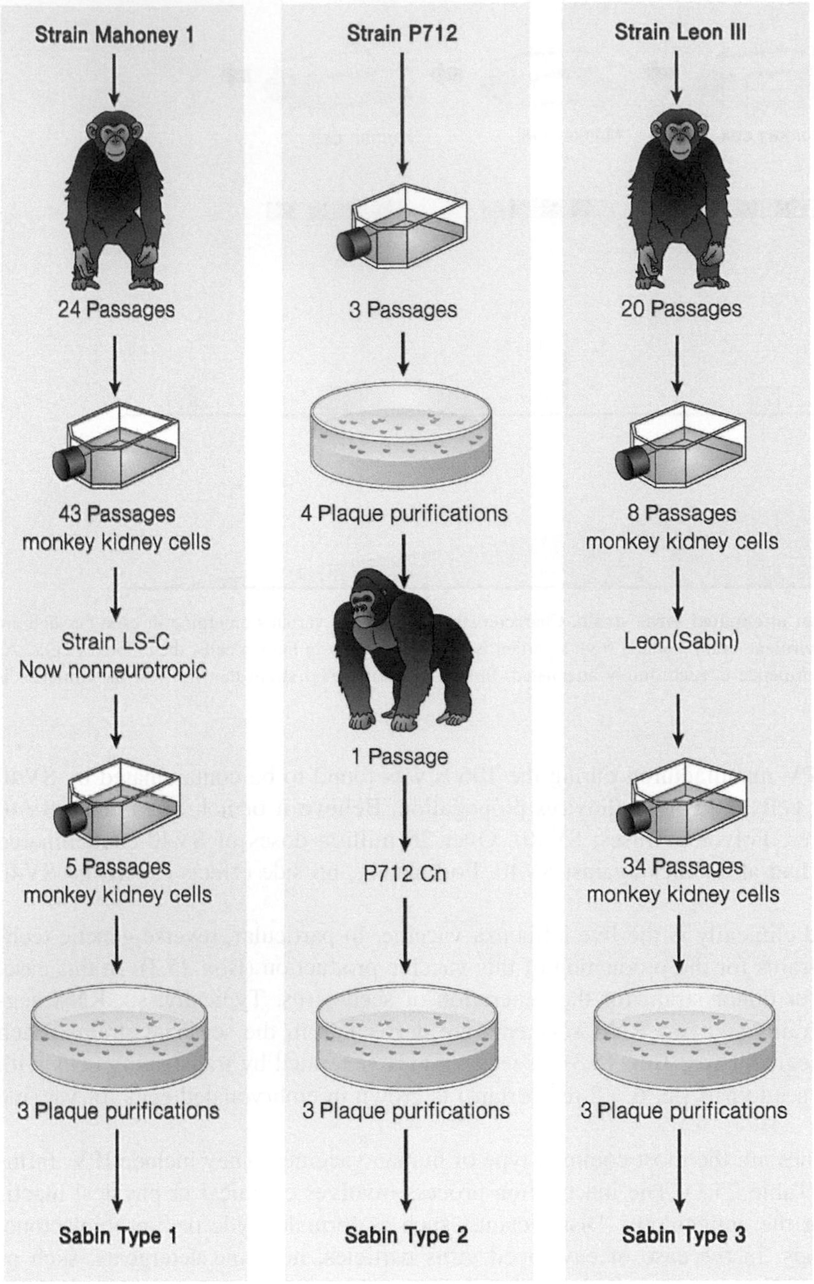

FIGURE 25.8 Three strains of Sabin's OPV. Production of Sabin's OPV vaccine strains was empirical. Viruses were grown using suboptimal conditions and in different host cells. The OPV is a mixture of all three attenuated poliovirus serotypes, Sabin 1, 2, and 3, and is therefore a trivalent vaccine.

February) of the WHO-recommended strains for the winter flu season. Three strains of flu are selected and chicken eggs are inoculated separately with each strain; these monovalent harvests are then combined to make the trivalent vaccine.

Subunit Vaccines: Recombinant DNA technology has allowed for the development of a new type of vaccine, the so-called subunit vaccines. Only a subset of viral antigens can constitute an effective vaccine, those that can induce neutralizing antibodies. Two kinds of subunit vaccine have been successfully developed. First, the HBV vaccine was developed by expressing the viral envelope glycoprotein (ie, HBsAg) in yeast via recombinant DNA technology (Fig. 25.11). HBsAg is the smaller isoform of the three HBsAg (see Fig. 18.10). It is believed that the particulate forms of HBsAg extracted from yeasts appropriately present the antigens to immune cells so that neutralizing antibodies are induced, as HBV virion particles do. Another subunit vaccine is the HPV vaccine that has been developed more recently (see Box 7.2). In fact, the monomeric form of the L1/L2 capsid protein of HPV could not elicit neutralizing

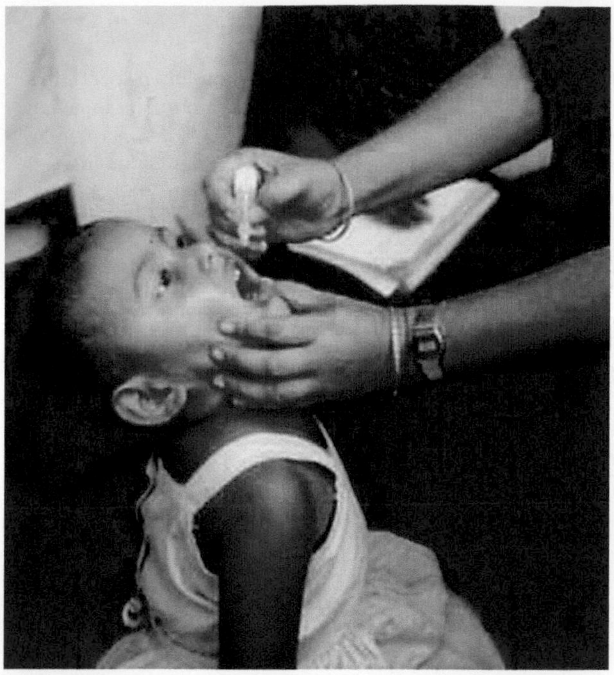

FIGURE 25.9 Sabin's OPV. *Sabin's OPV* vaccine is being given to a child.

antibodies, but the virus-like particle (VLP) of the L1/L2 capsid, which structurally mimics the HPV capsid, is used to induce the neutralizing antibody. This HPV vaccine is found to be a fairly effective vaccine.

Efforts to generate other subunit vaccines were not successful. Numerous attempts to make HIV and HCV vaccines have failed. Administration of the viral envelope glycoproteins of HIV gp120 or the HCV E1/E2 complex into an animal model has failed to protect the animals from challenge (ie, HIV or HCV). One of the reasons for the failure is that these viral antigens are not immunogenic. Another issue is that *IgA*, an immunoglobulin pertaining to mucosal infection by HIV and HCV, is not properly induced by the existing adjuvants. The development of novel *adjuvants*[6] that could induce IgA for the protection of mucosal infection by HIV and HCV represents an important unmet medical need (see below).

25.5 ADJUVANTS

In general, subunit vaccines are not highly immunogenic, because, perhaps, the viral antigens are expressed in isolation from other viral components. To boost the immune response by subunit vaccines, the vaccine preparation is supplemented with vaccine adjuvants. Adjuvants can enhance the immune response to the antigen in different ways: they can (1) extend the presence of antigen in the blood, (2) help absorb the antigen-presenting cells antigen, (3) activate macrophages and lymphocytes, and (4) support the production of cytokines. One of the most common adjuvants, *Freund's complete adjuvant (FCA)*,[7] has long been used for animal vaccines or veterinary vaccines and proven to be effective in stimulating T-cell immunity. Freund's adjuvant is a solution of antigen emulsified in mineral oil and used as an immunopotentiator (booster). The complete form is composed of inactivated and dried mycobacteria (usually *Mycobacterium tuberculosis*). FCA is effective in stimulating cell-mediated immunity and leads to a potentiation of T helper cells that stimulates the production of certain antibodies and effector T cells. However, FCA cannot be used for human vaccines due to its toxicity.

Immunization with purified protein antigens typically results in the induction of a modest antibody response with little or no T-cell response. Additionally, multiple immunizations may be required to elicit sufficient antibody responses. Developers may seek to include adjuvants in vaccine candidates to enhance the efficacy of weak antigens to induce appropriate immune responses. For instance, adjuvants enabling the use of lower vaccine doses will bring a sparing

6. **Adjuvant** An adjuvant (from Latin, *adiuvare:* to aid) is a pharmacological and/or immunological agent that modifies the effect of other agents.

7. **Freund's complete adjuvant (FCA)** Freund's adjuvant is a solution of antigen emulsified in mineral oil and used as an immunopotentiator (booster). The complete form, Freund's complete adjuvant (FCA) is composed of inactivated and dried mycobacteria (usually *M. tuberculosis*).

BOX 25.3 Live Influenza Vaccine

In addition to inactivated split vaccines, live influenza vaccine (ie, Flumist) has recently been developed. In many respects, this live vaccine is notable. First, live influenza vaccine is administered by intranasal spray, as opposed to the intramuscular injection of inactivated vaccines. The intranasal administration is a major advantage that improves patient compliance. More importantly, the live influenza vaccine is more effective than the inactivated vaccine in most patients, because it is delivered via the natural site of entry of the influenza virus and it produces a significantly stronger immune response than does the inactivated vaccine. Intriguingly, its manufacturing process involves the recombination of the RNA segments derived from two distinct strains. In other words, the generation of live influenza vaccine fully exploits reverse genetic technology (see Box 15.3) and the features of genetic reassortment of influenza viruses (see Fig. 15.14). To generate the vaccine strain, plasmids transfection is carried out. Specifically, six plasmids (ie, the PA, PB1, PB2, NP, M1, and NS1 genes) derived from the master donor virus and two plasmids (ie, HA and NA gene) derived from the seasonal flu strain are cotransfected into cells. These 6 + 2 reassorted viruses are used as seed viruses for vaccine production using embryonated eggs (see Fig. 25.10). The cold-adapted attenuated strains are employed as master donor viruses, which can grow well at 25°C but not at 37°C. This attenuated growth property enhances the vaccine's safety, because viral propagation in the lower respiratory tract is restricted. Three WHO-recommended strains for the coming winter flu season are chosen as seasonal flu strains. It is important to note that molecular virology technology built on basic science research is paying off in the development of effective live attenuated influenza vaccines.

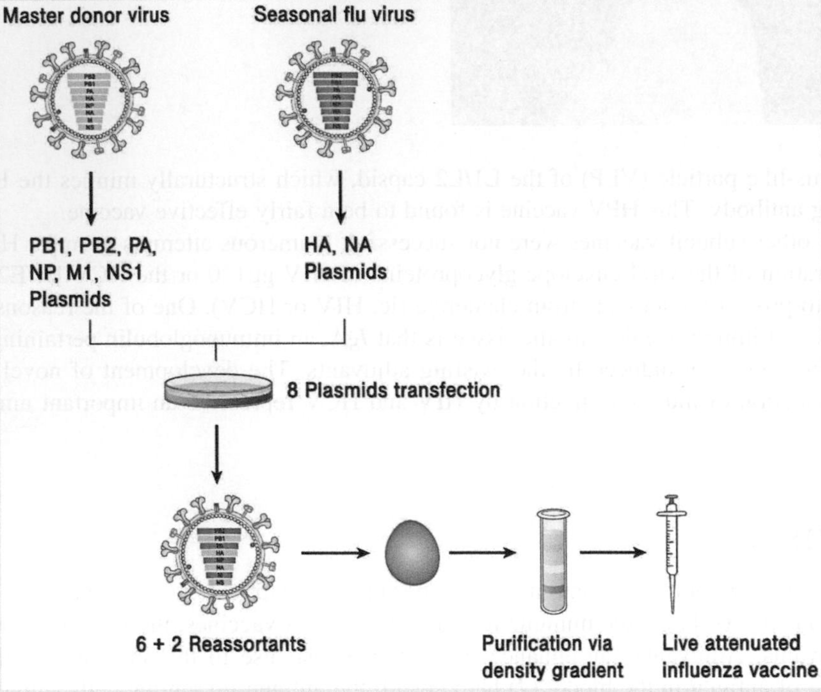

Steps involved in the generation of a live attenuated influenza vaccine.

effect (Fig. 25.12). In addition, the desired attributes of new adjuvants for future vaccines include: (1) the ability to elicit T-cell response (ie, T-cell vaccines for HIV and HCV infection, which are controlled by cellular immune response), (2) the ability to induce cytotoxic $CD8^+$ T cell (ie, therapeutic vaccines for HPV infection), and (3) antibody response broadening ability via expansion of B cell diversity (for influenza, HIV, and HPV infection).

For human vaccines, aluminum salt (or alum) is used in the HBV and HPV vaccines (Table 25.2). Although its immune stimulation effect is only modest, alum is the only vaccine adjuvant that was approved by the FDA for human use until recently. Alum functions as a delivery system by generating depots that trap antigens at the injection site, providing for slow release of the antigen and prolonged stimulation of the immune system. In addition, a direct immune-stimulatory effect of alum has recently been reported, in that alum directly stimulates dendritic cells (DCs), thereby inducing MHC class II and $CD8^+$ molecules on the DCs.

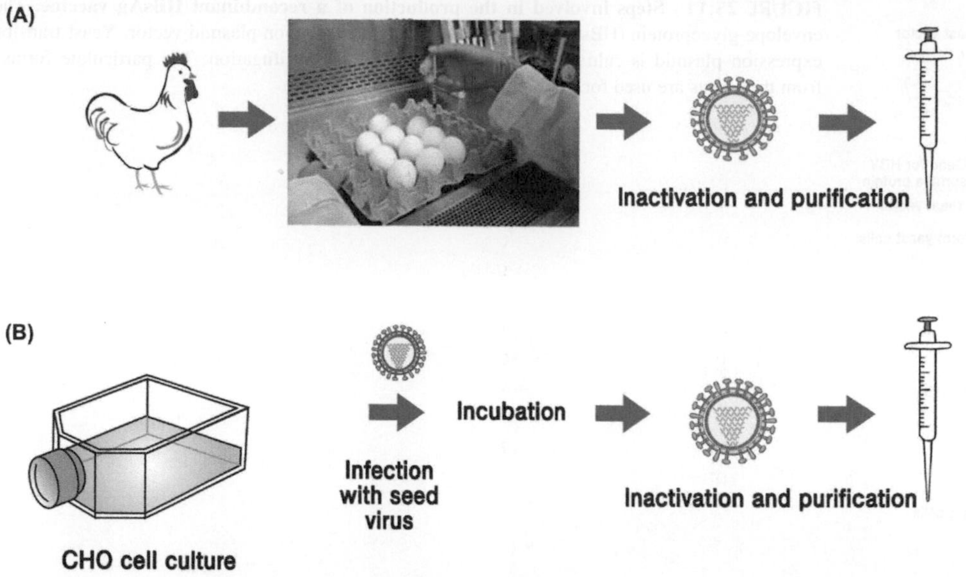

FIGURE 25.10 The manufacturing process for influenza vaccine. (A) Egg-based vaccines. Embryonated eggs are infected by injecting the virus into the allantoic cavity of the egg. The virus is purified from this allantoic fluid by methods such as filtration or centrifugation. The purified viruses are then inactivated (killed) with a disinfectant. The inactivated virus is treated with detergent to break up the virus particles and the viral envelope glycoproteins (ie, HA and NA) are enriched by biochemical fractionation. The final preparation is suspended in sterile phosphate buffered saline and ready for injection. One to two eggs are needed to make each dose of vaccine. (B) Cell-based vaccines. Chinese hamster ovary (CHO) cell is often used for the cultivation of influenza virus. Subsequent inactivation and biochemical enrichment of the viral envelope glycoproteins process are similar to those for the egg-based vaccines.

The need for more effective adjuvants became a serious concern in the midst of the 2009 new flu pandemic, because there was a shortage of appropriate antigen available. Implementation of adjuvant use was expected to reduce the amount of antigen needed by 5–10-fold. Two novel adjuvants under development were then approved under these circumstances: MF59 and AS03 (see Table 25.2). Both MF59 and AS03 are oil-in-water (O/W) emulsion-based adjuvants, with the oil being fish oil, squalene. The O/W emulsion is a reminiscent of FCA, which is a solution of antigen emulsified in mineral oil. Besides, AS04, which is essentially the alum fortified with monophosphoryl lipid A (MPL), has been approved for HBV and HPV vaccines (see Table 25.2). MPL is a nontoxic derivative of lipid A (Fig. 25.13), which is a lipid component of lipopolysaccharide (also called endotoxin), a ligand for TLR4 (see Fig 5.4).

Until recently, adjuvant technology has remained at the stage of "secret recipes." Adjuvants have been whimsically called the "dirty little secret" of vaccines in the scientific community. However, recent advances in our knowledge of innate immunity have changed the landscape of adjuvant research. Diverse types of TLR ligands are currently being explored as components of novel adjuvants, including CpG motif-containing oligonucleotides, which act as an agonist of TLR9 (see Fig 5.4). Certainly, the advent of potent adjuvants will not only help to improve the efficacy of existing vaccines but also facilitate the development of new vaccines.

25.6 PASSIVE IMMUNIZATION

As mentioned above, vaccination induces a host immune response via the administration of one or more antigens. In particular, humoral immunity (ie, antibody production) is the major component of the immune response indispensable for the protection from virus infection. Alternatively, ready-made antibodies can be used for prophylactic purposes as well. This type of immunity is termed *passive immunity*, as opposed to active immunity. In other words, passive immunity is the transfer of *active* humoral immunity, in the form of ready-made antibodies, from one individual to another. In fact, antibodies have been used for the prevention of virus infection. For instance, IgG is used for prophylaxis of respiratory syncytial virus (RSV) infection in high-risk infants (see Table 26.3). In addition, antibodies are used for the treatment of viral infections as well. For instance, IgG against rabies virus is used for postexposure prophylaxis and treatment of rabies virus infection (see Table 26.3).

On the other hand, passive immunity can occur naturally, such as when maternal antibodies are transferred to a fetus by its mother during pregnancy. This maternal transfer occurs around the third month of gestation. Immunoglobulin G

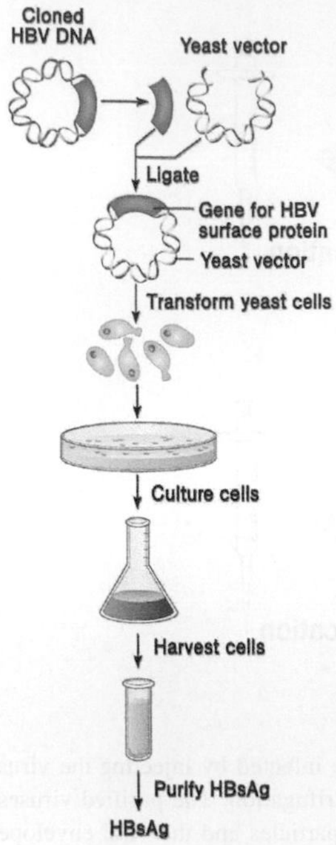

FIGURE 25.11 Steps involved in the production of a recombinant HBsAg vaccine. The gene for the viral envelope glycoprotein (HBsAg) is cloned into a yeast expression plasmid vector. Yeast transformed by the HBsAg expression plasmid is cultivated and then harvested by centrifugation. The particulate forms of HBsAg purified from the yeasts are used for vaccination.

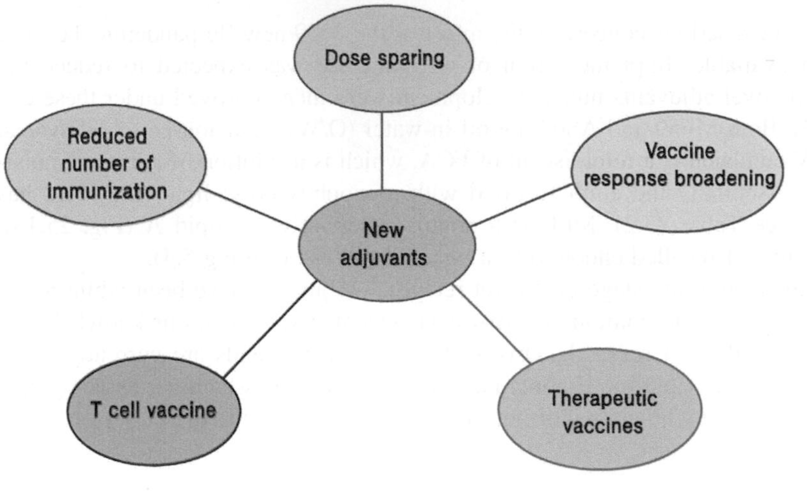

FIGURE 25.12 Potential benefits of adjuvants. Potential benefits of new adjuvants are as follows: (1) dose sparing effect to increase vaccine supply, (2) reduced number of immunizations, (3) broadening vaccine response to viruses with high mutation rate such as influenza virus and HIV, (4) to elicit T-cell response for treatment of HIV and HCV infection (ie, T-cell vaccine), and (5) to elicit cytotoxic T-cell response for treatment of HPV infection (ie, therapeutic vaccine).

is the only antibody isotype that can pass through the placenta. Passive immunity is also provided through the transfer of IgA antibodies found in breast milk, which are transferred to the gut of the infant and protect it against bacterial infections until the newborn can synthesize its own antibodies.

25.7 PERSPECTIVES

Despite the early success of the smallpox vaccine and polio vaccines, vaccines against the vast majority of human pathogenic viruses are not available yet. For instance, the development of vaccines for HIV and HCV infection has not been successful despite strenuous efforts. It is believed that cellular immunity, as opposed to humoral immunity, is primarily responsible for the protection of HIV and HCV infection. Nevertheless, recent success in developing vaccines for HPV has renewed interest in subunit vaccines (see Box 7.2). Effective vaccines elicit potent antibody responses to block

TABLE 25.2 Licensed Human Vaccine Adjuvants

Adjuvant Name	Composition	Mechanism	Target Virus (Brand)
Alum	Aluminum salt	Depot effect; slow release of antigens adsorbed to alum	HBV and HPV (Gardasil)
MF59	O/W (squalene) emulsion	Enhance antibody response	Influenza (Fluad)
AS03	O/W (squalene) + α-tocopherol	Enhance antibody response	Influenza (Pandemrix)
AS04	MPL(monophosphoryl lipid A) + aluminum salt	TLR4 ligand	HBV (Fendrix) and HPV (Cervarix)

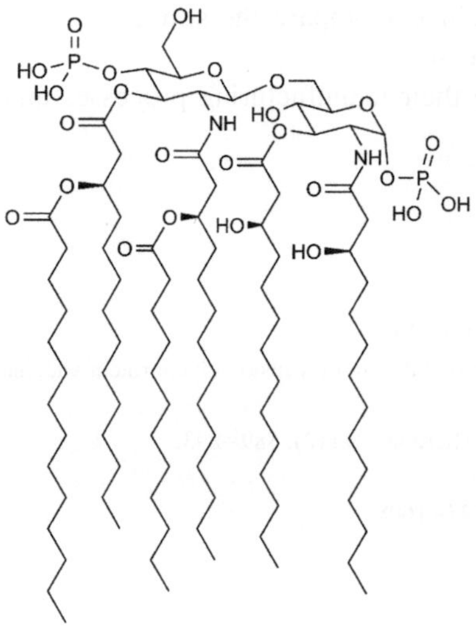

FIGURE 25.13 Chemical structure of lipid A. Lipid A consists of two glucosamine (carbohydrate/sugar) units with attached acyl chains (fatty acids), and normally contains one phosphate group on each carbohydrate. MPL has only one phosphate group.

the infection and/or clear pathogens. The protective antibodies are directed against invariant molecular components of the pathogens for all effective vaccines. This notion inspires the concept of a *universal influenza vaccine* that can provide protection from any seasonal flu strain. Identification of highly conserved epitopes in the HA trimer of the influenza virus supports the potential feasibility of a universal influenza vaccine. Innovative vaccines may require the contribution of seemingly unrelated disciplines such as structural biology and antibody technology. Intriguingly, the implementation of "structure biology" in antigen design is being considered, as demonstrated in work to produce a vaccine against RSV F glycoprotein (see Journal Club). In addition, the establishment of "*antibody engineering*" technology has made the passive immunization for prophylactic purposes more feasible. For instance, the generation of *humanized monoclonal antibodies* with neutralizing capability against human viruses, such as HIV and RSV, is now being explored for the treatment. It is clear that the contributions by scientists from multiple disciplines such as virology, immunology, and structure biology are essential for the development of successful HIV and HCV vaccines.

25.8 SUMMARY

- *Elements of vaccines*: The three elements of effective vaccines are efficacy, safety, and feasibility. Safety is most important, because vaccines are used to inoculate healthy people, particularly infants and children, for preventive purposes.
- *Polio vaccines*: Two types of vaccines are currently manufactured: live vaccines and killed vaccines. Poliovirus vaccines are available as both live and killed vaccines.
- *Live attenuated vaccine*: Live attenuated vaccines are generated by passage of virulent strains in cell lines other than the host cells.

- *Killed inactivated vaccine*: The inactivation process involves chemical or physical inactivation of viral infectivity without destroying antigenicity. Chemical treatments such as formaldehyde or Triton X-100 are used to inactivate virus particles.
- *Adjuvants*: Adjuvants may be added to a vaccine to modify the immune response by boosting it, giving a higher amount of antibodies and longer-lasting protection, thus minimizing the amount of foreign material that needs to be injected. Until recently, aluminum salt (or alum) has been the only vaccine adjuvant approved by the FDA for human use.
- *Passive immunization*: Passive immunization is the administration of ready-made antibodies, instead of antigens.

STUDY QUESTIONS

25.1 Two types of poliovirus vaccines are clinically available: killed and live vaccines. Compare the similarities and differences between the two vaccines with respect to their manufacturing processes.

25.2 Two types of influenza vaccines are clinically available: killed and live vaccines. Compare the similarities and differences between the two vaccines with respect to their manufacturing processes.

25.3 Two kinds of recombinant subunit vaccines are clinically available. Describe their manufacturing processes and their common characteristics.

SUGGESTED READING

Houser, K., Subbarao, K., 2015. Influenza vaccines: challenges and solutions. Cell Host Microbe. 17 (3), 295–300.

McLellan, J.S., Chen, M., Joyce, M.G., Sastry, M., Stewart-Jones, G.B., Yang, Y., et al., 2013. Structure-based design of a fusion glycoprotein vaccine for respiratory syncytial virus. Science. 342 (6158), 592–598.

Plotkin, S.A., Plotkin, S.L., 2011. The development of vaccines: how the past led to the future. Nat. Rev. Microbiol. 9 (12), 889–893.

Rappuoli, R., 2007. Bridging the knowledge gaps in vaccine design. Nat. Biotechnol. 25 (12), 1361–1366.

Reed, S.G., Orr, M.T., Fox, C.B., 2013. Key roles of adjuvants in modern vaccines. Nat. Med. 19 (12), 1597–1608.

JOURNAL CLUB

- McLellan, J.S., et al., 2013. Structure-based design of a fusion glycoprotein vaccine for respiratory syncytial virus. Science 342, 592–598.

  Highlight: For decades, researchers have hoped that structural biology would help them design better vaccines. This work provides convincing evidence that the structural approach can pay off. McLellan et al., have successfully utilized the prefusion state of respiratory syncytial virus (RSV) F glycoprotein that is targeted by extremely potent RSV-neutralizing antibodies. RSV is a leading cause of hospitalization for children under 5 years of age.

BOOK CLUB

- Dubos, R., 1998. Pasteur and Modern Science. ASM Press, Washington, DC.

  Highlight: This book provides a wonderfully captivating and insightful perspective into the life and scientific strength of Pasteur. The work takes the reader through the scientific and personal life of Pasteur as he progresses from school boy, to student, crystallographer, chemist, and biologist. For anyone who wants to know about the beginning of microbiology, immunology, virology, or modern medicine, this book is a must.

INTERNET RESOURCE

Global Polio Eradication Initiative: http://www.polioeradication.org/

Chapter 26

Antiviral Therapy

Chapter Outline

One of the important goals of virus research is to prevent and to control the viral infection. A vaccine protects against the virus infection; however, once infected, the treatment mainly relies on antiviral drugs and/or symptom treatment. Here, antiviral drugs and antibodies that are used to control viral infection are described. In particular, HIV drug development, an area which has been extensively pursued for past three decades, will be described. In addition, current efforts to develop new antivirals for HCV and influenza virus will be covered.

26.1 ANTIVIRAL DRUGS

In principle, antiviral drugs refer to any molecules that inhibit virus propagation; however, antivirals are generally limited to small organic molecules of less than 1 kDa in molecular weight (Table 26.1). Despite remarkable progress made in virology research, the number of antiviral drugs that are being used today is only 40 or so (Fig. 26.1). Moreover, the vast majority of them are anti-HIV drugs. Besides anti-HIV drugs, only a dozen of antiviral drugs have been developed. It is skimpy compared to the hundreds of antibiotics that had been made available.

One could consider a few reasons for the scarcity of antiviral drugs. Unlike antibiotics that control bacteria (prokaryotes) and fungi (lower eukaryotes), viruses rely on the host (higher eukaryotes) for their propagation. Therefore, a small molecule with antiviral activity is more likely to exhibit toxicity to the host cell due to the mechanistic similarities of the inhibition. In fact, many drug candidates with potent antiviral activity are frequently dropped due to toxicity during drug development. Second, since viruses inherently bear a high mutation rate, drug-resistant mutants are readily emerging. Emergence of the drug-resistant mutants frequently makes antiviral drugs obsolete.

26.2 MOLECULAR TARGETS OF ANTIVIRAL DRUGS

In principle, any stages or processes involved in a virus life cycle, from entry to exit, are amenable to antiviral intervention. Since stages in the virus life cycle can be viewed as a molecular interaction (ie, protein–protein interaction or protein–RNA interaction), a small molecule that could effectively interfere in the molecular interaction involved in the virus life cycle can be a candidate for antiviral drug. Historically, viral enzymes—in particular, viral DNA or RNA polymerases—have been the favorite. In addition, antivirals that block viral entry and exit are being exploited for

Molecular Virology of Human Pathogenic Viruses. DOI: http://dx.doi.org/10.1016/B978-0-12-800838-6.00026-6

TABLE 26.1 The Features of Representative Antiviral Drugs

Antiviral Drug	Virus	Target Molecule	Mode of Action	Chemical Structure
Acyclovir	HSV, VZV	Viral DNA polymerase	Chain terminator	Fig. 26.3
Boceprevir	HCV	NS3 serine protease	Inhibitor	Fig. 26.10
Foscarnet (PFA)	HSV	Viral DNA polymerase	Elongation inhibitor	–
Ganciclovir	HCMV, HSV	Viral DNA polymerase	Competitive inhibitor	–
Lamivudine (3TC)	HBV	RT (reverse transcriptase)	Chain terminator	Fig. 26.7
Oseltamivir	Influenza	Neuraminidase	Competitive inhibitor	Fig. 26.12
Ribavirin	RNA virus	Unknown	Unknown	–

Note: HBV, hepatitis B virus; HCV, hepatitis C virus; HCMV, human cytomegalovirus; HSV, Herpes simplex virus; VZV, Varicellar-Zoster virus.

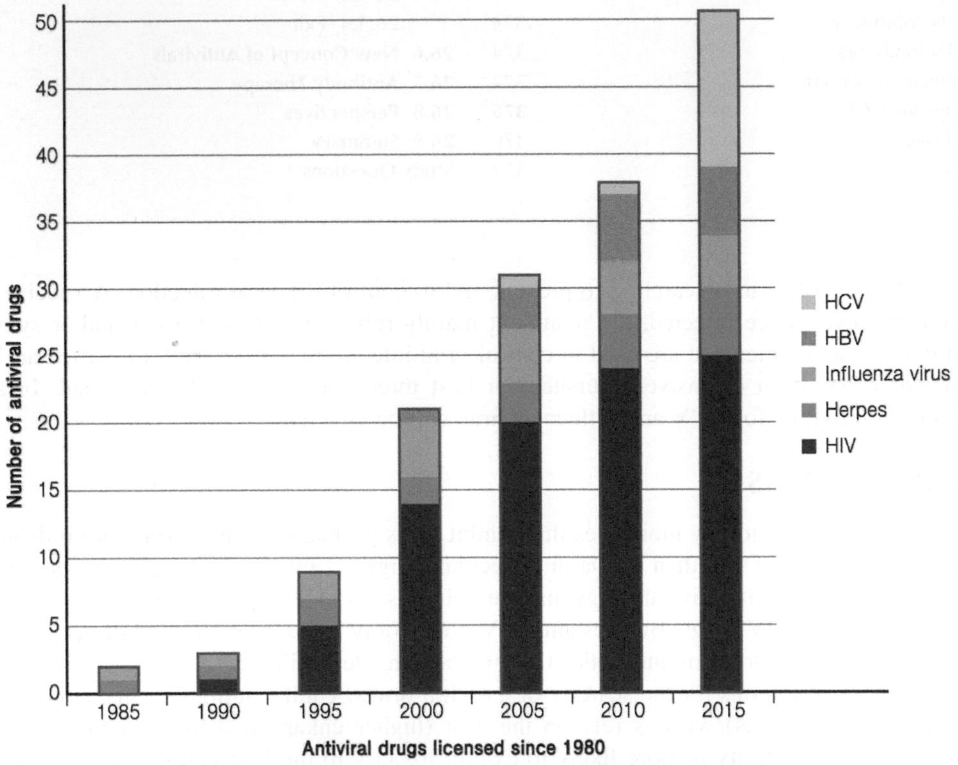

FIGURE 26.1 Antiviral drugs developed after FDA approval. In early 1980, only two antiviral drugs were developed: one for HSV-1 (ie, acyclovir) and another for influenza virus (ie, amantadine). The bulk of antiviral drugs, largely anti-HIV, have been developed since 1990.

human pathogenic viruses. Here, the potential antiviral targets will be described in the context of the virus life cycle, with an emphasis on future exploration.

Entry: Entry, the first step of virus infection, involves the recognition of the viral receptor by virions. Small molecules that interfere with the recognition of the receptor by virus particles could be an ideal antiviral drug, since the virus infection is preventable in the first place. In a case where the atomic details of viral capsids structure are available, the *structure-based drug design*[1] is possible. In fact, pleconaril is one that used to be under development for human

1. **Structure-based drug design** The inventive process of finding new medications based on knowledge of a biological target. The drug design that relies on the knowledge of the three-dimensional structure of the molecular target is known as the *rational drug design*.

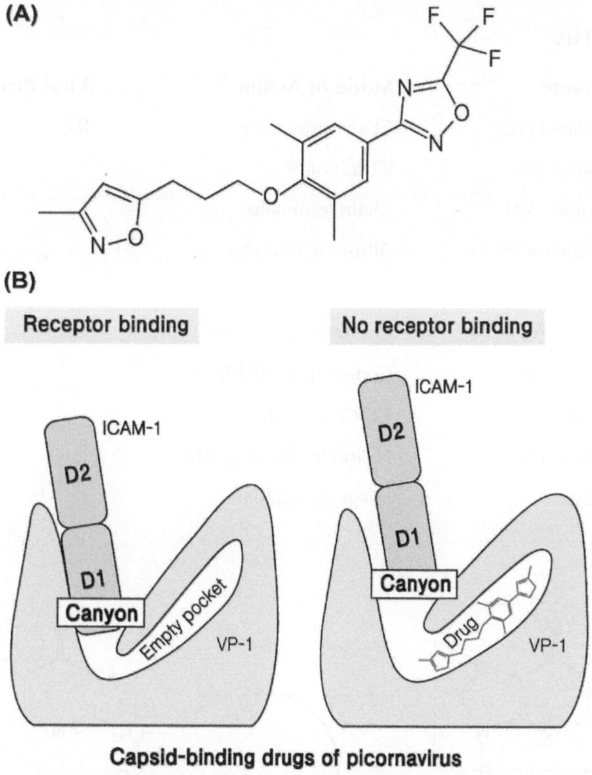

FIGURE 26.2 Capsid-binding drug of picornaviruses. (A) Chemical structure of pleconaril. Pleconaril is an antiviral drug currently being developed by Schering-Plough for common cold symptoms in patients exposed to picornavirus, including *Enterovirus* and *Rhinovirus*. (B) Cross-section view of the interaction between the ICAM receptor and VP1 of the capsid. (Left) In the absence of the drug, ICAM-1 binds to the floor of the canyon, inducing conformational changes that lead to the uncoating of the capsid and release of the viral RNA. (Right) The drug binds to the hydrophobic pocket, located beneath the floor of the canyon. The drug binding induces conformational changes, hampering the interaction with the receptor and subsequent uncoating of the capsid.

rhinovirus, a major cause of the common cold (Fig. 26.2). It was based on a rational design of an inhibitor to block the receptor-binding pocket known as a canyon on the surface of viral capsids. This canyon obscures the receptor-binding site from the immune system, and it was identified as a promising target for small-molecule drugs. Our understanding of these drugs is largely based on X-ray crystallographic studies of complexes between the viral capsids and drugs. The binding of the drug in the pocket alters the structure of the canyon floor, thereby preventing proper interaction of the virion with the cellular receptor. These capsid-binding drugs inhibit attachment and/or entry or even later steps of entry. Although an earlier submission of pleconaril for the approval of *Food and Drug Administration*[2] was rejected because of an unpredictable side effect, it demonstrated that the structure-based approach to develop the entry inhibitor is conceptually attractive.

The step after entry is receptor-mediated endocytosis, which involves membrane fusion between the endosomal membrane and the viral envelope (see Box 3.2). Envelope glycoproteins are the viral proteins that carry out the membrane fusion. For instance, *HA* (hemagglutinin) of influenza virus and *gp41* of HIV harbor the fusion peptide that triggers membrane fusion. Enfuvirtide is the anti-HIV drug that inhibits the gp41-mediated membrane fusion of HIV infection (Table 26.2).

Genome Replication: As stated above, viral DNA/RNA polymerases have been considered as the most attractive targets for antiviral purposes; therefore, they were extensively exploited as a target for antiviral drug discovery. As a result, many antiviral drugs available today are inhibitors of viral DNA/RNA polymerases (see Table 26.1).

Acyclovir deserves special attention, because it is the first generation of selective antiviral drugs. Since its discovery in 1977, acyclovir became the gold standard for the treatment of herpes viral infections, such as HSV-1, HSV-2, and

2. **Food and Drug Administration (FDA)** FDA is the authority that approves new drugs for medical use. In fact, the FDA is one of the United States federal executive departments.

TABLE 26.2 Antiviral Drugs for HIV

Class	Drug Name	Mode of Action	Viral Protein	Chemical Structure
NRTI	Lamivudine (3TC)	Chain terminator	RT	Fig. 26.7
	Tenofovir (TDF)	RT inhibitor		Box 23.2
	Zidovudine (AZT)	Chain terminator		Fig. 26.7
NNRTI	Nevirapine (NVP)	Allosteric inhibitor		Fig. 26.7
	Etravirine (EFV),	Allosteric inhibitor		Fig. 26.7
Protease inhibitor	Indinavir (IDV)	Competitive inhibitor	PR	–
	Saquinavir (SQV)	Competitive inhibitor		Fig. 26.8
Entry inhibitor	Maraviroc	CCR5 antagonist	–	Fig. 26.5
	Enfuvirtide (ENF)	Fusion inhibitor (gp41)	gp41	–
Integrase inhibitor	Raltegravir	Integrase inhibitor	IN	Fig. 26.5

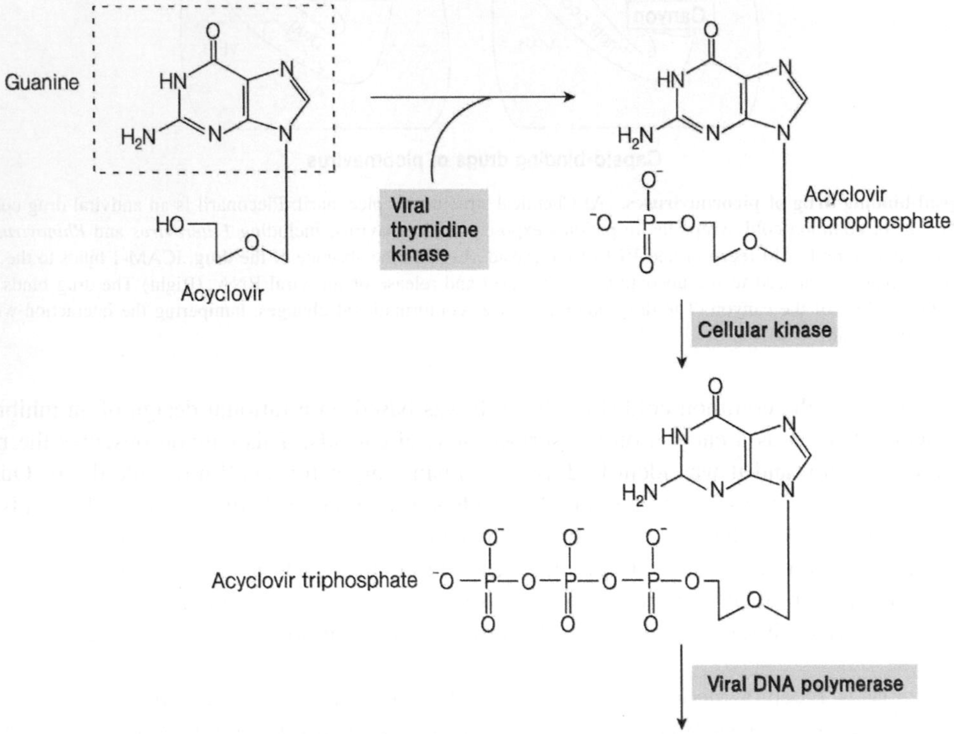

FIGURE 26.3 The conversion of acyclovir to a functional triphosphate form. The viral thymidine kinase converts acyclovir to acyclovir monophosphate. Next, the cellular pyrimidine kinase converts the monophosphate into acyclovir triphosphate. Acyclovir triphosphate, which lacks the ribose ring, acts as chain terminator for the viral DNA replication.

VZV. In fact, acyclovir, a nucleoside analog, is a *prodrug*[3] that becomes active only after the conversion to triphosphate form in the body (Fig. 26.3). Notably, acyclovir is converted into an active form (ie, acyclovir triphosphate) by two successive nucleoside kinases: first converted into acyclovir monophosphate by viral thymidine kinase, and then converted

3. **Prodrug** It refers to a medication that is initially administered to the body in an inactive (or less than fully active) form, and then becomes converted to its active form by the normal metabolic processes of the body. Prodrugs can be used to improve how the intended drug is absorbed, distributed, metabolized, and excreted (ADME).

into acyclovir triphosphate by cellular kinases. Thus, acyclovir triphosphate, which could potentially exhibit some cellular toxicity due to a possibility of being utilized as a substrate for the cellular DNA polymerase, is made only in the virus infected cells but not in uninfected cells, a feature that would reduce cellular toxicity.

Likewise, for HIV and HBV, the inhibitors of viral reverse transcriptase are the representative of the antiviral drugs. Most of these reverse transcriptase inhibitors are nucleoside analogs that act as chain terminators (see Table 26.1). Despite their potent antiviral effect, the emergence of drug-resistant variants is a major pitfall of these nucleoside analogs.

Viral Enzymes: Other viral enzymes such as proteases are also amenable to antiviral intervention. For instance, indivavir and saquinavir, the protease inhibitors of HIV PR (ie, aspartate protease) are the first-generation drugs of this category (see Table 26.2). The successful development of the HIV protease inhibitors has encouraged the exploitation of the NS3 serine protease of HCV as an antiviral target; such efforts were paid off by the development of the first direct-acting anti-HCV drugs in the market: boceprevir and telaprevir (see Table 26.1).

Capsid Assembly: The capsid assembly is also applicable to antiviral intervention. As capsid assembly involves the interactions between multimeric subunits, small molecules that interfere with the subunit interaction can be potent antiviral drugs. No such antiviral drug has been developed yet.

Release: Budding or the release of virions from infected cells can be antiviral targets. Blockade of the virions released from the infected cells by drugs is an effective strategy to stop viral spread to surrounding cells. Oseltamivir, an antiinfluenza drug, blocks the viral release from the infected cells (see Table 26.1). The *NA* protein (ie, neuraminidase) of influenza virus, an envelope glycoprotein, facilitates the virion release from cells by cleaving the sialic acid on the plasma membrane of infected cells (see Box 15.1 and Fig. 26.12).

26.3 ANTIVIRAL DRUGS FOR HIV

It is hard to describe the advances made in the antiviral drug discovery without stating that over two dozen anti-HIV drugs have been developed in the past 30 years. In other words, such productive endeavor established the milestones of drug discovery applicable not only to infectious diseases but also to other human diseases. In the past, drug discovery involved largely so-called blind screening, which relies on serendipity rather than a scientific reasoning (Box 26.1). Such an old-fashioned screening approach has been largely replaced by a structure-based rational drug design, which has been proven effective in the anti-HIV drug development. In particular, X-ray crystal structure of the viral proteins has guided the design of more effective inhibitors. Here, the targets, which are amenable to future antiviral intervention, as well as targets, which have been successfully exploited in the past, are described along with the viral life cycle.

26.3.1 Entry Inhibitor of HIV

The entry step of HIV has been well understood at the molecular level. It involves multiple engagements between two viral envelope proteins (ie, gp120 and gp41) and two viral receptors (i.e., CD4 and CCR5) (Fig. 26.4). Two HIV entry inhibitors have been recently developed. Enfuvirtide is an entry inhibitor that blocks membrane fusion, while maraviroc is an entry inhibitor that blocks CCR5 function (Fig. 26.5). Enfuvirtide is essentially a peptide composed of 30 amino acids that is a mimetic of one of four domains in gp41. Enfuvirtide blocks the HIV entry by interfering in an interdomain interaction within the gp41 trimer that precedes membrane fusion (Fig. 26.6). On the other hand, maraviroc binds to CCR5, thereby blocking gp120–CCR5 interaction. It is worthy of note that the molecule targeted by maraviroc is not a viral protein but a host protein, a feature that precludes the emergence of drug-resistant variants.

26.3.2 HIV RT Inhibitors

HIV RT is involved in viral reverse transcription, which converts the viral RNA into the DNA (see Fig. 26.4). HIV RT has been the most exploited antiviral drug target ever, resulting in the development of 12 drugs. HIV RT inhibitors can be divided into two classes depending on the mode of action: nucleoside RT inhibitors (*NRTIs*) and nonnucleoside RT inhibitors (*NNRTIs*). In fact, seven NRTIs and five NNRTIs are currently available. Zidovudine (AZT) and lamivudine (3TC), lacking the 3′OH group, act as a chain terminator of the viral reverse transcription (Fig. 26.7). These nucleoside analogs are prodrugs in that they act as an inhibitor only after conversion into a triphosphate form. On the other hand, NNRTIs, such as nevirapine (NVP) and efavirenz (ETR), bind to a region that is distinct from the dNTP-binding site on the viral RT protein. In other words, these NNRTIs act as allosteric inhibitors.

BOX 26.1 Drug Discovery

Historically, drugs were discovered through identifying an active ingredient from traditional remedies or through serendipitous discovery. These traditional methods have been largely replaced by modern drug discovery platform technologies: *high-throughput screening* and *rational drug design*. High-throughput screening (HTS), as its name implies, involves the screening of a large volume of chemical compounds, typically ~100,000 compounds, in a fully automated facility. HTS involves modern instruments such as robotics, liquid handling devices, and sensitive detectors. HTS allows a researcher to quickly conduct the testing of millions of compounds, which is termed the chemical library. Through this process one can rapidly identify active compounds, which modulate a particular target molecule. On the other hand, rational drug design is drug discovery that is based on the knowledge of a biological target. In particular, structure-based drug design relies on the knowledge of the three-dimensional structure of the target protein. It involves the design of small molecules that are complementary in shape and charge to the target protein with which they interact and therefore will bind to it.

Following the identification of "hit" compounds by the drug discovery technology, medicinal chemistry is involved for the optimization of those hits to increase the affinity, selectivity (to reduce the potential of side effects), efficacy/potency, metabolic stability (to increase the half-life), and oral bioavailability. Once a compound that fulfills all of these requirements has been identified as a "lead" compound, the process of drug development will begin.

Drug development is the process of bringing a new drug to the market once a lead compound has been identified through the process of drug discovery. It includes *preclinical research* (on animals) and *clinical trials* (on humans).

Preclinical trials: Preclinical research is to test the efficacy, pharmacokinetics, and the toxicity of the lead compounds through animal models. A compound that is selected as a "candidate" for clinical trials must meet certain criteria.

Clinical trials: Clinical trial is the step that involves human subjects. It is divided into three phases. Briefly, phases I and II are to test the toxicity, whereas phase III is to test the efficacy. Any drug candidates that exhibit toxicity in human, are immediately excluded from the drug development. Phase I is to test toxicity with a relatively small group of healthy volunteers (eg, 20 to 100). Biological safety, dosage, and side effects of a drug candidate are monitored. Phase II is to test the efficacy and side effects with a medium size group of patients (eg, 100 to 300). Once, the safety is ensured after phase II, phase III is conducted to test the efficacy with a larger group of patients (eg, 1000 to 3000). Here, the placebo group is included for comparison. If the efficacy is demonstrated in a phase III trial, an NDA (new drug application) can be submitted to the FDA for the final approval. The vast majority of drug development is terminated after phase I or phase II trials due to undesirable toxicity. Since human toxicity is unpredictable from the chemical structure in the first place, drug development inevitably suffers uncertainty.

Despite advances in technology and the understanding of biological systems, drug discovery is still a lengthy, "expensive, difficult, and inefficient process" with a low rate of new therapeutic discovery (~5%). In 2010, the research and development cost of each new drug was approximately USD 1.8 billion.

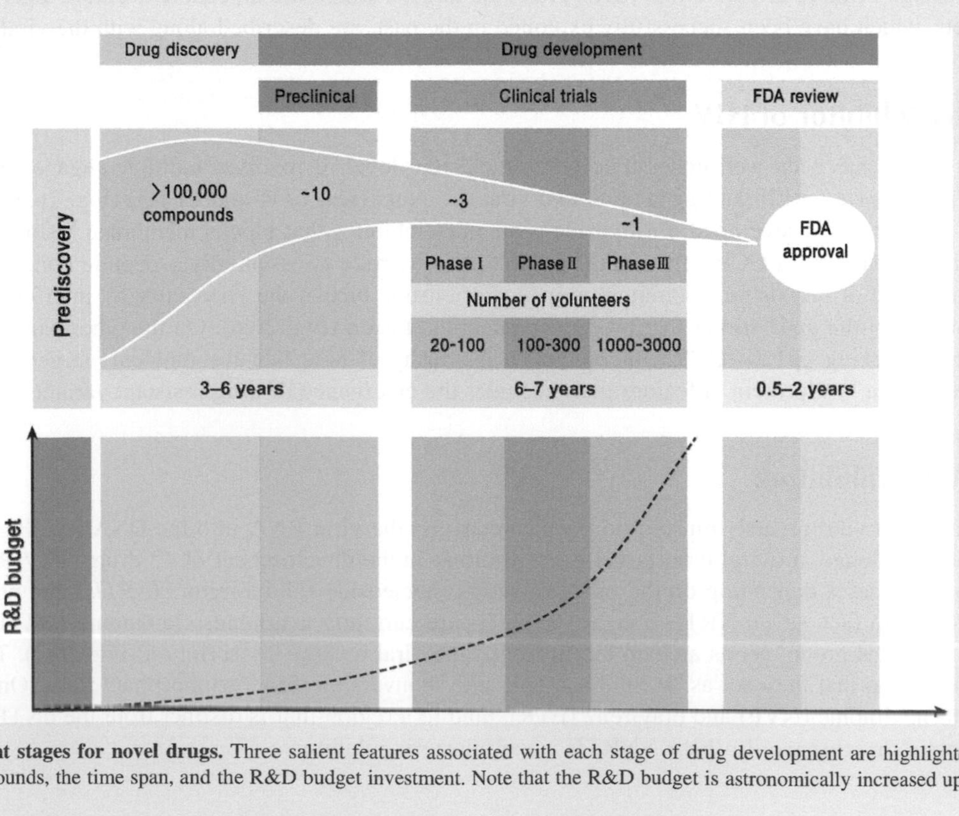

The development stages for novel drugs. Three salient features associated with each stage of drug development are highlighted including the number of compounds, the time span, and the R&D budget investment. Note that the R&D budget is astronomically increased upon human clinical trials.

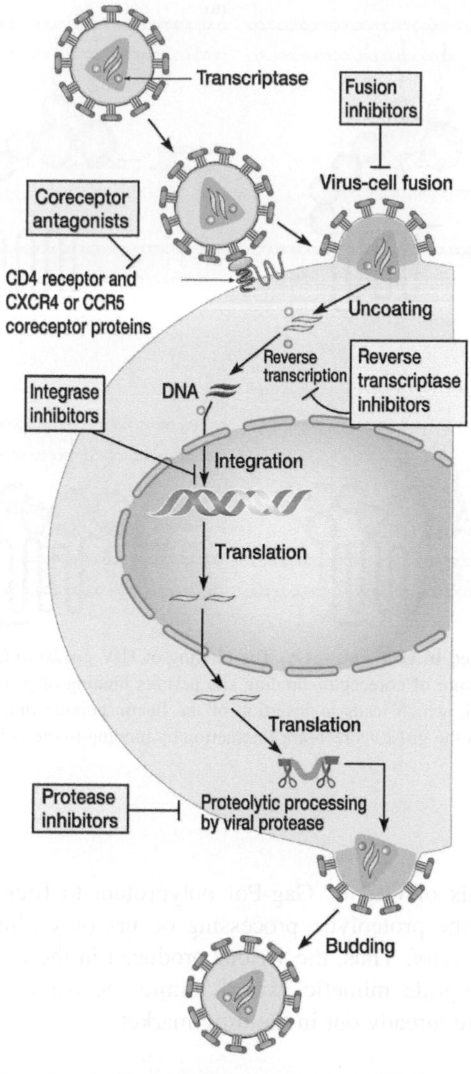

FIGURE 26.4 Antiviral drugs of HIV and their targets in the HIV life cycle. Anti-HIV drugs can be grouped into five categories: coreceptor antagonists, fusion inhibitors, reverse transcriptase inhibitors, integrase inhibitors, and protease inhibitors. The steps at which each anti-HIV drug inhibits are denoted in the context of HIV life cycle.

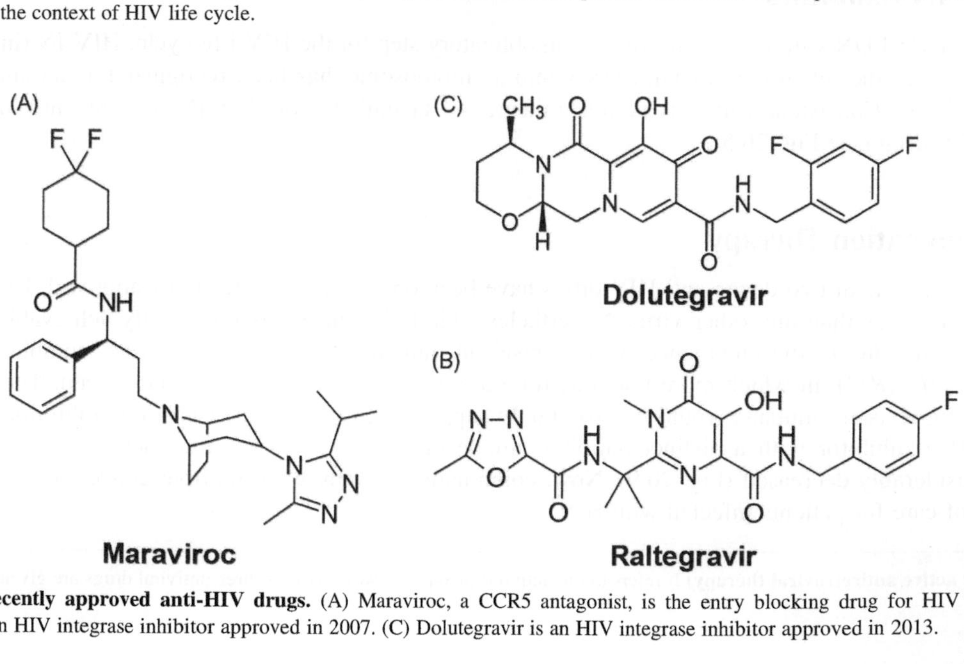

FIGURE 26.5 Recently approved anti-HIV drugs. (A) Maraviroc, a CCR5 antagonist, is the entry blocking drug for HIV approved in 2007. (B) Raltegravir is an HIV integrase inhibitor approved in 2007. (C) Dolutegravir is an HIV integrase inhibitor approved in 2013.

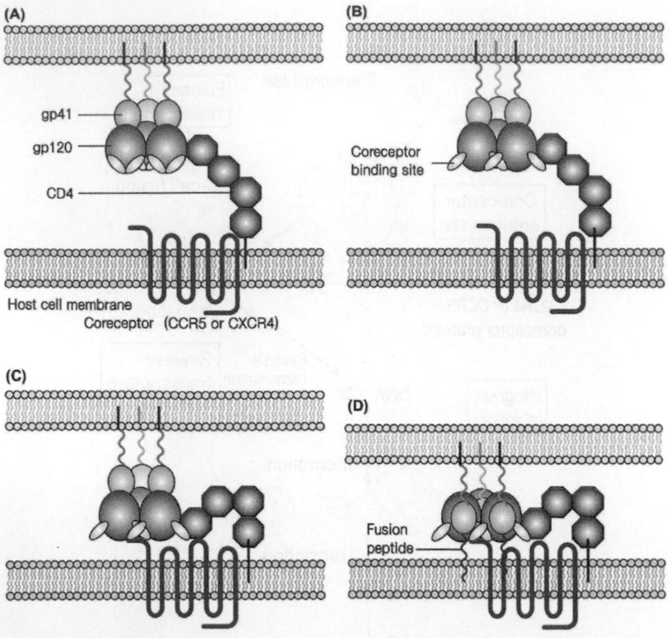

FIGURE 26.6 Molecular interactions engaged in HIV entry. (A) The binding of HIV gp120 to CD4 induces a conformational change in gp120, exposing the coreceptor binding site (B). Exposure of coreceptor binding site permits binding of gp120 to the coreceptor (C). Coreceptor binding of gp120 induces conformational changes in gp41, which leads to insertion of the fusion peptide into the host cell membrane and membrane fusion (D). Maraviroc, a coreceptor antagonist, inhibits the gp120/coreceptor interaction by binding to the coreceptor.

26.3.3 HIV PR Inhibitors

HIV PR is involved in the proteolysis of Gag or Gag-Pol polyprotein to four or six individual functional proteins, respectively (see Fig. 26.4). In fact, the proteolytic processing occurs only after the release of virions. This process termed "maturation" confers the infectivity. Thus, the virions produced in the presence of PR inhibitor lack infectivity. Intriguingly, the PR inhibitors are peptide mimetic, which mimic the peptide linkages of the Gag-Pol polyprotein (Fig. 26.8). Nine such PR inhibitors are already out in the drug market.

26.3.4 HIV IN Inhibitors

The integration of viral DNA into a chromosome is an obligatory step for the HIV life cycle. HIV IN (integrase), a viral enzyme that catalyzes the integration of viral DNA into a chromosome, has been recognized as an attractive antiviral target (see Fig. 26.4). Consistent with this notion, raltegravir, an inhibitor of HIV IN, was recently developed as an effective anti-HIV drug (see Fig. 26.5).

26.3.5 Combination Therapy

As stated above, more than two dozen anti-HIV drugs have been on the market after FDA approval. Undoubtedly, HIV has more antiviral drugs than any other virus. Nevertheless, viral clearance was not readily achievable by any one of these available drugs due to the emergence of drug-resistant mutants. To address this issue, combination therapy or cocktail therapy (*HAART*[4]), in which more than one drug are given together, has been introduced. The idea is that the emergence of drug-resistant mutants would be profoundly suppressed, as the leakiness of one inhibitor can be complemented by another inhibitor with a distinct inhibitory mode of action. Since its introduction in 1996, AIDS-related deaths have considerably decreased (Fig. 26.9). Now, combination therapy with multiple classes of antiretroviral drugs is the standard of care for patients infected with HIV.

4. **HAART (highly active antiretroviral therapy)** It refers to the antiviral therapy, in which two or three antiviral drugs are given together.

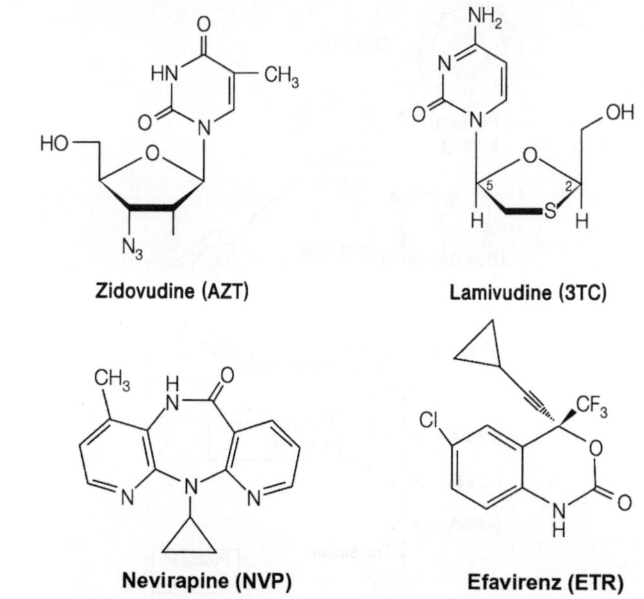

FIGURE 26.7 HIV RT inhibitors. Zidovudine (AZT) and lamivudine (3TC) are NRTIs that act as chain terminators of HIV RT following conversion into a triphosphate form in cells. Nevirapine (NVP) and efavirenz (ETR) are NNRTIs. Three-letter acronyms are shown in parenthesis.

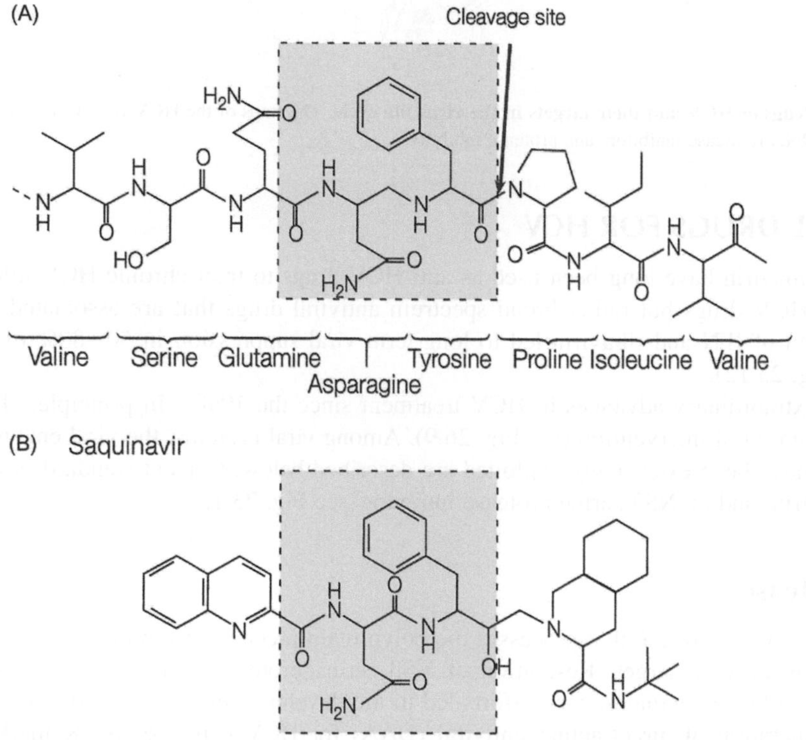

FIGURE 26.8 HIV PR inhibitors. (A) The natural substrate peptide of HIV PR protease. The cleavage site is denoted by a *red arrow*. (B) Chemical structure of saquinavir. HIV protease inhibitors are peptidomimetic, with the nonscissile hydroxyethylene [–CH2–CHOH–] bond instead of the readily hydrolyzable peptide [–CO–NH–] bond. The scaffold structure is highlighted by a *box*.

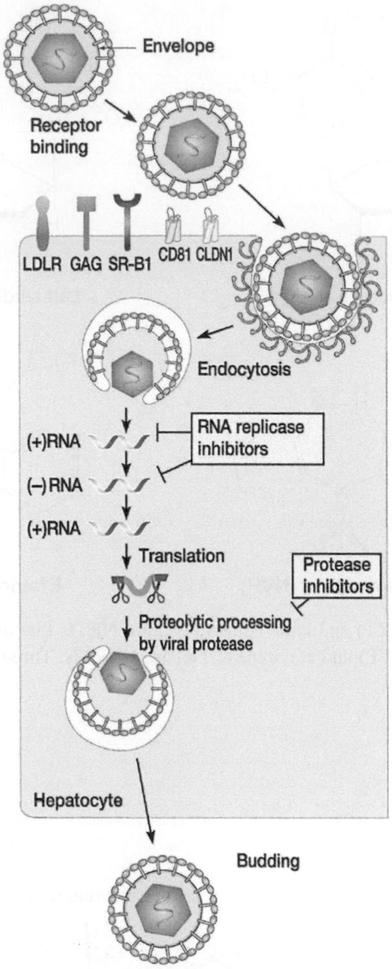

FIGURE 26.9 **Antiviral drugs of HCV and their targets in the virus life cycle.** The steps of the HCV life cycle that have been explored as antiviral targets are highlighted: RNA replicase inhibitors and protease inhibitors.

26.4 ANTIVIRAL DRUGS FOR HCV

Interferon (IFN) and ribavirin have long been used as anti-HCV drugs to treat chronic HCV infection. These drugs are not direct-acting anti-HCV drugs but rather broad spectrum antiviral drugs that are associated with an inevitable side effect. The combination of IFN and ribavirin led to long-term viral suppression in 50–80% of patients, depending on their genotype (see Fig. 23.12).

There have been extraordinary advances in HCV treatment since the 1990s. In principle, all steps in the HCV life cycle are amenable to antiviral intervention (see Fig. 26.9). Among viral proteins, the viral enzymes (ie, NS3 serine protease and RdRp) that have been extensively exploited are described below. Current standard of care involves pegylated IFN (*peg-IFN*[5]), ribavirin, and an NS3 serine protease inhibitor (see Fig. 23.13).

26.4.1 NS3 Protease

NS3 serine protease is a viral enzyme that processes the polyprotein precursor to multiple functional proteins. Hence, it represents an attractive antiviral target. Resolution of NS3 serine protease structure by X-ray crystallography made rational drug design feasible. Eventually, such efforts led to the development of two inhibitors (ie, boceprevir and telaprevir) as the first generation of direct-acting antiviral (DAA) for HCV (Fig. 26.10). Remarkably, triple therapy, in

5. **peg-IFN** It is an abbreviation for the pegylated interferon. Pegylation, the process of covalent attachment of polyethylene glycol (PEG) polymer chains, prolongs its circulatory time by reducing renal clearance, thereby improving the efficacy of drugs.

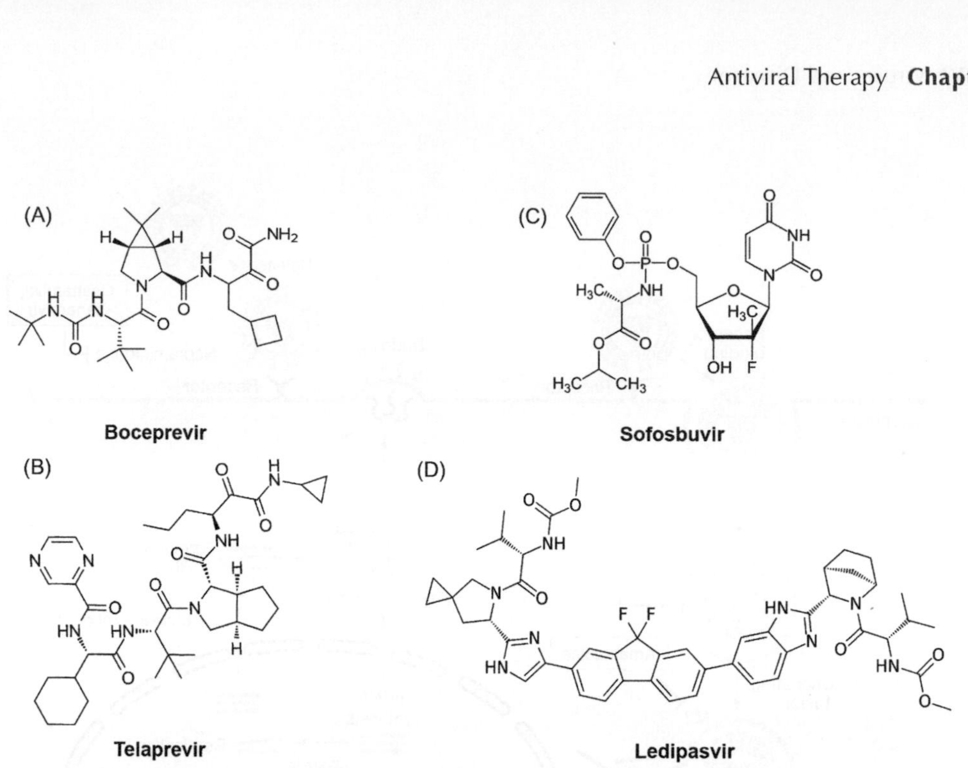

FIGURE 26.10 Direct-acting anti-HCV drugs. Chemical structures of direct-acting anti-HCV drugs (DAA) are presented. (A,B) Boceprevir and telaprevir, NS3 protease inhibitors, were approved by the FDA in 2011. (C) Sofosbuvir, an inhibitor of NS5B RdRp, was approved in 2013. (D) Ledipasvir, an inhibitor of NS5A, was approved in 2014.

which telaprevir is added to an IFN + ribavirin treatment regimen, led to a long-term viral suppression in 70–80% of patients, regardless of the viral genotype (see Fig. 23.12). Nonetheless, monotherapy by either of the NS3 inhibitors does not exhibit a therapeutic benefit.

26.4.2 NS5B RdRp

As stated above, historically, viral polymerase has been proven to be the favorite target for antiviral purposes. At the moment, many candidate inhibitors are under development. Sofosbuvir was approved by the FDA in 2013 (see Fig. 26.10). Preliminary work reported that monotherapy of sofosbuvir without IFN exhibits the long-term suppression of virus in nearly 100% of treated patients. It is a truly remarkable achievement.

26.4.3 NS5A

NS5A is a phosphoprotein with RNA binding ability, and is a component of the HCV replication complex. NS5A induces a double membrane vesicle, which serves as a platform for HCV RNA replication (see Fig. 12.6). Ledipasvir is a recently FDA-approved drug for the treatment of hepatitis C (see Fig. 26.10). The ledipasvir/sofosbuvir combination is a DAA agent that interferes with HCV replication and can be used to treat patients without PEG-IFN or ribavirin.

Overall, strenuous efforts made on the anti-HCV drug development are paying off with the development of four DAAs including boceprevir, telaprevir, sofosbuvir, and ledipasvir. Importantly, the ledipasvir/sofosbuvir combination represents an all-oral DAA regimen without needing PEG-IFN or ribavirin. In addition, more than 30 drugs are now under development. A few candidate drugs, at least, are expected to be approved by FDA within the next few years. Some experts are cautiously contending that "HCV infection is now a curable disease."

26.5 ANTIVIRAL DRUGS FOR INFLUENZA VIRUS

Although many steps in the influenza life cycle are amenable to therapeutic intervention, only a few have been exploited: M2 protein acting on entry and *NA* acting on release (Fig. 26.11). Besides these two viral proteins, other targets being explored will be described.

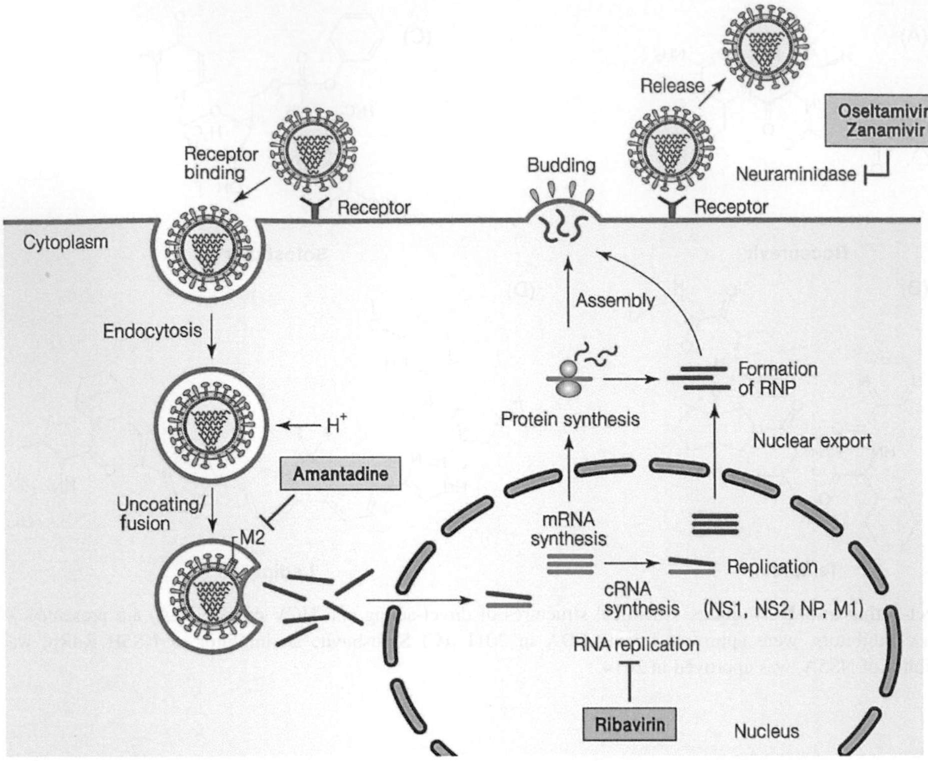

FIGURE 26.11 Antiviral targets of influenza virus and their targets in the virus life cycle. Currently available antiviral drugs are shown with their respective targets. The antiviral targets of influenza virus that can be further explored are presented in the context of the viral life cycle: receptor binding, uncoating/fusion, RNA replication, and assembly.

26.5.1 Entry

During viral entry, the intake of a proton via an M2 protein induces unfolding of the HA protein such that the fusion peptide is activated (see Fig. 26.11). Hence, the inhibition of the M2 ion channel will block the viral entry. Amantidine, an inhibitor of the M2 ion channel, has long been the only antiinfluenza drug available (Fig. 26.12). However, the rapid emergence of drug-resistant mutants to amantadine has limited its utility. Nonetheless, the M2 ion channel represents an attractive target that should be explored further.

26.5.2 RdRp

Unlike most other viruses, RdRp of influenza virus has not been explored yet for antiviral purposes. The RdRp is composed of 3 subunits (ie, PA, PB1, and PB2). In addition to RNA synthesis, cap-snatching by RdRp represents an attractive target, because it is a peculiar mechanism unique to influenza virus. At the moment, Favipiravir, a nucleoside analog inhibitor of RdRp, is the only drug candidate in this category, and is under clinical trial in Japan (see Fig. 26.12).

26.5.3 NS1

NS1 is a multifunctional protein that is associated with virulence. It suppresses host innate immune response by blocking IRF signaling and thereby IFN induction (see Fig. 15.13). Specifically, NS1 blocks the activation of RIG-I via its interaction with *TRIM25*.[6] It is possible that the blocking of NS1 could rescue the IFN induction, thereby suppressing viral infection. Hence, NS1 is an attractive antiviral target, which has not been explored.

6. **TRIM25** Refers to an ubiquitin E3 ligase that belongs to the tripartite motif family.

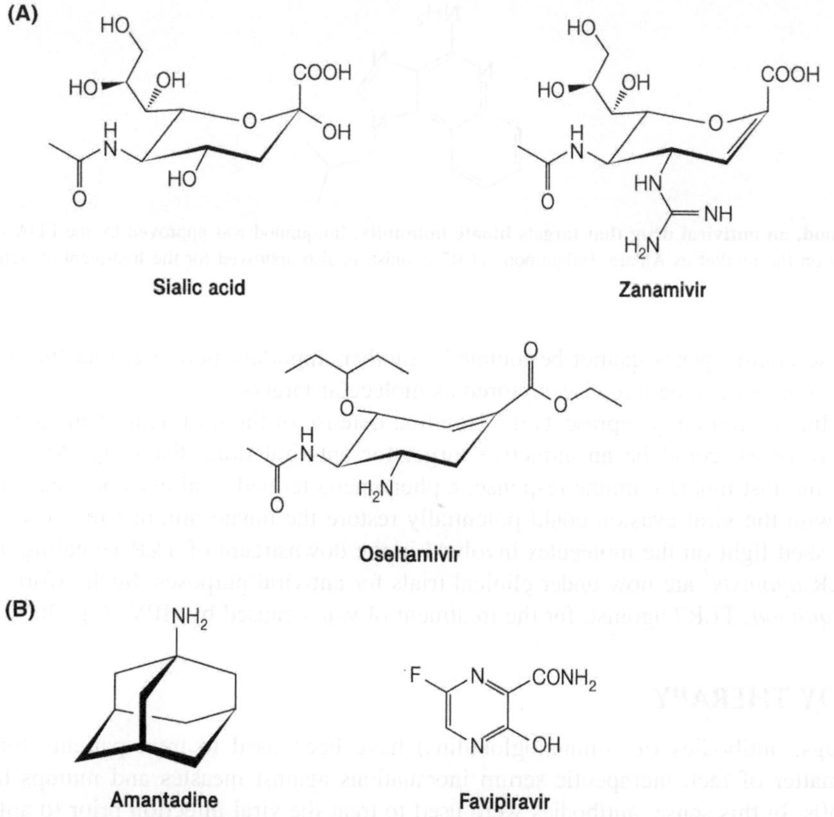

FIGURE 26.12 **Antiinfluenza drugs.** (A) Two neuraminidase inhibitors are shown in parallel with sialic acid. Note the striking structural similarity between sialic acid and the two sialic acid analog drugs. (B) Amantadine, the first antiinfluenza drug approved by the FDA in 1966, is an M2 ion channel inhibitor. Favipiravir, an RdRp inhibitor, is currently under clinical trial in Japan.

26.5.4 Exit

NA protein, an envelope glycoprotein, acts to facilitate the release of nascent virions. Its neuraminidase activity cleaves the sialic acid residue linked to glycans of the plasma membrane (see Fig. 15.5). Inhibition of neuraminidase activity will block the virion release from the infected cells (see Fig. 26.11). Two neuraminidase inhibitors have been developed: oseltamivir and zanamivir are the analogs of sialic acid (see Fig. 26.12). These drugs act as a competitive inhibitor of neuraminidase. The effectiveness of these drugs is in fact limited, since they need to be taken within 48 h after the first fever. The narrow window for their effectiveness is related to the mode of action in that the neuraminidase inhibitors do not suppress viral replication, but they block the viral spread. Once the virus spreads, these drugs are no longer effective.

Currently, only the neuraminidase inhibitors, with their short acting window, are available as an effective drug for the next pandemic. Certainly, new drugs that have a distinct mode of action represent the unmet medical needs.

26.6 NEW CONCEPT OF ANTIVIRALS

The antivirals that are available so far are inhibitors of viral proteins (ie, DNA/RNA polymerase, and proteases), which is, unfortunately, inevitably vulnerable to the emergence of drug-resistant mutants. What is an effective strategy to overcome the emergence of drug-resistant mutants?

Host Factors as Molecular Targets. Advances in molecular virology have laid a keystone for the development of novel antiviral drugs. Thanks to the introduction of siRNA technology, host factors essential for viral propagation have been revealed. In the case of HIV, more than 200 host factors that contribute to viral infection have been reported. Certainly, many, if not all, of these host factors can be explored as molecular targets of antiviral drug discovery. Importantly, in these cases, the emergence of drug-resistant mutants can be avoided, because the molecular targets are

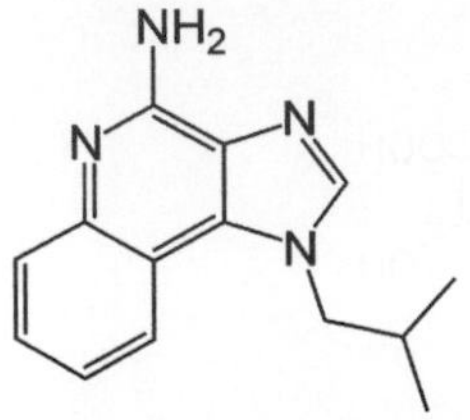

FIGURE 26.13 Imiquimod, an antiviral drug that targets innate immunity. Imiquimod was approved by the FDA in 2007 for the treatment of HPV-induced warts and is on the market as Aldara. Imiquimod, TLR7 agonist, is also approved for the treatment of actinic keratosis and basal cell carcinoma as well.

the host factors, whose coding genes cannot be mutated. Another important notion is that the interfaces between such host factors and viral proteins can be likewise explored as molecular targets.

Innate Immunity. Innate immunity represents the frontline defense of the host against invading pathogens. Thus, the host innate immune response could be an attractive target for antiviral drug discovery. Most viruses establish their infection by suppressing host innate immune response, a phenomena termed viral evasion (see Table 5.4). A small molecule that interferes with the viral evasion could potentially restore the innate immune response. Discovery of the toll-like receptors (TLR) shed light on the molecules involved in the downstream of TLR signaling as antiviral targets (see Fig. 5.19). A few TLR *agonists*[7] are now under clinical trials for antiviral purposes. Such efforts have been paid off by FDA approval of *imiquimod*, TLR7 agonist, for the treatment of warts caused by HPV (Fig. 26.13).

26.7 ANTIBODY THERAPY

Besides antiviral drugs, antibodies (ie, immunoglobulins) have been used to treat patients for therapeutic purposes (Table 26.3). As a matter of fact, therapeutic serum inoculations against measles and mumps have been practiced in Europe since the 1930s. In this sense, antibodies were used to treat the viral infection prior to antiviral drugs. Although antibodies are not as extensively used as antivirals these days, they are often a unique option under certain circumstances. For instance, for individuals bitten by rabid dogs, antibodies are a unique option for its treatment, because no antiviral drug for the *rabies*[8] virus is available. This treatment is termed "antibody therapy" or "*passive immunization*." Besides rabies, antibodies prepared from the vaccinated individuals are used as postexposure prophylaxis of HAV and HBV (see Table 26.3). Antibodies are also used to prevent severe Zoster.

What is the future prospect for antibody therapy for the viral disease? Recent advances in monoclonal antibody technology have made antibody therapy more feasible. As an evidence for this notion, the monoclonal antibody for *RSV*[9] is already developed for prophylaxis for high-risk infants. The cost for prophylatic RSV antibody (ie, Syngis) is 1000 USD per dose (see Table 26.3). Although it is unlikely that antibody therapy is widely used due to their prohibitive cost, it can be an option, where a vaccine is not yet available. In fact, such efforts are in progress to control HIV infection.

26.8 PERSPECTIVES

It is fair to say that antiviral drug development started only in the 1980s. Needless to say, the AIDS epidemic provided the motivation for the successful endeavor, resulting in the development of over two dozen anti-HIV drugs. Importantly, such endeavor has contributed to the establishment of structure-based drug design, which has become a standard drug discovery platform in general. It is hoped that the knowledge accumulated in anti-HIV drug discovery will facilitate the discovery of new antiviral drugs for other human viral infections, such as HBV and influenza virus, where effective antiviral therapy is not yet available. A payoff is that many human viral diseases are more or less under control, in the sense that either preventive vaccines or therapies are available. Nonetheless, numerous challenges are ahead of us, such as the emergence of new viruses (eg, Ebola virus outbreak). Knowledge obtained from basic virus

7. **Agonist** Refers to a chemical that binds to some receptor of a cell and triggers a response by that cell (cf., antagonist).
8. **Rabies** It refers to a viral disease that causes acute encephalitis in animals. Rabies virus belongs to a family Rhabdoviridae (see chapter: Rhabdovirus). The rabies is a Latin word for "madness."
9. **RSV (respiratory syncytial virus)** RSV belongs to a family Paramyxoviridae (see chapter: Other Negative-Strand RNA Viruses). It is a major cause of lower respiratory tract infections and hospital visits during infancy and childhood that causes worldwide 160,000 deaths per year.

TABLE 26.3 List of Antibodies Used for Prophylactic and Treatment Purpose

Antibody Type	Drug Name (Company)	Indication
Rabies IgG	IMOGAM Rabies (Aventis)	Postexposure prophylaxis and treatment against rabies virus infection
RSV IgG	Respigam (MedImmune)	RSV (respiratory syncytial virus) infection in children under 24 months of age
RSV Mab	Synagis (MedImmune)	Prophylaxis in high-risk infants
CMV IgG	Cytogam (MedImmune)	Prophylaxis of CMV (cytomegalovirus) associated diseases with organ transplantation
HBV IgG	HepaGamB (Cangene)	Treatment of acute exposure to HBV (hepatitis B virus)
VZV IgG	VariZIG (Cangene)	Prevents severe varicella zoster infection
Generic IgG	BayGam (Bayer)	Postexposure prophylaxis of HAV (hepatitis A virus)

research is expected to address these challenges in the near future. Lastly, it is important to note that the emergence of drug-resistant mutants represents a new challenge.

26.9 SUMMARY

- *Antiviral targets*: In principle, all steps in the virus life cycle ranging from entry to release can be explored as molecular targets for antiviral therapy, although viral enzymes are proven to be the most effective antiviral targets.
- *Drug-resistant mutants*: For most antiviral drugs, the rapid emergence of drug-resistant mutants limits their antiviral efficacy.
- *Multidrug therapy*: Multidrug therapy could overcome the emergence of drug-resistant mutants, as evidenced by HAART therapy.
- *Antibody therapy*: Besides antiviral drugs, antibodies (ie, immunoglobulins) have been used to treat patients for therapeutic purposes.

STUDY QUESTIONS

26.1 Acyclovir is an effective antiherpes drug with a low cytotoxicity. State reasons why?

26.2 Shown below is the chemical structure of lamivudine, a nucleoside analog drug for the treatment of chronic HBV infection. Please complete the derivative of lamivudine that inhibit the viral polymerase.

26.3 Suppose that you identified a new small organic molecule that potently inhibits influenza virus replication. How would you determine the molecular target of the antiviral compound?

SUGGESTED READING

Bartenschlager, R., Lohmann, V., Penin, F., 2013. The molecular and structural basis of advanced antiviral therapy for hepatitis C virus infection. Nat. Rev. Microbiol. 11 (7), 482–496.

De Clercq, E., 2006. Antiviral agents active against influenza A viruses. Nat. Rev. Drug Discov. 5 (12), 1015–1025.

Engelman, A., Cherepanov, P., 2012. The structural biology of HIV-1: mechanistic and therapeutic insights. Nat. Rev. Microbiol. 10 (4), 279–290.

Hennessy, E.J., Parker, A.E., O'Neill, L.A., 2010. Targeting Toll-like receptors: emerging therapeutics? Nat. Rev. Drug Discov. 9 (4), 293–307.

Marasco, W.A., Sui, J., 2007. The growth and potential of human antiviral monoclonal antibody therapeutics. Nat. Biotechnol. 25 (12), 1421–1434.

JOURNAL CLUB

- Kao, R.Y., et al., 2010. Identification of influenza A nucleoprotein as an antiviral target. Nat. Biotech. 28, 600–605.

  Highlight: Via high-throughput screening of over 1 million chemical compounds, a novel antiinfluenza drug, called "nucleozin," was identified. Intriguingly, nucleozin binds to viral NP protein, thereby preventing its nuclear import.

Answers to Study Questions

1.1. *Answer:*

Features	Biological Viruses	Computer Viruses
Dependence	Host organisms	Computer (PC)
Transmissibility	Yes (via contact or air)	Yes (via network)
Disease-causing	Yes	Yes (errors)
Variant	Yes (mutant)	Yes (variant)
Vaccine/therapy	Yes	Yes

1.2. *Answer: (1) Group IV: Positive-strand RNA viruses and group VI: Retroviruses. (2) In the case of positive-strand RNA viruses, the genome RNA is directly used as mRNA, while in the case of retroviruses, the genomic RNA is not used as mRNA, but converted to DNA genome via reverse transcription.*

1.3. *Answer: (1) Satellite virus: A satellite virus depends on other host viruses. Its dependency on other viruses stands out. (2) Viroids: The viroid is an "RNA only" agent, which does not encode any protein. The viroids are considered to be "infectious RNA" rather than viruses, since no proteins are expressed. (3) Prions: Prions are a "Protein only" agent, having no nucleic acid genome.*

2.1. *Answer: (1) Viral capsid protects the viral genome from degradation. (2) Viral capsid recognizes the receptor for entry, in the case of naked viruses. (3) Viral capsid delivers the viral genome to the site of genome replication.*

2.2. *Answer: (1)* $T = 16$ $(h = 4, k = 0)$. *(2)* $960 = 16 \times 60$.

2.3. *Answer: (1) Cryo-EM and X-ray crystallography. (2) Cryo-EM: The pro of cryo-EM is that it enables to obtain the image of the specimen in an intact state without fixing or staining. In addition, even dynamic states of virus particles could be examined by cryo-EM. The con of cryo-EM is that it is limited to particles that are symmetrical, rigid, and homogenous. X-ray crystallography: The pro of X-ray crystallography is that one can get [high resolution (atomic level)] structural information of a large complex particle. The con is that it can only be applied the particles that are crystallized.*

3.1. *Answer: (1) (see* Box 3.1*). (2) To validate the functionality of a newly identified "receptor," one needs to test whether the receptor could confer the susceptibility to infection to the otherwise unsusceptible cells. For instance, sodium taurocholate cotransporter (NTCP) is newly identified as an entry receptor for hepatitis B virus (HBV). Its functionality was validated by demonstrating that an otherwise HBV-unsusceptible HepG2 cell becomes susceptible to HBV infection, when NTCP is expressed (see Journal Club or Yan H, Zhong G, Xu G, He W, Jing Z, Gao Z, et al. Sodium taurocholate cotransporting polypeptide is a functional receptor for human hepatitis B and D virus. eLife. 2012;1: e00049).*

3.2. *Answer: Viruses overcome the first obstacle (ie, plasma membrane) by penetrating cells via either endocytosis or direct fusion. In contrast, the second obstacle is bypassed, as viruses enter the nucleus via nuclear pores.*

3.3. *Answer: (1) See* Fig. 3.10. *(2)*

Infection Type	Viral Gene Expression	Viral Genome Replication	Cytopathic Effect	Progeny Production
Lytic infection	yes	yes	Yes (cell lysis)	yes
Persistent infection	yes	yes	no	yes
Latent infection	limited	no	no	no
Transforming infection	limited	no	yes (tumor)	no
Abortive infection	limited	no	no	no

4.1. *Answer: (1) For the quantitation of antigen, the amount of antigens attached to the plate should be diluted to be limiting, whereas the amount of the primary antibody and the secondary antibody conjugate should be saturating. In this case, the measurement (ie, color change) is related to the amount of the antigens in specimens. (2) For the quantitation of antibody, however, the amount of the primary antibody should be diluted to be limiting, whereas the amount of the antigen attached to the plate and the secondary antibody conjugate should be saturating. In this case, the measurement (ie, color change) is related to the amount of the primary antibody in specimens.*

4.2. *Answer: (1) The polystyrene, of which the 96-well microtiter plate is made, serves as a solid support for ELISA. (2) A cover glass, onto which cells are grown, serves as a solid support for IFA. Nitrocellulose membrane serves as a solid support for immunoblotting.*

4.3. *Answer: (1) The viral replicon is critically important for viruses, where cell culture for viral propagation is undoable (eg, norovirus and hepatitis B virus until recently). By transfecting the replicon DNA into appropriate cells, the viral genome replication can be studied. (2) It has been undoable to do mutation analysis of RNA viruses. By introducing mutations in the replicon DNA, however, the genetic analysis of RNA viruses can be readily carried out (see* Box 12.3*). (3) The viral replicons enable us to carry out the complementation. In other words, the viral genome replication can be studied by cotransfecting two plasmids that are complementary to each other for the viral genome replication.*

5.1. *Answer: Similarities: Both TLR and RIG-I are similar in that (1) viral molecules are recognized by either by TLR or RIG-I, and (2) upon binding to the ligands, IRF signaling is commonly activated, and (3) IRF signaling culminates to induce interferon production. Differences: (1) receptor: TLR versus RIG-I/MDA5; (2) adaptor: TRIF or MyD88 versus MAVS; (3) cells: Immune cells (eg, dendritic cells) versus nonimmune cells (eg, epithelial cells).*

5.2. *Answer: (1) Ligand: PAMP versus IFN; (2) Receptor: TLR versus IFN receptor; (3) Transcription factor: IRF versus STAT; (4) Genes induced: IFN versus PKR or OAS.*

5.3. *Answer: See* Table 5.4.

6.1. *Answer: T-antigen facilitates the switch from early phase to late phase progressively via its binding to three T-antigen binding sites: (1) Its binding to the site I suppresses the early gene transcription; (2) Its binding to the site II triggers the viral genome replication; (3) Its binding to the site III transactivates the late gene transcription.*

6.2. *Answer: (1) Mutant A: This mutant will not produce the progeny virus in CV-1 cell, but will produce the progeny virus in COS cell, as the functional T-antigen is complemented by COS cell. (2) Mutant B: This mutant will not produce the progeny virus either in CV-1 cell or COS cell, since the T-antigen cannot bind to the defective origin.*

6.3. *Answer: Although humans are permissive host for SV40, human cells are technically nonpermissive for the Ori-defective mutants, since the mutants cannot support the viral genome replication. Thus, only early phase of life cycle is executed.*

7.1. *Answer: HPV genome replication is carried out in two distinct modes: (1) Episomal replication, and (2) Vegetative replication. Episomal replication, which occurs in the undifferentiated basal layer of the epithelium, executes the viral DNA synthesis in synchrony with cellular DNA synthesis, maintaining the viral DNA copy per cell constant. In other words, the viral genome replication is controlled by the host regulatory mechanism. In contrast, vegetative replication, which occurs in the differentiated suprabasal layer of the epithelium, executes the viral DNA synthesis not in the synchrony with cellular DNA synthesis or cell division. Thus, the viral DNA copy per cell is considerably increased.*

7.2. *Answer: HPV E6 protein binds to p53 and degrades it via the ubiquitin proteasome system. Since p53 is already inactivated in HPV infected cells, no selective pressure on p53 mutation is imposed. Therefore, p53 mutation is not found in HPV-associated carcinoma cells.*

7.3. *Answer: (1) Cervical carcinoma, (2) HPV E6 and E7, (3) Because p53 and Rb proteins, which are important regulators of apoptosis and cell cycle control, are already deregulated in HeLa cell.*

8.1. *Answer: (1) In the first round, the linear double-stranded DNA genome is used as a template, while the panhandle structure of the displaced single strand DNA is used as a template for the second round. Nonetheless, the terminal structure of both is identical in configuration. (2) The same set of viral and host proteins are used for both: Ad polymerase, DBP, pre-TP, NF-1, and Oct1.*

8.2. *Answer: (1) Cullin 2/5, Elongin B/C, E1B-55K, and E4 orf6. (2) p53 and MRN complex. The ubiquitin-mediated degradation of p53 and MRN complex contributes not only to the viral genome replication, but also to adenovirus-induced tumorigenesis.*

8.3. *Answer: (1) E1A, and E1B proteins. E1A releases E2F from the Rb-E2F complex via protein–protein interaction, leading to entry to the S phase in the cell cycle. E1B-55K downregulates the p53 and MRN complex, tumor suppressor proteins, while E1B-19K blocks apoptosis via interaction with Bax. (2) Adenovirus infection of a human cell, a permissive host, leads to cell lysis, while adenovirus infection of rodents' cells, a nonpermissive host, leads to tumor formation, instead.*

9.1. *Answer: (1) When HSV-1 DNA enters the nucleus, the viral genome is targeted to ND10 subnuclear structures, leading to a repressed viral chromosome. ICP0 (whether imported along with the genome during entry or newly made) is colocalized to the ND10 domain, and degrades the ND10 structure via its ubiquitin E3 ligase activity. Hence, WT HSV-1 manages to lead to lytic infection by disrupting intrinsic resistance. However, in mutant infected cells, the viral genome remains silent in the absence of ICP0. Therefore, the infection is aborted.*

9.2. *Answer:*

Modes	"θ" form Replication	Rolling-Circle Replication
Kinetics	1st phase of replication	2nd phase of replication
Extent	Trace	Bulk
Coupled to packing	Not	Coupled
Replication factors	Viral proteins: UL9 (DNA-binding protein), ICP8, UL30 (DNA polymerase), UL5/UL8/UL52 helicase/primase, UL42	Viral proteins: UL9 (DNA-binding protein), ICP8, UL30 (DNA polymerase), UL5/UL8/UL52 helicase/primase, UL42
Nick generation	Not applicable	Required
Cleavage factor	Not applicable	Required
Mode	Bidirectional DNA synthesis	Rolling-circle replication

9.3. *Answer: (1) No immune response is triggered by miRNAs, (2) The coding capacity for miRNA is minimal, as opposed to protein-coding genes.*

10.1. *Answer: (1) Refer to* Fig. 10.2A. *(2) These repeat elements fold into a hairpin structure that serves as a primer for the viral DNA synthesis, as shown in* Fig. 10.2B.

10.2. *Answer: (1) Refer to* Fig. 10.7. *(2) The linear DNA genome of poxvirus is flanked by inverted terminal repeat (ITR) sequences, which are believed to act as a telomere* (see Box 10.2).

11.1. *Answer: (1) Hypothesis: Like the VPg of the picornavirus the 5′-terminus linked viral protein may act as a protein primer for the viral genome replication. (2) Molecular Genetic Approach. One has to establish a "viral replicon" to perform a molecular genetic study. Once the replicon is generated, make a mutant in the replicon construct that harbors a mutation in the gene for the 5′ terminal viral protein (eg, substitution or small deletion), and test whether WT-like viral progeny is produced, when transfected into the appropriate cells. Since the protein primer is essential for viral RNA synthesis, no viral RNA would be detectable, if the 5′ terminal viral protein acts as a protein primer.*

11.2. *Answer: (1) VPg-free RNA: The cloverleaf structure at the 5′ NCR is sufficient to protect the viral genome RNA from degradation. Even in the absence of VPg, the RNA genome replication as well as translation will occur normally, and the viral progeny will be produced to a near WT level.* (2) Δ *cloverleaf structure: Translation will occur normally via IRES. However, the transit from translation to the genome replication will be aborted due to the lack of the cloverleaf structure that is instrumental for the transit. Perhaps, viral proteins continue to accumulate to a higher level, but no progeny virus will be produced. (3) Δ IRES: No viral protein synthesis will occur. (4) Δ CRE: Translation, and (−) RNA synthesis will occur normally. However, due to the lack of CRE, which is a template for the protein priming for (+) RNA synthesis, no (+) RNA synthesis will occur. No viral progeny will be produced. (5) Poly(A) tail-free RNA: Translation will occur normally but to a reduced level, since poly(A) tail facilitates translation. Importantly, (−) RNA synthesis will not occur, since the poly(A) tail serves as a template for the protein priming for (−) RNA synthesis.*

11.3. *Answer: Similarity: (1) $3D^{POL}$ executes the viral RNA synthesis for both (+) and (−) RNA synthesis. (2) Both are initiated by protein priming. Differences: (1) The template for (−) RNA synthesis is the poly(A) tail at the 3′ end of the RNA genome, while the template for (+) RNA synthesis is the CRE element in the middle of the RNA genome. (2) In the case of (−) RNA synthesis, only one molecule of RdRp engages per template, while multiple RdRp engage on template in a case of (+) RNA synthesis, leading to "asymmetric RNA synthesis."*

12.1. *Answer: (1) IRES element in 5′ NCR is essential, which facilitates cap-independent translation. (2) The X region in 3′ NCR is essential for the viral genome replication. However, the exact mechanism by which the X region contributes to the RNA genome replication remains unknown.*

12.2. *Answer: HCV NS3/4A serine protease cleaves two adaptor proteins essential for IRF signaling: (1) TRIF in TLR3 signaling, and (2) MAVS in RIG-I signaling.*

12.3. *Answer: One could test the hypothesis by making a fused cell between human hepatocyte and mouse hepatocytes. If fused cells allow HCV infection, it will be argued that no such restriction factors are present in the mouse hepatocyte. Instead, it can be concluded that the lack of positive factor in mouse cells is responsible for the nonpermissiveness of mouse hepatocytes. This was the case, as you can read at PLoS Pathogens 2012;8(12):e1003056.*

13.1. *Answer: (1) Hypothesis: The 5′ linked VPg may act as a protein primer for viral genome replication. Alternatively, it could facilitate the translation, by recruiting eIF4E, as is the case for calicivirus. (2) First, one could look for IRES element upstream of the ORF. If the novel virus does not harbor IRES element, it is likely that VPg is involved in translation initiation.*

13.2. *Answer: (1) Hypothesis: The negative-strand RNA will serve as a temple for both the positive-strand genomic and subgenomic RNA transcription. An internal promoter positioned downstream of the initiation site (5′ end of the subgenomic RNA or intergenic region between two ORFs) on the negative-strand RNA will be used for the synthesis of subgenomic RNA. (2) How would you test your hypothesis? A mutant harboring the deletion of the putative promoter would fail to synthesize the subgenomic RNA.*

14.1. *Answer: (1) The viral mRNA will be successfully transcribed off the (−)NC template by the associated L/P RdRp. Subsequently, the progeny virus will be produced. (2) The RNA genome replication off the introduced (+) NC template would not occur in the absence of excess N protein. Consequently, the progeny virus will be produced. (3) In an N protein expressing cell, the RNA genome replication would occur when (+) NC is introduced to the cell. Consequently, the progeny virus will be produced.*

14.2. *Answer: (1) It is possible that mutant N protein could not function as an antiterminator of transcription. Thus, the transit from transcription to the genome replication is impaired. VSV progeny cannot be produced, because the RNA genome replication may not occur at the nonpermissive temperature. (2) The viral genome replication will normally occur as wild-type does. Due to the defect in the M protein, the progeny virus will not be produced, as the budding process is impaired. (3) The progeny virus could be produced as wild-type does. However, due to the defect in the G protein, the reinfection to the surrounding cells would be impaired at nonpermissive temperature. Hence, the plaque size would be smaller than the wild-type.*

15.1. *Answer: (1) See* Fig. 15.2. *The vRNA could fold into a panhandle structure. (2) 5′ and 3′ end of viral mRNA contain nonviral sequences. 5′ cap RNA fragment is derived from cellular pre-mRNA, while 3′ poly (A) tail is synthesized by stuttering.*

15.2. *Answer: (1) NS1 functions as an IFN antagonist and blockage of cellular mRNA processing. At a nonpermissive temperature, in the absence of NS1 function, the rigorous IFN induction induced will block the virus replication and cellular gene expression would not be suppressed by the NS1. Thus, the virus protein expression would be suppressed. And the viral replication is fairly attenuated and may not form plaques. (2) NA acts to facilitate the virus release from infected cells. At a nonpermissive temperature, the virus replication would normally occur. However, in the absence of NA function, the virus would not be released from the infected cell.*

15.3. *Answer: Each year, a new strain of influenza virus with novel antigenicity emerges due to the high mutation rate (ie, antigenic drift). This new strain causes a seasonal flu epidemic, since antibodies generated by last year's vaccine cannot neutralize the new seasonal flu virus.*

16.1. *Answer: (1) Both are mononegavirales, (2) Both harbor noncoding RNA genes: Leader and trailer. (3) Both have an "intergenic region" between genes, and (4) Both have a U-tract at the end of genes.*

16.2. *Answer: (1) Both are multinegavirales, (2) Both get the RNA primer by cap-snatching, (3) Both form a panhandle structure by via intramolecular base-pairing, and (4) The N protein level serves as a switch from the RNA transcription to the RNA genome replication.*

16.3. *Answer: See* Box 16.1.

17.1. *Answer: (1) See* Fig. 17.5. *(2) Compared to the RNA genome, the U3 element is added to the upstream to 5′ R element, while the U5 element is added to the downstream of 3′ R. (3) The U3 in the 5′ LTR and the U5 in 3′ LTR are duplicated during reverse transcription, constituting LTR.*

17.2. *Answer: (1) Transcription from 5′ LTR will be prematurely terminated, since Tat is not expressed. Thus, no virus will be produced. (2) Doubly spliced RNA will be normally transcribed, and processed to express Tat and Nef. However, other viral proteins will not be expressed, since the unspliced and singly spliced RNAs will be trapped in the nucleus in the absence of Rev. Thus, HIV will not be produced. (3) Nef is mainly involved in the downregulation of CD4 and MHC class I, a function related to immune evasion. In addition, Nef enhances the virion release and infectivity. Thus, due to the absence of immune cells in transfection experiment, the immune evasion function may not be relevant. However, the virion release will be reduced in the absence of Nef. Thus, virus production will be somewhat reduced.*

17.3. *Answer: (1) Hypothesis: A host restriction factor present in the nonpermissive cell (but absent in permissive cells) limits the infection. In the case of WT, the restriction factor is counteracted by the viral protein that is missing in the mutant (eg, Vif). (2) Test: One could make a heterokaryon cells, a fusion cell, between permissive and nonpermissive cells, and test whether the HIV mutant can infect. If the mutant virus cannot infect the heterokaryon cell, the result would indicate that the nonpermissive phenotype is dominant in the fused cell, a finding that is consistent with the host restriction factor hypothesis.*

18.1. *Answer: (1) RC DNA, DL DNA, and SS DNA. RC DNA: The relaxed-circular DNA represents the mature product of viral reverse transcription and is also detected in released virions. DL DNA: A duplex linear DNA, that is made by in situ priming. DL DNA is not found in secreted virions. SS DNA: The single-strand DNA, the product of the first minus-strand DNA synthesis. SS DNA is not found in released virions. (2) cccDNA. Being an episomal DNA, it serves as a template for viral transcription.*

18.2. *Answer: (1) HBVP protein is linked to the 5′ terminus of the minus-strand, while an RNA fragment is linked to the 5′ terminus of the plus-strand DNA. (2) These non-DNA molecules served as primers for the synthesis of corresponding strands during viral reverse transcription.*

18.3. *Answer: (1) The pgRNA serves not only as mRNA for translation of C and P proteins but also serves as an RNA template for the viral genome synthesis. (2) Hypothesis: Transition from translation to genome replication is regulated by HBV P protein, in which the binding of the P protein to a 5′ stem-loop structure (epsilon) suppresses the translation, but concomitantly triggers the pgRNA encapsidation. (3) See related article [Virology 373:112–123(2008)].*

19.1. *Answer: The so-called self-inactivating retroviral vector* (SIN) *vector was developed. In this modified vector, the U3 region of the 3′ LTR in the plasmid vector (the Step (1)) was deliberately deleted so that the 5′ LTR promoter in target cells is inactivated (the Step (3)). As a result, only one RNA, instead of two, encoding the transgene is expressed (the Step (4)). In doing so, the full-genome length RNA, that expresses the no longer needed neomycin marker gene, is not expressed. It is notable that an in-depth understanding of the viral reverse transcription mechanism is aptly explored in the design of the SIN vector.*

Self-Inactivating Retroviral Vector

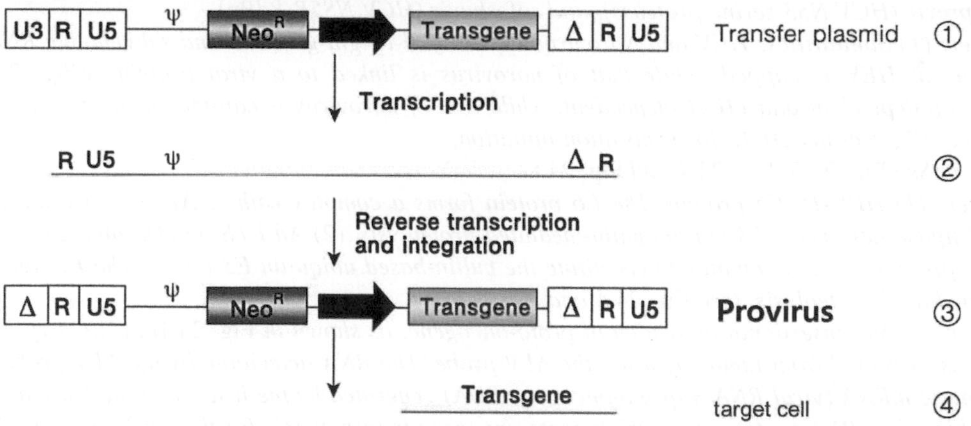

19.2. *Answer: (1) The retroviral DNA present in the packaging cell is present in the form of the transfer plasmid transfected, while the retroviral DNA present in the target cell is present in the form of provirus DNA integrated into chromosome (see* Fig. 19.6*). (2) The pseudotyping by VSV-G protein could widen the target cell range. The VSV-G protein pseudotyped lentivirus vector could infect almost all mammalian cells.*

19.3. *Answer: (1) Adenovirus vector induces a strong inflammatory response, which could lead to cell death. This feature makes the adenovirus vector more suitable for cancer gene therapy. (2) HEK293 cell constituents expresses E1 gene of adenovirus that can complement the first generation adenovirus vector lacking E1 gene. Likewise, AAV vector can be prepared by using HEK293 cell, as AAV needs adenovirus as a helper.*

20.1. *Answer: (1) Satellite virus is not related to host virus in terms of sequence homology. Satellite virus relies on helper virus for the propagation. (2) Satellite virus is morphologically distinct from helper virus. (3) Satellite virus often adversely affects the growth of helper virus.*

20.2. *Answer: (1) (1) RdRp for the RNA synthesis, (2) an endonuclease to cleave the multimeric RNA to a unit genome length, and (3) a RNA ligase to religate the RNA to a circle. (2) (1) Host RNA polymerase II is believed to replicate the viroid RNA, (2) and (3) Viroid-coded ribozyme confers both the endonuclease and ligase activity.*

20.3. *Answer: (1) No nucleic acid has been conclusively associated with prion infectivity; prion agent is resistant to ultraviolet radiation and nucleases. (2) The gene for prion protein is coded by the host chromosome, supporting that the host protein is responsible for the disease. (3) A knockout mouse lacking the PrP gene is resistant to prion infection, implicating that the protein–protein interaction is critical for the disease transmission.*

21.1. *Answer: (1) SARS-coronavirus: Human infection of bats' coronavirus. (2) Avian flu H5N1: Human infection of avian influenza virus. (3) 2009 H1N1 flu: Human infection of swine influenza virus.*

21.2. *Answer: (1) HIV: When HIV outbreaks occurred in the 1980s, its transmission appeared to be limited to homosexual males or intravenous drug users. It spread to most countries in a decade and the death toll rose to over 30 million. (2) WNV infection in U.S.A: The expansion of the mosquito vector's habitat to the Northern Hemisphere. (3) HAV infection in S. Korea: The decrease of the subclinically infected population due to improved sanitation.*

22.1. *Answer: (1) Higher mutation rate. (2) Structural features of HIV envelope spike. (3) Cell-to-cell spread. All above-described mechanisms allow the virus to evade the host's humoral immune response.*

22.2. *Answer: HIV reservoirs are the central obstacle to cure. HIV reservoirs represent the infected cell populations that enable the persistence of replication-competent HIV in patients treated with antiretroviral therapy regimens in the order of years. The HIV reservoir comprises both "latent HIV" and other as-yet incompletely defined sources of "persistent HIV" (eg, bone marrow hemopoietic stem cell). Latent HIV represents a quiescent, replication-competent provirus that exists within a long-lived population of resting T cells (ie, memory T cells) and that is capable of initiating new rounds of infection if therapy is interrupted.*

22.3. *Answer: See Journal club paper [N Engl J Med 370(10): 901–910, 2014].*

23.1. *Answer: (1) HBcAg, HBsAg. Anti-HBcAg. (2) anti-HCV antibodies (eg, core, E1, E2, and NS3, and NS5B). Note that anti-HBsAg antibodies are not detectable in hepatitis B carriers, while anti-HCV-E1, and -E2 antibodies are detectable in hepatitis C carriers. In other words, anti-HCV-E1, and -E2 antibodies do not effectively neutralize the HCV infection.*

23.2. *Answer: (1) HBV: Entecarvir (HBV RT), and tenofovir (HBV RT), (2) HCV: Boceprevir (HCV NS3 serine protease), telaprevir (HCV NS3 serine protease), and sofosbuvir (HCV NS5B RdRp).*

23.3. *Answer: (1) Similarities: HEV and Norovirus express full-length genomic and subgenomic RNAs. Differences: The 5' terminus of HEV is capped, while that of norovirus is linked to a viral protein, VPg. (2) Translation of HEV RNAs is cap-dependent and eIF4E-dependent, while that of norovirus is cap-independent and eIF4E-dependent. In norovirus, the VPg recruits eIF4E for translation initiation.*

24.1. *Answer: See* Fig. 24.3, Fig. 24.4 *and* Fig. 24.8.

24.2. *Answer: (1) HPV-16 E6 protein. The E6 protein forms a complex with E6AP to constitute the ubiquitin E3 ligase. The E3 ligase degrades p53 via ubiquitin-mediated proteolysis. (2) Ad E1B-55 kDa and E4 orf6 protein. These two adenoviral protein forms a complex to constitute the cullin-based ubiquitin E3 ligase. The E3 ligase degrades p53 via ubiquitin-mediated proteolysis. (see* Fig. 24.4 *and* Box 8.2*).*

24.3. *Answer: (1) Via insertional activation of proto-oncogene, as shown in* Fig. 24.16. *(2) Carry out Northern blot analysis of RNAs extracted from tumor by using the ALV probe. The RNA detectable by the ALV probe is likely to represent the chimeric mRNA (viral RNA + proto-oncogene RNA) generated by the insertional activation. Then, attempt to sequence the chimeric mRNA to determine which proto-oncogene is responsible for the viral oncogenesis.*

25.1. *Answer: (1) Inactivated polio vaccine (IPV) is the killed vaccine. Purified virus stock grown in animal cell culture is chemically inactivated by formaldehyde treatment. Intravenous administration is the disadvantage. (2) Oral polio vaccine (OPV) is the live attenuated vaccine. It is a combination of three attenuated poliovirus strains that have been attenuated by culturing in different animal cells. Its virulence is inactivated by serial passage of cultivation in different host cells. The oral administration is the advantage. The occasional emergence of revertants is the major concern.*

25.2. *Answer: (1) Similarities: Both killed inactivated and live attenuated vaccines are "split vaccines" that are produced by propagating in fertilized eggs. Both are trivalent or quadrivalent vaccines that are the combination of HA + NA antigens derived from three or four seasonal strains. (2) Differences: The killed inactivated vaccines are inactivated by formaldehyde following purification via density gradient. On the other hands, the seed strains of live attenuated vaccine are made by reverse genetics technology in that 8 plasmids transfection (ie, 6 from master strain + 2 from seasonal flu strain). The live vaccine is also purified via density gradient but not inactivated. The live flu vaccine is administered via nasal spray.*

25.3. *Answer: (1) HBV vaccine: HBsAg expression in yeast produces intracellular particles mimicking HBV subviral particles. Such HBsAg particles purified from yeast are immunogenic and used for HBV vaccine. (2) HPV vaccine: HPV capsid proteins (L1 + L2) of high-risk HPV genotypes (ie, HPV-16 and 18) are expressed in yeast or in insect cells as virus-like particle (VLP). Pros and cons: The advantage is the exclusion of potential contamination of unknown virus during the manufacturing process, because the viral antigens are expressed in heterologous systems such as yeasts. The drawback of subunit vaccine is the requirement of adjuvants to boost immune response.*

26.1. *Answer: Acyclovir is a nucleoside analog prodrug. It becomes active only after the conversion to the triphosphate form. The triphosphate form inhibits not only the viral DNA polymerase but also, to a lesser extent, cellular DNA polymerase, which could lead to cytotoxicity. Importantly, the first step of the conversion relies on the viral thymidine kinase, while the second step of the conversion relies on the host DNA polymerase. In other words, the conversion of acyclovir to a more toxic triphosphate form occurs only in the virus-infected cells, but not uninfected cells. The requirement of viral enzyme for the conversion confers the selectivity of acyclovir.*

26.2. *Answer: Lamivudine is a prodrug. It will act as a chain terminator following conversion to a lamivudine triphosphate in the cells.*

26.3. *Answer: (1) A time of addition experiment: This time of addition experiment allows you to determine the steps at which the hit compound acts by performing a time of addition experiment. One could determine how long the addition of a compound can be postponed before losing its antiviral activity. The target of an antiviral compound can be identified by comparing its relative position in the timescale to that of reference drugs. For instance, if the hit blocks entry, it will not inhibit viral replication, if it is added after the entry [see Nature Protocols 6:925–933 (2011)]. (2) The mapping of the viral genome, to which the drug-resistant mutants arise. Drug resistant mutants will emerge, if virus is propagated in the presence of the antiviral compound. Then, carry out nucleotide sequencing of the mutants and compare it with a wild-type. To prove that a certain mutation is responsible for the drug-resistant phenotype, one can make the same mutation in a wild-type and verify whether the drug-resistant phenotype is reproduced. In doing so, one could identify indirectly the target molecule of the antiviral compound.*

Glossary

Adjuvant An adjuvant (from Latin, *adiuvare:* to aid) is a pharmacological and/or immunological agent that modifies the effect of other agents.

Agonist It refers to a chemical that binds to some receptor of a cell and triggers a response by that cell. (cf., antagonist).

Amyloid It refers to the insoluble fibrous "protein aggregates" sharing specific structural traits.

Anaphylaxis It is a type of acute severe type I hypersensitivity allergic reaction. The term comes from the Greek word *ana* "against," and *phylaxis* "protection."

Antigenic drift It refers to a mechanism for variation that involves the accumulation of mutations within the genes that code for antibody-binding sites.

Antigenic shift It refers to the process by which two or more different strains of a virus combine to form a new subtype having a mixture of the surface antigens of the two or more original strains.

Antigen-presenting cell (APC) An immune cell that presents antigenic peptides (epitope) in the context of MHC to T lymphocytes. DC cell, macrophage, and B lymphocyte are said to be "professional APC."

APOBEC3G (apolipoprotein B editing enzyme) An RNA editing enzyme with cytidine deaminase activity.

Apoptosis It refers to the programmed cell death that may occur in multicellular organisms.

Arbovirus Arbovirus is a descriptive term applied to hundreds of predominantly RNA viruses that are transmitted by arthropods, notably mosquitoes and ticks. The word *arbovirus* is an acronym (Arthropod-Borne virus).

p14ARF (alternative reading frame) It refers to a tumor suppressor gene product, whose gene was discovered as an alternate reading frame (ARF) product of the CDKN2A locus, and is an inhibitor of Mdm2.

ATM (ataxia telangiectasia-mutated) A tumor suppressor gene identified as a mutated gene in ataxia telangiectasia. Ataxia telangiectasia is a rare, neurodegenerative, inherited disease causing severe disability. ATM, a serine/threonine protein kinase, is activated by double-strand break (DSB), acting as a sensor of the DNA damage.

ATR (ataxia telangiectasia and RAD3-related) ATM-related gene. It is a serine/threonine kinase that is involved in sensing DNA damage and activating the DNA damage checkpoint, leading to cell cycle arrest.

Bacteriophage A bacteriophage (informally, *phage*) is a virus that infects and replicates within bacteria. The term is derived from Greek word *phagein* "to eat."

Benign tumor A benign tumor is a mass of cells (tumor) that lacks the ability to invade neighboring tissue or metastasize.

Blood–brain barrier It refers to a highly selective permeability barrier that separates the circulating blood from the brain extracellular fluid (BECF) in the central nervous system (CNS).

Bovine spongiform encephalopathy (BSE) It is a fatal neurodegenerative disease (encephalopathy) in cattle that causes a spongy degeneration in brain and spinal cord.

Cancer Cancer is a group of diseases involving abnormal cell growth with the potential to invade or spread to other parts of the body. It is synonym of a malignant tumor.

Cap snatching It refers to a step in transcription process, which involves the cleavage of RNA fragments from the 5′ end of cellular mRNAs.

CAR (Coxsackie-Adenovirus Receptor) CAR is a type 1 membrane receptor with two Ig-like domain. CAR is an entry receptor for coxsackie type B3 as well as adenovirus type 5.

CARD (caspase recruitment domain) CARDs are "interaction motifs" found in a wide array of proteins, typically those involved in processes related to inflammation and apoptosis. These domains mediate the formation of larger protein complexes via direct interactions between individual CARDs.

Caveolins A family of integral membrane proteins which are the principal components of caveolae membranes.

CD4 A glycoprotein found on the surface of immune cells such as T helper cells, monocytes, macrophages, and dendritic cells. CD4 is best known as a cellular marker for T helper lymphocyte.

CD81 A ubiquitously expressed protein and a member of the tetraspanin superfamily.

Cell cycle Cell cycle is the series of events that take place in a cell leading to its division and duplication (replication) that produces two daughter cells.

Chemokine receptor It refers to cytokine receptors found on the surface of certain cells that interact with a type of cytokine called a chemokine. It belongs to the family of G-protein coupled receptor (GPCR) that has a 7 transmembrane (7TM) domain.

Cirrhosis Cirrhosis is a result of advanced liver disease. It is characterized by the replacement of liver tissue by fibrous connective tissue (scar tissue) and regenerative nodules. Cirrhosis (from Greek kirrhos "yellowish" and the common Greek suffix -sis meaning "condition").

***cis*-acting elements** It refers to the factors that are not diffusible (ie, the sequence element present in the plasmid).

Clathrin A protein that plays a major role in the formation of coated vesicles.

CPE (cytopathic effect) It refers to any pathological changes (or lesion) in the host cells that are caused by virus infection.

Cyclin-dependent kinases (Cdks) Cdks are a family of protein kinases first discovered for their role in regulating the cell cycle. Cdk binds a regulatory protein called a cyclin for the activation.

Cystic fibrosis Cystic fibrosis is a genetic disorder that affects the lungs. Affected individuals suffer difficulty breathing and coughing up sputum as a result of frequent lung infections. It is caused by the presence of mutations in both copies of the gene for the protein cystic fibrosis transmembrane conductance regulator (CFTR).

Cytotoxic T cell (CTL) A subset of T lymphocyte that expresses CD8 marker and is capable of killing target cells.

DCAF (DDB1-Cul4A-associated WD40 domain protein) It serves as a substrate receptor, linking DDB1 to Vpr. It is also called VprBP (Vpr binding protein).

DDB1 (DNA damage binding protein) A large subunit of DNA damage binding protein, which is a heterodimer composed of large (DDB1) and small subunit (DDB2). This protein functions in nucleotide excision repair.

DNA damage response (DDR) It refers to the signaling pathway that is induced by DNA damage (see Box 6.2).

Double membrane vesicle (DMV) It refers to double membrane structures formed as protrusions from the ER membrane into the cytosol, frequently connected to the ER membrane via a neck-like structure.

Dynamin A GTPase primarily involved in scission of newly formed vesicle during endocytosis.

E2F A transcription factor that was discovered as a DNA-binding protein of adenovirus E2 gene promoter. It is a representative S-phase specific transcription factor.

E6AP (E6-associated protein) A host factor first identified as HPV E6 binding protein. It is now classified as a member of HECT (homologous to E6AP carboxyl-terminus)-type ubiquitin E3 ligase family.

ECL (enhanced chemiluminescence) ECL is a common technique for a variety of detection assays in biology. An antibody that is conjugated to an enzyme (either HRP or AP) is used. The enzyme catalyzes the conversion of the enhanced chemiluminescent substrate into a sensitized reagent, which emits light. Enhanced chemiluminescence allows detection of minute quantities of a biomolecule.

ELISA (enzyme-linked immunosorbent assay) ELISA is the most frequently used diagnostic tool for virus detection that combines the exquisite specificity of antigen–antibody binding and the sensitivity of enzyme reaction.

Epidemics It refers to the rapid spread of infectious disease to a large number of people in a given population within a short period of time. The period is usually 2 weeks or less.

Epidemiology It refers to a discipline that studies the patterns, causes, and effects of disease conditions in defined populations. It is derived from Greek *epi*, meaning "upon," *demos*, meaning "people," and *logos*, meaning "study."

Epithelium A type of tissues that cover the surface of various organs in human body.

Episome A DNA that is stably present in the cell, excluding chromosomal DNA. The term "epi" is derived from Greek word for "above."

Epitope An epitope represents the part of an antigen that is recognized by the adaptive immune system, specifically by antibodies, B cells, or T cells.

Epstein-Barr virus (EBV) A human gamma-herpesvirus that is associated with Burkitt's lymphoma. The virus was named after the two scientists who discovered it: Dr Epstein and Dr Barr (see Box 9.4).

ESCRT (endosomal sorting complex required for transport) ESCRT machinery is made up of cytosolic protein complexes referred to as ESCRT-0, -I,-II, and -III. Together with a number of accessory proteins, these ESCRT complexes enable a unique mode of membrane remodeling that results in membranes bending/budding away from the cytoplasm.

Exocytosis The process in which a cell directs the contents of secretory vesicles out of the cell membrane into the extracellular space.

Extracellular matrix (ECM) The extracellular part of animal tissue that usually provides structural support to the animal cells. The extracellular matrix is the defining feature of connective tissue in animals, which is largely composed of fibrous protein, collagen, and glycoaminoglycan (GAG).

Fibrosis Fibrosis is the formation of excess fibrous connective tissue in an organ or tissue. Fibrosis can be used to describe the pathological state of excess deposition of fibrous tissue, as well as the process of connective tissue deposition in healing.

Food and drug administration (FDA) The FDA is the authority, which approves new prescription drugs, vaccines, and biopharmaceuticals for medical use. The FDA is one of the US federal executive departments.

Foot-and-mouth disease An animal disease that is caused by foot-and-mouth disease virus (FMDV), that belongs to the family Picornaviridae.

Freund's complete adjuvant (FCA) Freund's adjuvant is a solution of antigen emulsified in mineral oil and used as an immunopotentiator (booster). The complete form, Freund's Complete Adjuvant (FCA) is composed of inactivated and dried mycobacteria (usually *M. tuberculosis*).

Fusion peptide It refers to a hydrophobic peptide domain that triggers membrane fusion (see Box 3.2).

Genetic reassortment It refers to a kind of genetic recombination occurring during assembly of segmented genome.

Giant virus A novel virus found in ameba, which is bigger than any other known viruses (400 nm in diameter, 1200 kb in genome).

Glycan It refers to the carbohydrate moiety of glycoproteins.

HAART (highly active antiretroviral therapy) It refers to the antiviral therapy, in which two or three antiviral drugs are combined and given as a mixture.

Helper T-lymphocyte (Th) A subset of $CD4^+$ T lymphocyte that helps B or $CD8^+$ T-lymphocyte to proliferate and differentiate to effector cells.

Hemagglutinin (HA) Hemagglutinin refers to a substance that causes red blood cells to agglutinate or to clump together. This process is called hemagglutination.

Hematopoietic stem cells (HSC) It refers to the blood cells that give rise to all other blood cells: myeloid (monocytes, macrophage, etc.) and lymphoid lineages (B lymphocyte, T lymphocyte, and NK cell).

Heparin sulfate proteoglycan (HSPG) A kind of proteoglycans present abundantly on cell surface, in which heparin sulfate represents the glycan moiety.

Hepatitis It refers to a medical condition defined by the inflammation of liver and characterized by the presence of inflammatory cells in liver tissue. The term *hepa* is derived from the Greek word for "liver" and the term *titis* is derived from the Greek word for "inflammation."

Herd immunity It is a form of indirect protection from infectious disease that occurs when a large percentage of a population has become immune to an infection, thereby providing a measure of protection for individuals who are not immune.

HVEM (herpes virus entry mediator) It is a membrane protein that belongs to TNF-α superfamily.

IFI16 (interferon gamma-inducible protein 16) It is a type of pattern recognition receptor (PRR) that recognizes the DNA genome of DNA viruses. It is also known as a "nuclear DNA sensor."

Immunological memory It refers to the function of memory lymphocytes that are differentiated from the activated lymphocytes, and can swiftly respond to the antigens.

Immunological synapse An immunological synapse is the interface between an antigen-presenting cell or target cell and a lymphocyte such as an effector T cell, which is termed as an analogy to neural synapse.

Integrin Integrins are transmembrane proteins that mediate the attachment between a cell and its surroundings, such as other cells or the extracellular matrix (ECM).

Interferons Interferons are a group of cytokines, which trigger the induction of a broad array of antiviral proteins. Interferons are named for their ability to "interfere" with viral replication by protecting cells from virus infections.

Koch's postulates It refers to a set of criteria that Koch developed to establish a causative relationship between a microbe and a disease. Koch applied the postulates to describe the etiology of cholera and tuberculosis.

Late domain A peptide motif (four amino acids), that involves in the budding of enveloped viruses. It is composed of four amino acids such as "PTAP" or "PPXY" residues.

Lipid droplet (LD) Lipid-rich cellular organelles which regulate the storage and hydrolysis of neutral lipids. They are found largely in adipose tissue. They also serve as reservoirs of cholesterol and acyl-glycerols for membrane formation and maintenance.

Low density lipoprotein (LDL) It refers to a nanoparticle of 22 nm diameter composed of phospholipid, cholesterol, and apolipoproteins.

Macropinocytosis An endocytic mechanism normally involved in fluid uptake.

Malignant tumor A malignant tumor, also known as cancer, is a group of diseases involving abnormal cell growth with the potential to invade or spread to other parts of the body.

MAPK signaling (mitogen-activated protein kinase) A signal transduction pathway that plays a role in cell proliferation. Also known as the Ras-Raf-MEK-ERK pathway.

MAVS (mitochondrial antiviral signaling) It was discovered as a mitochondrial protein essential for antiviral signaling (ie, IFN induction). It is also termed as IPS (interferon promoter stimulator) or VISA or Cardif.

MDA5 (melanoma differentiation associated gene) It serves as an RNA sensor for the detection of viral RNAs and belongs to the DEAD-box RNA helicase family. It was first discovered as the melanoma differentiation associated gene, as the name implies.

MHC (major histocompatibility complex) MHC molecule displays an oligopeptide, called epitope, of antigens on cell surface of antigen-presenting cell (APC) such as DCs and macrophages.

MicroRNA A kind of short RNA (ie, 20–22 nt in length) found in eukaryotic cells, that regulates mRNA translation and mRNA stability.

MMR vaccine MMR vaccine is an immunization against measles, mumps, and rubella. It is a mixture of live attenuated viruses of the three diseases, administered via injection.

MOI (multiplicity of infection) It refers to as the number of virus particles imposed to one cell.

Monoclonal antibodies (Mab) It refers to monospecific antibodies that are made by identical immune cells that are all clones of a unique parent cell. Monoclonal antibodies have monovalent affinity, in that they bind to the same epitope.

MRN (Mre11-Rad50-Nbs1) A trimeric complex involved in DSB repair that senses the DNA damage (Box 8.1). It also exhibits endonuclease and exonuclease activity.

Mucosal immunity It refers to an immune response pertaining to the mucous membrane, which is distinct in that IgA, instead of IgG, is the type of immunoglobulin that acts.

Mucous membrane The mucous membrane (or mucosa) is the lining covered in epithelium, which is involved in absorption and secretion. It lines cavities that are exposed to the external environment and internal organs. The sticky and thick fluid secreted by the mucous membranes termed "mucus" plays a critical role in eliminating invading microbes.

Multivesicular bodies (MVBs) It refers to an intracellular structure that is generated by the inward vesiculation in late endosomes.

NCR (noncoding region) It refers to the region of the genome in which no protein is encoded.
NES (nuclear export signal) It refers to a peptide motif encoded by nuclear export proteins that is essential for nuclear export.
Neuraminidase An enzyme that cleaves a sialic acid of glycan.
NF-kB signaling A signal transduction pathway that plays a key role in regulating the immune response and cell survival.
Nonpermissive host A host that permits only early phase of the virus life cycle, but not late phase of virus life cycle or viral progeny production.
NTCP (sodium taurocholate cotransporting polypeptide) It is an integral membrane glycoprotein that participates in the enterohepatic circulation of bile acids. It is also an entry receptor necessary for hepatitis B virus infection.
Nuclear domain 10 (ND10) A small proteineous subnuclear structure, which is composed of multiple factors including PML (promyelocytic leukemia protein) and Sp100. It is also called PML nuclear bodies.
Nucleocapsid It refers a viral capsid that is associated with the viral genome.
2′5′ OAS (oligoadenylate synthetase) An enzyme that synthesizes the 2′5′ oligoadenylate (2–5A).
Oncogene An oncogene is a gene that has the potential to cause cancer. In tumor cells, the oncogenes are often mutated (activated) or expressed at high levels.
Opportunistic infection An opportunistic infection is an infection caused by pathogens, particularly "opportunistic pathogens" that usually do not cause disease in a healthy host.
Opportunistic pathogens Pathogens that usually do not cause disease in a healthy host, one with a healthy immune system, but do cause disease in compromised immune system.
Outbreak In epidemiology, an outbreak is an occurrence of disease greater than expected at a particular time and place. Outbreaks may also refer to epidemics which affect a particular region in a country or a group of countries.
p53 p53 is one of the most important as tumor suppressor gene. It was discovered as a T-antigen binding protein with a molecular mass of 53 kDa.
Packaging signal A sequence element in the viral genome that is essential for the genome packaging.
Pandemic An epidemiology term that refers to epidemic that affects multiple continents.
PCNA (proliferating cell nuclear antigen) PCNA was discovered as a nuclear antigen in dividing cell, as its name implies. It acts as a sliding clamp at the replication fork, leading to increase of the "processivity" of DNA polymerase.
peg-IFN It is an abbreviation for the pegylated interferon. Pegylation, the process of covalent attachment of polyethylene glycol (PEG) polymer chains, prolongs its circulatory time by reducing renal clearance, thereby improving the efficacy of drugs.
Phosphatidyl serine A kind of phospholipid, which is a major constituent of membrane lipid.
PKR (protein kinase RNA-activated) A serine/threonine kinase that is activated by double-stranded (dsRNA).
Plaque It refers to an area of empty hole in the monolayer of cells in plate, resulting from the cell lysis induced by virus infection.
Polarity It refers to the strandness of a single-strand DNA or RNA. The strand that corresponds to that of mRNA is defined as "plus" or "positive." The strand that is complementary to mRNA is defined as "minus" or "negative."
Poliomyelitis The term "poliomyelitis" is derived from Greek word for "gray"-*polio* + for "marrow"-*myelos* + "inflammation"-*titis*. The term implies paralytic poliomyelitis, resulted from destruction of motor neurons within the spinal cord in the CNS.
Polyprotein It refers to a large protein that is later processed to multiple functional proteins.
5′-pppRNA A characteristic nucleotide present at the 5′ end of the viral RNA genome such as influenza virus. Note that cellular mRNA has a cap structure at the 5′ end, while tRNA and rRNA have a monophosphate at the 5′ end.
Prion The word *prion*, is derived from the words *protein* and *infection*, in reference to a prion's ability to self-propagate and transmit its conformation to other prions.
Processivity It refers to the extent of DNA or RNA synthesis per engagement of RNA/DNA polymerase to template.
Prodrug It refers to a medication that is initially administered to the body in an inactive (or less than fully active) form, and then becomes converted to its active form by the normal metabolic processes of the body. Prodrugs can be used to improve how the intended drug is absorbed, distributed, metabolized, and excreted (ADME).
Protein-priming It refers to to RNA synthesis that is initiated by protein primer.
Protomer It refers to a structural subunit of poliovirus capsid. It is also called 5S structural unit, based on its sedimentation coefficient (ie, Svedberg sedimentation coefficient).
Proto-oncogene It refers to a normal gene that can become an oncogene due to mutations or increased expression.
Provirus It refers to the retroviral DNA integrated into the chromosomal DNA.
PRR (pattern recognition receptor) It refers to proteins expressed by cells of the innate immune system to identify pathogen-associated molecular patterns (PAMPs), which are associated with microbial pathogens. Toll-like receptor and RIG-I are representatives.
Pseudotyping It refers to the process of producing viruses or viral vectors in combination with foreign viral envelope proteins. The result is a pseudotyped virus particle. With this method, the foreign viral envelope proteins can be used to alter host tropism or an increased/decreased stability of the virus particles.
Quarantine Quarantine is used to separate and restrict the movement of people who may have been exposed to a communicable disease in order to monitor their health. The word came from an Italian word (17th-century Venetian) "quaranta," meaning 40. It is the number of days for which ships were required to be isolated before passengers and crew could go ashore during the Black Death plague epidemic.
Rabies It refers to a viral disease that causes acute encephalitis in animals. Rabies virus belongs to a family Rhabdoviridae (see chapter: Rhabdovirus). The rabies is a Latin word for "madness."
Rb (retinoblastoma) A tumor suppressor gene that was first identified as gene that was defective in retinoblastoma.

RdRp RNA-dependent RNA polymerase.

Regulatory T cells (Treg) A subset of T lymphocytes that suppress immune responses of other cells. This is an important "self-check" built in the immune system to prevent excessive reactions. Regulatory T cells come in many forms with the most well-understood being those that express CD4, CD25, and Foxp3.

Replication complex (RC) It refers to a subcellular site, where viral RNA replication is confined in HCV infected cells. It is also referred to as a "replication factory."

Replication origin (Ori) A *cis*-acting element where the DNA synthesis begins.

Replicon It refers to a plasmid construct that can induce the viral genome replication, when transfected into cells. A replicon in genetics is a region of DNA or RNA that replicates from a single origin of replication.

Replisome A complex molecular machine that carries out replication of DNA.

Reverse genetic A genetic method to examine phenotypic changes caused by experimental mutagenesis. This methodology is a reverse of classical genetic, which is to examine genetic changes associated with phenotype.

RIG-I (retinoic acid-inducible gene) It serves as an RNA sensor for the detection of viral RNAs and belongs to the DEAD-box RNA helicase family. It was first discovered as the retinoic acid-inducible gene, as the name implies.

RING (really interesting new gene) A protein motif found in RING finger family proteins.

RNase L (for latent) RNase that is activated by 2−5A. It is also called 2−5A-dependent ribonuclease.

RNP (ribonucleoprotein) complex It refers to a molecular complex that is composed of RNA and protein. vRNP refers to the viral nucleocapsid.

ROS (reactive oxygen species) It refers to chemically reactive molecules containing oxygen. Examples include oxygen ions and peroxides. ROS form as natural byproducts of the normal metabolism of oxygen and have important roles in cell signaling and homeostasis.

RSV (respiratory syncytial virus) RSV belongs to the family Paramyxoviridae (see chapter: Other Negative-Strand RNA Viruses).

SAMHD1 (SAM domain-and HD domain-containing protein 1) It is an enzyme that exhibits phosphohydrolase activity, converting nucleotide triphosphates to triphosphate and a nucleoside (ie, nucleotides without a phosphate group).

Satellite virus It refers to a virus, which depends on the host virus as a helper.

Scrapie It refers to a fatal, degenerative disease that affects the nervous systems of sheep and goats. The name scrapie is derived from one of the clinical signs of the condition, wherein affected animals will compulsively scrape off their fleeces against rocks, trees, or fences.

SDS−PAGE Polyacrylamide gel electrophoresis (PAGE) is a technique widely used to separate biological macromolecules, usually proteins or nucleic acids, according to their electrophoretic mobility. SDS−PAGE is a method to separate proteins following denaturation by SDS, a detergent.

Seasonal flu A flu epidemic that comes regularly during winter.

Sedimentation coefficient The unit of sedimentation coefficient represents the size of the particle that precipitates upon centrifugation. It is also called Svedberg number (S).

Sialic acid It is a generic term that refers to the N-acetyl neuraminic acid, an amino sugar, which is terminally linked to glycans on the cell membrane.

Smallpox A fatal infectious disease of human that is caused by a pox virus.

Split vaccine It refers to the vaccine preparation process, that involves fractionation (split) of immunogen (ie, virus particles).

Stuttering It refers to template-dependent polyadenylation that occurs in VSV as well as influenza virus (see Fig. 14.5).

Subgenomic RNA It refers to a viral RNA that is smaller than the full-length viral RNA.

Subtype It refers to a rank below "species" in taxonomic rank. According to rank-based classification, kingdoms are divided into phyla, and then, in turn, classes, orders, families, and into genera (singular: genus), and species.

Subviral agent It refers to virus-like transmissible agents, which do not comply with the classical definition of "virus."

SVR (sustained virologic response) SVR refers to "no detection of HCV RNA" in the serum at 24 weeks after cessation of antiviral therapy.

Syncytium A syncytium (pl., syncytia) refers to a multinucleated cell that can result from multiple cell fusions of uninuclear cells.

T-antigen It was discovered as a tumor-specific antigen (ie, tumor antigen) in SV40-induced tumors, thus named as "T-antigen."

TAP (transporter associated with antigen presentation) A membrane protein located in endoplasmic reticulum that functions to uptake peptides into lumen side.

T cell exhaustion A phenomena that leads to dysfunction of activated T lymphocyte.

$TCID_{50}$ (tissue culture infectious dose) It refers to the dilution fold of virus stock that could lead to CPE in 50% of wells seeded. It is a unit of virus titer.

TDP2 (5′-tyrosyl-DNA phosphodiesterase 2) An cellular enzyme that is known to cleave a tyrosine-DNA phosphodiester linkage found in topoisomerase II-DNA adducts.

Tetherin It is an interferon-induced membrane protein that inhibits the release of enveloped virus particles from infected cells.

Tight junction Tight junctions are the closely associated areas of two cells whose membranes join together forming a virtually impermeable barrier to fluid. They help to maintain the polarity of cells by preventing the lateral diffusion of integral membrane proteins between the apical and lateral/basal surfaces, allowing the maintenance of specialized functions of each surface.

Toll A gene that was discovered in *Drosophila* by Nusslein-Volhard (a Nobel laureate in 1995) in 1985. Toll was derived from her acclamation in German, "Das war ja toll!" (That is great!).

***trans*-acting factors** It refers to factors that can be diffusible (ie, proteins).

Transfection It refers to the process of experimentally introducing nucleic acids into cells. The term is often used for nonviral methods in eukaryotic cells. The word *transfection* is a blend of *trans-* and *infection*.

Transformation (malignant) Transformation is the process by which cells acquire the properties of cancer.

Transgene It refers to the genes of interest that are to be transferred or to be expressed in the target cell.

Trianglulation number It refers to the number of subunits that constitute a triangular facet of an icosahedral capsid.

TRIM (TRIpartite Motif) Tripartite motif represents three structural motifs composed of RING finger domain, B-box domain, and coiled-coil domain. TRIM family is a large protein family that constitutes over 50 members.

Tropism It refers to the cell specificity of viral infection. The term "tropism" is derived from the Greek word for "a turning"-*tropos*, indicating growth or turning movement of a biological organism.

Tumor Tumor, referred to as a neoplasm (from Ancient Greek—*neo*—"new" and *plasma* "formation, creation"), is an abnormal growth of tissue.

Tumor suppressor gene Tumor suppressor gene, or "antioncogene," is a gene that protects a cell from becoming cancerous.

VA (virus-associated) RNA A noncoding RNA transcribed by RNA polymerase III.

Vaccinia virus A pox virus strain that has been used to produce smallpox vaccine, as its name was derived from "vaccine."

Vector It refers to vehicles for the gene delivery. It could refer to either a plasmid DNA for the gene expression (ie, plasmid vector) or a recombinant virus.

Vector (epidemiology) In epidemiology, a vector is any agent (person, animal, or microorganism) that carries and transmits an infectious pathogen into another living organism.

Vhs (virion host shutof) A tegument protein of HSV-1, which blocks host immune response by degradation of cellular mRNAs via its endonuclease activity.

Viral quasispecies It refers to a group of closely related viruses which are distinguished by a high number of mutations.

Virome It refers to the collection of all the viruses (or viral genome) in a given animal.

VLP (virus-like particle) It refers to viral particles devoid of the viral genome.

vRNA (virion RNA) It refers to an RNA encapsidated inside a virion.

vRNP It refers to the viral nucleoprotein complex or nucleocapsid.

West Nile Virus (WNV) A mosquito-borne zoonotic arbovirus belonging to the genus *Flavivirus* in the family Flaviviridae.

Wnt signaling It refers to a signaling pathway, which has been implicated in colon cancer. The term "Wnt" is a combined word between "*Wingless*" and "*int*," as it was independently discovered as a gene associated with wingless phenotype of *Drosophila* and as a gene activated by the integration of a mouse retrovirus (Wingless/Int).

Zika virus A mosquito-borne flavivirus that was first discovered in 1947 in Zika forest in Uganda.

Zoonotic virus A virus that is transmitted between species (sometimes by a vector) from animals to human.

List of Credits

CHAPTER 1

Figure 1.1 http://en.wikipedia.org/wiki/Polio/
Figure 1.2A http://en.wikipedia.org/wiki/Dmitri_Ivanovsky and http://en.wikipedia.org/wiki/Martinus_Beijerinck
Figure 1.2B http://commons.wikimedia.org/wiki/Category:Martinus_Beijerinck/
Figure 1.9 Viralzone (http://viralzone.expasy.org/all_by_species/254.html)
Figure 1.10A http://en.wikipedia.org/wiki/Tobacco_mosaic_virus#
Figure 1.10B http://commons.wikimedia.org/wiki/File:Tobacco_mosaic_virus_symptoms_orchid.jpg#
Box 1.2 https://en.wikipedia.org/wiki/David_Baltimore
Box 1.3 Adapted from Duffy, S., et al., 2008. Nat. Rev. Genet. 9, 267–276, with permission.
Box 1.4 http://commons.wikimedia.org/wiki/File:Semper_Augustus_Tulip_17th_century.jpg#
Box 1.5A Adapted from La Scola, B., et al., 2008. The virophage as a unique parasite of the giant mimivirus. Nature 455, 100–104, with permission.
Box 1.5B Courtesy of Viralzone.

CHAPTER 2

Figure 2.1A Courtesy of Dr Fred Murphy, Sylvia Whitfield/CDC.
Figure 2.1B http://en.wikipedia.org/wiki/Poliovirus#
Figure 2.1C https://en.wikipedia.org/wiki/1918_flu_pandemic
Figure 2.1D http://commons.wikimedia.org/wiki/File:YellowFeverVirus.jpg/
Figure 2.3 http://commons.wikimedia.org/wiki/File:TMV_structure_full.png#
Figure 2.8A http://commons.wikimedia.org/wiki/File:Multiple_rotavirus_particles.jpg#
Figure 2.8B http://commons.wikimedia.org/wiki/File:HeLa-V.jpg#/media/File:HeLa-V.jpg
Figure 2.9 Courtesy of Michael Rossmann, Purdue University.
Figure 2.10 Courtesy of Michael Rossmann, Purdue University.
Figure 2.11 Courtesy of Carrillo-Tripp, M., Shepherd, C.M., Borelli, Ian A., Venkataraman, S., Lander, G., Natarajan, P., Johnson, J. E., Brooks, III, C. L., Reddy, V. S., 2009. VIPERdb^2: An enhanced and web API enabled relational database for structural virology. Nucleic Acid Res. 37, D436–D442.

CHAPTER 3

Figure 3.5 Adapted from Marsh, M., et al., 2006. Virus Entry: Open Sesame. Cell 124, 729–740, with permission.
Figure 3.6 Adapted from Sodeik, B., 2000. Mechanisms of viral transport in the cytoplasm. Trends Microbiol. 8, 465–472, with permission.
Box 3.2 Adapted from Vigant, F., et al., 2015. Broad-spectrum antivirals against viral fusion. Nat. Rev. Microbiol. 13, 426–437, with permission.
Box 3.3B Adapted from Freed, E., 2004. HIV-1 and the host cell: an intimate association. Trends Microbiol. 12, 170–177, with permission.
Box 3.4 Adapted from Sattentau, S., 2008. Avoiding the void: cell-to-cell spread of human viruses. Nat. Rev. Microbiol. 6, 815, with permission.

CHAPTER 4

Figure 4.2B Courtesy of EnCor Biotechnology Inc.
Figure 4.3 http://commons.wikimedia.org/wiki/File:Western_Blot_results_for_HIV_test.jpg#

Figure 4.5A http://en.wikipedia.org/wiki/Pcr
Figure 4.5B http://commons.wikimedia.org/wiki/File:PCR_masina_kasutamine.jpg#
Figure 4.12B Adapted from Figure 1.4 of Chapter 1, Introduction in Cann, A., 2016. Principles of Molecular Virology, sixth ed., p.14, with permission of Elsevier.
Figure 4.14 Courtesy of James Gathany/CDC.
Box 4.2 http://en.wikipedia.org/wiki/Dulbecco

CHAPTER 5

Figure 5.6 Adapted from Rehwinkel. L., et al., 2013. Targeting the viral Achilles's heel: recognition of 5'-triphosphate RNA in innate anti-viral defence. Curr. Opin. Microbiol. 16, 485–492, with permission.
Figure 5.8 Adapted from Sadler, A.J., et al. 2008. Interferon-inducible antiviral effectors. Nat. Rev. Immunol. 8, 559–568, with permission.
Figure 5.9 Adapted from Sadler, A.J., et al., 2008. Interferon-inducible antiviral effectors. Nat. Rev. Immunol. 8, 559–568, with permission.
Figure 5.14 Adapted from Villadangos, J.A., et al., 2007. Intrinsic and cooperative antigen-presenting functions of dendritic-cell subsets in vivo. Nat. Rev. Immunol. 7, 543, with permission.
Figure 5.17 Adapted from Marasco, W., et al., 2007. The growth and potential of human antiviral monoclonal antibody therapeutics. Nat. Biotechnol. 25, 1421–1434, with permission.
Box 5.1A Courtesy of Nobel Foundation.
Box 5.1B, C A photo from Nature 2003, Nature 423, 237, with permission.
Box 5.2 A photo from Interferon discovery and ferret flu, Interview by Alison Abbot, 2007. Nature 449, 126, with permission.
Box 5.3 http://en.wikipedia.org/wiki/Haematopoiesis

CHAPTER 6

Figure 6.1A Courtesy of Viralzone.
Figure 6.1B Courtesy of Dr Erskine Palmer/CDC.
Figure 6.8 Adapted from Sowd, G.A., et al., 2012. A wolf in sheep's clothing: SV40 co-opts host genome maintenance proteins to replicate viral DNA. PLoS Pathog. 8:e1002994.
Figure 6.10 Adapted from De Caprio, J., et al., 1998. SV40 large tumor antigens forms a specific complex with the product of the retinoblastoma susceptibility gene. Cell 54, 275–283, with permission.
Box 6.1 Adapted from Moore, S., Chang, Y., 2014. The conundrum of causality in tumor virology: the cases of KSHV and MCV. Semin. Cancer Biol. 26, 4–12, with permission.
Box 6.2A Adapted from Cimprich, K., et al., 2008. ATR: an essential regulator of genome integrity. Nat. Rev. MCB 9, 616, with permission.

CHAPTER 7

Figure 7.1A Courtesy of Viralzone.
Figure 7.1B http://commons.wikimedia.org/wiki/File:Papilloma_Virus_(HPV)_EM.jpg#
Figure 7.3 Adapted from Moody, C.A. et al., 2010. Nature Review Cancer 10, 550–560, with permission.
Figure 7.6 Adapted from You, J., et al., 2004. Interaction of the bovine papillomavirus E2 protein with Brd4 tethers the viral DNA to host mitotic chromosomes. Cell 117, 349, with permission.
Box 7.1 Courtesy of the Nobel Foundation 2008.
Box 7.2 http://commons.wikimedia.org/wiki/File:Gardasil_vaccine_and_box_new.jpg#
Box 7.4A http://en.wikipedia.org/wiki/File:Henrietta_Lacks_(1920-1951).jpg#
Box 7.4B http://en.wikipedia.org/wiki/HeLa#/media/File:HeLa-III.jpg
Box 7.5 http://commons.wikimedia.org/wiki/File:Cases_of_HPV_cancers_graph.png#

CHAPTER 8

Figure 8.1B Adapted from Nemerow, G., et al., 2009. Insights into adenovirus host cell interactions from structural studies. Virology 384, 380–388, with permission.
Figure 8.5 Adapted from Nemerow, G., et al., 2009. Insights into adenovirus host cell interactions from structural studies. Virology 384, 380–388, with permission.
Box 8.1 Adapted from Stracker, T., et al., 2009. Adenovirus oncoproteins inactivates the Mre11-Rad50-NBS1 DNA repair complex. Nature 418, 348–352, with permission.
Box 8.3 http://commons.wikimedia.org/wiki/File:HEK_293_cells_grown_in_tissue_culture_medium.jpg#

CHAPTER 9

Figure 9.1A Courtesy of CDC.
Figure 9.1B http://en.wikipedia.org/wiki/Aciclovir#/media/File:Guanosine-acyclovir-comparison.png
Figure 9.2 Courtesy of Dr Prashant Desai, The Johns Hopkins University.
Figure 9.11 Adapted from Johnson, D., 2011. Herpesviruses remodel host membranes for virus egress. Nat. Rev. Microbiol. 9, 382–394, with permission.

CHAPTER 10

Figure 10.1A Courtesy of Viralzone.
Figure 10.1B Courtesy of Dr Cornelia Büchen-Osmond, Columbia University.
Figure 10.5 Courtesy of CDC.
Figure 10.6A Courtesy of Viralzone.
Figure 10.6B http://en.wikipedia.org/wiki/Smallpox#
Box 10.1 https://en.wikipedia.org/wiki/Myxomatosis#

CHAPTER 11

Figure 11.1 Courtesy of Viralzone.

CHAPTER 12

Figure 12.1A Courtesy of Viralzone.
Figure 12.1B Courtesy of Center for the Study of Hepatitis C, The Rockefeller University.
Figure 12.5 Adapted from Lindenbach, B., et al., 2005. Unraveling hepatitis C virus replication from genome to function. Nature 436, 933, with permission.
Figure 12.6 Adapted from Lindenbach, B., 2011. Understanding how hepatitis C virus builds its unctuous home. Cell Host & Microbe 9, 1–2, with permission.
Figure 12.7 Adapted from Lemon, S., et al., 2012. Gastroenterology 142, 1274–1278, with permission.
Box 12.1 http://en.wikipedia.org/wiki/Dengue_fever#/media/File:Dengue06.png

CHAPTER 13

Figure 13.1A Courtesy of Viralzone.
Figure 13.1B http://en.wikipedia.org/wiki/Norovirus#/media/File:Norwalk.jpg
Figure 13.2 Courtesy of Viralzone.
Figure 13.3 Courtesy of Viralzone.
Figure 13.4 Courtesy of Viralzone.
Figure 13.5A Adapted from Millet, J., et al., 2015. Host cell proteases: critical determinants of coronavirus tropism. Virus Res. 202, 120–134, with permission.
Figure 13.5B Courtesy of Dr Fred Murphy/CDC.
Figure 13.6 Courtesy of Viralzone.

CHAPTER 14

Figure 14.1B Courtesy of CDC.
Figure 14.2 Courtesy of Viralzone.
Figure 14.6 Adapted from Jayakar, H., et al., 2004. Rhabdovirus assembly and budding. Virus Res. 106, 117−132, with permission.

CHAPTER 15

Figure 15.1A Adapted from Medina, R., et al., 2011. Influenza Z viruses: new research developments. Nat. Rev. Microbiol. 9, 590, with permission.
Figure 15.1B http://commons.wikimedia.org/wiki/File:H1N1_navbox.jpg#
Box 15.1 https://en.wikipedia.org/wiki/Oseltamivir
Box 15.2 Adapted from De Clercq, E., 2006. Antiviral agents active against influenza A viruses. Nat. Rev. Drug Discov. 5, 1015, with permission.
Box 15.4A Courtesy of James Gathany/CDC.
Box 15.4B https://en.wikipedia.org/wiki/1918_flu_pandemic

CHAPTER 16

Figure 16.1 Courtesy of Viralzone.
Figure 16.2 Courtesy of Viralzone.
Figure 16.4A Courtesy of Viralzone.
Figure 16.4B Courtesy of Dr Cynthia Goldsmith/CDC.
Figure 16.5 Courtesy of Viralzone.
Figure 16.6A Courtesy of Viralzone.
Figure 16.6B http://en.wikipedia.org/wiki/Hantavirus_hemorrhagic_fever_with_renal_syndrome/
Figure 16.9A Courtesy of Viralzone.
Figure 16.9B Courtesy of CDC.
Figure 16.10 Courtesy of Viralzone.
Figure 16.11A Courtesy of Viralzone.
Figure 16.11B Courtesy of CDC.
Figure 16.12 Courtesy of Viralzone.
Box 16.1 Adapted from Dong, H., et al., 2008. Flavivirus methyltransferase: a novel antiviral target. Antiviral Res. 80, 1−10, with permission.

CHAPTER 17

Figure 17.4 Adapted from Hatziioannou, T., Evans, D.T., 2012. Animal models for HIV/AIDS research. Nat. Rev. Microbiol. 10, 852, with permission.
Figure 17.10 Adapted from Freed, E., 2015. HIV-1 assembly, release and maturation. Nat. Rev. Microbiol. 13, 484−496, with permission.
Figure 17.11 Courtesy of Drs A. Harrison and P. Feorino/CDC.
Figure 17.13 Adapted from Rice, A.P., 2010. The HIV-1 Tat team gets biggers. Cell Host Microbe 7, 179−181, with permission.
Figure 17.15 Adapted from Simon, V., et al., 2015. Intrinsic host restrictions to HIV-1 and mechanisms of viral escape. Nat. Immunol 16, 546−553, with permission.
Figure 17.17 Adapted from Simon, V., et al., 2015. Intrinsic host restrictions to HIV-1 and mechanisms of viral escape. Nat. Immunol. 16, 546−553, with permission.
Box 17.1 Courtesy of The Nobel Foundation 1975.
Box 17.3 Courtesy of The Nobel Foundation 2008.

CHAPTER 18

Figure 18.10 Adapted from Baumert, T., et al., 2014. Entry of hepatitis B and C viruses-recent progress and future impact. Curr. Opin. Virol. 4, 58−65, with permission.

CHAPTER 20

Figure 20.1 Courtesy of Dr Cornelia Büchen-Osmond, Columbia University.
Figure 20.2 Courtesy of Insect Image, University of Georgia.
Figure 20.5 Adapted from Flores, R., et al., 1997. Viroids: the noncoding genomes. Semin. Virol. 8, 65–73, with permission.
Figure 20.6A http://en.wikipedia.org/wiki/Prion#/media/File:Histology_bse.jpg
Figure 20.6B Courtesy of Dr Rita P.-Y. Chen, Academia Sinica, Taiwan.
Figure 20.7 Courtesy of The Nobel Foundation 1976.
Figure 20.8 https://en.wikipedia.org/wiki/Bovine_spongiform_encephalopathy
Figure 20.9 Adapted from Figure 8.6 of Chapter 8 Subviral agents: Genome without Viruses, Viruses without Genomes, in Cann, A., 2016. Principles of Molecular Virology, sixth ed. p. 274, with permission of Elsevier.
Figure 20.11 Adapted from Figure 8.10 of Chapter 8 Subviral agents: Genome without Viruses, Viruses without Genomes, in Cann, A., 2016. Principles of Molecular Virology, sixth ed., p. 277, with permission of Elsevier.
Figure 20.13 Adapted from Figure 8.8 of Chapter 8 Subviral agents: Genome without Viruses, Viruses without Genomes, in Cann, A., 2016. Principles of Molecular Virology, sixth ed. p. 275, with permission of Elsevier.
Box 20.1 Courtesy of The Nobel Foundation.

CHAPTER 21

Figure 21.5 https://en.wikipedia.org/wiki/Ebola_virus_epidemic_in_West_Africa/
Figure 21.7 https://en.wikipedia.org/wiki/West_Nile_virus_in_the_United_States
Figure 21.8 http://en.wikipedia.org/wiki/1993_Four_Corners_hantavirus_outbreak#/
Figure 21.9 http://en.wikipedia.org/wiki/Henipavirus#/media/File:Pteropus_vampyrus2.jpg
Figure 21.10 http://en.wikipedia.org/wiki/File:Sars_Cases_and_Deaths.pdf
Figure 21.11A Courtesy of Drs Maureen Metcalfe, Cynthia Goldsmith, and AzaibiTamin, CDC.
Figure 21.12 Adapted from Wang, T., et al., 2009. Unraveling of the mystery of swine influenza virus. Cell 137, 983, with permission of Elsevier.
Figure 21.13 http://en.wikipedia.org/wiki/Influenza_A_virus_subtype_H5N1#/
Figure 21.14 http://en.wikipedia.org/wiki/Influenza_A_virus_subtype_H5N1#/
Figure 21.15 http://commons.wikimedia.org/wiki/File:Measles_incidence_and_vaccination_England/

CHAPTER 22

Figure 22.1 http://en.wikipedia.org/wiki/Epidemiology_of_HIV/AIDS/
Figure 22.2 http://aidsinfo.unaids.org/
Figure 22.3 https://en.wikipedia.org/wiki/Epidemiology_of_HIV/AIDS
Figure 22.4 http://en.wikipedia.org/wiki/HIV/AIDS#/media/File:Hiv-timecourse.png
Figure 22.5 https://en.wikipedia.org/wiki/Subtypes_of_HIV#HIV-2
Figure 22.6 http://en.wikipedia.org/wiki/Structure_and_genome_of_HIV
Figure 22.7 http://commons.wikimedia.org/wiki/File:Stribild_bottle_Dutch_labeling.jpg
Box 22.1 Adapted from Feinberg, M.B., et al., 2012. Born this way: Understanding the immunological basis of effective HIV control. Nat. Immunol. 13, 632–634, with permission.

CHAPTER 23

Figure 23.2 http://commons.wikimedia.org/wiki/File:HAV_prevalence_2005.svg#
Figure 23.3 http://commons.wikimedia.org/wiki/File:HAV_Infection.png#
Figure 23.4 http://en.wikipedia.org/wiki/Hepatitis_B#/media/File:HBV_prevalence_2005.png
Figure 23.5 Adapted from Rehermann, B., 2005. Immunology of hepatitis B virus and hepatitis C virus infection. Nat. Rev. Immunol. 5, 215–229, with permission.
Figure 23.7 Adapted from Rehermann, B., 2005. Immunology of hepatitis B virus and hepatitis C virus infection. Nat. Rev. Immunol. 5, 215–229, with permission.
Figure 23.8 Adapted from Farazi, P., et al., 2006. Hepatocellular carcinoma pathogenesis: from genes to environment. Nat. Rev. Cancer. 6, 674–687, with permission.
Figure 23.9 http://en.wikipedia.org/wiki/Hepatitis_C

Figure 23.10 Adapted from Rehermann, B., 2005. Immunology of hepatitis B virus and hepatitis C virus infection. Nat. Rev. Immunol. 5, 215–229, with permission.

Figure 23.14 Adapted from Panda, S.K., et al., 2007. Hepatitis E virus. Rev. Med. Virol. 17, 151–180, with permission.

Figure 23.15 Courtesy of Viralzone.

Box 23.1 Adapted from Protzer, U., et al., 2012. Living in the liver: hepatic infections. Nat. Rev. Immunol. 12, 201–213, with permission.

CHAPTER 24

Box 24.1 Courtesy of The Nobel Foundation.

Box 24.2 Adapted from Moore, P.S., Chang, Y., 2010. Why do viruses cause cancer? Highlights of the first century of human tumor virology. Nat. Rev. Cancer 10, 878, with permission.

CHAPTER 25

Figure 25.1 https://en.wikipedia.org/wiki/Smallpox_vaccine#

Figure 25.2 Adapted from Figure 5.13A of Chapter 5 Vaccines and Vaccination, in Harper, D.R., 2012, Viruses: Biology, Applications, Control, first ed. p.125, with permission of Garland Science.

Figure 25.3 Adapted from Minor, P., 2004. Polio eradication, cessation of vaccination and re-emergence of disease. Nat. Rev. Microbiol. 2, 473–482, with permission.

Figure 25.5 http://commons.wikimedia.org/wiki/File:Ped-O-Jet-TearyChild-crop.jpg#

Figure 25.8 Adapted from Minor, P., 2004. Polio eradication, cessation of vaccination and re-emergence of disease. Nat. Rev. Microbiol. 2, 473–482, with permission.

Figure 25.9 http://en.wikipedia.org/wiki/Polio_vaccine#Oral_vaccine

Figure 25.12 Adapted from Reed, S., et al., 2013. Key roles of adjuvants in modern vaccines. Nat. Med. 19, 1597–1608, with permission.

Figure 25.13 http://en.wikipedia.org/wiki/Lipid_A

Box 25.1 http://en.wikipedia.org/wiki/Louis_Pasteur.

Box 25.2 http://en.wikipedia.org/wiki/Jonas_Salk and http://en.wikipedia.org/wiki/Albert_Sabin

CHAPTER 26

Figure 26.2A http://en.wikipedia.org/wiki/Pleconaril

Figure 26.2B Adapted from De Palma, A., et al., 2008. Selective inhibitors of picornavirus replication. Med. Res. Rev. 28, 823–884, with permission.

Figure 26.4 Adapted from De Clercq, E., 2007. The design of drugs for HIV and HCV. Nat. Rev. Drug Discov. 6, 1001, with permission.

Figure 26.5 http://en.wikipedia.org/wiki/Maraviroc, http://en.wikipedia.org/wiki/Raltegravir and http://en.wikipedia.org/wiki/Dolutegravir

Figure 26.6 Adapted from De Clercq, E., 2007. The design of drugs for HIV and HCV. Nat. Rev. Drug Discov. 6, 1001, with permission.

Figure 26.8 Adapted from Figure 6.4 of Chapter 6 Antiviral Drugs, in Harper, D.R., 2012. Viruses: Biology, Applications, Control, first ed. p.137, with permission of Garland Science.

Figure 26.9 Adapted from De Clercq, E., 2007. The design of drugs for HIV and HCV. Nat. Rev. Drug Discov. 6, 1001, with permission.

Figure 26.10 http://en.wikipedia.org/wiki/Boceprevir, http://en.wikipedia.org/wiki/Telaprevir, http://en.wikipedia.org/wiki/Sofosbuvir, and https://en.wikipedia.org/wiki/Ledipasvir

Figure 26.11 Adapted from Itzstein, A., 2007. The war against influenza: discovery and development of sialidase inhibitors. Nat. Rev. Drug Discov. 6, 967, with permission.

Figure 26.12 https://en.wikipedia.org/wiki/Oseltamivir, http://en.wikipedia.org/wiki/Amantadine, and http://en.wikipedia.org/wiki/Favipiravir

Figure 26.13 http://en.wikipedia.org/wiki/Imiquimod

List of Figures

List of Tables

Index

Note: Page numbers followed by "*b*," "*f*," and "*t*" refer to boxes, figures, and tables, respectively.

I

N

O

P

Q

R

U

V

Printed and bound by CPI Group (UK) Ltd, Croydon, CR0 4YY
03/10/2024
01040325-0019